Operator Theory: Advances and Applications
Vol. 139

Vladimir Müller

Spectral Theory of Linear Operators

and Spectral Systems in Banach Algebras

Second edition

Springer Basel AG

Author:

Vladimir Müller
Institute of Mathematics
Czech Academy of Sciences
Zitna 25
115 67 Praha 1
Czech Republic
e-mail: muller@math.cas.cz

2000 Mathematics Subject Classification 47Axx; 47B48

Library of Congress Control Number: 2007929011

Bibliographic information published by Die Deutsche Bibliothek
Die Deutsche Bibliothek lists this publication in the Deutsche Nationalbibliografie; detailed bibliographic data is available in the Internet at <http://dnb.ddb.de>.

ISBN 978-3-7643-8264-3

Originally published by Birkhäuser Verlag Basel, Switzerland 2007
Printed on acid-free paper produced from chlorine-free pulp. TCF ∞
Cover design: Heinz Hiltbrunner, Basel

ISBN 978-3-7643-8264-3 ISBN 978-3-7643-8265-0 (eBook)
DOI 10.1007/978-3-7643-8265-0

9 8 7 6 5 4 3 2 1 www.birkhauser.ch

Contents

Preface

Spectral theory is an important part of functional analysis. It has numerous applications in many parts of mathematics and physics including matrix theory, function theory, complex analysis, differential and integral equations, control theory and quantum physics.

In recent years, spectral theory has witnessed an explosive development. There are many types of spectra, both for one or several commuting operators, with important applications, for example the approximate point spectrum, Taylor spectrum, local spectrum, essential spectrum, etc.

The present monograph is an attempt to organize the available material most of which exists only in the form of research papers scattered throughout the literature. The aim is to present a survey of results concerning various types of spectra in a unified, axiomatic way.

The central unifying notion is that of a regularity, which in a Banach algebra is a subset of elements that are considered to be "nice". A regularity R in a Banach algebra $\mathcal{A}$ defines the corresponding spectrum $\sigma_R(a) = \{\lambda \in \mathbb{C} : a - \lambda \notin R\}$ in the same way as the ordinary spectrum is defined by means of invertible elements, $\sigma(a) = \{\lambda \in \mathbb{C} : a - \lambda \notin \operatorname{Inv}(\mathcal{A})\}$.

Axioms of a regularity are chosen in such a way that there are many natural interesting classes satisfying them. At the same time they are strong enough for non-trivial consequences, for example the spectral mapping theorem.

Spectra of n-tuples of commuting elements of a Banach algebra are described similarly by means of a notion of joint regularity. This notion is closely related to the axiomatic spectral theory of Żelazko and Słodkowski.

The book is organized in five chapters. The first chapter contains spectral theory in Banach algebras which form a natural frame for spectral theory of operators.

In the second chapter the spectral theory of Banach algebras is applied to operators. Of particular interest are regular functions – operator-valued functions whose ranges (kernels) behave continuously. Applied to the function $z \mapsto T - z$ where T is a fixed operator, this gives rise to the important class of Kato operators and the corresponding Kato spectrum (studied in the literature under many names, e.g., semi-regular operators, Apostol spectrum etc.).

The third chapter gives a survey of results concerning various types of essential spectra, Fredholm and Browder operators etc.

The next chapter concentrates on the Taylor spectrum, which is by many experts considered to be the proper generalization of the ordinary spectrum of single operators. The most important property of the Taylor spectrum is the existence of the functional calculus for functions analytic on a neighbourhood of the Taylor spectrum. We present the Taylor functional calculus in an elementary way, without the use of sheaf theory or cohomological methods.

Further we generalize the concept of regular functions. We introduce and study operator-valued functions that admit finite-dimensional discontinuities of the kernel and range. This is closely related with stability results for the index of complexes of Banach spaces.

The last chapter is concentrated on the study of orbits of operators. By an orbit of an operator T we mean a sequence $\{T^n x : n = 0, 1, \dots\}$ where x is a fixed vector. Similarly, a weak orbit is a sequence of the form $\{\langle T^n x, x^* \rangle : n = 0, 1, \dots\}$ where $x \in X$ and $x^* \in X^*$ are fixed, and a polynomial orbit is a set $\{p(T)x : p \text{ polynomial}\}$. These notions, which originated in the theory of dynamical systems, are closely related to the invariant subspace problem. We investigate these notions by means of the essential approximate point spectrum.

All results are presented in an elementary way. We assume only a basic knowledge of functional analysis, topology and complex analysis. Moreover, basic notions and results from the theory of Banach spaces, analytic and smooth vector-valued functions and semi-continuous set-valued functions are given in the Appendix.

The author would like to express his gratitude to many experts in the field who influenced him in various ways. In particular, he would like to thank V. Pták for his earlier guidance and later interest in the subject, and A. Sołtysiak and J. Zemánek who read parts of the manuscript and made various comments. The author is further indebted to V. Kordula, W. Żelazko, M. Mbekhta, F.-H. Vasilescu, C. Ambrozie, E. Albrecht, F. Leon and many others for cooperation and useful discussions over the years. Finally, the author would like to acknowledge that this book was written while he was partially supported by grant No. 201/00/0208 of the Grant Agency of the Czech Republic.

Prague
November 2001

V. M.

Preface to the Second Edition

Since this book was written several years ago, further progress has been made in some parts of the theory. I use the opportunity to include some of the new results, improve the arguments in other places, and also to correct some unfortunate errors and misprints that appeared in the first edition.

My sincere thanks are due to A. Sołtysiak, J. Bračič and J. Vršovský who contributed to the improvement of the text. The work was supported by grant No. 201/06/0128 of GA ČR.

Prague
April 2007

V. M.

Chapter I

Banach Algebras

In this chapter we study spectral theory in Banach algebras. Basic concepts and classical results are summarized in the first two sections. In the subsequent sections we study the approximate point spectrum, which is one of the most important examples of a spectrum in Banach algebras. The approximate point spectrum is closely related with the notions of removable and non-removable ideals.

The axiomatic theory of spectrum is introduced in Sections 6 and 7. This enables us to study various types of spectra, both of single elements and commuting n-tuples, in a unified way.

All algebras considered here are complex and unital. The field of complex numbers will be denoted by $\mathbb{C}$.

1 Basic Concepts

This section contains basic definitions and results from the theory of Banach algebras. For more details see the monograph [BD] or some other textbook about Banach algebras (e.g., [Ric], [Zel6], [Pal]).

Definition 1. An *algebra* $\mathcal{A}$ is a complex linear space $\mathcal{A}$ together with a multiplication mapping $(x,y) \mapsto xy$ from $\mathcal{A} \times \mathcal{A}$ into $\mathcal{A}$ which satisfies the following conditions (for all $x, y, z \in \mathcal{A}, \alpha \in \mathbb{C}$):

(i) $(xy)z = x(yz)$;

(ii) $x(y+z) = xy + xz, (x+y)z = xz + yz$;

(iii) $(\alpha x)y = \alpha(xy) = x(\alpha y)$;

(iv) there exists a *unit element* $e \in \mathcal{A}$ such that $e \neq 0$ and $ex = xe = x$ for all $x \in \mathcal{A}$.

It is easy to show that the unit element is determined uniquely. Indeed, if e' is another unit element, then $e = ee' = e'$. The unit element of an algebra $\mathcal{A}$

will be denoted by $1_{\mathcal{A}}$, (or simply 1 when no confusion can arise). Similarly, for a complex number α, the symbol α also denotes the algebra element $\alpha \cdot 1_{\mathcal{A}}$.

Definition 2. Let $\mathcal{A}$ be an algebra. An *algebra seminorm* in $\mathcal{A}$ is a function $\|\cdot\|$: $\mathcal{A} \to \langle 0, \infty)$, $x \mapsto \|x\|$ satisfying (for all $x, y \in \mathcal{A}, \alpha \in \mathbb{C}$):

(i) $\|\alpha x\| = |\alpha| \cdot \|x\|$;
(ii) $\|x + y\| \leq \|x\| + \|y\|$;
(iii) $\|xy\| \leq \|x\| \cdot \|y\|$;
(iv) $\|1_{\mathcal{A}}\| = 1$.

An *algebra norm* in $\mathcal{A}$ is an algebra seminorm such that

(v) if $\|x\| = 0$, then $x = 0$.

Definition 3. A *normed algebra* is a pair $(\mathcal{A}, \|\cdot\|)$, where $\mathcal{A}$ is an algebra and $\|\cdot\|$ is an algebra norm in $\mathcal{A}$. A *Banach algebra* is a normed algebra that is complete in the topology defined by the norm (in other words, $(\mathcal{A}, \|\cdot\|)$ considered as a linear space is a Banach space).

Examples 4. There are many examples of Banach algebras that appear naturally in functional analysis.

(i) For the purpose of this monograph the most important example of a Banach algebra is the algebra $\mathcal{B}(X)$ of all (bounded linear) operators acting on a Banach space X, $\dim X \geq 1$, with naturally defined algebraic operations and with the *operator-norm* $\|T\| = \sup\{\|Tx\| : x \in X, \|x\| = 1\}$. The unit element in $\mathcal{B}(X)$ is the *identity operator* I defined by $Ix = x \quad (x \in X)$.
In particular, if $\dim X = n < \infty$, then $\mathcal{B}(X)$ can be identified with the algebra of all $n \times n$ complex matrices.

(ii) Let K be a non-empty compact space. Then the algebra $C(K)$ of all continuous complex-valued functions on K with the sup-norm $\|f\| = \sup\{|f(z)| : z \in K\}$ is a Banach algebra.

(iii) Let E be a non-empty set. The set of all bounded complex-valued functions defined on E with pointwise algebraic operations and the sup-norm is a Banach algebra.
Similarly, let L^∞ be the set of all bounded measurable complex-valued functions defined on the real line (as usually, we identify two functions that differ only on a set of Lebesgue measure zero). Then L^∞ with pointwise algebraic operations and the usual L^∞ norm $\|f\|_\infty = \operatorname{ess\,sup}\{|f(t)| : t \in \mathbb{R}\}$ is a Banach algebra.

(iv) Let $\mathbb{D}$ be the open unit disc in the complex plane. Denote by H^∞ the algebra of all functions that are analytic and bounded on $\mathbb{D}$.
The *disc algebra* $A(\mathbb{D})$ is the algebra of all functions that are continuous on $\overline{\mathbb{D}}$ and analytic on $\mathbb{D}$. Both H^∞ and $A(\mathbb{D})$ with the sup-norm are important examples of Banach algebras.

(v) Let S be a semigroup with unit and let $\ell^1(S)$ be the set of all functions $f : S \to \mathbb{C}$ satisfying $\|f\| = \sum_{s\in S} |f(s)| < \infty$. Then $\ell^1(S)$ with the multiplication defined by $(fg)(s) = \sum_{t_1t_2=s} f(t_1)g(t_2)$ is a Banach algebra.
More generally, if $\alpha : S \to (0,\infty)$ is a submultiplicative function and $\alpha(1_S) = 1$, then $\{f : S \to \mathbb{C} : \sum |f(s)|\alpha(s) < \infty\}$ is a Banach algebra with the norm $\|f\| = \sum |f(s)|\alpha(s)$ and the above-defined multiplication.

(vi) Let L^1 be the set of all integrable functions $f : \mathbb{R} \to \mathbb{C}$. Define the multiplication and norm in L^1 by

$$(f * g)(s) = \int_{-\infty}^{\infty} f(t)g(s-t)\,\mathrm{d}t$$

and

$$\|f\| = \int_{-\infty}^{\infty} |f(t)|\,\mathrm{d}t.$$

Then L^1 satisfies all axioms of Banach algebras except of the existence of the unit element. The unitization (see C.1.1) $L^1(\mathbb{R}) \oplus \mathbb{C}$ with the norm $\|f \oplus \lambda\| = \|f\| + |\lambda|$ and multiplication $(f \oplus \lambda)\cdot(g \oplus \mu) = f * g + \lambda g + \mu f \oplus \lambda\mu$ is a Banach algebra.

Remark 5. Sometimes little bit different definitions of Banach algebras are used. Frequently the existence of the unit element is not assumed or condition (iii) of Definition 2 is replaced by a weaker condition of continuity of the multiplication. However, it is possible to reduce these more general definitions of Banach algebras to the present definition, see C.1.1 and C.1.3. Many results get a more natural formulation in this way and the proofs are not obscured by technical difficulties.

Definition 6. Let $\mathcal{A}, \mathcal{B}$ be algebras. A linear mapping $\rho : \mathcal{A} \to \mathcal{B}$ is called a *homomorphism* if $\rho(xy) = \rho(x)\rho(y)$ for all $x, y \in \mathcal{A}$ and $\rho(1_{\mathcal{A}}) = 1_{\mathcal{B}}$.

Let $\mathcal{A}$ and $\mathcal{B}$ be normed algebras. A homomorphism $\rho : \mathcal{A} \to \mathcal{B}$ is continuous if $\|\rho\| := \sup\{\|\rho(x)\| : x \in \mathcal{A}, \|x\| = 1\} < \infty$. A continuous homomorphism ρ satisfying $\inf\{\|\rho(x)\| : x \in \mathcal{A}, \|x\| = 1\} > 0$ is called an *isomorphism*. A homomorphism ρ is called *isometrical* if $\|\rho(x)\| = \|x\|$ for all $x \in \mathcal{A}$.

A subset M of an algebra $\mathcal{A}$ is called a *subalgebra* if it is closed under the algebraic operations (i.e., M is a linear subspace of $\mathcal{A}$, $1_{\mathcal{A}} \in M$ and $x, y \in M \Rightarrow xy \in M$).

If $\mathcal{A}$ is a closed subalgebra of a Banach algebra $\mathcal{B}$, then $\mathcal{A}$ with the restricted norm is again a Banach algebra.

Each normed algebra $\mathcal{A}$ has a *completion* – the uniquely determined (up to an isometrical isomorphism) Banach algebra $\mathcal{B}$ such that $\mathcal{A}$ is a dense subalgebra of $\mathcal{B}$.

Remark 7. Let $\mathcal{A}$ be a Banach algebra. For $a \in \mathcal{A}$ define the operator $L_a : \mathcal{A} \to \mathcal{A}$ by $L_a x = ax$. It is easy to verify that the mapping $a \mapsto L_a$ is an isometrical homomorphism $\mathcal{A} \to \mathcal{B}(\mathcal{A})$. If we identify $\mathcal{A}$ with the image of this homomorphism, then we can consider $\mathcal{A}$ as a closed subalgebra of $\mathcal{B}(\mathcal{A})$.

This simple but important construction enables us often to generalize results from operator theory to Banach algebras.

Definition 8. Let $\mathcal{A}$ be a Banach algebra. A set $J \subset \mathcal{A}$ is called a *left* (*right*) *ideal* in $\mathcal{A}$ if J is a subspace of $\mathcal{A}$ and $ax \in J$ ($xa \in J$) for all $x \in J$, $a \in \mathcal{A}$. J is a *two-sided ideal* in $\mathcal{A}$ if J is both a left and right ideal in $\mathcal{A}$. An ideal $J \subset \mathcal{A}$ (left, right or two-sided) is called *proper* if $J \neq \mathcal{A}$. Equivalently, J is proper if and only if $1_{\mathcal{A}} \notin J$.

Let $\rho : \mathcal{A} \to \mathcal{B}$ be a continuous homomorphism from a Banach algebra $\mathcal{A}$ to a Banach algebra $\mathcal{B}$. It is easy to see that $\operatorname{Ker} \rho = \{x \in A : \rho(x) = 0\}$ is a closed two-sided ideal in $\mathcal{A}$.

Conversely, if $J \subset \mathcal{A}$ is a closed proper two-sided ideal in $\mathcal{A}$, then we can define a multiplication in the quotient space $\mathcal{A}/J$ by $(x+J)(y+J) = xy+J$. The space $\mathcal{A}/J$ then becomes a Banach algebra with the unit $1_{\mathcal{A}}+J$. For the canonical homomorphism $\pi : \mathcal{A} \to \mathcal{A}/J$ defined by $\pi x = x+J \quad (x \in \mathcal{A})$ we have $\operatorname{Ker} \pi = J$.

Invertible elements

Definition 9. Let x, y be elements of an algebra $\mathcal{A}$. Then y is called a *left* (*right*) *inverse* of x if $yx = 1$ ($xy = 1$). If y is both a left and right inverse of x, then it is called an *inverse* of x. If x has a left inverse y and a right inverse z, then $y = z$. Indeed, we have $y = y(xz) = (yx)z = z$. In particular, an element has at most one inverse.

An element of $\mathcal{A}$ for which there exists an inverse (left inverse, right inverse) will be called *invertible (left invertible, right invertible)*. The unique inverse of an invertible element x will be denoted by x^{-1}. The set of all invertible elements in an algebra $\mathcal{A}$ will be denoted by $\operatorname{Inv}(\mathcal{A})$. Similarly, the set of all left (right) invertible elements in $\mathcal{A}$ will be denoted by $\operatorname{Inv}_l(\mathcal{A})$ and $\operatorname{Inv}_r(\mathcal{A})$,

$$\operatorname{Inv}_l(\mathcal{A}) = \{x \in \mathcal{A} : \text{there exists } y \in \mathcal{A} \text{ such that } yx = 1\},$$

$$\operatorname{Inv}_r(\mathcal{A}) = \{x \in \mathcal{A} : \text{there exists } y \in \mathcal{A} \text{ such that } xy = 1\}.$$

Obviously, $\operatorname{Inv}(\mathcal{A}) = \operatorname{Inv}_l(\mathcal{A}) \cap \operatorname{Inv}_r(\mathcal{A})$.

It is easy to see that an element $a \in \mathcal{A}$ is left (right) invertible if and only if there is no proper left (right) ideal containing a.

Remark 10. The left and right properties in Banach algebras are perfectly symmetrical. The simplest way how to give an exact meaning to this statement is to consider the following construction: for a Banach algebra $\mathcal{A}$ consider the reversed multiplication $a \odot b = ba$ for all $a, b \in \mathcal{A}$. In this way we obtain the Banach algebra $\operatorname{rev} \mathcal{A}$, and the left ideals, left inverses etc. in $\mathcal{A}$ correspond to the right objects in the algebra rev $\mathcal{A}$. Using this construction, each one-sided result implies immediately the corresponding symmetrical result.

Theorem 11. *Let $\mathcal{A}$ be a Banach algebra. Then:*

(i) *if $a \in \mathcal{A}$, $\|a\| < 1$, then $1 - a \in \operatorname{Inv}(\mathcal{A})$;*
(ii) *the sets $\operatorname{Inv}_l(\mathcal{A})$, $\operatorname{Inv}_r(\mathcal{A})$ and $\operatorname{Inv}(\mathcal{A})$ are open;*
(iii) *the mapping $x \mapsto x^{-1}$ is continuous in $\operatorname{Inv}(\mathcal{A})$.*

Proof. (i) If $\|a\| < 1$, then $\|a^j\| \le \|a\|^j$ for all j, so the series $\sum_{j=0}^{\infty} a^j$ is convergent in $\mathcal{A}$ and

$$(1-a)\sum_{j=0}^{\infty} a^j = \Big(\sum_{j=0}^{\infty} a^j\Big)(1-a) = \lim_{k\to\infty}\Big(\sum_{j=0}^{k} a^j - \sum_{j=0}^{k} a^{j+1}\Big) = \lim_{k\to\infty}(1-a^{k+1}) = 1.$$

(ii) Let $yx = 1$ and let $\|u\| < \|y\|^{-1}$. Then $y(x+u) = 1 + yu$, where $\|yu\| \le \|y\| \cdot \|u\| < 1$. By (i), $y(x+u)$ is invertible and $\big(y(x+u)\big)^{-1} y(x+u) = 1$. Thus $x + u \in \operatorname{Inv}_l(\mathcal{A})$. Hence $\operatorname{Inv}_l(\mathcal{A})$ is an open set.

Similarly, $\operatorname{Inv}_r(\mathcal{A})$ and $\operatorname{Inv}(\mathcal{A}) = \operatorname{Inv}_l(\mathcal{A}) \cap \operatorname{Inv}_r(\mathcal{A})$ are open subsets of $\mathcal{A}$.

(iii) Let $a \in \operatorname{Inv}(\mathcal{A})$ and let $\|x\| < \|a^{-1}\|^{-1}$. Then the series $\sum_{i=0}^{\infty}(xa^{-1})^i$ is convergent in $\mathcal{A}$ and one can check directly that

$$(a-x)^{-1} = a^{-1}\sum_{i=0}^{\infty}(xa^{-1})^i.$$

Thus

$$\begin{aligned}\|(a-x)^{-1} - a^{-1}\| &= \Big\|a^{-1}\sum_{i=1}^{\infty}(xa^{-1})^i\Big\| \\ &\le \|a^{-1}\| \cdot \sum_{i=1}^{\infty}\|x\|^i \cdot \|a^{-1}\|^i = \frac{\|x\| \cdot \|a^{-1}\|^2}{1 - \|x\| \cdot \|a^{-1}\|},\end{aligned}$$

and so $(a-x)^{-1} \to a^{-1}$ for $\|x\| \to 0$. □

Lemma 12. *Let x, x_n, y_n $(n = 1, 2, \dots)$ be elements of a Banach algebra $\mathcal{A}$, $x_n \to x$ and $\sup_n\{\|y_n\|\} < \infty$. If $y_n x_n = 1$ for all n, then $x \in \operatorname{Inv}_l(\mathcal{A})$.*

If $x_n y_n = 1$ for all n, then $x \in \operatorname{Inv}_r(\mathcal{A})$.

Proof. Choose n such that $\|x - x_n\| < (\sup\|y_n\|)^{-1}$. Then $y_n x = y_n x_n + y_n(x - x_n) = 1 + y_n(x - x_n)$, where $\|y_n(x - x_n)\| < 1$, and so $y_n x \in \operatorname{Inv}(\mathcal{A})$. Thus $(y_n x)^{-1} y_n x = 1$ and $x \in \operatorname{Inv}_l(\mathcal{A})$.

The second statement can be proved similarly. □

Definition 13. An element a of a Banach algebra $\mathcal{A}$ is called a *left (right) divisor of zero* if $ax = 0$ $(xa = 0)$ for some non-zero element $x \in \mathcal{A}$.

An element a of a Banach algebra $\mathcal{A}$ is called a *left topological divisor of zero* if $\inf\{\|ax\| : x \in A, \|x\| = 1\} = 0$. Similarly, a is a *right topological divisor of zero* if $\inf\{\|xa\| : x \in A, \|x\| = 1\} = 0$.

Topological divisors of zero are closely related to non-invertible elements.

Theorem 14. *Let a be an element of a Banach algebra $\mathcal{A}$. Then:*

(i) *if a is left (right) invertible, then a is not a left (right) topological divisor of zero;*

(ii) *if a is invertible, then a is neither a left nor a right topological divisor of zero;*

(iii) *if $a \in \partial\,\mathrm{Inv}_l(\mathcal{A})$ (the topological boundary of $\mathrm{Inv}_l(\mathcal{A})$), then a is a right topological divisor of zero;*

(iv) *if $a \in \partial\,\mathrm{Inv}(\mathcal{A})$, then a is both a left and right topological divisor of zero.*

Proof. (i) Suppose that b is a left inverse of a, $ba = 1$. For $x \in \mathcal{A}$, $\|x\| = 1$ we have $1 = \|x\| = \|bax\| \le \|b\| \cdot \|ax\|$, so $\inf\{\|ax\| : x \in \mathcal{A}, \|x\| = 1\} \ge \|b\|^{-1} > 0$ and a is not a left topological divisor of zero.

The right version can be proved similarly.

This implies also (ii).

(iii) Let $a \in \partial\,\mathrm{Inv}_l(\mathcal{A})$. Then there exist $a_n, b_n \in \mathcal{A}$ such that $\lim_{n\to\infty} a_n = a$ and $b_n a_n = 1$ for all n. By Lemma 12, $\lim \|b_n\| = \infty$. Set $c_n = \frac{b_n}{\|b_n\|}$. Then $\|c_n\| = 1$ for every n and

$$\|c_n a\| = \left\| \frac{b_n}{\|b_n\|}(a_n + (a - a_n)) \right\| \le \frac{1}{\|b_n\|} + \|a - a_n\|,$$

so $\lim_{n\to\infty} \|c_n a\| = 0$ and a is a right topological divisor of zero.

(iv) Let $a \in \partial\,\mathrm{Inv}(\mathcal{A})$. Then there exist $a_n \in \mathrm{Inv}(\mathcal{A})$ with $a_n \to a \quad (n \to \infty)$. By Lemma 12, $\lim_{n\to\infty} \|a_n^{-1}\| = \infty$. Set $c_n = \frac{a_n^{-1}}{\|a_n^{-1}\|}$. Then $\|c_n\| = 1$ and, as in (iii), one can get easily that $\|c_n a\| \to 0$ and $\|a c_n\| \to 0$. Thus a is both a left and right topological divisor of zero. □

Spectrum and spectral radius

Definition 15. Let a be an element of a Banach algebra $\mathcal{A}$. The *spectrum* of a in $\mathcal{A}$ is the set of all complex numbers λ such that $a - \lambda$ is not invertible in $\mathcal{A}$. The spectrum of a in $\mathcal{A}$ will be denoted by $\sigma^{\mathcal{A}}(a)$, or $\sigma(a)$ if the algebra is clear from the context.

By Theorem 11, $\sigma(a)$ is a closed subset of $\mathbb{C}$. The function $\lambda \mapsto (a - \lambda)^{-1}$ defined in the open set $\mathbb{C} \setminus \sigma(a)$ is called the *resolvent* of a.

Theorem 16. *Let a be an element of a Banach algebra $\mathcal{A}$. Then the resolvent $\lambda \mapsto (a - \lambda)^{-1}$ is analytic in $\mathbb{C} \setminus \sigma(a)$.*

Proof. For $\lambda, \mu \notin \sigma(a)$ we have

$$\begin{aligned}(a-\mu)^{-1}-(a-\lambda)^{-1} &= (a-\mu)^{-1}\big((a-\lambda)-(a-\mu)\big)(a-\lambda)^{-1}\\ &= (\mu-\lambda)(a-\mu)^{-1}(a-\lambda)^{-1},\end{aligned}$$

and so

$$\lim_{\mu\to\lambda}\frac{(a-\mu)^{-1}-(a-\lambda)^{-1}}{\mu-\lambda}=(a-\lambda)^{-2}.$$

Thus the function $\lambda \mapsto (a-\lambda)^{-1}$ is analytic in $\mathbb{C}\setminus\sigma(a)$. □

The following theorem is one of the most important results in the theory of Banach algebras.

Theorem 17. *Let x be an element of a Banach algebra $\mathcal{A}$. Then $\sigma(x)$ is a non-empty compact set.*

Proof. Let $\lambda \in \mathbb{C}$, $|\lambda| > \|x\|$. Then the series $\sum_{j=0}^{\infty}\frac{x^j}{\lambda^{j+1}}$ is convergent in $\mathcal{A}$ and

$$(x-\lambda)\sum_{j=0}^{\infty}\frac{-x^j}{\lambda^{j+1}} = -\sum_{j=1}^{\infty}\frac{x^j}{\lambda^j}+\sum_{j=0}^{\infty}\frac{x^j}{\lambda^j}=1.$$

Similarly $\left(\sum_{j=0}^{\infty}\frac{-x^j}{\lambda^{j+1}}\right)(x-\lambda)=1$, and so $\lambda\notin\sigma(x)$. Thus $\sigma(x)$ is bounded and hence compact.

Suppose on the contrary that $\sigma(x)=\emptyset$. Consider the function $f:\mathbb{C}\to\mathcal{A}$ defined by $f(\lambda)=(x-\lambda)^{-1}$. By Theorem 16, f is an entire function. For $|\lambda| > \|x\|$ we have $f(\lambda)=\sum_{j=0}^{\infty}\frac{-x^j}{\lambda^{j+1}}$, and so $\|f(\lambda)\|\le\sum_{j=0}^{\infty}\frac{\|x\|^j}{|\lambda|^{j+1}}=\frac{1}{|\lambda|-\|x\|}$. Thus $f(\lambda)\to 0$ for $\lambda\to\infty$. By the Liouville theorem, $f(\lambda)=0$ for each $\lambda\in\mathbb{C}$. This is a contradiction, since $f(\lambda)$ is invertible for each λ. □

Remark 18. Let T be an operator on a finite-dimensional Banach space X (i.e., T is a square matrix). Then $\sigma(T)$ is finite and consists of eigenvalues of T.

Since the eigenvalues of a matrix are precisely the roots of its characteristic polynomial, the non-emptiness of $\sigma(T)$ is equivalent to the "fundamental theorem of algebra" that each complex polynomial has a root. This illustrates how deep is the previous theorem, and also that operators on finite-dimensional spaces are far from being trivial.

Corollary 19. (Gelfand, Mazur) *Let $\mathcal{A}$ be a Banach algebra such that every non-zero element of $\mathcal{A}$ is invertible (i.e., $\mathcal{A}$ is a field). Then $\mathcal{A}$ consists of scalar multiples of the identity, $\mathcal{A}=\{\lambda\cdot 1_{\mathcal{A}}:\lambda\in\mathbb{C}\}$. Thus $\mathcal{A}$ is isometrically isomorphic to the field of complex numbers $\mathbb{C}$.*

Proof. For every $x\in\mathcal{A}$ there exists $\lambda\in\sigma(x)$ such that $x-\lambda\cdot 1_{\mathcal{A}}\notin\operatorname{Inv}(\mathcal{A})$. Thus $x=\lambda\cdot 1_{\mathcal{A}}$. □

Definition 20. The *spectral radius* $r(x)$ of an element $x \in \mathcal{A}$ is the number

$$r(x) = \max\{|\lambda| : \lambda \in \sigma(x)\}.$$

Lemma 21. *Let $s_1, s_2, \dots$ be non-negative real numbers. Then:*

(i) *if $s_{n+m} \le s_n \cdot s_m$ for all $m, n \in \mathbb{N}$, then the limit $\lim_{n\to\infty} s_n^{1/n}$ exists and is equal to $\inf_n s_n^{1/n}$;*

(ii) *if $s_n > 0$ and $s_{n+m} \ge s_n \cdot s_m$ for all $m, n \in \mathbb{N}$, then the limit $\lim_{n\to\infty} s_n^{1/n}$ exists and is equal to $\sup_n s_n^{1/n}$.*

Proof. (i) Write $t = \inf_n s_n^{1/n}$ and let $\varepsilon > 0$. Fix k such that $s_k^{1/k} < t + \varepsilon$. Any number $n \ge k$ can be expressed in the form $n = n_1 k + r$, where $0 \le r \le k-1$ and $n_1 \ge 1$. Then

$$s_n \le s_r \cdot s_k^{n_1} \le \max\{1, s_1, s_2, \dots, s_{k-1}\} \cdot (t+\varepsilon)^{kn_1}$$

and

$$s_n^{1/n} \le \max\{1, s_1, s_2, \dots, s_{k-1}\}^{1/n} \cdot (t+\varepsilon)^{kn_1/n} \to t + \varepsilon$$

as $n \to \infty$, since $kn_1/n \to 1$. Thus $\limsup_{n\to\infty} s_n^{1/n} \le t + \varepsilon$ and, since ε was arbitrary, we have $\lim_{n\to\infty} s_n^{1/n} = t = \inf_n s_n^{1/n}$.

(ii) The second statement can be reduced to (i) by considering the numbers s_n^{-1}. □

Theorem 22. (spectral radius formula) *Let a be an element of a Banach algebra $\mathcal{A}$. Then*

$$r(a) = \lim_{n\to\infty} \|a^n\|^{1/n} = \inf_n \|a^n\|^{1/n}.$$

Proof. Since $\|a^{m+n}\| \le \|a^m\| \cdot \|a^n\|$ for all m, n, by the previous lemma the limit $\lim \|a^n\|^{1/n}$ exists and is equal to the infimum.

Let λ be a complex number with $|\lambda| > \lim \|a^n\|^{1/n}$. Then $\sum_{n=0}^{\infty} \frac{a^n}{\lambda^{n+1}}$ converges and it is easy to verify that $(a-\lambda)^{-1} = \sum_{n=0}^{\infty} \frac{-a^n}{\lambda^{n+1}}$. Consequently, $r(a) \le \lim \|a^n\|^{1/n}$.

It remains to show that $\lim \|a^n\|^{1/n} \le r(a)$. Consider the function $f(\lambda) = (1 - \lambda a)^{-1}$.

For $\lambda \neq 0$ we have $f(\lambda) = \lambda^{-1}(\lambda^{-1} - a)^{-1}$. Clearly f is analytic in $\{\lambda \in \mathbb{C} : 0 < |\lambda| < r(a)^{-1}\}$ and continuous in $\{\lambda \in \mathbb{C} : |\lambda| < r(a)^{-1}\}$. So f is analytic in $\{\lambda \in \mathbb{C} : |\lambda| < r(a)^{-1}\}$ (if $r(a) = 0$, then f is analytic in $\mathbb{C}$). For $|\lambda| < \|a\|^{-1}$ we can write $f(\lambda) = (1 - \lambda a)^{-1} = \sum_{n=0}^{\infty} a^n \lambda^n$. Therefore we have $f(\lambda) = \sum_{n=0}^{\infty} a^n \lambda^n$ for all λ, $|\lambda| < r(a)^{-1}$.

For the radius of convergence of the power series $\sum a^n \lambda^n$ we have (see Theorem A.2.1)

$$\liminf \|a^n\|^{-1/n} \ge r(a)^{-1},$$

and so

$$r(a) \ge \limsup \|a^n\|^{1/n} = \lim \|a^n\|^{1/n}.$$ □

Definition 23. Let M be a subset of a Banach algebra $\mathcal{A}$. The *commutant* of M is defined by $M' = \{a \in \mathcal{A} : am = ma \ (m \in M)\}$. We write M'' instead of $(M')'$ for the *second commutant* of M. If $xy = yx$ for all $x, y \in \mathcal{A}$, then $\mathcal{A}$ is called commutative.

Lemma 24. *Let M, N be subsets of a Banach algebra $\mathcal{A}$. Then:*

(i) *M' is a closed subalgebra of $\mathcal{A}$;*

(ii) *if $M \subset N$, then $M' \supset N'$ and $M'' \subset N''$;*

(iii) *$M \subset M''$ and $M' = M'''$;*

(iv) *if M consists of mutually commuting elements, then $M \subset M'' \subset M'$ and M'' is a commutative Banach algebra.*

Proof. The first three statements are clear.

To see (iv), note first that $M \subset M'$. Therefore $M'' \subset M' = M'''$, which means that M'' is a commutative algebra. □

Lemma 25. *Let $a \in \mathcal{A}$ and let $\lambda \in \mathbb{C} \setminus \sigma(a)$. Then $(a - \lambda)^{-1} \in \{a\}''$. In particular, $\sigma(a)$ equals to the spectrum of a in the commutative Banach algebra $\{a\}''$.*

Proof. Let $b \in \mathcal{A}$ and $ab = ba$. Then $(a - \lambda)b = b(a - \lambda)$ and, by multiplicating this equality from both sides by $(a - \lambda)^{-1}$, we get $b(a - \lambda)^{-1} = (a - \lambda)^{-1}b$. Thus $(a - \lambda)^{-1} \in \{a\}''$. □

Theorem 26. *Let $\mathcal{A}, \mathcal{B}$ be Banach algebras, $\rho : \mathcal{A} \to \mathcal{B}$ a homomorphism and let $x \in \mathcal{A}$. Then $\sigma^{\mathcal{B}}(\rho(x)) \subset \sigma^{\mathcal{A}}(x)$.*

Proof. Let $\lambda \in \mathbb{C} \setminus \sigma^{\mathcal{A}}(x)$ and let $y = (x - \lambda)^{-1} \in \mathcal{A}$. Then $(\rho(x) - \lambda) \cdot \rho(y) = \rho(x - \lambda) \cdot \rho(y) = \rho(1_{\mathcal{A}}) = 1_{\mathcal{B}}$ and similarly $\rho(y) \cdot (\rho(x) - \lambda) = 1_{\mathcal{B}}$. □

Theorem 27. *Let $\mathcal{A}$ be a subalgebra of a Banach algebra $\mathcal{B}$ and let $x \in \mathcal{A}$. Then:*

(i) *if x is a left (right) topological divisor of zero in $\mathcal{A}$, then x is a left (right) topological divisor of zero in $\mathcal{B}$;*

(ii) *$\partial\sigma^{\mathcal{A}}(x) \subset \sigma^{\mathcal{B}}(x) \subset \sigma^{\mathcal{A}}(x)$.*

Proof. (i) We have

$$\inf\{\|xb\| : b \in \mathcal{B}, \|b\| = 1\} \leq \inf\{\|xa\| : a \in \mathcal{A}, \|a\| = 1\} = 0.$$

(ii) If $\lambda \in \partial\sigma^{\mathcal{A}}(x)$, then $x - \lambda$ is a left topological divisor of zero in $\mathcal{A}$, and so, by (i), $\lambda \in \sigma^{\mathcal{B}}(x)$.

The inclusion $\sigma^{\mathcal{B}}(x) \subset \sigma^{\mathcal{A}}(x)$ follows from the previous theorem. □

Consequently, $\sigma^{\mathcal{A}}(x)$ is obtained by filling in some holes in $\sigma^{\mathcal{B}}(x)$.

In the algebra $\mathcal{B}(X)$ we have additional information.

Theorem 28. *Let X be a Banach space, $\dim X \geq 1$ and let $T \in \mathcal{B}(X)$. Then:*

(i) *T is invertible if and only if T is one-to-one and onto;*
(ii) *if $\lambda \in \partial\sigma(T)$, then $(T-\lambda)X \neq X$ and $\inf\{\|(T-\lambda)x\| : x \in X, \|x\|=1\} = 0$.*

Proof. (i) Follows from the open mapping theorem.

(ii) By Theorem 14, there exist operators $S_n \in \mathcal{B}(X)$ $(n \in \mathbb{N})$ such that $\|S_n\| = 1$ and $\|S_n(T-\lambda)\| \to 0$. For each n there exists $x_n \in X$ with $\|x_n\| = 1$ and $\|S_n x_n\| \geq 1/2$. Suppose on the contrary that $T-\lambda$ is onto. By the open mapping theorem, there exists $k > 0$ such that $(T-\lambda)B_X \supset kB_X$ where B_X denotes the closed unit ball in X. Thus there exists $y_n \in X$ such that $(T-\lambda)y_n = x_n$ and $\|y_n\| \leq k^{-1}$. Hence

$$\|S_n(T-\lambda)\| \geq \left\| S_n(T-\lambda)\frac{y_n}{\|y_n\|} \right\| = \frac{1}{\|y_n\|}\|S_n x_n\| \geq \frac{k}{2},$$

a contradiction with the assumption that $\|S_n(T-\lambda)\| \to 0$.

Similarly, there exist operators $R_n \in \mathcal{B}(X)$ $(n \in \mathbb{N})$ such that $\|R_n\| = 1$ and $\|(T-\lambda)R_n\| \to 0$. There exist vectors $x_n \in X$ with $\|x_n\| = 1$ and $\|R_n x_n\| \geq 1/2$. Set $y_n = \frac{R_n x_n}{\|R_n x_n\|}$. Then $\|y_n\| = 1$ and

$$\|(T-\lambda)y_n\| = \frac{\|(T-\lambda)R_n x_n\|}{\|R_n x_n\|} \leq 2\|(T-\lambda)R_n\| \to 0.$$

Hence $\inf\{\|(T-\lambda)x\| : x \in X, \|x\| = 1\} = 0$. □

Theorem 29. *Let $a, b \in \mathcal{A}$ and let λ be a non-zero complex number. Then $ab-\lambda$ is left (right) invertible if and only if $ba-\lambda$ is left (right) invertible.*

Proof. Let $c \in \mathcal{A}$, $c(ab-\lambda) = 1$. Then

$$\begin{aligned}(-\lambda^{-1} + \lambda^{-1}bca)(ba-\lambda) &= -\lambda^{-1}ba + 1 + \lambda^{-1}bcaba - bca \\ &= 1 - \lambda^{-1}ba + \lambda^{-1}bc(ab-\lambda)a = 1.\end{aligned}$$

Similarly, if $(ba-\lambda)d = 1$ for some $d \in \mathcal{A}$, then

$$(ab-\lambda)(-\lambda^{-1} + \lambda^{-1}adb) = 1.$$ □

Corollary 30. *Let x, y be elements of a Banach algebra $\mathcal{A}$. Then*

$$\sigma(xy) \setminus \{0\} = \sigma(yx) \setminus \{0\}.$$

In general, the spectrum and the spectral radius in a Banach algebra do not behave continuously, see C.1.14. However, they are always upper semicontinuous (for definitions and basic properties of semicontinuous set-valued functions see Appendix A.4). Moreover, we prove later in Section 6 that the set of all discontinuity points of the spectrum is a set of the first category.

Theorem 31. (upper semicontinuity of the spectrum) *Let $\mathcal{A}$ be a Banach algebra, $x \in \mathcal{A}$, let U be an open neighbourhood of $\sigma(x)$. Then there exists $\varepsilon > 0$ such that $\sigma(y) \subset U$ for all $y \in \mathcal{A}$ with $\|y - x\| < \varepsilon$. In particular, the function $x \mapsto r(x)$ is upper semicontinuous.*

Proof. Suppose on the contrary that for every n there exist $x_n \in \mathcal{A}$ and $\lambda_n \in \sigma(x_n) \setminus U$ such that $\|x_n - x\| < 1/n$. Then $|\lambda_n| \leq \|x_n\| \leq \|x\| + 1$, and so there exists a subsequence of (λ_n) converging to some $\lambda \in \mathbb{C} \setminus U$. Since $x_n - \lambda_n \notin \mathrm{Inv}(\mathcal{A})$, we have $x - \lambda \notin \mathrm{Inv}(\mathcal{A})$ by Theorem 11. Thus $\lambda \in \sigma(x)$ and $\lambda \notin U$, which is a contradiction with the assumption that U is a neighbourhood of $\sigma(x)$. □

Equivalent norms

Two norms $\|\cdot\|$ and $\|\cdot\|'$ on a vector space X are called *equivalent* if there exists a positive constant k such that

$$k^{-1}\|x\| \leq \|x\|' \leq k\|x\|$$

for all $x \in X$.

Theorem 32. *Let $(\mathcal{A}, \|\cdot\|)$ be a Banach algebra and let $S \subset \mathcal{A}$ be a bounded semigroup. Then there exists an equivalent algebra norm $\|\cdot\|'$ on $\mathcal{A}$ such that $\|s\| \leq 1$ for every $s \in S$.*

More precisely, there is such a norm $\|\cdot\|'$ satisfying

$$k^{-1}\|a\| \leq \|a\|' \leq k\|a\|$$

for all $a \in \mathcal{A}$, where $k = \sup\{1, \|s\| : s \in S\}$.

Proof. Without loss of generality we can assume that $1 \in S$. Thus $k := \sup\{\|s\| : s \in S\} \geq 1$.

For $a \in \mathcal{A}$ define $q(a) = \sup\{\|sa\| : s \in S\}$. Since $1 \in S$, we have $q(a) \geq \|a\|$. Thus

$$\|a\| \leq q(a) \leq k\|a\|$$

for all $a \in \mathcal{A}$.

Clearly, q is a norm. We have $q(1) = k$ and $q(s) \leq k$ for every $s \in S$.

For $a \in \mathcal{A}$ and $s \in S$ we have

$$q(sa) = \sup\{\|s'sa\| : s' \in S\} \leq q(a).$$

Further, for $a_1, a_2 \in \mathcal{A}$ we have

$$\begin{aligned} q(a_1a_2) &= \sup\{\|sa_1a_2\| : s \in S\} \leq \sup\{\|sa_1\| \cdot \|a_2\| : s \in S\} \\ &\leq q(a_1)\|a_2\| \leq q(a_1)q(a_2). \end{aligned}$$

Define now

$$\|a\|' = \sup\{q(ax) : x \in \mathcal{A}, q(x) \le 1\} = \sup\Big\{\frac{q(ax)}{q(x)} : x \ne 0\Big\}.$$

Since $q(ax) \le q(a)q(x)$, we have $\|a\|' \le q(a) \le k\|a\|$ and $\|a\|' \ge \frac{q(a)}{q(1)} \ge k^{-1}\|a\|$. Hence

$$k^{-1}\|a\| \le \|a\|' \le k\|a\|$$

for every $a \in \mathcal{A}$.

Clearly $\|\cdot\|'$ is a norm and $\|1\|' = 1$.

Let $a_1, a_2 \in \mathcal{A}$. We show that $\|a_1a_2\|' \le \|a_1\|'\|a_2\|'$. This is clear if $a_1a_2 = 0$. If $a_1a_2 \ne 0$, then

$$\begin{aligned}\|a_1a_2\|' &= \sup\Big\{\frac{q(a_1a_2x)}{q(x)} : x \ne 0\Big\} = \sup\Big\{\frac{q(a_1a_2x)}{q(x)} : x \ne 0, q(a_1a_2x) \ne 0\Big\}\\ &= \sup\Big\{\frac{q(a_1a_2x)}{q(a_2x)} \cdot \frac{q(a_2x)}{q(x)} : q(a_1a_2x) \ne 0\Big\} \le \|a_1\|'\|a_2\|'.\end{aligned}$$

Finally, for $s \in S$ we have

$$\|s\|' = \sup\Big\{\frac{q(sx)}{q(x)} : x \ne 0\Big\} \le 1. \qquad \square$$

Corollary 33. *Let $(\mathcal{A}, \|\cdot\|)$ be a Banach algebra and let $x \in \mathcal{A}$. Then*

$$r(x) = \inf\{\|x\|' : \|\cdot\|' \text{ is an equivalent algebra norm on } \mathcal{A}\}.$$

Proof. Since $r(x)$ does not depend on the choice of an equivalent algebra norm, we have $r(x) \le \|x\|'$ for every equivalent algebra norm $\|\cdot\|'$.

Conversely, let $\varepsilon > 0$. Consider the semigroup

$$S = \left\{\left(\frac{x}{r(x) + \varepsilon}\right)^n : n = 0, 1, \dots\right\}.$$

Then S is a bounded semigroup and, by Theorem 32, there exists an equivalent algebra norm $\|\cdot\|'$ on $\mathcal{A}$ such that $\|s\|' \le 1$ for each $s \in S$. In particular, $\|x\|' \le r(x) + \varepsilon$. This completes the proof. $\square$

Functional calculus

Let $p(z) = \sum_{i=0}^n \alpha_i z^i$ be a polynomial with coefficients $\alpha_i \in \mathbb{C}$. For $x \in \mathcal{A}$ we write $p(x) = \sum_{i=0}^n \alpha_i x^i$. It is clear that the mapping $p \mapsto p(x)$ is a homomorphism from the algebra of all polynomials to $\mathcal{A}$. The spectra of x and $p(x)$ and related in the following way:

Theorem 34. (spectral mapping theorem) *Let x be an element of a Banach algebra $\mathcal{A}$ and let p be a polynomial. Then $\sigma(p(x)) = p(\sigma(x))$.*

Proof. The equality is clear if p is a constant polynomial. Suppose p is non-constant and let $\lambda \in \mathbb{C}$. Then we can write $p(z) - \lambda = \beta(z - \alpha_1) \cdots (z - \alpha_n)$ for some $\beta, \alpha_1, \ldots, \alpha_n \in \mathbb{C}, \beta \neq 0, n \geq 1$. Clearly, $p(x) - \lambda = \beta(x - \alpha_1) \cdots (x - \alpha_n)$ and $p(x) - \lambda$ is non-invertible if and only if at least one of the factors $x - \alpha_i$ is non-invertible, i.e., if $\alpha_i \in \sigma(x)$ for some i. Thus $\lambda \in \sigma(p(x))$ if and only if $p(z) - \lambda = 0$ for some $z \in \sigma(x)$. Hence $\sigma(p(x)) = p(\sigma(x))$. □

In Banach algebras we can substitute an element $a \in \mathcal{A}$ not only to polynomials but also to functions analytic on a neighbourhood of the spectrum $\sigma(a)$.

If f is a function analytic on a disc $\{z : |z| < R\}$ where $R > r(a)$ and $f(z) = \sum_{i=0}^{\infty} \alpha_i z^i$ is the Taylor expansion of f, then the series $\sum_{i=0}^{\infty} \alpha_i a^i$ converges in $\mathcal{A}$ to an element denoted by $f(a)$.

If f is a function analytic only on a neighbourhood U of $\sigma(a)$, then we can define $f(a)$ by means of a Cauchy integral. We define

$$f(a) = \frac{1}{2\pi i} \int_{\Gamma} f(z)(z - a)^{-1} \mathrm{d}z,$$

where Γ is a contour surrounding $\sigma(a)$ in U, see Appendix A.2. The integral is well defined since the mapping $z \mapsto (z - a)^{-1}$ is continuous on Γ by Theorem 11. By the Cauchy formula, the integral does not depend on the choice of Γ. The definition coincides with the previous definition for polynomials:

Proposition 35. *Let a be an element of a Banach algebra $\mathcal{A}$, let Γ be a contour surrounding $\sigma(a)$. Let $p(z) = \sum_{j=0}^{n} \alpha_j z^j$ be a polynomial with complex coefficients α_j. Then*

$$\frac{1}{2\pi i} \int_{\Gamma} p(z)(z - a)^{-1} \mathrm{d}z = \sum_{j=0}^{n} \alpha_j a^j.$$

Proof. It is sufficient to show that

$$\frac{1}{2\pi i} \int_{\Gamma} z^k (z - a)^{-1} \mathrm{d}z = a^k$$

for all $k \geq 0$. For $R > r(a)$ we have

$$\frac{1}{2\pi i} \int_{\Gamma} z^k (z - a)^{-1} \mathrm{d}z = \frac{1}{2\pi i} \int_{|z|=R} z^k (z - a)^{-1} \mathrm{d}z = \frac{1}{2\pi i} \int_{|z|=R} z^k \sum_{j=0}^{\infty} \frac{a^j}{z^{j+1}} \mathrm{d}z$$

$$= \frac{1}{2\pi i} \int_{|z|=R} \left(z^{k-1} + a z^{k-2} + \cdots + a^{k-2} z + a^{k-1} + \sum_{j=0}^{\infty} \frac{a^{k+j}}{z^{j+1}} \right) \mathrm{d}z = a^k,$$

by the residue theorem. □

Proposition 36. *Let $a \in \mathcal{A}$, let Γ be a contour surrounding $\sigma(a)$ and let $\frac{p(z)}{q(z)}$ be a rational function such that no zero of q is surrounded by Γ. Then*

$$\frac{1}{2\pi i}\int_\Gamma \frac{p(z)}{q(z)}(z-a)^{-1}\mathrm{d}z = p(a)q(a)^{-1}$$

(note that $q(a)^{-1}$ exists by Theorem 34).

Proof. We first prove that for $k = 0, 1, \dots$ and for $\lambda \in \mathbb{C}$ not surrounded by Γ we have

$$\frac{1}{2\pi i}\int_\Gamma \frac{1}{(z-\lambda)^k}(z-a)^{-1}\mathrm{d}z = (a-\lambda)^{-k}. \tag{1}$$

For $k = 0$ this was proved in Proposition 35. Suppose that (1) is true for some $k \geq 0$. We have

$$\begin{aligned}(z-a)^{-1} - (\lambda-a)^{-1} &= (z-a)^{-1}\big((\lambda-a)-(z-a)\big)(\lambda-a)^{-1}\\ &= (\lambda-z)(z-a)^{-1}(\lambda-a)^{-1}.\end{aligned}$$

Thus $(z-a)^{-1} = (\lambda-a)^{-1} + (a-\lambda)^{-1}(z-\lambda)(z-a)^{-1}$ and

$$\begin{aligned}&\frac{1}{2\pi i}\int_\Gamma \frac{1}{(z-\lambda)^{k+1}}(z-a)^{-1}\mathrm{d}z\\ &= \frac{1}{2\pi i}\int_\Gamma \frac{1}{(z-\lambda)^{k+1}}(\lambda-a)^{-1}\mathrm{d}z + \frac{(a-\lambda)^{-1}}{2\pi i}\int_\Gamma \frac{1}{(z-\lambda)^k}(z-a)^{-1}\mathrm{d}z\\ &= (a-\lambda)^{-(k+1)}\end{aligned}$$

by the induction assumption (the first integral is equal to 0 since the function $z \mapsto \frac{1}{(z-\lambda)^{k+1}}$ is analytic inside Γ).

Let now $\frac{p(z)}{q(z)}$ be an arbitrary rational function, let $\lambda_1, \dots, \lambda_n$ be the roots of q of multiplicities $k_1, \dots, k_n$. Then $\frac{p(z)}{q(z)}$ can be expressed as

$$\frac{p(z)}{q(z)} = p_1(z) + \sum_{j=1}^n \sum_{s=1}^{k_j} \frac{c_{j,s}}{(z-\lambda_j)^s}$$

for some polynomial p_1 and complex numbers $c_{j,s}$. It is easy to verify that

$$p(a)q(a)^{-1} = p_1(a) + \sum_{j=1}^n \sum_{s=1}^{k_j} c_{j,s}(a-\lambda_j)^{-s},$$

and, by (1), we have

$$\frac{1}{2\pi i}\int_\Gamma \frac{p(z)}{q(z)}(z-a)^{-1}\mathrm{d}z = p_1(a) + \sum_{j=1}^n \sum_{s=1}^{k_j} c_{j,s}(a-\lambda_j)^{-s} = p(a)q(a)^{-1}. \qquad \square$$

For a non-empty compact set $K \subset \mathbb{C}$ denote by H_K the set of all functions analytic on a neighbourhood of K. We identify two such functions if they coincide on a neighbourhood of K. Thus, more precisely, H_K is the algebra of all germs of functions analytic on a neighbourhood of K.

Theorem 37 (functional calculus). *Let a be an element of a Banach algebra $\mathcal{A}$. Then there exists a homomorphism $f \mapsto f(a)$ from the algebra $H_{\sigma(a)}$ into $\mathcal{A}$ with the following properties:*

(i) *if $f(z) = \sum_{i=0}^n \alpha_i z^i$ is a polynomial with complex coefficients α_i, then $f(a) = \sum_{i=0}^n \alpha_i a^i$;*
(ii) *$f(a) \in \{a\}''$ for each f;*
(iii) *if U is a neighbourhood of $\sigma(a)$, f, f_k are analytic on U and $f_k \to f$ uniformly on U, then $f_k(a) \to f(a)$;*
(iv) *$\sigma\big(f(a)\big) = f\big(\sigma(a)\big)$.*

Properties (i) *and* (iii) *determine this homomorphism uniquely.*

Proof. Define

$$f(a) = \frac{1}{2\pi i}\int_\Gamma f(z)(z-a)^{-1}\mathrm{d}z,$$

where Γ is a contour surrounding $\sigma(a)$ in the domain of definition of f. The linearity of the mapping $f \mapsto f(a)$ is clear and (i) was proved in Proposition 35.

(ii) follows directly from the definition since $\{a\}''$ is a closed algebra containing $(z-a)^{-1}$ for every $z \in \Gamma$.

To prove (iii), we can replace Γ by a contour Γ' surrounding $\sigma(a)$ in U. Then $f_k(z) \to f(z)$ uniformly on Γ' and (iii) is clear.

By Proposition 36, $(f_1 f_2)(a) = f_1(a)f_2(a)$ if f_1, f_2 are rational functions with poles outside $\sigma(a)$. By the Runge theorem, any $f \in H_{\sigma(a)}$ can be approximated uniformly on some neighbourhood of $\sigma(a)$ by rational functions, so we conclude that the mapping $f \mapsto f(a)$ is multiplicative. Since property (i) determines $f(a)$ uniquely for rational functions f, we see that properties (i) and (iii) determine the functional calculus uniquely.

It remains to prove (iv). If $\lambda \notin f(\sigma(a))$, then $g(z) = (f(z)-\lambda)^{-1}$ is a function analytic on a neighbourhood of $\sigma(a)$. Thus $(f(a)-\lambda)g(a) = 1$ and $\lambda \notin \sigma(f(a))$.

Conversely, if $\lambda \in f(\sigma(a))$, then there exists $z_0 \in \sigma(a)$ with $f(z_0) = \lambda$ and $f(z) - \lambda = (z - z_0)g(z)$ for some function $g \in H_{\sigma(a)}$. Then $f(a) - \lambda = (a - z_0)g(a)$ and, since $a - z_0 \notin \operatorname{Inv}(\mathcal{A})$, we have $f(a) - \lambda \notin \operatorname{Inv}(\mathcal{A})$. Hence $\lambda \in \sigma(f(a))$. □

We mention at least one important corollary of the functional calculus for the algebra of operators:

Corollary 38. *Let T be an operator on a Banach space X, $\dim X \geq 1$. Suppose that U_1, U_2 are disjoint open subsets of $\mathbb{C}$ such that $\sigma(T) \subset U_1 \cup U_2$. Then there* *exist closed subspaces* $X_1, X_2 \subset X$ *such that* $X = X_1 \oplus X_2$, $TX_i \subset X_i$ *and* $\sigma(T|X_i) \subset U_i \quad (i = 1, 2)$.

Proof. Let $f : U_1 \cup U_2 \to \mathbb{C}$ be defined by $f|U_1 = 1$, $f|U_2 = 0$. Since $f^2 = f$, the operator $P = f(T)$ is a projection. Set $X_1 = PX$ and $X_2 = (I - P)X$. Clearly, X_1 and X_2 are invariant with respect to T and $X = X_1 \oplus X_2$. Write $T_i = T|X_i \quad (i = 1, 2)$. We show that $\sigma(T_1) \subset U_1$. Let $\lambda \notin U_1$ and define $g : U_1 \cup U_2 \to \mathbb{C}$ by $g(z) = (z - \lambda)^{-1} \quad (z \in U_1)$, $g|U_2 = 0$. Then $(z - \lambda)gf = f$, and so $(T - \lambda)g(T)P = P$. We have $g(T)X_1 = g(T)PX = Pg(T)X \subset X_1$, and so the restriction of the last equality to X_1 gives $(T_1 - \lambda)g(T)|X_1 = I_{X_1}$, where I_{X_1} denotes the identity operator on X_1. Thus $\lambda \notin \sigma(T_1)$ and $\sigma(T_1) \subset U_1$. Similarly, $\sigma(T_2) \subset U_2$. □

The spaces X_1, X_2 given in the preceding corollary are called the *spectral subspaces* of T corresponding to U_1 and U_2, respectively.

Radical

We finish this section with the basic properties of a radical.

Definition 39. Let $\mathcal{A}$ be a Banach algebra, let $J \subset \mathcal{A}$ be a left ideal. We say that J is a *maximal left ideal* if J is proper and if the only proper left ideal containing J is J itself.

Similarly we define maximal right ideals.

Theorem 40. *Let* $\mathcal{A}$ *be a Banach algebra. Then:*

(i) *the closure of a proper left ideal is a proper left ideal;*

(ii) *every proper left ideal is contained in a maximal left ideal;*

(iii) *every maximal left ideal is closed.*

Proof. (i) If $J \subset \mathcal{A}$ is a proper left ideal, then $\mathrm{Inv}(\mathcal{A}) \cap J = \emptyset$. By Theorem 11, $\mathrm{dist}\{1_{\mathcal{A}}, J\} \geq 1$, and so $1_{\mathcal{A}} \notin \overline{J}$. Hence $\overline{J}$ is proper.

(ii) is an easy application of the Zorn lemma and (iii) follows from (i). □

Theorem 41. *Let* $\mathcal{A}$ *be a Banach algebra. The following sets are identical:*

(i) *the intersection of all maximal left ideals in* $\mathcal{A}$*;*

(ii) *the intersection of all maximal right ideals in* $\mathcal{A}$*;*

(iii) *the set of all* $x \in \mathcal{A}$ *such that* $1 - ax$ *is invertible for every* $a \in \mathcal{A}$*;*

(iv) *the set of all* $x \in \mathcal{A}$ *such that* $1 - xa$ *is invertible for every* $a \in \mathcal{A}$.

Proof. By Theorem 29, the sets described in (iii) and (iv) are equal. It is sufficient to show the equivalence of (i) and (iii) since the equivalence of (ii) and (iv) can be proved similarly.

Suppose that $1-ax$ is invertible for all $a \in \mathcal{A}$ and let J be a maximal left ideal such that $x \notin J$. Then $J+\mathcal{A}x$ is a left ideal containing J, and so $J+\mathcal{A}x=\mathcal{A}$. Thus there exists $a \in \mathcal{A}$ such that $1-ax \in J$. Since $1-ax \in \operatorname{Inv}(\mathcal{A})$, we conclude that $1 \in J$, a contradiction.

In the opposite direction, let x be in the intersection of all maximal left ideals of $\mathcal{A}$. Suppose that there exists $a \in \mathcal{A}$ such that $1-ax$ is not invertible. Thus $\sigma(ax) \neq \{0\}$ and let $\lambda \in \sigma(ax)$ satisfy $|\lambda| = r(ax) > 0$. By Theorem 14, $\lambda - ax$ is a left topological divisor of zero, and so $\lambda - ax$ is not left invertible. Consequently, $\mathcal{A}(\lambda - ax)$ is a proper left ideal, and so there exists a maximal left ideal $J \supset \mathcal{A}(\lambda - ax)$. We have $\lambda - ax \in J$ and $x \in J$, and so $1 = \lambda^{-1}(\lambda - ax) + \lambda^{-1}ax \in J$, a contradiction. □

Definition 42. The set of all x with properties (i)–(iv) of the previous theorem is called the *radical* of $\mathcal{A}$ and denoted by $\operatorname{rad}\mathcal{A}$. Evidently, $\operatorname{rad}\mathcal{A}$ is a closed two-sided ideal of $\mathcal{A}$.

An algebra $\mathcal{A}$ is called *semisimple* if $\operatorname{rad}\mathcal{A} = \{0\}$.

Theorem 43. *Let $\mathcal{A}$ be a Banach algebra. Then:*

(i) *$\mathcal{A}/\operatorname{rad}\mathcal{A}$ is semisimple;*

(ii) *an element $x \in \mathcal{A}$ is invertible in $\mathcal{A}$ if and only if $x + \operatorname{rad}\mathcal{A}$ is invertible in $\mathcal{A}/\operatorname{rad}\mathcal{A}$;*

(iii) *if $x \in \operatorname{rad}\mathcal{A}$, then $\sigma(x) = \{0\}$.*

Proof. (i) Denote by $\rho : \mathcal{A} \to \mathcal{A}/\operatorname{rad}\mathcal{A}$ the canonical projection. If $x \in \mathcal{A}$, $x \notin \operatorname{rad}\mathcal{A}$, then there exists a maximal left ideal J with $x \notin J$. Since $\operatorname{rad}\mathcal{A} \subset J$, it is easy to check that $J + \operatorname{rad}\mathcal{A} = \rho(J)$ is a maximal left ideal in $\mathcal{A}/\operatorname{rad}\mathcal{A}$ and $\rho(x) = x + \operatorname{rad}\mathcal{A} \notin \rho(J)$. Thus $x + \operatorname{rad}\mathcal{A} \notin \operatorname{rad}(\mathcal{A}/\operatorname{rad}\mathcal{A})$.

Since x was an arbitrary element in $\mathcal{A}\setminus\operatorname{rad}\mathcal{A}$, the algebra $\mathcal{A}/\operatorname{rad}\mathcal{A}$ is semisimple.

(ii) If $x \in \operatorname{Inv}(\mathcal{A})$, then $\rho(x) \in \operatorname{Inv}(\mathcal{A}/\operatorname{rad}\mathcal{A})$ by Theorem 26. Conversely, if $\rho(x) \in \operatorname{Inv}(\mathcal{A}/\operatorname{rad}\mathcal{A})$, then there exists $y \in \mathcal{A}$ such that $xy \in 1+\operatorname{rad}\mathcal{A}$, $yx \in 1+\operatorname{rad}\mathcal{A}$. By Theorem 41 (iii) and (iv), the elements $1+1(xy-1) = xy$ and $1+1(yx-1) = yx$ are invertible. Hence $x \in \operatorname{Inv}(\mathcal{A})$.

(iii) Let $x \in \operatorname{rad}\mathcal{A}$ and $\lambda \neq 0$. Then $\lambda - x = \lambda(1-\lambda^{-1}x)$, which is invertible by Theorem 41 (iii). □

Theorem 44. *$\mathcal{B}(X)$ is semisimple for every Banach space X with $\dim X \geq 1$.*

Proof. Let $T \in \mathcal{B}(X)$, $T \neq 0$. Then $Tx \neq 0$ for some non-zero $x \in X$. Choose $g \in X^*$ such that $g(Tx) = 1$ and define $S \in \mathcal{B}(X)$ by $Sy = g(y)\cdot x \quad (y \in X)$. Then $STx = x$, and so $1 \in \sigma(ST)$. By Theorem 41, $T \notin \operatorname{rad}(\mathcal{B}(X))$. Hence $\mathcal{B}(X)$ is semisimple. □

2 Commutative Banach algebras

In this section we give a survey of basic results of the theory of commutative Banach algebras.

The most important example of a commutative Banach algebra is the algebra $C(K)$ of all complex-valued continuous functions defined on a non-empty compact space K with the sup-norm $\|f\| = \sup\{|f(z)| : z \in K\}$. For further examples see 1.4 (iii), (iv), (vi).

In commutative algebras the notions of left, right and two-sided ideals coincide. In the same way, the notions of maximal left (right) ideals and left (right) topological divisors of zero coincide. We are going to speak only about ideals, maximal ideals and topological divisors of zero.

Theorem 1. *Every maximal ideal in a commutative Banach algebra is closed and of codimension* 1.

Proof. Let J be a maximal ideal in $\mathcal{A}$. By Theorem 1.40, J is closed. Furthermore, $\mathcal{A}/J$ is a commutative Banach algebra with no non-trivial ideals, and so every non-zero element of $\mathcal{A}/J$ is invertible. By Corollary 1.19, $\dim \mathcal{A}/J = 1$. □

Definition 2. Let $\mathcal{A}$ be a commutative Banach algebra. A linear functional $\varphi : \mathcal{A} \to \mathbb{C}$ is called *multiplicative*if $\varphi(1_{\mathcal{A}}) = 1$ and $\varphi(xy) = \varphi(x)\varphi(y)$ for all $x, y \in \mathcal{A}$ (in other words, φ is a homomorphism).

Multiplicative functionals are in 1-1 correspondence with maximal ideals.

Theorem 3. *Let $\mathcal{A}$ be a commutative Banach algebra. Then:*

(i) *if φ is a multiplicative functional on $\mathcal{A}$, then* $\operatorname{Ker} \varphi$ *is a maximal ideal;*

(ii) *if $J \subset \mathcal{A}$ is a maximal ideal in $\mathcal{A}$, then $\mathcal{A} = \{x + \lambda \cdot 1_{\mathcal{A}} : x \in J, \lambda \in \mathbb{C}\}$ and the mapping $\varphi : \mathcal{A} \to \mathbb{C}$ defined by $\varphi(x + \lambda \cdot 1_{\mathcal{A}}) = \lambda$ is a multiplicative functional. Clearly,* $\operatorname{Ker} \varphi = J$.

Proof. An easy verification. □

The set of all multiplicative functionals on a commutative Banach algebra $\mathcal{A}$ will be denoted by $\mathcal{M}(\mathcal{A})$. As the multiplicative functionals are in 1-1 correspondence with the maximal ideals, multiplicative functionals are frequently identified with the corresponding maximal ideals (we are not going to use this convention). From this reason $\mathcal{M}(\mathcal{A})$ is usually called the *maximal ideal space*.

Theorem 4. *Let x be an element of a commutative Banach algebra $\mathcal{A}$. Then:*

(i) *if $\varphi \in \mathcal{M}(\mathcal{A})$, then $\varphi(x) \in \sigma(x)$;*

(ii) *if $\lambda \in \sigma(x)$, then there exists $\varphi \in \mathcal{M}(\mathcal{A})$ such that $\varphi(x) = \lambda$;*

(iii) *an element $x \in \mathcal{A}$ is invertible if and only if $\varphi(x) \neq 0$ for every $\varphi \in \mathcal{M}(\mathcal{A})$.*

Proof. (i) If $\varphi \in \mathcal{M}(\mathcal{A})$, then $x - \varphi(x) \cdot 1_{\mathcal{A}} \in \operatorname{Ker} \varphi$. Since $\operatorname{Ker} \varphi$ is a proper ideal, the element $x - \varphi(x) \cdot 1_{\mathcal{A}}$ is not invertible, and so $\varphi(x) \in \sigma(x)$.

(ii) If $\lambda \in \sigma(x)$, then $x - \lambda$ is contained in a proper ideal, and so there exists a maximal ideal containing $x - \lambda$. The corresponding multiplicative functional φ then satisfies $\varphi(x - \lambda) = 0$, and so $\varphi(x) = \lambda$.

(iii) Follows from (i) and (ii). □

Corollary 5. *Let $\mathcal{A}$ be a commutative Banach algebra. Then $\mathcal{M}(\mathcal{A}) \neq \emptyset$.*

Theorem 6. *Let φ be a multiplicative functional on a commutative Banach algebra $\mathcal{A}$. Then φ is continuous and $\|\varphi\| = 1$.*

Proof. For $x \in \mathcal{A}$ we have $\varphi(x) \in \sigma(x)$, and so $|\varphi(x)| \leq r(x) \leq \|x\|$. Thus $\|\varphi\| \leq 1$, and since $\varphi(1_{\mathcal{A}}) = 1$, we have $\|\varphi\| = 1$. □

We consider the topology of pointwise convergence of multiplicative functionals on $M(\mathcal{A})$. The base of open neighbourhoods of a multiplicative functional $\varphi \in \mathcal{M}(\mathcal{A})$ is formed by the sets

$$U_{x_1,\dots,x_n,\varepsilon} = \{\psi \in \mathcal{M}(\mathcal{A}) : |\psi(x_i) - \varphi(x_i)| < \varepsilon, i = 1, \dots, n\},$$

where $x_1, \dots, x_n \in \mathcal{A}, \varepsilon > 0$.

Theorem 7. *$\mathcal{M}(\mathcal{A})$ with the above-defined topology is a non-empty compact Hausdorff space.*

Proof. $\mathcal{M}(\mathcal{A})$ is non-empty by Corollary 5 and Hausdorff by definition. Since the closed unit ball $B_{\mathcal{A}^*}$ of $\mathcal{A}^*$ is compact in the w^*-topology and $\mathcal{M}(\mathcal{A})$ is a w^*-closed subset of $B_{\mathcal{A}^*}$, we conclude that $\mathcal{M}(\mathcal{A})$ is compact. □

Consider the Banach algebra $C(\mathcal{M}(\mathcal{A}))$ of all complex-valued continuous functions on the compact space $\mathcal{M}(\mathcal{A})$ with the sup-norm.

Definition 8. The mapping $G : \mathcal{A} \to C(\mathcal{M}(\mathcal{A}))$ defined by

$$G(a)(\varphi) = \varphi(a) \qquad (a \in \mathcal{A}, \varphi \in \mathcal{M}(\mathcal{A}))$$

is called the *Gelfand transform*.

The Gelfand transform has the following properties:

Theorem 9. *Let $\mathcal{A}$ be a commutative Banach algebra. Then:*

(i) *$G : \mathcal{A} \to C(\mathcal{M}(\mathcal{A}))$ is a continuous homomorphism, $\|G\| = 1$;*
(ii) *$\|G(a)\| = r(a) \quad (a \in \mathcal{A})$;*
(iii) *$G(a)(\mathcal{M}(\mathcal{A})) = \sigma(a) \quad (a \in \mathcal{A})$;*
(iv) *$G(a) = 0 \iff \sigma(a) = \{0\} \iff a \in \operatorname{rad} \mathcal{A}$.*

Proof. (iii) We have $\{G(a)(\varphi) : \varphi \in \mathcal{M}(\mathcal{A})\} = \{\varphi(a) : \varphi \in \mathcal{M}(\mathcal{A})\} = \sigma(a)$.

(ii) By (iii), $\|G(a)\| = \sup\{|G(a)(\varphi)| : \varphi \in \mathcal{M}(\mathcal{A})\} = \sup\{|\lambda| : \lambda \in \sigma(a)\} = r(a)$.

(i) By (ii), $\|G\| \le 1$. Since $\|G(1_{\mathcal{A}})\| = r(1_{\mathcal{A}}) = 1$, we have $\|G\| = 1$.

(iv) The first equivalence follows from (ii). An element a belongs to the radical if and only if it is contained in every maximal ideal, i.e., if $\varphi(a) = 0$ for every $\varphi \in \mathcal{M}(\mathcal{A})$. This means that $\sigma(a) = \{0\}$. □

The spectrum in commutative Banach algebras has a number of nice properties. The following result means the continuity of the spectrum.

Theorem 10. *Let $\mathcal{A}$ be a commutative Banach algebra, $x, x_k \in \mathcal{A}$ $(k \in \mathbb{N})$, $x_k \to x$. Then $\lambda \in \sigma(x)$ if and only if there exist points $\lambda_k \in \sigma(x_k)$ $(k = 1, 2, \dots)$ such that $\lambda = \lim_{k\to\infty} \lambda_k$.*

Proof. If $\lambda \in \sigma(x)$, then there exists $\varphi \in \mathcal{M}(\mathcal{A})$ with $\varphi(x) = \lambda$. Set $\lambda_k = \varphi(x_k)$. Then $\lambda_k \in \sigma(x_k)$ and $\lambda_k \to \lambda$.

Conversely, let $\lambda_k \in \sigma(x_k)$ and $\lambda_k \to \lambda$. For each k there exists $\varphi_k \in \mathcal{M}(\mathcal{A})$ with $\varphi_k(x_k) = \lambda_k$. Set $\mu_k = \varphi_k(x) \in \sigma(x)$. Then

$$|\lambda - \mu_k| \le |\lambda - \lambda_k| + |\lambda_k - \mu_k| = |\lambda - \lambda_k| + |\varphi_k(x_k) - \varphi_k(x)| \le |\lambda - \lambda_k| + \|x_k - x\| \to 0.$$

Thus $\mu_k \to \lambda$ and $\lambda \in \sigma(x)$. □

Theorem 11. *Let $\mathcal{A}$ be a Banach algebra, $x, y \in A$, $xy = yx$. Then:*

(i) *$\sigma(xy) \subset \sigma(x) \cdot \sigma(y)$ and $\sigma(x + y) \subset \sigma(x) + \sigma(y)$;*

(ii) *$r(xy) \le r(x) \cdot r(y)$ and $r(x + y) \le r(x) + r(y)$.*

In particular, the spectral radius in a commutative Banach algebra is an algebra seminorm.

Proof. Clearly, (ii) is a consequence of (i).

To prove (ii), suppose first that $\mathcal{A}$ is a commutative Banach algebra. Then

$$\begin{aligned}\sigma(xy) &= \{\varphi(xy) : \varphi \in \mathcal{M}(\mathcal{A})\} = \{\varphi(x)\varphi(y) : \varphi \in \mathcal{M}(\mathcal{A})\} \\ &\subset \{\varphi(x)\psi(y) : \varphi, \psi \in \mathcal{M}(\mathcal{A})\} = \sigma(x) \cdot \sigma(y).\end{aligned}$$

In the same way, $\sigma(x + y) \subset \sigma(x) + \sigma(y)$.

In general, write $\mathcal{A}_0 = \{x, y\}''$. By Lemma 1.24, $\mathcal{A}_0$ is a commutative Banach algebra and it is easy to check that $\sigma^{\mathcal{A}_0}(x) = \sigma^{\mathcal{A}}(x)$, $\sigma^{\mathcal{A}_0}(y) = \sigma^{\mathcal{A}}(y)$, $\sigma^{\mathcal{A}_0}(xy) = \sigma^{\mathcal{A}}(xy)$ and $\sigma^{\mathcal{A}_0}(x+y) = \sigma^{\mathcal{A}}(x+y)$. Thus the result follows from the corresponding inclusions for the commutative Banach algebra $\mathcal{A}_0$. □

The following result characterizes multiplicative functionals.

Theorem 12. (Gleason-Kahane-Żelazko) *Let $\mathcal{A}$ be a commutative Banach algebra and let $\varphi : \mathcal{A} \to \mathbb{C}$ be a linear functional. Then φ is multiplicative if and only if $\varphi(x) \in \sigma(x)$ for every $x \in \mathcal{A}$.*

Proof. If $\varphi \in \mathcal{M}(\mathcal{A})$, then $\varphi(x) \in \sigma(x)$ for every $x \in \mathcal{A}$ by Theorem 4.

Suppose that φ is a linear functional satisfying $\varphi(x) \in \sigma(x)$ for all $x \in \mathcal{A}$. Then $\varphi(1_{\mathcal{A}}) = 1$. First, we prove the implication $\varphi(a) = 0 \Rightarrow \varphi(a^2) = 0$ for all $a \in \mathcal{A}$.

Let $\varphi(a) = 0$, $n \geq 2$ and denote by p the polynomial $p(\lambda) = \varphi((\lambda - a)^n)$. Let $\lambda_1, \dots, \lambda_n$ be the roots of p. Since $0 = p(\lambda_i) = \varphi((\lambda_i - a)^n)$, we have $(\lambda_i - a)^n \notin \operatorname{Inv}(\mathcal{A})$, and so $\lambda_i - a \notin \operatorname{Inv}(\mathcal{A})$. Thus $\lambda_i \in \sigma(a)$ for $i = 1, \dots, n$. We can write

$$\begin{aligned} p(\lambda) = (\lambda - \lambda_1) \cdots (\lambda - \lambda_n) &= \lambda^n - \lambda^{n-1} \sum_{i=0}^{n} \lambda_i + \lambda^{n-2} \sum_{i \neq j} \lambda_i \lambda_j + \cdots \\ &= \lambda^n - \lambda^{n-1} n\varphi(a) + \lambda^{n-2} \frac{n(n-1)}{2} \varphi(a^2) + \cdots . \end{aligned}$$

Thus we have $\sum \lambda_i = n\varphi(a) = 0$ and $\sum_{i \neq j} \lambda_i \lambda_j = \frac{n(n-1)}{2} \varphi(a^2)$. We have $0 = \left(\sum \lambda_i\right)^2 = \sum \lambda_i^2 + 2 \sum_{i \neq j} \lambda_i \lambda_j$, and so

$$|\varphi(a^2)| = \frac{|2 \sum_{i \neq j} \lambda_i \lambda_j|}{n(n-1)} = \frac{\sum \lambda_i^2}{n(n-1)} \leq \frac{n(r(a))^2}{n(n-1)} = \frac{(r(a))^2}{n-1}.$$

Letting $n \to \infty$ yields $\varphi(a^2) = 0$.

Let $a \in \mathcal{A}$. Then $\varphi\big(a - \varphi(a) \cdot 1_{\mathcal{A}}\big) = 0$, and so $\varphi\big(a^2 - 2a\varphi(a) + (\varphi(a))^2\big) = 0$, which implies $\varphi(a^2) = (\varphi(a))^2$.

Consequently, for $x, y \in \mathcal{A}$ we have

$$\begin{aligned} &(\varphi(x))^2 + 2\varphi(x)\varphi(y) + \big(\varphi(y)\big)^2 = \big(\varphi(x) + \varphi(y)\big)^2 = \big(\varphi(x+y)\big)^2 \\ &= \varphi((x+y)^2) = \varphi(x^2) + 2\varphi(xy) + \varphi(y^2), \end{aligned}$$

and so $\varphi(xy) = \varphi(x)\varphi(y)$. □

The Gelfand transform commutes with the functional calculus which was introduced in the previous section.

Theorem 13. *Let $\mathcal{A}$ be a commutative Banach algebra, $x \in \mathcal{A}$ and let f be a function analytic on a neighbourhood of $\sigma(x)$. Then:*

(i) *$\varphi(f(x)) = f(\varphi(x))$ for all $\varphi \in \mathcal{M}(\mathcal{A})$;*

(ii) *$G(f(x)) = f(G(x))$.*

Proof. (i) We have $f(x) = \frac{1}{2\pi i}\int_\Gamma f(z)(z-x)^{-1}\mathrm{d}z$, where Γ is a contour surrounding $\sigma(x)$. Since the integral is defined as a limit of Riemann's sums and φ is continuous and multiplicative, we have

$$\varphi(f(x)) = \frac{1}{2\pi i}\int_\Gamma f(z)\big(z-\varphi(x)\big)^{-1}\mathrm{d}z = f(\varphi(x)).$$

The second statement follows from the definition of the Gelfand transform. □

In commutative Banach algebras it is possible to introduce the notion of spectrum for n-tuples of elements.

Definition 14. Let $\mathcal{A}$ be a commutative Banach algebra, $x_1,\dots,x_n \in \mathcal{A}$. The *spectrum* $\sigma(x_1,\dots,x_n)$ is the set

$$\sigma(x_1,\dots,x_n) = \big\{(\varphi(x_1)\dots\varphi(x_n)) : \varphi \in \mathcal{M}(\mathcal{A})\big\}.$$

Theorem 15. *Let $x_1,\dots,x_n$ be elements of a commutative Banach algebra $\mathcal{A}$. Then:*

(i) *The spectrum $\sigma(x_1,\dots,x_n)$ is a non-empty compact subset of $\mathbb{C}^n$;*
(ii) *$\lambda \in \sigma(x_1,\dots,x_n)$ if and only if the ideal $(x_1-\lambda_1)\mathcal{A}+\dots+(x_n-\lambda_n)\mathcal{A}$ is proper, i.e., if $1_\mathcal{A} \notin (x_1-\lambda_1)\mathcal{A}+\dots+(x_n-\lambda_n)\mathcal{A}$;*
(iii) *if $m<n$, then $\sigma(x_1,\dots,x_m) = P\sigma(x_1,\dots,x_n)$, where $P:\mathbb{C}^n\to\mathbb{C}^m$ is the natural projection onto the first m coordinates.*

Proof. (i) Since $\mathcal{M}(\mathcal{A}) \neq \emptyset$, the spectrum $\sigma(x_1,\dots,x_n)$ is also non-empty. The mapping $\varphi \mapsto (\varphi(x_1,\dots,\varphi(x_n))$ from $\mathcal{M}(\mathcal{A})$ onto $\sigma(x_1,\dots,x_n)$ is continuous, and so $\sigma(x_1,\dots,x_n)$ is compact.

(ii) If $(\lambda_1,\dots,\lambda_n) \in \sigma(x_1,\dots,x_n)$, then there exists $\varphi \in \mathcal{M}(\mathcal{A})$ such that $\varphi(x_i) = \lambda_i \quad (i=1,\dots,n)$. Then $(x_1-\lambda_1)\mathcal{A}+\dots+(x_n-\lambda_n)\mathcal{A}$ is contained in $\operatorname{Ker}\varphi$, and so it is a proper ideal.

Conversely, if $(x_1-\lambda_1)\mathcal{A}+\dots+(x_n-\lambda_n)\mathcal{A}$ is a proper ideal, then it is contained in a maximal ideal. The corresponding multiplicative functional φ satisfies $\varphi(x_i)=\lambda_i \quad (i=1,\dots,n)$, and so $(\lambda_1,\dots,\lambda_n) \in \sigma(x_1,\dots,x_n)$.

(iii) Clear. □

Theorem 16. *Let $\mathcal{A}$ be a commutative Banach algebra with a finite number of generators $x_1,\dots,x_n$ (i.e., $\mathcal{A}$ is the smallest closed algebra containing the elements $x_1,\dots,x_n$). Then $\mathcal{M}(\mathcal{A})$ is homeomorphic to $\sigma(x_1,\dots,x_n)$.*

Proof. Consider the mapping $\Psi : \mathcal{M}(\mathcal{A}) \to \sigma(x_1,\dots,x_n)$ defined by

$$\Psi(\varphi) = (\varphi(x_1),\dots,\varphi(x_n)) \qquad (\varphi \in \mathcal{M}(\mathcal{A})).$$

Clearly, Ψ is continuous and onto. It is sufficient to show that Ψ is one-to-one. If $\varphi_1,\varphi_2 \in \mathcal{M}(\mathcal{A})$ with $\Psi(\varphi_1)=\Psi(\varphi_2)$, then $\varphi_1(x_i)=\varphi_2(x_i)$ for all $i=1,\dots,n$. Since $\mathcal{A}$ is generated by $x_1,\dots,x_n$, we have $\varphi_1=\varphi_2$. □

Denote by $\mathcal{P}(n)$ the set of all complex polynomials in n variables. If $p \in \mathcal{P}(n)$ and $x_1, \dots, x_n \in \mathcal{A}$, then we define $p(x_1, \dots, x_n)$ in the natural way.

Definition 17. Let K be a non-empty compact subset of $\mathbb{C}^n$. The *polynomially convex hull* of K is the set

$$\widehat{K} = \Big\{ z \in \mathbb{C}^n : |p(z)| \le \max\{|p(w)| : w \in K\} \text{ for every polynomial } p \in \mathcal{P}(n) \Big\}.$$

A set K is called *polynomially convex* if $\widehat{K} = K$.

In the case $n = 1$ there is a simple characterization of polynomially convex sets: a non-empty compact subset $K \subset \mathbb{C}$ is polynomially convex if and only if $\mathbb{C} \setminus K$ is a connected set. Thus the polynomially convex hull of a compact subset K of $\mathbb{C}$ is the union of K with the bounded components of $\mathbb{C} \setminus K$.

Theorem 18. *Let $\mathcal{A}$ be a commutative Banach algebra with a finite number of generators $x_1, \dots, x_n$. Then $\sigma(x_1, \dots, x_n)$ is a polynomially convex subset of $\mathbb{C}^n$.*

Proof. Fix $\lambda = (\lambda_1, \dots, \lambda_n)$ in the polynomially convex hull of $\sigma(x_1, \dots, x_n)$. Then

$$\begin{aligned} |p(\lambda)| &\le \sup\big\{|p(z_1, \dots, z_n)| : z_1, \dots, z_n \in \sigma(x_1, \dots, x_n)\big\} \\ &= \sup\big\{|p(\varphi(x_1), \dots, \varphi(x_n))| : \varphi \in \mathcal{M}(\mathcal{A})\big\} \\ &= \sup\big\{|\varphi(p(x_1, \dots, x_n))| : \varphi \in \mathcal{M}(\mathcal{A})\big\} = r(p(x_1, \dots, x_n)) \\ &\le \|p(x_1, \dots, x_n)\| \end{aligned}$$

for each polynomial $p \in \mathcal{P}(n)$. Set $\mathcal{A}_0 = \big\{p(x_1, \dots, x_n) : p \in \mathcal{P}(n)\big\}$ and let $\psi : \mathcal{A}_0 \to \mathbb{C}$ be defined by $\psi : p(x_1, \dots, x_n) \mapsto p(\lambda)$. Since $|\psi(y)| \le \|y\|$ for all $y \in \mathcal{A}_0$, the definition of ψ is correct and ψ can be uniquely extended to a multiplicative functional (denoted also by ψ) on $\overline{\mathcal{A}_0} = \mathcal{A}$. Thus $(\lambda_1, \dots, \lambda_n) = (\psi(x_1), \dots, \psi(x_n)) \in \sigma(x_1, \dots, x_n)$, and so $\sigma(x_1, \dots, x_n)$ is polynomially convex. □

In particular, if $\mathcal{A}$ is generated by a single element x, then $\sigma^{\mathcal{A}}(x)$ has no holes ($\mathbb{C} \setminus \sigma^{\mathcal{A}}(x)$ is connected).

Let $\mathcal{A}$ be a Banach algebra and $x_1, \dots, x_n \in \mathcal{A}$. We denote by $\langle x_1, \dots, x_n \rangle$ the closed subalgebra generated by the elements $x_1, \dots, x_n$. By definition, $\langle x_1, \dots, x_n \rangle$ contains the unit of $\mathcal{A}$.

Examples 19.

(i) Let K be a non-empty compact Hausdorff space and $C(K)$ the algebra of all continuous functions on K with the sup-norm. It is not difficult to show that the multiplicative functionals on $C(K)$ are precisely the evaluations $\mathcal{E}_\lambda$ $(\lambda \in K)$ defined by $\mathcal{E}_\lambda(f) = f(\lambda)$. Thus $\mathcal{M}(C(K))$ is homeomorphic to K and the Gelfand transform is the identical mapping.
For $f \in C(K)$, we have $\sigma^{C(K)}(f) = f(K)$.

(ii) If K is a non-empty compact subset of $\mathbb{C}^n$ and $\mathcal{P}(K)$ the uniform closure on K of all polynomials, then $\mathcal{M}(\mathcal{P}(K))$ can be identified with the polynomially convex hull $\widehat{K}$.
This applies also to the disc algebra $A(\mathbb{D})$, see Example 1.4 (iv), since $A(\mathbb{D}) = \mathcal{P}(\overline{\mathbb{D}})$.

(iii) Consider the Banach algebra H^∞ of all bounded analytic functions on the open unit disc. The maximal ideal space $\mathcal{M}(H^\infty)$ is quite large and complicated. The celebrated corona theorem of Carleson says: if $f_1, \dots, f_n \in H^\infty$ and $\inf_{z \in \mathbb{D}} \sum_{i=1}^n |f_i(z)| > 0$, then $\sum f_i g_i = 1$ for some $g_i \in H^\infty$. This can be reformulated as follows: the set of all evaluations $\mathcal{E}_\lambda \quad (\lambda \in \mathbb{D})$ is dense in $\mathcal{M}(H^\infty)$.
Thus $\sigma^{H^\infty}(f) = \overline{f(\mathbb{D})}$ for $f \in H^\infty$.

(iv) Let $\mathbb{T} = \{z \in \mathbb{C} : |z| = 1\}$ be the unit torus and let $\mathcal{A}$ be the algebra of all continuous functions on $\mathbb{T}$ with absolutely convergent Fourier series (i.e., $f(z) = \sum_{-\infty}^{\infty} \alpha_i z^i$ where $\sum |\alpha_i| < \infty$). It is easy to see that $\mathcal{M}(\mathcal{A})$ coincides with $\mathbb{T}$. Thus the Gelfand theory gives a very simple proof of the following Wiener theorem: if $f \in \mathcal{A}$ and $f(z) \neq 0 \quad (z \in \mathbb{T})$, then $1/f \in \mathcal{A}$.

(v) Multiplicative functionals on L^1 (see Example 1.4 (vi)) are of the form

$$f \mapsto \int_{-\infty}^{\infty} f(x) e^{itx} \, dx$$

where $t \in \mathbb{R}$. Thus $\mathcal{M}(L^1 \oplus \mathbb{C})$ coincides with the one-point compactification of $\mathbb{R}$. The Gelfand transform is closely connected with the Fourier transform.

In commutative Banach algebras it is possible to extend the functional calculus to analytic functions of n-variables. Here we formulate the result without proof since it will be an easy consequence of the more general Taylor functional calculus that will be discussed later. Recall that for a non-empty compact subset of $K \subset \mathbb{C}^n$ we denote by H_K the algebra of all functions analytic on a neighbourhood of K (more precisely, H_K is the algebra of all germs of functions analytic on a neighbourhood of K).

Theorem 20. *Let $\mathcal{A}$ be a commutative Banach algebra. To each finite family $a = (a_1, \dots, a_n)$ of elements of $\mathcal{A}$ and each function $f \in H_{\sigma(a)}$ it is possible to assign an element $f(a_1, \dots, a_n) \in \mathcal{A}$ such that the following conditions are satisfied:*

(i) *if $p(z_1, \dots, z_n) = \sum_{\alpha \in \mathbb{Z}_+^n} c_\alpha z_1^{\alpha_1} \cdots z_n^{\alpha_n}$ is a polynomial in n variables with complex coefficients c_α, then $p(a) = \sum_{\alpha \in \mathbb{Z}_+^n} c_\alpha a_1^{\alpha_1} \cdots a_n^{\alpha_n}$;*

(ii) *the mapping $f \mapsto f(a_1, \dots, a_n)$ is an algebra homomorphism from the algebra $H_{\sigma(a)}$ to $\mathcal{A}$;*

(iii) *if $\varphi \in \mathcal{M}(\mathcal{A})$ and $f \in H_{\sigma(a)}$, then*

$$\varphi\big(f(a_1, \dots, a_n)\big) = f\big(\varphi(a_1), \dots, \varphi(a_n)\big);$$

(iv) $\sigma\big(f(a_1,\dots,a_n)\big) = f\big(\sigma(a_1,\dots,a_n)\big)$;

(v) *if* U *is a neighbourhood of* $\sigma(a_1,\dots,a_n)$, f, f_k $(k \in \mathbb{N})$ *are functions analytic on* U *and* f_k *converge to* f *uniformly on* U, *then*

$$f_k(a_1,\dots,a_n) \to f(a_1,\dots,a_n).$$

We finish this section with the basic properties of the Shilov boundary.

Let K be a compact Hausdorff space. For a non-empty subset $M \subset K$ and a continuous function $f : K \to \mathbb{C}$ write $\|f\|_M = \sup\{|f(z)| : z \in M\}$.

Definition 21. Let K be a non-empty compact Hausdorff space, let $\mathcal{A} \subset C(K)$ be an algebra of continuous functions containing the constant functions and separating the points of K. The *Shilov boundary* $\Gamma(K,\mathcal{A})$ is the subset of all points $x \in K$ with the following property: for every neighbourhood U of x there exists a function $f \in \mathcal{A}$ such that $\|f\|_U > \|f\|_{K\setminus U}$.

Theorem 22. *Let* K *be a non-empty compact Hausdorff space, let* $\mathcal{A} \subset C(K)$ *be an algebra of continuous functions containing the constant functions and separating the points of* K. *Then:*

(i) $\Gamma(K,\mathcal{A})$ *is a closed subset of* K;

(ii) $\|f\|_{\Gamma(K,\mathcal{A})} = \|f\|_K$ *for every* $f \in \mathcal{A}$;

(iii) *if* F *is a closed subset of* K *with the property that* $\|f\|_K = \|f\|_F$ *for all* $f \in \mathcal{A}$, *then* $F \supset \Gamma(K,\mathcal{A})$.

Proof. (i) follows directly from the definition.

(iii) For $f \in \mathcal{A}$ write $S(f) = \{x \in K : |f(x)| = \|f\|_K\}$. Clearly, $S(f)$ is a non-empty compact subset of K for every $f \in \mathcal{A}$. We say that a closed subset $F \subset K$ is maximizing if $\|f\|_F = \|f\|_K$ for every $f \in \mathcal{A}$. Obviously, F is maximizing if and only if $S(f) \cap F \neq \emptyset$ for every $f \in \mathcal{A}$.

Let F be a maximizing set and $x \in \Gamma(K,\mathcal{A})$. For every neighbourhood U of x there exists $f \in \mathcal{A}$ with $S(f) \subset U$, and so $F \cap U \neq \emptyset$. Since U was an arbitrary neighbourhood of x and F is closed, we conclude that $x \in F$. This proves (iii).

(ii) Denote by $\mathcal{F}$ the family of all maximizing sets ordered by inclusion. If $\{F_\alpha\}_\alpha$ is a totally ordered subset of $\mathcal{F}$, then $F = \bigcap_\alpha F_\alpha$ is also a maximizing set. Indeed, for every $f \in \mathcal{A}$, $S(f) \cap F_\alpha$ is a totally ordered family of non-empty compact subsets of K, and so $F \cap S(f) = \bigcap_\alpha (S(f) \cap F_\alpha) \neq \emptyset$. Thus $F = \bigcap_\alpha F_\alpha$ is maximizing. By the Zorn lemma there exists a minimal maximizing set F_0.

By (iii), $\Gamma(K,\mathcal{A}) \subset F_0$. Conversely, let $x \in F_0$ and let U be an open neighbourhood of x in K. For every $y \in K \setminus U$ there exists $f_y \in \mathcal{A}$ with $f_y(x) = 0$ and $f_y(y) = 1$. Let $U_y = \{z \in K : |f_y(z)| > 1/2\}$. Since the set $K \setminus U$ is compact, we can find finitely many points $y_1,\dots,y_n \in K\setminus U$ with $\bigcup U_{y_i} \supset K\setminus U$, and functions $f_1 = f_{y_1},\dots,f_n = f_{y_n} \in \mathcal{A}$ such that

$$U_0 := \{z \in K : |f_i(z)| < 1/2 \quad (i = 1,\dots,n)\} \subset U.$$

Since F_0 is a minimal maximizing set, there exists $f \in \mathcal{A}$ such that $\|f\|_{F_0\setminus U_0} < \|f\|_{F_0} = \|f\|_K$ (otherwise $F_0 \setminus U_0$ would be a maximizing set smaller than F_0). We may assume that $\|f\|_K = 1$ and, by replacing f by a suitable power f^m if necessary, $\|f\|_{F_0\setminus U_0} < \frac{1}{2\max\{\|f_1\|_K,\dots,\|f_n\|_K\}}$. For $i = 1,\dots,n$ we have

$$\|ff_i\|_K = \max\{\|ff_i\|_{F_0\setminus U_0}, \|ff_i\|_{F_0\cap U_0}\} < 1/2.$$

Let $y \in F_0$ be a point with the property that $|f(y)| = \|f\|_K = 1$. Then $|f_i(y)| = |(ff_i)(y)| < 1/2 \quad (i = 1,\dots,n)$, and so $y \in U_0$. Thus $S(f) \subset U_0 \subset U$ and $x \in \Gamma(K,\mathcal{A})$.

Hence $F_0 = \Gamma(K,\mathcal{A})$ and $\Gamma(K,\mathcal{A})$ is a maximizing set. □

Let $\mathcal{A}$ be a commutative Banach algebra. The *Shilov boundary* of $\mathcal{A}$ is the set $\Gamma(\mathcal{A}) = \Gamma(\mathcal{M}(\mathcal{A}), G(\mathcal{A}))$.

The *Shilov boundary of a commutative Banach algebra* $\mathcal{A}$ has the following properties:

Corollary 23. *Let $\mathcal{A}$ be a commutative Banach algebra. Then:*

(i) *$\Gamma(\mathcal{A})$ is a non-empty closed subset of $\mathcal{M}(\mathcal{A})$;*

(ii) $\max\{|\varphi(a)| : \varphi \in \Gamma(\mathcal{A})\} = \max\{|\varphi(a)| : \varphi \in \mathcal{M}(\mathcal{A})\} = r(a)$ *for every $a \in \mathcal{A}$;*

(iii) *if F is a closed subset of $M(\mathcal{A})$ satisfying* $\max\{|\varphi(a)| : \varphi \in F\} = r(a)$ *for every $a \in \mathcal{A}$, then $F \supset \Gamma(\mathcal{A})$;*

(iv) *if $\varphi \in \mathcal{M}(\mathcal{A})$, then $\varphi \in \Gamma(\mathcal{A})$ if and only if for every neighbourhood U of φ there exists $a \in \mathcal{A}$ such that*

$$\sup\{|\psi(a)| : \psi \in U\} > \sup\{|\psi(a)| : \psi \in \mathcal{M}(\mathcal{A}) \setminus U\}.$$

3 Approximate point spectrum in commutative Banach algebras

In this section we introduce and study the approximate point spectrum in commutative Banach algebras. This is, apart from the ordinary spectrum, the most important example of a spectrum.

All algebras in this section will be commutative.

Let $x_1,\dots,x_n$ be elements of a commutative Banach algebra $\mathcal{A}$. We will write

$$d^{\mathcal{A}}(x_1,\dots,x_n) = \inf\left\{\sum_{i=1}^{n} \|x_i z\| : z \in \mathcal{A}, \|z\| = 1\right\}.$$

If no confusion can arise we write simply $d(x_1,\dots,x_n)$ instead of $d^{\mathcal{A}}(x_1,\dots,x_n)$.

Clearly, an element $x_1 \in \mathcal{A}$ is a topological divisor of zero in $\mathcal{A}$ if and only if $d^{\mathcal{A}}(x_1) = 0$.

Lemma 1. *Let $x_1, \dots, x_n, y_1, \dots, y_n$ be elements of a commutative Banach algebra $\mathcal{A}$. Then:*

(i) $d(x_1, \dots, x_n) \le \sum_{i=1}^n \|x_i\|$;

(ii) $\big|d(x_1, \dots, x_n) - d(y_1, \dots, y_n)\big| \le \sum_{i=1}^{n} \|x_i - y_i\|$;

(iii) *the function $d : \mathcal{A}^n \to \langle 0, \infty)$ is continuous, where $\mathcal{A}^n$ is considered with the product topology;*

(iv) $d(x_1)d(y_1) \le d(x_1y_1) \le \|x_1\|d(y_1)$;

(v) *if x_1 is invertible, then $d(x_1) = \|x^{-1}\|^{-1}$.*

Proof. (i)–(iv) Clear.
(v) For $y \in \mathcal{A}$ we have $\|y\| = \|x_1^{-1}x_1y\| \le \|x_1^{-1}\| \cdot \|x_1y\|$, and so $d(x_1) \ge \|x_1^{-1}\|^{-1}$. On the other hand, we have $\|x_1 \cdot x_1^{-1}\| = \|1_{\mathcal{A}}\| = 1$, and so $d(x_1) \le \|x_1^{-1}\|^{-1}$. □

Definition 2. Let M be a subset of a commutative Banach algebra $\mathcal{A}$. We say that M consists of *joint topological divisors of zero* if $d(x_1, \dots, x_n) = 0$ for every finite subset $\{x_1, \dots, x_n\} \subset M$.

Theorem 3. *A set $M \subset A$ consists of joint topological divisors of zero if and only if there exists a net $(z_\alpha) \subset \mathcal{A}$ such that $\|z_\alpha\| = 1$ for all α and $\lim_\alpha z_\alpha x = 0$ for each $x \in M$.*

Proof. If there exists such a net, then clearly M consists of joint topological divisors of zero.

Suppose on the contrary that M consists of joint topological divisors of zero. For each finite subset $F \subset M$ and each $k \in \mathbb{N}$ there exists an element $z_{F,k} \in \mathcal{A}$ such that $\|z_{F,k}\| = 1$ and $\sum_{x \in F} \|z_{F,k}x\| \le k^{-1}$. Consider the order $(F, k) \le (F', k')$ if and only if $F \subset F'$ and $k \le k'$. Then the net $(z_{F,k})_{F,k}$ satisfies the conditions of the theorem. □

Theorem 4. *Let (z_α) be a net of elements of a commutative Banach algebra A such that $\|z_\alpha\| = 1$ for every α. Then the set $M = \{x \in A : \lim_\alpha xz_\alpha = 0\}$ is a closed ideal.*

Proof. It is clear that M is an ideal. Let $x \in \overline{M}$. For every $\varepsilon > 0$ there exists $y \in M$ such that $\|x - y\| < \frac{\varepsilon}{2}$ and α_0 such that $\|yz_\alpha\| < \frac{\varepsilon}{2}$ for all $\alpha \ge \alpha_0$. Then

$$\|xz_\alpha\| \le \|(x - y)z_\alpha\| + \|yz_\alpha\| < \varepsilon$$

for all $\alpha \ge \alpha_0$. Hence $x \in M$ and M is closed. □

Corollary 5. *Let $M \subset \mathcal{A}$ consist of joint topological divisors of zero. Then the smallest closed ideal containing M consists of joint topological divisors of zero.*

The following construction is very important in the study of topological divisors of zero.

Let $\mathcal{A}$ be a commutative Banach algebra. Denote by $\ell^\infty(\mathcal{A})$ the set of all bounded sequences of elements of $\mathcal{A}$. If we consider the pointwise algebraic operations and the norm $\|(a_j)_{j=1}^\infty\| = \sup_j \|a_j\|$, then $\ell^\infty(\mathcal{A})$ becomes a commutative Banach algebra. Denote by $c_0(\mathcal{A})$ the set of all sequences (a_j) with $\lim_{j\to\infty} a_j = 0$. Clearly, $c_0(\mathcal{A})$ is a closed ideal in $\ell^\infty(\mathcal{A})$. Denote by $Q(\mathcal{A})$ the quotient algebra $\ell^\infty(\mathcal{A})/c_0(\mathcal{A})$.

The elements of $Q(\mathcal{A})$ can be considered as bounded sequences of elements of $\mathcal{A}$ where we identify two sequences (a_j) and (a'_j) whenever they satisfy $\lim_{j\to\infty}\|a_j - a'_j\| = 0$. With this convention $\|(a_j)\|_{Q(\mathcal{A})} = \limsup_{j\to\infty}\|a_j\|$ and $Q(\mathcal{A})$ contains $\mathcal{A}$ as a subalgebra of constant sequences.

The most useful property of the algebra $Q(\mathcal{A})$ is that topological divisors of zero in $\mathcal{A}$ become divisors of zero in $Q(\mathcal{A})$.

Theorem 6. *Let $\mathcal{A}$ be a commutative Banach algebra and let $a \in \mathcal{A}$. Then:*

(i) *$d^{\mathcal{A}}(a) = d^{Q(\mathcal{A})}(a)$ for all $a \in \mathcal{A}$;*

(ii) *if $a_1, \dots, a_n \in \mathcal{A}$ and $d^{\mathcal{A}}(a_1, \dots, a_n) = 0$, then there exists $\tilde{b} \in Q(\mathcal{A})$ of norm 1 such that $\tilde{b} \cdot a_i = 0$ for $i = 1, \dots, n$.*

Proof. (i) Clearly, $d^{Q(\mathcal{A})}(a) \le d^{\mathcal{A}}(a)$. Conversely, we have $\|ab\| \ge d^{\mathcal{A}}(a)\cdot\|b\|$ for all $b \in \mathcal{A}$. Let $\tilde{b} = (b_j) \in Q(\mathcal{A})$. Then

$$\|a\tilde{b}\|_{Q(\mathcal{A})} = \limsup_{j\to\infty}\|ab_j\| \ge \limsup_{j\to\infty}\big(d^{\mathcal{A}}(a)\cdot\|b_j\|\big) = d^{\mathcal{A}}(a)\cdot\|\tilde{b}\|_{Q(\mathcal{A})}.$$

Thus $d^{Q(\mathcal{A})}(a) \ge d^{\mathcal{A}}(a)$.

(ii) If $d^{\mathcal{A}}(a_1, \dots, a_n) = 0$, then there exists a sequence $\tilde{b} = (b_j)$ of elements of $\mathcal{A}$ with $\|b_j\| = 1 \quad (j = 1, 2, \dots)$ and $\lim_{j\to\infty}\|a_i b_j\| = 0$ for $i = 1, \dots, n$. Then $\|\tilde{b}\|_{Q(\mathcal{A})} = 1$ and $a_i\tilde{b} = 0 \quad (i = 1, \dots, n)$. □

Theorem 7. *Let $x_0, x_1, \dots, x_n$ be elements of a commutative Banach algebra $\mathcal{A}$ satisfying $d(x_1, \dots, x_n) = 0$. Then there exists $\lambda \in \mathbb{C}$ such that*

$$d(x_0 - \lambda, x_1, \dots, x_n) = 0.$$

Proof. Regard $\mathcal{A}$ as a subalgebra of the algebra $Q(\mathcal{A})$ constructed above. Set

$$J = \big\{\tilde{a} = (a_i) \in Q(\mathcal{A}) : x_r\tilde{a} = 0 \quad (r = 1, \dots, n)\big\}.$$

Then J is a closed ideal in $Q(\mathcal{A})$ and, by the preceding lemma, $J \ne \{0\}$. Define the operator $T : J \to J$ by $T\big((a_i)\big) = (x_0 a_i)$.

Let $\lambda \in \partial\sigma^{\mathcal{B}(J)}(T)$. Then, by Theorem 1.28, there exist $\tilde{u}_j \in J$ $(j \in \mathbb{N})$ such that $\|\tilde{u}_j\|_{Q(\mathcal{A})} = 1$ and $\lim_{j\to\infty}\|(T-\lambda)\tilde{u}_j\|_{Q(\mathcal{A})} = \lim_{j\to\infty}\|(x_0-\lambda)\tilde{u}_j\|_{Q(\mathcal{A})} = 0$.

For every $k \in \mathbb{N}$ there exists $j \in \mathbb{N}$ such that $\|(x_0 - \lambda)\tilde{u}_j\|_{Q(\mathcal{A})} < k^{-1}$. As $\tilde{u}_j \in J$, we can find an element z_k of the sequence $\tilde{u}_j$ such that $\|x_r z_k\| < k^{-1}$ $(r = 1, \dots, n)$, $\|(x_0 - \lambda)z_k\| < k^{-1}$ and $1 - k^{-1} < \|z_k\| < 1 + k^{-1}$. If we consider the sequence $\left(\frac{z_k}{\|z_k\|}\right)$, it is easy to see that $d(x_0 - \lambda, x_1, \dots, x_n) = 0$. $\square$

Let $\mathcal{A}$ be a commutative Banach algebra. Denote by $\text{ł}(\mathcal{A})$ the set of all ideals in $\mathcal{A}$ consisting of joint topological divisors of zero.

Corollary 8. *Let $\mathcal{A}$ be a commutative Banach algebra and let $I \in \text{ł}(\mathcal{A})$. Then there exists a maximal ideal J such that $J \in \text{ł}(\mathcal{A})$ and $J \supset I$.*

Proof. It follows easily from the Zorn lemma that there exists an ideal J in $\text{ł}(\mathcal{A})$ containing I that is maximal in this class with respect to the inclusion. It is sufficient to show that $\operatorname{codim} J = 1$.

If $\operatorname{codim} J \geq 2$, then there exists $x \in \mathcal{A}$ such that $x \notin J + \mathbb{C} \cdot 1_{\mathcal{A}}$.

For every finite subset $F = \{x_1, \dots, x_n\} \subset J$ set

$$C_F = \big\{\lambda \in \mathbb{C} : d(x - \lambda, x_1, \dots, x_n) = 0\big\}.$$

By the previous theorem, we have $C_F \neq \emptyset$. Moreover, C_F is compact and $C_F \cap C_{F'} \supset C_{F \cup F'} \neq \emptyset$ for all finite subsets $F, F' \subset J$. Hence the system $\{C_F\}$ has the finite intersection property and there exists $\lambda \in \bigcap_F C_F$.

Let J' be the ideal generated by J and $x - \lambda$. Then $J \subset J'$, $J \neq J'$ and, by Corollary 5, we have $J' \in \text{ł}(\mathcal{A})$, which is a contradiction. $\square$

The set of all multiplicative functionals $\varphi \in \mathcal{M}(\mathcal{A})$ with $\operatorname{Ker} \varphi \in \text{ł}(\mathcal{A})$ is called the *cortex* of $\mathcal{A}$ and denoted by $\operatorname{cor} \mathcal{A}$.

Definition 9. Let $x_1, \dots, x_n$ be elements of a commutative Banach algebra $\mathcal{A}$. The *approximate point spectrum* $\tau(x_1, \dots, x_n)$ is defined by

$$\tau(x_1, \dots, x_n) = \big\{(\lambda_1, \dots, \lambda_n) \in \mathbb{C}^n : d(x_1 - \lambda_1, \dots, x_n - \lambda_n) = 0\big\}.$$

The approximate point spectrum has similar properties as the spectrum σ defined in the previous section.

Theorem 10. *Let $x = (x_1, \dots, x_n)$ be an n-tuple of elements of a commutative Banach algebra $\mathcal{A}$. Then:*

(i) $\tau(x) = \{(\varphi(x_1), \dots, \varphi(x_n)) : \varphi \in \operatorname{cor} \mathcal{A}\}$;

(ii) $\partial\sigma(x_1) \subset \tau(x_1) \subset \sigma(x_1)$;

(iii) *$\operatorname{cor} \mathcal{A}$ is a non-empty compact subset of $\mathcal{M}(\mathcal{A})$ and $\tau(x)$ is a non-empty compact subsets of $\mathbb{C}^n$;*

(iv) *if $m \leq n$ and $P : \mathbb{C}^n \to \mathbb{C}^m$ is the canonical projection onto the first m coordinates, then $\tau(x_1, \dots, x_m) = P\tau(x_1, \dots, x_n)$;*

(v) *if* $f = (f_1, \dots, f_m)$ *is an* m*-tuple of functions analytic in a neighbourhood of* $\sigma(x)$, *then* $\tau(f(x)) = f(\tau(x))$;

(vi) *if* $\mathcal{B}$ *is a closed subalgebra of* $\mathcal{A}$ *and* $x_1, \dots, x_n \in \mathcal{B}$, *then*

$$\tau^{\mathcal{B}}(x) \subset \tau^{\mathcal{A}}(x) \subset \sigma^{\mathcal{A}}(x) \subset \sigma^{\mathcal{B}}(x);$$

(vii) *the polynomially convex hulls of* $\tau(x)$ *and* $\sigma(x)$ *coincide, i.e.,*

$$\widehat{\tau}(x) = \widehat{\sigma}(x) = \sigma^{\langle x \rangle}(x).$$

Proof. (i) If $\varphi \in \operatorname{cor} \mathcal{A}$, then $x_i - \varphi(x_i) \in \operatorname{Ker} \varphi \quad (i = 1, \dots, n)$, and so $(\varphi(x_1), \dots, \varphi(x_n)) \in \tau(x_1, \dots, x_n)$. Conversely, if $(\lambda_1, \dots, \lambda_n) \in \tau(x_1, \dots, x_n)$, then $d(x_1 - \lambda_1, \dots, x_n - \lambda_n) = 0$ and, by Corollary 8, there is a $\varphi \in \operatorname{cor} \mathcal{A}$ such that $x_i - \lambda_i \in \operatorname{Ker} \varphi$ and $\varphi(x_i) = \lambda_i \quad (i = 1, \dots, n)$.

(ii) follows from Theorem 1.14.

(iii) Since $\tau(x_1) \supset \partial\sigma(x_1)$, the set $\tau(x_1)$ is non-empty. By (i), $\operatorname{cor} \mathcal{A} \neq \emptyset$ and also $\tau(x_1, \dots, x_n) \neq \emptyset$.

We show that $\operatorname{cor} \mathcal{A}$ is a closed subset of $\mathcal{M}(\mathcal{A})$. Let φ_α be a net of elements of $\operatorname{cor} \mathcal{A}$ and $\varphi_\alpha \to \varphi \in \mathcal{M}(\mathcal{A})$. Let $y_1, \dots, y_m \in \operatorname{Ker} \varphi$ and $\varepsilon > 0$. Then $|\varphi_\alpha(y_j)| < \varepsilon/2$ for all j and all α sufficiently large. Further, $y_j - \varphi_\alpha(y_j) \cdot 1_{\mathcal{A}} \in \operatorname{Ker} \varphi_\alpha$ and there is a $u \in \mathcal{A}$ with $\|u\| = 1$ and $\|(y_j - \varphi_\alpha(y_j))u\| < \varepsilon/2 \quad (j = 1, \dots, m)$. Then $\|y_j u\| < \varepsilon \quad (j = 1, \dots, m)$ and, consequently, $d(y_1, \dots, y_m) = 0$. Hence $\varphi \in \operatorname{cor} \mathcal{A}$ and $\operatorname{cor} \mathcal{A}$ is compact.

Consider the mapping $\varphi \mapsto (\varphi(x_1), \dots, \varphi(x_n))$ from $\operatorname{cor} \mathcal{A}$ onto $\tau(x_1, \dots, x_n)$. The mapping is continuous and therefore $\tau(x_1, \dots, x_n)$ is compact.

(iv) follows from (i).

(v) By Theorem 2.20, we have

$$f(\tau(x)) = \{f(\varphi(x)) : \varphi \in \operatorname{cor} \mathcal{A}\} = \{\varphi(f(x)) : \varphi \in \operatorname{cor} \mathcal{A}\} = \tau(f(x)).$$

(vi) The first inclusion follows from the inequality

$$d^{\mathcal{A}}(x_1 - \lambda_1, \dots, x_n - \lambda_n) \le d^{\mathcal{B}}(x_1 - \lambda_1, \dots, x_n - \lambda_n)$$

for all $(\lambda_1, \dots, \lambda_n) \in \mathbb{C}^n$.

The second inclusion is clear. To show the third inclusion, let $(\lambda_1, \dots, \lambda_n) \in \sigma^{\mathcal{A}}(x_1, \dots, x_n)$. Then there exists a multiplicative functional $\varphi \in \mathcal{M}(\mathcal{A})$ such that $\varphi(x_i) = \lambda_i \ (i = 1, \dots, n)$. The restriction $\psi = \varphi|\mathcal{B} \in \mathcal{M}(\mathcal{B})$ has the same property, and so $(\lambda_1, \dots, \lambda_n) \in \sigma^{\mathcal{B}}(x_1, \dots, x_n)$.

(vii) For each polynomial p we have

$$\begin{aligned} \max\{|p(\lambda)| : \lambda \in \sigma(x)\} &= \max\{|\mu| : \mu \in \sigma(p(x))\} \\ &= \max\{|\mu| : \mu \in \tau(p(x))\} = \max\{|p(\lambda)| : \lambda \in \tau(x)\}. \end{aligned}$$

Hence $\widehat{\tau}(x) = \widehat{\sigma}(x)$.

By Theorem 2.18, $\sigma^{\langle x\rangle}(x)$ is polynomially convex. By (vi), we have $\tau^{\langle x\rangle}(x) \subset \tau^{\mathcal{A}}(x) \subset \sigma^{\mathcal{A}}(x) \subset \sigma^{\langle x\rangle}(x)$ and $\widehat{\tau}^{\langle x\rangle}(x) = \widehat{\sigma}^{\langle x\rangle}(x)$. Hence $\widehat{\sigma}^{\mathcal{A}}(x) = \sigma^{\langle x\rangle}(x)$. □

If we replace the norm by the spectral radius in the definition of ideals consisting of joint topological divisors of zero, then we get similar results.

Let $\mathcal{A}$ be a commutative Banach algebra. Denote by $\gamma(\mathcal{A})$ the set of all ideals J in $\mathcal{A}$ such that

$$\inf\Big\{\sum_{i=1}^{n} r(x_i y) : y \in \mathcal{A}, r(y) = 1\Big\} = 0$$

for every finite subset $\{x_1, \dots, x_n\} \subset J$.

This notion is closely related to the Shilov boundary $\Gamma(\mathcal{A})$ of the algebra $\mathcal{A}$.

Theorem 11. *Let φ be a multiplicative functional on a commutative Banach algebra $\mathcal{A}$. Then $\varphi \in \Gamma(\mathcal{A})$ if and only if* $\operatorname{Ker}\varphi \in \gamma(\mathcal{A})$.

Proof. Let $\varphi \in \Gamma(\mathcal{A})$, $x_1, \dots, x_n \in \operatorname{Ker}\varphi$ and $\varepsilon > 0$.

Consider the neighbourhood

$$U = \big\{\psi \in \mathcal{M}(\mathcal{A}) : |\psi(x_i)| < \varepsilon \quad (i = 1, \dots, n)\big\}$$

of φ in $\mathcal{M}(\mathcal{A})$. By Corollary 2.23, there exists $y \in \mathcal{A}$ such that $r(y) = 1$ and

$$\sup\big\{|\psi(y)| : \psi \in \mathcal{M}(\mathcal{A}) \setminus U\big\} < 1.$$

For a suitable power $z = y^k$ we have $r(z) = 1$ and

$$\sup\big\{|\psi(z)| : \psi \in \mathcal{M}(\mathcal{A}) \setminus U\big\} < \varepsilon.$$

Then

$$\begin{aligned} r(x_i z) &= \max\big\{\sup\{|\psi(x_i z)| : \psi \in U\}, \sup\{|\psi(x_i z)| : \psi \in \mathcal{M}(\mathcal{A}) \setminus U\}\big\} \\ &\le \varepsilon \cdot \max\big\{1, \|x_1\|, \dots, \|x_n\|\big\}. \end{aligned}$$

Hence $\sum_{i=1}^{n} r(x_i z) \le n\varepsilon \cdot \max\big\{1, \|x_1\|, \dots, \|x_n\|\big\}$ and, consequently, $\operatorname{Ker}\varphi \in \gamma(\mathcal{A})$.

In the opposite direction, let $\operatorname{Ker}\varphi \in \gamma(\mathcal{A})$. Let $x_1, \dots, x_n \in \mathcal{A}$, $\varepsilon > 0$ and let

$$U = \big\{\psi \in \mathcal{M}(\mathcal{A}) : |\psi(x_i) - \varphi(x_i)| < \varepsilon \ \ (i = 1, \dots, n)\big\}.$$

Then $y_i := x_i - \varphi(x_i) \cdot 1_{\mathcal{A}} \in \operatorname{Ker}\varphi \quad (i = 1, \dots, n)$, and so there exists $z \in \mathcal{A}$ with $r(z) = 1$ and $\sum_{i=1}^{n} r(z y_i) < \varepsilon/2$.

If $\psi \in \mathcal{M}(\mathcal{A}) \setminus U$, then there exists $i \in \{1, \dots, n\}$ with $|\psi(x_i) - \varphi(x_i)| \ge \varepsilon$. Then $|\psi(y_i)| = |\psi(x_i) - \varphi(x_i)| \ge \varepsilon$, and so

$$|\psi(z)| = \frac{|\psi(z y_i)|}{|\psi(y_i)|} \le \frac{r(z y_i)}{\varepsilon} < \frac{1}{2}.$$

Hence

$$1 = r(z) = \sup\{|\psi(z)| : \psi \in U\} > \frac{1}{2} \geq \sup\{|\psi(z)| : \psi \in \mathcal{M}(\mathcal{A}) \setminus U\},$$

and so $\varphi \in \Gamma(\mathcal{A})$. □

Lemma 12. *Let J be an ideal in a commutative Banach algebra $\mathcal{A}$. Then $J \in \gamma(\mathcal{A})$ if and only if there exists a net $(z_\alpha)_\alpha$ of elements of $\mathcal{A}$ such that $r(z_\alpha) = 1$ for all α and $\lim_\alpha r(xz_\alpha) = 0$ for all $x \in J$.*

Proof. Let $J \in \gamma(\mathcal{A})$. For every finite set $F \subset J$ and every $n \in \mathbb{N}$ find $z_{F,n} \in \mathcal{A}$ with $r(z_{F,n}) = 1$ and $r(z_{F,n}x) < 1/n \quad (x \in F)$. As in Theorem 3 we can show that $(z_{F,n})_{F,n}$ is the required net.

The opposite implication is clear. □

Theorem 13. *Let I be an ideal in a commutative Banach algebra $\mathcal{A}$. Then $I \in \gamma(\mathcal{A})$ if and only if there exists a multiplicative functional $\varphi \in \Gamma(\mathcal{A})$ such that $I \subset \operatorname{Ker} \varphi$.*

Proof. If $\varphi \in \Gamma(\mathcal{A})$ and $I \subset \operatorname{Ker}\varphi$, then $\operatorname{Ker}\varphi \in \gamma(\mathcal{A})$ by Theorem 11. Consequently, $I \in \gamma(\mathcal{A})$.

For the converse, let $I \in \gamma(\mathcal{A})$. By Theorem 2.11, the spectral radius is an algebra norm in the algebra $\mathcal{A}/\operatorname{rad}\mathcal{A}$. Let $\mathcal{B}$ be the completion of the algebra $(\mathcal{A}/\operatorname{rad}\mathcal{A}, r(\cdot))$ and let $\rho = \rho_2\rho_1 : \mathcal{A} \to \mathcal{B}$ where $\rho_1 : \mathcal{A} \to \mathcal{A}/\operatorname{rad}\mathcal{A}$ is the canonical projection and $\rho_2 : \mathcal{A}/\operatorname{rad}\mathcal{A} \to \mathcal{B}$ the natural embedding. Clearly, ρ is a continuous homomorphisms with dense range and we have $\|\rho(x)\|_{\mathcal{B}} = r(x)$ for each $x \in \mathcal{A}$. Using the spectral radius formula and the continuity of the spectral radius in the commutative Banach algebra $\mathcal{B}$ we see that the norm in $\mathcal{B}$ coincides with the spectral radius. In particular, $\gamma(\mathcal{B}) = \natural(\mathcal{B})$.

Using Corollary 5 we obtain that $\rho(I)$ is contained in an ideal in $\gamma(\mathcal{B}) = \natural(\mathcal{B})$. By Corollary 8, there exists a maximal ideal $J_1 \in \natural(\mathcal{B})$ such that $J_1 \supset \rho(I)$. Set $J = \rho^{-1}(J_1)$. Then $I \subset J$, and J is a maximal ideal in $\mathcal{A}$ since it is of codimension 1. Furthermore, $J_1 \in \gamma(\mathcal{B})$, and so there exists a net $(z_\alpha) \subset \mathcal{B}$ with $r(z_\alpha) = 1$ and $r(z_\alpha u) \to 0$ for all $u \in J_1$. Since $\rho(\mathcal{A})$ is dense in $\mathcal{B}$, we can choose $(z_\alpha) \subset \rho(\mathcal{A})$. Since ρ preserves the spectral radius, we can see that $J = \rho^{-1}(J_1) \subset \gamma(\mathcal{A})$. By Theorem 11, $J = \operatorname{Ker}\varphi$ for some $\varphi \in \Gamma(\mathcal{A})$, and so $I \subset \operatorname{Ker}\varphi$. □

Lemma 14. *Let $x_1, \dots, x_n, y$ be elements of a commutative Banach algebra $\mathcal{A}$. Then*

$$\sum_{i=1}^{n} r(x_i y) \geq d(x_1, \dots, x_n) \cdot r(y).$$

Proof. The inequality is clear if $d(x_1, \dots, x_n) = 0$.

Suppose that $c := d(x_1, \dots, x_n) > 0$. Let $\mathcal{B}$ be the set of all formal power series in variables $t_1, \dots, t_n$ of the form $\sum_{\alpha \in \mathbb{Z}_+^n} c_\alpha t^\alpha$ satisfying $\sum_{\alpha \in \mathbb{Z}_+^n} \|c_\alpha\| < \infty$, where $\alpha = (\alpha_1, \dots, \alpha_n), c_\alpha \in \mathcal{A}$ and $t^\alpha = t_1^{\alpha_1} \cdots t_n^{\alpha_n} \quad (\alpha \in \mathbb{Z}_+^n)$. Together with the naturally defined algebraic operations and the norm

$$\Big\| \sum_{\alpha \in \mathbb{Z}_+^n} c_\alpha t^\alpha \Big\| = \sum_{\alpha \in \mathbb{Z}_+^n} \|c_\alpha\|,$$

$\mathcal{B}$ is a commutative Banach algebra containing $\mathcal{A}$ as a subalgebra of constants. Set $u = \sum_{i=1}^n x_i t_i$. Then $\|uy\| \geq c \cdot \|y\|$ for all $y \in \mathcal{A}$. We can show easily by induction that

$$\sum_{\substack{\alpha \in \mathbb{Z}_+^n \\ |\alpha| = k}} \frac{k!}{\alpha_1! \cdots \alpha_n!} \|x^\alpha y\| \geq c^k \|y\|,$$

and so

$$\|u^k y\| = \Big\| \sum_{\substack{\alpha \in \mathbb{Z}_+^n \\ |\alpha| = k}} \frac{k!}{\alpha_1! \cdots \alpha_n!} x^\alpha t^\alpha y \Big\| = \sum_{\substack{\alpha \in \mathbb{Z}_+^n \\ |\alpha| = k}} \frac{k!}{\alpha_1! \cdots \alpha_n!} \|x^\alpha y\| \geq c^k \|y\|$$

for all $y \in \mathcal{A}, k \in \mathbb{N}$. Thus $\|u^k y^k\| \geq c^k \|y^k\|$ and $r(uy) \geq c\, r(y)$ for all $y \in \mathcal{A}$. Hence

$$c \cdot r(y) \leq r(uy) \leq \sum_{i=1}^n r(x_i t_i y) \leq \sum_{i=1}^n r(x_i y)$$

for every $y \in \mathcal{A}$, since $r(t_i) = 1 \quad (i = 1, \dots, n)$ and the spectral radius in $\mathcal{B}$ is subadditive and submultiplicative. □

Corollary 15. *Let $\mathcal{A}$ be a commutative Banach algebra. Then $\gamma(\mathcal{A}) \subset \text{ł}(\mathcal{A})$ and $\Gamma(\mathcal{A}) \subset \operatorname{cor} \mathcal{A}$.*

Let $\mathcal{A}$ be a commutative Banach algebra. We have already studied the spectrum

$$\sigma(x_1, \dots, x_n) = \big\{(\varphi(x_1), \dots, \varphi(x_n)) : \varphi \in \mathcal{M}(\mathcal{A})\big\}$$

and the approximate point spectrum

$$\tau(x_1, \dots, x_n) = \big\{(\varphi(x_1), \dots, \varphi(x_n)) : \varphi \in \operatorname{cor} \mathcal{A}\big\}.$$

Another important closed subset of $\mathcal{M}(\mathcal{A})$ is the Shilov boundary $\Gamma(\mathcal{A})$. We define the *Shilov spectrum* σ_Γ for $(x_1, \dots, x_n) \in \mathcal{A}^n$ by

$$\sigma_\Gamma(x_1, \dots, x_n) = \big\{(\varphi(x_1), \dots, \varphi(x_n)) : \varphi \in \Gamma(\mathcal{A})\big\}.$$

It is easy to see that σ_Γ has similar properties as the spectrum σ and the approximate point spectrum τ. Later we formulate a result of this kind for general spectral systems.

Theorem 16. *Let $x = (x_1, \dots, x_n)$ be an n-tuple of elements of a commutative Banach algebra $\mathcal{A}$. Then:*

(i) *$\sigma_\Gamma(x)$ is a non-empty compact subset of $\mathbb{C}^n$ and $\sigma_\Gamma(x) \subset \tau(x) \subset \sigma(x)$;*

(ii) *if $f = (f_1, \dots, f_m)$ is an m-tuple of functions analytic in a neighbourhood of $\sigma(x)$, then*

$$\sigma_\Gamma\big(f(x)\big) = f\big(\sigma_\Gamma(x)\big);$$

(iii) $\max\{|\lambda| : \lambda \in \sigma_\Gamma(x_1)\} = r(x_1)$ *for all $x_1 \in \mathcal{A}$;*

(iv) *the polynomially convex hulls of $\sigma_\Gamma(x)$ and $\sigma(x)$ coincide, i.e.,*

$$\widehat{\sigma}_\Gamma(x) = \widehat{\sigma}(x) = \sigma^{\langle x\rangle}(x).$$

In particular, $\partial\sigma(x_1) \subset \sigma_\Gamma(x_1) \subset \sigma(x_1)$.

4 Permanently singular elements and removability of spectrum

Let $\mathcal{A}, \mathcal{B}$ be Banach algebras. We say that $\mathcal{B}$ is an *extension* of $\mathcal{A}$ if there exists an isometrical homomorphism $\rho : \mathcal{A} \to \mathcal{B}$. If we identify $\mathcal{A}$ with the image $\rho(\mathcal{A})$ we can consider $\mathcal{A}$ as a closed subalgebra of $\mathcal{B}$ and write simply $\mathcal{A} \subset \mathcal{B}$.

In fact, if the homomorphism ρ is only isomorphic, then we can renorm $\mathcal{B}$ such that ρ becomes an isometrical embedding.

Theorem 1. *Let $(\mathcal{B}, \|\cdot\|)$ be a Banach algebra and $\mathcal{A} \subset \mathcal{B}$ a subalgebra. Let $k \geq 1$ and let $|\cdot|$ be an algebra norm on $\mathcal{A}$ satisfying*

$$k^{-1}|a| \leq \|a\| \leq k|a| \qquad (a \in \mathcal{A}).$$

Then there exists an algebra norm $|||\cdot|||$ on $\mathcal{B}$ equivalent to $\|\cdot\|$ such that $|||a||| = |a|$ for every $a \in \mathcal{A}$.

Proof. Consider the bounded semigroup $S = \{a \in \mathcal{A} : |a| \leq 1\}$. By Theorem 1.32, there exists an algebra norm $\|\cdot\|'$ on $\mathcal{B}$ such that

$$k^{-1}\|b\| \leq \|b\|' \leq k\|b\|$$

for every $b \in \mathcal{B}$ and $\|a\|' \leq |a| \quad (a \in \mathcal{A})$.

Define the new norm $|||\cdot|||$ on $\mathcal{B}$ by

$$|||b||| = \inf\big\{|a| + k^2\|b-a\|' : a \in \mathcal{A}\big\}.$$

For $a \in \mathcal{A}$ we have $|||a||| \leq |a|$ and

$$\begin{aligned} |||a||| = \inf\big\{|a_1| + k^2\|a-a_1\|' : a_1 \in \mathcal{A}\big\} &\geq \inf_{a_1\in\mathcal{A}}\big(|a_1| + k\|a-a_1\|\big) \\ &\geq \inf_{a_1\in\mathcal{A}}\big(|a_1| + |a-a_1|\big) \geq |a|. \end{aligned}$$

Hence $|||a||| = |a|$ for every $a \in \mathcal{A}$.

For $b \in \mathcal{B}$ we have $|||b||| \le k^2\|b\|' \le k^3\|b\|$ and

$$|||b||| = \inf\{|a| + k^2\|b-a\|' : a \in \mathcal{A}\} \ge \inf_{a\in\mathcal{A}}\bigl(k^{-1}\|a\| + k\|b-a\|\bigr) \ge k^{-1}\|b\|.$$

Hence $|||\cdot|||$ is a norm on $\mathcal{B}$ equivalent to $\|\cdot\|$.

It remains to show the submultiplicativity of $|||\cdot|||$. We have

$$\begin{aligned}
|||b_1|||\cdot|||b_2||| &= \inf_{a_1,a_2\in\mathcal{A}}\Bigl((|a_1| + k^2\|b_1-a_1\|')\cdot(|a_2| + k^2\|b_2-a_2\|')\Bigr)\\
&\ge \inf_{a_1,a_2\in\mathcal{A}}\Bigl(|a_1|\cdot|a_2| + k^2\bigl(\|a_1\|'\cdot\|b_2-a_2\|' + \|b_1-a_1\|'\cdot\|a_2\|'\\
&\qquad + \|b_1-a_1\|'\cdot\|b_2-a_2\|'\bigr)\Bigr)\\
&\ge \inf_{a_1,a_2\in\mathcal{A}}\Bigl(|a_1a_2| + k^2\|b_1b_2 - a_1a_1\|'\Bigr) \ge |||b_1b_2|||.
\end{aligned}$$

□

By Theorem 1.27 (ii), a topological divisor of zero is singular (= non-invertible) in any extension $\mathcal{B} \supset \mathcal{A}$. For commutative Banach algebras the opposite statement is also true.

Definition 2. An element x in a commutative Banach algebra $\mathcal{A}$ is called *permanently singular* if it is singular in each commutative Banach algebra $\mathcal{B} \supset \mathcal{A}$.

Theorem 3. *Let x be an element of a commutative Banach algebra $\mathcal{A}$. Then x is permanently singular if and only if it is a topological divisor of zero.*

Proof. Let $x \in \mathcal{A}$ be a topological divisor of zero and let $\mathcal{B}$ be a commutative extension of $\mathcal{A}$. Then there exists a sequence $(u_k)_{k=1}^{\infty} \subset \mathcal{A}$, $\|u_k\| = 1$ $(k = 1, 2, \dots)$ such that $\lim_{k\to\infty} u_k x = 0$. Suppose on the contrary that x is invertible in $\mathcal{B}$, so there exists $y \in \mathcal{B}$ such that $xy = 1$. Then $1 = \|u_k\| = \|u_k xy\| \le \|u_k x\|\cdot\|y\| \to 0$, a contradiction.

In the opposite direction, suppose that $x \in \mathcal{A}$ is not a topological divisor of zero. Let $q = (d^{\mathcal{A}}(x))^{-1}$, so $\|a\| \le q\|ax\|$ for all $a \in \mathcal{A}$. Consider the algebra $\mathcal{C}$ of all power series $\sum_{i=0}^{\infty} a_i b^i$ with coefficients $a_i \in \mathcal{A}$ in one variable b such that

$$\Bigl\|\sum_{i=0}^{\infty} a_i b^i\Bigr\| = \sum_{i=0}^{\infty} \|a_i\| q^i < \infty.$$

With the multiplication given by

$$\left(\sum_{i=0}^{\infty} a_i b^i\right)\cdot\left(\sum_{j=0}^{\infty} a'_j b^j\right) = \sum_{k=0}^{\infty} b^k \left(\sum_{i+j=k} a_i a'_j\right),$$

$\mathcal{C}$ is a commutative Banach algebra containing $\mathcal{A}$ as a subalgebra of constants. Let J be the closed ideal generated by the element $1 - xb$ and set $\mathcal{B} = \mathcal{C}/J$.

Let $\rho : \mathcal{A} \to \mathcal{B}$ be the composition of the embedding $\mathcal{A} \to \mathcal{C}$ and the canonical homomorphism $\mathcal{C} \to \mathcal{B} = \mathcal{C}/J$. Then

$$\rho(x) \cdot (b + J) = (x + J)(b + J) = 1_{\mathcal{A}} + J = 1_{\mathcal{B}},$$

and so it is sufficient to show that ρ is an isometry, i.e., that for each $a \in \mathcal{A}$ we have $\|a\|_{\mathcal{A}} = \|\rho(a)\|_{\mathcal{B}}$.

Obviously, $\|\rho(a)\|_{\mathcal{B}} = \inf_{c \in \mathcal{C}} \|a + (1 - xb)c\|_{\mathcal{C}} \leq \|a\|_{\mathcal{A}}$.

Conversely, let $c = \sum_{i=0}^{\infty} a_i b^i \in \mathcal{C}$. So $\sum_{i=0}^{\infty} \|a_i\| q^i < \infty$ and

$$\begin{aligned}
\|a + (1 - xb)c\|_{\mathcal{C}} &= \left\| (a + a_0) + \sum_{i=1}^{\infty} b^i (a_i - a_{i-1}x) \right\|_{\mathcal{C}} \\
&= \|a + a_0\| + \sum_{i=1}^{\infty} \|a_i - a_{i-1}x\| q^i \geq \|a\| - \|a_0\| + \sum_{i=1}^{\infty} q^i \big(\|a_{i-1}x\| - \|a_i\|\big) \\
&\geq \|a\| - \|a_0\| + \sum_{i=1}^{\infty} \big(q^{i-1}\|a_{i-1}\| - q^i\|a_i\|\big) = \lim_{k\to\infty} \big(\|a\| - q^k\|a_k\|\big) = \|a\|.
\end{aligned}$$

Hence ρ is an isometry and $\mathcal{B}$ is the required extension of $\mathcal{A}$. □

Corollary 4. *Let x be an element of a commutative Banach algebra $\mathcal{A}$. Then*

$$\tau^{\mathcal{A}}(x) = \bigcap_{\mathcal{B} \supset \mathcal{A}} \sigma^{\mathcal{B}}(x)$$

(the intersection of all subsets of $\mathbb{C}$ that are of the form $\sigma^{\mathcal{B}}(x)$ for some commutative extension $\mathcal{B} \supset \mathcal{A}$).

In fact, a stronger result is true – there is a single extension $\mathcal{B} \supset \mathcal{A}$ such that $\tau^{\mathcal{A}}(a) = \sigma^{\mathcal{B}}(x)$. To show this we need several lemmas.

Lemma 5. *Let $\mathcal{A}$ be a commutative Banach algebra, $x \in \mathcal{A}$, let G be an open connected subset of $\mathbb{C} \setminus \tau(x)$, let U be a non-empty open subset of G. Suppose that $f : G \to \mathcal{A}$ and $g : U \to \mathcal{A}$ are analytic functions satisfying*

$$f(z) = (x - z)g(z) \qquad (z \in U). \tag{1}$$

Then it is possible to extend g to an analytic function on G (it is clear that the extension, denoted also by g, satisfies (1) for all $z \in G$).

Proof. Denote by G_0 the set of all $z \in G$ such that there exists an analytic solution of (1) in a neighbourhood of z. Since for $z \in G$ the value of $g(z)$ is determined by (1) uniquely, it is sufficient to show that $G_0 = G$.

Let $\lambda \in G$ and let $\varphi : \langle 0, 1\rangle \to G$ be a continuous function with $\varphi(0) \in U \subset G_0$ and $\varphi(1) = \lambda$. Let $Z = \{\varphi(t) : t \in \langle 0, 1\rangle\}$ and let r be a positive constant satisfying $\operatorname{dist}\{Z, \mathbb{C} \setminus G\} > r$ and $\inf_{z \in Z} d(x - z) \geq r$. Let $M = \max\{\|f(z)\| :$

dist$\{z, Z\} \le r\}$. Let $\mu \in G_0 \cap Z$. Then there exist a neighbourhood U_1 of μ and an analytic function $g : U_1 \to \mathcal{A}$ such that $U_1 \subset G$, $f(z) = (x - z)g(z) \quad (z \in U_1)$, and $f(z) = \sum_{i=0}^{\infty} f_i(z-\mu)^i$, $g(z) = \sum_{i=0}^{\infty} g_i(z-\mu)^i \quad (z \in U_1)$ for some coefficients $f_i, g_i \in \mathcal{A} \quad (i = 0, 1, \ldots)$. By the Cauchy formulas, $\|f_i\| \le \frac{M}{r^i} \quad (i = 0, 1, \ldots)$. Furthermore,

$$\sum_{i=0}^{\infty} f_i(z-\mu)^i = f(z) = (x-z)g(z) = \big((x-\mu) - (z-\mu)\big) \sum_{i=0}^{\infty} g_i(z-\mu)^i$$

$$= (x-\mu)g_0 + \sum_{i=1}^{\infty} (z-\mu)^i \big((x-\mu)g_i - g_{i-1}\big)$$

for all $z \in U_1$. Thus $f_0 = (x-\mu)g_0$ and $(x-\mu)g_i = g_{i-1} + f_i \quad (i \ge 1)$. Hence $\|g_0\| \le d(x-\mu)^{-1}\|f_0\| \le r^{-1}M$ and $\|g_i\| \le d(x-\mu)^{-1}(\|g_{i-1}\| + \|f_i\|) \le r^{-1}(\|g_{i-1}\| + Mr^{-i})$. It is easy to show by induction that $\|g_i\| \le (i+1)Mr^{-(i+1)} \quad (i = 1, 2, \ldots)$. Thus the series $\sum_{i=0}^{\infty} g_i(z-\mu)^i$ converges for $|z - \mu| < r$, and so $\{z : |z - \mu| < r\} \subset G_0$.

Let $t_0 = \sup\{t \in \langle 0, 1\rangle : \varphi(s) \in G_0$ for every $s, 0 \le s \le t\}$. It is easy to check that $t_0 = 1$, and so $\lambda \in G_0$. Hence $G_0 = G$. □

Lemma 6. *Let a, x be elements of a commutative Banach algebra $\mathcal{A}$, let U be an open neighbourhood of $\tau(x)$ and let $g : U \to \mathcal{A}$ be an analytic function satisfying $a = (x - z)g(z) \quad (z \in U)$. Then $a = 0$.*

Proof. Let G be a component of $\mathbb{C} \setminus \tau(x)$. Obviously, $U \cap G \ne \emptyset$. By Lemma 5, there is an analytic function $g_G : G \to \mathcal{A}$ satisfying $a = (x - z)g_G(z) \quad (z \in G)$. Since the function g_G is uniquely determined on G, it coincides with g on $G \cap U$. Hence the function g can be extended to an entire function (denoted also by g) satisfying $a = (x-z)g(z) \quad (z \in \mathbb{C})$. For $|z| > \|x\|$ we have $g(z) = (x-z)^{-1}a$. Thus $g(z) \to 0$ as $z \to \infty$. By the Liouville theorem, $g = 0$, and so $a = (x - z)g(z) = 0$. □

Let $\mathcal{A}$ be a commutative Banach algebra and let U be an open subset of the complex plane. Denote by $H^\infty(U, \mathcal{A})$ the algebra of all bounded analytic functions $f : U \to \mathcal{A}$ with the norm $\|f\|_U = \sup\{\|f(z)\| : z \in U\}$. Clearly, $H^\infty(U, \mathcal{A})$ is a commutative Banach algebra.

Lemma 7. *Let $\mathcal{A}$ be a commutative Banach algebra, $x \in \mathcal{A}$, let U be an open neighbourhood of $\tau(x)$, Then there exists a constant $k > 0$ such that $\|a\| \le k \cdot \|g\|_U$ whenever $a \in \mathcal{A}$, $f, g \in H^\infty(U, \mathcal{A})$ and $a = g(z) + (x - z)f(z) \quad (z \in U)$.*

Proof. Without loss of generality we can assume that U is bounded. Suppose on the contrary that there exist $a_n \in \mathcal{A}, f_n, g_n \in H^\infty(U, \mathcal{A}) \quad (n = 1, 2, \ldots)$ such that

$$a_n = g_n(z) + (x - z)f_n(z) \qquad (z \in U, n = 1, 2, \ldots),$$

$\|a_n\| = 1$, and $\|g_n\|_U \to 0 \quad (n \to \infty)$.

Consider the algebra $Q(\mathcal{A}) = \ell^\infty(\mathcal{A})/c_0(\mathcal{A})$ defined in Section 3. The elements of $Q(\mathcal{A})$ are bounded sequences of elements of $\mathcal{A}$ with the norm $\|(u_n)\|_{Q(\mathcal{A})} = \limsup_{n\to\infty} \|u_n\|$; we identify sequences (u_n) and (v_n) in $Q(\mathcal{A})$ if $\lim_{i\to\infty} \|u_i - v_i\| = 0$.

Write $\tilde{a} = (a_n) \in Q(\mathcal{A})$. Then $\|\tilde{a}\|_{Q(\mathcal{A})} = \limsup_{i\to\infty} \|a_i\| = 1$. Define $\tilde{g} : U \to Q(\mathcal{A})$ by $\tilde{g}(z) = (g_n(z))_n$. Since $\|g_n\|_U \to 0$ we have $\tilde{g} = 0$. Let $Z = \{z \in U : \operatorname{dist}\{z, \tau(x)\} \geq 1/2 \operatorname{dist}\{\tau(x), \mathbb{C} \setminus U\}$. The continuity of the function $z \mapsto d(x-z)$ implies that $d(x-z) \geq r \quad (z \in Z)$ for some positive r. Thus, using the maximum modulus principle,

$$\begin{aligned}\sup_n \|f_n(z)\|_U &= \sup_n \|f_n(z)\|_Z \leq r^{-1} \sup_n \|(x-z)f_n(z)\|_Z \\ &= r^{-1} \sup_n \|a_n - g_n(z)\|_Z < \infty.\end{aligned}$$

Define $\tilde{f} : U \to Q(\mathcal{A})$ by $\tilde{f}(z) = (f_n(z))$. Let $\lambda \in U$, $0 < s < \operatorname{dist}\{\lambda, \mathbb{C} \setminus U\}$, $M = \sup_n \|f_n\|_U$ and let

$$f_n(z) = \sum_{i=0}^{\infty} f_{n,i}(z-\lambda)^i \qquad \big(|z-\lambda| < \operatorname{dist}\{\lambda, \mathbb{C} \setminus U\}\big).$$

Then $\|f_{n,i}\| \leq \frac{M}{s^i}$, and so $\tilde{f}_i := (f_{n,i})_n \in Q(\mathcal{A})$ for all $i \geq 0$. Furthermore, $\tilde{f}(z) = \sum_{i=0}^{\infty} \tilde{f}_{n,i}(z-\lambda)^i \quad (|z-\lambda| < s)$. Thus $\tilde{f} : U \to Q(\mathcal{A})$ is a bounded analytic function. Further, $\tilde{a} = (x-z)\tilde{f}(z) \quad (z \in U)$ and $U \supset \tau^{\mathcal{A}}(x) = \tau^{Q(\mathcal{A})}(x)$. By the previous lemma, we have $\tilde{a} = 0$, which is a contradiction. □

Proposition 8. *Let $\mathcal{A}$ be a commutative Banach algebra, $x \in \mathcal{A}$, let U be an open neighbourhood of $\tau(x)$. Then there exists a commutative Banach algebra $\mathcal{B} \supset \mathcal{A}$ such that $\sigma^{\mathcal{B}}(x) \subset \overline{U}$.*

Proof. Consider the Banach algebra $H^\infty(U, \mathcal{A})$ of all bounded analytic functions $f : U \to \mathcal{A}$. Let $J \subset H^\infty(U, \mathcal{A})$ be the closed ideal generated by the function $z \mapsto x - z \cdot 1_{\mathcal{A}}$ and let $\mathcal{B} = H^\infty(U, \mathcal{A})/J$. Let $\rho : \mathcal{A} \to \mathcal{B}$ be the composition of the natural embedding $\mathcal{A} \to H^\infty(U, \mathcal{A})$ and the canonical projection $H^\infty(U, \mathcal{A}) \to \mathcal{B}$. Clearly, $\|\rho(a)\|_{\mathcal{B}} \leq \|a\|_{\mathcal{A}} \quad (a \in \mathcal{A})$. On the other hand,

$$\|\rho(a)\|_{\mathcal{B}} = \inf_{f \in H^\infty(U,\mathcal{A})} \big\|a - (x-z)f\big\|_U \geq k^{-1}\|a\| \qquad (a \in \mathcal{A}),$$

where k is the constant from the previous lemma. Thus ρ is an isomorphism.

Let $\lambda \notin \overline{U}$. Then $z \mapsto (z-\lambda)^{-1} 1_{\mathcal{A}} \in H^\infty(U, \mathcal{A})$ and

$$(z-\lambda)^{-1}(x-\lambda) = (z-\lambda)^{-1}\big((z-\lambda) + (x-z)\big) = 1 + (x-z)(z-\lambda)^{-1} \in 1_{\mathcal{A}} + J,$$

and so $\rho(x) - \lambda$ is invertible in $\mathcal{B}$. Thus $\sigma^{\mathcal{B}}(\rho(x)) \subset \overline{U}$.

By Theorem 1, it is possible to renorm $\mathcal{B}$ such that it becomes an extension of $\mathcal{A}$. □

Let $\mathcal{A}$ be a commutative Banach algebra and let $x \in \mathcal{A}$. We construct by transfinite induction algebras $\mathcal{A}_\alpha \quad (\alpha \le \omega_1)$ by $\mathcal{A}_0 = \mathcal{A}$, $\mathcal{A}_{\alpha+1} = Q(\mathcal{A}_\alpha)$ and for a limit ordinal α let $\mathcal{A}_\alpha$ be the completion of $\bigcup_{\beta<\alpha} \mathcal{A}_\beta$. Here ω_1 is the first uncountable ordinal. Note that we consider each $\mathcal{A}_\alpha$ to be a subalgebra of $\mathcal{A}_{\alpha+1} = Q(\mathcal{A}_\alpha)$, so we have $\mathcal{A}_\alpha \subset \mathcal{A}_\beta$ whenever $\alpha \le \beta$.

Set $\mathcal{A}' = \mathcal{A}_{\omega_1}$. Note that $\mathcal{A}' = \bigcup_{\alpha<\omega_1} \mathcal{A}_\alpha$. Indeed, let (x_n) be a Cauchy sequence in $\bigcup_{\alpha<\omega_1} \mathcal{A}_\alpha$. Then $x_n \in \mathcal{A}_{\alpha_n}$ for some $\alpha_n < \omega_1$. Let $\alpha = \sup \alpha_n$. Then the sequence (x_n) is convergent in $\mathcal{A}_\alpha \subset \mathcal{A}'$.

By Theorem 3.6, it is easy to show that $d^{\mathcal{A}_\alpha}(a) = d^{\mathcal{A}}(a)$ for all $a \in \mathcal{A}$ and $\alpha \le \omega_1$. Consequently, $\tau^{\mathcal{A}'}(x) = \tau^{\mathcal{A}}(x)$.

Lemma 9. *Let $\mathcal{A}$ be a commutative Banach algebra, $x \in \mathcal{A}$, let $U_1, U_2, \dots$ be open subsets of $\mathbb{C}$ such that $U_1 \supset U_2 \supset \cdots \supset \tau(x)$. Let $\mathcal{A}'$ be the algebra constructed above. Then there exist numbers $1 \le k_1 \le k_2 \le \cdots$ with the following property: if $n \in \mathbb{N}$, $a \in \mathcal{A}'$, $a \ne 0$, $f \in H^\infty(U_n, \mathcal{A}')$, $g_i \in H^\infty(U_i, \mathcal{A}') \quad (i = 1, \dots, n)$ and*

$$a = \sum_{i=1}^{n} g_i(z) + (x - z) f(z) \qquad (z \in U_n),$$

then $\|a\| < \sum_{i=1}^{n} k_i \|g_i\|_{U_i}$.

Proof. We find the numbers k_i inductively. The existence of k_1 was proved in Lemma 7.

Suppose that the statement of Lemma 9 is true for n, so there are positive constants $k_1, \dots, k_n$ such that

$$\|a\| < \sum_{i=1}^{n} k_i \|g_i\|_{U_i} \tag{2}$$

whenever $a = (x - z) f(z) + \sum_{i=1}^{n} g_i(z) \quad (z \in U_n)$ for some non-zero $a \in \mathcal{A}'$ and bounded analytic functions $f : U_n \to \mathcal{A}'$ and $g_i : U_i \to \mathcal{A}' \quad (i = 1, \dots, n)$. We prove (2) for $n + 1$. Suppose on the contrary that there is no constant k_{n+1} for which (2) were true. Then there exist elements $a^{(r)} \in \mathcal{A}'$ of norm 1 and analytic functions $f^{(r)} : U_{n+1} \to \mathcal{A}'$, $g_i^{(r)} : U_i \to \mathcal{A}' \quad (i = 1, \dots, n+1, r \in \mathbb{N})$ such that

$$a^{(r)} = (x - z) f^{(r)}(z) + \sum_{i=1}^{n} g_i^{(r)}(z) \qquad (z \in U_{n+1})$$

and

$$\sum_{i=1}^{n} k_i \|g_i^{(r)}\|_{U_i} + r \cdot \|g_{n+1}^{(r)}\|_{U_{n+1}} \le 1.$$

In particular, $\|g_{n+1}^{(r)}\|_{U_{n+1}} \le r^{-1}$ and $\|g_i^{(r)}\|_{U_i} \le k_i^{-1} \quad (i = 1, \dots, n, r \in \mathbb{N})$.

Passing to a subsequence if necessary, we can assume that the sequences $(\|g_1^{(r)}\|_{U_1})_{r=1}^\infty, \dots, (\|g_n^{(r)}\|_{U_n})_{r=1}^\infty$ are convergent.

It is easy to see that there is an ordinal $\beta < \omega_1$ such that $a^{(r)} \in \mathcal{A}_\beta$ and the functions $f^{(r)}, g_i^{(r)}$ are $\mathcal{A}_\beta$-valued for all r and i (to see this, note that a continuous function h is $\mathcal{A}_\beta$-valued if $h(z) \in \mathcal{A}_\beta$ for all z in a given countable set dense in the domain of definition of h).

Let $\tilde{a} = (a^{(r)})$. Then $\tilde{a}$ can be considered as element of $Q(\mathcal{A}_\beta) = \mathcal{A}_{\beta+1} \subset \mathcal{A}'$ with $\|\tilde{a}\| = \limsup_{r\to\infty} \|a^{(r)}\| = 1$. In the same way we define functions $\tilde{f} : U_{n+1} \to Q(\mathcal{A}_\beta)$ and $\tilde{g}_i : U_i \to Q(\mathcal{A}_\beta)$ $(i = 1, \ldots, n+1)$ by $\tilde{f}(z) = (f^{(r)}(z))_{r=1}^\infty$ and $\tilde{g}_i(z) = (g_i^{(r)}(z))_{r=1}^\infty$. We have

$$\|\tilde{g}_i\|_{U_i} = \sup_{z\in U_i} \limsup_{r\to\infty} \|g_i^{(r)}(z)\| \le k_i^{-1} \qquad (i = 1, \ldots, n)$$

and

$$\|\tilde{g}_{n+1}\|_{U_{n+1}} = \sup_{z\in U_{n+1}} \limsup_{r\to\infty} \|g_{n+1}^{(r)}(z)\| \le \limsup_{r\to\infty} r^{-1} = 0,$$

so $\tilde{g}_{n+1} = 0$. As in the proof of Lemma 7 we can show that functions $f^{(r)}$ are uniformly bounded on U_{n+1}, and so $\tilde{g}_1, \ldots, \tilde{g}_n, \tilde{f}$ are bounded analytic functions. We have $\tilde{a} = \sum_{i=1}^n \tilde{g}_i(z) + (x - z)\tilde{f}(z)$ $(z \in U_{n+1})$. By Lemma 5, $\tilde{f}$ can be extended to U_n. Since $Q(\mathcal{A}_\beta) = \mathcal{A}_{\beta+1} \subset \mathcal{A}'$, by the induction assumption we have

$$\begin{aligned} 1 = \|\tilde{a}\| &< \sum_{i=1}^n k_i \|\tilde{g}_i\|_{U_i} = \sum_{i=1}^n k_i \sup_{z\in U_i} \limsup_{r\to\infty} \|g_i^{(r)}(z)\| \\ &\le \sum_{i=1}^n k_i \limsup_{r\to\infty} \|g_i^{(r)}\|_{U_i} = \lim_{r\to\infty} \sum_{i=1}^n k_i \|g_i^{(r)}\|_{U_i} \le 1, \end{aligned}$$

a contradiction. □

Theorem 10. *Let x be an element of a commutative Banach algebra $\mathcal{A}$. Then there exists a commutative extension $\mathcal{B} \supset \mathcal{A}$ such that $\tau^{\mathcal{A}}(x) = \sigma^{\mathcal{B}}(x)$.*

Proof. For $i = 1, 2, \ldots$ set $U_i = \{z \in \mathbb{C} : \operatorname{dist}\{z, \tau^{\mathcal{A}}(x)\} < 1/i\}$. Then U_i are open subsets of $\mathbb{C}$, $U_1 \supset U_2 \supset \cdots$ and $\bigcap_{i=1}^\infty U_i = \tau^{\mathcal{A}}(x)$. Let $k_1, k_2, \ldots$ be the positive numbers constructed in the previous lemma. Let $\mathcal{C}$ be the algebra of all $\mathcal{A}$-valued functions analytic on a neighbourhood of $\tau^{\mathcal{A}}(x)$ (more precisely, the algebra of all germs of analytic functions). For $h \in \mathcal{C}$ define $|||h|||$ by

$$|||h||| = \inf\Big\{\|c\|_{\mathcal{A}} + \sum_{i=1}^n k_i \|g_i\|_{U_i}\Big\},$$

where the infimum is taken over all $n \in \mathbb{N}$, $c \in \mathcal{A}$ and over all bounded analytic functions $g_i : U_i \to \mathcal{A}$ $(i = 1, \ldots, n)$ such that, for some $f \in H^\infty(U_n, \mathcal{A})$,

$$h(z) = c + \sum_{i=1}^n g_i(z) + (x - z)f(z) \qquad (z \in U_n).$$

We identify elements of $\mathcal{A}$ with the corresponding constant functions; so $\mathcal{A} \subset \mathcal{C}$.

Let $a \in \mathcal{A}$. We have $|||a||| \le \|a\|_{\mathcal{A}}$ and, by the preceding lemma,

$$|||a||| = \inf\Big\{\|c\|_{\mathcal{A}} + \sum_{i=1}^{n} k_i\|g_i\|_{U_i} : a = c + \sum_{i=1}^{n} g_i(z) + (x-z)f(z)\Big\}$$
$$\ge \inf_{c\in\mathcal{A}}\{\|c\|_{\mathcal{A}} + \|a-c\|_{\mathcal{A}}\} \ge \|a\|_{\mathcal{A}}.$$

Thus $|||a||| = \|a\|_{\mathcal{A}}$ for all $a \in \mathcal{A}$.

It is clear that $|||h+h'||| \le |||h||| + |||h'|||$ and $|||\alpha h||| = |\alpha| \cdot |||h|||$ for all $h, h' \in \mathcal{C}$, $\alpha \in \mathbb{C}$.

We show that $|||\cdot|||$ is submultiplicative. Let $h, h' \in \mathcal{C}$ and $\varepsilon > 0$. By the definition of $|||\cdot|||$, there are $c, c' \in \mathcal{A}$, $n, m \in \mathbb{N}$ and functions $g_i \in H^\infty(U_i, \mathcal{A})$ $(i = 1, \dots, n)$, $f \in H^\infty(U_n, \mathcal{A})$, $g'_j \in H^\infty(U_j, \mathcal{A})$ $(j = 1, \dots, m)$, $f' \in H^\infty(U_m, \mathcal{A})$ such that

$$h(z) = c + \sum_{i=1}^{n} g_i(z) + (x-z)f(z) \qquad (z \in U_n),$$

$$h'(z) = c' + \sum_{j=1}^{m} g'_j(z) + (x-z)f'(z) \qquad (z \in U_m),$$

$$|||h||| \le \varepsilon + \|c\|_{\mathcal{A}} + \sum_{i=1}^{n} k_i\|g_i\|_{U_i}$$

and

$$|||h'||| \le \varepsilon + \|c'\|_{\mathcal{A}} + \sum_{j=1}^{m} k_j\|g'_j\|_{U_j}.$$

For some function $u \in H^\infty(U_{\max\{n,m\}}, \mathcal{A})$ we have

$$(hh')(z) = cc' + \sum_{i=1}^{n} c'g_i(z) + \sum_{j=1}^{m} cg'_j(z) + \sum_{i=1}^{n}\sum_{j=1}^{m} g_i(z)g'_j(z) + (x-z)u(z)$$

for $z \in U_{\max\{n,m\}}$. By definition,

$$|||hh'||| \le \|c\| \cdot \|c'\| + \sum_{i=1}^{n} k_i\|c'\| \cdot \|g_i\|_{U_i} + \sum_{j=1}^{m} k_j\|c\| \cdot \|g'_j\|_{U_j}$$
$$+ \sum_{i=1}^{n}\sum_{j=1}^{m} k_{\max\{i,j\}}\|g_ig'_j\|_{U_{\max\{i,j\}}}$$
$$\le \Big\{\|c\| + \sum_{i=1}^{n} k_i\|g_i\|_{U_i}\Big\} \cdot \Big\{\|c'\| + \sum_{j=1}^{m} k_j\|g'_j\|_{U_j}\Big\} \le (|||h||| + \varepsilon)(|||h'||| + \varepsilon).$$

Since $\varepsilon > 0$ was arbitrary, we have $|||hh'||| \le |||h||| \cdot |||h'|||$.

Let $J = \{h \in \mathcal{C} : |||h||| = 0\}$. It is clear that J is an ideal in $\mathcal{C}$. Denote by $\mathcal{B}$ the completion of $\mathcal{C}/J$. Then $\mathcal{B}$ is an extension of $\mathcal{A}$. Furthermore, $(x - z)f \in J$ for each $f \in \mathcal{C}$. Let $\lambda \in \mathbb{C} \setminus \tau^{\mathcal{A}}(x)$. Then

$$(z-\lambda)^{-1}(x-\lambda) = (z-\lambda)^{-1}\big((x-z)+(z-\lambda)\big) = 1 + (x-z)(z-\lambda)^{-1} \in 1 + J,$$

and so $x - \lambda$ is invertible in $\mathcal{B}$. Hence $\sigma^{\mathcal{B}}(x) \subset \tau^{\mathcal{A}}(x)$ and the opposite inclusion is clear. □

The following theorem is an analogue of the spectral radius formula.

Theorem 11. *Let x be an element of a commutative Banach algebra $\mathcal{A}$. Then the limit $\lim_{k\to\infty} d(x^k)^{1/k}$ exists and*

$$\lim_{k\to\infty} d(x^k)^{1/k} = \sup_{k\in\mathbb{N}} d(x^k)^{1/k} = \min\{|\lambda| : \lambda \in \tau(x)\}. \tag{3}$$

Proof. If $0 \in \tau(x)$, then $0 \in \tau(x^k)$ for all k by the spectral mapping theorem, and both sides of (3) are equal to 0.

Suppose that $0 \notin \tau(x)$. By Lemma 3.1, $d(x^{k+l}) \geq d(x^k) \cdot d(x^l)$ for all $k, l \in \mathbb{N}$. By Lemma 1.21, the limit $\lim_{k\to\infty} d(x^k)^{1/k}$ exists and is equal to the supremum.

Let $\lambda \in \mathbb{C}$, $|\lambda| < d(x)$. Then $d(x - \lambda) \geq d(x) - |\lambda| > 0$, and so $\lambda \notin \tau(x)$. Thus $d(x) \leq \min\{|\lambda| : \lambda \in \tau(x)\}$. Similarly,

$$d(x^k) \leq \min\{|\mu| : \mu \in \tau(x^k)\} = \min\{|\lambda| : \lambda \in \tau(x)\}^k$$

for each $k \in \mathbb{N}$. Hence

$$\lim_{k\to\infty} d(x^k)^{1/k} \leq \min\{|\lambda| : \lambda \in \tau(x)\}.$$

To prove the opposite inequality, let $\mathcal{B} \supset \mathcal{A}$ be a commutative extension of the algebra $\mathcal{A}$ satisfying $\sigma^{\mathcal{B}}(x) = \tau^{\mathcal{A}}(x)$. Then x is invertible in $\mathcal{B}$, and so

$$\min\{|\lambda| : \lambda \in \tau^{\mathcal{A}}(x)\} = \min\{|\lambda| : \lambda \in \sigma^{\mathcal{B}}(x)\} = r(x^{-1})^{-1}$$
$$= \lim_{k\to\infty} \|x^{-k}\|^{-1/k} = \lim_{k\to\infty} d^{\mathcal{B}}(x^k)^{1/k} \leq \lim_{k\to\infty} d^{\mathcal{A}}(x^k)^{1/k}.$$ □

5 Non-removable ideals

The notion of permanently singular elements can be generalized to ideals.

Definition 1. An ideal I in a commutative Banach algebra $\mathcal{A}$ is called *non-removable* if in every commutative Banach algebra $\mathcal{B} \supset \mathcal{A}$ there exists a proper ideal $J \supset I$.

Non-removable ideals are closely related to permanently singular elements. Clearly, $x \in \mathcal{A}$ is permanently singular if and only if the ideal $x\mathcal{A}$ is non-removable.

It is easy to see that an ideal consisting of joint topological divisors of zero is non-removable.

Our goal will be to show that the opposite statement is also true, so there is an analogy with the characterization of permanently singular elements.

Theorem 2. *Let $u_1, \dots, u_n$ be elements of a commutative Banach algebra $\mathcal{A}$ such that $d(u_1, \dots, u_n) > 0$. Then there exist a commutative Banach algebra $\mathcal{B} \supset \mathcal{A}$ and elements $b_1, \dots, b_n \in \mathcal{B}$ such that $\sum_{i=1}^n u_i b_i = 1$.*

Proof. For $n = 1$, the result was proved in Theorem 4.3. Assume that $n \geq 2$. The extension $\mathcal{B} \supset \mathcal{A}$ will be constructed in a similar way.

We may assume that $\|u_1\| = \dots = \|u_n\| = 1$. Let $q = (d(u_1, \dots, u_n))^{-1}$, so $\|x\| \leq q \sum_{i=1}^n \|xu_i\|$ for every $x \in \mathcal{A}$. In particular, for $x = 1_{\mathcal{A}}$ we get $qn \geq 1$.

Set $R = 8(n-1)^2(nq)^{2n+1}$. Let $\mathcal{C}$ be the ℓ^1-algebra over $\mathcal{A}$ and adjoined elements $b_1, \dots, b_n$ such that $\|b_1\| = \dots = \|b_n\| = R$. More precisely, the elements of $\mathcal{C}$ are power series $\sum_{\alpha \in \mathbb{Z}_+^n} a_\alpha b_1^{\alpha_1} \cdots b_n^{\alpha_n}$ in n variables $b_1, \dots, b_n$ with coefficients $a_\alpha \in \mathcal{A}$ $\quad (\alpha = (\alpha_1, \dots, \alpha_n) \in \mathbb{Z}_+^n)$ such that

$$\Big\| \sum_{\alpha \in \mathbb{Z}_+^n} a_\alpha b_1^{\alpha_1} \cdots b_n^{\alpha_n} \Big\| = \sum_{\alpha \in \mathbb{Z}_+^n} \|a_\alpha\| R^{|\alpha|} < \infty.$$

The multiplication in $\mathcal{C}$ is defined by

$$\Big(\sum_{\alpha \in \mathbb{Z}_+^n} a_\alpha b^\alpha \Big) \cdot \Big(\sum_{\beta \in \mathbb{Z}_+^n} a'_\beta b^\beta \Big) = \sum_{\gamma \in \mathbb{Z}_+^n} \Big(\sum_{\alpha+\beta=\gamma} a_\alpha a'_\beta \Big) b^\gamma,$$

where b^α stands for $b_1^{\alpha_1} \cdots b_n^{\alpha_n}$.

It is clear that $\mathcal{C}$ is a commutative Banach algebra containing $\mathcal{A}$ as a subalgebra of constants.

Let $z = 1 - u_1 b_1 - \dots - u_n b_n$ and let J be the closed ideal in $\mathcal{C}$ generated by z.

Set $\mathcal{B} = \mathcal{C}/J$. Let $\rho : \mathcal{A} \to \mathcal{B}$ be the composition of the embedding $\mathcal{A} \to \mathcal{C}$ and the canonical homomorphism $\mathcal{C} \to \mathcal{B}$. Then $\rho(u_1) \cdot (b_1 + J) + \dots + \rho(u_n) \cdot (b_n + J) = 1_{\mathcal{B}}$, so it is sufficient to show that ρ is an isometry.

Obviously, $\|\rho(a)\|_{\mathcal{B}} \leq \|a\|_{\mathcal{A}}$ for all $a \in \mathcal{A}$. Suppose on the contrary that there exists $a \in \mathcal{A}$ such that

$$\|a\|_{\mathcal{A}} > \|\rho(a)\|_{\mathcal{B}} = \inf_{c \in \mathcal{C}} \|a + cz\|_{\mathcal{C}}.$$

Thus there exists $c = \sum_{\alpha \in \mathbb{Z}_+^n} a_\alpha b^\alpha$ such that

$$\|a\| > \Big\| a + z \sum_{\alpha \in \mathbb{Z}_+^n} a_\alpha b^\alpha \Big\|_{\mathcal{C}}.$$

Since the polynomials are dense in $\mathcal{C}$, we can assume that only finitely many coefficients a_α are non-zero and

$$\begin{aligned}\|a\| &> \Big\|a + z \sum_{\alpha\in\mathbb{Z}_+^n} a_\alpha b^\alpha\Big\|_{\mathcal{C}} = \Big\|a + (1 - u_1b_1 + \cdots + u_nb_n) \sum_{\alpha\in\mathbb{Z}_+^n} a_\alpha b^\alpha\Big\|_{\mathcal{C}} \\ &= \Big\|a + a_0 + \sum_{\substack{\alpha\in\mathbb{Z}_+^n\\|\alpha|\geq 1}} f_\alpha b^\alpha\Big\|_{\mathcal{C}} = \|a + a_0\| + \sum_{\substack{\alpha\in\mathbb{Z}_+^n\\|\alpha|\geq 1}} \|f_\alpha\| \cdot R^{|\alpha|} \\ &\geq \|a\| - \|a_0\| + \sum_{\substack{\alpha\in\mathbb{Z}_+^n\\|\alpha|\geq 1}} \|f_\alpha\| \cdot R^{|\alpha|},\end{aligned}$$

where

$$f_\alpha = a_\alpha - \sum_{\substack{1\leq t\leq n\\ \alpha_t\neq 0}} a_{\alpha_1,\ldots,\alpha_t-1,\ldots,\alpha_n} u_t \qquad (\alpha \in \mathbb{Z}_+^n, |\alpha| \geq 1). \tag{1}$$

Thus

$$\|a_0\| > \sum_{\substack{\alpha\in\mathbb{Z}_+^n\\|\alpha|\geq 1}} \|f_\alpha\| R^{|\alpha|}.$$

Without loss of generality we can assume that $\|a_0\| = 1$. For $k \in \mathbb{N}$ write

$$S_k = \sum_{\substack{\alpha\in\mathbb{Z}_+^n\\|\alpha|=k}} \|f_\alpha\| \cdot R^{|\alpha|},$$

so $\sum_{k=1}^{\infty} S_k < 1$. We will show that this is impossible.

The idea of the proof is to express a_α in terms of f_γ's (which are small in the norm) and a_γ's with $|\gamma| > |\alpha|$ (which are equal to zero for $|\gamma|$ large enough since only finitely many of the elements a_γ are non-zero).

First, we need some technical lemmas.

We will use the standard multi-index notation, see Appendix A.2. For $j, n \geq 1$ and $\alpha \in \mathbb{Z}_+^n$ write

$$m_{j,\alpha} = \binom{|\alpha| + j - 1}{j - 1} \cdot \frac{|\alpha|!}{\alpha!}.$$

Note that $m_{1,\alpha} = \frac{|\alpha|!}{\alpha!}$, and so $m_{j,\alpha} = \binom{|\alpha|+j-1}{j-1} \cdot m_{1,\alpha}$.

Lemma 3. *If $\alpha \in \mathbb{Z}_+^n$ and $|\alpha| \geq 1$, then*

$$m_{1,\alpha} = \sum_{\substack{\beta\leq\alpha\\|\beta|=|\alpha|-1}} m_{1,\beta}.$$

Proof. We have

$$\alpha! \sum_{\substack{\beta\le\alpha \\ |\beta|=|\alpha|-1}} m_{1,\beta} = \alpha_1!\cdots\alpha_n! \sum_{\substack{1\le t\le n \\ \alpha_t\neq 0}} \frac{(|\alpha|-1)!}{\alpha_1!\cdots(\alpha_t-1)!\cdots\alpha_n!}$$

$$= \sum_{t=1}^{n} \alpha_t(|\alpha|-1)! = |\alpha|!,$$

which implies the statement of Lemma 3. □

Lemma 4. *For all $\alpha \in \mathbb{Z}_+^n$ and $s \ge 0$ we have*

$$\sum_{\substack{\alpha\in\mathbb{Z}_+^n \\ |\alpha|=s}} m_{1,\alpha} = n^s.$$

Proof. The proof follows from the identity

$$(x_1+\cdots+x_n)^s = \sum_{\substack{\alpha\in\mathbb{Z}_+^n \\ |\alpha|=s}} m_{1,\alpha} x_1^{\alpha_1}\cdots x_n^{\alpha_n}$$

by substituting $x_1 = \cdots = x_n = 1$. □

Note that for $n = 2$, the statement of Lemma 4 reduces to the well-known identity $\sum_{k=0}^{n} \binom{s}{k} = 2^s$.

Lemma 5. *For $j \ge 1$ and $\alpha \in \mathbb{Z}_+^n$ we have*

$$m_{j,\alpha} = \sum_{\substack{\beta\in\mathbb{Z}_+^n \\ \beta\le\alpha}} m_{1,\beta} m_{j-1,\alpha-\beta}.$$

Proof. The number $m_{1,\alpha}$ is the number of ways to order n elements $x_1, \ldots, x_n$ into a sequence of length $|\alpha|$ in which every element x_t occurs exactly α_t times (permutations with repetition). The number $m_{j,\alpha}$ is the number of ways to divide these sequences into j (possibly empty) subsequences (combinations with repetition), i.e., in how many ways it is possible to form j sequences $s_1, \ldots, s_j$ from the elements $x_1, \ldots, x_n$ such that x_t occurs in them altogether α_t times (sequences s_1, s_2 and s_2, s_1 are counted twice).

The right-hand side is the same number obtained in another way: for $\beta \le \alpha$, $m_{1,\beta} m_{j-1,\alpha-\beta}$ is the number of ways how to form these subsequences in such a way that x_t occurs in the first subsequence exactly β_t-times ($t = 1, \ldots, n$). □

For $\alpha = (\alpha_1, \ldots, \alpha_n) \in \mathbb{Z}_+^n$ we write $\alpha' = (\alpha_1, \ldots, \alpha_{n-1})$. Similarly, we write $u' = (u_1, \ldots, u_{n-1})$ and $u'^{\alpha'} = u_1^{\alpha_1} \cdots u_{n-1}^{\alpha_{n-1}}$.

Lemma 6. *Let* $k, j \geq 1$, $\alpha \in \mathbb{Z}_+^n$, $|\alpha| = k$. *Then*

$$a_\alpha u_n^{k+j} = \sum_{\beta \leq \alpha'} (-1)^{|\beta|} m_{j\beta} a_{\alpha'-\beta,\alpha_n+|\beta|+j} u'^{\beta} u_n^{k-|\beta|}$$
$$+ \sum_{p=1}^{j} \sum_{\beta \leq \alpha'} (-1)^{|\beta|+1} m_{p,\beta} f_{\alpha'-\beta,\alpha_n+|\beta|+p} u'^{\beta} u_n^{k+j-p-|\beta|}.$$

Proof. Note that all terms $a_\gamma u^\delta$ and $f_\gamma u^\delta$ which appear in the statement of Lemma 6 satisfy $\gamma + \delta = (\alpha', \alpha_n + k + j)$. In order to simplify the notation, for $\gamma = (\gamma', \gamma_n) \in \mathbb{Z}_+^n$ with $\gamma \leq (\alpha', \alpha_n + k + j)$ write

$$\begin{aligned} d_\gamma &= a_\gamma u'^{\alpha'-\gamma'} u_n^{\alpha_n+k+j-\gamma_n}, \\ g_\gamma &= f_\gamma u'^{\alpha'-\gamma'} u_n^{\alpha_n+k+j-\gamma_n}. \end{aligned} \tag{2}$$

Using this notation, the statement of Lemma 6 can be rewritten as

$$d_\alpha = \sum_{\beta \leq \alpha'} (-1)^{|\beta|} m_{j,\beta} d_{\alpha'-\beta,\alpha_n+|\beta|+j} + \sum_{p=1}^{j} \sum_{\beta \leq \alpha'} (-1)^{|\beta|+1} m_{p,\beta} g_{\alpha'-\beta,\alpha_n+|\beta|+p}. \tag{3}$$

We prove (3) by induction on j.

For $j = 1$, (3) becomes

$$d_\alpha = \sum_{\beta \leq \alpha'} (-1)^{|\beta|} m_{1,\beta} \big(d_{\alpha'-\beta,\alpha_n+|\beta|+1} - g_{\alpha'-\beta,\alpha_n+|\beta|+1}\big). \tag{4}$$

By (1) and (2) we have

$$g_\delta = d_\delta - \sum_{\substack{\gamma \leq \delta \\ |\gamma| = |\delta|-1}} d_\gamma$$

for every δ. Thus (4) can be rewritten as

$$d_\alpha = \sum_{\beta \leq \alpha'} (-1)^{|\beta|} m_{1,\beta} \sum_{\substack{\gamma \leq (\alpha'-\beta,\alpha_n+|\beta|+1) \\ |\gamma| = |\alpha|}} d_\gamma. \tag{5}$$

The condition $\gamma_n \leq \alpha_n + |\beta| + 1$ means $|\gamma| - |\gamma'| \leq |\alpha| - |\alpha'| + |\beta| + 1$, and so $|\beta| \geq |\alpha' - \gamma'| - 1$. Thus (5) becomes

$$d_\alpha = \sum_{\substack{|\gamma| = |\alpha| \\ \gamma' \leq \alpha'}} d_\gamma \left(\sum_{\substack{\beta \leq \alpha' - \gamma' \\ |\beta| \geq |\alpha'-\gamma'|-1}} (-1)^{|\beta|} m_{1,\beta} \right). \tag{6}$$

By Lemma 3, the expression in the parenthesis is equal to 0 whenever $\gamma' \neq \alpha'$. This proves (6), and also (3) for $j = 1$.

Suppose that (3) holds for some $j \geq 1$. We prove it for $j+1$. Substituting (4) into the induction assumption (3) yields

$$\begin{aligned} d_\alpha = \sum_{\beta \leq \alpha'} (-1)^{|\beta|} m_{j,\beta} & \sum_{\gamma \leq \alpha'-\beta} (-1)^{|\gamma|} m_{1,\gamma} \\ & \cdot \left(d_{\alpha'-\beta-\gamma,\alpha_n+|\beta|+j+|\gamma|+1} - g_{\alpha'-\beta-\gamma,\alpha_n+|\beta|+j+|\gamma|+1}\right) \\ & + \sum_{p=1}^{j} \sum_{\beta \leq \alpha'} (-1)^{|\beta|+1} m_{p,\beta} g_{\alpha'-\beta,\alpha_n+|\beta|+p}. \end{aligned}$$

Using Lemma 5, the first sum can be rewritten as

$$\begin{aligned} & \sum_{\beta \leq \alpha'} \sum_{\gamma \leq \alpha'-\beta} (-1)^{|\beta+\gamma|} m_{j\beta} m_{1\gamma} \cdot \\ & \qquad \left(d_{\alpha'-\beta-\gamma,\alpha_n+|\beta|+|\gamma|+j+1} - g_{\alpha'-\beta-\gamma,\alpha_n+|\beta|+|\gamma|+j+1}\right) \\ & = \sum_{\delta \leq \alpha'} \Big(\sum_{\beta+\gamma=\delta} m_{j\beta} m_{1\gamma} \Big) (-1)^{|\delta|} \left(d_{\alpha'-\delta,\alpha_n+|\delta|+j+1} - g_{\alpha'-\delta,\alpha_n+|\delta|+j+1}\right) \\ & = \sum_{\delta \leq \alpha'} m_{j+1,\delta} (-1)^{|\delta|} \left(d_{\alpha'-\delta,\alpha_n+|\delta|+j+1} - g_{\alpha'-\delta,\alpha_n+|\delta|+j+1}\right). \end{aligned}$$

Thus (replacing δ by β) we have

$$\begin{aligned} d_\alpha = & \sum_{\beta \leq \alpha'} (-1)^{|\beta|} m_{j+1,\beta} \left(d_{\alpha'-\beta,\alpha_n+|\beta|+j+1} - g_{\alpha'-\beta,\alpha_n+|\beta|+j+1}\right) \\ & + \sum_{p=1}^{j} \sum_{\beta \leq \alpha'} (-1)^{|\beta|+1} m_{p,\beta} g_{\alpha'-\beta,\alpha_n+|\beta|+p}, \end{aligned}$$

which is (3) for $j+1$. This finishes the proof of Lemma 6. □

For $k, j \in \mathbb{N}$ set

$$s_{k,j} = \max_{\substack{\alpha \in \mathbb{Z}_+^n \\ |\alpha|=k}} \max_{\substack{\beta \in \mathbb{Z}_+^n \\ |\beta|=j}} \|a_\alpha u^\beta\|.$$

Corollary 7. *If $k, j \geq 1$, $\alpha, \gamma \in \mathbb{Z}_+^n$ and $|\alpha| = k$, then*

$$\|a_\alpha u_n^{k+j} u^\gamma\| \leq 2^{k+j}(n-1)^k s_{k+j,|\gamma|+k} + 2^k (n-1)^k R^{-k-1} \sum_{p=k+1}^{k+j} S_p.$$

Proof. By the previous lemma, we have

$$\begin{aligned} \|a_\alpha u_n^{k+j} u^\gamma\| \leq & \sum_{\beta \leq \alpha'} m_{j,\beta} \left\| a_{\alpha'-\beta,\alpha_n+|\beta|+j} u'^\beta u_n^{k-|\beta|} u^\gamma \right\| \\ & + \sum_{p=1}^{j} \sum_{\beta \leq \alpha'} m_{p,\beta} \|f_{\alpha'-\beta,\alpha_n+|\beta|+p}\|. \end{aligned}$$

We have

$$m_{p,\beta} = \binom{|\beta|+p-1}{p-1} m_{1,\beta} \le 2^{|\beta|+p-1} \cdot (n-1)^{|\beta|} \le 2^{k+p-1} \cdot (n-1)^k$$

and

$$\sum_{\beta \le \alpha'} m_{j,\beta} \le \sum_{\beta \le \alpha'} m_{j,\beta} m_{1,\alpha'-\beta} = m_{j+1,\alpha'} \le 2^{k+j}(n-1)^k.$$

Thus

$$\|a_\alpha u_n^{k+j} u^\gamma\| \le 2^{k+j}(n-1)^k s_{k+j,|\gamma|+k} + 2^k(n-1)^k \sum_{p=1}^{j} \sum_{\beta \le \alpha'} 2^{p-1} \|f_{\alpha'-\beta,\alpha_n+|\beta|+p}\|.$$

Furthermore,

$$\begin{aligned}
&\sum_{p=1}^{j} \sum_{\beta \le \alpha'} 2^{p-1} \|f_{\alpha'-\beta,\alpha_n+|\beta|+p}\| \\
&= \frac{1}{R^{k+1}} \sum_{p=1}^{j} \frac{2^{p-1}}{R^{p-1}} \sum_{\beta \le \alpha'} \|f_{\alpha'-\beta,\alpha_n+|\beta|+p}\| R^{k+p} \\
&\le \frac{1}{R^{k+1}} \sum_{p=1}^{j} \frac{2^{p-1}}{R^{p-1}} S_{k+p} \le \frac{1}{R^{k+1}} \sum_{p=k+1}^{k+j} S_p,
\end{aligned}$$

which gives the statement of Corollary 7. □

For $k \in \mathbb{N}$ set $r_k = s_{k,nk}$.

Corollary 8. *If* $k \in \mathbb{N}$, *then*

$$r_k \le R_1^k r_{2k} + \frac{1}{RR_1^k} \sum_{p=k+1}^{2k} S_p,$$

where $R_1 = 4(n-1)(nq)^{n+1}$.

Proof. For each $x \in \mathcal{A}$ we have

$$\|x\| \le q \cdot \sum_{t=1}^{n} \|xu_t\| \le qn \max\{\|xu_t\| : 1 \le t \le n\}.$$

It is easy to show by induction on j that

$$\|x\| \le (qn)^j \max\{\|xu^\gamma\| : \gamma \in \mathbb{Z}_+^n, |\gamma| = j\}.$$

Let $\alpha, \beta \in \mathbb{Z}_+^n$, $|\alpha| = k$, $|\beta| = nk$. Then

$$\|a_\alpha u^\beta\| \le (qn)^{nk} \max\{\|a_\alpha u^{\beta+\gamma}\| : \gamma \in \mathbb{Z}_+^n, |\gamma| = nk\}.$$

Since $|\beta+\gamma| = 2nk$, there exists an index t, $1 \le t \le n$ such that $(\beta+\gamma)_t \ge 2k$. Since the situation is symmetrical in indices, we can apply Corollary 7.

Thus, for $j = k$,

$$\begin{aligned}\|a_\alpha u^\beta\| &\le (qn)^{nk}\Big(2^{2k}(n-1)^k s_{2k,2nk-k} + 2^k(n-1)^k R^{-k-1}\sum_{p=k+1}^{2k} S_p\Big)\\ &\le (qn)^{nk}\Big(4^k(n-1)^k(qn)^k r_{2k} + 2^k(n-1)^k R^{-k-1}\sum_{p=k+1}^{2k} S_p\Big)\\ &= R_1^k r_{2k} + \frac{1}{R}\cdot\frac{(qn)^{nk}2^k(n-1)^k}{8^k(n-1)^{2k}(qn)^{2nk+k}}\sum_{p=k+1}^{2k} S_p = R_1^k r_{2k} + \frac{1}{RR_1^k}\sum_{p=k+1}^{2k} S_p. \qquad \square\end{aligned}$$

Continuation of the proof of Theorem 2:
We prove by induction on j that

$$R_1 r_1 \le R_1^{2^j} r_{2^j} + \frac{1}{R}\sum_{p=2}^{2^j} S_p. \tag{7}$$

For $j = 1$ this follows from Corollary 8. If the statement is true for some $j \ge 1$, then, by the induction assumption and Corollary 8,

$$R_1 r_1 \le R_1^{2^j}\Big(R_1^{2^j} r_{2^{j+1}} + \frac{1}{RR_1^{2^j}}\sum_{p=2^j+1}^{2^{j+1}} S_p\Big) + \frac{1}{R}\sum_{p=2}^{2^j} S_p = R_1^{2^{j+1}} r_{2^{j+1}} + \frac{1}{R}\sum_{p=2}^{2^{j+1}} S_p,$$

which gives (7) for $j+1$.

Since $r_k = 0$ for k sufficiently large, we have $R_1 r_1 \le \frac{1}{R}\sum_{p=2}^{\infty} S_p < \frac{1}{R}$ and $r_1 < \frac{1}{RR_1}$.

Furthermore, we have

$$\begin{aligned}\|a_0\| &\le q\sum_{t=1}^n \|a_0 u_t\| \le q\sum_{t=1}^n\big(\|a_{\underbrace{0,\dots,0}_{t-1},1,0,\dots,0}\| + \|a_0 u_t - a_{\underbrace{0,\dots,0}_{t-1},1,0,\dots,0}\|\big)\\ &\le qns_{1,0} + q\sum_{t=1}^n \|f_{\underbrace{0,\dots,0}_{t-1},1,0,\dots,0}\| \le (qn)^{n+1} r_1 + \frac{qS_1}{R} \le \frac{(qn)^{n+1}}{RR_1} + \frac{q}{R} < 1,\end{aligned}$$

which is a contradiction. $\square$

Theorem 9. *An ideal in a commutative Banach algebra is non-removable if and only if it consists of joint topological divisors of zero.*

Proof. Let I be an ideal in a Banach algebra $\mathcal{A}$ consisting of joint topological divisors of zero. Then there exists a net $\{z_\alpha\} \subset \mathcal{A}$ of elements of norm 1 such that $z_\alpha x \to 0$ for every $x \in I$. Let $\mathcal{B} \supset \mathcal{A}$ be a commutative extension and let

$J = \{x_1b_1 + \cdots + x_nb_n : n \in \mathbb{N}, x_1, \ldots, x_n \in I, b_1, \ldots, b_n \in \mathcal{B}\}$ be the smallest ideal in $\mathcal{B}$ containing I. Suppose on the contrary that J is not proper, and so $1_{\mathcal{B}} = x_1b_1 + \cdots + x_nb_n$ for some $x_1, \ldots, x_n \in I, b_1, \ldots, b_n \in \mathcal{B}$. Then

$$1 = \|z_\alpha\| = \left\| \sum_{i=1}^{n} z_\alpha x_i b_i \right\| \le \sum_{i=1}^{n} \|z_\alpha x_i\| \cdot \|b_i\| \to 0,$$

which is a contradiction. Hence I is non-removable.

Conversely, let $I \notin \mathfrak{k}(\mathcal{A})$. It means that there are elements $u_1, \ldots, u_n \in I$ such that $d(u_1, \ldots, u_n) > 0$.

Let $\mathcal{B} \supset \mathcal{A}$ be the extension constructed in Theorem 2. Clearly, I is contained in no proper ideal in $\mathcal{B}$, and so I is a removable ideal. □

Corollary 10. *Let φ be a multiplicative functional on a commutative Banach algebra $\mathcal{A}$. The following statements are equivalent:*

(i) $\varphi \in \operatorname{cor} \mathcal{A}$;

(ii) *for every commutative Banach algebra $\mathcal{B} \supset \mathcal{A}$ there exists a multiplicative functional $\psi \in \mathcal{M}(\mathcal{B})$ such that $\varphi = \psi|\mathcal{A}$;*

(iii) *for every commutative Banach algebra $\mathcal{B} \supset \mathcal{A}$ there exists a multiplicative functional $\psi \in \operatorname{cor} \mathcal{B}$ such that $\varphi = \psi|\mathcal{A}$.*

Proof. (i) ⇒ (iii): Let $\varphi \in \operatorname{cor} \mathcal{A}$ and $\mathcal{B} \supset \mathcal{A}$. Since $\operatorname{Ker} \varphi \in \mathfrak{k}(\mathcal{A})$, there exists a net $(z_\alpha)_\alpha$ of elements of $\mathcal{A}$ of norm 1 such that $xz_\alpha \to 0$ for every $x \in \operatorname{Ker} \varphi$. Let $I = \{y \in \mathcal{B} : yz_\alpha \to 0\}$. Then $I \supset \operatorname{Ker} \varphi$ and $I \in \mathfrak{k}(\mathcal{B})$, so there exists a maximal ideal $J \in \mathfrak{k}(\mathcal{B})$ containing I. The corresponding multiplicative functional $\psi \in \operatorname{cor} \mathcal{B}$ satisfies $\psi|\mathcal{A} = \varphi$.

(iii) ⇒ (ii): Clear.

(ii) ⇒ (i): If $\mathcal{B} \supset \mathcal{A}$ and $\psi \in \mathcal{M}(\mathcal{B})$ extends φ, then $\operatorname{Ker} \varphi \subset \operatorname{Ker} \psi$. Hence $\operatorname{Ker} \varphi$ is a non-removable ideal, and $\varphi \in \operatorname{cor} \mathcal{A}$. □

Corollary 11. *Let $\mathcal{A}$ be a commutative Banach algebra, $\varphi \in \Gamma(\mathcal{A})$ and let $\mathcal{B}$ be a commutative extension of $\mathcal{A}$. Then there exists $\psi \in \Gamma(\mathcal{B})$ such that $\varphi = \psi|\mathcal{A}$.*

Proof. By Theorem 3.11, $\operatorname{Ker}(\varphi) \in \gamma(\mathcal{A})$. By Lemma 3.12, there exists a net $(z_\lambda)_\alpha \subset \mathcal{A}$ such that $r(z_\alpha) = 1$ for every α and $\lim_\alpha r(xz_\alpha) = 0$ for all $x \in \operatorname{Ker} \varphi$. Let $I = \{b \in \mathcal{B} : \lim_\alpha r(bz_\alpha) = 0\}$. Then $I \in \gamma(\mathcal{B})$ and there exists a multiplicative functional $\psi \in \Gamma(\mathcal{B})$ such that $\operatorname{Ker} \varphi \subset I \subset \operatorname{Ker} \psi$. Thus ψ is the required extension of φ. □

The following result is an easy consequence of Theorem 2.

Theorem 12. *Let $x_1, \ldots, x_n$ be elements of a commutative Banach algebra $\mathcal{A}$. Then*

$$\tau^{\mathcal{A}}(x_1, \ldots, x_n) = \bigcap_{\mathcal{B} \supset \mathcal{A}} \sigma^{\mathcal{B}}(x_1, \ldots, x_n),$$

(the intersection of all subsets of $\mathbb{C}^n$ that are of the form $\sigma^{\mathcal{B}}(x_1, \ldots, x_n)$ for some commutative extension $\mathcal{B} \supset \mathcal{A}$).

6 Axiomatic theory of spectrum

In Section 1 we studied the basic properties of the set-valued function $a \mapsto \sigma(a)$ defined on a Banach algebra $\mathcal{A}$. Since there are many naturally defined set-valued functions with similar properties, in this section we introduce an axiomatic spectral theory.

The ordinary spectrum of an element $a \in \mathcal{A}$ is defined by means of invertible elements, $\sigma(a) = \{\lambda : a - \lambda \notin \mathrm{Inv}(\mathcal{A})\}$. In the same way we define a generalized spectrum by means of "regular elements".

Definition 1. Let $\mathcal{A}$ be a Banach algebra. A non-empty subset R of $\mathcal{A}$ is called a *regularity* if it satisfies the following conditions:

(i) if $a \in \mathcal{A}$ and $n \in \mathbb{N}$, then $a \in R \Leftrightarrow a^n \in R$;

(ii) if a, b, c, d are mutually commuting elements of $\mathcal{A}$ and $ac + bd = 1_{\mathcal{A}}$, then

$$ab \in R \Leftrightarrow a \in R \text{ and } b \in R.$$

A regularity $R \subset \mathcal{A}$ assigns to each $a \in \mathcal{A}$ a subset of $\mathbb{C}$ defined by

$$\sigma_R(a) = \{\lambda \in \mathbb{C} : a - \lambda \notin R\}.$$

This mapping will be called the *spectrum corresponding to the regularity R*.

Proposition 2. *Let R be a regularity in a Banach algebra $\mathcal{A}$. Then:*

(i) $1_{\mathcal{A}} \in R$;

(ii) $\mathrm{Inv}(\mathcal{A}) \subset R$;

(iii) *If $a, b \in \mathcal{A}, ab = ba$ and $a \in \mathrm{Inv}(\mathcal{A})$, then $ab \in R \Leftrightarrow b \in R$.*
In particular, if $a \in R$ and $\lambda \in \mathbb{C}, \lambda \neq 0$, then $\lambda a \in R$;

(iv) $\sigma_R(a) \subset \sigma(a)$ *for all* $a \in \mathcal{A}$;

(v) $\sigma_R(a - \lambda) = \sigma_R(a) - \lambda$ *for all* $a \in \mathcal{A}, \lambda \in \mathbb{C}$.

Proof. (i) Choose $b \in R$. We have $1_{\mathcal{A}} \cdot 1_{\mathcal{A}} + b \cdot 0 = 1_{\mathcal{A}}$ and $1_{\mathcal{A}} \cdot b \in R$. Thus $1_{\mathcal{A}} \in R$.
(ii) Let $c \in \mathrm{Inv}(\mathcal{A})$. Then $c \cdot c^{-1} + c^{-1} \cdot 0 = 1_{\mathcal{A}}$ and $c \cdot c^{-1} \in R$ by (i). Thus $c \in R$.
(iii) We have $a \cdot a^{-1} + b \cdot 0 = 1_{\mathcal{A}}$, so we can apply property (ii) of Definition 1.
(iv) and (v) are clear. □

In general, $\sigma_R(a)$ is neither compact nor non-empty.

Theorem 3. *Let $\mathcal{A}$ be a Banach algebra. Then:*

(i) *if $(R_\alpha)_\alpha$ is a family of regularities in $\mathcal{A}$, then $R = \bigcap_\alpha R_\alpha$ is a regularity. The corresponding spectrum satisfies*

$$\sigma_R(a) = \bigcup_\alpha \sigma_{R_\alpha}(a).$$

(ii) *if* (R_α) *is a directed system (i.e., for all* α, β *there exists* γ *such that* $R_\gamma \supset R_\alpha \cup R_\beta$*) of regularities in* $\mathcal{A}$*, then* $R' = \bigcup_\alpha R_\alpha$ *is a regularity. The corresponding spectrum satisfies*

$$\sigma_{R'}(a) = \bigcap_\alpha \sigma_{R_\alpha}(a);$$

(iii) *if* J *is a closed two-sided ideal in* $\mathcal{A}$*,* $\pi : \mathcal{A} \to \mathcal{A}/J$ *the canonical projection and* R *a regularity in* $\mathcal{A}/J$*, then* $\pi^{-1}R$ *is a regularity in* $\mathcal{A}$*.*

Proof. Clear. □

In many cases it is possible to verify the axioms of a regularity using the following criterion:

Theorem 4. *Let* R *be a non-empty subset of a Banach algebra* $\mathcal{A}$ *satisfying*

$$ab \in R \Leftrightarrow a \in R \quad \text{and} \quad b \in R \tag{P1}$$

for all commuting elements $a, b \in \mathcal{A}$*. Then* R *is a regularity.*

Proof. Clear. □

Examples 5. Let $\mathcal{A}$ be a Banach algebra. The following sets are regularities since they satisfy (P1).

(i) $R_1 = \mathcal{A}$; the corresponding spectrum is empty for every $a \in \mathcal{A}$.
(ii) $R_2 = \mathrm{Inv}(\mathcal{A})$; this gives the ordinary spectrum $\sigma_R(a)$.
(iii) The sets $R_3 = \mathrm{Inv}_l(\mathcal{A})$ and $R_4 = \mathrm{Inv}_r(\mathcal{A})$ of all left (right) invertible elements of $\mathcal{A}$.
(iv) The sets R_5 and R_6 all elements of $\mathcal{A}$ which are not left (right) topological divisors of zero.
(v) The sets R_7 and R_8 of all elements that are not left (right) divisors of zero.

Definition 6. The spectra corresponding to the regularities R_3–R_8 are:
the *left spectrum*

$$\sigma_l(a) = \{\lambda \in \mathbb{C} : a - \lambda \neq \mathrm{Inv}_l(\mathcal{A})\} = \{\lambda \in \mathbb{C} : 1_\mathcal{A} \notin \mathcal{A}(a-\lambda)\};$$

the *right spectrum*

$$\sigma_r(a) = \{\lambda \in \mathbb{C} : a - \lambda \neq \mathrm{Inv}_r(\mathcal{A})\} = \{\lambda \in \mathbb{C} : 1_\mathcal{A} \notin (a-\lambda)\mathcal{A}\};$$

the *left approximate point spectrum*

$$\tau_l(a) = \{\lambda \in \mathbb{C} : a - \lambda \text{ is a left topological divisor of zero}\};$$

the *right approximate point spectrum*

$$\tau_r(a) = \{\lambda \in \mathbb{C} : a - \lambda \text{ is a right topological divisor of zero}\};$$

the *left point spectrum* $\pi_l(a) = \{\lambda \in \mathbb{C} : a - \lambda \text{ is a left divisor of zero}\}$ and the *right point spectrum* $\pi_r(a) = \{\lambda \in \mathbb{C} : a - \lambda \text{ is a right divisor of zero}\}$.

Obviously, $\pi_l(a) \subset \tau_l(a) \subset \sigma_l(a)$, $\pi_r(a) \subset \tau_r(a) \subset \sigma_r(a)$ and $\partial\sigma(a) \subset \tau_l(a) \cap \tau_r(a)$. Theorems 1.14 and 1.17 imply that $\sigma_l, \sigma_r, \tau_l$ and τ_r are always closed and non-empty. The point spectra π_l and π_r can be both empty and non-closed.

Further examples of regularities will be given later.

Every spectrum defined by a regularity satisfies the spectral mapping theorem:

Theorem 7. (spectral mapping theorem) *Let R be a regularity in a Banach algebra $\mathcal{A}$ and let σ_R be the corresponding spectrum. Then*

$$\sigma_R(f(a)) = f(\sigma_R(a))$$

for every $a \in \mathcal{A}$ and every function f analytic on a neighbourhood of $\sigma(a)$ which is non-constant on each component of its domain of definition.

Proof. Let $\mu \in \mathbb{C}$. It is sufficient to show that

$$\mu \notin \sigma_R(f(a)) \Leftrightarrow \mu \notin f(\sigma_R(a)). \tag{1}$$

Since $f - \mu$ has only a finite number of zeros $\lambda_1, \dots, \lambda_n$ in $\sigma(a)$, it can be written as $f(z) - \mu = (z - \lambda_1)^{k_1} \cdots (z - \lambda_n)^{k_n} \cdot g(z)$, where g is a function analytic on a neighbourhood of $\sigma(a)$ and $g(z) \neq 0$ for $z \in \sigma(a)$. Then $f(a) - \mu = (a - \lambda_1)^{k_1} \cdots (a - \lambda_n)^{k_n} \cdot g(a)$ and $g(a)$ is invertible by the spectral mapping theorem for the ordinary spectrum.

Thus (1) is equivalent to

$$f(a) - \mu \in R \Leftrightarrow a - \lambda_i \in R \qquad (i = 1, \dots, n). \tag{2}$$

Since $g(a)$ is invertible, by Proposition 2 (iii) and Definition 1 (ii), this is equivalent to

$$(a - \lambda_1)^{k_1} \cdots (a - \lambda_n)^{k_n} \in R \Leftrightarrow (a - \lambda_i)^{k_i} \in R \qquad (i = 1, \dots, n). \tag{3}$$

Since for all relatively prime polynomials p, q there exist polynomials p_1, q_1 such that $pp_1 + qq_1 = 1$, we have $p(a)p_1(a) + q(a)q_1(a) = 1_{\mathcal{A}}$ and it is possible to apply property (ii) of Definition 1 inductively to get (3). This completes the proof. □

We will see later that the assumption that f is non-constant on each component is really necessary. However, in many cases this condition can be left out. The question whether the spectral mapping theorem is true for all analytic functions is closely related to the non-emptiness of the spectrum (clearly the spectral mapping theorem for constant functions cannot be true if $\sigma_R(x) = \emptyset$ for some $x \in \mathcal{A}$ and $0 \notin R$. We give a simple criterion (in the most interesting case of the algebra $\mathcal{B}(X)$) which is usually easy to verify.

Let X be a Banach space, let R be a regularity in $\mathcal{B}(X)$ and let $X = X_1 \oplus X_2$. Let us write $R_1 = \{T_1 \in \mathcal{B}(X_1) : T_1 \oplus I \in R\}$ and $R_2 = \{T_2 \in \mathcal{B}(X_2) : I \oplus T_2 \in R\}$. If $X_i \neq \{0\}$, then R_i is a regularity in $\mathcal{B}(X_i)$ $(i = 1, 2)$. Indeed, to see condition (ii) of Definition 1 (e.g., for R_1), note that if $A_1C_1+B_1D_1 = I_{X_1}$ for some commuting $A_1, B_1, C_1, D_1, \in \mathcal{B}(X_1)$, then

$$(A_1 \oplus I)(C_1 \oplus I) + (B_1 \oplus I)(D_1 \oplus 0) = I_X.$$

If $T_1 \in \mathcal{B}(X_1)$ and $T_2 \in \mathcal{B}(X_2)$, then

$$T_1 \oplus T_2 \in R \Leftrightarrow T_1 \in R_1 \quad \text{and} \quad T_2 \in R_2.$$

Indeed, this follows from the observation that

$$(T_1 \oplus I)(0 \oplus I) + (I \oplus T_2)(I \oplus 0) = I_X.$$

Note that in all examples given above the regularities R_1, R_2 are defined in a canonical way: for example, if $R = \{T \in \mathcal{B}(X) : T \text{ is left invertible}\}$, then we have $R_1 = \{T_1 \in \mathcal{B}(X_1) : T_1 \text{ is left invertible}\}$ and $R_2 = \{T_2 \in \mathcal{B}(X_2) : T_2 \text{ is left invertible}\}$.

Theorem 8. *Let X be a Banach space, let R be a regularity in $\mathcal{B}(X)$ and let σ_R be the corresponding spectrum. Suppose that for all pairs of complementary subspaces X_1, X_2, $X = X_1 \oplus X_2$ such that $R_1 = \{S_1 \in \mathcal{B}(X_1) : S_1 \oplus I \in R\} \neq \mathcal{B}(X_1)$ the corresponding spectrum $\sigma_{R_1}(T_1) = \{\lambda : T_1 - \lambda \notin R_1\}$ is non-empty for every $T_1 \in \mathcal{B}(X_1)$. Then $\sigma_R(f(T)) = f(\sigma_R(T))$ for every $T \in \mathcal{B}(X)$ and every function f analytic on a neighbourhood of $\sigma(T)$.*

Proof. Let $\mu \in \mathbb{C}$. It is sufficient to show that

$$\mu \notin \sigma_R(f(T)) \Longleftrightarrow \mu \notin f(\sigma_R(T)).$$

Let U_1, U_2 be open subsets of the domain of definition of f such that $U_1 \cup U_2 \supset \sigma(T)$, $f|U_1$ is identically equal to μ and $(f-\mu)|U_2$ can be written as $(f-\mu)|U_2 = p(z)g(z)$, where p is a non-zero polynomial and g is analytic and has no zeros in $U_2 \cap \sigma(T)$. Let X_1, X_2 be the spectral subspaces corresponding to U_1 and U_2, i.e., $X = X_1 \oplus X_2$, $T = T_1 \oplus T_2$ where $T_i = T|X_i$ and $\sigma(T_i) \subset U_i \quad (i = 1, 2)$, see Corollary 1.38.

Let $R_1 \subset \mathcal{B}(X_1)$ and $R_2 \subset \mathcal{B}(X_2)$ be the regularities defined above and let $\sigma_{R_1}, \sigma_{R_2}$ be the corresponding spectra. It is clear that $\sigma_R(T) = \sigma_{R_1}(T_1) \cup \sigma_{R_2}(T_2)$. The following statements are equivalent:

$\mu \notin \sigma_R(f(T))$; $\quad f(T) - \mu \in R$;
$0 \in R_1$ and $f(T_2) - \mu I_{X_2} \in R_2$; $\quad R_1 = \mathcal{B}(X_1)$ and $p(T_2) \in R_2$;
$\sigma_{R_1}(T_1) = \emptyset$ and $0 \notin p(\sigma_{R_2}(T_2))$; $\quad \mu \notin f(\sigma_{R_1}(T_1) \cup \sigma_{R_2}(T_2))$;
$\mu \notin f(\sigma_R(T))$. □

We are now going to study the continuity properties of spectra.

For basic definitions and properties of semicontinuous set-valued mappings we refer to Appendix A.4.

Let R be a regularity in a Banach algebra $\mathcal{A}$ and let σ_R be the corresponding spectrum. We consider the following properties of R (or σ_R):

(P2) "Upper semicontinuity of σ_R": if $a_n, a \in \mathcal{A}$, $a_n \to a$, $\lambda_n \in \sigma_R(a_n)$ and $\lambda_n \to \lambda$, then $\lambda \in \sigma_R(a)$.

(P3) "Upper semicontinuity on commuting elements": if $a_n, a \in \mathcal{A}$, $a_n \to a$, $a_n a = a a_n$ for every n, $\lambda_n \in \sigma_R(a_n)$ and $\lambda_n \to \lambda$, then $\lambda \in \sigma_R(a)$.

(P4) "Continuity on commuting elements": if $a_n, a \in \mathcal{A}$, $a_n \to a$ and $a_n a = a a_n$ for every n, then $\lambda \in \sigma_R(a)$ if and only if there exists a sequence $\lambda_n \in \sigma_R(a_n)$ such that $\lambda_n \to \lambda$.

Evidently, either (P2) or (P4) implies (P3). If σ_R satisfies (P3), then, by considering a constant sequence $a_n = a$, the spectrum $\sigma_R(a)$ is closed for every $a \in \mathcal{A}$.

Proposition 9. *Let R be a regularity in a Banach algebra $\mathcal{A}$, let σ_R be the corresponding spectrum. The following conditions are equivalent:*

(i) *(P2);*

(ii) *$\sigma_R(a)$ is closed for every $a \in \mathcal{A}$ and the mapping $a \mapsto \sigma_R(a)$ is upper semicontinuous;*

(iii) *R is an open subset of $\mathcal{A}$.*

Proof. (iii) $\Rightarrow$ (i): Let $a_n, a \in \mathcal{A}$, $a_n \to a$, $\lambda_n \in \sigma_R(a_n)$ and $\lambda_n \to \lambda$. Then $a_n - \lambda_n \notin R$. Since $\mathcal{A} \setminus R$ is closed, we conclude that $a - \lambda \notin R$. Hence $\lambda \in \sigma_R(a)$.

(i) $\Rightarrow$ (iii): We prove that $\mathcal{A} \setminus R$ is closed. Let $a_n \in \mathcal{A} \setminus R$, $a_n \to a$. Then $0 \in \sigma_R(a_n)$ for each n. From (i) we conclude that $0 \in \sigma_R(a)$. Hence $a \in \mathcal{A} \setminus R$.

(i) $\Leftrightarrow$(ii) follows from Theorem A.4.3. □

Proposition 10. *Let R be a regularity in a Banach algebra $\mathcal{A}$ and let σ_R be the corresponding spectrum. The following conditions are equivalent:*

(i) *(P3);*

(ii) *$\sigma_R(a)$ is closed for every $a \in \mathcal{A}$, and for each neighbourhood U of $\sigma_R(a)$ there exists $\varepsilon > 0$ such that $\sigma_R(a+u) \subset U$ whenever $u \in \mathcal{A}$, $au = ua$ and $\|u\| < \varepsilon$;*

(iii) *If $a \in R$, then there exists $\varepsilon > 0$ such that $u \in \mathcal{A}$, $ua = au$ and $\|u\| < \varepsilon$ implies $a + u \in R$.*

Proof. Analogous to Proposition 9. □

Recall that for two sets $L, M \subset \mathbb{C}$ we write $\Delta(L, M) = \sup_{l \in L} \operatorname{dist}\{l, M\}$. The Hausdorff distance is defined by $\widehat{\Delta}(L, M) = \max\{\Delta(L, M), \Delta(M, L)\}$.

Proposition 11. *Let R be a regularity in a Banach algebra $\mathcal{A}$, let σ_R be the corresponding spectrum.*

(i) *Suppose that, for every $a \in \mathcal{A}$,*

$$\inf\{|z| : z \in \mathbb{C}, a - z \notin R\} = \inf\{\|u\| : u \in \mathcal{A}, ub = bu, a - u \notin R\}.$$

Then $\widehat{\Delta}(\sigma_R(a), \sigma_R(b)) \leq \|a - b\|$ for all commuting $a, b \in \mathcal{A}$.

(ii) *If $\sigma_R(a)$ is closed for every $a \in \mathcal{A}$ and $\Delta(\sigma_R(a), \sigma_R(b)) \leq \|a - b\|$ for all commuting $a, b \in \mathcal{A}$, then σ_R satisfies (P4).*

Proof. (i) Let $a, b \in \mathcal{A}$, $ab = ba$ and let $\lambda \in \sigma_R(a)$. We prove $\operatorname{dist}\{\lambda, \sigma_R(b)\} \leq \|a - b\|$. This is clear if $\lambda \in \sigma_R(b)$. If $\lambda \notin \sigma_R(b)$, then

$$\begin{aligned}\|a - b\| &= \|(a - \lambda) - (b - \lambda)\| \\ &\geq \inf\{\|u\| : u \in \mathcal{A}, u(b - \lambda) = (b - \lambda)u, (b - \lambda) + u \notin R\} \\ &= \inf\{|z| : z \in \sigma_R(b - \lambda)\} = \operatorname{dist}\{0, \sigma_R(b - \lambda)\} = \operatorname{dist}\{\lambda, \sigma_R(b)\}.\end{aligned}$$

Thus

$$\Delta(\sigma_R(a), \sigma_R(b)) = \sup_{\lambda \in \sigma_R(a)} \operatorname{dist}\{\lambda, \sigma_R(b)\} \leq \|a - b\|$$

and, by symmetry, $\widehat{\Delta}(\sigma_R(a), \sigma_R(b)) \leq \|a - b\|$.

(ii) Let $a_n a = a a_n$, $a_n \to a$, $\lambda_n \in \sigma_R(a_n)$ and $\lambda_n \to \lambda$. Then, for each n, there exists $\mu_n \in \sigma_R(a)$ with $|\mu_n - \lambda_n| \leq \|a_n - a\|$. Clearly, $\mu_n \to \lambda$, and so $\lambda \in \sigma_R(a)$ since $\sigma_R(a)$ is closed. This proves the upper semicontinuity.

If we restrict σ_R to the set $\{a, a_1, a_2, \dots\}$, then the lower semicontinuity follows from Theorem A.4.4. □

In general, the left and right point spectra are not closed, and so they do not satisfy (P3) (and therefore neither (P2) nor (P4)). On the other hand $\sigma_l, \sigma_r, \tau_l$ and τ_r are defined by open regularities, so these spectra satisfy (P2). They satisfy also (P4) (we prove a more general result in the next section).

The upper semicontinuity on commuting elements enables us to weaken the axioms of regularity.

Theorem 12. *Let R be a non-empty subset of a Banach algebra $\mathcal{A}$ satisfying* (P3) *and*

(i) *if $a \in R$ and $n \in \mathbb{N}$, then $a^n \in R$,*

(ii) *if a,b,c,d, are mutually commuting elements of $\mathcal{A}$ and $ac + bd = 1_{\mathcal{A}}$, then $ab \in R \Leftrightarrow a \in R$ and $b \in R$.*

Then R is a regularity.

Proof. It is sufficient to show the implication $a^n \in R \Rightarrow a \in R \quad (n \geq 2)$. By (P3), $a^n - \mu a = a(a^{n-1} - \mu) \in R$ for some non-zero complex number μ. Since

$$(a^{n-1} - \mu) \cdot (-\mu^{-1}) + a(\mu^{-1} a^{n-2}) = 1_{\mathcal{A}},$$

we have $a \in R$ by (ii). □

Let R be a regularity in a Banach algebra $\mathcal{A}$ and σ_R the corresponding spectrum. We say that σ_R is *spectral-radius-preserving* if $\sigma_R(a)$ is closed and $\max\{|\lambda| : \lambda \in \sigma_R(a)\} = r(a)$ for each $a \in \mathcal{A}$.

Theorem 13. *Let R be a regularity in a Banach algebra $\mathcal{A}$ such that the corresponding spectrum σ_R is spectral-radius-preserving. Then $\partial\sigma(a) \subset \sigma_R(a)$ for all $a \in \mathcal{A}$.*

Proof. Suppose on the contrary that $\lambda_0 \in \partial\sigma(a)$ and there exists $\varepsilon > 0$ such that $\{z : |z - \lambda_0| < \varepsilon\} \cap \sigma_R(a) = \emptyset$. Choose $\lambda_1 \in \mathbb{C} \setminus \sigma(a)$ with $|\lambda_0 - \lambda_1| < \varepsilon/2$. Consider the function $f(z) = (z - \lambda_1)^{-1}$. Then

$$\begin{aligned}\operatorname{dist}\{\lambda_1, \sigma_R(a)\}^{-1} &= \sup\{|f(z)| : z \in \sigma_R(a)\} = \sup\{|z| : z \in \sigma_R(f(a))\} \\ &= r(f(a)) = \max\{|f(z)| : z \in \sigma(a)\} \geq \frac{1}{|\lambda_0 - \lambda_1|} > \frac{2}{\varepsilon}.\end{aligned}$$

Thus there exists $\lambda_2 \in \sigma_R(a)$ with $|\lambda_2 - \lambda_1| < \varepsilon/2$, and so $|\lambda_2 - \lambda_0| < \varepsilon$, a contradiction. □

Upper semicontinuous spectra have always many continuity points.

Theorem 14. *Let R be a regularity in a Banach algebra $\mathcal{A}$ satisfying property* (P2) *(upper semicontinuity). Then the set of all discontinuity points of the set-valued function $a \mapsto \sigma_R(a)$ is of the first category.*

Proof. For every complex number λ let $f_\lambda : \mathcal{A} \to \langle 0, \infty\rangle$ be the function defined by $f_\lambda(a) = \operatorname{dist}\{\lambda, \sigma_R(a)\}$ (if $\sigma_R(a) = \emptyset$, then we set $f_\lambda(a) = \infty$). The result follows from the following three statements:

(a) f_λ is lower semicontinuous for every λ.

(b) The mapping $a \mapsto \sigma_R(a)$ is continuous at x if and only if f_λ is continuous at x for every λ from a dense subset of $\mathbb{C}$.

(c) The set of all discontinuity points of a lower semicontinuous function $f : \mathcal{A} \to \langle 0, \infty\rangle$ is of the first category.

(a) Let $a \in \mathcal{A}$, $\lambda \in \mathbb{C}$. We must show that $f_\lambda(a) \leq \liminf_{x \to a} f_\lambda(x)$. This is clear if $f_\lambda(a) = 0$. Suppose that $f_\lambda(a) > 0$ and choose r, $0 < r < f_\lambda(a) = \operatorname{dist}\{\lambda, \sigma_R(a)\}$. Then $\sigma_R(a) \subset \mathbb{C} \setminus \{z : |z - \lambda| \leq r\}$, so there exists $\delta > 0$ such that $\sigma_R(x) \subset \mathbb{C} \setminus \{z : |z - \lambda| \leq r\}$ for every $x \in \mathcal{A}$ with $\|x - a\| < \delta$. For x in this neighbourhood we have $f_\lambda(x) \geq r$ and $\liminf_{x \to a} f_\lambda(x) \geq r$. Since $r < f_\lambda(a)$ was arbitrary, we have $\liminf_{x \to a} f_\lambda(x) \geq f_\lambda(a)$ and f_λ is lower semicontinuous.

(b) Suppose that $x \mapsto \sigma_R(x)$ is continuous at $a \in \mathcal{A}$. Let $\lambda \in \mathbb{C}$. We show that f_λ is continuous at a. By (a), it is sufficient to show that $f_\lambda(a) \geq \limsup_{x\to a} f_\lambda(x)$. This is clear if $f_\lambda(a) = \infty$, i.e., if $\sigma_R(a) = \emptyset$. Suppose that $\sigma_R(a) \neq \emptyset$. Let $\varepsilon > 0$. By lower semicontinuity of σ_R, there exists $\delta > 0$ such that $\operatorname{dist}\{\lambda, \sigma_R(x)\} \leq f_\lambda(a)+\varepsilon$ whenever $\|x-a\| < \delta$. Thus $\limsup_{x\to a} f_\lambda(x) \leq f_\lambda(a)+\varepsilon$ and, since ε was arbitrary, we have $\limsup_{x\to a} f_\lambda(x) \leq f_\lambda(a)$. Hence f_λ is continuous at a.

Conversely, let $x \mapsto \sigma_R(x)$ be discontinuous at a. Since σ_R is upper semicontinuous, there exist an open set $U \subset \mathbb{C}$ and points $x_1, x_2, \dots \in \mathcal{A}$ such that $U \cap \sigma_R(a) \neq \emptyset$, $x_n \to a$ and $U \cap \sigma_R(x_n) = \emptyset$ for every n. Choose $\lambda_0 \in U \cap \sigma_R(a)$ and let r be a positive number with $\{z : \|z - \lambda_0\| < r\} \subset U$. Choose λ in the dense subset of $\mathbb{C}$ with $|\lambda - \lambda_0| < r/3$. Then $f_\lambda(a) = \operatorname{dist}\{\lambda, \sigma_R(a)\} < r/3$ and $f_\lambda(x_n) > 2r/3$ for every n. Hence f_λ is not continuous at a.

(c) Let $f : \mathcal{A} \to \langle 0, \infty\rangle$ be a lower semicontinuous function. Suppose f is not continuous at a point $a \in \mathcal{A}$. Then $f(a) < \limsup_{x\to a} f(x)$. Find a rational number r such that $f(a) < r < \limsup_{x\to a} f(x)$. Then $f^{-1}(\langle 0, r\rangle)$ is a closed subset of $\mathcal{A}$ and $a \in \partial f^{-1}(\langle 0, r\rangle)$. Hence the set of all discontinuity points of f is contained in $\bigcup\{\partial f^{-1}(\langle 0, r\rangle) : r \text{ rational}\}$, which is of the first category. □

7 Spectral systems

In Section 3 we studied the spectrum, approximate point spectrum and the Shilov spectrum defined for n-tuples of elements of a commutative Banach algebra. In this section we introduce and study general spectra defined for commuting n-tuples in a (in general non-commutative) Banach algebra.

Let $\mathcal{A}$ be a Banach algebra. Denote by $c_n(\mathcal{A})$ the set of all n-tuples of mutually commuting elements of $\mathcal{A}$ and set $c(\mathcal{A}) = \bigcup_{n=1}^\infty c_n(\mathcal{A})$.

Definition 1. Let R be a subset of $c(\mathcal{A})$, $R = \bigcup_{n=1}^\infty R_n$ where $R_n \subset c_n(\mathcal{A})$. The set R is called a *joint regularity* if it satisfies the following conditions (for all $n \geq 1$):

(i) if $(x_1, \dots, x_n, y_1, \dots, y_n) \in c(\mathcal{A})$ and $\sum_{i=1}^n x_i y_i = 1_{\mathcal{A}}$, then $(x_1, \dots, x_n) \in R_n$;

(ii) if $(x_1, \dots, x_n) \in R_n$ and x_{n+1} commutes with x_i $(1 \leq i \leq n)$, then $(x_1, \dots, x_n, x_{n+1}) \in R_{n+1}$;

(iii) if $(x_0, x_1, \dots, x_n) \in c(\mathcal{A})$ and $(x_0 - \lambda, x_1, \dots, x_n) \in R_{n+1}$ for every $\lambda \in \mathbb{C}$, then $(x_1, \dots, x_n) \in R_n$.

Proposition 2. *Let $R = \bigcup_{i=1}^\infty R_n$ be a joint regularity in a Banach algebra $\mathcal{A}$ and let $(x_1, \dots, x_n) \in R_n$. Then $(x_{\pi(1)}, \dots, x_{\pi(n)}) \in R_n$ for every permutation π of the set $\{1, \dots, n\}$.*

Proof. By (ii), we have $(x_1, \dots, x_n, x_{\pi(1)}, \dots, x_{\pi(n)}) \in R_{2n}$. Let $\lambda \in \mathbb{C}$, $\lambda \neq 0$. Then x_1 appears in the n-tuple $x_{\pi(1)}, \dots, x_{\pi(n)}$ and $-\lambda^{-1}(x_1 - \lambda) + 0x_2 + \cdots +$

$0x_n + \lambda^{-1}x_1 = 1_{\mathcal{A}}$. Thus, by Definition 1 (i),

$$(x_1 - \lambda, x_2, \dots, x_n, x_{\pi(1)}, \dots, x_{\pi(n)}) \in R_{2n}.$$

By (iii), we conclude that $(x_2, \dots, x_n, x_{\pi(1)}, \dots, x_{\pi(n)}) \in R_{2n-1}$. By repeating this argument we get $(x_3, \dots, x_n, x_{\pi(1)}, \dots, x_{\pi(n)}) \in R_{2n-2}$, and finally, $(x_{\pi(1)}, \dots, x_{\pi(n)}) \in R_n$. □

Corollary 3. *Let R be a joint regularity in a Banach algebra $\mathcal{A}$, let $n \geq 2$ and $1 \leq k \leq n$. Then:*

(i) *if $(x_1, \dots, x_n) \in c_n(\mathcal{A})$ and $(x_1, \dots, x_{k-1}, x_{k+1}, \dots, x_n) \in R_{n-1}$, then*

$$(x_1, \dots, x_n) \in R_n;$$

(ii) *if $(x_1, \dots, x_{k-1}, x_k - \lambda, x_{k+1}, \dots, x_n) \in R_n$ for all $\lambda \in \mathbb{C}$, then*

$$(x_1, \dots, x_{k-1}, x_{k+1}, \dots, x_n) \in R_{n-1}.$$

Definition 4. Let $\mathcal{A}$ be a Banach algebra. A *spectral system* is a mapping which assigns to every finite family $x_1, \dots, x_n$ of mutually commuting elements of $\mathcal{A}$ a subset $\tilde{\sigma}(x_1, \dots, x_n) \subset \mathbb{C}^n$ that satisfies the following conditions:

(i) if $x = (x_1, \dots, x_n) \in c_n(\mathcal{A})$ and $\langle x \rangle$ is the algebra generated by $x_1, \dots, x_n$, then $\tilde{\sigma}(x) \subset \sigma^{\langle x \rangle}(x)$;

(ii) (projection property) if $1 \leq m \leq n$, $1 \leq i_1 < i_2 < \cdots < i_m \leq n$ and $(x_1, \dots, x_n) \in c_n(\mathcal{A})$, then $\tilde{\sigma}(x_{i_1}, \dots, x_{i_m}) = Q\tilde{\sigma}(x_1, \dots, x_n)$ where $Q : \mathbb{C}^n \to \mathbb{C}^m$ is the projection defined by $Q(\lambda_1, \dots, \lambda_n) = (\lambda_{i_1}, \dots, \lambda_{i_m})$.

Spectral systems are in one-to-one correspondence with the joint regularities.

Theorem 5. *Let $\mathcal{A}$ be a Banach algebra. Then:*

(i) *if $R \subset c(\mathcal{A})$ is a joint regularity, then σ_R defined by*

$$\sigma_R(x_1, \dots, x_n) = \{(\lambda_1, \dots, \lambda_n) \in \mathbb{C}^n : (x_1 - \lambda_1, \dots, x_n - \lambda_n) \notin R_n\}$$

is a spectral system;

(ii) *if $\tilde{\sigma}$ is a spectral system, then $R = \bigcup_{n=1}^{\infty} R_n$ defined by*

$$R_n = \{(x_1, \dots, x_n) \in c_n(\mathcal{A}) : (0, \dots, 0) \notin \tilde{\sigma}(x_1, \dots, x_n)\}$$

is a joint regularity.

Proof. (i) If $(x_1, \dots, x_n) \in c_n(\mathcal{A})$ and $\lambda = (\lambda_1, \dots, \lambda_n) \notin \sigma^{\langle x \rangle}(x)$, then $\sum_{i=1}^{n}(x_i - \lambda_i)y_i = 1_{\mathcal{A}}$ for some $y_1, \dots, y_n \in \langle x \rangle$. Then, by Definition 1 (i), $(x_1 - \lambda_1, \dots, x_n - \lambda_n) \in R_n$, and so $(\lambda_1, \dots, \lambda_n) \notin \sigma_R(x_1, \dots, x_n)$.

Let $1 \le k \le n$ and let $P : \mathbb{C}^n \to \mathbb{C}^{n-1}$ be the projection defined by $P(\lambda_1, \dots, \lambda_n) = (\lambda_1, \dots, \lambda_{k-1}, \lambda_{k+1}, \dots, \lambda_n)$. By Corollary 3, we get

$$\sigma_R(x_1, \dots, x_{k-1}, x_{k+1}, \dots, x_n) = P\sigma_R(x_1, \dots, x_n).$$

The projection property can be obtained by a repeated use of this observation.

(ii) Conversely, let $\tilde{\sigma}$ be a spectral system and let R be defined by $R = \bigcup_{n=0}^{\infty} R_n$, where $R_n = \{(x_1, \dots, x_n) \in c_n(\mathcal{A}) : (0, \dots, 0) \notin \tilde{\sigma}(x_1, \dots, x_n)\}$. Then R is a joint regularity. Indeed, axioms (ii) and (iii) of Definition 1 follow immediately from the projection property. Let $(x_1, \dots, x_n, y_1, \dots, y_n) \in c_{2n}(\mathcal{A})$, $\sum_{i=1}^n x_i y_i = 1_{\mathcal{A}}$. Then

$$\underbrace{(0, \dots, 0)}_{n} \notin \sigma^{\langle x_1, \dots, x_n, y_1, \dots, y_n \rangle}(x_1, \dots, x_n),$$

and so, by the projection property of the spectrum in the commutative Banach algebra $\langle x_1, \dots, x_n, y_1, \dots, y_n \rangle$,

$$(0, \dots, 0, \lambda_1, \dots, \lambda_n) \notin \sigma^{\langle x_1, \dots, x_n, y_1, \dots, y_n \rangle}(x_1, \dots, x_n, y_1, \dots, y_n)$$

for all $(\lambda_1, \dots, \lambda_n) \in \mathbb{C}^n$. Thus $(0, \dots, 0, \lambda_1, \dots, \lambda_n) \notin \tilde{\sigma}(x_1, \dots, x_n, y_1, \dots, y_n)$ and, by the projection property, $(0, \dots, 0) \notin \tilde{\sigma}(x_1, \dots, x_n)$. Hence $(x_1, \dots, x_n) \in R_n$. □

Proposition 6. *Let $R \subset c(\mathcal{A})$ be a joint regularity and σ_R the corresponding spectral system. The following conditions are equivalent:*

(i) $\{0\} \notin R_1$;

(ii) $\sigma_R(x_1, \dots, x_n)$ *is non-empty for all* $(x_1, \dots, x_n) \in c(\mathcal{A})$;

(iii) *there exist* $(x_1, \dots, x_n) \in c(\mathcal{A})$ *such that* $\sigma_R(x_1, \dots, x_n) \neq \emptyset$;

(iv) $R \neq c(\mathcal{A})$.

Proof. (i) ⇒ (ii): Let $x = (x_1, \dots, x_n) \in c(\mathcal{A})$. We have $0 \in \sigma_R(0)$ and so, by the projection property, there exists $\lambda \in \mathbb{C}^n$ such that $(0, \lambda) \in \sigma_R(0, x)$. Again by the projection property, $\lambda \in \sigma_R(x)$.

The implications (ii) ⇒ (iii) and (iii) ⇒ (iv) are trivial.

(iv) ⇒ (i): Let $(x_1, \dots, x_n) \in c(\mathcal{A})$ and suppose that $0 \in R_1$. Then $\lambda \cdot 1_{\mathcal{A}} \in R_1$ for every $\lambda \in \mathbb{C}$, and so $(\lambda, x_1, \dots, x_n) \in R$ for all $\lambda \in \mathbb{C}$. Consequently, $(x_1, \dots, x_n) \in R$. □

An arbitrary spectral system possesses the spectral mapping property.

Theorem 7. *Let $\mathcal{A}$ be a Banach algebra, $\tilde{\sigma}$ a spectral system in $\mathcal{A}$ and suppose that $x = (x_1, \dots, x_n) \in c(\mathcal{A})$. Let $f = (f_1, \dots, f_m)$ be an m-tuple of functions analytic in a neighbourhood of $\sigma^{\langle x \rangle}(x)$. Then*

$$\tilde{\sigma}(f(x)) = f(\tilde{\sigma}(x)), \qquad \text{where} \qquad f(x) = (f_1(x), \dots, f_m(x)).$$

Proof. We have

$$\begin{aligned}\tilde{\sigma}(x, f(x)) &\subset \sigma^{\langle x, f(x)\rangle}(x, f(x)) = \sigma^{\langle x\rangle}(x, f(x)) \\ &= \{(\varphi(x), \varphi(f(x))) : \varphi \in \mathcal{M}(\langle x\rangle)\} = \{(z, f(z)) : z \in \sigma^{\langle x\rangle}(x)\}.\end{aligned}$$

The following statements are equivalent:

$w \in f(\tilde{\sigma}(x))$;

there exists $z \in \tilde{\sigma}(x)$ such that $w = f(z)$;

there exists $z \in \tilde{\sigma}(x)$ such that $(z, w) \in \tilde{\sigma}(x, f(x))$;

$w \in \tilde{\sigma}(f(x))$. □

Theorem 8. *Let R be a joint regularity in a Banach algebra $\mathcal{A}$. Then $R_1 = \{x_1 \in \mathcal{A} : \{x_1\} \in R\}$ is a regularity satisfying property (P1).*

Proof. Let $a, b \in \mathcal{A}$, $ab = ba$. If $a, b \in R_1$, then

$$\begin{aligned}\sigma_R(ab) = \{\lambda\mu : (\lambda, \mu) \in \sigma_R(a, b)\} &\subset \{\lambda\mu : \lambda \in \sigma_R(a), \mu \in \sigma_R(b)\} \\ &\subset \{\lambda\mu : \lambda, \mu \neq 0\} \subset \mathbb{C} \setminus \{0\},\end{aligned}$$

and so $ab \in R_1$.

Conversely, if $a \notin R_1$, then $0 \in \sigma_R(a)$ and there exists $\lambda \in \mathbb{C}$ such that $(0, \lambda) \in \sigma_R(a, b)$. By Theorem 7, $0 = 0 \cdot \lambda \in \sigma_R(ab)$ and $ab \notin R_1$. Similarly, $b \notin R_1 \Rightarrow ab \notin R_1$.

This proves (P1) and also that R_1 is a regularity. □

Remark 9. Condition (P1) provides a criterion whether a spectrum given for single elements can be extended to commuting n-tuples, cf. C.7.4. Questions of this type appear frequently in spectral theory.

Theorem 10. *Let $\tilde{\sigma}$ be a spectral system in a Banach algebra $\mathcal{A}$. Then*

$$(x_1, \dots, x_n) \mapsto \overline{\tilde{\sigma}(x_1, \dots, x_n)}$$

is also a spectral system.

Proof. Obviously, $\overline{\tilde{\sigma}(x)} \subset \sigma^{\langle x\rangle}(x)$ for every $x \in c(\mathcal{A})$.

Let $1 \leq i_1 < \cdots < i_m \leq n$ and let $Q : \mathbb{C}^n \to \mathbb{C}^m$ be the projection defined by $Q(\lambda_1, \dots, \lambda_n) = (\lambda_{i_1}, \dots, \lambda_{i_m})$. Since $\tilde{\sigma}(x_{i_1}, \dots, x_{i_m}) = Q\tilde{\sigma}(x_1, \dots, x_n)$, we have $Q\big(\overline{\tilde{\sigma}(x_1, \dots, x_n)}\big) \subset \overline{\tilde{\sigma}(x_{i_1}, \dots, x_{i_m})}$.

The equality follows from the fact that $Q\big(\overline{\tilde{\sigma}(x_1, \dots, x_n)}\big)$ is compact and contains $\tilde{\sigma}(x_{i_1}, \dots, x_{i_m})$. □

A spectral system $\tilde{\sigma}$ will be called *compact-valued* if $\tilde{\sigma}(x_1, \dots, x_n)$ is a compact subset of $\mathbb{C}^n$ for all $(x_1, \dots, x_n) \in c(\mathcal{A})$. The most important spectral systems satisfy this property.

Theorem 11. *Let $\mathcal{A}$ be a Banach algebra. Then:*

(i) *if $(R_\alpha)_\alpha$ is a family of joint regularities in $\mathcal{A}$, then $R = \cap_\alpha R_\alpha$ is a joint regularity. The corresponding spectral system satisfies*

$$\sigma_R(a) = \bigcup_\alpha \sigma_{R_\alpha}(a) \qquad (a \in c(\mathcal{A}));$$

(ii) *if (R_α) is a directed system (i.e., for all α, β there exists γ such that $R_\gamma \supset R_\alpha \cup R_\beta$) of joint regularities in $\mathcal{A}$ such that the corresponding spectral systems are compact-valued, then $R' = \bigcup_\alpha R_\alpha$ is a joint regularity. The corresponding spectral system satisfies*

$$\sigma_{R'}(a) = \bigcap_\alpha \sigma_{R_\alpha} \qquad (a \in c(\mathcal{A}));$$

(iii) *if J is a closed two-sided ideal in $\mathcal{A}$, $\pi : \mathcal{A} \to \mathcal{A}/J$ the canonical projection and R a joint regularity in $\mathcal{A}/J$, then $\pi^{-1}R = \{a \in c(\mathcal{A}) : \pi(a) \in R\}$ is a joint regularity in $\mathcal{A}$.*

Proof. (i) and (iii) are clear.

(ii) The first two axioms of Definition 1 are clear. To prove the third axiom, let $(x_0, x_1, \dots, x_n) \in c(\mathcal{A})$ and suppose that for each $\lambda \in \mathbb{C}$ there exists α such that $(x_0 - \lambda, x_1, \dots, x_n) \in R_\alpha$. Thus $(x - \mu, x_1, \dots, x_n) \in R_\alpha$ for all μ in a certain neighbourhood of λ. From the compactness of $\sigma^{\langle x_0, x_1, \dots, x_n\rangle}(x_0)$ and the fact that (R_α) is a directed system we conclude that there is a β such that $(x_0 - \mu, x_1, \dots, x_n) \in R_\beta$ for all $\mu \in \mathbb{C}$. Hence $(x_1, \dots, x_n) \in R_\beta \subset R$. □

In the commutative case it is possible to describe all compact-valued spectral systems easily. They are in 1-1 correspondence with compact subsets of $\mathcal{M}(\mathcal{A})$. For $x = (x_1, \dots, x_n) \in c_n(\mathcal{A})$ and $\varphi \in \mathcal{M}(\mathcal{A})$ write $\varphi(x) = \big(\varphi(x_1), \dots, \varphi(x_n)\big)$.

Theorem 12. *Let A be a commutative Banach algebra. Then:*

(i) *if K is a compact subset of $\mathcal{M}(\mathcal{A})$, then the mapping*

$$x \in c(\mathcal{A}) \mapsto \tilde{\sigma}(x) = \{\varphi(x) : \varphi \in K\}$$

is a compact-valued spectral system;

(ii) *if K_1, K_2 are compact subset of $\mathcal{M}(\mathcal{A})$ and $K_1 \neq K_2$, then there exists a finite family $x \in c(\mathcal{A})$ such that*

$$\{\varphi(x) : \varphi \in K_1\} \neq \{\varphi(x) : \varphi \in K_2\};$$

(iii) *if $\tilde{\sigma}$ is a compact-valued spectral system, then*

$$K(\tilde{\sigma}) = \big\{\varphi \in \mathcal{M}(\mathcal{A}) : \varphi(x) \in \tilde{\sigma}(x) \text{ for all } x \in c(\mathcal{A})\big\}$$

is a compact subset of $\mathcal{M}(\mathcal{A})$;

(iv) *if $\tilde{\sigma}$ is a spectral system in $\mathcal{A}$, then $\tilde{\sigma}(x) \subset \sigma^{\mathcal{A}}(x)$ for all $x = (x_1, \dots, x_n) \in c(\mathcal{A})$;*

(v) *if $\tilde{\sigma}$ is a compact-valued spectral system, $x = (x_1, \dots, x_n) \in c_n(\mathcal{A})$ and $\lambda = (\lambda_1, \dots, \lambda_n) \in \tilde{\sigma}(x)$, then there exists $\varphi \in K(\tilde{\sigma})$ such that $\varphi(x) = \lambda$.*

Proof. The proof of (i) is clear.

(ii) For $x \in c(\mathcal{A})$ write

$$\tilde{\sigma}_1(x) = \{\varphi(x) : \varphi \in K_1\} \quad \text{and} \quad \tilde{\sigma}_2(x) = \{\varphi(x) : \varphi \in K_2\}.$$

Suppose on the contrary that $\tilde{\sigma}_1(x) = \tilde{\sigma}_2(x)$ for every $x \in c(\mathcal{A})$.

Let $\varphi \in K_1$. For $x \in c(\mathcal{A})$ set $M_x = \{\psi \in K_2 : \psi(x) = \varphi(x)\}$. By assumption, $\varphi(x) \in \tilde{\sigma}_1(x) = \tilde{\sigma}_2(x)$, and so M_x is a non-empty compact subset of $\mathcal{M}(\mathcal{A})$. Since $M_x \cap M_y = M_{(x,y)}$ for all $x, y \in c(\mathcal{A})$, the system $\{M_x\}_{x \in c(\mathcal{A})}$ has the intersection property and we have $\bigcap_{x \in c(\mathcal{A})} M_x \neq \emptyset$. Let $\psi \in \bigcap_{x \in c(\mathcal{A})} M_x$. Then $\psi \in K_2$ and $\psi(x_1) = \varphi(x_1)$ for every $x_1 \in \mathcal{A}$. Thus $\psi = \varphi$, $\varphi \in K_2$ and $K_1 \subset K_2$.

By symmetry, we get $K_1 = K_2$, which is a contradiction.

(iii) Clearly, $\{\varphi \in \mathcal{M}(\mathcal{A}) : \varphi(x) \in \tilde{\sigma}(x)\}$ is a compact set for every $x \in c(\mathcal{A})$, and so

$$K(\tilde{\sigma}) = \bigcap_{x \in c(\mathcal{A})} \{\varphi \in \mathcal{M}(A) : \varphi(x) \in \tilde{\sigma}(x)\}$$

is a compact subset of $\mathcal{M}(\mathcal{A})$.

(iv) Let $x = (x_1, \dots, x_n) \in c(\mathcal{A})$. Suppose on the contrary that there is a $\lambda = (\lambda_1, \dots, \lambda_n) \in \tilde{\sigma}(x) \setminus \sigma^{\mathcal{A}}(x)$. Then there are elements $y_1, \dots, y_n \in \mathcal{A}$ such that $\sum_{i=1}^n (x_i - \lambda_i) y_i = 1$. Write $y = (y_1, \dots, y_n)$. By the projection property for $\tilde{\sigma}$, there is a $\mu \in \mathbb{C}^n$ such that $(\lambda, \mu) \in \tilde{\sigma}(x, y) \subset \sigma^{\langle x,y \rangle}(x, y)$, a contradiction.

(v) Let $x \in c(A)$ and $\lambda \in \tilde{\sigma}(x)$. For every $y \in c(\mathcal{A})$ set $M_y = \{\varphi \in \mathcal{M}(A) : \varphi(x) = \lambda, \varphi(y) \in \tilde{\sigma}(y)\}$. By the projection property, there exists μ such that $(\lambda, \mu) \in \tilde{\sigma}(x, y) \subset \sigma^{\mathcal{A}}(x, y)$. Hence $M_y \neq 0$ and it is clearly a compact set. Furthermore,

$$M_y \cap M_z \supset M_{(y,z)} \neq 0 \qquad (y, z \in c(\mathcal{A})),$$

so the system $\{M_y\}_{y \in c(\mathcal{A})}$ has the intersection property and $\bigcap_{y \in c(\mathcal{A})} M_y \neq \emptyset$. Let $\varphi \in \bigcap_{y \in c(\mathcal{A})} M_y$. Then $\varphi(x) = \lambda$ and $\varphi(y) \in \tilde{\sigma}(y)$ for every $y \in c(\mathcal{A})$. □

Corollary 13. *Let $\mathcal{A}$ be a commutative Banach algebra. Then the mapping*

$$\tilde{\sigma} \mapsto K(\tilde{\sigma}) = \{\varphi \in \mathcal{M}(\mathcal{A}) : \varphi(x) \in \tilde{\sigma}(x) \text{ for every } x \in c(\mathcal{A})\}$$

is a one-to-one correspondence between the compact-valued spectral systems and compact subsets of $\mathcal{M}(\mathcal{A})$.

For all $(x_1, \dots, x_n) \in c(\mathcal{A})$, $\tilde{\sigma}(x_1, \dots, x_n) = \{(\varphi(x_1), \dots, \varphi(x_n)) : \varphi \in K(\tilde{\sigma})\}$.

Any spectral system behaves continuously (even uniformly continuously) on commuting elements. For the definition of Hausdorff distance $\widehat{\Delta}$ see Definition A.4.2.

Theorem 14. *Let $\tilde{\sigma}$ be a spectral system in a Banach algebra $\mathcal{A}$ and let $R_1 = \{x_1 \in \mathcal{A} : 0 \notin \tilde{\sigma}(x_1)\}$. Then:*

(i) *if $a, u \in \mathcal{A}$, $au = ua$, $a \in R_1$ and*

$$\inf\{|z| : z \in \tilde{\sigma}(a)\} > \sup\{|z| : z \in \tilde{\sigma}(u)\},$$

then $a + u \in R_1$;

(ii) *if $a \in \mathcal{A}$, then*

$$\operatorname{dist}\{0, \tilde{\sigma}(a)\} = \inf\big\{\|u\| : u \in \mathcal{A}, ua = au, a + u \notin R_1\big\};$$

(iii) *if $a, b \in \mathcal{A}$ are commuting elements, then*

$$\widehat{\Delta}(\tilde{\sigma}(a), \tilde{\sigma}(b)) \le \|a - b\|.$$

Proof. (i) Denote by $\mathcal{A}_0$ the commutative Banach algebra generated by a and u. By Theorem 12, there exists a compact subset $K \subset \mathcal{M}(\mathcal{A}_0)$ such that $\overline{\tilde{\sigma}(y)} = \{\varphi(y) : \varphi \in K\}$ for all $y \in \mathcal{A}_0$. In particular, $\overline{\tilde{\sigma}(a+u)} = \{\varphi(a+u) : \varphi \in K\}$. For $\varphi \in K$ we have

$$|\varphi(a+u)| \ge |\varphi(a)| - |\varphi(u)| \ge \inf\{|\psi(a)| : \psi \in K\} - \sup\{|\psi(u)| : \psi \in K\} > 0.$$

Thus $0 \notin \overline{\tilde{\sigma}(a+u)}$ and $a + u \in R_1$.

(ii) Clearly, $\operatorname{dist}\{0, \tilde{\sigma}(a)\} \ge \inf\big\{\|u\| : u \in \mathcal{A}, ua = au, a + u \notin R_1\big\}$.

To prove the opposite inequality, let $u \in \mathcal{A}$, $ua = au$ and $\|u\| < \operatorname{dist}\{0, \tilde{\sigma}(a)\}$. Then

$$\sup\{|z| : z \in \tilde{\sigma}(u)\} \le \|u\| < \operatorname{dist}\{0, \tilde{\sigma}(a)\} = \inf\{|z| : z \in \tilde{\sigma}(a)\},$$

and so $a + u \in R_1$ by (i).

The third statement follows from Proposition 6.11. □

Corollary 15. *Let $\tilde{\sigma}$ be a compact-valued spectral system in a Banach algebra $\mathcal{A}$. Then $R_1 = \{x \in \mathcal{A} : 0 \notin \tilde{\sigma}(a)\}$ is a regularity satisfying property (P4) (continuity on commuting elements).*

Theorem 16. *Let $\tilde{\sigma}$ be a spectral system in a Banach algebra $\mathcal{A}$ and let $R_1 = \{a \in \mathcal{A} : 0 \notin \tilde{\sigma}(a)\}$. Let $a \in R_1$ and let u be a quasinilpotent (i.e., $r(u) = 0$) commuting with a. Then $a + u \in R_1$.*

Proof. Since $\tilde{\sigma}(a, u) \subset \tilde{\sigma}(a) \times \tilde{\sigma}(u) = \tilde{\sigma}(a) \times \{0\}$, the projection property of $\tilde{\sigma}$ gives that $\tilde{\sigma}(a, u) = \tilde{\sigma}(a) \times \{0\}$.

By the spectral mapping property (Theorem 7), we have $\tilde{\sigma}(a+u) = \big\{\lambda + \mu : (\lambda, \mu) \in \tilde{\sigma}(a, u)\big\} = \tilde{\sigma}(a)$. □

Definition 17. We say that a spectral system $\tilde{\sigma}$ in a Banach algebra $\mathcal{A}$ is upper semicontinuous, if $\tilde{\sigma}(a)$ is closed for all $a \in c(\mathcal{A})$ and the mapping $(a_1, \dots, a_n) \in c_n(\mathcal{A}) \mapsto \tilde{\sigma}(a_1, \dots, a_n)$ is upper semicontinuous for all $n \in \mathbb{N}$.

Theorem 18. *Let $R = \bigcup_{n=1}^{\infty} R_n \subset c_n(\mathcal{A})$ be a joint regularity in a Banach algebra $\mathcal{A}$. The following conditions are equivalent:*

(i) *R_n is open in $c_n(\mathcal{A})$ for all $n \in \mathbb{N}$,*

(ii) *the corresponding spectral system is upper semicontinuous.*

Proof. Straightforward (cf. Appendix A.4). □

By definition, each upper semicontinuous spectral system is compact-valued.

In commutative Banach algebras, a spectral system is compact-valued if and only if it is upper semicontinuous (and if and only if it is continuous).

Let A be a commutative algebra and $a_1, \dots, a_n \in \mathcal{A}$. Let $I_{\mathcal{A}}(a_1, \dots, a_n)$ denote the smallest ideal in $\mathcal{A}$ containing the elements $a_1, \dots, a_n$. It is easy to see that $I_{\mathcal{A}}(a_1, \dots, a_n) = \left\{\sum_{i=1}^{n} a_i b_i : b_1, \dots, b_n \in \mathcal{A}\right\}$.

Proposition 19. *Let K be a compact Hausdorff space and $\mathcal{A}$ a subalgebra of $C(K)$. Let $n \in \mathbb{N}$, $f_1, \dots, f_n, g \in \mathcal{A}$ and suppose that $I_{\mathcal{A}}(f_1, \dots, f_n)$ consists of functions which achieve zero on K. Then there exists $\mu \in \mathbb{C}$ such that $I_{\mathcal{A}}(f_1, \dots, f_n, g - \mu)$ consists of functions which achieve zero on K.*

Proof. Suppose on the contrary that for each $\mu \in \mathbb{C}$ there are functions $h_1^{(\mu)}, \dots, h_{n+1}^{(\mu)} \in \mathcal{A}$ such that

$$\sum_{i=1}^{n} h_i^{(\mu)} f_i + h_{n+1}^{(\mu)}(g - \mu)$$

does not vanish on K. Clearly $g(K)$ is a compact subset of the complex plane. Let Δ be the polynomially convex hull of $g(K)$.

For $m \in \mathbb{N}$ let U_m be the set of all $\mu \in \Delta$ such that there are functions $h_1^{(\mu)}, \dots, h_{n+1}^{(\mu)} \in \mathcal{A}$ with $\|h_{n+1}^{(\mu)}\| < m$ and

$$\inf_{z \in K} \Big| \sum_{i=1}^{n} h_i^{(\mu)}(z) f_i(z) + h_{n+1}^{(\mu)}(z)\big(g(z) - \mu\big) \Big| > 1. \tag{1}$$

Clearly, the sets U_m are open, $U_1 \subset U_2 \subset \cdots$ and $\bigcup_{m=1}^{\infty} U_m = \Delta$. Therefore there exists $m_0 \in \mathbb{N}$ such that $U_{m_0} = \Delta$, i.e., for each $\mu \in \Delta$ there are functions $h_1^{(\mu)}, \dots, h_{n+1}^{(\mu)} \in \mathcal{A}$ satisfying (1) and $\|h_{n+1}^{(\mu)}\| < m_0$.

Find a finite subset $\{\nu_1, \dots, \nu_k\} \subset \Delta$ such that $\operatorname{dist}\{\mu, \{\nu_1, \dots, \nu_k\}\} < \frac{1}{m_0}$ for each $\mu \in \Delta$. For $j = 1, \dots, k$ let

$$u_j = \sum_{i=1}^{n} h_i^{(\nu_j)} f_i + h_{n+1}^{(\nu_j)}(g - \nu_j).$$

Then $u_j \in \mathcal{A}$. Since $\inf_{z \in K} |u_j(z)| > 1$, we have $\|u_j^{-1}\| < 1$.

Let $\mathcal{B}$ be the smallest subalgebra of $C(K)$ containing $\mathcal{A}$ and the functions $u_1^{-1}, \dots, u_k^{-1}$. Let $I = I_{\mathcal{B}}(f_1, \dots, f_n)$. Every function $v \in I$ can be expressed as

$$v = \sum_{i=1}^{n} \sum_{\alpha \in \mathbb{Z}_+^n} f_i u^{-\alpha} b_{i,\alpha}, \tag{2}$$

where the second sum is finite, $b_{i,\alpha} \in \mathcal{A}$ for all i and α, and u^α stands for $u_1^{\alpha_1} \cdots u_k^{\alpha_k}$.

For v given by (2) find $\beta \in \mathbb{Z}_+^k$ such that $\beta \geq \alpha$ whenever $b_{i,\alpha} \neq 0$ for some i. Then

$$vu^\beta = \sum_{i=1}^{n} \sum_{\alpha \in \mathbb{Z}_+^k} f_i u^{\beta - \alpha} b_{i,\alpha} \in I_{\mathcal{A}}(f_1, \dots, f_n).$$

By assumption, vu^β vanishes on K, and so does v. Thus $\overline{I}$ is a closed ideal in the Banach algebra $\overline{\mathcal{B}}$ consisting of functions singular in $\overline{\mathcal{B}}$, and so $\overline{I} \neq \overline{\mathcal{B}}$. Therefore there exists a multiplicative functional $\varphi \in \mathcal{M}(\overline{\mathcal{B}})$ with $\overline{I} \subset \operatorname{Ker} \varphi$.

Set $\mu_0 = \varphi(g)$. Then $\mu_0 \in \sigma^{\overline{\mathcal{B}}}(g) \subset \widehat{\sigma}^{C(K)}(g) = \widehat{g(K)} = \Delta$. Therefore there exists $j \in \{1, \dots, k\}$ such that $|\mu_0 - \nu_j| < \frac{1}{m_0}$. Then

$$\begin{aligned} w &:= \sum_{i=1}^{n} h_i^{(\nu_j)} f_i + h_{n+1}^{(\nu_j)}(g - \mu_0) = \sum_{i=1}^{n} h_i^{(\nu_j)} f_i + h_{n+1}^{(\nu_j)}(g - \nu_j) + h_{n+1}^{(\nu_j)}(\nu_j - \mu_0) \\ &= u_j + h_{n+1}^{(\nu_j)}(\nu_j - \mu_0). \end{aligned}$$

We have $\|u_j^{-1}\| < 1$ and

$$\|wu_j^{-1} - 1\| = \|h_{n+1}^{(\nu_j)}(\nu_j - \mu_0)u_j^{-1}\| < 1.$$

Thus wu_j^{-1} is invertible in $\overline{\mathcal{B}}$ and so $w \in \operatorname{Inv}(\overline{\mathcal{B}})$. This is a contradiction with the fact that $\varphi(w) = 0$. This completes the proof. □

Theorem 20. *Let $\mathcal{A}$ be a commutative Banach algebra. Let $R_1 \subset \mathcal{A}$ be an open regularity satisfying (P1). Then the following conditions are equivalent:*

- (i) *there is a joint regularity R in $\mathcal{A}$ such that $R_1 = R \cap c_1(\mathcal{A})$ and the corresponding spectral system is compact-valued;*
- (ii) *for every $x \in \mathcal{A} \setminus R_1$ there exists $\varphi \in \mathcal{M}(\mathcal{A})$ such that $x \in \operatorname{Ker} \varphi \subset \mathcal{A} \setminus R_1$;*
- (iii) *if $I \subset \mathcal{A}$ is an ideal satisfying $I \subset \mathcal{A} \setminus R_1$, then there exists $\varphi \in \mathcal{M}(\mathcal{A})$ such that $I \subset \operatorname{Ker} \varphi \subset \mathcal{A} \setminus R_1$;*
- (iv) *if $a_1, \dots, a_n \in \mathcal{A}$, $I_{\mathcal{A}}(a_1, \dots, a_n) \subset \mathcal{A} \setminus R_1$ and $b \in \mathcal{A}$, then there exists $\mu \in \mathbb{C}$ such that $I_{\mathcal{A}}(a_1, \dots, a_n, b - \mu) \subset \mathcal{A} \setminus R_1$.*

Proof. (iii) ⇒ (ii): Let $x \in \mathcal{A} \setminus R_1$. Since R_1 satisfies (P1), we have $I_{\mathcal{A}}(x) = \{xa : a \in \mathcal{A}\} \subset \mathcal{A} \setminus R_1$. By (iii), there exists $\varphi \in \mathcal{M}(\mathcal{A})$ such that $x \in \operatorname{Ker} \varphi \subset \mathcal{A} \setminus R_1$.

(ii) $\Rightarrow$ (i): Let $K = \{\varphi \in \mathcal{M}(\mathcal{A}) : \operatorname{Ker}\varphi \subset \mathcal{A} \setminus R_1\}$. Since

$$K = \bigcap_{x \in \mathcal{A}} \{\varphi \in \mathcal{M}(\mathcal{A}) : \varphi(x) \in \sigma_{R_1}(x)\},$$

K is a compact subset of $\mathcal{M}(\mathcal{A})$.

Let $x \in \mathcal{A}$. By (ii), we have $x \notin R_1$ if and only if there exists $\varphi \in K$ with $\varphi(x) = 0$. Hence $\sigma_{R_1}(x) = \{\varphi(x) : \varphi \in K\}$ for each $x \in \mathcal{A}$, and so the joint regularity corresponding to K extends R_1.

(iv) $\Rightarrow$ (iii): Let I be an ideal satisfying $I \subset \mathcal{A} \setminus R_1$. Consider the set of all ideals in $\mathcal{A}$ which are contained in $\mathcal{A} \setminus R_1$ ordered by the inclusion. By the Zorn lemma, there exists an ideal $J \supset I$ which is maximal among those contained in $\mathcal{A} \setminus R_1$. It is sufficient to show that $\operatorname{codim} J = 1$ since the corresponding multiplicative functional will satisfy the required properties.

Suppose on the contrary that $\operatorname{codim} J \geq 2$. Let $b \in \mathcal{A}$ satisfy $b \notin J + \mathbb{C} \cdot 1_{\mathcal{A}}$. For a finite subset $F = \{a_1, \dots, a_n\}$ of J let

$$M_F = \{\mu \in \mathbb{C} : I_{\mathcal{A}}(a_1, \dots, a_n, b - \mu) \subset \mathcal{A} \setminus R_1.$$

By assumption, $M_F \neq \emptyset$. Clearly $M_F \subset \sigma^{\mathcal{A}}(b)$, and so M_F is bounded. If $\mu_k \in M_F \quad (k \in \mathbb{N})$ and $\mu_k \to \mu$, then every element of $I_{\mathcal{A}}(a_1, \dots, a_n, b - \mu)$ can be written as a limit of elements of $I_{\mathcal{A}}(a_1, \dots, a_n, b - \mu_k) \subset \mathcal{A} \setminus R_1$. Since $\mathcal{A} \setminus R_1$ is closed, $\mu \in M_F$, and so M_F is compact. Furthermore, $M_F M_{F \cup F'} M_{F'} \supset M_{F \cap F'} \neq \emptyset$, and so the system $(M_F)_F$ has finite intersection property. Thus there exists $\mu \in \bigcap_F M_F$. It means that the ideal generated by J and $g - \mu$ is contained in $\mathcal{A} \setminus R_1$ and is strictly greater than J. This contradicts to the assumption that J was maximal in this class of ideals and proves (iii).

(i) $\Rightarrow$ (iv): Let $K \subset \mathcal{M}(\mathcal{A})$ be a compact set satisfying $x \in \mathcal{A} \setminus R_1$ if and only if there exists $\varphi \in K$ with $\varphi(x) = 0$. Let $a_1, \dots, a_n \in \mathcal{A}$, $I_{\mathcal{A}}(a_1, \dots, a_n) \subset \mathcal{A} \setminus R_1$ and let $b \in \mathcal{A}$. Consider the algebra $G(\mathcal{A})|K = \{G(x)|K : x \in \mathcal{A}\}$ and the functions $G(a_1)|K, \dots, G(a_n)|K, G(b)|K \in C(\mathcal{M}(\mathcal{A}))$. By Proposition 19, there exists $\mu \in \mathbb{C}$ such that $I_{\mathcal{A}}(a_1, \dots, a_n, b - \mu) \subset \mathcal{A} \setminus R_1$. □

As an application of the previous theorem, consider the regularity consisting of all elements that are not topological divisors of zero. We get the following analogy of Corollary 3.8.

Corollary 21. *Let $\mathcal{A}$ be a commutative Banach algebra, let $I \subset \mathcal{A}$ be an ideal consisting of topological divisors of zero. Then there exists a maximal ideal $J \supset I$ consisting of topological divisors of zero.*

Proof. Follows from the implication (i) $\Rightarrow$ (iii) of Theorem 20. □

Theorem 22. *Let $\mathcal{A}$ be a Banach algebra and let σ_1, σ_2 be compact-valued spectral systems in $\mathcal{A}$ satisfying $\sigma_1(x) \subset \sigma_2(x)$ and*

$$\max\{|\lambda| : \lambda \in \sigma_1(x)\} = \max\{|\lambda| : \lambda \in \sigma_2(x)\}$$

for all $x \in \mathcal{A}$. *Then* $\Gamma(\sigma_2(a_1,\dots,a_n),\mathcal{P}(n)) \subset \sigma_1(a_1,\dots,a_n)$ *for all commuting* n*-tuples* $a_1,\dots,a_n \in \mathcal{A}$.

In particular, the polynomially convex hulls of the sets $\sigma_1(a_1,\dots,a_n)$ *and* $\sigma_2(a_1,\dots,a_n)$ *coincide.*

Similarly, the closed convex hulls of $\sigma_1(a_1,\dots,a_n)$ *and* $\sigma_2(a_1,\dots,a_n)$ *coincide.*

Proof. Let $\lambda \in \Gamma(\sigma_2(a_1,\dots,a_n),\mathcal{P}(n))$ and let U be a neighbourhood of λ in $\mathbb{C}^n$. By the definition of the Shilov boundary, there exists a polynomial $p \in \mathcal{P}(n)$ such that

$$\sup\{|p(z)| : z \in U \cap \sigma_2(a_1,\dots,a_n)\} > \sup\{|p(z)| : z \in \sigma_2(a_1,\dots,a_n) \setminus U\}.$$

Write $y = p(a_1,\dots,a_n)$. Then

$$\begin{aligned}\max\{|p(z)| : z \in \sigma_1(a_1,\dots,a_n)\} &= \max\{|\mu| : \mu \in \sigma_1(y)\}\\ = \max\{|\mu| : \mu \in \sigma_2(y)\} &= \max\{|p(z)| : z \in \sigma_2(a_1,\dots,a_n)\}.\end{aligned}$$

Thus $U \cap \sigma_1(a_1,\dots,a_n) \neq \emptyset$. Since U was an arbitrary neighbourhood of λ and $\sigma_1(a_1,\dots,a_n)$ is closed, $\lambda \in \sigma_1(a_1,\dots,a_n)$. Consequently, $\widehat{\sigma}_1(a_1,\dots,a_n) = \widehat{\sigma}_2(a_1,\dots,a_n)$.

By considering the linear polynomials, we obtain in the same way that the closed convex hulls of $\sigma_1(a_1,\dots,a_n)$ and $\sigma_2(a_1,\dots,a_n)$ coincide. □

We say that a spectral system $\tilde{\sigma}$ in a Banach algebra $\mathcal{A}$ is *spectral-radius-preserving* if $\tilde{\sigma}(a)$ is closed for all $a \in c(\mathcal{A})$ and $\max\{|\lambda| : \lambda \in \tilde{\sigma}(a_1)\} = r(a)$ for each $a_1 \in \mathcal{A}$.

In Section 3 we studied already three spectral-radius-preserving spectral systems in commutative Banach algebras: the spectrum σ, the approximate point spectrum τ and the Shilov spectrum σ_Γ.

From the properties of the Shilov boundary it follows easily that σ_Γ is the smallest spectral system with this property:

Theorem 23. *Let* $\tilde{\sigma}$ *be a spectral-radius-preserving spectral system in a commutative Banach algebra* $\mathcal{A}$. *Then* $\sigma_\Gamma(a_1,\dots,a_n) \subset \tilde{\sigma}(a_1,\dots,a_n)$ *for all commuting* n*-tuples* $a_1,\dots,a_n \in \mathcal{A}$.

8 Basic spectral systems in Banach algebras

We now introduce the basic spectral systems in non-commutative Banach algebras.

Definition 1. Let $\mathcal{A}$ be a Banach algebra, $x = (x_1,\dots,x_n) \in c(\mathcal{A})$. We define the *left spectrum* $\sigma_l(x)$ as the set of all $\lambda = (\lambda_1,\dots,\lambda_n) \in \mathbb{C}^n$ such that

$$\mathcal{A}(x_1 - \lambda_1) + \cdots + \mathcal{A}(x_n - \lambda_n) \neq \mathcal{A}.$$

Equivalently,

$$\lambda \in \sigma_l(x) \Longleftrightarrow 1_{\mathcal{A}} \notin \mathcal{A}(x_1-\lambda_1)+\cdots+\mathcal{A}(x_n-\lambda_n).$$

The *right spectrum* $\sigma_r(x)$ is defined analogously:

$$\begin{aligned}\lambda \in \sigma_r(x) \quad &\Longleftrightarrow \quad (x_1-\lambda_1)\mathcal{A}+\cdots+(x_n-\lambda_n)\mathcal{A} \neq \mathcal{A}\\ &\Longleftrightarrow \quad 1_{\mathcal{A}} \notin (x_1-\lambda_1)\mathcal{A}+\cdots+(x_n-\lambda_n)\mathcal{A}.\end{aligned}$$

The *Harte spectrum* $\sigma(x)$ is defined as the union of the left and right spectrum, $\sigma_H(x)=\sigma_l(x)\cup\sigma_r(x) \quad (x\in c(\mathcal{A}))$.

For single elements the Harte spectrum coincides with the ordinary spectrum and Definition 1 coincides with Definition 6.6. If $\mathcal{A}$ is a commutative Banach algebra, then $\sigma_l(x_1,\dots,x_n)=\sigma_r(x_1,\dots,x_n)=\sigma(x_1,\dots,x_n)$.

Proposition 2. *Let $\mathcal{A}$ be a Banach algebra, $(x_0,x_1,\dots,x_n)\in c(\mathcal{A})$ and let $(0,\dots,0,)\in\sigma_l(x_1,\dots,x_n)$. Then there exists $\lambda\in\mathbb{C}$ such that $(\lambda,0,\dots,0)\in\sigma_l(x_0,x_1,\dots,x_n)$.*

Proof. Let $J=\overline{\mathcal{A}x_1+\cdots+\mathcal{A}x_n}$. Clearly, J is a proper left ideal. Consider the Banach space $X=\mathcal{A}/J$ and the operator $T:X\to X$ defined by $T(a+J)=ax_0+J \quad (a+J\in X)$. The definition is correct since $Jx_0\subset J$. Choose $\lambda\in\partial\sigma^{\mathcal{B}(X)}(T)$. By Theorem 1.28, $(T-\lambda)X\neq X$. Thus $J+\mathcal{A}(x_0-\lambda)\neq\mathcal{A}$, which finishes the proof. □

Theorem 3. *Let $\mathcal{A}$ be a Banach algebra. Then σ_l,σ_r and σ_H are upper semicontinuous spectral systems.*

Proof. For $n\geq 1$ let $R_n=\{(x_1,\dots,x_n)\in c_n(\mathcal{A}):1_{\mathcal{A}}\in x_1\mathcal{A}+\cdots+x_n\mathcal{A}\}$. Clearly, $R=\bigcup_{n=1}^{\infty}$ is a joint regularity (the only non-trivial part was proved in Proposition 2) and σ_l is the corresponding spectral system. Further, R_n is open in $c_n(\mathcal{A})$, and so σ_l is upper semicontinuous.

For the right spectrum the result follows by symmetry and for the Harte spectrum $\sigma_H=\sigma_l\cup\sigma_r$ by Theorem 7.11 (i). □

Definition 4. For $x=(x_1,\dots,x_n)\in c(\mathcal{A})$ let

$$d_l(x)=\inf\left\{\sum_{j=1}^{n}\|x_ju\| : u\in\mathcal{A}, \|u\|=1\right\}$$

and

$$d_r(x)=\inf\left\{\sum_{j=1}^{n}\|ux_j\| : u\in\mathcal{A}, \|u\|=1\right\}.$$

The *left* and *right approximate spectra* are defined by

$$\tau_l(x)=\{\lambda\in\mathbb{C}^n : d_l(x-\lambda)=0\} \quad \text{and} \quad \tau_r(x)=\{\lambda\in\mathbb{C}^n : d_r(x-\lambda)=0\}.$$

It is easy to see that $\tau_l(x) \subset \sigma_l(x)$ and $\tau_r(x) \subset \sigma_r(x)$. For single elements $\tau_l(x_1) = \{\lambda_1 \in \mathbb{C} : x_1 - \lambda_1$ is a left topological divisor of $0\}$ and $\tau_r(x_1) = \{\lambda_1 \in \mathbb{C} : x_1 - \lambda_1$ is a right topological divisor of $0\}$. For commutative Banach algebras

$$\tau_l(x_1, \dots, x_n) = \tau_r(x_1, \dots, x_n) = \tau(x_1, \dots, x_n).$$

The next lemma enables us to reduce problems for non-commutative Banach algebras to the commutative case.

Lemma 5. *Let $\mathcal{A}$ be a closed commutative subalgebra of a Banach algebra $\mathcal{B}$. Then there exists a commutative Banach algebra $\mathcal{C}$ containing $\mathcal{A}$ as subalgebra such that $d_l^{\mathcal{B}}(x_1, \dots, x_n) = d^{\mathcal{C}}(x_1, \dots, x_n)$ for every $(x_1, \dots, x_n) \in \mathcal{A}^n$.*

Proof. Let $\mathcal{C} = \mathcal{A} \times \mathcal{B} = \{(a, b) : a \in \mathcal{A}, b \in \mathcal{B}\}$. Define in $\mathcal{C}$ algebraic operations (for all $a, a' \in \mathcal{A}, b, b' \in \mathcal{B}, \alpha \in \mathbb{C}$) by

$$(a, b) + (a', b') = (a + a', b + b'),$$
$$\alpha\,(a, b) = (\alpha a, \alpha b),$$
$$(a, b) \cdot (a', b') = (aa', ab' + a'b)$$

and the norm $\|(a, b)\| = \|a\| + \|b\|$. It is easy to show that $\mathcal{C}$ is a commutative Banach algebra with the unit element $(1_{\mathcal{A}}, 0)$. If we identify $a \in \mathcal{A}$ with $(a, 0) \in \mathcal{C}$, then $\mathcal{A}$ is a subalgebra of $\mathcal{C}$. Let $(x_1, \dots, x_n) \in \mathcal{A}^n$. Then

$$d^{\mathcal{C}}\big((x_1, 0), \dots, (x_n, 0)\big) = \inf\left\{ \sum_{i=1}^{n} \|(x_i, 0)(a, b)\| : a, b \in \mathcal{A}, \|a\| + \|b\| = 1 \right\}$$
$$= \inf\left\{ \sum_{i=1}^{n} (\|x_i a\| + \|x_i b\|) : a, b \in \mathcal{A}, \|a\| + \|b\| = 1 \right\} = d_l^{\mathcal{B}}(x_1, \dots, x_n). \qquad \square$$

Corollary 6. *Let $\mathcal{A}$ be a Banach algebra, $(x_0, x_1, \dots, x_n) \in c(\mathcal{A})$, and let $d_l(x_1, \dots, x_n) = 0$. Then there exists $\lambda \in \mathbb{C}$ such that $d_l(x_0 - \lambda, x_1, \dots, x_n) = 0$.*

Proof. The statement follows from the previous lemma and the corresponding result 3.7 for commutative Banach algebras (in fact, the proof of Theorem 3.7 works without any change in the non-commutative case, too). $\square$

Corollary 7. *The left and right approximate point spectra are upper semicontinuous spectral systems.*

Proof. Let $R_n = \{x_1, \dots, x_n) \in c_n(\mathcal{A}) : d_l(x_1, \dots, x_n) > 0\}$. Using Corollary 6 we can see that $R = \bigcup_{n=1}^{\infty} R_n$ is a joint regularity. Furthermore, R_n is an open subset of $c_n(\mathcal{A})$, and so τ_l is an upper semicontinuous spectral system.

The statement for τ_r follows by symmetry. $\square$

Thus all results for general spectral system apply for $\sigma_l, \sigma_r, \sigma_H, \tau_l$ and τ_r. In particular we have:

Theorem 8. *Let $\mathcal{A}$ be a Banach algebra. Let $\tilde{\sigma}$ stand for any of $\sigma_l, \sigma_r, \sigma_H, \tau_l, \tau_r$. Then, for all $x = (x_1, \dots, x_n) \in c_n(\mathcal{A})$ we have:*

(i) *$\tilde{\sigma}(x)$ is a non-empty compact subset of $\mathbb{C}^n$, $\tau_l(x) \subset \sigma_l(x)$, $\tau_r(x) \subset \sigma_r(x)$, $\sigma_H(x) = \sigma_l(x) \cup \sigma_r(x)$;*

(ii) *if $(x_0, x_1, \dots, x_n) \in c_{n+1}(\mathcal{A})$, then $\tilde{\sigma}(x_1, \dots, x_n) = P\tilde{\sigma}(x_0, x_1, \dots, x_n)$ where $P : \mathbb{C}^{n+1} \to \mathbb{C}^n$ is the projection onto the last n coordinates;*

(iii) *if $f = (f_1, \dots, f_m)$ is an m-tuple of functions analytic in a neighbourhood of $\sigma^{\langle x \rangle}(x_1, \dots, x_n)$, then*
$$f(\tilde{\sigma}(x) = \tilde{\sigma}(f(x));$$

(iv) *$\partial\sigma(x_1) \subset \tau_l(x_1) \cap \tau_r(x_1)$ for all $x_1 \in \mathcal{A}$;*

(v) *$\Gamma(\sigma^{\langle x \rangle}(x), \mathcal{P}(n)) \subset \tau_l(x) \cap \tau_r(x)$;*

(vi) *the polynomially convex hull of $\tilde{\sigma}(x)$ is equal to $\sigma^{\langle x \rangle}(x)$.*

Theorem 9. *Let x be an element of a Banach algebra $\mathcal{B}$. Then*
$$\lim_{k\to\infty} d_l(x^k)^{1/k} = \sup_{k\in\mathbf{N}} d_l(x^k)^{1/k} = \min\{|\lambda| : \lambda \in \tau_l(x)\}.$$

Similarly,
$$\lim_{k\to\infty} d_r(x^k)^{1/k} = \sup_{k\in\mathbf{N}} d_r(x^k)^{1/k} = \min\{|\lambda| : \lambda \in \tau_r(x)\}.$$

Proof. Let $\mathcal{A} = \langle x \rangle$ and let $\mathcal{C}$ be the commutative Banach algebra constructed in Lemma 5.

Since $d_l^{\mathcal{B}}(y) = d^{\mathcal{C}}\big((y, 0)\big)$ and $\tau_l^{\mathcal{B}}(y) = \tau^{\mathcal{C}}(y, 0)$ for every $y \in \langle x \rangle$, the proof of the first statement follows from Theorem 4.11 for the algebra $\mathcal{C}$.

The second statement follows by symmetry. □

In Section 1 we have proved that $\sigma(ab) \setminus \{0\} = \sigma(ba) \setminus \{0\}$ for all elements a, b in a Banach algebra $\mathcal{A}$. Similar relations are also true for other spectra.

Proposition 10. *Let a, b be elements of a Banach algebra $\mathcal{A}$, let $\lambda \in \mathbb{C}$, $\lambda \neq 0$. Then:*

(i) *$ab - \lambda$ is a left (right) divisor of zero if and only if $ba - \lambda$ is a left (right) divisor of zero;*

(ii) *$ab - \lambda$ is a left (right) topological divisor of zero if and only if $ba - \lambda$ is a left (right) topological divisor of zero.*

Proof. (i) Let $ab-\lambda$ be a left divisor of zero. Let $x \in \mathcal{A}$, $x \neq 0$ and $(ab-\lambda)x = 0$. Then $abx = \lambda x \neq 0$, and so $bx \neq 0$. We have

$$(ba-\lambda)bx = b(ab-\lambda)x = 0,$$

and so $ba-\lambda$ is a left divisor of zero.

The converse implication and the statement for right divisors of zero follow by symmetry.

(ii) Let $ab-\lambda$ be a left topological divisor of zero. Let $x_n \in \mathcal{A}$, $\|x_n\| = 1 \quad (n = 1, 2, \dots)$ and $(ab-\lambda)x_n \to 0$. Then $\lim_{n\to\infty} \|abx_n\| = \lim_{n\to\infty} \|\lambda x_n\| = |\lambda| > 0$, and so $\liminf_{n\to\infty} \|bx_n\| > 0$. Further,

$$(ba-\lambda)bx_n = b(ab-\lambda)x_n \to 0,$$

and so $ba-\lambda$ is a left topological divisor of zero.

The rest follows by symmetry. □

Corollary 11. *Let a, b be elements of a Banach algebra $\mathcal{A}$. Then $\tilde{\sigma}(ab) \setminus \{0\} = \tilde{\sigma}(ba) \setminus \{0\}$, where $\tilde{\sigma}$ stands for any of σ_l, σ_r, τ_l, τ_r, π_l, π_r.*

Proof. For σ_l and σ_r the statement follows from Theorem 1.29, for the remaining spectra from the previous proposition. □

Comments on Chapter I

By C.i.j we denote the jth comment on Section i.

C.1.1. Let $\mathcal{A}$ be an algebra without unit, i.e., $\mathcal{A}$ satisfies only axioms (i)–(iii) of Definition 1.1. Then it is possible to define the *unitization* $\mathcal{A}_1$ of $\mathcal{A}$ by $\mathcal{A}_1 = \{\alpha + a : \alpha \in \mathbb{C}, a \in \mathcal{A}\}$, with the algebraic operations

$$\begin{aligned}(\alpha + a) + (\beta + b) &= (\alpha + \beta) + (a + b),\\ (\alpha + a) \cdot (\beta + b) &= \alpha\beta + (\alpha b + \beta a + ab),\\ \alpha \cdot (\beta + b) &= \alpha\beta + \alpha b\end{aligned}$$

for all $a, b \in \mathcal{A}, \alpha, \beta \in \mathbb{C}$. Then $\mathcal{A}_1$ is an algebra with the unit element $1 + 0_{\mathcal{A}}$.

If $\|\cdot\|$ is an algebra norm on $\mathcal{A}$, then $\|\alpha + a\| = |\alpha| + \|a\|$ defines an algebra norm on $\mathcal{A}_1$, and $\mathcal{A}_1$ contains an isometrical copy of $\mathcal{A}$. In this way it is possible to define the spectrum of elements of an algebra without unit as the spectrum in its unitization.

An alternative approach uses the concept of quasi-inverses, see [BD].

C.1.2. Let $\mathcal{A}$ be a *real algebra*, i.e., suppose that $\mathcal{A}$ satisfies all axioms of Definition 1.1 with the complex field replaced by the field of real numbers. Then it is possible to define the *complexification* $\mathcal{A}_c = \{a+ib : a, b \in \mathcal{A}\}$ with the algebraic operations defined by

$$\begin{aligned}(a+ib)+(c+id) &= (a+c)+i(b+d),\\ (a+ib)\cdot(c+id) &= (ac-bd)+i(ad+bc),\\ (\alpha+i\beta)\cdot(a+ib) &= (\alpha a-\beta b)+i(\alpha b+\beta a)\end{aligned}$$

for all $a,b,c,d \in \mathcal{A}$, $\alpha,\beta \in \mathbb{R}$. Then $\mathcal{A}_c$ is a (complex) algebra with the unit element $1_{\mathcal{A}} + i\cdot 0_{\mathcal{A}}$.

If $\|\cdot\|$ is a (real) algebra norm in $\mathcal{A}$, then it is possible to extend it to $\mathcal{A}_c$ by

$$\|a+ib\|_{\mathcal{A}_c} = \inf\Big\{\sum_{j=1}^{n} |\lambda_j|\cdot\|a_j\| : n \in \mathbb{N}, \lambda_j \in \mathbb{C}, a_j \in \mathcal{A} \text{ and } \sum_{j=1}^{n}\lambda_j a_j = a+ib\Big\}.$$

In this way it is possible to define the spectrum of an element a of a real algebra $\mathcal{A}$ as the spectrum of $a + i\cdot 0$ in $\mathcal{A}_c$. Note that the spectrum is still complex in this case.

C.1.3. Let $\mathcal{A}$ be an algebra (in the sense of Definition 1.1) and let $\|\cdot\|$ be a norm on $\mathcal{A}$ which makes of $\mathcal{A}$ a Banach space such that the multiplication is (jointly) continuous ($a_n \to a, b_n \to b$ implies $a_n b_n \to ab$). Thus $\|\cdot\|$ satisfies conditions (i), (ii) and (v) of Definition 1.2 and condition (iii) is replaced by a weaker condition

(iii′): there exists $k > 0$ such that $\|xy\| \le k\|x\|\cdot\|y\| \qquad (x, y \in \mathcal{A})$.

Define a new norm $|||\cdot|||$ on $\mathcal{A}$ by $|||a||| = \sup\{\|ax\| : x \in \mathcal{A}, \|x\| \le 1\}$. It is easy to check that $|||\cdot|||$ is equivalent to the original norm $\|\cdot\|$, and $|||\cdot|||$ satisfies all conditions of Definition 1.2.

Thus the definition of Banach algebras which uses the continuity of the multiplication (this definition is natural in the context of topological algebras) is essentially equivalent to Definition 1.3. Also, the axiom $\|1_{\mathcal{A}}\| = 1$ is not essential.

The condition (iii′) of joint continuity of multiplication can be replaced by a seemingly weaker assumption that the multiplication is only separately continuous, i.e., that the operators $L_a, R_a : \mathcal{A} \to \mathcal{A}$ defined by $L_a x = ax$, $R_a x = xa$ are continuous for every $a \in \mathcal{A}$. Indeed, an easy application of the Banach-Steinhaus theorem gives the joint continuity of the multiplication.

C.1.4. The most important generalizations of Banach algebras are locally convex algebras and m-convex algebras, see [Mi], [Zel1], [Zel4].

A *locally convex algebra* $\mathcal{A}$ is a locally convex space $\mathcal{A}$ together with a jointly continuous multiplication which makes of $\mathcal{A}$ an algebra. Equivalently, the topology of $\mathcal{A}$ can be given by a system of seminorms $\{\|\cdot\|_\alpha : \alpha \in \Lambda\}$ satisfying:

(i) the system $\{\|\cdot\|_\alpha : \alpha \in \Lambda\}$ is directed, i.e., for all $\alpha, \beta \in \Lambda$ there exists $\gamma \in \Lambda$ such that

$$\max\{\|x\|_\alpha, \|x\|_\beta\} \le \|x\|_\gamma \qquad (x \in \mathcal{A}), \tag{1}$$

(ii) for every $\alpha \in \Lambda$ there exists $\beta \in \Lambda$ such that

$$\|xy\|_\alpha \le \|x\|_\beta \cdot \|y\|_\beta \qquad (x, y \in \mathcal{A}).$$

If $\mathcal{A}$ is a metrizable locally convex algebra, then its topology can be given by means of a sequence $\{\|\cdot\|_n : n \in \mathbb{N}\}$ of seminorms satisfying (for all $n \in \mathbb{N}$, $x, y \in \mathcal{A}$)

$$\|x\|_n \le \|x\|_{n+1}$$

and

$$\|xy\|_n \le \|x\|_{n+1} \cdot \|y\|_{n+1}.$$

Complete metrizable locally convex algebras are sometimes called B_0*-algebras.*

C.1.5. A locally convex algebra $\mathcal{A}$ is called *m-convex* (multiplicatively convex) if its topology can be given by means of a system of seminorms $\{\|\cdot\|_\alpha : \alpha \in \Lambda\}$ satisfying (1) and

$$\|xy\|_\alpha \le \|x\|_\alpha \cdot \|y\|_\alpha \qquad (\alpha \in \Lambda, x, y \in \mathcal{A}).$$

Then, for each $\alpha \in \Lambda$, $J_\alpha = \{x \in \mathcal{A} : \|x\|_\alpha = 0\}$ is a closed two-sided ideal and $(\mathcal{A}/J_\alpha, \|\cdot\|_\alpha)$ is a normed algebra. In this way it is easy to show that complete m-convex algebras are projective limits of Banach algebras. This enables us to generalize many properties of Banach algebras to complete m-convex algebras.

Complete metrizable m-convex algebras are called *Fréchet algebras.*

C.1.6. The openness of the set of all invertible elements in a topological algebra is sometimes called property Q. Obviously, property Q implies the compactness of the spectrum.

An example of a Fréchet algebra where the invertible elements are not open is the algebra of all entire functions with the topology given by a sequence of seminorms $\|f\|_n = \max_{|z|\le n} |f(z)| \quad (n = 1, 2, \dots)$.

C.1.7. The mapping $x \mapsto x^{-1}$ is continuous for every m-convex algebra. For locally convex (even B_0) algebras this is not true. An example is the algebra $L^\omega = \bigcap_{p\ge1} L^p(0,1)$ with the topology given by seminorms of $L^n \quad (n = 1, 2, \dots)$, see [Ar1], [Zel1].

For B_0-algebras the continuity of the mapping $x \mapsto x^{-1}$ is equivalent to the condition that the set $\mathrm{Inv}(\mathcal{A})$ is G_δ, see [Zel1].

C.1.8. Let $\mathcal{A}$ be a Banach algebra and let $a \in \partial\,\mathrm{Inv}(\mathcal{A})$. By Theorem 1.14, there is a sequence (x_n) of norm 1 elements of $\mathcal{A}$ such that both $x_n a \to 0$ and $a x_n \to 0$.

On the other hand, it is possible (example of P.G. Dixon, see [BD, p. 13]) that an element $b \in \mathcal{A}$ is both left and right topological divisor of zero and yet there is no sequence (u_n) of norm 1 elements of $\mathcal{A}$ with both $u_n b \to 0$ and $b u_n \to 0$.

C.1.9. By Theorem 1.14, $\partial\sigma_l(a) \subset \tau_r(a)$. On the other hand, the inclusion $\partial\sigma_l(a) \subset \tau_l(a)$ is not true in general, see [Pi1, pp. 367–368], [Bu1, Example 1.11] or [Mü20].

C.1.10. The basic properties of the spectrum including the spectral radius formula were proved by Gelfand [Ge]. The Gelfand-Mazur theorem (Corollary 1.19) appeared in [Ma] and [Ge] (the Mazur's proof was not published from technical reasons).

C.1.11. The spectrum of an element a in an algebra $\mathcal{A}$ can be defined in the same way as in Banach algebras by $\sigma(a) = \{\lambda \in \mathbb{C} : a - \lambda \text{ is not invertible}\}$.

Let $\mathcal{A}$ be a normed algebra. Considering its completion gives easily that the spectrum in normed algebras is always non-empty. In general, however, it is neither bounded nor closed (consider for example the algebra of all polynomials $p(z) = \sum_i c_i z^i$ with the norm $\|p\| = \sum_i |c_i|$ and the polynomial $p(z) = z$, or the algebra of all "trigonometrical polynomials" $\sum_{i\in\mathbb{Z}} c_i z^i$ with the same norm; all sums are finite).

Similarly, it is easy to see that the spectrum in m-convex algebras is always non-empty.

The spectrum in locally convex algebras (even in B_0-algebras) can be empty. However, if we define the set $\Sigma(a) = \sigma(a) \cup \sigma'(a) \cup \sigma^\infty(a) \subset \mathbb{C} \cup \{\infty\}$, where $\sigma'(a) = \{\lambda : z \mapsto (a - z)^{-1} \text{ is discontinuous at } \lambda\}$ and

$$\sigma^\infty(a) = \begin{cases} \emptyset & z \mapsto (1 - za)^{-1} \text{ is continuous at } 0, \\ \{\infty\} & \text{otherwise,} \end{cases}$$

then $\Sigma(a) \neq \emptyset$ for all elements a in a locally convex algebra, see [Zel4]. For similar concepts of spectrum in locally convex algebras see [Wa], [All1] and [Neu1], [Neu2].

C.1.12. By the Gelfand-Mazur theorem, the complex numbers are the only Banach algebra where all non-zero elements are invertible. For real Banach algebras the situation is more complicated. There are three such examples: the real numbers, complex numbers and the field of quaternions, see [Ric, p. 40] or [BD, p. 73].

C.1.13. By the Gelfand-Mazur theorem and Theorem 1.14, a Banach algebra either possesses non-zero one-sided topological divisors of zero or it is isometrically isomorphic to the complex field. An analogous statement fails even for Fréchet algebras.

However, each m-convex algebra different from the complex field possesses generalized topological divisors of zero, i.e., there is a neighbourhood U of zero such that $0 \in \overline{(\mathcal{A} \setminus U)^2}$. On the other hand, this is no longer true for locally convex algebras [AZ].

C.1.14. In general neither the spectrum nor the spectral radius in a Banach algebra is continuous.

Let H be the Hilbert space with an orthonormal basis $\{e_i : i \in \mathbb{Z}\}$. Let $T, T_n \in \mathcal{B}(H)$ be the bilateral weighted shifts defined by $T_n e_i = e_{i+1} \quad (i \neq 0)$, $T_n e_0 = n^{-1} e_1$, $T e_0 = 0$ and $T e_i = e_{i+1} \quad (i \neq 0)$. It is easy to see that $\|T_n - T\| \to 0$, $\sigma(T) = \overline{\mathbb{D}}$ and $\sigma(T_n) = \mathbb{T}$ for all n, hence the spectrum is discontinuous.

In the previous example the spectral radius behaves continuously. An example of discontinuous spectral radius was given by Kakutani, see [Ric, pp. 282–283].

Let K be the Hilbert space with an orthonormal basis $\{e_1, e_2, \dots\}$. Let $T \in \mathcal{B}(K)$ be the weighted shift defined by $Te_n = w_n e_{n+1}$, where $w_n = 2^{-k}$ if n can be written as $n = 2^k \cdot m$ with m odd. It is a matter of routine to verify that $r(T) > 0$ but T is a limit of nilpotent operators: define T_s by $T_s e_n = w_n e_{n+1}$ if $w_n \geq 2^{-s}$, and $T_s e_n = 0$ otherwise. Then $T_s^{2^{s+1}} = 0$, and so $r(T_s) = 0$ for all s.

C.1.15. An example of a Banach algebra with continuous spectral radius but discontinuous spectrum was given by Apostol [Ap4]. In [Mü1] it was given an example that the spectrum (and the spectral radius) in a Banach algebra can be discontinuous even on straight lines. For details and further examples see [Au1].

C.1.16. The upper semicontinuity of spectrum in Banach algebras was proved in [New]. Although the spectral radius and the spectrum are in general discontinuous, there are plenty of continuity points. By Theorem 6.14, the set of all continuity points is dense and residual.

Let a be an element of a Banach algebra and let U_1, U_2 be disjoint open subsets such that $\sigma(a) \subset U_1 \cup U_2$ and $\sigma(a) \cap U_1 \neq \emptyset$. Then $U_1 \cap \sigma(b) \neq \emptyset$ for all b sufficiently close to a [New]. Indeed, let $f \equiv 1$ on U_1 and $f \equiv 0$ on U_2. Then $f(b)$ is defined for all b in a neighbourhood of a, $f(b)^2 = f(b)$ and $\lim_{b \to a} f(b) = f(a) \neq 0$. Consequently, $\sigma(b) \cap U_1 \neq \emptyset$.

In particular, the spectrum is continuous at a if the spectrum of a is totally disconnected, [New].

For a survey of results concerning the continuity of spectrum and the spectral radius see [Bu2].

C.1.17. Let G be an open subset of the complex plane and let $f : G \to \mathcal{A}$ be an analytic function. Although the function $z \in G \to r(f(z))$ can be discontinuous, by [Ve1] it is always subharmonic (a function $f : G \to \mathbb{R}$ is *subharmonic* if it is upper semicontinuous and $f(z) \leq \frac{1}{2\pi} \int_0^{2\pi} f(z + re^{it})\, \mathrm{d}t$ whenever G contains the ball $\{w : |z - w| \leq r\}$). Moreover, the function $z \mapsto \log r(f(z))$ is also subharmonic [Ve2].

The subharmonicity of these functions has many interesting consequences, see [Au1]. For example, it implies the maximum principle for the spectral radius. Furthermore, $r(f(z)) = \limsup_{w \to z} r(f(w))$ for all $z \in G$.

C.1.18. The functional calculus for one operator was first used by Riesz [Ri1] for construction of the spectral projections of a compact operator. The general spectral mapping theorem was proved by Dunford [Du1].

C.1.19. Various types of radical were studied intensely in the ring theory; the radical defined in Definition 1.42 is due to Jacobson [Ja]. For a detailed information about the history of radical see [Pal].

C.2.1. The basic results concerning commutative Banach algebras are due to Gelfand [Ge]. It is possible to formulate the Gelfand theory for algebras without unit. The maximal ideal space for a non-unital Banach algebra $\mathcal{A}$ is only locally compact; the maximal ideal space of the corresponding unitization $\mathcal{A}_1$ (see C.1.1) is the one-point compactification of $\mathcal{M}(\mathcal{A})$.

C.2.2. By Theorem 2.6, every multiplicative functional on a commutative Banach algebra is automatically continuous. A much more general result of this type was proved by B.E. Johnson [J]:

Let $\mathcal{A}, \mathcal{B}$ be Banach algebras, $\mathcal{B}$ semisimple and let $\varphi : \mathcal{A} \to \mathcal{B}$ be a surjective homomorphism. Then φ is continuous. In particular, all algebra norms on a semisimple Banach algebra are equivalent.

This result was further generalized by Aupetit [Au4], Theorem 5.5.1:

Let $\mathcal{A}, \mathcal{B}$ be Banach algebras, $\mathcal{B}$ semisimple, let $\varphi : \mathcal{A} \to \mathcal{B}$ be a surjective linear mapping satisfying $r(\varphi(x)) \le r(x)$ for all $x \in \mathcal{A}$. Then φ is continuous.

It is an open problem whether it is sufficient to assume in the above results that φ has only dense range. For details on automatic continuity see [Sin], [Dal].

C.2.3. Let $\mathcal{A}$ be a commutative Banach algebra. It is easy to see that the Gelfand transform is an isometry if and only if $r(a) = \|a\|$ for all $a \in \mathcal{A}$. Equivalently, $\|a^2\| = \|a\|^2 \quad (a \in \mathcal{A})$. Algebras with this property are called function algebras.

Another possible definition is that function algebras are closed subalgebras of $C(X)$ that contain the constant functions and separate the points of X, where X is a compact Hausdorff space.

C.2.4. It is easy to see that every commutative m-convex algebra has at least one continuous multiplicative functional. This is not true if the algebra is not m-convex. The algebra L^ω defined in C.1.7 is an example of a B_0-algebra without multiplicative functionals, see [Zel1].

C.2.5. It is a longstanding open problem (see Michael [Mi]) whether each multiplicative functional on a commutative Fréchet algebra is automatically continuous. This is true for finitely generated Fréchet algebras [Ar3]. The Michael problem is not true if we drop the assumption of metrizability. An example of a complete commutative m-convex algebra with a discontinuous multiplicative functional is the algebra $C(T_{\omega_1})$ of all continuous functions with the compact open topology on the compact space T_{ω_1}, where ω_1 is the first uncountable ordinal, and $T_{\omega_1} = \{\alpha : \alpha < \omega_1\}$ with the order topology.

C.2.6. Denote by Q the set of all *quasinilpotent* elements of a Banach algebra $\mathcal{A}$, $Q = \{x \in \mathcal{A} : r(x) = 0\}$. By Theorem 2.9, $Q = \operatorname{rad} \mathcal{A}$ for commutative Banach algebras.

For non-commutative Banach algebras this is no longer true. An extreme example was given by [Di]: there exists an infinite-dimensional semisimple Banach algebra such that the closure of the nilpotent elements is an ideal of codimension 1.

Theorem 2.9 implies that a subalgebra of a semisimple commutative Banach algebra is also semisimple. For non-commutative Banach algebras this is not true: consider the semisimple algebra of all $n \times n$ matrices and its subalgebra of upper triangular matrices.

The radical of a non-commutative Banach algebra can be characterized as follows [Ze1], [Ze3], [Ze4]:

Theorem. *$a \in \operatorname{rad} \mathcal{A}$ if and only if $a + Q \subset Q$.*

C.2.7. It is easy to see that the spectral radius and the spectrum are continuous in Banach algebras that are commutative modulo the radical. An example of a semisimple non-commutative Banach algebra with continuous spectrum is the algebra of all $n \times n$ matrices.

In [Au2], see also [PZ], [Ze2], the following result was proved:

Theorem. *Let $\mathcal{A}$ be a Banach algebra. The following properties are equivalent:*

(i) *the spectral radius is uniformly continuous on $\mathcal{A}$;*
(ii) $r(x+y) \leq r(x) + r(y) \qquad (x, y \in \mathcal{A})$;
(iii) $r(xy) \leq r(x) \cdot r(y) \qquad (x, y \in \mathcal{A})$;
(iv) *the algebra $\mathcal{A} / \operatorname{rad} \mathcal{A}$ is commutative.*

C.2.8. The Gleason-Kahane-Żelazko theorem was proved in [Gle] and [KZ]. The same statement is true also for non-commutative Banach algebras, see [Zel2].

The following stronger version of the Gleason-Kahane-Żelazko theorem was proved in [KS]:

Theorem. *Let $\mathcal{A}$ be a commutative Banach algebra and let $\varphi : \mathcal{A} \to \mathbb{C}$ be an arbitrary function. Suppose that $\varphi(0) = 0$ and $\varphi(x) - \varphi(y) \in \sigma(x - y)$ for all $x, y \in \mathcal{A}$. Then φ is a multiplicative functional.*

C.2.9. The Gelfand theory can be generalized to Banach algebras satisfying a polynomial identity (PI-algebras). An important role for PI-algebras play the standard identities $e_n(x_1, \ldots, x_n) = 0$ for all $x_1, \ldots, x_n$ in the algebra, where

$$e_n(x_1, \ldots, x_n) = \sum_{\pi} (-1)^{\operatorname{sign} \pi} x_{\pi(1)} \cdots x_{\pi(n)}$$

and the sum is taken over all permutations π of the set $\{1, \ldots, n\}$. If an algebra satisfies any polynomial identity, then it satisfies some standard identity.

Note that $e_2(x_1, x_2) = x_1 x_2 - x_2 x_1$, and so commutative Banach algebras are PI. By [AL], the algebra of all $n \times n$ matrices satisfies the standard identity $e_{2n}(A_1, \ldots, A_{2n}) = 0$ for all matrices $A_1, \ldots, A_{2n}$. Thus the PI-algebras include also the matrix algebras.

For PI-algebras it is possible to develop the Gelfand theory (with multiplicative functionals replaced by finite-dimensional representations), see [Kr].

C.2.10. The multivariable functional calculus in commutative Banach algebras was constructed by Arens, Calderon [AC], Waelbroeck [Wa] and Shilov [Sh2]. For simpler proofs see [BD] or [Ga].

The calculus is uniquely determined by properties (i), (ii), (iii) and (v) of Theorem 2.20, see [Zam]. In Section 30 we show a weaker unicity property.

C.2.11. The concept of the Shilov boundary is due to Shilov [Sh1].

It is easy to see that the Shilov boundary of the algebra $C(K)$ is equal to its maximal ideal space, $\Gamma(C(K)) = \mathcal{M}(C(K)) \sim K$.

The maximal ideal space of the disc algebra $A(D)$ can be identified with $\overline{D}$; the Shilov boundary then coincides with the topological boundary of D.

C.3.1. The construction of the algebra $Q(\mathcal{A})$ is attributed in [Ha5] to Berberian [Ber] and Quigley, see [Ric], page 25. It was simplified by Wolff, see [ChD].

C.3.2. Theorem 3.7 and its consequences (Corollary 3.8 and the projection property for the approximate point spectrum) were proved in [Sl1] and [ChD].

C.3.3. The cortex $\operatorname{cor}\mathcal{A}$ was defined by Arens [Ar4] as the set of all elements of $\mathcal{M}(\mathcal{A})$ that admit an extension to a multiplicative functional on any commutative extension $\mathcal{B} \supset \mathcal{A}$. The present definition is equivalent, see Section 5.

C.3.4. The inclusion $\Gamma(\mathcal{A}) \subset \operatorname{cor}\mathcal{A}$ was proved in [Zel5], cf. also [Sh1].

The class $\gamma(\mathcal{A})$ and the characterization of the Shilov boundary given in Theorem 3.11 are new, but the ideas are present already in [Zel5]. The definition of the spectrum σ_Γ is also new.

C.3.5. The analogue of the Gleason-Kahane-Żelazko theorem for the cortex is not true as the following example of Żelazko shows. Let $\mathcal{A}$ be the Banach algebra of all functions analytic on the unit ball $B = \{(\lambda,\mu) \in \mathbb{C}^2 : |\lambda|^2 + |\mu|^2 < 1\}$ and continuous in $\overline{B}$. For $(\lambda,\mu) \in \overline{B}$ let $\mathcal{E}_{\lambda,\mu}$ be the evaluation functional on $\mathcal{A}$ defined by $\mathcal{E}_{\lambda,\mu}(f) = f(\lambda,\mu)$. Then $\Gamma(\mathcal{A}) = \{\mathcal{E}_{\lambda,\mu} : (\lambda,\mu) \in \partial B\}$ and

$$\mathcal{E}_{0,0}(f) = f(0,0) \in \big\{f(\lambda,\mu) : (\lambda,\mu) \in \partial B\big\} = \big\{\varphi(f) : \varphi \in \Gamma(\mathcal{A})\big\} \subset \tau(f)$$

for all $f \in \mathcal{A}$. On the other hand, it is easy to see that $\mathcal{E}_{0,0} \notin \operatorname{cor}\mathcal{A}$ since $z_1, z_2 \in \operatorname{Ker}\mathcal{E}_{0,0}$ and (z_1, z_2) are not joint topological divisors of zero since $\|z_1 f\| + \|z_2 f\| \geq \|f\|$ for all $f \in \mathcal{A}$.

C.4.1. Theorem 4.1 was proved for commutative Banach algebras in [Li] and for general Banach algebras in [FM].

The characterization of permanently singular elements (Theorem 4.3) is due to Arens [Ar2].

C.4.2. If F is a finite subset of a commutative Banach algebra $\mathcal{A}$ such that no $f \in F$ is a topological divisor of zero, then their product in not a topological divisor of zero either. Using Theorem 4.3 for this product, we see that there exists a commutative extension $\mathcal{B} \supset \mathcal{A}$ such that all $f \in F$ are invertible in $\mathcal{B}$.

This result was generalized by Bollobás [Bo1] for countable sets, i.e., it is always possible to adjoin inverses to a countably many elements of a commutative Banach algebra that are not permanently singular.

In general, this is not possible for uncountable sets, see [Bo1] or [Mü5].

C.4.3. The statement analogous to Theorem 4.3 is not true in non-commutative Banach algebras. By [Mü10], there exists a non-commutative Banach algebra $\mathcal{A}$ and an element $a \in \mathcal{A}$ such that $\|ax\| \geq \|x\| \quad (x \in \mathcal{A})$ (i.e., a is not a left topological divisor of zero) but a is left invertible in no Banach algebra $\mathcal{B} \supset \mathcal{A}$.

C.4.4. By [Mü8], there exists a (non-commutative) Banach algebra $\mathcal{A}$, an element $a \in \mathcal{A}$ and two extensions $\mathcal{B}_1 \supset \mathcal{A}$ and $\mathcal{B}_2 \supset \mathcal{A}$ such that a is left invertible in $\mathcal{B}_1$, right invertible in $\mathcal{B}_2$ but a is invertible (both left and right) in no extension $\mathcal{B} \supset \mathcal{A}$.

C.4.5. Theorem 4.10 was proved by Read [Re2] who gave a positive answer to a problem posed by Bollobás [Bo3].

C.4.6. If x, y are elements of a commutative Banach algebra $\mathcal{A}$, then in general there is no extension $\mathcal{B} \supset \mathcal{A}$ such that both $\sigma^{\mathcal{B}}(x) = \tau^{\mathcal{A}}(x)$ and $\sigma^{\mathcal{B}}(y) = \tau^{\mathcal{A}}(y)$, see [Mü5], [Re5].

Consequently, by the spectral mapping theorem for both σ and τ, there is no extension $\mathcal{B} \supset \mathcal{A}$ where $\sigma^{\mathcal{B}}(x, y) = \tau^{\mathcal{A}}(x, y)$. On the other hand, we always have $\tau^{\mathcal{A}}(x, y) = \bigcap_{\mathcal{B} \supset \mathcal{A}} \sigma^{\mathcal{B}}(x, y)$ by Theorem 5.12.

C.4.7. The "inverse" spectral radius formula (Theorem 4.11) was proved in [MZ] (using the sheaf theory) and in [Mü4] (by combinatorial methods). The present proof is based on the above-mentioned result of Read (Theorem 4.10). Note that for the proof of Theorem 4.11 it is sufficient to use only the simpler result of Proposition 4.8.

C.5.1. Removable and non-removable ideals were introduced and studied by Arens, [Ar2], [Ar4], [Ar5] and further by Żelazko [Zel3], [Zel5]. The problem whether an ideal in a commutative Banach algebra is non-removable if and only if it consists of joint topological divisors of zero was raised in [Ar4]. A positive answer was given in [Mü2]. The present proof follows the line of [Mü2] but the estimates are essentially improved here.

C.5.2. Let $u_1, \dots, u_n$ be elements of a commutative Banach algebra $\mathcal{A}$ and c a positive constant such that $\sum_{i=1}^{n} \|u_i x\| \geq c \cdot \|x\| \quad (x \in \mathcal{A})$. An interesting open problem is what are the smallest norms of elements b_i in a commutative extension $\mathcal{B} \supset \mathcal{A}$ for which $\sum_{i=1}^{n} u_i b_1 = 1$.

By Theorem 4.3, if $n = 1$, then it is possible to take $\|b_1\| = c^{-1}$ (which is also the smallest possible norm). For $n \geq 2$ the situation is much more complicated. In the simplest non-trivial case of $n = 2, c = 1$ the present proof gives the existence of elements $b_1, b_2 \in \mathcal{B} \supset \mathcal{A}$ with $b_1u_1 + b_2u_2 = 1$ and $\|b_i\| \leq 2^8$ (the original proof only gave $\|b_i\| = 2^{17}$).

On the other hand, rather surprisingly, in general it is not possible to find an extension $\mathcal{B}$ and $b_1, b_2 \in \mathcal{B}$ with $\|b_i\| \leq 1 \quad (i = 1, 2)$, see [Bo2]. Thus there is still an enormous gap between the upper and lower estimates.

C.5.3. Let $I_1, I_2, \dots$ be a countable family of removable ideals in a commutative Banach algebra $\mathcal{A}$. Then there exists a commutative extension $\mathcal{B} \supset \mathcal{A}$ in which all ideals $I_1, I_2, \dots$ are removed (= neither of them is contained in a proper ideal in the extension), see [Mü6]. This generalizes the corresponding result of Bollobás [Bo1] for non-permanently singular elements, cf. C.4.2.

C.5.4. A stronger version of Theorem 4.3 (with essentially the same proof) says: if u, v are elements of a commutative Banach algebra $\mathcal{A}$ and c is a positive constant such that $\|ux\| \geq c \cdot \|vx\| \quad (x \in \mathcal{A})$, then there exists a commutative extension $\mathcal{B} \supset \mathcal{A}$ and $b \in \mathcal{B}$ such that $ub = v$. The analogous result is not true for n-tuples [Mü3]: it is possible to have $v, u_1, u_2 \in \mathcal{A}$ with $\|u_1x\| + \|u_2x\| \geq \|vx\| \quad (x \in \mathcal{A})$ and yet there is no extension $\mathcal{B} \supset \mathcal{A}$ and $b_1, b_2 \in \mathcal{B}$ such that $u_1b_1 + u_2b_2 = v$.

C.6.1. The notion of regularity and the corresponding axiomatic spectral theory was presented in [KM2].

For other related approaches see [GL] and [Rn].

C.6.2. Theorem 6.14 for the ordinary spectrum in the algebra of operators on a Hilbert space was noted in [CM]. In general form it was proved in [Rn], Proposition 3, see also [LvS]. The argument is based on a classical result of Kuratowski, cf. [Au4], p. 50.

C.7.1. The definition of spectral systems for n-tuples of commuting elements (Definition 7.4) is a slightly modified concept of Słodkowski and Żelazko [SZ1] and Żelazko [Zel7], see also Curto [Cu4].

C.7.2. By [SZ1], it is possible to replace condition (i) of Definition 7.4 ($\sigma_R(x) \subset \sigma^{\langle x \rangle}(x)$ for all n-tuples x of commuting elements) by the same condition only for commuting triples of elements.

Using the Kowalski-Słodkowski theorem (see C.2.8) it is even sufficient to require this only for commuting pairs. On the other hand, it is not sufficient to require in Definition 7.4 the condition $\tilde{\sigma}(x) \subset \sigma(x)$ for single elements x only; an example is the product $\tilde{\sigma}(x_1, \dots, x_n) = \prod_i \sigma(x_i)$.

C.7.3. Let $\tilde{\sigma}$ be a compact-valued spectral system in a Banach algebra $\mathcal{A}$. It is possible to extend $\tilde{\sigma}$ to infinite commuting subsets of $\mathcal{A}$. Indeed, let $\tilde{\sigma}((x_\alpha)_{\alpha \in \Lambda})$ be the set of all $(\lambda_\alpha)_{\alpha \in \Lambda}$ such that $(\lambda_{\alpha_1}, \dots, \lambda_{\alpha_n}) \in \tilde{\sigma}(x_{\alpha_1}, \dots, x_{\alpha_n})$ for all finite subsets $\{\alpha_1, \dots, \alpha_n\} \subset \Lambda$, see [SZ1].

A compactness argument gives that $\tilde{\sigma}((x_\alpha)_{\alpha\in\Lambda})$ is non-empty for all commuting families $(x_\alpha)_{\alpha\in\Lambda}$.

C.7.4. An interesting problem is to characterize those regularities that can be extended to joint regularities (equivalently, which spectra $\tilde{\sigma}$ defined for single elements can be extended to n-tuples of commuting elements).

Necessary conditions are property (P1), stability of the spectrum under commuting quasinilpotent perturbations (Theorem 7.16), and the property that $\widehat{\Delta}(\tilde{\sigma}(a), \tilde{\sigma}(b)) \le \|a-b\|$ for all commuting a, b (uniform continuity on commuting elements, see Theorem 7.14).

C.7.5. By [SZ2], a complex-valued function φ defined on a Banach algebra $\mathcal{A}$ is called a *semicharacter* if its restriction to any commutative subalgebra of $\mathcal{A}$ is a multiplicative functional (sometimes also called a character).

Any set K of semicharacters on a Banach algebra $\mathcal{A}$ defines a spectral system by

$$\sigma_K(a_1, \dots, a_n) = \{(\varphi(a_1), \dots, \varphi(a_n)) : \varphi \in K\}.$$

On the other hand, it is not possible to describe all spectral system in this way since there are Banach algebras (for example the algebra of all 3×3 matrices) without semicharacters.

Clearly, each multiplicative functional is a semicharacter. An example of a discontinuous semicharacter can be found in the algebra of all 2×2 matrices [SZ2]. An example of a continuous semicharacter which is not a multiplicative functional was found in [KM4].

It is an open problem whether a uniformly continuous semicharacter is already automatically a multiplicative functional.

C.7.6. Proposition 7.20 and Corollary 7.22 were proved in [Waw].

C.8.1. The one-sided and one-sided approximate point spectra for n-tuples of elements in a Banach algebra were introduced and studied by Harte [Ha1] and [Ha2].

C.8.2. The one-sided and one-sided approximate point spectra can be defined, using exactly the same definitions, for non-commuting n tuples of Banach algebra elements, see Harte [Ha1]. However, in this case the spectrum can be empty. The simplest example are the matrices

$$A_1 = \begin{pmatrix} 0 & 1 \\ 0 & 0 \end{pmatrix} \qquad A_2 = \begin{pmatrix} 0 & 0 \\ 1 & 0 \end{pmatrix}$$

in the Banach algebra of 2×2 matrices. It is easy to verify that $A_1^2 = A_2^2 = 0$ and $A_1A_2 + A_2A_1 = I$. Thus $\sigma_l(A_1, A_2) = \sigma_r(A_1, A_2) = \emptyset$.

Similarly, the projection property is not satisfied since $\sigma(A_1) \neq \emptyset$. In general, for non-commuting n-tuples there is only one inclusion.

C.8.3. Let $\mathcal{A}$ be a Banach algebra. By [FS1], $\sigma_l(a_1, \ldots, a_n) \neq \emptyset$ for all (non-commuting) n-tuples $a_1, \ldots, a_n \in \mathcal{A}^n$ if and only if there exists a multiplicative functional on $\mathcal{A}$.

Further, $\tau_l(a_1, \ldots, a_n) \neq \emptyset$ for all n-tuples $a_1, \ldots, a_n \in \mathcal{A}^n$ if and only if there exists a multiplicative functional on $\mathcal{A}$ whose kernel consists of joint left topological divisors of zero, see [So3].

Similar statements are true also for the right spectra.

C.8.4. By [FS2], a Banach algebra $\mathcal{A}$ is commutative modulo its radical if and only if $\sigma_l(a_1, \ldots, a_n) \subset \sigma_r(a_1, \ldots, a_n)$ for all n-tuples $a_1, \ldots, a_n \in \mathcal{A}$.

C.8.5. By Theorem 8.9, it is possible to calculate the distance of 0 to the left (right) approximate point spectrum of a Banach algebra element.

An interesting problem is to obtain a similar formula for the distance of 0 to the left (right) spectrum. In [Ze5] it was conjectured that

$$\operatorname{dist}\{0, \sigma_l(a)\} = \sup\{r(b)^{-1} : ba = 1\}. \tag{2}$$

If a is invertible, then this reduces to the spectral radius formula. Also, (2) is true in the algebra $\mathcal{B}(H)$ of all bounded operators on a Hilbert space H, see [BM].

In general, it is easy to see that the inequality $\geq$ in (2) is always true; the opposite inequality is an open problem. By [Ze5], the conjecture is also equivalent to another interesting problem:

For each compact subset K of the disc $\{z \in \mathbb{C} : |z| < \sup\{r(b)^{-1} : ba = 1\}\}$ there exists an analytic $\mathcal{A}$-valued function g defined on a neighbourhood U of K such that

$$\begin{aligned} g(z)(a-z) &= 1 \\ g(z) - g(w) &= (z-w)g(z)g(w) \end{aligned}$$

for all $z, w \in U$ (so-called left resolvent of a), cf. C.13.2.

Note that the resolvent $z \mapsto (a-z)^{-1}$ defined on the complement of $\sigma(a)$ satisfies the resolvent identity $(a-z)^{-1} - (a-w)^{-1} = (z-w)(a-z)^{-1}(a-w)^{-1}$, cf. Theorem 1.16.

Chapter II

Operators

In this chapter we study the Banach algebra of all operators acting on a Banach space. All Banach spaces are assumed to be complex and non-trivial, of dimension at least 1. By an operator we always mean a bounded linear mapping between two Banach spaces.

Let X, Y be Banach spaces. Denote by $\mathcal{B}(X, Y)$ the set of all operators from X to Y. Write for short $\mathcal{B}(X) = \mathcal{B}(X, X)$. For $T \in \mathcal{B}(X, Y)$ define the operator norm $\|T\| = \sup\{\|Tx\| : x \in X, \|x\| \leq 1\}$. It is clear that $\mathcal{B}(X, Y)$ with this norm becomes a Banach space and $\mathcal{B}(X)$ is a Banach algebra. The unit element in $\mathcal{B}(X)$ is the identity operator denoted by I_X (or simply I if no confusion can arise).

For basic results and notations from operator theory see Appendix A.1.

9 Spectrum of operators

In this section we reformulate the results from Section 8 for the algebra $\mathcal{B}(X)$ of all operators acting on a Banach space X.

We start with the left and right point spectra π_l and π_r.

Theorem 1. *Let T be an operator acting on a Banach space X and let $\lambda \in \mathbb{C}$. Then:*

(i) *$\lambda \in \pi_l(T)$ if and only if λ is an eigenvalue of T (i.e., $\operatorname{Ker}(T - \lambda) \neq \{0\}$);*

(ii) *$\lambda \in \pi_r(T)$ if and only if $\overline{(T - \lambda)X} \neq X$.*

Proof. (i) If $\lambda \in \pi_l(T)$, then $(T - \lambda)S = 0$ for some $S \in \mathcal{B}(X)$, $S \neq 0$. Let $x \in X$ satisfy $Sx \neq 0$. Then $(T - \lambda)Sx = 0$, and so Sx is an eigenvector of T.

Conversely, let $(T - \lambda)x = 0$ for some non-zero $x \in X$. Let $f \in X^*$ satisfy $f(x) = 1$ and define $S \in \mathcal{B}(X)$ by $Sy = f(y)x \quad (y \in X)$. Then $Sx = x$; so $S \neq 0$ and $(T - \lambda)S = 0$.

(ii) Let $S(T-\lambda) = 0$ for some $S \in \mathcal{B}(X)$, $S \neq 0$. Then $\overline{(T-\lambda)X} \subset \operatorname{Ker} S \neq X$.

Conversely, if $\overline{(T-\lambda)X} \neq X$, then there exists a non-zero functional $f \in X^*$ such that $f|\overline{(T-\lambda)X} = 0$. Let $x \in X$ be any non-zero vector and define $S \in \mathcal{B}(X)$ by $Sy = f(y) \cdot x$. Then $S \neq 0$ and $S(T-\lambda) = 0$. □

Corollary 2. *Let $T \in \mathcal{B}(X)$. Then $\pi_l(T) \subset \pi_r(T^*)$ and $\pi_r(T) = \pi_l(T^*)$.*

Proof. Follows from Theorem 1 and Appendix A.1.15. □

In general, $\pi_l(T)$ is not equal to $\pi_r(T^*)$, since it is possible that T is an injective operator such that $\operatorname{Ran} T^*$ is not dense (in general, $\operatorname{Ran} T^*$ is only w^*-dense). As an example, consider the operator $(a_n) \mapsto (n^{-1}a_n)$ acting in ℓ^1.

The equality $\pi_l(T) = \pi_r(T^*)$ is true for operators in reflexive Banach spaces.

Characterizations similar to those in Theorem 1 are also true for the approximate point spectra. We start with the definition of two quantities connected with an operator. Recall that B_X denotes the closed unit ball in X.

Definition 3. Let X, Y be Banach spaces and let $T \in \mathcal{B}(X,Y)$. We define the *injectivity modulus* of T (sometimes also called the minimum modulus) by

$$j(T) = \inf\{\|Tx\| : x \in X, \|x\| = 1\}$$

and the *surjectivity modulus* by

$$k(T) = \sup\{r \geq 0 : TB_X \supset r \cdot B_Y\}.$$

We say that T is *bounded below* if $j(T) > 0$.

Clearly, $j(T) \leq \|T\|$ and $k(T) \leq \|T\|$ since $TB_X \subset \|T\| \cdot B_Y$. Furthermore,

$$j(T) = \inf\left\{\frac{\|Tx\|}{\|x\|} : x \in X, x \neq 0\right\}.$$

Theorem 4. *An operator $T \in \mathcal{B}(X,Y)$ is bounded below if and only if it is one-to-one and* $\operatorname{Ran} T$ *is closed. T is onto if and only if $k(T) > 0$.*

Proof. If T is onto, then $k(T) > 0$ by the open mapping theorem. If T is not onto, then $k(T) = 0$ by definition.

If T is one-to-one and $\operatorname{Ran} T$ is closed, then $j(T) > 0$ by the open mapping theorem. Conversely, if $j(T) > 0$, then T is one-to-one and $T : X \to \operatorname{Ran} T$ is an isomorphism. Thus $\operatorname{Ran} T$ is complete and therefore closed in Y. □

Theorem 5. *Let $T \in \mathcal{B}(X,Y)$. Then T is bounded below if and only if T^* is onto. T is onto if and only if T^* is bounded below.*

Proof. Follows from Corollary A.1.15 and Theorem A.1.16. □

Theorem 6. *Let X, Y, Z be Banach spaces, $T \in \mathcal{B}(X, Y)$ and $S \in \mathcal{B}(Y, Z)$. Then:*

(i) $j(ST) \leq \|S\| \cdot j(T)$;
(ii) $j(ST) \geq j(S) \cdot j(T)$;
(iii) $k(ST) \leq k(S) \cdot \|T\|$;
(iv) $k(ST) \geq k(S) \cdot k(T)$.

Proof. (i) $j(ST) = \inf\left\{\frac{\|STx\|}{\|x\|} : x \in X, x \neq 0\right\} \leq \|S\| \cdot \inf\left\{\frac{\|Tx\|}{\|x\|} : x \in X, x \neq 0\right\} = \|S\| \cdot j(T)$.

(ii) For $x \in X$ we have $\|STx\| \geq j(S) \cdot \|Tx\| \geq j(S)j(T)\|x\|$, and so $j(ST) \geq j(S)j(T)$.

(iii) The statement is clear if $k(ST) = 0$. Suppose that ST is onto, and so $T \neq 0$. Let $z \in Z$, $\|z\| < \frac{k(ST)}{\|T\|}$. Then there exists $x \in X$ such that $STx = z$ and $\|x\| \leq \|T\|^{-1}$. Then $y = Tx \in B_Y$ and $Sy = z$. Thus $k(S) \geq \frac{k(ST)}{\|T\|}$, which gives (iii).

(iv) The statement is clear if either $k(T)$ or $k(S)$ is equal to 0. Let both T and S be onto and let $\varepsilon > 0$, $\varepsilon < \min\{k(S), k(T)\}$. We have

$$(ST)B_X \supset S\big((1-\varepsilon)k(T)B_Y\big) = (1-\varepsilon)k(T)SB_Y \supset (1-\varepsilon)^2 k(T)k(S)B_X.$$

Thus $k(ST) \geq (1-\varepsilon)^2 k(S)k(T)$. Letting $\varepsilon \to 0$ gives $k(ST) \geq k(S)k(T)$. □

Theorem 7. *If $T \in \mathcal{B}(X, Y)$ is bijective, then $j(T) = \|T^{-1}\|^{-1} = k(T)$.*

Proof. We have

$$\begin{aligned} j(T) &= \inf\left\{\frac{\|Tx\|}{\|x\|} : x \in X, x \neq 0\right\} = \left(\sup\left\{\frac{\|x\|}{\|Tx\|} : x \in X, x \neq 0\right\}\right)^{-1} \\ &= \left(\sup\left\{\frac{\|T^{-1}y\|}{\|y\|} : y \in Y, y \neq 0\right\}\right)^{-1} = \|T^{-1}\|^{-1}. \end{aligned}$$

Let $c > 0$. The following statements are equivalent:

$$c < k(T); \quad c^{-1}TB_X \supset B_Y;$$
$$c^{-1}B_X \supset T^{-1}B_Y; \quad \|T^{-1}\| \leq c^{-1}; \quad \|T^{-1}\|^{-1} \geq c.$$

Hence $k(T) = \sup\{c > 0 : c < k(T)\} = \|T^{-1}\|^{-1}$. □

Theorem 8. *Let $T \in \mathcal{B}(X, Y)$. Then $j(T) = k(T^*)$ and $k(T) = j(T^*)$.*

Proof. (a) By Theorem 5, $j(T) = 0$ if and only if $k(T^*) = 0$. If $j(T) > 0$, then $\operatorname{Ran} T$ is closed and $T = JT_0$, where $T_0 : X \to \operatorname{Ran} T$ is induced by T and $J : \operatorname{Ran} T \to Y$ is the natural embedding. Clearly, $j(T) = j(T_0) = \|T_0^{-1}\|^{-1} =$

$\|T_0^{*-1}\|^{-1} = k(T_0^*)$. Further, $T^* = T_0^* J^*$ where $J^* : Y^* \to (\operatorname{Ran} T)^*$ assigns to each functional $f \in Y^*$ the restriction $f|\operatorname{Ran} T$, cf. A.1.19. By the Hahn-Banach theorem, $J^* B_{Y^*} = B_{(\operatorname{Ran} T)^*}$. Thus $k(J^*) = 1 = \|J^*\|$, $k(T^*) = k(T_0^*)$ by Theorem 6, and $k(T^*) = j(T)$.

(b) By Theorem 5, $k(T) = 0$ if and only if $j(T^*) = 0$. Suppose that $k(T) > 0$; so T is onto. Then $T = T_0 Q$, where $Q : X \to X/\operatorname{Ker} T$ is the canonical projection and $T_0 : X/\operatorname{Ker} T \to Y$ is one-to-one and onto. We have $k(T_0)k(Q) \le k(T) \le k(T_0)\|Q\|$. Since $\|Q\| = 1 = k(Q)$, we have $k(T) = k(T_0) = \|T_0^{-1}\|^{-1} = \|(T_0^*)^{-1}\|^{-1} = j(T_0^*)$. Further, $T^* = Q^* T_0^*$ where Q^* is the natural embedding of $(X/\operatorname{Ker} T)^* = (\operatorname{Ker} T)^\perp$ into X^*, see A.1.20. Therefore $j(T^*) = j(T_0^*) = k(T)$. □

Proposition 9. Let $T, S \in \mathcal{B}(X,Y)$. Then $|j(T) - j(S)| \le \|T - S\|$ and $|k(T) - k(S)| \le \|T - S\|$. In particular, the injectivity and surjectivity moduli are continuous.

Proof. Let $x \in X$, $\|x\| = 1$. Then $\|Sx\| \ge \|Tx\| - \|(T-S)x\| \ge j(T) - \|T - S\|$, and so $j(T) - j(S) \le \|T - S\|$. The symmetry implies the first statement of Proposition 9 and the second one follows from Theorem 8. □

Proposition 10. *Let $T \in \mathcal{B}(X,Y)$. Then:*

(i) $j(T) = \sup\{r > 0 : T - S$ *is bounded below for all* $S \in \mathcal{B}(X,Y), \|S\| < r\}$;

(ii) $k(T) = \sup\{r > 0 : T - S$ *is onto for all* $S \in \mathcal{B}(X,Y), \|S\| < r\}$.

Proof. (i) Let $S \in \mathcal{B}(X,Y)$, $\|S\| < j(T)$. Then $j(T - S) \ge j(T) - \|S\| > 0$, and so $T - S$ is bounded below.

Conversely, let $\varepsilon > 0$. Then there exists $x_0 \in X$ of norm 1 such that $\|Tx_0\| < j(T) + \varepsilon$. Let $f \in X^*$ satisfy $\|f\| = 1 = f(x_0)$. Define $S \in \mathcal{B}(X,Y)$ by $Sx = f(x) \cdot Tx_0$. Then $\|S\| = \|Tx_0\| < j(T) + \varepsilon$ and $(T - S)x_0 = 0$. Hence $T - S$ is not one-to-one.

(ii) The statement is clear if $k(T) = 0$. Suppose that $k(T) > 0$.

Let $S \in \mathcal{B}(X,Y)$, $\|S\| < k(T)$. Then $k(T - S) \ge k(T) - \|S\| > 0$, and so $T - S$ is onto.

Conversely, let $\varepsilon > 0$ and let $y \in Y$ satisfy $\|y\| < k(T) + \varepsilon$ and $y \notin TB_X$. Choose $x_0 \in X$ such that $Tx_0 = y$. It is clear that $\operatorname{dist}\{x_0, \operatorname{Ker} T\} \ge 1$. Let $f \in (\operatorname{Ker} T)^\perp$ satisfy $\|f\| = 1$ and $f(x_0) = \operatorname{dist}\{x_0, \operatorname{Ker} T\} \ge 1$. Define S by $Sx = f(x) f(x_0)^{-1} Tx_0$. Then $\|S\| = \|Tx_0\| f(x_0)^{-1} < k(T) + \varepsilon$. We show that $y \notin \operatorname{Ran}(T - S)$. We have $(T - S)x_0 = 0$, and so $\operatorname{Ran}(T - S) = T \operatorname{Ker} f$. Suppose on the contrary that there is an $x \in \operatorname{Ker} f$ such that $y = Tx$. Then $x - x_0 \in \operatorname{Ker} T \subset \operatorname{Ker} f$ and $0 = f(x) = f(x_0) \ge 1$, a contradiction. □

Recall that for an n-tuple $(A_1, \dots, A_n)$ of operators in a Banach space X we write

$$d_l^{\mathcal{B}(X)}(A_1, \dots, A_n) = \inf\left\{\sum_{i=1}^n \|A_i S\| : S \in \mathcal{B}(X), \|S\| \le 1\right\}$$

and

$$d_r^{\mathcal{B}(X)}(A_1,\dots,A_n) = \inf\Big\{\sum_{i=1}^n \|SA_i\| : S \in \mathcal{B}(X), \|S\| \le 1\Big\}.$$

Theorem 11. *Let $A = (A_1,\dots,A_n)$ be an n-tuple of mutually commuting operators on a Banach space X. Then:*

(i) $d_l^{\mathcal{B}(X)}(A) = \inf\Big\{\sum_{i=1}^n \|A_i x\| : x \in X, \|x\| = 1\Big\}$;

(ii) $d_r^{\mathcal{B}(X)}(A) = \sup\{r : A_1B_X + \cdots + A_rB_X \supset rB_X\}$.

Proof. (i) Let $x \in X$, $\|x\| = 1$ and let $f \in X^*$ satisfy $f(x) = 1 = \|f\|$. Define $S \in \mathcal{B}(X)$ by $Sz = f(z)\cdot x \quad (z \in X)$. Then $\|S\| = 1$ and

$$\sum_{i=1}^n \|A_iS\| = \sum_{i=1}^n \sup\{\|A_iSz\| : z \in X, \|z\| = 1\} =$$
$$= \sum_{i=1}^n \sup\{|f(z)|\cdot\|A_ix\| : z \in X, \|z\| = 1\} \le \sum_{i=1}^n \|A_ix\|.$$

Thus

$$d_l(A) \le \inf\left\{\sum_{i=1}^n \|A_ix\| : x \in X, \|x\| = 1\right\}.$$

To show the opposite inequality, let $S \in \mathcal{B}(X)$, $\|S\| = 1$. For each $\varepsilon > 0$ there exists $y \in X$, $\|y\| = 1$ such that $\|Sy\| \ge 1-\varepsilon$. Then $\frac{Sy}{\|Sy\|}$ is of norm 1 and

$$\inf\left\{\sum_{i=1}^n \|A_ix\| : x \in X, \|x\| = 1\right\} \le \sum_{i=1}^n \frac{\|A_iSy\|}{\|Sy\|} \le (1-\varepsilon)^{-1}\sum_{i=1}^n \|A_iS\|.$$

Since S and ε were arbitrary, we have

$$\inf\left\{\sum_{i=1}^n \|A_ix\| : x \in X, \|x\| = 1\right\} \le d_l(A).$$

(ii) Suppose that $A_1B_X+\cdots+A_nB_X \supset rB_X$. Let $S : X \to X$ be an operator of norm 1 and $\varepsilon > 0$. Then there exists $y \in X$, $\|y\| = r$ such that $\|Sy\| > (1-\varepsilon)r$ and $x_1,\dots,x_n \in B_X$ such that $y = \sum_{i=1}^n A_ix_i$.

We have

$$\sum_{i=1}^n \|SA_i\| \ge \sum_{i=1}^n \|SA_ix_i\| \ge \|Sy\| > (1-\varepsilon)r.$$

Thus

$$d_r(A) \ge \sup\{r : A_1B_X + \cdots + A_nB_X \supset rB_X\}.$$

Conversely, let $x \in X$, $\|x\| = 1$ and $x \notin r^{-1} \cdot \overline{(A_1 B_X + \cdots + A_n B_X)}$. For every $\varepsilon > 0$ we have $x \notin (r+\varepsilon)^{-1}\overline{A_1 B_X + \cdots + A_n B_X}$. By the Hahn-Banach theorem A.1.3, there exists a functional $f \in X^*$ such that

$$f(x) = 1 > (r+\varepsilon)^{-1} \sup\left\{\left|f\left(\sum_{i=1}^n A_i x_i\right)\right| : \|x_1\| = \cdots = \|x_n\| = 1\right\}.$$

Define $S \in \mathcal{B}(X)$ by $Sz = f(z) \cdot x \quad (z \in X)$. Then $\|S\| = \|f\| \geq 1$ and

$$\begin{aligned}\sum_{i=1}^n \|SA_i\| &= \sup\left\{\sum_{i=1}^n \|SA_i x_i\| : x_i \in X, \|x_1\| = \cdots = \|x_n\| = 1\right\}\\ &= \sup\left\{\sum_{i=1}^n |f(A_i x_i)| : \|x_1\| = \cdots = \|x_n\| = 1\right\}\\ &= \sup\left\{\left|f\left(\sum_{i=1}^n A_i x_i\right)\right| : \|x_1\| = \cdots = \|x_n\| = 1\right\} < r + \varepsilon.\end{aligned}$$

Thus

$$d_r(A) \leq \sup\{r : A_1 B_X + \cdots + A_n B_X \supset r B_X\}.$$ □

Remark 12. The symmetry between the left and right spectrum is rather lost in the algebra $\mathcal{B}(X)$. The left point spectrum of an operator is much more important than the right point spectrum. Therefore in operator theory the left point spectrum is usually called just the *point spectrum*; the right point spectrum is rather the point spectrum of the adjoint operator.

Similarly, the left approximate point spectrum of an operator is much more important than the right approximate point spectrum. Therefore the left approximate point spectrum in $\mathcal{B}(X)$ is usually called only the *approximate point spectrum*. The right approximate point spectrum is called the *surjective spectrum* (sometimes also the defect spectrum).

The point, approximate point, and surjective spectrum of a commuting n-tuple $A = (A_1, \dots, A_n)$ of operators on a Banach space X are denoted by $\sigma_p(A)$, $\sigma_\pi(A)$ and $\sigma_\delta(A)$, respectively.

The notation is not consistent with the notation used in the Banach algebra theory, but since it is generally accepted and is also quite convenient (it is not necessary to remember which spectrum is left and which is right), we are going to use it.

Corollary 13. *Let $A = (A_1, \dots, A_n)$ be an n-tuple of mutually commuting operators on a Banach space X. Then*

$$\sigma_p(A) = \left\{\lambda \in \mathbb{C}^n : \bigcap_{i=1}^n \operatorname{Ker}(A_i - \lambda_i) \neq \{0\}\right\},$$

$$\sigma_\pi(A) = \left\{\lambda \in \mathbb{C}^n : \inf\left\{\sum_{i=1}^n \|(A_i - \lambda_i)x\| : x \in X, \|x\| = 1\right\} = 0\right\},$$

$$\sigma_\delta(A) = \left\{\lambda \in \mathbb{C}^n : (A_1 - \lambda_1)X + \cdots + (A_n - \lambda_n)X \neq X\right\}.$$

In particular, for a single operator $T \in \mathcal{B}(X)$ we have $\sigma_p(T) = \{\mu \in \mathbb{C} : T - \mu$ is not one-to-one$\}$, $\sigma_\pi(T) = \{\mu \in \mathbb{C} : T - \mu$ is not bounded below$\}$ and $\sigma_\delta(T) = \{\mu \in \mathbb{C} : T - \mu$ is not onto$\}$.

Let $A = (A_1, \ldots, A_n)$ be a mutually commuting n-tuple of operators on a Banach space X. Denote by X_1^n and X_∞^n the Banach space $\underbrace{X \oplus \cdots \oplus X}_{n}$ with the norms

$$\|x_1 \oplus \cdots \oplus x_n\|_1 = \sum_{i=1}^n \|x_i\|,$$

and

$$\|x_1 \oplus \cdots \oplus x_n\|_\infty = \max\{\|x_i\| : i = 1, \ldots, n\},$$

respectively.

Denote by $\delta_A : X \to X_1^n$ the operator defined by

$$\delta_A x = A_1 x \oplus \cdots \oplus A_n x \qquad (x \in X).$$

By Theorem 11, we have $d_l(A) = j(\delta_A)$. Similarly, $d_r(A) = k(\eta_A)$ where $\eta_A = X_\infty^n \to X$ is defined by

$$\eta_A(x_1 \oplus \cdots \oplus x_n) = \sum_{i=1}^n A_i x_i \qquad (x_1, \ldots, x_n \in X).$$

Corollary 14. *Let $A = (A_1, \ldots, A_n)$ be an n-tuple of mutually commuting operators on a Banach space X. Write $A^* = (A_1^*, \ldots, A_n^*) \in \mathcal{B}(X^*)$. Then $\sigma_\pi(A^*) = \sigma_\delta(A)$ and $\sigma_\delta(A^*) = \sigma_\pi(A)$*

Proof. Let $\delta_A : X \to X_1^n$ and let $\eta_A : X_\infty^n \to X$ be the operators defined above. Similarly define operators $\delta_{A^*} : X^* \to (X^*)_1^n$ and $\eta_{A^*} : (X^*)_\infty^n \to X^*$. Clearly, $\delta_{A^*} = (\eta_A)^*$ and $\eta_{A^*} = (\delta_A)^*$. Thus

$$d_l^{\mathcal{B}(X^*)}(A^*) = j(\delta_{A^*}) = k(\eta_A) = d_r^{\mathcal{B}(X)}(A)$$

and

$$d_r^{\mathcal{B}(X^*)}(A^*) = k(\eta_{A^*}) = j(\delta_A) = d_l^{\mathcal{B}(X)}(A).$$

Consequently, $\sigma_\pi(A^*) = \sigma_\delta(A)$ and $\sigma_\delta(A^*) = \sigma_\pi(A)$. □

Remark 15. Note that there are two different conventions of taking adjoints. We are using the Banach space convection: for an operator T on a Banach space X and $\alpha \in \mathbb{C}$ we have $(T - \alpha I_X)^* = T^* - \alpha I_{X^*}$.

If X is a Hilbert space, then X^* is usually identified with X and with this (Hilbert space) convention $(T-\alpha I)^* = T^* = \bar{\alpha} I$. In this notation we have rather $\sigma_\pi(T^*) = \{\bar{z} : z \in \sigma_\delta(T)\}$ and a similar change must be done in all duality results.

Theorem 16. Let X, Y be Banach spaces and $T \in \mathcal{B}(X,Y)$. Then there exists and operator $S : Y \to X$ satisfying $ST = I_X$ if and only if T is bounded below and $\operatorname{Ran} T$ is a complemented subspace of Y.

Similarly, there exists an operator $S : Y \to X$ with $TS = I_Y$ if and only if T is onto and $\operatorname{Ker} T$ is a complemented subspace of X.

Proof. Let $S : Y \to X$, $ST = I_X$. Then $(TS)^2 = TS$ and $\operatorname{Ran} T \supset \operatorname{Ran} TS \supset \operatorname{Ran} TST = \operatorname{Ran} T$. Hence TS is a projection onto $\operatorname{Ran} T$ and $\operatorname{Ran} T$ is a complemented subspace of Y. In particular, $\operatorname{Ran} T = \operatorname{Ker}(I - TS)$, which is closed. Clearly, T is one-to-one and hence bounded below.

Conversely, let T be bounded below and let $P \in \mathcal{B}(X)$ be a projection onto $\operatorname{Ran} T$. Let $S_0 : \operatorname{Ran} T \to X$ be the inverse of T. Set $S = S_0 P$. Then $ST = I_X$.

Suppose now that $TS = I_Y$ for some $S : Y \to X$. Then T is onto and $(ST)^2 = ST$ is a projection satisfying $\operatorname{Ker}(ST) = \operatorname{Ker} T$. Hence $\operatorname{Ker} T$ is a complemented subspace of X.

Conversely, let T be onto and let $X = \operatorname{Ker} T \oplus M$ for some closed subspace $M \subset X$. The restriction $T|M : M \to Y$ is one-to-one and onto; let S_0 be its inverse. Let $J : M \to X$ be the natural embedding and $S = JS_0$. Then $TS = I_Y$. □

Theorem 17. *Let $A = (A_1, \dots, A_n)$ be an n-tuple of mutually commuting operators on a Banach space X. Then:*

(i) *$(0,\dots,0) \notin \sigma_l(A)$ if and only if $j(\delta_A) > 0$ and the space $\delta_A X$ is complemented;*

(ii) *$(0,\dots,0) \notin \sigma_r(A)$ if and only if $A_1X + \cdots + A_nX = X$ and $\operatorname{Ker} \eta_A$ is complemented.*

Proof. (i) Let $j(\delta_A) > 0$ and let $\delta_A X$ be a complemented subspace of X_1^n. By the preceding theorem, there exists an operator $S : X_1^n \to X$ such that $S\delta_A = I_X$. If we express S in the matrix form as $S = (B_1, \dots, B_n)$, we have $B_1A_1 + \cdots + B_nA_n = I_X$. Hence $0 \notin \sigma_l(A)$.

Conversely, if $\sum B_iA_i = I_X$ for some $B_i \in \mathcal{B}(X)$, then $S\delta_A = I_X$ for $S : X_1^n \to X$ defined by $S(x_1 \oplus \cdots \oplus x_n) = \sum B_i x_i$. By the preceding theorem, δ_A is bounded below and $\operatorname{Ran} \delta_A$ complemented.

The second statement can be shown similarly. □

Corollary 18. *Let $A = (A_1, \dots, A_n)$ be an n-tuple of mutually commuting operators on a Hilbert space H. Then*

$$\sigma_l(A) = \sigma_\pi(A) \qquad \text{and} \qquad \sigma_r(A) = \sigma_\delta(A).$$

Proof. The spaces H_1^n, H_∞^n are isomorphic to a Hilbert space, so every closed subspace is complemented. □

Example 19. The preceding result is not true for Banach spaces. An example is based on the well-known fact that the space c_0 is not complemented in ℓ^∞.

Let $X = c_0 \oplus \ell^\infty$ and let $T \in \mathcal{B}(X)$ be defined by

$$T\big((x_1, x_2, \dots) \oplus (y_1, y_2, \dots)\big) = (0, 0, \dots) \oplus (x_1, y_1, x_2, y_2, \dots).$$

It is easy to see that T is an isometry and $\operatorname{Ran} T$ is not complemented in X. Thus $0 \in \sigma_l(T) \setminus \sigma_\pi(T)$.

A similar example can be constructed to show that in general $\sigma_r \neq \sigma_\delta$.

Theorem 20. *Let T be an operator in a Banach space X. Then*

$$\sigma(T) = \sigma_p(T) \cup \sigma_\delta(T) = \sigma_p(T^*) \cup \sigma_\pi(T).$$

Proof. Evidently, $\sigma_p(T) \cup \sigma_\delta(T) \subset \sigma(T)$. If $\lambda \notin \sigma_p(T) \cup \sigma_\delta(T)$, then $T - \lambda$ is both one-to-one and onto. Thus $T - \lambda$ is invertible by the open mapping theorem.

The second equality can be proved similarly. □

Remark 21. The equality $\sigma(a) = \tau_l(a) \cup \pi_r(a) = \tau_r(a) \cup \pi_l(a)$, which is true for operators by Theorem 20, is not true for Banach algebras.

An easy example is the algebra of all formal power series $\sum_{i=0}^\infty \alpha_i x^i$ with complex coefficients α_i such that $\|\sum \alpha_i x^i\| = \sum |\alpha_i| < \infty$. It is easy to check that $\sigma(x) = \{z \in \mathbb{C} : |z| \leq 1\}$, $\tau_l(x) = \tau_r(x) = \{z \in \mathbb{C} : |z| = 1\}$, and so even $\sigma(x) \neq \tau_l(x) \cup \tau_r(x)$.

Note that for any Banach algebra element x we have a weaker relation $\sigma(x) = \sigma_l(x) \cup \pi_r(X) = \sigma_r(x) \cup \pi_l(x)$. Indeed, suppose that $yx = 1$ and x is not a right divisor of zero. Then $(xy - 1)x = 0$, and so $xy = 1$. Hence x is invertible.

Examples 22. (i) Let H be a Hilbert space with an orthonormal basis $\{e_i\}_{i \geq 0}$. The *weighted unilateral shift* with weights $w_i \geq 0$ is the operator on H defined by $Te_i = w_i e_{i+1}$. Its adjoint (satisfying $T^* e_0 = 0$ and $T^* e_{i+1} = w_i e_i$ for $i \geq 0$) is the *weighted backward shift*. If H has an orthonormal basis $\{e_i\}_{i \in \mathbb{Z}}$ and T is defined by $Te_i = w_i e_{i+1} \quad (i \in \mathbb{Z})$, then T is called the *weighted bilateral shift* with weights $w_i > 0$.

Weighted shifts are an important source of examples and counterexamples.

For weighted shifts all spectra are circularly symmetrical since cT is unitarily equivalent to T for $|c| = 1$.

Let T be a weighted unilateral shift. Then

$$\begin{aligned}
\|T\| &= \sup w_i, \\
j(T) &= \inf w_i, \\
r(T) &= \lim_n \sup_i (w_i \cdots w_{i+n-1})^{1/n}, \\
\sigma(T) &= \{z \in \mathbb{C} : |z| \leq r(T)\}, \\
\sigma_\pi(T) &= \{z \in \mathbb{C} : m(T) \leq |z| \leq r(T)\},
\end{aligned}$$

where $m(T) = \lim_n \inf_i (w_i \cdots w_{i+n-1})^{1/n}$. Furthermore,

$$\{0\} \cup \{z \in \mathbb{C} : |z| < r'(T)\} \subset \sigma_p(T^*) \subset \{z \in \mathbb{C} : |z| \le r'(T)\}$$

where $r'(T) = \liminf (w_0 \cdots w_{n-1})^{1/n}$.

For bilateral weighted shifts there are similar formulas but little bit more complicated. For example, $\sigma(T) = \{z \in \mathbb{C} : m(T) \le |z| \le r(T)\}$ and $\sigma_\pi(T)$ can be in general a union of two annuli. For details see [Shl].

(ii) Weighted shifts play an important role in analysis. Let H^2 be the Hardy space of all functions analytic on the open unit disc satisfying

$$\|f\|_2^2 = \frac{1}{2\pi} \sup_{0<r<1} \int_0^{2\pi} |f(re^{it})|^2 \, \mathrm{d}t < \infty.$$

Then H^2 is a Hilbert space and the operator of multiplication by the variable z is a unilateral shift (with weights equal to 1). Many deep results in analysis can be formulated in the language of operator theory, see [Ni].

Similarly, weighted shifts can be considered as multiplication operators in various spaces of analytic functions (e.g., Bergman spaces), see [Shl].

(iii) An important example of commuting n-tuples of operators are *multishifts* and *weighted multishifts.* Let H be Hilbert space with an orthogonal basis e_α $(\alpha \in M)$, where M is a "translation invariant" subset of $\mathbb{Z}^n$, (i.e., $m \in M, 1 \le j \le n \Rightarrow m + \varepsilon_j \in M$ where $\varepsilon_j = (\underbrace{0, \dots, 0}_{j-1}, 1, 0, \dots, 0)$). Define operators $S_1, \dots, S_n \in \mathcal{B}(H)$ by

$$S_j e_m = e_{m+\varepsilon_j} \qquad (m \in M).$$

Then $(S_1, \dots, S_n)$ is a commuting n-tuple of operators.

More generally, we can define $S_j e_m = w_{m,j} e_{m+\varepsilon_j}$ where $w_{m,j}$ $(m \in M, j = 1, \dots, n)$ are positive weights such that $w_{m,i} w_{m+\varepsilon_i,j} = w_{m,j} w_{m+\varepsilon_j,i}$ for all m, i, j.

The most important examples of the index set M are $\mathbb{Z}^n$ and $\mathbb{Z}_+^n$ but there are many other possibilities (e.g., for $n = 2$ the sets $\{(i,j) : i + j \ge 0\}$ or $\{(i,j) :$ either $i \ge 0$ or $j \ge 0\}$).

Weighted multishifts can be frequently interpreted as n-tuples of multiplications by variables $z_1, \dots, z_n$ in various spaces of analytic functions of n variables (e.g., in the Hardy and Bergman spaces over the unit polydisc, unit ball etc.).

(iv) An operator T on a Hilbert space H is called *selfadjoint* if $T^* = T$ and *normal* if $TT^* = T^*T$. A normal operator is selfadjoint if and only if its spectrum is contained in the real line.

The structure of a normal operator is described by the spectral decomposition

$$T = \int z \, \mathrm{d}E(z)$$

where E is the spectral measure of T and $\operatorname{supp} E = \sigma(T)$. Normal operators satisfy $\|T\| = r(T)$ and $\sigma(T) = \sigma_\pi(T) = \sigma_\delta(T)$.

Commuting n-tuples of normal operators have similar properties. If $T = (T_1, \dots, T_n)$ is an n-tuple of mutually commuting normal operators, then $T_i T_j^* = T_j^* T_i \quad (i, j = 1, \dots, n)$ by the Fuglede-Putnam theorem and $\sigma_\pi(T) = \sigma_\delta(T) = \sigma(T)$. Furthermore, it is possible to express T as

$$T_i = \int_{\sigma(T)} z_i \, \mathrm{d}E(z) \qquad (i = 1, \dots, n)$$

where E is the joint spectral measure of $(T_1, \dots, T_n)$.

(v) Let H be a Hilbert space with an orthonormal basis $\{e_i\}_{i \geq 0}$; let $(c_i)_{i \geq 0}$ be a bounded sequence of complex numbers. Let $T \in \mathcal{B}(H)$ be the diagonal operator defined by $Te_i = c_i e_i$. Then $\|T\| = \sup\{|c_i| : i \geq 0\} = r(T)$, $\sigma_p(T) = \{c_i : i \geq 0\}$ and $\sigma(T) = \sigma_\pi(T) = \sigma_\delta(T) = \{c_i : i \geq 0\}^-$.

Theorem 23. *Let T be an operator on a Banach space X. Then there exists a Banach space Y containing X as a closed subspace and an operator $S \in \mathcal{B}(Y)$ such that $S|X = T$ and $\sigma^{\mathcal{B}(Y)}(S) = \sigma_\pi^{\mathcal{B}(X)}(T)$ (clearly, $\sigma_\pi(T) \subset \sigma_\pi(S) \subset \sigma(S)$ whenever $S \in \mathcal{B}(Y), S|X = T$).*

Proof. Let $\mathcal{A}$ be a closed commutative subalgebra of $\mathcal{B}(X)$ containing T.

Set $\mathcal{B} = \mathcal{A} \oplus X$. Define the norm and multiplication in $\mathcal{B}$ by $\|A \oplus x\| = \|A\| + \|x\|$ and $(A \oplus x)(A' \oplus x') = AA' \oplus (Ax' + A'x) \quad (A, A' \in \mathcal{A}, x, x' \in X)$. Then $\mathcal{B}$ is a commutative Banach algebra and $A \mapsto A \oplus 0 \quad (A \in \mathcal{A})$ is an isometrical embedding $\mathcal{A} \to \mathcal{B}$.

Let $\lambda \in \mathbb{C}$. It is easy to show (cf. Lemma 8.5) that

$$d_l^{\mathcal{B}}((T - \lambda) \oplus 0) = d_l^{\mathcal{B}(X)}(T - \lambda),$$

and so

$$\tau^{\mathcal{B}}(T \oplus 0) = \sigma_\pi^{\mathcal{B}(X)}(T).$$

By Theorem 4.10, there exists a commutative Banach algebra $\mathcal{C} \supset \mathcal{B}$ such that

$$\sigma^{\mathcal{C}}(T \oplus 0) = \tau^{\mathcal{B}}(T \oplus 0) = \sigma_\pi(T).$$

Consider the operator $S : \mathcal{C} \to \mathcal{C}$ defined by $Sc = (T \oplus 0)c \quad (c \in \mathcal{C})$. Then

$$\sigma^{\mathcal{B}(\mathcal{C})}(S) \subset \sigma^{\mathcal{C}}(T \oplus 0) = \sigma_\pi(T).$$

For $x \in X$ we have

$$S(0 \oplus x) = (T \oplus 0)(0 \oplus x) = 0 \oplus Tx.$$

If we identify $x \in X$ with $0 \oplus x \in \mathcal{B} \subset \mathcal{C}$, then $S|X = T$ and $\sigma(S) = \sigma_\pi(T)$. □

Theorem 24. *Let $T = (T_1, \dots, T_n)$ be an n-tuple of mutually commuting operators on a Banach space X, let $0 \notin \sigma_\pi(T)$. Then there exist a Banach space $Y \supset X$ and commuting operators $S_1, \dots, S_n, V_1, \dots, V_n \in \mathcal{B}(Y)^n$ such that $S_i|X = T_i$ $(i = 1, \dots, n)$ and $\sum_{i=1}^n S_i V_i = I_Y$.*

Proof. We use the construction from the previous theorem and Theorem 5.12. □

Theorem 25. *Let $T \in \mathcal{B}(X)$. Then*

$$\operatorname{dist}\{0, \sigma_\pi(T)\} = \lim_{n\to\infty} j(T^n)^{1/n} \quad \text{and} \quad \operatorname{dist}\{0, \sigma_\delta(T)\} = \lim_{n\to\infty} k(T^n)^{1/n}.$$

Proof. Since $d_l^{\mathcal{B}(X)}(T^n) = j(T^n)$ and $d_r^{\mathcal{B}(X)}(T^n) = k(T^n)$, the statement is a reformulation of Theorem 8.9. □

The theory of Banach algebras and operator theory are closely related. Obviously, every result for Banach algebras holds also for operators. On the other hand, it is possible to generalize many results for operators to general Banach algebras in the following way:

Let $\mathcal{A}$ be a Banach algebra and let $b \in \mathcal{A}$. Define operators L_b, R_b: $\mathcal{A} \to \mathcal{A}$ by $L_b x = bx$, $R_b x = xb$ $(x \in \mathcal{A})$. For an n-tuple $a = (a_1, \dots, a_n) \in c(\mathcal{A})$ write $L_a = (L_{a_1}, \dots, L_{a_n})$, $R_a = (R_{a_1}, \dots, R_{a_n}) \in c\big(\mathcal{B}(\mathcal{A})\big)$. The spectrum of $a \in c(\mathcal{A})$ in the algebra $\mathcal{A}$ and the spectra of L_a, R_a in $\mathcal{B}(\mathcal{A})$ are closely related.

Theorem 26. *Let $\mathcal{A}$ be a Banach algebra and $a \in c(\mathcal{A})$. Then:*

(i) $\tau_l^{\mathcal{A}}(a) = \sigma_\pi^{\mathcal{B}(\mathcal{A})}(L_a)$, $\tau_r^{\mathcal{A}}(a) = \sigma_\pi^{\mathcal{B}(\mathcal{A})}(R_a)$;

(ii) $\sigma_r^{\mathcal{A}}(a) = \sigma_\delta^{\mathcal{B}(\mathcal{A})}(L_a) = \sigma_r(L_a)$, $\sigma_l^{\mathcal{A}}(a) = \sigma_\delta^{\mathcal{B}(\mathcal{A})}(R_a) = \sigma_r(R_a)$.

Proof. (i) By Corollary 13, we have

$$\begin{aligned}\sigma_\pi^{\mathcal{B}(\mathcal{A})}(L_a) &= \Big\{\lambda \in \mathbb{C}^n : \inf\Big\{\sum_{i=1}^n \|L_{a_i-\lambda_i}x\| : x \in \mathcal{A}, \|x\| = 1\Big\} = 0\Big\} \\ &= \Big\{\lambda \in \mathbb{C}^n : \inf\Big\{\sum_{i=1}^n \|(a_i - \lambda_i)x\| : x \in \mathcal{A}, \|x\| = 1\Big\} = 0\Big\} = \tau_l^{\mathcal{A}}(a).\end{aligned}$$

Similarly, $\sigma_\pi^{\mathcal{B}(\mathcal{A})}(R_a) = \tau_r^{\mathcal{A}}(a)$.

(ii) We have

$$\begin{aligned}\sigma_\delta^{\mathcal{B}(\mathcal{A})}(L_a) &= \Big\{\lambda \in \mathbb{C}^n : L_{a_1-\lambda_1}\mathcal{A} + \dots + L_{a_n-\lambda_n}\mathcal{A} \neq \mathcal{A}\Big\} \\ &= \Big\{\lambda \in \mathbb{C}^n : (a_1 - \lambda_1)\mathcal{A} + \dots + (a_n - \lambda_n)\mathcal{A} \not\ni 1\Big\} = \sigma_r^{\mathcal{A}}(a).\end{aligned}$$

Since $a \in \mathcal{A} \mapsto L_a \in \mathcal{B}(\mathcal{A})$ is an isometrical embedding, we have $\sigma_r^{\mathcal{B}(\mathcal{A})}(L_a) \subset \sigma_r^{\mathcal{A}}(a) = \sigma_\delta^{\mathcal{B}(\mathcal{A})}(L_a) \subset \sigma_r^{\mathcal{B}(\mathcal{A})}(L_a)$, which proves the first statement of (ii).

The second statement can be proved analogously. □

10 Operators with closed range

Operators with closed range play an important role in operator theory. The following quantity is closely connected with the closeness of the range.

Let X, Y be Banach spaces and let $T : X \to Y$ be a non-zero operator. The *reduced minimum modulus* of T is defined by

$$\gamma(T) = \inf\{\|Tx\| : x \in X, \operatorname{dist}\{x, \operatorname{Ker} T\} = 1\}.$$

Formally we set $\gamma(0) = \infty$.

If T is one-to-one, then clearly $\gamma(T) = j(T)$.

An operator $T : X \to Y$ defines naturally the operator $T_0 : X/\operatorname{Ker} T \to \overline{\operatorname{Ran} T}$ by $T_0(x + \operatorname{Ker} T) = Tx$.

Lemma 1. $\gamma(T) = j(T_0)$.

Proof. We have

$$\begin{aligned} j(T_0) &= \inf\{\|T_0(x + \operatorname{Ker} T)\| : \|x + \operatorname{Ker} T\|_{X/\operatorname{Ker} T} = 1\} \\ &= \inf\{\|Tx\| : \operatorname{dist}\{x, \operatorname{Ker} T\} = 1\} = \gamma(T). \end{aligned}$$ □

The notion of the reduced minimum modulus is motivated by the following characterization:

Theorem 2. *Let $T : X \to Y$ be an operator. Then $\operatorname{Ran} T$ is closed if and only if $\gamma(T) > 0$.*

Proof. The statement is clear for $T = 0$. If $T \neq 0$, then $\operatorname{Ran} T = \operatorname{Ran} T_0$ and $\operatorname{Ran} T_0$ is closed if and only if $j(T_0) > 0$. □

Theorem 3. $\gamma(T) = \gamma(T^*)$ *for every operator $T : X \to Y$.*

Proof. By Theorem 2, $\gamma(T) = 0$ if and only if $\gamma(T^*) = 0$.

Let $\gamma(T) > 0$; so $\operatorname{Ran} T$ is closed. We have $T = JT_0Q$, where $Q : X \to X/\operatorname{Ker} T$ is the canonical projection, $T_0 : X/\operatorname{Ker} T \to \operatorname{Ran} T$ is one-to-one and onto and $J : \operatorname{Ran} T \to Y$ is the natural embedding. The corresponding decomposition for T^* is $T^* = Q^* T_0^* J^*$. We have $\gamma(T) = j(T_0) = \|T_0^{-1}\|^{-1} = \|T_0^{*-1}\|^{-1} = j(T_0^*) = \gamma(T^*)$. □

Corollary 4. *Let $T \in \mathcal{B}(X, Y)$. Then*

$$\gamma(T) = \sup\{c \geq 0 : TB_X \supset c \cdot (B_Y \cap \operatorname{Ran} T)\}. \tag{1}$$

In particular, if T is onto, then $\gamma(T) = k(T)$.

Proof. Let $T = JT_0Q$ be the canonical decomposition of T as above. Suppose first that $\operatorname{Ran} T$ is closed. Then $\gamma(T) = j(T_0) = \|T_0^{-1}\|^{-1} = k(T_0)$. Further,

$k(T_0)k(Q) \le k(T_0Q) \le k(T_0)\|Q\|$ and $k(Q) = 1 = \|Q\|$; so

$$\begin{aligned} k(T_0) = k(T_0Q) &= \sup\{c \ge 0 : T_0QB_X \supset c \cdot (B_Y \cap \operatorname{Ran} T)\} \\ &= \sup\{c \ge 0 : TB_X \supset c(B_Y \cap \operatorname{Ran} T)\}. \end{aligned}$$

If $\operatorname{Ran} T$ is not closed, then $\gamma(T) = 0$. We show that the right-hand side of (1) is also equal to 0. Suppose on the contrary that there is a $c > 0$ such that $TB_X \supset c \cdot (B_Y \cap \operatorname{Ran} T)$. Then T induces an isomorphism from $X/\operatorname{Ker} T$ onto $\operatorname{Ran} T$. Thus $\operatorname{Ran} T$ is complete and hence closed, a contradiction. □

The reduced minimum modulus γ is closely connected with the gap function:

Definition 5. Let M, L be subspaces of a Banach space X. Define

$$\delta(M, L) = \sup_{\substack{x \in M \\ \|x\| \le 1}} \operatorname{dist}\{x, L\}.$$

The *gap* $\widehat{\delta}(M, L)$ is defined by $\widehat{\delta}(M, L) = \max\{\delta(M, L), \delta(L, M)\}$.

The gap measures the "distance" between two subspaces. Clearly,

$$0 \le \delta(M, L) \le 1, \quad \delta(M, L) = \delta(\overline{M}, \overline{L}) \quad \text{and} \quad \widehat{\delta}(M, L) = \widehat{\delta}(\overline{M}, \overline{L}).$$

For closed subspaces M and L we have $\delta(M, L) = 0$ if and only if $M \subset L$ and $\widehat{\delta}(M, L) = 0 \Leftrightarrow M = L$.

The following result is a kind of triangular inequality.

Lemma 6. *Let M_1, M_2, M_3 be closed subspaces of a Banach space X. Then*

$$\delta(M_1, M_3) \le \delta(M_1, M_2) + \delta(M_2, M_3) + \delta(M_1, M_2)\delta(M_2, M_3).$$

Proof. Let $x \in M_1$, $\|x\| < 1$. Then there exists $y \in M_2$ with $\|x - y\| \le \delta(M_1, M_2)$. Clearly, $\|y\| \le \|x\| + \|x - y\| < 1 + \delta(M_1, M_2)$ and there exists $z \in M_3$ with $\|y - z\| \le (1 + \delta(M_1, M_2))\delta(M_2, M_3)$. Hence

$$\begin{aligned} \operatorname{dist}\{x, M_3\} &\le \|x - z\| \le \|x - y\| + \|y - z\| \\ &\le \delta(M_1, M_2) + \delta(M_2, M_3) + \delta(M_1, M_2)\delta(M_2, M_3). \end{aligned}$$ □

Lemma 7. *Let M be a closed subspace of X, $x \in X$ and $f \in X^*$. Then:*

(i) $\operatorname{dist}\{x, M\} = \sup\{|\langle x, g\rangle| : g \in M^\perp, \|g\| \le 1\}$;
(ii) $\operatorname{dist}\{f, M^\perp\} = \sup\{|\langle m, f\rangle| : m \in M, \|m\| \le 1\}$.

Proof. (i) Let $Q : X \to X/M$ be the canonical projection. By Theorem A.1.20, $Q^* : (X/M)^* \to X^*$ is an isometry with range $M^\perp$. Then

$$\begin{aligned} \operatorname{dist}\{x, M\} = \|Qx\| &= \sup\{|\langle Qx, h\rangle| : h \in (X/M)^*, \|h\| \le 1\} \\ &= \sup\{|\langle x, Q^*h\rangle| : h \in (X/M)^*, \|h\| \le 1\} \\ &= \sup\{|\langle x, g\rangle| : g \in M^\perp, \|g\| \le 1\}. \end{aligned}$$

(ii) Let $J : M \to X$ be the natural embedding. By Theorem A.1.19, for $f \in X^*$ we have $J^*f = f|M$ and $\|J^*f\| = \|f + M^\perp\|_{X^*/M^\perp}$. Thus

$$\begin{aligned}\operatorname{dist}\{f, M^\perp\} &= \|f + M^\perp\|_{X^*/M^\perp} = \|f|M\|_{M^*}\\ &= \sup\{|\langle m, f\rangle| : m \in M, \|m\| \le 1\}.\end{aligned}$$ □

Theorem 8. *Let M, L be closed subspaces of X. Then $\delta(M, L) = \delta(L^\perp, M^\perp)$ and $\widehat{\delta}(M, L) = \widehat{\delta}(L^\perp, M^\perp)$.*

Proof. We have

$$\begin{aligned}\delta(M, L) &= \sup_{x \in B_M} \operatorname{dist}\{x, L\} = \sup_{x \in B_M} \sup_{g \in B_{L^\perp}} |\langle x, g\rangle|\\ &= \sup_{g \in B_{L^\perp}} \sup_{x \in B_M} |\langle x, g\rangle| = \sup_{g \in B_{L^\perp}} \operatorname{dist}\{g, M^\perp\} = \delta(L^\perp, M^\perp).\end{aligned}$$ □

The following result is intuitively clear but the proof is surprisingly deep; it uses essentially the Borsuk antipodal theorem A.1.26.

Lemma 9. *Let M and L be subspaces of a finite-dimensional Banach space X such that $\dim M > \dim L$. Then there exists $m \in M$ such that $\|m\| = 1 = \operatorname{dist}\{m, L\}$.*

Proof. Suppose first that X is strictly convex, i.e., $\|\frac{u+v}{2}\| < 1$ for all $u, v \in X$, $\|u\| = \|v\| = 1$ and $u \ne v$.

Let $S_M = \{m \in M : \|m\| = 1\}$ be the unit sphere in M. For $m \in S_M$ let $d(m) = \operatorname{dist}\{m, L\}$. If $l \in L$, $\|l\| > 2$, then $\|m - l\| \ge \|l\| - \|m\| > 1 \ge d(m)$; so $d(m) = \inf\{\|m - l\| : l \in L, \|l\| \le 2\}$. From the compactness of the ball $\{l \in L : \|l\| \le 2\}$ it follows that there is a vector $g(m) \in L$ nearest to m, $\|m - g(m)\| = d(m)$.

Moreover, $g(m)$ is determined uniquely by this property. Indeed, if $l, l' \in L$, $l \ne l'$ and $\|m - l\| = \|m - l'\| = d(m)$, then $\left\|m - \frac{l+l'}{2}\right\| = \left\|\frac{(m-l)+(m-l')}{2}\right\| < d(m)$, a contradiction. Consequently, $g(-m) = -g(m)$.

We show that g is continuous. Suppose on the contrary that there is a sequence (m_k) in S_M converging to an $m \in S_M$ such that $g(m_k) \not\to g(m)$. Passing to a subsequence if necessary, we may assume that $g(m_k) \to l$ for some $l \in L$, $l \ne g(m)$.

Set $\varepsilon = \|m - l\| - \|m - g(m)\|$. By assumption, $\varepsilon > 0$. Choose k such that $\|m_k - m\| < \varepsilon/3$ and $\|g(m_k) - l\| < \varepsilon/3$. Then

$$\begin{aligned}\|m_k - g(m)\| &\le \|m_k - m\| + \|m - g(m)\| < \varepsilon/3 + \|m - l\| - \varepsilon\\ &\le \|m - l\| - \|m - m_k\| - \|l - g(m_k)\| \le \|m_k - g(m_k)\|,\end{aligned}$$

which is a contradiction with the definition of $g(m_k)$. Hence $g : S_M \to L$ is continuous.

By the Borsuk antipodal theorem, there exists $m \in M$ with $\|m\| = 1$ and $g(m) = 0$. Thus $\operatorname{dist}\{m, L\} = \|m - g(m)\| = \|m\| = 1$.

In the general case choose any Hilbert norm $|||\cdot|||$ on X and define strictly convex norms $\|\cdot\|_n = \|\cdot\| + \frac{1}{n}|||\cdot|||$ for $n \in \mathbb{N}$. Let $m_n \in M$ be vectors satisfying $\|m_n\|_n = 1 = \operatorname{dist}_n\{m_n, L\}$, where dist_n means the distance in the sense of $\|\cdot\|_n$. Evidently, $\|m_n\| \le 1$ for all n. Passing to a subsequence if necessary, we can assume that (m_n) is a convergent sequence. Its limit m clearly satisfies the conditions required. □

Corollary 10. *Let M, L be subspaces of a Banach space X. If $\delta(M, L) < 1$, then $\dim M \le \dim L$. If $\widehat{\delta}(M, L) < 1$, then $\dim M = \dim L$.*

Proof. The first statement is clear if $\dim L = \infty$. Suppose on the contrary that $\delta(M, L) < 1$, $\dim L < \infty$ and $\dim M > \dim L$. Choose a subspace $M_0 \subset M$ with $\dim M_0 = \dim L + 1$. By the previous lemma for the finite-dimensional space $M_0 + L$, there exists $m \in M_0 \subset M$ with $\|m\| = 1 = \operatorname{dist}\{m, L\}$, which is a contradiction with the assumption that $\delta(M, L) < 1$.

The second statement follows from the first one. □

Note that the dimension of a subspace is either finite or it is equal to ∞ (we do not distinguish different infinite cardinalities).

In most of the applications it is sufficient to replace the preceding result by the following statement which can be proved quite elementarily: if $L \subset X$, then there exists $\varepsilon > 0$ such that $\delta(M, L) < \varepsilon \Rightarrow \dim M \le \dim L$.

Complemented subspaces are also stable under small perturbations.

Theorem 11. *Let M be a complemented subspace of a Banach space X. Let $P \in \mathcal{B}(X)$ be a projection with $\operatorname{Ran} P = M$ and let $M' \subset X$ be a closed subspace satisfying $\widehat{\delta}(M, M') < \frac{1}{\|P\|+1}$. Then M' is also complemented.*

More precisely, there exists a projection $Q \in \mathcal{B}(X)$ satisfying $\operatorname{Ran} Q = M'$ and $\operatorname{Ker} Q = \operatorname{Ker} P$.

Proof. Write $L = \operatorname{Ker} P$. We show first that $M' \cap L = \{0\}$.

Suppose on the contrary that $x \in M' \cap L$, $\|x\| = 1$. Then there exists $y \in M$ with $\|x - y\| < \frac{1}{\|P\|+1}$. Therefore $\|Py\| = \|y\| \ge \|x\| - \|x - y\| \ge 1 - \frac{1}{\|P\|+1} = \frac{\|P\|}{\|P\|+1}$. On the other hand, $\|Py\| = \|P(y - x)\| \le \|P\| \cdot \|y - x\| < \frac{\|P\|}{\|P\|+1}$, a contradiction. Hence $M' \cap L = \{0\}$.

We show now that $M' + L = X$. Let $x_0 \in X$, $\|x_0\| = 1$. Set $l_0 = (I - P)x_0$. There exists $m_0' \in M'$ with $\|m_0' - Px_0\| < \frac{\|Px_0\|}{\|P\|+1} \le \frac{\|P\|}{\|P\|+1}$. Set $x_1 = x_0 - (m_0' + l_0)$. Then $\|x_1\| = \|Px_0 - m_0'\| \le \frac{\|P\|}{\|P\|+1}$. We have $\|l_0\| \le \|P\| + 1$ and $\|m_0'\| \le \|Px_0\| + \|m_0' - Px_0\| \le \|P\| + \frac{\|P\|}{\|P\|+1}$.

By induction we can construct elements $m'_n \in M'$, $l_n \in L$ and $x_n := x_{n-1} - (m'_{n-1} + l_{n-1}) \in X$ such that $\|x_n\| \le \left(\frac{\|P\|}{\|P\|+1}\right)^n$, $l_n = (I-P)x_n$, $\|l_n\| \le (\|P\| + 1)\|x_n\|$ and $\|m'_n\| \le \left(\|P\| + \frac{\|P\|}{\|P\|+1}\right) \cdot \|x_n\|$.

Set $m' = \sum_{n=0}^{\infty} m_n$ and $l = \sum_{n=0}^{\infty} l_n$. Clearly these series are convergent, $m' \in M'$ and $l \in L$. Further $x_0 = m' + l$. Hence $M' \oplus L = X$. □

Lemma 12. *Let $T, T' \in \mathcal{B}(X,Y)$ and let $\operatorname{Ran} T$ be closed. Then:*

(i) $\delta(\operatorname{Ker} T', \operatorname{Ker} T) \le \gamma(T)^{-1}\|T - T'\|$;

(ii) $\delta(\operatorname{Ran} T, \operatorname{Ran} T') \le \gamma(T)^{-1}\|T - T'\|$.

Proof. Let s be a positive number, $s < \gamma(T)$.

(i) Let $x \in \operatorname{Ker} T'$, $\|x\| = 1$. Then $\|Tx\| = \|(T - T')x\| \le \|T - T'\|$. There exists $x_1 \in X$ such that $Tx_1 = Tx$ and $\|x_1\| \le s^{-1}\|Tx\| \le s^{-1}\|T - T'\|$. Since $x - x_1 \in \operatorname{Ker} T$, we have

$$\operatorname{dist}\{x, \operatorname{Ker} T\} \le \|x - (x - x_1)\| = \|x_1\| \le s^{-1}\|T - T'\|,$$

and so $\delta(\operatorname{Ker} T', \operatorname{Ker} T) \le s^{-1}\|T - T'\|$.

Hence $\delta(\operatorname{Ker} T', \operatorname{Ker} T) \le \gamma(T)^{-1}\|T - T'\|$.

(ii) Let $y \in \operatorname{Ran} T$, $\|y\| = 1$. Then there exists $x \in X$ such that $Tx = y$ and $\|x\| \le s^{-1}$. So

$$\operatorname{dist}\{y, \operatorname{Ran} T'\} \le \|y - T'x\| = \|(T - T')x\| \le s^{-1}\|T - T'\|$$

and $\delta(\operatorname{Ran} T, \operatorname{Ran} T') \le s^{-1}\|T - T'\|$.

Hence $\delta(\operatorname{Ran} T, \operatorname{Ran} T') \le \gamma(T)^{-1}\|T - T'\|$. □

Lemma 13. *Let $T, T' \in \mathcal{B}(X,Y)$ and $\delta(\operatorname{Ker} T, \operatorname{Ker} T') < 1/2$. Then*

$$\gamma(T) \le \frac{\|T - T'\| + \gamma(T')}{1 - 2\delta(\operatorname{Ker} T, \operatorname{Ker} T')}.$$

Proof. If $T' = 0$, then the inequality is clear. If $T = 0$, then the assumption $\delta(\operatorname{Ker} T, \operatorname{Ker} T') < 1/2$ implies that $T' = 0$; so the inequality is also clear. Suppose that $T \ne 0$, $T' \ne 0$. Let ε be a positive number, $\varepsilon < 1 - 2\delta(\operatorname{Ker} T, \operatorname{Ker} T')$. Find $x' \in X \setminus \operatorname{Ker} T'$ such that $\frac{\|T'x'\|}{\operatorname{dist}\{x', \operatorname{Ker} T'\}} \le (1+\varepsilon)\gamma(T')$. Find $x \in X$ such that $T'x = T'x'$ and $\|x\| \le (1+\varepsilon)\operatorname{dist}\{x', \operatorname{Ker} T'\} = (1+\varepsilon)\operatorname{dist}\{x, \operatorname{Ker} T'\}$. Obviously, $x \ne 0$. We have

$$\|T'x\| \le (1+\varepsilon)\gamma(T')\operatorname{dist}\{x, \operatorname{Ker} T'\} \le (1+\varepsilon)\|x\|\gamma(T').$$

Furthermore,

$$\operatorname{dist}\{x, \operatorname{Ker} T\} = \inf\big\{\|x - u\| : u \in \operatorname{Ker} T, \|u\| < 2\|x\|\big\}.$$

For every $u \in \operatorname{Ker} T$ with $\|u\| < 2\|x\|$ there exists $z \in \operatorname{Ker} T'$ such that

$$\|u - z\| \le 2\|x\|\delta(\operatorname{Ker} T, \operatorname{Ker} T').$$

Then $\|x - u\| \ge \|x - z\| - \|u - z\| \ge \operatorname{dist}\{x, \operatorname{Ker} T'\} - 2\|x\|\delta\big(\operatorname{Ker} T, \operatorname{Ker} T'\big)$. Thus $\operatorname{dist}\{x, \operatorname{Ker} T\} \ge \operatorname{dist}\{x, \operatorname{Ker} T'\} - 2\|x\|\delta(\operatorname{Ker} T, \operatorname{Ker} T')$ and

$$\gamma(T) \le \frac{\|Tx\|}{\operatorname{dist}\{x, \operatorname{Ker} T\}} \le \frac{\|(T - T')x\| + \|T'x\|}{\operatorname{dist}\{x, \operatorname{Ker} T'\} - 2\|x\|\delta(\operatorname{Ker} T, \operatorname{Ker} T')}$$

$$\le \frac{\|T - T'\| \cdot \|x\| + (1+\varepsilon)\|x\|\gamma(T')}{(1+\varepsilon)^{-1}\|x\| - 2\|x\|\delta(\operatorname{Ker} T, \operatorname{Ker} T')} = \frac{\|T - T'\| + (1+\varepsilon)\gamma(T')}{(1+\varepsilon)^{-1} - 2\delta(\operatorname{Ker} T, \operatorname{Ker} T')}.$$

Letting $\varepsilon \to 0$ gives the required inequality. □

Theorem 14. *Let $c \ge 0$. Then the set $\{T \in \mathcal{B}(X, Y) : \gamma(T) \ge c\}$ is closed.*

Proof. The statement is clear for $c = 0$. Suppose that $c > 0$, $\gamma(T_n) \ge c$ and $\|T - T_n\| \to 0$. We prove that $\gamma(T) \ge c$. By Theorem 3, it is sufficient to show that $\gamma(T^*) \ge c$.

Let $y^* \in Y^*$ and $0 < \varepsilon < c$. For each n find $y_n^* \in Y^*$ such that $T_n^* y_n^* = T_n^* y^*$ and $\|y_n^*\| < (c-\varepsilon)^{-1}\|T_n^* y^*\|$. Since closed balls in Y^* are w^*-compact, there exists a w^*-accumulation point u^* of the sequence $\{y_n^*\}$.

We show that $T^*u^* = T^*y^*$. Let $x \in X$. Then there exists a subsequence $(y_{n_k}^*)$ such that $\langle Tx, y_{n_k}^* \rangle \to \langle Tx, u^* \rangle$. We have

$$\langle x, T^*u^* \rangle = \langle Tx, u^* \rangle = \lim_{k\to\infty} \langle Tx, y_{n_k}^* \rangle = \lim_{k\to\infty} \langle T_{n_k} x, y_{n_k}^* \rangle$$

$$= \lim_{k\to\infty} \langle x, T_{n_k}^* y^* \rangle = \lim_{k\to\infty} \langle T_{n_k} x, y^* \rangle = \langle Tx, y^* \rangle = \langle x, T^* y^* \rangle.$$

Hence $T^*u^* = T^*y^*$ and

$$\|u^*\| \le \limsup_n \|y_n^*\| \le (c-\varepsilon)^{-1}\|T^*y^*\|.$$

Thus $\gamma(T^*) \ge c - \varepsilon$. Letting $\varepsilon \to 0$ gives $\gamma(T) = \gamma(T^*) \ge c$. □

Corollary 15. *The function $\gamma : \mathcal{B}(X, Y) \to \langle 0, \infty \rangle$ is upper semicontinuous.*

Example 16. In general, the function γ is not continuous. Let $T_n = \begin{pmatrix} 1 & 0 \\ 0 & 1/n \end{pmatrix}$ and $T = \begin{pmatrix} 1 & 0 \\ 0 & 0 \end{pmatrix}$. Then $\gamma(T_n) = 1/n$, $T_n \to T$ and $\gamma(T) = 1$.

Theorem 17. *Let X, Y be Banach spaces, $T, T_k \in \mathcal{B}(X, Y)$ $(k = 1, 2, \dots)$, suppose that T has closed range and let $\lim_{k\to\infty} \|T_k - T\| = 0$. Then the following statements are equivalent:*

(i) $\gamma(T) = \lim_{k\to\infty} \gamma(T_k)$;

(ii) $\liminf_{k\to\infty} \gamma(T_k) > 0$;

(iii) $\lim_{k\to\infty} \delta(\operatorname{Ker} T, \operatorname{Ker} T_k) = 0$;

(iv) $\lim_{k\to\infty} \widehat{\delta}(\operatorname{Ker} T, \operatorname{Ker} T_k) = 0$;

(v) $\lim_{k\to\infty} \delta(\operatorname{Ran} T_k, \operatorname{Ran} T) = 0$;

(vi) $\lim_{k\to\infty} \widehat{\delta}(\operatorname{Ran} T_k, \operatorname{Ran} T) = 0$.

Proof. We have $\lim\limits_{k\to\infty} \delta(\operatorname{Ker} T_k, \operatorname{Ker} T) = 0$ and $\lim\limits_{k\to\infty} \delta(\operatorname{Ran} T, \operatorname{Ran} T_k) = 0$ by Lemma 12. This means that (iii) $\Leftrightarrow$ (iv) and (v) $\Leftrightarrow$ (vi).

The implication (i) $\Rightarrow$ (ii) is clear.

(ii) $\Rightarrow$ (iii): There exists a positive constant s such that $\gamma(T_k) \geq s$ for all k large enough. By Lemma 12, we have $\delta(\operatorname{Ker} T, \operatorname{Ker} T_k) \leq \gamma(T_k)^{-1}\|T_k - T\| \leq s^{-1}\|T_k - T\|$, and so $\lim_{k\to\infty} \delta(\operatorname{Ker} T, \operatorname{Ker} T_k) = 0$.

(iii) $\Rightarrow$ (i): For k large enough we have $\delta(\operatorname{Ker} T, \operatorname{Ker} T_k) < 1/2$. By Lemma 13, $\gamma(T_k) \geq \big(1-2\delta(\operatorname{Ker} T, \operatorname{Ker} T_k)\big)\gamma(T) - \|T_k - T\|$, and so $\liminf_{k\to\infty} \gamma(T_k) \geq \gamma(T) > 0$. Since $\gamma(T) \geq \limsup_{k\to\infty} \gamma(T_k)$ by Corollary 15, we have $\gamma(T) = \lim_{k\to\infty} \gamma(T_k)$.

(v) $\Leftrightarrow$ (i): The following statements are equivalent:
$\lim_{k\to\infty} \delta(\operatorname{Ran} T_k, \operatorname{Ran} T) = 0$;
$\lim_{k\to\infty} \delta\big((\operatorname{Ran} T)^\perp, (\operatorname{Ran} T_k)^\perp\big) = 0$;
$\lim_{k\to\infty} \delta(\operatorname{Ker} T^*, \operatorname{Ker} T_k^*) = 0$;
$\lim_{k\to\infty} \gamma(T_k^*) = \gamma(T^*)$;
$\lim_{k\to\infty} \gamma(T_k) = \gamma(T)$. □

Corollary 18. Let $T, T_k \in \mathcal{B}(X) \quad (k = 1, 2, \ldots)$, let

$$\|T_k - T\| \to 0 \quad \text{and} \quad \delta(\operatorname{Ker} T, \operatorname{Ker} T_k) \to 0.$$

Then $\gamma(T_k) \to \gamma(T)$.

Proof. If $\gamma(T) = 0$, then the upper semicontinuity of γ implies that $\gamma(T_k) \to \gamma(T)$.

If $\gamma(T) > 0$, then $\operatorname{Ran} T$ is closed and the statement follows from the previous theorem. □

Corollary 19. *Let $T, T_k \in \mathcal{B}(X, Y) \quad (k = 1, 2, \ldots)$, $\lim_{k\to\infty} \|T_k - T\| = 0$ and let $\limsup_{k\to\infty} \gamma(T_k) > 0$. Suppose that $y \in Y$ and $y_k \in \operatorname{Ran} T_k$ such that $\lim_{k\to\infty} y_k = y$. Then $y \in \operatorname{Ran} T$.*

Proof. Without loss of generality we can assume that $\inf_{k\to\infty} \gamma(T_k) > 0$. By Corollary 15, we have $\gamma(T) \geq \lim_{k\to\infty} \gamma(T_k) > 0$. Thus $\operatorname{Ran} T$ is closed and $\delta(\operatorname{Ran} T_k, \operatorname{Ran} T) \to 0$. For every k find a vector $y'_k \in \operatorname{Ran} T$ such that $\|y_k - y'_k\| \leq \|y_k\| \cdot \big(\delta(\operatorname{Ran} T_k, \operatorname{Ran} T) + k^{-1}\big)$. So

$$\begin{aligned} \operatorname{dist}\{y, \operatorname{Ran} T\} &\leq \|y - y'_k\| \leq \|y - y_k\| + \|y_k - y'_k\| \\ &\leq \|y - y_k\| + \|y_k\| \cdot \big(\delta(\operatorname{Ran} T_k, \operatorname{Ran} T) + k^{-1}\big) \to 0 \end{aligned}$$

as $k \to \infty$. Since $\operatorname{Ran} T$ is closed, we have $y \in \operatorname{Ran} T$. □

Corollary 20. *Let $\varepsilon > 0$. Then the function $T \mapsto \gamma(T)$ is continuous on the set $\{T \in \mathcal{B}(X,Y) : \gamma(T) \geq \varepsilon\}$.*

Definition 21. Let M be a metric space, let $T : M \to \mathcal{B}(X,Y)$ be a norm-continuous function and let $w \in M$. We say that the function T is *regular at* w if the range of $T(w)$ is closed and the function $z \mapsto \gamma(T(z))$ is continuous at w. We say that T is regular in M if T is regular at each point $w \in M$.

The following theorem is an easy consequence of Theorem 17.

Theorem 22. *Let M be a metric space, let $T : M \to \mathcal{B}(X,Y)$ be a norm-continuous function and let $w \in M$. Suppose that $\operatorname{Ran} T(w)$ is closed. Then the following statements are equivalent:*

(i) *T is regular at w;*

(ii) $\liminf_{z \to w} \gamma(T(z)) > 0$;

(iii) $\lim_{z \to w} \delta(\operatorname{Ker} T(w), \operatorname{Ker} T(z)) = 0$;

(iv) $\lim_{z \to w} \widehat{\delta}(\operatorname{Ker} T(w), \operatorname{Ker} T(z)) = 0$;

(v) $\lim_{z \to w} \delta(\operatorname{Ran} T(z), \operatorname{Ran} T(w)) = 0$;

(vi) $\lim_{z \to w} \widehat{\delta}(\operatorname{Ran} T(z), \operatorname{Ran} T(w)) = 0$.

Corollary 23. *A norm-continuous function $T : M \to \mathcal{B}(X,Y)$ is regular at $w \in M$ if and only if the function $z \in M \mapsto T(z)^* \in \mathcal{B}(Y^*, X^*)$ is regular at w. The set $\{w \in M : T$ is regular at $w\}$ is open.*

Examples 24. (i) Let T be an operator on a finite-dimensional Banach space X. Then

$$\operatorname{Ker}(T - z) = \{0\} \Longleftrightarrow z \notin \sigma(T) \Longleftrightarrow \operatorname{Ran}(T - z) = X.$$

Thus the function $z \mapsto T - z$ is regular on the set $\mathbb{C} \setminus \sigma(T)$ and it is not regular at the points of spectrum.

For $\lambda \in \sigma(T)$ both $\operatorname{Ker}(T - z)$ and $\operatorname{Ran}(T - z)$ are discontinuous: $\operatorname{Ker}(T - \lambda)$ is "larger" than $\operatorname{Ker}(T - z)$ and $\operatorname{Ran}(T - \lambda)$ is "smaller" than $\operatorname{Ran}(T - z)$ for z close to λ.

By Theorem 22, this is a general fact for norm-continuous functions with closed range: continuity of the range is equivalent to the continuity of the kernel.

(ii) Let M be a metric space, let $T : M \to \mathcal{B}(X,Y)$ be a norm-continuous function, $w \in M$ and suppose that $T(w)$ is bounded below. Then $T(z)$ is bounded below in a certain neighbourhood U of w, so $\operatorname{Ker} T(z) = \{0\}$ for $z \in U$ and T is regular at w.

(iii) Similarly, if $T(w)$ is onto, then there is a neighbourhood U of w such that $T(z)$ is onto for $z \in U$. So $\operatorname{Ran} T(z) = Y$ for $z \in U$ and T is regular at w.

(iv) Let X, Y, Z be Banach spaces, let $T : M \to \mathcal{B}(X,Y)$ and $S : M \to \mathcal{B}(Y,Z)$ be norm-continuous functions satisfying $S(z)T(z) = 0$ for $z \in M$. Let

$w \in M$, and suppose that $\operatorname{Ker} S(w) = \operatorname{Ran} T(w)$ and $\operatorname{Ran} S(w)$ is closed. Then $\lim_{z\to w} \delta(\operatorname{Ran} T(w), \operatorname{Ran} T(z)) = 0$ and $\lim_{z\to w} \delta(\operatorname{Ker} S(z), \operatorname{Ker} S(w)) = 0$ by Lemma 12. Thus

$$\delta(\operatorname{Ran} T(z), \operatorname{Ran} T(w)) = \delta(\operatorname{Ran} T(z), \operatorname{Ker} S(w)) \leq \delta(\operatorname{Ker} S(z), \operatorname{Ker} S(w)) \to 0.$$

Similarly, $\delta(\operatorname{Ker} S(w), \operatorname{Ker} S(z)) \leq \delta(\operatorname{Ran} T(w), \operatorname{Ran} T(z)) \to 0$. Hence T and S are regular at w.

In fact, (ii) and (iii) are particular cases of (iv) for either $X = 0$ or $Z = 0$.

If the function T is not only continuous but also Lipschitz, then we have a better information about the behaviour of the function $z \mapsto \gamma(T(z))$.

Theorem 25. *Let M be a convex subset of a Banach space Z, let $k > 0$ and let $T : M \to \mathcal{B}(X, Y)$ be a regular function satisfying $\|T(z) - T(w)\| \leq k \cdot |z - w|$ $(z, w \in M)$. Then $|\gamma(T(z)) - \gamma(T(w))| \leq 3k \cdot |z - w|$ for all $z, w \in M$.*

Proof. Fix $z, w \in M$ and $\varepsilon > 0$. Let $S : \langle 0, 1\rangle \to \mathcal{B}(X, Y)$ be defined by $S(t) = T(z + t(w - z))$. So $S(0) = T(z)$, $S(1) = T(w)$ and $\|S(t) - S(s)\| \leq k|t - s| \cdot \|z - w\|$.

We first prove that for every $t \in \langle 0, 1\rangle$ there exists $\eta_t > 0$ such that

$$\gamma(S(t)) - \gamma(S(s)) \leq (3 + \varepsilon)k|t - s| \cdot \|z - w\| \qquad (s \in U_t),$$

where $U_t = \{s \in \langle 0, 1\rangle : |s - t| < \eta_t\}$. To see this, take η_t such that $\frac{\gamma(S(t))}{\gamma(S(s))} < 1 + \varepsilon/2$ $(s \in U_t)$ and $\eta_t < \frac{1}{2k\|z-w\|} \cdot \min\{\gamma(S(u)) : u \in \langle 0, 1\rangle\}$. For $s \in U_t$ we have

$$\|S(t) - S(s)\| < \eta_t k\|z - w\| < \frac{1}{2} \min\{\gamma(S(u)) : u \in \langle 0, 1\rangle\}.$$

So, by Lemma 12, $\delta(\operatorname{Ker} S(t), \operatorname{Ker} S(s)) < 1/2$. By Lemma 13,

$$\gamma(S(t)) \leq \frac{\|S(t) - S(s)\| + \gamma(S(s))}{1 - 2\delta(\operatorname{Ker} S(t), \operatorname{Ker} S(s))},$$

and so

$$\begin{aligned}
\gamma(S(t)) - \gamma(S(s)) &\leq \|S(t) - S(s)\| + 2\gamma(S(t))\delta(\operatorname{Ker} S(t), \operatorname{Ker} S(s)) \\
&\leq \|S(t) - S(s)\| + 2\frac{\gamma(S(t))}{\gamma(S(s))} \cdot \|S(t) - S(s)\| \leq (3 + \varepsilon)\|S(t) - S(s)\| \\
&\leq (3 + \varepsilon)k|t - s| \cdot \|z - w\|.
\end{aligned}$$

Let $t_0 = \max\{s \in \langle 0, 1\rangle : \gamma(S(0)) - \gamma(S(s)) \leq (3 + \varepsilon)ks\|z - w\|\}$. If $t_0 < 1$, then for t_1 with $t_0 < t_1 < t_0 + \eta_{t_0}$ we have $\gamma(S(0)) - \gamma(S(t_1)) \leq \gamma(S(0)) - \gamma(S(t_0)) + \gamma(S(t_0)) - \gamma(S(t_1)) \leq (3 + \varepsilon)kt_0\|z - w\| + (3 + \varepsilon)k(t_1 - t_0)\|z - w\| = (3 + \varepsilon)kt_1\|z - w\|$, a contradiction with the maximality of t_0. Thus $t_0 = 1$ and $\gamma(T(z)) - \gamma(T(w)) = \gamma(S(0)) - \gamma(S(1)) \leq (3 + \varepsilon)k\|z - w\|$. Letting $\varepsilon \to 0$ gives $\gamma(T(z)) - \gamma(T(w)) \leq 3k\|z - w\|$.

The statement of the theorem now follows by symmetry. □

11 Factorization of vector-valued functions

In this section we study in more details regular and analytic regular functions.

Theorem 1. *Let M be a metric space, let $T : M \to \mathcal{B}(X,Y)$ be a regular function and let $f : M \to Y$ be a continuous function satisfying $f(z) \in \operatorname{Ran} T(z) \quad (z \in M)$. Let $\varepsilon > 0$. Then there exists a continuous function $g : M \to X$ satisfying $T(z)g(z) = f(z)$ and $\|g(z)\| \le \|f(z)\| \cdot \gamma(T(z))^{-1} + \varepsilon$ for all $z \in M$.*

Moreover, if $w_0 \in M$ and $x_0 \in X$ satisfy $T(w_0)x_0 = f(w_0)$ and $\|x_0\| < \|f(w_0)\| \cdot \gamma(T(w_0))^{-1} + \varepsilon$, then it is possible to choose $g : M \to X$ in such a way that $g(w_0) = x_0$.

Proof. Note that the function $z \mapsto \gamma(T(z))$ is continuous and positive in M.

We first prove an approximate version of the theorem.

Claim A. Let $T : M \to \mathcal{B}(X,Y)$ be regular, $f : M \to Y$ continuous, let $f(z) \in \operatorname{Ran} T(z)$ for all $z \in M$, let $\varepsilon > 0$, $w_0 \in M$, $x_0 \in X$, $T(w_0)x_0 = f(w_0)$ and $\|x_0\| < \|f(w_0)\| \cdot \gamma(T(w_0))^{-1} + \varepsilon$. Then there exists a continuous function $h : M \to X$ such that $h(w_0) = x_0$, $\|T(z)h(z) - f(z)\| \le \varepsilon \cdot \gamma(T(z))$ and $\|h(z)\| \le \|f(z)\| \cdot \gamma(T(z))^{-1} + \varepsilon$ for all $z \in M$.

Proof of Claim A. For every $w \in M$ find a vector $x_w \in X$ with $T(w)x_w = f(w)$ and $\|x_w\| < \|f(w)\|\gamma(T(w))^{-1} + \varepsilon$; for $w = w_0$ set $x_{w_0} = x_0$. Choose a neighbourhood U_w of w such that $\|T(z)x_w - f(z)\| < \varepsilon \cdot \gamma(T(z))$ and $\|x_w\| < \|f(z)\|\gamma(T(z))^{-1} + \varepsilon$ for all $z \in U_w$.

For $w \ne w_0$ we can also assume that $w_0 \notin U_w$. Then $\{U_w : w \in M\}$ is an open cover of M.

Let $\{\varphi_\alpha\}$ be a partition of unity subordinate to this cover, i.e., $\varphi_\alpha : M \to \langle 0, 1 \rangle$ are continuous functions, for each α there exists a point $w_\alpha \in M$ such that $\operatorname{supp} \varphi_\alpha \subset U_{w_\alpha}$, for each $z \in M$ there is a neighbourhood V of z such that $\operatorname{supp} \varphi_\alpha \cap V \ne \emptyset$ for only a finite number of α's, and $\sum_\alpha \varphi_\alpha(z) = 1 \quad (z \in M)$, see A.3.1.

Set $h(z) = \sum_\alpha \varphi_\alpha(z) x_{w_\alpha}$. The sum is well defined, the function $h(z) : M \to X$ is continuous and

$$\|T(z)h(z) - f(z)\| = \left\| \sum_{\alpha : z \in \operatorname{supp} \varphi_\alpha} T(z)\varphi_\alpha(z)x_{w_\alpha} - f(z) \right\|$$

$$= \left\| \sum_{\alpha : z \in \operatorname{supp} \varphi_\alpha} \varphi_\alpha(z)\bigl(T(z)x_{w_\alpha} - f(z)\bigr) \right\| \le \sum_\alpha \varphi_\alpha(z) \cdot \varepsilon \cdot \gamma(T(z)) = \varepsilon \cdot \gamma(T(z))$$

for all $z \in M$.

Furthermore,

$$h(w_0) = \sum_{w_0 \in \operatorname{supp} \varphi_\alpha} \varphi_\alpha(w_0) x_{w_\alpha} = x_0$$

and

$$\|h(z)\| = \Big\| \sum_{\alpha : z \in \operatorname{supp} \varphi_\alpha} \varphi_\alpha(z) x_{w_\alpha} \Big\|$$
$$\leq \Big\| \sum_{z \in U_{x_\alpha}} \varphi_\alpha(z) \big(\|f(z)\| \gamma(T(z))^{-1} + \varepsilon \big) \Big\| = \|f(z)\| \gamma(T(z))^{-1} + \varepsilon.$$

Proof of Theorem 1. Using A, we can find a continuous function $h_1 : M \to X$ such that $\|T(z)h_1(z) - f(z)\| \leq \varepsilon/3 \cdot \gamma(T(z))$, $\|h_1(z)\| \leq \|f(z)\| \cdot \gamma(T(z))^{-1} + \varepsilon/3 \quad (z \in M)$ and $h_1(w_0) = x_0$.

Set $f_1(z) = f(z) - T(z)h_1(z)$. So $f_1(z) \in \operatorname{Ran} T(z)$, $\|f_1(z)\| \leq \gamma(T(z)) \cdot \varepsilon/3$ for all $z \in M$, and $f_1(w_0) = 0$. For $i = 2, 3, \dots$ we construct inductively functions $h_i, f_i : M \to X$ such that $\|f_{i-1}(z) - T(z)h_i(z)\| \leq \varepsilon 3^{-i} \gamma(T(z))$,

$$\|h_i(z)\| \leq \|f_{i-1}(z)\| \cdot \gamma(T(z))^{-1} + \frac{\varepsilon}{3^i},$$

$h_i(w_0) = 0$ and $f_i(z) := f_{i-1}(z) - T(z)h_i(z) \in \operatorname{Ran} T(z)$ for all $z \in M$. Thus $\|h_i(z)\| \leq 3^{-i+1}\varepsilon + 3^{-i}\varepsilon$.

Set $g(z) = \sum_{i=1}^\infty h_i(z)$. The sum converges uniformly in M, and so $g : M \to X$ is a continuous function. We have $g(w_0) = h_1(w_0) = x_0$,

$$\|g(z)\| \leq \sum_{i=1}^\infty \|h_i(z)\| \leq \|f(z)\| \cdot \gamma(T(z))^{-1} + \frac{\varepsilon}{3} + \sum_{i=2}^\infty \Big(\frac{\varepsilon}{3^{i-1}} + \frac{\varepsilon}{3^i} \Big)$$
$$= \|f(z)\| \cdot \gamma(T(z))^{-1} + \varepsilon,$$

and

$$T(z)g(z) = \sum_{i=1}^\infty T(z)h_i(z) = f(z) - f_1(z) + \sum_{i=2}^\infty \big(f_{i-1}(z) - f_i(z)\big) = f(z)$$

for all $z \in M$. □

An alternative proof of Theorem 1 can be done using the Michael selection theorem, see A.4.5. In fact, the method of the present proof can be used to prove also the Michael selection theorem.

Lemma 2. *Let X, Y, Z be Banach spaces, let $T : X \to Y$ and $S : Y \to Z$ be operators satisfying $ST = 0$. The following statements are equivalent:*

(i) $\operatorname{Ran} T = \operatorname{Ker} S$ *and* $\operatorname{Ran} S$ *is closed;*

(ii) $\operatorname{Ran} S^* = \operatorname{Ker} T^*$ *and* $\operatorname{Ran} T^*$ *is closed.*

Proof. (i) ⇒ (ii): It is clear that $\operatorname{Ran} T$ is closed, and so is $\operatorname{Ran} T^*$. We have $T^*S^* = 0$, and so $\operatorname{Ran} S^* \subset \operatorname{Ker} T^*$.

Let $y^* \in \operatorname{Ker} T^*$. For all $x \in X$ we have $\langle Tx, y^*\rangle = \langle x, T^*y^*\rangle = 0$. Define a functional $z_1^* \in (\operatorname{Ran} S)^*$ by

$$\langle Sy, z_1^*\rangle = \langle y, y^*\rangle \qquad (y \in Y).$$

If $Sy = 0$, then $y \in \operatorname{Ran} T$ and $\langle y, y^*\rangle = 0$, so z_1^* is well defined.

We show that z_1^* is continuous. Let $0 < s < \gamma(S)$ and $z \in \operatorname{Ran} S$, $\|z\| = 1$. Then there exists $y \in Y$ with $\|y\| \le s^{-1}$ and $Sy = z$. We have

$$|\langle z, z_1^*\rangle| = |\langle Sy, z_1^*\rangle| = |\langle y, y^*\rangle| \le \|y\| \cdot \|y^*\| \le s^{-1}\|y^*\|.$$

Hence z_1^* is continuous. By the Hahn-Banach theorem, we can extend z_1^* to a functional $z^* \in Z^*$ with the same norm. Furthermore, for all $y \in Y$ we have $\langle y, S^*z^*\rangle = \langle Sy, z^*\rangle = \langle y, y^*\rangle$, and so $y^* = S^*z^* \in \operatorname{Ran} S^*$. Hence $\operatorname{Ker} T^* = \operatorname{Ran} S^*$.

(ii) $\Rightarrow$ (i): $\operatorname{Ran} S^*$ is closed, and so is $\operatorname{Ran} S$. Similarly, $\operatorname{Ran} T$ is closed. By the previous implication, $\operatorname{Ran} T^{**} = \operatorname{Ker} S^{**}$. Hence, by Corollary A.1.17,

$$\operatorname{Ran} T = \operatorname{Ran} T^{**} \cap Y = \operatorname{Ker} S^{**} \cap Y = \operatorname{Ker} S. \qquad \square$$

The following lemma is useful in many situations:

Lemma 3. *Let X, Y, Z be Banach spaces, let $T : X \to Y$ and $S : Y \to Z$ be operators satisfying $\operatorname{Ran} T = \operatorname{Ker} S$ and let $\operatorname{Ran} S$ be closed. Let $0 < c < 1$. Then there exists $\varepsilon > 0$ such $\operatorname{Ran} T' = \operatorname{Ker} S'$, $\gamma(T') > c \cdot \gamma(T)$ and $\gamma(S') > c \cdot \gamma(S)$ for all pairs of operators $T' : X \to Y$ and $S' : Y \to Z$ satisfying $\|T' - T\| < \varepsilon$, $\|S' - S\| < \varepsilon$ and $S'T' = 0$.*

Proof. Consider the metric space

$$\Lambda = \{(T', S') : T' \in \mathcal{B}(X, Y), S' \in \mathcal{B}(Y, Z), S'T' = 0\}$$

with the metric $\operatorname{dist}\{(T', S), (T'', S'')\} = \max\{\|T' - T''\|, \|S' - S''\|\}$. Let $Q_1 : \Lambda \to \mathcal{B}(X, Y)$, $Q_2 : \Lambda \to \mathcal{B}(Y, Z)$ be defined by $Q_1(T', S') = T'$, $Q_2(T', S') = S'$.

By Example 10.24 (iv), the functions Q_1, Q_2 are regular at (T, S), and so they are regular in a certain neighbourhood of (T, S). In particular, there is a positive ε' such that $(T', S') \in \Lambda$, $\|T' - T\| < \varepsilon'$ and $\|S' - S\| < \varepsilon'$ implies that $\gamma(T') > c \cdot \gamma(T)$ and $\gamma(S') > c \cdot \gamma(S)$.

Let $\varepsilon = \min\{\varepsilon', \frac{1}{3}\gamma(T), \frac{1}{3}\gamma(S)\}$, $\|T' - T\| < \varepsilon$ and $\|S' - S\| < \varepsilon$. By Lemma 10.12, we have $\delta(\operatorname{Ker} S', \operatorname{Ker} S) < 1/3$ and $\delta(\operatorname{Ran} T, \operatorname{Ran} T') < 1/3$. Furthermore,

$$\begin{aligned}
&\delta(\operatorname{Ker} S', \operatorname{Ran} T') \\
&\quad \le \delta(\operatorname{Ker} S', \operatorname{Ker} S) + \delta(\operatorname{Ker} S, \operatorname{Ran} T') + \delta(\operatorname{Ker} S', \operatorname{Ker} S) \cdot \delta(\operatorname{Ker} S, \operatorname{Ran} T') \\
&\quad \le 1/3 + 4/3\ \delta(\operatorname{Ker} S, \operatorname{Ran} T') = 1/3 + 4/3\ \delta(\operatorname{Ran} T, \operatorname{Ran} T') < 1.
\end{aligned}$$

Since $\operatorname{Ran} T' \subset \operatorname{Ker} S'$, we conclude that $\operatorname{Ran} T' = \operatorname{Ker} S'$. $\square$

Theorem 4. *Let M be a metric space, let $T : M \to \mathcal{B}(X,Y)$ be a continuous function such that $\operatorname{Ran} T(z)$ is closed for all $z \in M$. The following statements are equivalent:*

(i) *T is regular in M;*

(ii) *there exist a Banach space Z and a continuous function $S : M \to \mathcal{B}(Z,X)$ such that $\operatorname{Ran} S(z) = \operatorname{Ker} T(z) \quad (z \in M)$;*

(iii) *there exist a Banach space Z' and a continuous function $V : M \to \mathcal{B}(Y,Z')$ such that $\operatorname{Ran} T(z) = \operatorname{Ker} V(z)$ and $\operatorname{Ran} V(z)$ is closed for all $z \in M$.*

Proof. By Example 10.24 (iv), either (ii) or (iii) implies (i).

(i) $\Rightarrow$ (ii): Let $(g_\alpha)_{\alpha\in\Lambda}$ be the set of all continuous functions $g_\alpha : M \to X$ such that $T(z)g_\alpha(z) = 0$ and $\|g_\alpha(z)\| \le 1 \quad (z \in M)$. By Theorem 1, for all $w \in M$ and $x \in \operatorname{Ker} T(w)$ with $\|x\| < 1$ there exists $\alpha \in \Lambda$ with $g_\alpha(w) = x$.

Let Z be the ℓ^1 space over the set Λ, i.e., Z is the set of all functions $c : \Lambda \to \mathbb{C}$ with $\|c\| = \sum_{\alpha\in\Lambda} |c(\alpha)| < \infty$ and define $S(z) : Z \to X$ by $S(z)(c) = \sum_{\alpha\in\Lambda} c(\alpha) \cdot g_\alpha(z) \quad (z \in M)$. It is clear that

$$\|S(z)c\| \le \sum_{\alpha\in\Lambda} |c(\alpha)| \cdot \|g_\alpha(z)\| \le \sum_{\alpha\in\Lambda} |c(\alpha)| = \|c\|,$$

and so $\|S(z)\| \le 1$. Further, $\operatorname{Ran} S(z) = \operatorname{Ker} T(z)$ for all $z \in M$.

(i) $\Rightarrow$ (iii): The function $T^* : M \to \mathcal{B}(Y^*, X^*)$ defined by $T^*(z) = (T(z))^*$ is regular in M, and so, by (ii), there exist a Banach space Z and a continuous function $S : M \to \mathcal{B}(Z, Y^*)$ such that $\operatorname{Ran} S(z) = \operatorname{Ker} T^*(z) \quad (z \in M)$. By Lemma 2, $\operatorname{Ran}(T(z))^{**} = \operatorname{Ker}(S(z))^* \quad (z \in M)$. Let $V : M \to \mathcal{B}(Y, Z^*)$ be the restriction $V(z) = (S(z))^*|Y$. By Corollary A.1.17,

$$\operatorname{Ran} T(z) = Y \cap \operatorname{Ran}(T(z))^{**} = Y \cap \operatorname{Ker}(S(z))^* = \operatorname{Ker} V(z) \quad (z \in M).$$

It remains to show that $\operatorname{Ran} V(z)$ is closed.

Fix $z \in M$. Write $E = Y/{}^\perp \operatorname{Ran} S(z)$ and let $Q : Y \to E$ be the canonical projection. By A.1.11, E^* is isometrically isomorphic to $({}^\perp \operatorname{Ran} S(z))^\perp = \operatorname{Ran} S(z)$, since $\operatorname{Ran} S(z) = \operatorname{Ker} T(z)^*$, and therefore it is w^*-closed. Furthermore, Q^* is an isometrical embedding of E^* into Y^*. Thus we can write $S(z) = Q^* S_0$ for the operator $S_0 : Z \to E^*$ induced by $S(z)$. It is clear that S_0 is onto. Consequently, $S(z)^* : Y^{**} \to Z^*$ can be written as $S(z)^* = S_0^* Q^{**}$. Then we have $\operatorname{Ran} V(z) = S_0^* Q^{**} Y = S_0^* Q Y = S_0^* E$. Since S_0^* is bounded below, we conclude that $\operatorname{Ran} V(z)$ is closed. □

In the following we study analytic regular functions.

For $w = (w_1, \dots, w_n) \in \mathbb{C}^n$ and $r > 0$ we write

$$\Delta(w,r) = \{(z_1, \dots, z_n) \in \mathbb{C}^n : |z_i - w_i| < r \quad (i = 1, \dots, n\}$$

and

$$\overline{\Delta}(w,r) = \{(z_1, \dots, z_n) \in \mathbb{C}^n : |z_i - w_i| \le r \quad (i = 1, \dots, n\}.$$

For simplicity we start with functions of one variable.

Theorem 5. *Let G be an open subset of $\mathbb{C}$, let $T : G \to \mathcal{B}(X,Y)$ and $f : G \to Y$ be analytic functions, let $w, w_k \in G \quad (k = 1,2\dots)$, $\lim_{k\to\infty} w_k = w$, $\limsup_{k\to\infty} \gamma(T(w_k)) > 0$ and $f(w_k) \in \operatorname{Ran} T(w_k) \quad (k = 1,2,\dots)$. Then there exist a neighbourhood U of w and an analytic function $g : U \to X$ such that $T(z)g(z) = f(z) \quad (z \in U)$.*

More precisely, if r and s are positive numbers such that $\overline{\Delta}(w,r) \subset G$, $s < \min\{1, \gamma(T(w))\}$ and $M = \max_{z\in\overline{\Delta}(w,r)} \max\{1, \|T(z)\|, \|f(z)\|\}$, then we can take $U = \Delta(w,r')$ where $r' = \frac{rs}{2M}$. If $x \in X$ satisfies $T(w)x = f(w)$, then there exists an analytic function $g : U \to X$ such that $g(w) = x, T(z)g(z) = f(z)$ and $\|g(z)-x\| \le \frac{|z-w|}{r'-|z-w|} \max\{1, \|x\|\}$ for all $z \in U$.

Proof. Without loss of generality we can assume that $w = 0$. Then $\gamma(T(0)) \ge \limsup_{k\to\infty} \gamma(T(w_k)) > 0$.

Let r, s, M and x satisfy the conditions of the theorem. Let $T(z) = \sum_{i=0}^{\infty} T_i z^i$ and $f(z) = \sum_{i=0}^{\infty} f_i z^i$ be the Taylor expansions of T and f about 0. Then $\|T_i\| \le \frac{M}{r^i}$ and $\|f_i\| \le \frac{M}{r^i} \quad (i = 0,1,\dots)$. Set $r' = \frac{rs}{2M}$ and $U = \Delta(0,r')$.

To find the required analytic function $g : U \to X$, it is sufficient to construct its coefficients $g_i \in X \quad (i = 0,1,\dots)$ such that $g_0 = x$,

$$f_m = \sum_{i=0}^{m} T_{m-i} g_i \tag{1}$$

and $\|g_i\| \le r'^{-i} \max\{1, \|x\|\}$ for all $i \ge 0$. Indeed, if $g_i \quad (i = 0,1,\dots)$ satisfy (1), then $g(z) = \sum_{i=0}^{\infty} g_i z^i$ is convergent in U, $g(0) = x$, $T(z)g(z) = f(z)$ and

$$\|g(z) - x\| = \left\|\sum_{i=1}^{\infty} g_i z^i\right\| \le \sum_{i=1}^{\infty} \|g_i\| \cdot |z|^i \le \frac{|z|}{r' - |z|} \max\{1, \|x\|\}.$$

We construct the vectors $g_i \in X \quad (i = 0,1,\dots)$ satisfying (1) inductively. Suppose that we have already found elements $g_0 = x, g_1, \dots, g_{m-1} \in X$ with properties (1). For k large enough we have $|w_k| < r' \le r$. Set

$$y_k = f(w_k) - \sum_{i=0}^{m-1} w_k^i T(w_k) g_i \in \operatorname{Ran} T(w_k)$$

and

$$y_k' = \sum_{j=0}^{m} f_j w_k^j - \sum_{i=0}^{m-1} w_k^i \sum_{l=0}^{m-i} w_k^l T_l g_i.$$

Clearly, y'_k is an approximation of y_k and

$$\|y_k - y'_k\| = \left\| \sum_{j=m+1}^{\infty} f_j w_k^j - \sum_{i=0}^{m-1} w_k^i \Big(\sum_{l=m-i+1}^{\infty} w_k^l T_l \Big) g_i \right\|$$

$$\leq \sum_{j=m+1}^{\infty} \frac{M}{r^j} |w_k^j| + \sum_{i=0}^{m-1} \sum_{l=m-i+1}^{\infty} |w_k|^{i+l} \|g_i\| \cdot \frac{M}{r^l} \leq c \cdot |w_k^{m+1}|,$$

where c is a constant independent of k.

Set $u = f_m - \sum_{j=0}^{m-1} T_{m-j} g_j$. Using the induction assumption, we have

$$w_k^m u = w_k^m \left(f_m - \sum_{j=0}^{m-1} T_{m-j} g_j \right) + \sum_{i=0}^{m-1} w_k^i \left(f_i - \sum_{j=0}^{i} T_{i-j} g_j \right) = y'_k.$$

Furthermore, $\frac{y_k}{w_k^m} \in \operatorname{Ran} T(w_k)$ and

$$\lim_{k\to\infty} \frac{y_k}{w_k^m} = \lim_{k\to\infty} \frac{y_k - y'_k}{w_k^m} + \lim_{k\to\infty} \frac{y'_k}{w_k^m} = u.$$

By Corollary 10.19, $u \in \operatorname{Ran} T(0)$. Thus there exists a vector $g_m \in X$ with $T(0) g_m = u = f_m - \sum_{j=0}^{m-1} T_{m-j} g_j$ and

$$\|g_m\| \leq s^{-1} \|u\| \leq s^{-1} \Big(\frac{M}{r^m} + \sum_{j=0}^{m-1} \frac{M}{r^{m-j}} \frac{1}{r'^j} \max\{1, \|x\|\} \Big)$$

$$\leq \max\{1, \|x\|\} \cdot \frac{M}{s} \Big(\frac{1}{r^m} + \sum_{j=0}^{m-1} \frac{2^j M^j}{r^{m-j} r^j s^j} \Big)$$

$$\leq \max\{1, \|x\|\} \cdot \frac{M^m}{r^m s^m} \Big(1 + \sum_{j=0}^{m-1} 2^j \Big) = r'^{-m} \cdot \max\{1, \|x\|\}.$$

This finishes the induction step and also the proof. □

Corollary 6. *Let G be an open subset of $\mathbb{C}$, $w \in G$ and let $T : G \to \mathcal{B}(X, Y)$ be an analytic function. Then:*

(i) *the limit $\lim_{z\to w} \gamma(T(z))$ exists;*

(ii) *T is regular at w if and only if $\lim_{z\to w} \gamma(T(z)) > 0$.*

Proof. Both statements are trivial if $\limsup_{z\to w} \gamma(T(z)) = 0$.

If $\limsup_{z\to w} \gamma(T(z)) > 0$, then there exists a sequence (w_k) converging to w such that $\limsup_k \gamma(T(w_k)) > 0$. Thus $\gamma(T(w)) > 0$.

We show that T is regular at w. Let $f : G \to Y$ be identically equal to 0 and let U be the neighbourhood of w constructed in the previous theorem. For $x \in \operatorname{Ker} T(w), \|x\| = 1$ we can find an analytic function $g : U \to X$ such that $g(w) = x, T(z)g(z) = 0 \quad (z \in U)$ and $\|g(z) - x\| = \|g(z) - g(w)\| \le c(z) \cdot |z - w|$, where $c(z)$ is a constant independent of x and $\limsup_{z \to w} c(z) < \infty$. Since $g(z) \in \operatorname{Ker} T(z)$, we have $\operatorname{dist}\{x, \operatorname{Ker} T(z)\} \le c(z)|z - w|$, and so

$$\lim_{z \to w} \delta\big(\operatorname{Ker} T(w), \operatorname{Ker} T(z)\big) = 0.$$

Hence T is regular at w and $\lim_{z \to w} \gamma(T(z)) = \gamma(T(w))$. □

Corollary 7. *Let G be an open subset of $\mathbb{C}$ and let $T : G \to \mathcal{B}(X, Y)$ be an analytic function. Then the set of all $w \in G$ such that $\operatorname{Ran} T(w)$ is closed and T is not regular at w is at most countable.*

Proof. For $k = 1, 2, \dots$ denote by M_k the set of all $w \in G$ such that $\gamma\big(T(w)\big) \ge 1/k$ and T is not regular at w. The previous theorem implies that every $w \in M_k$ is an isolated point of M_k, so the set M_k is at most countable. The same is clearly true for the union $\bigcup_{k=1}^{\infty} M_k$. □

Corollary 8. *Let G be an open connected subset of $\mathbb{C}$, let $T : G \to \mathcal{B}(X, Y)$ be an analytic function regular in G and let $f : G \to Y$ be an analytic function. Let $w, w_k \in G \quad (k = 1, 2, \dots), \lim_{k \to \infty} w_k = w$ and $f(w_k) \in \operatorname{Ran} T(w_k)$ for all $k \in \mathbb{N}$. Then $f(z) \in \operatorname{Ran} T(z)$ for all $z \in G$.*

Proof. Let $u \in G$. Connect w and u by a continuous curve $\psi : \langle 0, 1 \rangle \to G$ such that $\psi(0) = w$ and $\psi(1) = u$. Since $z \mapsto \gamma(T(z))$ is a continuous positive function in G, it is bounded below on the curve $\{\psi(t) : 0 \le t \le 1\}$.

Let $M = \big\{t \in \langle 0, 1 \rangle : f(\psi(t)) \in \operatorname{Ran} T(\psi(t))$ for all $s, 0 \le s \le t\big\}$ and $t_0 = \sup\{t : t \in M\}$. Theorem 5 implies that $t_0 > 0$ and also, using a standard argument, one can see that $t_0 = 1$. Thus $f(u) \in \operatorname{Ran} T(u)$. □

Similar results can be proved for analytic functions of n variables.

Theorem 9. *Let G be an open subset of $C^n, w \in G$, let $T : G \to \mathcal{B}(X, Y)$ be an analytic function. Suppose that T satisfies the following condition: if $w_k \in G$, $w_k \to w$, $y_k \in \operatorname{Ran} T(w_k)$ and $y_k \to y$, then $y \in \operatorname{Ran} T(w)$ (in particular, this condition is satisfied if T is regular at w).*

Let $f : G \to Y$ be an analytic function satisfying $f(z) \in \operatorname{Ran} T(z) \quad (z \in G)$. Then there exists a neighbourhood U of w such that, for each $x \in X$ with $T(w)x = f(w)$, there exists an analytic function $g : U \to X$ satisfying $T(z)g(z) = f(z) \quad (z \in U)$ and $g(w) = x$.

Proof. The argument is similar to the proof of Theorem 5. We can assume that $w = 0$.

Considering the constant sequence $w_k = 0$ gives that $\operatorname{Ran} T(0)$ is closed. Let r, s and M satisfy $\overline{\Delta}(0,r) \subset G$, $r \le 1$, $0 < s < \min\{1, \gamma(T(0))\}$ and $M = \max_{z\in\overline{\Delta}(0,r)} \max\{1, \|T(z)\|, \|f(z)\|\}$. Set $U = \Delta(0, \frac{rs}{4Mn})$.

For $z \in \Delta(0,r)$ we can express $T(z)$ and $f(z)$ as power series $T(z) = \sum_{\alpha\in\mathbb{Z}_+^n} T_\alpha z^\alpha$ and $f(z) = \sum_{\alpha\in\mathbb{Z}_+} f_\alpha z^\alpha$, where $f_\alpha \in Y, T_\alpha \in \mathcal{B}(X,Y)$, $\|f_\alpha\| \le \frac{M}{r^{|\alpha|}}$ and $\|T_\alpha\| \le \frac{M}{r^{|\alpha|}}$ for all $\alpha \in \mathbb{Z}_+^n$.

We define a new order $\prec$ on $\mathbb{Z}_+^n$. If $\beta, \gamma \in \mathbb{Z}_+^n$, then write $\beta \prec \gamma$ if either $|\beta| < |\gamma|$, or $|\beta| = |\gamma|$ and there exists $j, 1 \le j \le n$ such that $\beta_i = \gamma_i \quad (1 \le i < j)$ and $\beta_j < \gamma_j$ (i.e., $\prec$ is the lexicographic order on each set $\{\alpha \in \mathbb{Z}_+^n, |\alpha| = \text{const}\}$).

Let $x \in X$ satisfy $T(0)x = f(0)$.

Let $g_0 = x$. We construct inductively (with respect to the order $\prec$) points $g_\alpha \in X \quad (\alpha \in \mathbb{Z}_+^n)$ such that

$$f_\alpha = \sum_{\beta\le\alpha} T_{\alpha-\beta} g_\beta \tag{2}$$

and $\|g_\alpha\| \le \left(\frac{4Mn}{rs}\right)^{|\alpha|} \max\{1, \|x\|\}$.

Let $\alpha \in \mathbb{Z}_+^n, |\alpha| = m \ge 1$ and suppose that we have already found points $g_{\alpha'} \in X \quad (\alpha' \prec \alpha)$ satisfying (2).

For $\varepsilon > 0$ set $w_\varepsilon = \left(\varepsilon^{1+\frac{1}{m+1}}, \varepsilon^{1+\frac{1}{(m+1)^2}}, \dots, \varepsilon^{1+\frac{1}{(m+1)^n}}\right)$. It is clear that $w_\varepsilon \in G$ for all ε small enough. Set $y_\varepsilon = f(w_\varepsilon) - \sum_{\gamma\prec\alpha} w_\varepsilon^\gamma T(w_\varepsilon) g_\gamma$ and let y'_ε be its approximation

$$y'_\varepsilon = \sum_{|\beta|\le m} f_\beta w_\varepsilon^\beta - \sum_{\gamma\prec\alpha} w_\varepsilon^\gamma \sum_{|\delta|\le m-|\gamma|} w_\varepsilon^\delta T_\delta g_\gamma.$$

Obviously, $y_\varepsilon \in \operatorname{Ran} T(w_\varepsilon)$.

We have $\operatorname{card}\{\beta \in \mathbb{Z}_+^n : |\beta| = j\} = \binom{n+j-1}{j}$. For $j \ge n$ we have $\binom{n+j-1}{j} \le \binom{2j}{j} \le 2^{2j} = 4^j$. For $2 \le j < n$ we have $\binom{n+j-1}{j} \le (n+j-1)^j \le (2n)^j$. Thus $\operatorname{card}\{\beta \in \mathbb{Z}_+^n : |\beta| = j\} \le (2n)^j$ for all j.

For all ε small enough we have

$$\begin{aligned}
\|y_\varepsilon - y'_\varepsilon\| &= \left\| \sum_{|\beta|\ge m+1} f_\beta w_\varepsilon^\beta - \sum_{\gamma\prec\alpha} \sum_{|\delta|\ge m-|\gamma|+1} w_\varepsilon^{\delta+\gamma} T_\delta g_\gamma \right\| \\
&\le \sum_{|\beta|\ge m+1} \frac{M}{r^{|\beta|}} \varepsilon^{|\beta|} + \sum_{\gamma\prec\alpha} \sum_{|\delta|\ge m-|\gamma|+1} \varepsilon^{|\delta+\gamma|} \frac{M}{r^{|\delta|}} \cdot \|g_\gamma\| \\
&\le \sum_{k=m+1}^{\infty} M r^{-k} \varepsilon^k (2n)^k + \sum_{\gamma\prec\alpha} \sum_{k=m-|\gamma|+1}^{\infty} \varepsilon^{|\gamma|+k} M r^{-k} (2n)^k \|g_\gamma\| \\
&\le c \cdot \varepsilon^{m+1},
\end{aligned}$$

where c is a constant independent of ε.

Set $u = f_\alpha - \sum_{\substack{\gamma\le\alpha\\ \gamma\ne\alpha}} T_{\alpha-\gamma} g_\gamma$.

Using the induction assumption we have

$$y'_\varepsilon = \sum_{|\beta|\le m} w_\varepsilon^\beta \left(f_\beta - \sum_{\substack{\gamma\le\beta\\ \gamma\prec\alpha}} T_{\beta-\gamma} g_\gamma \right) = w_\varepsilon^\alpha u + \sum_{\substack{|\beta|=m\\ \beta\succ\alpha}} w_\varepsilon^\beta \left(f_\beta - \sum_{\substack{\gamma\le\beta\\ \gamma\prec\alpha}} T_{\beta-\gamma} g_\gamma \right).$$

If $|\beta| = m = |\alpha|, \beta \succ \alpha$, then there exists $j \le n-1$ such that $\beta_j > \alpha_j$ and $\beta_i = \alpha_i \quad (i = 1, \dots, j-1)$. So

$$\left|\frac{w_\varepsilon^\beta}{w_\varepsilon^\alpha}\right| = \varepsilon^{\frac{\beta_j-\alpha_j}{(m+1)^j}} \cdot \varepsilon^{\sum_{i=j+1}^n \frac{\beta_i-\alpha_i}{(m+1)^i}} \le \varepsilon^{\frac{1}{(m+1)^j} - \frac{m}{(m+1)^{j+1}}} \le \varepsilon^{\frac{1}{(m+1)^{j+1}}} \le \varepsilon^b,$$

where $b = (m+1)^{-n}$. Similarly,

$$\left|\frac{\varepsilon^{m+1}}{w_\varepsilon^\alpha}\right| = \varepsilon^{1-\sum_{i=1}^n \frac{\alpha_i}{(m+1)^i}} \le \varepsilon^{\frac{1}{m+1}} \le \varepsilon^b.$$

So

$$\begin{aligned} \left\| \frac{y_\varepsilon}{w_\varepsilon^\alpha} - u \right\| &\le \left\| \frac{y_\varepsilon - y'_\varepsilon}{w_\varepsilon^\alpha} \right\| + \left\| \frac{y'_\varepsilon}{w_\varepsilon^\alpha} - u \right\| \\ &\le c \cdot \varepsilon^b + \sum_{\substack{|\beta|=m\\ \beta\succ\alpha}} \frac{w_\varepsilon^\beta}{w_\varepsilon^\alpha} \left\| f_\beta - \sum_{\substack{\gamma\le\beta\\ \gamma\prec\alpha}} T_{\beta-\gamma} g_\gamma \right\| \le c'\varepsilon^b, \end{aligned}$$

where c' is a constant independent of ε.

Since $\frac{y_\varepsilon}{w_\varepsilon^\alpha} \in \operatorname{Ran} T(w_\varepsilon)$ and $\lim_{\varepsilon\to 0_+} \frac{y_\varepsilon}{w_\varepsilon^\alpha} = u$, we conclude that $u \in \operatorname{Ran} T(0)$.

Thus there exists $g_\alpha \in X$ such that $T_0 g_\alpha = u = f_\alpha - \sum_{\substack{\gamma\le\alpha\\ \gamma\ne\alpha}} T_{\alpha-\gamma} g_\gamma$, i.e., $f_\alpha = \sum_{\gamma\le\alpha} T_{\alpha-\gamma} g_\gamma$.

We can choose the point g_α in such a way that

$$\begin{aligned} \|g_\alpha\| \le s^{-1}\|u\| &\le s^{-1} \left(\frac{M}{r^{|\alpha|}} + \sum_{\substack{\gamma\le\alpha\\ \gamma\ne\alpha}} \frac{M}{r^{|\alpha|-|\gamma|}} \cdot \left(\frac{4Mn}{rs}\right)^{|\gamma|} \cdot \max\{1, \|x\|\} \right) \\ &\le \left(\frac{M}{rs}\right)^{|\alpha|} \max\{1, \|x\|\} \left(1 + \sum_{\substack{\gamma\le\alpha\\ \gamma\ne\alpha}} n^{|\gamma|} 4^{|\gamma|} \right) \\ &\le \left(\frac{M}{rs}\right)^m \max\{1, \|x\|\} \left(1 + \sum_{i=0}^{m-1} n^i 4^i \cdot (2n)^{m-i} \right) \le \left(\frac{4Mn}{rs}\right)^m \max\{1, \|x\|\} \end{aligned}$$

(we used the estimate $\operatorname{card}\{\gamma \le \alpha : |\gamma| = i\} = \operatorname{card}\{\gamma \le \alpha : |\gamma| = m-i\} \le (2n)^{m-i}$).

Define now $g(z) = \sum_{\alpha\in\mathbb{Z}_+^n} g_\alpha z^\alpha$. This series converges for all $z \in U = \Delta(0, \frac{rs}{4Mn})$.

For $z \in U$ we have

$$T(z)g(z) = \sum_{\beta\in\mathbb{Z}_+^n} T_\beta z^\beta \left(\sum_{\gamma\in\mathbb{Z}_+^n} g_\gamma z^\gamma\right) = \sum_{\alpha\in\mathbb{Z}_+^n} z^\alpha \left(\sum_{\gamma\le\alpha} T_{\alpha-\gamma} g_\gamma\right) = \sum_{\alpha\in\mathbb{Z}_+^n} f_\alpha z^\alpha = f(z).$$

□

Theorem 10. *Let G be an open subset of $\mathbb{C}^n$, $w \in G$ and let $T : G \to \mathcal{B}(X,Y)$ be an analytic function such that $\operatorname{Ran} T(w)$ is closed. The following statements are equivalent:*

(i) *T is regular at w;*

(ii) *there exist a neighbourhood U of w, a Banach space Z and an analytic function $S : U \to \mathcal{B}(Z,X)$ such that $\operatorname{Ran} S(z) = \operatorname{Ker} T(z) \quad (z \in U)$, (i.e., the sequence $Z \xrightarrow{S(z)} X \xrightarrow{T(z)} Y$ is exact for $z \in U$);*

(iii) *there exist a neighbourhood U of w, a Banach space Z' and an analytic function $S' : U \to \mathcal{B}(Y,Z')$ such that $\operatorname{Ran} T(z) = \operatorname{Ker} S'(z)$ (i.e., the sequence $X \xrightarrow{T(z)} Y \xrightarrow{S'(z)} Z'$ is exact) and $\operatorname{Ran} S'(z)$ is closed for all $z \in U$.*

(iv) *there exist a neighbourhood U of w, a Banach space Z' and an analytic function $S' : U \to \mathcal{B}(Y,Z')$ such that $\operatorname{Ran} T(z) = \operatorname{Ker} S'(z) \quad (z \in U)$.*

Proof. Clearly, (ii) implies (i) by Example 10.24 (iv).

(i) ⇒ (ii): Let T be regular at w. By Theorem 9, there exists a neighbourhood U_1 of w and a system $\{g_\alpha\}_{\alpha\in\Lambda}$ of analytic functions, $g_\alpha : U_1 \to X$, such that $\{g_\alpha(w), \alpha \in \Lambda\} = \operatorname{Ker} T(w)$ and $T(z)g_\alpha(z) = 0 \quad (z \in U_1, \alpha \in \Lambda)$. Let U_2 be a neighbourhood of w such that $\overline{U_2} \subset U_1$. Let $Z = \ell^1(\Lambda)$, i.e., Z is the space of all complex valued functions $c : \Lambda \to X$ such that $\|c\| := \sum_{\alpha\in\Lambda} \|c(\alpha)\| < \infty$. For $z \in U_2$, define $S(z) : Z \to X$ by

$$S(z)(c) = \sum_{\alpha\in\Lambda} \frac{c(\alpha)g_\alpha(z)}{\max\{1, \|g_\alpha\|_{\overline{U_2}}\}}.$$

Then $S : U_2 \to \mathcal{B}(Z,X)$ is an analytic function, $T(z)S(z) = 0 \quad (z \in U_2)$ and $\operatorname{Ran} S(w) = \operatorname{Ker} T(w)$. By Lemma 3, there exists a neighbourhood U of w such that $\operatorname{Ran} S(z) = \operatorname{Ker} T(z)$ for all $z \in U$.

The implication (i) ⇒ (iii) can be proved as in Theorem 4 using the duality argument.

The implication (iii) ⇒ (iv) is trivial.

(iv) ⇒ (ii): If $w_k \in G$, $w_k \to w$, $y_k \in \operatorname{Ran} T(w_k)$ and $y_k \to y$, then $S(w_k)y_k = 0$ for all k and $\|S(w)y\| = \|S(w)y - S(w_k)y_k\| \le \|S(w)(y - y_k)\| + \|(S(w) - S(w_k))y_k\| \to 0$. Hence $y \in \operatorname{Ker} S(w) = \operatorname{Ran} T(w)$. Thus the conditions of Theorem 9 are satisfied and we can prove (ii) as in the implication (i) ⇒ (ii). □

Let $G \subset \mathbb{C}^n$ be an open set. A subset M of G is called *analytic* if for every $w \in \mathbb{C}^n$ there exists a neighbourhood U of w and a family $\{f_\alpha\}$ of analytic scalar-valued functions defined in U such that $M \cap U = \{z : f_\alpha(z) = 0 \text{ for all } \alpha\}$.

Corollary 11. *Let G be an open subset of $\mathbb{C}^n$, let $T : G \to \mathcal{B}(X,Y)$ and $f : G \to Y$ be analytic functions and let T be regular in G. Then the set $\{z \in G : f(z) \in \operatorname{Ran} T(z)\}$ is analytic.*

Proof. Let $w \in G$ and let U be a neighbourhood of w, M a Banach space and $S : U \to \mathcal{B}(Y,M)$ an analytic function satisfying $\operatorname{Ran} T(z) = \operatorname{Ker} S(z) \quad (z \in U)$, see Theorem 10. For $z \in U$ we have $f(z) \in \operatorname{Ran} T(z)$ if and only if $S(z)f(z) = 0$, which is equivalent to the condition $\langle S(z)f(z), m^* \rangle = 0$ for all $m^* \in M^*$. Thus $\{z \in G : f(z) \in \operatorname{Ran} T(z)\}$ is an analytic set. □

A typical problem of complex analysis is to construct from local solutions (Theorem 9) a global analytic solution. The global version is also true. We state the results without proof since it involves rather advanced techniques of complex analysis (cf. C.11.2). Note that the next theorem applies in particular to any domain in $\mathbb{C}$.

Theorem 12. *Let G be a domain of holomorphy in $\mathbb{C}^n$, let $T : G \to \mathcal{B}(X,Y)$ be an analytic function regular in G. Then:*

(i) *for every analytic function $f : G \to Y$ satisfying $f(z) \in \operatorname{Ran} T(z)$ for each $z \in G$ there exists an analytic function $g : G \to X$ such that $T(z)g(z) = f(z) \quad (z \in G)$;*

(ii) *there exist a Banach space Z and a regular analytic function $S : G \to \mathcal{B}(Z,X)$ such that $\operatorname{Ran} S(z) = \operatorname{Ker} T(z) \quad (z \in G)$;*

(iii) *there exist a Banach space Z' and a regular analytic function $S' : G \to \mathcal{B}(Y,Z')$ such that $\operatorname{Ran} T(z) = \operatorname{Ker} S'(z)$ for all $z \in G$.*

Corollary 13. *Let $G \subset \mathbb{C}^n$ be a domain of holomorphy, let $T : G \to \mathcal{B}(X,Y)$ and $f : G \to Y$ be analytic functions, and let $T(z)$ be onto for all $z \in G$. Then there exists an analytic function $g : G \to X$ such that $T(z)g(z) = f(z) \quad (z \in G)$.*

Proof. If $T(z)$ is onto for every $z \in G$, then T is regular in G. The rest follows from Theorem 12. □

Corollary 14. *Let $\mathcal{A}$ be a Banach algebra, let $G \subset \mathbb{C}$ be an open set and let $a : G \to \mathcal{A}$ be an analytic function such that $a(z)$ is right invertible for every $z \in G$. Then there exists an analytic function $b : G \to \mathcal{A}$ such that $a(z)b(z) = 1 \quad (z \in G)$.*

Proof. Consider the analytic function $z \mapsto L_{a(z)} \in \mathcal{B}(\mathcal{A})$. Since $L_{a(z)}$ is onto for all $z \in G$, by the previous theorem there exists an analytic function $b : G \to \mathcal{A}$ such that $a(z)b(z) = L_{a(z)}b(z) = 1 \quad (z \in G)$. □

Using the algebra with the reversed multiplication we can formulate the analogous result for left inverses.

12 Kato operators

In this section we study the regularity of the function $z \mapsto T - z$ where T is an operator on a Banach space X.

We start with a simple purely algebraic lemma.

Lemma 1. *Let $T \in \mathcal{B}(X)$, $k, n \in \mathbb{N}, k < n$. Suppose that* $\operatorname{Ker} T^k \subset \operatorname{Ran} T^{n-k}$. *Then* $\operatorname{Ker} T^j \subset \operatorname{Ran} T^{n-j}$ *for all* $j, 1 \le j < n$.

Proof. It is sufficient to show two implications:

(i) if $1 \le k \le n-2$ and $\operatorname{Ker} T^k \subset \operatorname{Ran} T^{n-k}$, then $\operatorname{Ker} T^{k+1} \subset \operatorname{Ran} T^{n-k-1}$;

(ii) if $2 \le k \le n-1$ and $\operatorname{Ker} T^k \subset \operatorname{Ran} T^{n-k}$, then $\operatorname{Ker} T^{k-1} \subset \operatorname{Ran} T^{n-k+1}$.

To prove (i), suppose that $1 \le k \le n-2$ and $\operatorname{Ker} T^k \subset \operatorname{Ran} T^{n-k}$. Let $x \in \operatorname{Ker} T^{k+1}$. Then $Tx \in \operatorname{Ker} T^k \subset \operatorname{Ran} T^{n-k}$, so $Tx = T^{n-k}y$ for some $y \in X$. Thus

$$x - T^{n-k-1}y \in \operatorname{Ker} T \subset \operatorname{Ker} T^k \subset \operatorname{Ran} T^{n-k} \subset \operatorname{Ran} T^{n-k-1},$$

and so $x \in \operatorname{Ran} T^{n-k-1}$.

To prove (ii), let $2 \le k \le n-1$, $\operatorname{Ker} T^k \subset \operatorname{Ran} T^{n-k}$ and let $x \in \operatorname{Ker} T^{k-1}$. Then $x \in \operatorname{Ker} T^k \subset \operatorname{Ran} T^{n-k}$, and so $x = T^{n-k}y$ for some $y \in X$. Since $T^k T^{n-k-1} y = T^{k-1}x = 0$, we have $T^{n-k-1}y \in \operatorname{Ker} T^k \subset \operatorname{Ran} T^{n-k}$ and $T^{n-k-1}y = T^{n-k}z$ for some $z \in X$. Thus $x = T^{n-k}y = T(T^{n-k-1}y) = T^{n-k+1}z \in \operatorname{Ran} T^{n-k+1}$. □

For $T \in \mathcal{B}(X)$ write $R^\infty(T) = \bigcap_{n=1}^\infty \operatorname{Ran} T^n$ and $N^\infty(T) = \bigcup_{n=1}^\infty \operatorname{Ker} T^n$. Clearly both $R^\infty(T)$ and $N^\infty(T)$ are linear subspaces of X but in general neither $R^\infty(T)$ nor $N^\infty(T)$ is closed.

Theorem 2. *Let $T \in \mathcal{B}(X)$ and let* $\operatorname{Ran} T$ *be closed. The following conditions are equivalent:*

(i) *the function $z \mapsto \gamma(T - z)$ is continuous at 0;*

(ii) $\limsup_{z\to 0} \gamma(T - z) > 0$;

(iii) $\lim_{z\to 0} \hat{\delta}(\operatorname{Ker} T, \operatorname{Ker}(T - z)) = 0$;

(iv) $\lim_{z\to 0} \hat{\delta}(\operatorname{Ran}(T - z), \operatorname{Ran} T) = 0$;

(v) $\operatorname{Ker} T \subset R^\infty(T)$;

(vi) $N^\infty(T) \subset \operatorname{Ran} T$;

(vii) $N^\infty(T) \subset R^\infty(T)$;

(viii) $\operatorname{Ker} T \subset \overline{R^\infty(T)}$.

Moreover, if any of the previous conditions is satisfied, then $\operatorname{Ran} T^k$ *is closed for all* $k \in \mathbb{N}$.

Proof. The equivalence of the first four condition is true in general, see Theorem 10.17 and Corollary 11.6.

The equivalence of (v), (vi) and (vii) follows from the previous lemma and the implication (v) $\Rightarrow$ (viii) is obvious.

(i) $\Rightarrow$ (v): The function $z \mapsto \gamma(T-z)$ is regular at 0. Let $x \in \operatorname{Ker} T$. By Theorem 11.5, there exist a neighbourhood U of 0 and an analytic function $f : U \to X$ such that $f(0) = x$ and $(T-z)f(z) = 0 \quad (z \in U)$. Let

$$f(z) = \sum_{i=0}^{\infty} x_i z^i \qquad (z \in U)$$

be the Taylor expansion of f. Then $x_0 = x$ and $x_i = Tx_{i+1} \quad (i = 0, 1, \dots)$. Thus $x = x_0 = Tx_1 = T^2 x_2 = \cdots$, and so $x \in R^\infty(T)$.

(vi) $\Rightarrow$ (iii): Let s be a positive number, $s < \gamma(T)$ and let $x \in \operatorname{Ker} T \subset \operatorname{Ran} T$, $\|x\| = 1$. We construct inductively a sequence $x_0 = x, x_1, x_2, \dots$ of points of X such that $Tx_{i+1} = x_i$ and $\|x_{i+1}\| \le s^{-1}\|x_i\|$ for all i (clearly, for every i we have $x_i \in N^\infty(T) \subset \operatorname{Ran} T$). Thus $\|x_i\| \le s^{-i} \quad (i \in \mathbb{N})$. For $|z| < s$ define $f(z) = \sum_{i=0}^{\infty} x_i z^i$. Clearly, $(T-z)f(z) = 0$ and

$$\begin{aligned}\operatorname{dist}\{x, \operatorname{Ker}(T-z)\} \le \|x - f(z)\| = \left\|\sum_{i=1}^{\infty} x_i z^i\right\| \le \sum_{i=1}^{\infty} \|x_i\| \cdot |z^i| \\ \le \sum_{i=1}^{\infty} \left(\frac{|z|}{s}\right)^i = \frac{|z|}{s - |z|}.\end{aligned}$$

Thus $\delta(\operatorname{Ker} T, \operatorname{Ker}(T-z)) \le \frac{|z|}{s-|z|}$ and $\lim_{z\to 0} \delta(\operatorname{Ker} T, \operatorname{Ker}(T-z)) = 0$. Since $\lim_{z\to 0} \delta(\operatorname{Ker}(T-z), \operatorname{Ker} T) = 0$ is true in general by Lemma 10.12, we have (iii).

(viii) $\Rightarrow$ (v): We prove by induction on k that $\operatorname{Ran} T^k$ is closed for all $k \ge 1$. This is assumed for $k = 1$.

Suppose that $k \ge 1$, $\operatorname{Ran} T^k$ is closed and $\operatorname{Ker} T \subset \overline{R^\infty(T)} \subset \operatorname{Ran} T^k$.

Let $u \in \overline{\operatorname{Ran} T^{k+1}}$. By the induction assumption, $u \in \operatorname{Ran} T^k$, and so $u = T^k v$ for some $v \in X$. Furthermore, there are vectors $v_j \in X \quad (j = 1, 2, \dots)$ such that $T^{k+1} v_j \to u \quad (j \to \infty)$. Thus $T(T^k v_j - T^{k-1} v) \to 0$. Consider the operator $\widehat{T} : X/\operatorname{Ker} T \to \operatorname{Ran} T$ induced by T. It is clear that $\widehat{T}$ is bounded below and $\widehat{T}(T^k v_j - T^{k-1} v + \operatorname{Ker} T) \to 0$, so $T^k v_j - T^{k-1} v + \operatorname{Ker} T \to 0 \quad (j \to \infty)$ in the quotient space $X/\operatorname{Ker} T$. Thus there exist vectors $w_j \in \operatorname{Ker} T \subset \operatorname{Ran} T^k$ such that $T^k v_j + w_j \to T^{k-1} v$. Since $\operatorname{Ran} T^k$ is closed, we have $T^{k-1} v \in \operatorname{Ran} T^k$. Hence $u = T^k v \in \operatorname{Ran} T^{k+1}$ and $\operatorname{Ran} T^{k+1}$ is closed.

Thus $R^\infty(T) = \bigcap_{k+1}^{\infty} \operatorname{Ran} T^k$ is also closed and (viii) implies (v). □

Definition 3. Let $T \in \mathcal{B}(X)$. We say that T is *Kato* if $\operatorname{Ran} T$ is closed and T satisfies any of the (equivalent) conditions of the previous theorem.

Corollary 4. *$T \in \mathcal{B}(X)$ is Kato if and only if T^* is Kato. If T is Kato, then $T-z$ is Kato for all z in a neighbourhood of 0. Moreover,* $\dim \operatorname{Ker} T = \lim_{z\to 0} \dim \operatorname{Ker}(T-z)$ *and* $\operatorname{codim} \operatorname{Ran} T = \lim_{z\to 0} \operatorname{codim} \operatorname{Ran}(T-z)$.

Proof. See Corollary 10.23. The last statement follows from Corollary 10.10. □

Example 5. Any operator that is either onto or bounded below is Kato. In particular, the isometrical shift S on a Hilbert space H is Kato. Note that in this case $R^\infty(S) = \{0\} = N^\infty(S)$.

Similarly, S^* is also Kato and $R^\infty(S^*) = H = \overline{N^\infty(S)}$.

The direct sum $S \oplus S^*$ is an example of a Kato operator that is neither onto nor bounded below.

Proposition 6. *Let $T, S \in \mathcal{B}(X)$, $TS = ST$. If TS is Kato, then both T and S are Kato.*

Proof. It is sufficient to show that T is Kato. We have $\operatorname{Ker} T^n \subset \operatorname{Ker}(TS)^n \subset \operatorname{Ran}(TS) \subset \operatorname{Ran} T$ for all n, and so $N^\infty(T) \subset \operatorname{Ran} T$.

It remains to show that $\operatorname{Ran} T$ is closed. Let $x_k \in X$ and $Tx_k \to v$ for some $v \in X$. Then $STx_k \to Sv$, and so $Sv = STu$ for some $u \in X$. Thus $v - Tu \in \operatorname{Ker} S \subset \operatorname{Ker}(TS) \subset \operatorname{Ran}(TS) \subset \operatorname{Ran} T$, and so $v \in \operatorname{Ran} T$. □

Theorem 7. *Let $T \in \mathcal{B}(X)$. The following conditions are equivalent:*

(i) *T is Kato;*

(ii) *T^n is Kato for all $n \in \mathbb{N}$;*

(iii) *T^n is Kato for some $n \in \mathbb{N}$.*

Proof. (ii) ⇒ (iii): Clear.

(iii) ⇒ (i): The implication follows from the preceding proposition.

(i) ⇒ (ii): Since $\operatorname{Ker} T \subset \operatorname{Ker} T^2 \subset \cdots$, we have $N^\infty(T^n) = N^\infty(T)$ for all $n \in \mathbb{N}$. Similarly, $R^\infty(T^n) = R^\infty(T)$. Thus $N^\infty(T^n) \subset R^\infty(T^n)$.

Moreover, $\operatorname{Ran} T^n$ is closed by Theorem 2, and so T^n is Kato for each n. □

To show that the set of all Kato operators is a regularity, we need the following lemma, which will also be useful later.

Lemma 8. *Let A, B, C, D be mutually commuting operators on X such that $AC + BD = I$. Then:*

(i) *for every n there are $C_n, D_n \in \mathcal{B}(X)$ such that A^n, B^n, C_n, D_n are mutually commuting and $A^nC_n + B^nD_n = I$;*

(ii) $\operatorname{Ran}(A^nB^n) = \operatorname{Ran} A^n \cap \operatorname{Ran} B^n$ *and* $\operatorname{Ker}(A^nB^n) = \operatorname{Ker} A^n + \operatorname{Ker} B^n$ *for each n. Consequently, $R^\infty(AB) = R^\infty(A) \cap R^\infty(B)$ and $N^\infty(AB) = N^\infty(A) + N^\infty(B)$;*

(iii) *$N^\infty(A) \subset R^\infty(B)$ and $N^\infty(B) \subset R^\infty(A)$;*

(iv) $\operatorname{Ran}(A^nB^n)$ *is closed if and only if* $\operatorname{Ran} A^n$ *and* $\operatorname{Ran} B^n$ *are closed.*

Proof. (i) We have

$$I = (AC+BD)^{2n-1} = \sum_{i=0}^{2n-1} \binom{2n-1}{i} A^i C^i B^{2n-1-i} D^{2n-1-i} = A^n C_n + B^n D_n$$

for some $C_n, D_n \in \mathcal{B}(X)$ commuting with A^n, B^n.

(ii) Clearly, $\operatorname{Ran}(AB) \subset \operatorname{Ran} A \cap \operatorname{Ran} B$. If $x \in \operatorname{Ran} A \cap \operatorname{Ran} B, x = Au = Bv$ for some $u, v \in X$, then set $w = Cv + Du$. Then

$$Bw = BCv + BDu = Cx + u - ACu = u,$$

and so $x = Au = ABw \in \operatorname{Ran}(AB)$. Thus $\operatorname{Ran}(AB) = \operatorname{Ran} A \cap \operatorname{Ran} B$.

By (i), we have $\operatorname{Ran}(A^n B^n) = \operatorname{Ran} A^n \cap \operatorname{Ran} B^n$ for all n and

$$R^\infty(AB) = \bigcap_n \operatorname{Ran}(A^n B^n) = \bigcap_n (\operatorname{Ran} A^n \cap \operatorname{Ran} B^n) = R^\infty(A) \cap R^\infty(B).$$

Similarly, $\operatorname{Ker} A + \operatorname{Ker} B \subset \operatorname{Ker}(AB)$. If $x \in \operatorname{Ker}(AB)$, then $x = ACx + BDx$, where $ACx \in \operatorname{Ker} B$ and $BDx \in \operatorname{Ker} A$. Thus $\operatorname{Ker}(AB) = \operatorname{Ker} A + \operatorname{Ker} B$ and, by (i), $\operatorname{Ker}(A^n B^n) = \operatorname{Ker} A^n + \operatorname{Ker} B^n$. Hence

$$N^\infty(AB) = \bigcup_n \operatorname{Ker}(A^n B^n) = \bigcup_n (\operatorname{Ker} A^n + \operatorname{Ker} B^n) = N^\infty(A) + N^\infty(B).$$

(iii) If $x \in \operatorname{Ker} A$, then $x = BDx \in \operatorname{Ran} B$. Thus $\operatorname{Ker} A \subset \operatorname{Ran} B$ and, by (i), $\operatorname{Ker} A^n \subset \operatorname{Ran} B^n$ for all n. If $m \geq n$, then $\operatorname{Ker} A^n \subset \operatorname{Ker} A^m \subset \operatorname{Ran} B^m$, so $\operatorname{Ker} A^n \subset R^\infty(B)$. Consequently, $N^\infty(A) \subset R^\infty(B)$. The inclusion $N^\infty(B) \subset R^\infty(A)$ follows by symmetry.

(iv) If $\operatorname{Ran} A^n$ and $\operatorname{Ran} B^n$ are closed, then clearly $\operatorname{Ran}(A^n B^n) = \operatorname{Ran} A^n \cap \operatorname{Ran} B^n$ is closed.

Suppose that $\operatorname{Ran}(A^n B^n)$ is closed and let (x_k) be a sequence of elements of X such that $A^n x_k \to v \in X$. Then $A^n B^n x_k \to B^n v$, and so $B^n v = A^n B^n u$ for some $u \in X$. Thus $v - A^n u \in \operatorname{Ker} B^n \subset \operatorname{Ran} A^n$, and so $v \in \operatorname{Ran} A^n$. Hence $\operatorname{Ran} A^n$ is closed. □

Theorem 9. *The set of all Kato operators is a regularity.*

Proof. The first axiom of regularities was proved in Theorem 7 and one implication of the second axiom follows from Proposition 6. Thus it is sufficient to show that if $A, B, C, D \in \mathcal{B}(X)$ are mutually commuting operators, $AC + BD = I$ and A, B are Kato, then AB is Kato. By Lemma 8, $\operatorname{Ran}(AB) = \operatorname{Ran} A \cap \operatorname{Ran} B$, and so $\operatorname{Ran}(AB)$ is closed. Since A, B are Kato, we have $\operatorname{Ker} A \subset R^\infty(A)$ and $\operatorname{Ker} B \subset R^\infty(B)$. By Lemma 8, $\operatorname{Ker} A \subset R^\infty(B)$ and $\operatorname{Ker} B \subset R^\infty(A)$. Thus

$$\operatorname{Ker}(AB) = \operatorname{Ker} A + \operatorname{Ker} B \subset R^\infty(A) \cap R^\infty(B) = R^\infty(AB).$$

Hence AB is Kato. □

Definition 10. For $T \in \mathcal{B}(X)$ denote by $\sigma_K(T) = \{\lambda \in \mathbb{C} : T - \lambda \text{ is not Kato}\}$ the *Kato spectrum* corresponding to the regularity of all Kato operators.

Theorem 11. *Let $T \in \mathcal{B}(X)$. Then:*

(i) *$\sigma_K(T)$ is a non-empty compact subset of the complex plane;*

(ii) *$\partial\sigma(T) \subset \sigma_K(T) \subset \sigma_\pi(T) \cap \sigma_\delta(T) \subset \sigma(T)$;*

(iii) *$\sigma_K(f(T)) = f(\sigma_K(T))$ for every function f analytic on a neighbourhood of $\sigma(T)$;*

(iv) *$\sigma_K(T^*) = \sigma_K(T)$.*

Proof. (i) and (ii): The set of all regularity points of the function $z \mapsto T - z$ is open, so $\sigma_K(T)$ is closed. Since the operators that are bounded below or onto are Kato, we have $\sigma_K(T) \subset \sigma_\pi(T) \cap \sigma_\delta(T)$. To show that $\sigma_K(T)$ is non-empty, it is sufficient to prove $\partial\sigma(T) \subset \sigma_K(T)$. Suppose on the contrary that there exists $\lambda \in \partial\sigma(T)$ and $\lambda \notin \sigma_K(T)$. By Corollary 4, $\dim \operatorname{Ker}(T - \lambda) = 0 = \operatorname{codim} \operatorname{Ran}(T - \lambda)$. Thus $T - \lambda$ is invertible, a contradiction.

(iii) If $X = X_1 \oplus X_2$ for closed subspaces X_1, X_2 of X and $T_1 \in \mathcal{B}(X_1)$, $T_2 \in \mathcal{B}(X_2)$, then $\sigma_K(T_1 \oplus T_2) = \sigma_K(T_1) \cup \sigma_K(T_2)$. By (i), $\sigma_K(T_1) \neq \emptyset$ whenever $X_1 \neq \{0\}$, and so the spectral mapping theorem follows from Theorem 6.8.

(iv) is clear from Corollary 4. □

Theorem 12. *Let $T \in \mathcal{B}(X)$ and let λ be a complex number. Then the limit $\lim_{z\to\lambda} \gamma(T - z)$ exists and $\lambda \in \sigma_K(T)$ if and only if $\lim_{z\to\lambda} \gamma(T - z) = 0$.*

Proof. See Corollary 11.6.

Theorem 13. *The set $\{\lambda \in \sigma_K(T) : \operatorname{Ran}(T - \lambda) \text{ is closed}\}$ is at most countable.*

Proof. See Corollary 11.7.

Examples 14. (i) By the preceding theorem, the set

$$\{\lambda \in \sigma_K(T) : \operatorname{Ran}(T - \lambda) \text{ is closed}\}$$

is at most countable. The following example shows that this set can contain a convergent sequence.

Let H be a separable infinite-dimensional Hilbert space, let $T \in \mathcal{B}(H \oplus H)$ be given in the matrix form by

$$T = \begin{pmatrix} D & I \\ 0 & 0 \end{pmatrix},$$

where D (with respect to an orthonormal basis) is a diagonal operator, $D = \operatorname{diag}(1, 1/2, 1/3, \dots)$. Note that $\operatorname{Ran}(T - z) = H \oplus H$ for $z \notin \{0\} \cup \{1/n : n \in \mathbb{N}\}$, $\operatorname{codim} \operatorname{Ran}(T - 1/n) = 1$ and $\operatorname{Ran} T = H \oplus 0$. Thus $\operatorname{Ran}(T - \lambda)$ is closed for all λ and the Kato spectrum $\sigma_K(T) = \{0\} \cup \{1/n : n \in \mathbb{N}\}$ contains a convergent sequence.

(ii) The product of two commuting Kato operators need not be Kato in general. Thus the set of all Kato operators does not satisfy condition (P1) of Section 6. In particular, it is not possible to extend reasonably the Kato spectrum to n-tuples of commuting operators.

Let H be a Hilbert space with an orthonormal basis $\{e_{i,j} : i,j \in \mathbb{Z}, ij \le 0\}$. Define $T \in \mathcal{B}(H)$ by $Te_{i,j} = 0$ if $i = 0, j > 0$ and $Te_{i,j} = e_{i+1,j}$ otherwise.

Similarly, let $Se_{i,j} = 0$ if $j = 0, i > 0$ and $Se_{i,j} = e_{i,j+1}$ otherwise. It is easy to verify that $TS = ST$, both T and S are Kato but their product TS is not.

(iii) The set of all Kato operators is not open, so the Kato spectrum is not upper semicontinuous.

Let H be a Hilbert space with an orthonormal basis $\{e_{i,j} : i \ge 1, j \ge 0\}$. Let $T \in \mathcal{B}(H)$ be defined by $Te_{i,j} = e_{i,j+1}$, i.e., T is a direct sum of countably many isometrical shifts. Let $S \in \mathcal{B}(H)$ be defined by $Se_{i,0} = i^{-1}e_{i,0}$ and $Se_{i,j} = 0$ for $j \ge 1$.

For each $\varepsilon > 0$, the range of $T + \varepsilon S$ is non-closed. So T is Kato while $T + \varepsilon S$ is not.

Theorem 15. *Let $T \in \mathcal{B}(X)$ be a Kato operator. Then:*

(i) *$R^\infty(T)$ is closed;*
(ii) *if $x \in X$ and $Tx \in R^\infty(T)$, then $x \in R^\infty(T)$;*
(iii) *$TR^\infty(T) = R^\infty(T)$.*

Proof. (i) Since $R^\infty(T) = \bigcap_n \operatorname{Ran} T^n$ and $\operatorname{Ran} T^n$ is closed for all n by Theorem 2, we have (i).

(ii) Let $n \in \mathbb{N}$. Then $Tx = T^{n+1}y$ for some $y \in X$. Thus $x - T^n y \in \operatorname{Ker} T \subset \operatorname{Ran} T^n$, and so $x \in \operatorname{Ran} T^n$. Since n was arbitrary, we have $x \in R^\infty(T)$.

(iii) Clearly, $TR^\infty(T) \subset R^\infty(T)$. If $x \in R^\infty(T)$, then $x = Ty$ for some $y \in X$. By (ii), $y \in R^\infty(T)$, and so $R^\infty(T) = TR^\infty(T)$. □

Remark 16. The space $N^\infty(T)$ need not be closed even for Kato operators. The simplest example is the backward shift in a separable Hilbert space.

Theorem 17. *Let T be a Kato operator on a Banach space X. Then:*

(i) *$R^\infty(T^*) = N^\infty(T)^\perp$;*
(ii) *$R^\infty(T) = {}^\perp N^\infty(T^*)$;*
(iii) *$\overline{N^\infty(T)} = {}^\perp R^\infty(T^*)$;*
(iv) *$\overline{N^\infty(T^*)}^{w^*} = R^\infty(T)^\perp$.*

Proof. (i) We have

$$N^\infty(T)^\perp = \Big(\bigcup_{n=1}^\infty \operatorname{Ker} T^n\Big)^\perp = \bigcap_{n=1}^\infty (\operatorname{Ker} T^n)^\perp = \bigcap_{n=1}^\infty \operatorname{Ran} T^{*n} = R^\infty(T^*).$$

(iii) ${}^\perp R^\infty(T^*) = {}^\perp\Big(N^\infty(T)^\perp\Big) = \overline{N^\infty(T)}.$

(ii) and (iv) can be proved similarly using A.1.11 and A.1.12. □

Theorem 18. *Let $T \in \mathcal{B}(X)$ be Kato and let G be an open connected subset of $\mathbb{C} \setminus \sigma_K(T)$ containing 0. Then:*

(i) $R^\infty(T) = \bigcap_{z\in G} \operatorname{Ran}(T-z)$;

(ii) $\overline{N^\infty(T)} = \bigvee_{z\in G} \operatorname{Ker}(T-z)$.

Proof. (i) Suppose that $x \in \bigcap_{z\in G} \operatorname{Ran}(T-z)$. By Theorem 11.5, there exist a neighbourhood U of 0 and an analytic function $g : U \to X$ satisfying $(T-z)g(z) = x \quad (z \in U)$. Let $g(z) = \sum_{i=0}^\infty x_i z^i$ be the Taylor expansion of g about 0. The equality $(T-z)g(z) = x$ implies that $Tx_0 = x$ and $Tx_i = x_{i-1}$ for all $i \geq 1$. Thus $x = Tx_0 = T^2x_1 = T^3x_2 = \cdots$, and so $x \in R^\infty(T)$.

In the opposite direction, let $x \in R^\infty(T)$. Let s be a positive number, $s < \gamma(T)$. By Theorem 15, we can find inductively a sequence of points $x_k \in R^\infty(T)$ such that $Tx_0 = x$, $Tx_k = x_{k-1} \quad (k = 1, 2, \dots)$ and $\|x_k\| \leq s^{-1}\|x_{k-1}\|$. Then the series $g(z) = \sum_{i=0}^\infty x_k z^k$ converges for $|z| < s$ and $(T-z)g(z) = x$. Thus $x \in \operatorname{Ran}(T-z)$ for all $|z| < s$, and so for all $z \in G$, by Corollary 11.8.

(ii) We have

$$\overline{N^\infty(T)} = {}^\perp(R^\infty(T^*)) = {}^\perp\Big(\bigcap_{z\in G} \operatorname{Ran}(T^*-z)\Big)$$
$$\supset \bigvee_{z\in G} {}^\perp \operatorname{Ran}(T^*-z) = \bigvee_{z\in G} \operatorname{Ker}(T-z).$$

Conversely, let $n \in \mathbb{N}$ and $x \in \operatorname{Ker} T^n$. Let s be a positive number such that $s < \gamma(T)$ and $\{z \in \mathbb{C} : |z| < s\} \subset G$. Set $x_n = x$, $x_{n-k} = T^k x \quad (k = 1, 2, \dots, n-1)$. Since $x \in \operatorname{Ker} T^n \subset R^\infty(T)$, we can find inductively points $x_{n+1}, x_{n+2}, \dots \in R^\infty(T)$ such that $Tx_{i+1} = x_i$ and $\|x_{i+1}\| \leq s^{-1}\|x_i\|$. Set $g(z) = \sum_{i=1}^\infty x_i z^i$. This series converges for $|z| < s$ and $(T-z)g(z) = 0$. Thus $g(z) \in \operatorname{Ker}(T-z)$ and, by the Cauchy formula,

$$x = x_n = \frac{1}{2\pi i}\int_{|z|=s/2} \frac{g(z)}{z^{n+1}}\,\mathrm{d}z \in \bigvee_{z\in G} \operatorname{Ker}(T-z).$$ □

Corollary 19. *Let $T \in \mathcal{B}(X)$ and let G be a component of $\mathbb{C} \setminus \sigma_K(T)$. Then:*

(i) *the mapping $\lambda \mapsto R^\infty(T-\lambda)$ is constant on G;*
(ii) *the mapping $\lambda \mapsto \overline{N^\infty(T-\lambda)}$ is constant on G;*
(iii) *for all $\lambda \in G$ and $x \in \operatorname{Ker}(T-\lambda)$ there exists an analytic function $f: G \to X$ such that $f(\lambda) = x$ and $(T-z)f(z) = 0 \quad (z \in G)$.*

Proof. The first two statements follow from the previous theorem, the constants are $\bigcap_{z\in G} \operatorname{Ran}(T-z)$ and $\bigvee_{z\in G} \operatorname{Ker}(T-z)$, respectively.

(iii) follows from Theorem 11.12. □

In fact, if $T \in \mathcal{B}(X)$ is Kato, then $R^\infty(T+S) = R^\infty(T)$ and $\overline{N^\infty(T+S)} = \overline{N^\infty(T)}$ for all S commuting with T such that $\|S\|$ is small enough. We postpone the proof to Section 21.

Theorem 20. *Let $T \in \mathcal{B}(X)$ be an operator with closed range. The following conditions are equivalent:*

(i) *T is Kato;*
(ii) *$\operatorname{Ran} T \supset \bigcap_{z\neq 0} \overline{\operatorname{Ran}(T-z)}$;*
(iii) *$\operatorname{Ker} T \subset \bigvee_{z\neq 0} \operatorname{Ker}(T-z)$.*

Proof. (i) ⇒ (ii) and (i) ⇒ (iii): Let G be the connected component of $\mathbb{C} \setminus \sigma_K(T)$ containing 0. Let G' be an open subset of G such that $0 \notin G'$ and let $w \in G'$. Then

$$\operatorname{Ran} T \supset R^\infty(T) = R^\infty(T-w) = \bigcap_{z\in G'}^{\infty} \operatorname{Ran}(T-z) \supset \bigcap_{z\neq 0} \overline{\operatorname{Ran}(T-z)}$$

and

$$\operatorname{Ker} T \subset \overline{N^\infty(T)} = \overline{N^\infty(T-w)} = \bigvee_{z\in G'} \operatorname{Ker}(T-z) \subset \bigvee_{z\neq 0} \operatorname{Ker}(T-z).$$

(iii) ⇒ (i): Let $\lambda \neq 0$ and $x \in \operatorname{Ker}(T-\lambda)$. Then $Tx = \lambda x$ and $x = \frac{T^n x}{\lambda^n} \in \operatorname{Ran} T^n$, so $x \in R^\infty(T)$. Thus $\operatorname{Ker} T \subset \bigvee_{z\neq 0} \operatorname{Ker}(T-z) \subset \overline{R^\infty(T)}$ and T is Kato.

(ii) ⇒ (i): Let $x \in \operatorname{Ker} T^n$ and $z \neq 0$. Then

$$(T-z)(T^{n-1} + zT^{n-2} + \cdots + z^{n-1})x = T^n x - z^n z = -z^n x,$$

and so $x \in \operatorname{Ran}(T-z)$. Thus $N^\infty(T) \subset \bigcap_{z\neq 0} \operatorname{Ran}(T-z) \subset \operatorname{Ran} T$ and T is Kato. □

Theorem 21. *Let $T \in \mathcal{B}(X)$. Then T is Kato if and only if there exists a closed subspace $M \subset X$ invariant with respect to T such that $T|M$ is onto and the operator $\widehat{T}: X/M \to X/M$ induced by T is bounded below.*

For the space M it is possible to take $R^\infty(T)$.

Proof. Let T be Kato and set $M = R^\infty(T)$. Then M is closed. By Theorem 15, $TM = M$ and the operator $\widehat{T} : X/M \to X/M$ induced by T is one-to-one.

Moreover, $\operatorname{Ran} T$ is closed and $M \subset \operatorname{Ran} T$. Then $\operatorname{Ran} \widehat{T} = \operatorname{Ran} T + M$, which is is closed. Indeed, let $x, x_n \in X$ and $Tx_n + M \to x + M$ in X/M. There are $m_n \in M$ such that $Tx_n + m_n \to x$. Thus $x \in \operatorname{Ran} T$ and $x + M \in \operatorname{Ran} \widehat{T}$. Hence $\widehat{T}$ is bounded below.

Conversely, let M be a subspace of X with the required properties. The condition $TM = M$ implies that $M \subset R^\infty(T)$. If $Tx = 0$, then $\widehat{T}(x + M) = 0$ and the injectivity of $\widehat{T}$ implies that $x \in M$. Thus $\operatorname{Ker} T \subset M \subset R^\infty(T)$.

It remains to prove that T has closed range. Let $Q : X \to X/M$ be the canonical projection. We show that $\operatorname{Ran} T = Q^{-1} \operatorname{Ran} \widehat{T}$. If $y \in \operatorname{Ran} T, y = Tx$ for some $x \in X$, then $Qy = Tx + M = \widehat{T}(x + M) \in \operatorname{Ran} \widehat{T}$, and so $\operatorname{Ran} T \subset Q^{-1} \operatorname{Ran} \widehat{T}$. If $y \in X$ and $Qy \in \operatorname{Ran} \widehat{T}$, then $y + M = Tx + M$ for some $x \in X$, and so $y \in Tx + M \subset \operatorname{Ran} T$, since $M \subset \operatorname{Ran} T$. Thus $\operatorname{Ran} T = Q^{-1} \operatorname{Ran} \widehat{T}$, which is closed, since $\operatorname{Ran} \widehat{T}$ is closed and Q continuous. □

Corollary 22. *The regularity of all Kato operators satisfies* (P4), *so σ_K is continuous on commuting elements.*

Proof. Let $T \in \mathcal{B}(X)$ be Kato. Let $\varepsilon = \inf\{|z| : T - z \text{ is not Kato}\}$ and $M = R^\infty(T)$. Since $R^\infty(T - \lambda) = M$ for $|\lambda| < \varepsilon$, we have $(T - \lambda)M = M$ and the induced operator $\widehat{T - \lambda} : X/M \longrightarrow X/M$ is bounded below.

If $UT = TU$ and $\|U\| < \varepsilon$, then $UM \subset M$ and we can define the operator $\widehat{U} : X/M \to X/M$ induced by U. Clearly, $\|\widehat{U}\| \le \|U\| < \varepsilon$. By Theorem 7.14 (ii) for the spectral systems σ_π and σ_δ, we conclude that $(T + U)M = M$ and $\widehat{T} + \widehat{U}$ is bounded below. By Theorem 21, $T + U$ is Kato.

By Theorem 6.11, the regularity of all Kato operators satisfies property (P4). □

Corollary 23. *Let $T, Q \in \mathcal{B}(X)$, $TQ = QT$, let T be Kato and Q quasinilpotent. Then $T + Q$ is Kato.*

Proof. Let $M = R^\infty(T)$. Then $QM \subset M$. Let $\widehat{T}$ and $\widehat{Q}$ be the operators acting in X/M induced by T and Q, respectively. By the spectral radius formula, $Q_1 = Q|M$ is a quasinilpotent operator commuting with $T_1 = T|M$ and $\widehat{Q}$ is a quasinilpotent operator commuting with $\widehat{T}$. Theorem 7.16 for the spectral systems σ_π and σ_δ implies that $T_1 + Q_1$ is onto and $\widehat{T} + \widehat{Q}$ is bounded below. Consequently, $T + Q$ is Kato. □

Theorem 24. *Let $T \in \mathcal{B}(X)$ be Kato. Then $\gamma(T^{m+n}) \ge \gamma(T^m) \cdot \gamma(T^n)$ for all $m, n \in \mathbb{N}$.*

Proof. Fix $\varepsilon > 0$ and let $x \in \operatorname{Ran} T^{n+m}$. Then there exists $y \in X$ such that $T^m y = x$ and $\|y\| \le (\gamma(T^m) - \varepsilon)^{-1} \|x\|$. Further, $x = T^{n+m} z$ for some $z \in X$.

Thus $T^n z - y \in \operatorname{Ker} T^m \subset \operatorname{Ran} T^n$, and so $y \in \operatorname{Ran} T^n$. There exists $u \in X$ such that $T^n u = y$ and

$$\|u\| \le (\gamma(T^n) - \varepsilon)^{-1}\|y\| \le (\gamma(T^n) - \varepsilon)^{-1}(\gamma(T^m) - \varepsilon)^{-1}\|x\|.$$

Hence $\gamma(T^{n+m}) \ge (\gamma(T^n) - \varepsilon)(\gamma(T^m) - \varepsilon)$. Letting $\varepsilon \to 0$ yields $\gamma(T^{n+m}) \ge \gamma(T^n) \cdot \gamma(T^m)$. □

Lemma 25. *Let $T \in \mathcal{B}(X)$ be Kato. Suppose that T is neither onto nor bounded below. Let $M = R^\infty(T)$, let $T_1 : M \to M$ and $\widehat{T} : X/M \to X/M$ be the operators induced by T. Then the limit $\lim_{n\to\infty} \gamma(T^n)^{1/n}$ exists and*

$$\lim_{n\to\infty} \gamma(T^n)^{1/n} = \min\Big\{\lim_{n\to\infty} k(T_1^n)^{1/n}, \lim_{n\to\infty} j(\widehat{T}^n)^{1/n}\Big\}.$$

Proof. By the preceding theorem and Lemma 1.21, the limit $\lim_{n\to\infty} \gamma(T^n)^{1/n}$ exists. By Theorem 21, T_1 is onto and $\widehat{T}$ is bounded below, so the limits on the right-hand side exist by Theorem 9.25.

Since $\operatorname{Ker} T^n \subset R^\infty(T) = M$, we have $\operatorname{Ker} T_1^n = \operatorname{Ker} T^n$. Thus

$$\begin{aligned} k(T_1^n) = \gamma(T_1^n) &= \inf\left\{\frac{\|T_1^n x\|}{\operatorname{dist}\{x, \operatorname{Ker} T_1^n\}} : x \in M \setminus \operatorname{Ker} T_1^n\right\} \\ &= \inf\left\{\frac{\|T^n x\|}{\operatorname{dist}\{x, \operatorname{Ker} T^n\}} : x \in M \setminus \operatorname{Ker} T^n\right\} \ge \gamma(T^n). \end{aligned}$$

Since $TM = M$, we have

$$\begin{aligned} j(\widehat{T}^n) &= \inf\left\{\frac{\|\widehat{T}^n(x+M)\|}{\|x+M\|} : x \notin M\right\} = \inf\left\{\frac{\|T^n x + M\|}{\operatorname{dist}\{x, M\}} : x \notin M\right\} \\ &= \inf\left\{\frac{\|T^n x + T^n m\|}{\operatorname{dist}\{x, M\}} : x \notin M, m \in M\right\} = \inf\left\{\frac{\|T^n y\|}{\operatorname{dist}\{y, M\}} : y \notin M\right\} \\ &\ge \inf\left\{\frac{\|T^n y\|}{\operatorname{dist}\{y, \operatorname{Ker} T^n\}} : y \notin M\right\} \ge \gamma(T^n). \end{aligned}$$

Thus $\gamma(T^n) \le \min\{k(T_1^n), j(\widehat{T}^n)\}$ and

$$\lim_{n\to\infty} \gamma(T^n)^{1/n} \le \min\Big\{\lim_{n\to\infty} k(T_1^n)^{1/n}, \lim_{n\to\infty} j(\widehat{T}^n)^{1/n}\Big\}.$$

To prove the opposite inequality, let

$$0 < s < \min\Big\{\lim_{n\to\infty} k(T_1^n)^{1/n}, \lim_{n\to\infty} j(\widehat{T}^n)^{1/n}\Big\}.$$

We prove that $\lim_{n\to\infty} \gamma(T^n)^{1/n} \ge s$.

Let $n \ge 1$, $x = x_0 \in \operatorname{Ran} T^n$, $\|x\| = 1$ and let $\varepsilon > 0$. Then $x + M \in \operatorname{Ran} \widehat{T}^n$ and

$$\|\widehat{T}^{-i}(x+M)\| \le j(\widehat{T}^i)^{-1}\|x+M\| \le j(\widehat{T}^i)^{-1} \qquad (i = 1, \dots, n).$$

Thus there exist vectors $x_i \in \widehat{T}^{-i}(x+M)$ such that

$$\|x_i\| \le j(\widehat{T}^i)^{-1}(1+\varepsilon) \qquad (i=1,\dots,n).$$

For $i=0,\dots,n-1$ write $m_i = Tx_{i+1} - x_i$. Then

$$\|m_i\| \le \|T\|\cdot\|x_{i+1}\| + \|x_i\| \le (1+\varepsilon)\big(\|T\| j(\widehat{T}^{i+1})^{-1} + j(\widehat{T}^i)^{-1}\big)$$

for all $i=0,\dots,n-1$. Further, $\widehat{T}^i(m_i+M) = T^{i+1}x_{i+1} - T^i x_i + M = M$, so $m_i \in M$ for each i. We have

$$\sum_{i=0}^{n-1} T^i m_i = (T^n x_n - T^{n-1}x_{n-1}) + (T^{n-1}x_{n-1} - T^{n-2}x_{n-2}) + \cdots$$
$$\cdots + (Tx_1 - x_0) = T^n x_n - x.$$

Since $T_1 : M \to M$ is onto, there exist vectors $m_i' \in M$ such that $T^{n-i}m_i' = m_i$ and $\|m_i'\| \le (1+\varepsilon)k(T_1^{n-i})^{-1}\|m_i\|$. Thus

$$T^n\Big(x_n - \sum_{i=0}^{n-1} m_i'\Big) = T^n x_n - \sum_{i=0}^{n-1} T^i m_i = x$$

and

$$\Big\|x_n - \sum_{i=0}^{n-1} m_i'\Big\| \le (1+\varepsilon)j(\widehat{T}^n)^{-1} + \sum_{i=0}^{n-1}(1+\varepsilon)^2 k(T_1^{n-i})^{-1}\Big(\|T\| j(\widehat{T}^{i+1})^{-1} + j(\widehat{T}^i)^{-1}\Big).$$

Thus

$$\gamma(T^n)^{-1} \le (1+\varepsilon)j(\widehat{T}^n)^{-1} + \sum_{i=0}^{n-1}(1+\varepsilon)^2 k(T_1^{n-i})^{-1}\Big(\|T\| j(\widehat{T}^{i+1})^{-1} + j(\widehat{T}^i)^{-1}\Big).$$

Find n_0 such that $k(T_1^i) \ge s^i$ and $j(\widehat{T}^i) \ge s^i$ for all $i \ge n_0$. Set

$$K = \max_{1\le i\le n_0+1} \max\big\{k(T_1^i)^{-1}, j(\widehat{T}^i)^{-1}, s^{-i}\big\}.$$

For n large enough we have

$$\begin{aligned}
\gamma(T^n)^{-1} &\le (1+\varepsilon)^2\Big(s^{-n} + \sum_{i=0}^{n_0-1} s^{i-n}\big(\|T\|\cdot K + K\big) \\
&\quad + \sum_{i=n_0}^{n-n_0-1} s^{i-n}\big(\|T\| s^{-i-1} + s^{-i}\big) + \sum_{i=n-n_0}^{n-1} K\big(\|T\| s^{-i-1} + s^{-i}\big)\Big) \\
&\le (1+\varepsilon)^2 s^{n_0-n}\Big(K + 2n_0 K(\|T\|\cdot K + K) + (n-2n_0)(K\cdot\|T\| + K)\Big) \\
&\le (1+\varepsilon)^2 s^{n_0-n} n \cdot K',
\end{aligned}$$

where K' is a constant independent of n. Hence

$$\lim_{n\to\infty} \gamma(T^n)^{1/n} \geq \lim_{n\to\infty} s^{\frac{n-n_0}{n}} = s.$$

Letting

$$s \to \min\Big\{\lim_{n\to\infty} k(T_1^n)^{1/n}, \lim_{n\to\infty} j(\widehat{T}^n)^{1/n}\Big\}$$

yields the required equality. □

Theorem 26. *Let $T \in B(X)$ be Kato. Then*

$$\operatorname{dist}\{0, \sigma_K(T)\} = \lim_{n\to\infty} \gamma(T^n)^{1/n} = \sup_n \gamma(T^n)^{1/n}.$$

Proof. By Theorem 24 and Lemma 1.21, the limit $\lim \gamma(T^n)^{1/n}$ exists and equals to the supremum.

Set $r = \operatorname{dist}\{0, \sigma_K(T)\}$. Suppose first that T is neither onto nor bounded below. Let $M = R^\infty(T)$, let $T_1 = T|M$ and $\widehat{T} : X/M \to X/M$ be the operators induced by T. If λ is a complex number satisfying

$$|\lambda| < \lim_{n\to\infty} \gamma(T^n)^{1/n} = \min\Big\{\lim_{n\to\infty} k(T_1^n)^{1/n}, \lim_{n\to\infty} j(\widehat{T}^n)^{1/n}\Big\},$$

then $T_1 - \lambda$ is onto and $\widehat{T} - \lambda$ is bounded below. Thus $T - \lambda$ is Kato by Theorem 21 and $\lim_{n\to\infty} \gamma(T^n)^{1/n} \leq r$.

Conversely, by Corollary 19, $R^\infty(T - \lambda) = M$ for $|\lambda| < r$. If $|\lambda| < r$, then $(T - \lambda)M = M$ and $\widehat{T} - \lambda = \widehat{T - \lambda} : X/M \to X/M$ is bounded below. Thus $\lim_{n\to\infty} k(T_1^n)^{1/n} \geq r$ and $\lim_{n\to\infty} j(\widehat{T}^n)^{1/n} \geq r$. Hence $\lim_{n\to\infty} \gamma(T^n)^{1/n} \geq r$ by Lemma 25.

If T is bounded below, then $\lim \gamma(T^n)^{1/n} = \operatorname{dist}\{0, \sigma_\pi(T)\} \leq \operatorname{dist}\{0, \sigma_K(T)\}$ by Theorem 9.25. Moreover, $\overline{N^\infty(T - z)} = \overline{N^\infty(T)} = \{0\}$ for all z with $|z| < r$, and so $\operatorname{dist}\{0, \sigma_\pi(T)\} = \operatorname{dist}\{0, \sigma_K(T)\}$.

Similarly, if T is onto, then $\lim \gamma(T^n)^{1/n} = \operatorname{dist}\{0, \sigma_\delta(T)\} = \operatorname{dist}\{0, \sigma_K(T)\}$. □

Theorem 27. *Let $T \in \mathcal{B}(X)$ be Kato. Then*

$$\gamma(T) - \gamma(T - z) \leq 3|z|$$

for all $z \in \mathbb{C}$.

Proof. The inequality is trivial if $|z| \geq \gamma(T)$. By the previous theorem, the function $z \mapsto T - z$ is regular in $\{z : |z| < \gamma(T)\}$. Thus the inequality for $|z| < \gamma(T)$ follows from Theorem 10.25. □

Proposition 28. *Let $T, S \in \mathcal{B}(X)$, $\lambda \in \mathbb{C}$, $\lambda \neq 0$ and $k \in \mathbb{N}$. Then:*

(i) $\operatorname{Ker}(ST-\lambda)^k = S\operatorname{Ker}(TS-\lambda)^k$;

(ii) $\operatorname{Ran}(ST-\lambda)^k = T^{-1}\operatorname{Ran}(TS-\lambda)^k$;

(iii) $\dim\operatorname{Ker}(ST-\lambda)^k = \dim\operatorname{Ker}(TS-\lambda)^k$;

(iv) $\dim\operatorname{Ran}(ST-\lambda)^k = \dim\operatorname{Ran}(TS-\lambda)^k$.

Proof. We have $(ST-\lambda)S = S(TS-\lambda)$ and so $(ST-\lambda)^k S = S(TS-\lambda)^k$ for each $k \geq 0$.

(i) If $x \in \operatorname{Ker}(TS-\lambda)^k$, then $(ST-\lambda)^k Sx = S(TS-\lambda)^k x = 0$. Hence $S\operatorname{Ker}(TS-\lambda)^k \subset \operatorname{Ker}(ST-\lambda)^k$ for each k.

We prove the opposite inclusion by induction on k. This is clear for $k = 0$.

Suppose that $S\operatorname{Ker}(TS-\lambda)^k = \operatorname{Ker}(ST-\lambda)^k$ and let $x \in \operatorname{Ker}(ST-\lambda)^{k+1}$. Then $(TS-\lambda)^{k+1}Tx = T(ST-\lambda)^{k+1}x = 0$. So $Tx \in \operatorname{Ker}(TS-\lambda)^{k+1}$ and $STx \in S\operatorname{Ker}(TS-\lambda)^{k+1}$. By induction assumption, $(ST-\lambda)x \in \operatorname{Ker}(ST-\lambda)^k = S\operatorname{Ker}(TS-\lambda)^k \subset S\operatorname{Ker}(TS-\lambda)^{k+1}$, and so

$$x = -\lambda^{-1}(ST-\lambda)x + \lambda^{-1}STx \in S\operatorname{Ker}(TS-\lambda)^{k+1}.$$

Hence $\operatorname{Ker}(ST-\lambda)^{k+1}S = S\operatorname{Ker}(TS-\lambda)^{k+1}$.

(iii) Since $\operatorname{Ker}(ST-\lambda)^k = S\operatorname{Ker}(TS-\lambda)^k$, we have $\dim\operatorname{Ker}(ST-\lambda)^k \leq \dim\operatorname{Ker}(TS-\lambda)^k$. The equality follows by symmetry.

(ii) Let $x \in \operatorname{Ran}(ST-\lambda)^k$, i.e., $x = (ST-\lambda)^k y$ for some $y \in X$. Then $Tx = T(ST-\lambda)^k y = (TS-\lambda)^k Ty \in \operatorname{Ran}(TS-\lambda)^k$. Hence $\operatorname{Ran}(ST-\lambda)^k \subset T^{-1}\operatorname{Ran}(TS-\lambda)^k$ for all k.

We prove the opposite inclusion by induction on k. This is clear for $k = 0$. Suppose that $\operatorname{Ran}(ST-\lambda)^k = T^{-1}\operatorname{Ran}(TS-\lambda)^k$ and let $x \in T^{-1}\operatorname{Ran}(TS-\lambda)^{k+1}$. Then $Tx = (TS-\lambda)^{k+1}y$ for some $y \in X$ and $STx = S(TS-\lambda)^{k+1}y = (ST-\lambda)^{k+1}Sy \in \operatorname{Ran}(ST-\lambda)^{k+1}$.

By the induction assumption, $x \in T^{-1}\operatorname{Ran}(TS-\lambda)^{k+1} \subset T^{-1}\operatorname{Ran}(TS-\lambda)^k = \operatorname{Ran}(ST-\lambda)^k$, and so $x = -\lambda^{-1}(ST-\lambda)x + \lambda^{-1}STx \in \operatorname{Ran}(ST-\lambda)^{k+1}$. Hence $\operatorname{Ran}(ST-\lambda)^{k+1} = T^{-1}\operatorname{Ran}(TS-\lambda)^{k+1}$.

(iv) Since $\operatorname{Ran}(ST-\lambda)^k = T^{-1}\operatorname{Ran}(TS-\lambda)^k$, we have $\operatorname{codim}\operatorname{Ran}(ST-\lambda)^k \leq \operatorname{codim}\operatorname{Ran}(TS-\lambda)^k$. The equality follows by symmetry. ☐

Proposition 29. *Let $T, S \in \mathcal{B}(X)$, $\lambda \in \mathbb{C}$, $\lambda \neq 0$ and $k \in \mathbb{N}$. Then:*

(i) $\operatorname{Ran}(TS-\lambda)^k$ *is closed if and only if* $\operatorname{Ran}(ST-\lambda)^k$ *is closed;*

(ii) $\operatorname{Ker}(TS-\lambda) \subset R^\infty(TS-\lambda)$ *if and only if* $\operatorname{Ker}(ST-\lambda) \subset R^\infty(ST-\lambda)$.

Proof. (i) Suppose that $\operatorname{Ran}(ST-\lambda)^k$ is closed. Let $x_n \in \operatorname{Ran}(TS-\lambda)^k$, $x_n = (TS-\lambda)^k y_n$ for some $y_n \in X$ $\quad (n = 1, 2, \dots)$ and $x_n \to x$.

Then $Sx_n \to Sx$, where $Sx_n = S(TS-\lambda)^k y_n = (ST-\lambda)^k Sy_n \in \operatorname{Ran}(ST-\lambda)^k$, and so $Sx = (ST-\lambda)^k y$ for some $y \in X$. We have $TSx = T(ST-\lambda)^k y =$

$(TS-\lambda)^k Ty \in \operatorname{Ran}(TS-\lambda)^k$. Therefore $(TS)^j x \in \operatorname{Ran}(TS-\lambda)^k$ for all $j \geq 1$. We have $(TS-\lambda)^k x = (-\lambda)^k x + \sum_{j=1}^{k} \binom{k}{j} (TS)^j (-\lambda)^{k-j} x$, and so $x \in \operatorname{Ran}(TS-\lambda)^k$.

(ii) Suppose that $\operatorname{Ker}(TS-\lambda) \subset R^\infty(TS-\lambda)$. It is sufficient to show that $\operatorname{Ker}(ST-\lambda) \subset \operatorname{Ran}(ST-\lambda)^k$ foe each $k \in \mathbb{N}$.

We have $TS \operatorname{Ker}(TS-\lambda) \subset \operatorname{Ker}(TS-\lambda) \subset \operatorname{Ran}(TS-\lambda)^k$, and so $\operatorname{Ker}(ST-\lambda) = S \operatorname{Ker}(TS-\lambda) \subset T^{-1} \operatorname{Ran}(TS-\lambda)^k = \operatorname{Ran}(ST-\lambda)^k$. □

Corollary 30. Let $T, S \in \mathcal{B}(X)$. Then $\sigma_K(TS) \setminus \{0\} = \sigma_K(ST) \setminus \{0\}$.

13 General inverses and Saphar operators

Let X, Y be Banach spaces and $T \in \mathcal{B}(X,Y)$ an operator. An operator $S : Y \to X$ is called a *generalized inverse* of T if $TST = T$ and $STS = S$.

It is easy to see that if $S : Y \to X$ is a one-sided inverse of T (i.e., either $TS = I_Y$ or $ST = I_X$), then S is a generalized inverse of T.

Proposition 1. *Let X, Y be Banach spaces, let $T : X \to Y$ be an operator. The following conditions are equivalent:*

(i) *T has a generalized inverse;*

(ii) *there exists an operator $S : Y \to X$ such that $TST = T$;*

(iii) *$\operatorname{Ran} T$ is closed and both $\operatorname{Ker} T$ and $\operatorname{Ran} T$ are complemented subspaces of X and Y, respectively.*

Proof. (ii) ⇒ (i): Let $TST = T$ for some operator $S : Y \to X$. Set $S' = STS$. It is easy to check that $TS'T = T$ and $S'TS' = S'$.

(i) ⇒ (iii): Let $TST = T$ and $STS = S$. Then $(TS)^2 = TS$ and $\operatorname{Ran} T \supset \operatorname{Ran}(TS) \supset \operatorname{Ran}(TST) = \operatorname{Ran} T$, so TS is a projection onto $\operatorname{Ran} T$.

Similarly, ST is a projection with $\operatorname{Ker}(ST) = \operatorname{Ker} T$.

(iii) ⇒ (ii): Let $X = \operatorname{Ker} T \oplus M$ and let $P \in \mathcal{B}(Y)$ be a projection onto $\operatorname{Ran} T$. Then $T|M : M \to \operatorname{Ran} T$ is one-to-one and onto. Set $S = (T|M)^{-1} P$. Then $TST = T(T|M)^{-1} PT = T$. □

Corollary 2. *Let $T \in \mathcal{B}(H_1, H_2)$ where H_1, H_2 are Hilbert spaces. Then T has a generalized inverse if and only if $\operatorname{Ran} T$ is closed.*

The difference between the left spectrum and the approximate point spectrum (the right spectrum and the defect spectrum) consists in the requirement that certain subspace should be complemented. This is a typical situation; frequently spectra in $\mathcal{B}(X)$ appear in pairs that differ in this way.

In this section we study the "complemented" version of the Kato operators.

Definition 3. An operator $T \in \mathcal{B}(X)$ is called *Saphar* if T is Kato and has a generalized inverse.

Equivalently, T is Saphar if and only if T has a generalized inverse and $\operatorname{Ker} T \subset R^\infty(T)$.

Obviously, in Hilbert spaces the Saphar operators coincide with the Kato operators.

Our aim is to show that the set of all Saphar operators is a regularity. This will be an immediate consequence of the following three lemmas, which are of independent interest.

Lemma 4. *Let $A \in \mathcal{B}(X)$ be a Saphar operator, let $S \in \mathcal{B}(X)$ satisfy $ASA = A$ and let $n \in \mathbb{N}$. Then $A^n S^n A^n = A^n$. In particular, A^n is a Saphar operator.*

Proof. Let $S \in \mathcal{B}(X)$ satisfy $ASA = A$. We prove by induction on n that $A^n S^n A^n = A^n$.

Suppose that $n \geq 1$ and $A^n S^n A^n = A^n$. Then

$$A^{n+1} S^{n+1} A^{n+1} = A\big(A^n S^n (SA - I) + A^n S^n\big) A^n.$$

Since $A^n S^n A^n = A^n$ and $ASA = A$, we can check easily (cf. the proof of Proposition 1) that $A^n S^n$ is a projection onto $\operatorname{Ran} A^n$ and $I - SA$ is a projection onto $\operatorname{Ker} A \subset \operatorname{Ran} A^n$. Thus

$$A^{n+1} S^{n+1} A^{n+1} = A\big((SA - I) + A^n S^n\big) A^n = A \cdot A^n S^n A^n = A^{n+1}.$$

Hence A^{n+1} is a Saphar operator. □

Lemma 5. *Let A, B, C, D be mutually commuting operators on a Banach space X satisfying $AC + BD = I$. Then AB has a generalized inverse if and only if both A and B have generalized inverses.*

Proof. Suppose that $ASA = A$ and $BTB = B$ for some $S, T \in \mathcal{B}(X)$. Then

$$\begin{aligned} ABTSAB &= ABT(CA + BD)SAB = ABTCASAB + ABTBDSAB \\ &= ABTCAB + ABDSAB = ABT(I - BD)B + A(I - AC)SAB \\ &= ABTB - ABTBDB + ASAB - ACASAB \\ &= AB - ABDB + AB - ACAB = 2AB - A(BD + CA)B = AB. \end{aligned}$$

For the converse, let $ABZAB = AB$ for some $Z \in \mathcal{B}(X)$. Then

$$\begin{aligned} A\big(C + BZ(I - AC)\big)A &= ACA + ABZA - ABZACA \\ &= ACA + ABZA(I - CA) \\ &= ACA + ABZABD = ACA + ABD = A, \end{aligned}$$

and similarly, $B\big(D + AZ(I - BD)\big)B = B$. □

Lemma 6. *Let $T \in \mathcal{B}(X)$ be a Saphar operator. Then there exists $\varepsilon > 0$ such that $T - U$ has a generalized inverse for every operator $U \in \mathcal{B}(X)$ commuting with T such that $\|U\| < \varepsilon$.*

More precisely, if T is Kato, $TST = T$, $UT = TU$ and $\|U\| < \|S\|^{-1}$, then $(T - U)S(I - US)^{-1}(T - U) = T - U$.

Proof. Let $TST = T$, $UT = TU$ and $\|U\| < \|S\|^{-1}$.

We first prove by induction on n that $U(SU)^n \operatorname{Ker} T \subset \operatorname{Ker} T^{n+1}$. This is clear for $n = 0$. Suppose that $U(SU)^{n-1} \operatorname{Ker} T \subset \operatorname{Ker} T^n \subset \operatorname{Ran} T$ and let $z \in \operatorname{Ker} T$. Then $U(SU)^{n-1}z = Tv$ for some $v \in X$, and

$$T^{n+1}U(SU)^n z = T^{n+1}USTv = T^n UTSTv = T^n UTv = UT^n U(SU)^{n-1}z = 0$$

by the induction assumption. Thus $U(SU)^n \operatorname{Ker} T \subset \operatorname{Ker} T^{n+1}$ for all n.

Since $I - ST$ is a projection onto $\operatorname{Ker} T$, we have

$$U(SU)^n(I - ST)X \subset \operatorname{Ker} T^{n+1} \subset \operatorname{Ran} T \quad (n \geq 0),$$

and so

$$(I - TS)U(SU)^n(I - ST) = 0 \quad (n \geq 0).$$

Then

$$\begin{aligned}
(T-U)S(I-US)^{-1}(T-U) &= (T-U)S\sum_{i=0}^{\infty}(US)^i(T-U) \\
&= TST - UST - TSU + TSUST \\
&\quad + \sum_{i=0}^{\infty}(TS(US)^{i+2}T - US(US)^{i+1}T - TS(US)^{i+1}U + US(US)^i U) \\
&= T - UST - TSU + TSUST + \sum_{i=0}^{\infty}(I-TS)(US)^{i+1}U(I-ST) \\
&= T - U + (I-TS)U(I-ST) = T - U.
\end{aligned}$$

Hence $T - U$ has a generalized inverse. □

Theorem 7. *The set of all Saphar operators in X is a regularity satisfying* (P3) (*upper semi-continuity on commuting elements*).

Proof. The proof follows immediately from Theorem 6.12 and the preceding three lemmas. □

For $T \in \mathcal{B}(X)$ denote by $\sigma_{\mathrm{Sap}}(T) = \{\lambda \in \mathbb{C} : T - \lambda \text{ is not Saphar}\}$ the corresponding spectrum. Property (P3) implies that $\sigma_{\mathrm{Sap}}(T)$ is always a compact set. Further, $\partial\sigma(T) \subset \sigma_K(T) \subset \sigma_{\mathrm{Sap}}(T)$, and so $\sigma_{\mathrm{Sap}}(T)$ is non-empty.

If $X = X_1 \oplus X_2$ is a decomposition of X and $T_1 \in \mathcal{B}(X_1)$, then it is easy to verify that T_1 is Saphar if and only if $T_1 \oplus I_{X_2}$ is Saphar. By Theorem 6.8, this gives the spectral mapping property.

Corollary 8. *Let $T \in \mathcal{B}(X)$. Then $\sigma_{\mathrm{Sap}}(f(T)) = f(\sigma_{\mathrm{Sap}}(T))$ for every function f analytic on a neighbourhood of $\sigma(T)$.*

If T is a Saphar operator, then $T - z$ is Saphar, and therefore has a generalized inverse, for every z in a certain neighbourhood of 0. In fact, the generalized inverse can be chosen in such a way that it depends analytically on z.

Theorem 9. *Let $T \in \mathcal{B}(X)$. The following conditions are equivalent:*

(i) *T is Saphar;*

(ii) *there exist a neighbourhood $U \subset \mathbb{C}$ of 0 and an analytic function $S : U \to \mathcal{B}(X)$ such that $(T-z)S(z)(T-z) = T-z$.*

More precisely, it is possible to take $S(z) = S(I-zS)^{-1} = \sum_{i=0}^{\infty} S^{i+1}z^i$ for $|z| < \|S\|^{-1}$, where S is a generalized inverse of T.

Proof. (i) $\Rightarrow$ (ii): Let $S \in \mathcal{B}(X)$ satisfy $TST = T$. For $|z| < \|S\|^{-1}$ set $S(z) = S(I-zS)^{-1}$. By Lemma 6, $(T-z)S(z)(T-z) = T-z$ for all $|z| < \|S\|^{-1}$.

(ii) $\Rightarrow$ (i): Let $S(z)$ be a function analytic on a neighbourhood U of 0 satisfying $(T-z)S(z)(T-z) = T-z \quad (z \in U)$. It is sufficient to show that $\operatorname{Ker} T \subset R^\infty(T)$. Let $x \in \operatorname{Ker} T$. Since $TS(0)T = T$, $I - S(0)T$ is a projection onto $\operatorname{Ker} T$. Set $g(z) = (I - S(z)(T-z))x \quad (z \in U)$. Clearly, $(T-z)g(z) = 0 \quad (z \in U)$ and $g(0) = x$. Let $g(z) = \sum_{i=0}^{\infty} x_i z^i$ be the Taylor expansion of g about 0. The above relations give $x_0 = x$ and $Tx_i = x_{i-1} \quad (i \geq 1)$. Thus $x = x_0 = Tx_1 = T^2x_2 = \cdots$, and so $x \in R^\infty(T)$. □

The next theorem shows that it is possible to find a global analytic general inverse of $T - z$.

Theorem 10. *Let $T \in \mathcal{B}(X)$. Let $G = \{z \in \mathbf{C} : T - z$ is Saphar$\}$. Then there exists an analytic function $S : G \to \mathcal{B}(X)$ such that*

$$(T-z)S(z)(T-z) = T-z$$

and

$$S(z)(T-z)S(z) = S(z) \qquad (z \in G).$$

Proof. The set G is open by Theorem 9. For $z \in G$ let $\Phi(z) : \mathcal{B}(X) \to \mathcal{B}(X)$ be the operator defined by

$$\Phi(z)A = (T-z)A(T-z) \qquad (A \in \mathcal{B}(X)).$$

Evidently, $\Phi : G \to \mathcal{B}(\mathcal{B}(X))$ is an analytic function and $T - z \in \operatorname{Ran} \Phi(z)$ for all $z \in G$.

We show that Φ is regular at each point $\lambda \in G$. By Theorem 9, there is a neighbourhood U of λ and an analytic function $S_1 : U \to \mathcal{B}(X)$ such that $(T-z)S_1(z)(T-z) = T-z \quad (z \in U)$. For $z \in U$ let $\Psi(z) : \mathcal{B}(X) \to \mathcal{B}(X)$ be the operator defined by $\Psi(z)A = (T-z)S_1(z)AS_1(z)(T-z) - A$. It is clear that Ψ is analytic and $\Psi(z)\Phi(z) = 0 \quad (z \in U)$. If $A \in \operatorname{Ker} \Psi(z)$, then $A = (T-z)S_1(z)AS_1(z)(T-z) \in \operatorname{Ran} \Phi(z)$ for all $z \in U$.

Thus $\operatorname{Ran} \Phi(z) = \operatorname{Ker} \Psi(z)$ and Φ is regular in U by Theorem 11.10.

Hence Φ is regular in G. By Theorem 11.12, there exists an analytic function $S_2 : G \to \mathcal{B}(X)$ such that $\Phi(z)S_2(z) = T - z$, i.e., $(T-z)S_2(z)(T-z) = T-z$ for $z \in G$. Set

$$S(z) = S_2(z)(T-z)S_2(z) \qquad (z \in G).$$

Then

$$(T-z)S(z)(T-z) = (T-z)S_2(z)(T-z)S_2(z)(T-z) = T-z$$

and

$$\begin{aligned} S(z)(T-z)S(z) &= S_2(z)(T-z)S_2(z)(T-z)S_2(z)(T-z)S_2(z) \\ &= S_2(z)(T-z)S_2(z) = S(z) \end{aligned}$$

for all $z \in G$. □

Proposition 11. *Let $T, S \in \mathcal{B}(X)$, $\lambda \in \mathbb{C}$, $\lambda \neq 0$. Then $TS - \lambda$ has a generalized inverse if and only is $ST - \lambda$ has a generalized inverse.*

Proof. Let $A \in \mathcal{B}(X)$ satisfy $(TS-\lambda)A(TS-\lambda) = TS - \lambda$. Set $B = -\lambda^{-1}I + \lambda^{-1}SAT$. Then

$$\begin{aligned} (ST-\lambda)B(ST-\lambda) &= \lambda^{-1}(ST-\lambda)SAT(ST-\lambda) - \lambda^{-1}(ST-\lambda)^2 \\ &= \lambda^{-1}S(TS-\lambda)A(TS-\lambda)T - \lambda^{-1}(ST-\lambda)^2 \\ &= \lambda^{-1}S(TS-\lambda)T - \lambda^{-1}(ST)^2 + 2ST - \lambda = ST - \lambda. \end{aligned}$$ □

Corollary 12. Let $T, S \in \mathcal{B}(X)$. Then $\sigma_{\mathrm{Sap}}(ST) \setminus \{0\} = \sigma_{\mathrm{Sap}}(TS) \setminus \{0\}$.

Theorem 9 enables us to extend the notion of Saphar spectrum for elements in a Banach algebra.

Let $\mathcal{A}$ be a Banach algebra, $a, b \in \mathcal{A}$. As in the case of operators, we say that b is a generalized inverse of a if $aba = a$ and $bab = b$.

Definition 13. An element a in a Banach algebra $\mathcal{A}$ is called Saphar if there exist a neighbourhood U of 0 in $\mathbb{C}$ and an analytic function $f : U \to \mathcal{A}$ such that $(a-z)f(z)(a-z) = a - z \quad (z \in U)$.

For the study of Saphar elements in a Banach algebra $\mathcal{A}$ it is useful to consider the operator of left multiplication $L_a : \mathcal{A} \to \mathcal{A}$ defined by $L_a x = ax \quad (x \in \mathcal{A})$. Clearly, if $a \in \mathcal{A}$ is Saphar, then L_a is also Saphar (as an element of $\mathcal{B}(\mathcal{A})$). The opposite is not true: let $\mathcal{A}$ be the algebra of all power series $\sum_{i=0}^{\infty} \alpha_i a^i$ with complex coefficients α_i such that $\|\sum_{i=0}^{\infty} \alpha_i a^i\| = \sum |\alpha_i| < \infty$. Then L_a is bounded below and $\operatorname{Ran} L_a$ is complemented in $\mathcal{A}$, since it is of codimension 1. Thus L_a is Saphar. On the other hand, $aba \in a^2\mathcal{A}$ for every $b \in \mathcal{A}$, so a has no generalized inverse and thus it is not Saphar.

We have the following characterization:

Theorem 14. *An element $a \in \mathcal{A}$ is Saphar if and only if it has a generalized inverse and $\operatorname{Ker} L_a \subset R^\infty(L_a)$.*

Proof. If $a \in \mathcal{A}$ is Saphar, then it has a generalized inverse and L_a is Saphar, so $\operatorname{Ker} L_a \subset R^\infty(L_a)$.

Conversely, let a have a generalized inverse $b \in \mathcal{A}$. Then L_b is a generalized inverse of L_a and L_a is Saphar. By Theorem 9, $\sum_{i=0}^{\infty} L_b^{i+1} z^i$ is an analytically dependant generalized inverse of $L_a - z$ for $|z| < \|L_b\|^{-1} = \|b\|^{-1}$. Note that $\sum_{i=0}^{\infty} L_b^{i+1} z^i = L_{b(z)}$ where $b(z) = \sum_{i=0}^{\infty} b^{i+1} z^i$, and so $b(z)$ is a generalized inverse of $a - z$ depending analytically on z. □

Theorem 15. *The set of all Saphar elements in a Banach algebra $\mathcal{A}$ is a regularity satisfying* (P3) (*upper semicontinuity on commuting elements*).

Proof. Recall that $a \in \mathcal{A}$ is Saphar if and only if it has a generalized inverse and L_a is Kato. The Kato operators form a regularity satisfying (P3), see Theorem 12.9. To prove that the set of all Saphar elements satisfies the conditions of Theorem 6.12 it is sufficient to show:

(a) a is Saphar, $n \in \mathbb{N} \Rightarrow a^n$ has a generalized inverse;

(b) if a, b, c, d are mutually commuting elements of $\mathcal{A}$ such that $ac + bd = 1$, then

$$ab \text{ has a generalized inverse } \iff a, b \text{ have generalized inverses;}$$

(c) if $a \in \mathcal{A}$ is Saphar, then there exists $\varepsilon > 0$ such that $u \in \mathcal{A}$, $ua = au$ and $\|u\| < \varepsilon$ implies that $a - u$ has a generalized inverse.

Note that an element $x \in \mathcal{A}$ has a generalized inverse if and only if L_x has a generalized inverse which belongs to the set $\{L_y : y \in \mathcal{A}\}$. This is really the case if we apply Lemmas 4, 5 and 6 to the operators of left multiplication. In this way we obtain (a), (b) and (c), which finishes the proof. □

It is easy to see that Proposition 11 and Corollary 12 are also true for elements of a Banach algebra.

In commutative Banach algebras the class of Saphar elements coincides with the invertible elements.

Theorem 16. *Let $\mathcal{A}$ be a commutative Banach algebra. An element $a \in \mathcal{A}$ is Saphar if and only if it is invertible.*

Proof. Clearly, invertible elements are Saphar.

Let $a \in \mathcal{A}$ be Saphar and let $b(z) = \sum_{i=0}^{\infty} b_i z^i$ be a function analytic on a neighbourhood of 0 satisfying $(a-z)b(z)(a-z) = a-z$. Comparing the coefficients at z we get $-1 = a^2 b_1 - 2ab_0 = a(ab_1 - 2b_0)$. Hence a is invertible. □

14 Local spectrum

Further important examples of regularities are provided by local spectra.

Let x be a vector in a Banach space X. Denote by $R_x(X)$ the set of all operators $T \in \mathcal{B}(X)$ for which there exists a neighbourhood U of 0 in $\mathbb{C}$ and an analytic function $f : U \to X$ such that $(T - z)f(z) = x \quad (z \in U)$.

If $f(z) = \sum_{i=0}^{\infty} x_{i+1} z^i$ is the Taylor expansion of f in a neighbourhood of 0, then $(T-z)f(z) = Tx_1 + \sum_{i=1}^{\infty} z^i(Tx_{i+1} - x_i)$, so $Tx_{i+1} = x_i \quad (i = 1, 2, \dots)$ and $Tx_1 = x$. Thus $T \in R_x(X)$ if and only if there exist vectors $x_1, x_2, \dots \in X$ such that $Tx_i = x_{i-1} \quad (i = 1, 2, \dots)$, where $x_0 = x$, and $\sup_{i\geq 1} \|x_i\|^{1/i} < \infty$.

We start with the following lemma which is a refined formulation of Lemma 12.8 (ii).

Lemma 1. *Let A, B, C, D be mutually commuting operators on a Banach space X such that $AC + BD = I$ and let $u, v \in X$ satisfy $Au = Bv$. Then there exists $w \in X$ such that $Aw = v$, $Bw = u$ and $\|w\| \leq (\|C\| + \|D\|) \cdot \max\{\|u\|, \|v\|\}$.*

Proof. Set $w = Du + Cv$. Then $\|w\| \leq (\|C\| + \|D\|) \cdot \max\{\|u\|, \|v\|\}$, $Aw = ADu + ACv = ADu + (I - BD)v = DAu + v - DBv = v$ and similarly, $Bw = BDu + BCv = u - ACu + BCv = u$. □

Theorem 2. *Let x be a vector in a Banach space X. Then $R_x(X)$ is a regularity. Moreover, for each $T \in R_x(X)$ there exists $\varepsilon > 0$ such that $T + U \in R_x(X)$ for all $U \in \mathcal{B}(X)$ commuting with T such that $\|U\| < \varepsilon$.*

Thus $R_x(X)$ satisfies (P3) (*upper semicontinuity on commuting elements*).

Proof. Clearly $I \in R_x(X)$, and so $R_x(X)$ is non-empty.

Suppose that $T \in R_x(X)$ and let n be a positive integer. Write $x_0 = x$ and let $x_i \in X$ satisfy $Tx_i = x_{i-1} \quad (i = 1, 2, \dots)$ and $\sup_{i\geq 1} \|x_i\|^{1/i} < \infty$. Set $y_i = x_{ni} \quad (i = 0, 1, \dots)$. Then $T^n y_i = y_{i-1} \quad (i = 1, 2, \dots)$, $y_0 = x$ and

$$\sup_{i\geq 1} \|y_i\|^{1/i} \leq \left(\sup_{i\geq 1} \|x_i\|^{1/i}\right)^n < \infty.$$

Thus $T^n \in R_x(X)$.

Let $AB = BA \in R_x(X)$. Let $x_i \in X$ satisfy $ABx_i = x_{i-1} \quad (i = 1, 2, \dots)$ with $x_0 = x$ and let $\sup_{i\geq 1} \|x_i\|^{1/i} < \infty$. Set $y_i = B^i x_i$. Then $y_0 = x$, $Ay_i = AB^i x_i = B^{i-1} x_{i-1} = y_{i-1} \quad (i = 1, 2, \dots)$ and

$$\sup_{i\geq 1} \|y_i\|^{1/i} \leq \|B\| \cdot \sup_{i\geq 1} \|x_i\|^{1/i} < \infty.$$

Thus $A \in R_x(X)$ and similarly $B \in R_x(X)$. In particular, $T^n \in R_x(X)$ implies $T \in R_x(X)$.

Let A, B, C, D be mutually commuting operators with $AC + BD = I$ and let $A, B \in R_x$. Let $x_{i,0}, x_{0,j} \in X \quad (i, j = 1, 2, \dots)$ satisfy $Ax_{i,0} = x_{i-1,0}$, $Bx_{0,j} = x_{0,j-1} \quad (i, j \geq 1)$, where $x_{0,0} = x$ and $\|x_{i,0}\| \leq k^i$, $\|x_{0,j}\| \leq k^j$ for some k, $1 \leq k < \infty$. Using Lemma 1 inductively, we can construct elements $x_{i,j} = Dx_{i-1,j} + Cx_{i,j-1}$ such that $Ax_{i,j} = x_{i-1,j}$, $Bx_{i,j} = x_{i,j-1}$ and

$$\|x_{i,j}\| \leq (\|C\| + \|D\|) \cdot \max\{\|x_{i,j-1}\|, \|x_{i-1,j}\|\} \qquad (i, j \geq 1).$$

It is easy to show by induction that $\|x_{i,j}\| \le \left(k \cdot \max\{1, \|C\| + \|D\|\}\right)^{i+j} \|x\|$. Set $y_i = x_{i,i}$. Then $ABy_i = y_{i-1}$, $y_0 = x$ and $\sup \|y_i\|^{1/i} < \infty$, so $AB \in R_x$. Hence R_x is a regularity.

To prove property (P3), let $T \in R_x(X)$, let $x_i \in X$ satisfy $Tx_i = x_{i-1}$ $(i = 1, 2, \dots)$, $x_0 = x$ and $\sup_{i \ge 1} \|x_i\|^{1/i} = k < \infty$. Let $U \in \mathcal{B}(X), UT = TU$ and $\|U\| < k^{-1}$. Set $g(\lambda) = \sum_{i=0}^{\infty} (\lambda - U)^i x_{i+1}$. This series is convergent for $|\lambda| < k^{-1} - \|U\|$ and we have

$$(T + U - \lambda) g(\lambda) = Tx_1 + \sum_{i=1}^{\infty} T(\lambda - U)^i x_{i+1} - \sum_{i=0}^{\infty} (\lambda - U)^{i+1} x_{i+1} = Tx_1 = x.$$

Thus $T + U \in R_x(X)$. □

Denote by γ_x the spectrum corresponding to the regularity $R_x(X)$. Since $R_x(X)$ satisfies $(P3)$, $\gamma_x(T)$ is always closed. Obviously, $\gamma_x(T) \subset \sigma(T)$.

Remark 3. The standard notation is $\gamma_T(x)$. For our approach, however, the notation $\gamma_x(T)$ seems to be more natural. The set $\gamma_x(T)$ is called the *local spectrum* of T at x.

Corollary 4. *Let x be a vector in a Banach space X, let $T \in \mathcal{B}(X)$. Then*

$$\gamma_x(f(T)) = f(\gamma_x(T))$$

for each function f analytic on a neighbourhood of $\sigma(T)$ which is non-constant on every component of its domain of definition.

Examples 5. (i) $\gamma_x(T)$ can be empty.

Let H be a Hilbert space with an orthonormal basis $\{e_i : i \ge 0\}$, let S be the unilateral shift $(Se_i = e_{i+1})$ and let S^* be its adjoint, $S^* e_i = e_{i-1}$ $(i \ge 1)$, $S^* e_0 = 0$. Clearly, $S^* S = I$. Let $x = \sum_{i=0}^{\infty} 2^{-i} e_i$. Then $S^* x = \frac{x}{2}$. For $|z| > 1/2$ set $f(z) = -\sum_{i=0}^{\infty} \frac{S^{*i} x}{z^{i+1}}$. Clearly,

$$(S^* - z) f(z) = -\sum_{i=0}^{\infty} \frac{S^{*i+1} x}{z^{i+1}} + \sum_{i=0}^{\infty} \frac{S^{*i} x}{z^i} = x \qquad (|z| > 1/2).$$

Further, set $g(z) = \sum_{i=0}^{\infty} S^{i+1} x z^i$. This sum is convergent for $|z| < 1$ and

$$(S^* - z) g(z) = \sum_{i=0}^{\infty} S^i x z^i - \sum_{i=0}^{\infty} S^{i+1} x z^{i+1} = x.$$

Hence $\gamma_x(S^*) = \emptyset$.

(ii) The previous example also shows that in general it is not possible to find a global analytic solution of the equation $(S^* - z) h(z) = x$ $\;(z \notin \gamma_x(T))$.

Suppose on the contrary that there exists an entire function h satisfying $(S^* - z)h(z) = x \quad (z \in \mathbb{C})$. Since for $|z| > 1 = r(S^*)$ the function h is uniquely determined by $h(z) = (S^* - z)^{-1}x$ and $(S^* - z)^{-1}x \to 0$ as $z \to \infty$, the Liouville theorem gives $h \equiv 0$. So $x = 0$, a contradiction.

(iii) The assumption in Corollary 4 that f is non-constant on each component is really necessary, since $\gamma_x(T)$ might be empty and $\gamma_x(I) \neq \emptyset$.

(iv) $R_x(X)$ does not satisfy (P2) (upper semicontinuity of γ_x). To see this, consider a 2-dimensional space X with a basis x, y, and let

$$T = \begin{pmatrix} 1 & 0 \\ 0 & 0 \end{pmatrix}.$$

Then $T \in R_x(X)$ and

$$\begin{pmatrix} 1 & 0 \\ \varepsilon & 0 \end{pmatrix} \notin R_x(X)$$

for each $\varepsilon > 0$. Hence $R_x(X)$ is not open, and so γ_x does not satisfy (P2).

(v) $R_x(X)$ does not satisfy (P4) (continuity on commuting elements).

Consider the operator S^* and the vector x from (i). Then $\frac{S^*}{n} \to 0$ and $\gamma_x(\frac{S^*}{n}) = \emptyset$ for all n. On the other hand, $\gamma_x(0) = \{0\} \neq \emptyset$.

(vi) R_x does not satisfy (P1):

Let H be a Hilbert space with an orthonormal basis $\{e_{i,j} : i \geq 0 \text{ or } j \geq 0\}$ and let the operators $A, B \in \mathcal{B}(H)$ be defined by $Ae_{i,j} = e_{i+1,j}$, $Be_{i,j} = e_{i,j+1}$. It is clear that $AB = BA$. Let $x = e_{0,0}$. It is easy to see that $A, B \in R_x$ and $AB \notin R_x$, since $x \notin \operatorname{Ran}(AB)$.

Consider now the subset $R(X) \subset \mathcal{B}(X)$ defined by: $T \notin R(X)$ if and only if there exists a function $f : U \longrightarrow X$ analytic on a neighbourhood U of 0 such that f is not identically equal to 0 and $(T - z)f(z) = 0 \quad (z \in U)$.

As above, it is easy to see that $T \notin R(X)$ if and only if there exist vectors $x_i \in X \quad (i = 1, 2, \dots)$ not all of them equal to 0 such that $Tx_i = x_{i-1} \quad (i = 1, 2, \dots)$, where $x_0 = 0$ and $\sup_{i \geq 1} \|x_i\|^{1/i} < \infty$. We can assume that $x_1 \neq 0$.

Evidently, if $T \notin R(X)$, then there is an open neighbourhood of 0 consisting of eigenvalues of T. The simplest example of an operator $T \notin R(X)$ is the backward shift.

Theorem 6. *$R(X)$ is a regularity.*

Proof. The set $R(X)$ is non-empty, since $I \in R(X)$.

Let $A, B \in \mathcal{B}(X), AB = BA \notin R(X)$. We prove that either $A \notin R(X)$ or $B \notin R(X)$. Let $x_i \in X$ satisfy $ABx_i = x_{i-1} \quad (i = 1, 2, \dots)$, where $x_0 = 0$, $x_1 \neq 0$ and $\sup_{i \geq 1} \|x_i\|^{1/i} < \infty$. Set $u_i = B^i x_i \quad (i = 0, 1, \dots)$. Then $u_0 = 0$, $Au_i = u_{i-1}$ for all $i \geq 1$ and $\sup_{i \geq 1} \|u_i\|^{1/i} < \infty$. If $u_1 \neq 0$, then $A \notin R(X)$.

Suppose on the contrary that $u_1 = Bx_1 = 0$. Set $v_0 = 0, v_i = A^{i-1}x_i$ for all $i \in \mathbb{N}$. Then $Bv_i = v_{i-1} \quad (i = 1, 2, \dots)$, $\sup_{i\geq 1} \|v_i\|^{1/i} < \infty$ and $v_1 = x_1 \neq 0$. Thus $B \notin R(X)$. Hence $A, B \in R(X), AB = BA$ implies $AB \in R(X)$.

In particular, $A \in R(X) \Rightarrow A^n \in R(X)$ for all n.

Let $A \notin R(X)$ and let $x_i \in X$ satisfy the conditions required. Then $y_i = x_{ni}$ satisfy all the required conditions for A^n, and so $A^n \notin R(X)$. Hence $A^n \in R(X) \Rightarrow A \in R(X)$.

Suppose that A, B, C, D are mutually commuting operators satisfying $AC + BD = I$ and $A \notin R(X)$. Let $x_{i,0} \in X$ satisfy $Ax_{i,0} = x_{i-1,0}$ for all $i = 1, 2, \dots$, $x_{0,0} = 0$, $x_{1,0} \neq 0$ and $\sup_{i\geq 1} \|x_{i,0}\|^{1/i} < \infty$.

Set $x_{0,j} = 0 \quad (j = 0, 1, \dots)$. Using Lemma 1 inductively, we construct vectors $x_{i,j} = Dx_{i-1,j} + Cx_{i,j-1} \quad (i, j \in \mathbb{N})$. As in the proof of Theorem 2 we can show that $ABx_{i,i} = x_{i-1,j-1} \quad (i \geq 1)$, $x_{1,1} \neq 0$ and $\sup_i \|x_{i,i}\|^{1/i} < \infty$. Thus $AB \notin R(X)$ and $AB \in R(X) \Rightarrow A, B \in R(X)$.

Hence $R(X)$ is a regularity. □

Denote by $S_0(T)$ the spectrum of T corresponding to the regularity $R(X)$. Clearly $S_0(T)$ is the union of all open subsets $U \subset \mathbb{C}$ for which there is an analytic function $f : U \to X$ not identically equal to 0 satisfying $(T - z)f(z) = 0 \quad (z \in U)$. Obviously, $S_0(T)$ is contained in the point spectrum of T. In general, $S_0(T)$ is not closed (on the contrary, it is always open), and so $R(X)$ cannot satisfy (P2), (P3) or (P4). Neither does $R(X)$ satisfy (P1). To see this, let X be a separable Hilbert space, $A = 0$ and let B be the backward shift. It is easy to see that $0 = AB \in R(X)$ and $B \notin R(X)$.

The closure of $S_0(T)$ is called the *analytic residuum* of T and denoted by $S(T)$. An operator T is said to have the *single value extension property* (SVEP) if $S_0(T)$ is empty.

Corollary 7. *Let $T \in \mathcal{B}(X)$ and let f be a function analytic on a neighbourhood of $\sigma(T)$ which is non-constant on each component of its domain of definition. Then*

$$S_0(f(T)) = f(S_0(T)) \qquad \text{and} \qquad S(f(T)) = f(S(T)).$$

The set $S(T) \cup \gamma_x(T)$ will be denoted by $\sigma_x(T)$; this set is also called the *local spectrum* (the standard notation is again rather $\sigma_T(x)$ instead of $\sigma_x(T)$).

Proposition 8. *Let $T \in \mathcal{B}(X), x \in X, x \neq 0$. Then $S_0(T) \cup \gamma_x(T) \neq \emptyset$. In particular, $\sigma_x(T) \neq \emptyset$.*

Proof. Suppose on the contrary that $S_0(T) \cup \gamma_x(T) = \emptyset$. Then for every $w \in \mathbb{C}$ there exists a neighbourhood U_w of w and an analytic function $f_w : U_w \to X$ such that $(T - z)f_w(z) = x \quad (z \in U_w)$. Since $S_0(T) = \emptyset$, functions f_w and $f_{w'}$ coincide on $U_w \cap U_{w'} \quad (w, w' \in \mathbb{C})$, so we have an entire function $f : \mathbb{C} \to X$ such that $(T - z)f(z) = x \quad (\lambda \in \mathbb{C})$.

For $|z| > r(T)$ we have $f(z) = (T - z)^{-1}x$, so $\lim_{z\to\infty} f(\lambda) = 0$. By the Liouville theorem, $f \equiv 0$, and so $x = 0$, a contradiction. □

Theorem 9. *Let $T \in \mathcal{B}(X), x \in X, x \neq 0$ and let f be a locally non-constant function analytic on a neighbourhood of $\sigma(T)$. Then*

$$S_0(f(T)) \cup \gamma_x(f(T)) = f(S_0(T)) \cup f(\gamma_x(T)) \qquad \text{and} \qquad \sigma_x(f(T)) = f(\sigma_x(T)).$$

Proof. Since $R_x(X) \cap R(X)$ is a regularity, we have

$$S_0(f(T)) \cup \gamma_x(f(T)) = f(S_0(T)) \cup f(\gamma_x(T)).$$

Taking closures on both sides we get

$$\sigma_x(f(T)) = \overline{f(S_0(T))} \cup f(\gamma_x(T)) = f(\overline{S_0(T)}) \cup f(\gamma_x(T)) = f(\sigma_x(T)). \qquad \square$$

Example 10. (i) The assumption that f is locally non-constant is really necessary. This does not contradict to Theorem 6.8 since $\sigma_x(T)$ is non-empty only for $x \neq 0$; if $x = 0$, then $\sigma_x(T)$ can be either empty or non-empty.

For an example, let S^* be the backward shift on a separable Hilbert space H and let u be a non-zero vector in H. Consider $T = S^* \oplus 2I \in \mathcal{B}(H \oplus H)$ and $x = 0 \oplus u$. It is easy to verify that $S(T) = \{z \in \mathbb{C} : |z| \leq 1\}$ and $\gamma_x(T) = \{2\}$. Let $f = 0$ on a neighbourhood of $\{z : |z| \leq 1\}$ and $f = 1$ on a neighbourhood of $\{2\}$. Then $f(\sigma_x(T)) = \{0\} \cup \{2\}$. Further, $f(T) = 0 \oplus 2I$, $S(f(T)) = \emptyset$ and $\sigma_x(f(T)) = \gamma_x(f(T)) = \{2\}$. Hence $\sigma_x(f(T)) \neq f(\sigma_x(T))$.

(ii) The local spectrum has a natural meaning for normal operators. Let $T = \int z \, \mathrm{d}E(z)$ be a normal operator on a Hilbert space H, let $x \in H$. Then T has SVEP and $\sigma_x(T)$ is the support of the scalar measure $\|E(\cdot)x\|^2 = \langle E(\cdot)x, x\rangle$.

Proposition 11. Let $T, U \in \mathcal{B}(X)$. Then $S_0(TU) = S_0(UT)$ and $S(TU) = S(UT)$.

Proof. Let $\lambda \in \mathbb{C}$, $\lambda \neq 0$. Suppose that $\lambda \in S_0(TU)$. Then there are $x_1, x_2, \dots \in X$ such that $x_1 \neq 0$, $(TU-\lambda)x_i = x_{i-1}$ $(i \geq 2)$, $(TU-\lambda)x_1 = 0$ and $\sup \|x_i\|^{1/i} < \infty$.

Set $y_i = Ux_i \quad (i = 1, 2, \dots)$. For $i \geq 2$ we have $(UT-\lambda)y_i = (UT-\lambda)Ux_i = U(TU-\lambda)x_i = Ux_{i-1} = y_{i-1}$ and $(UT-\lambda)y_1 = U(TU-\lambda)x_1 = 0$. Further, $TUx_1 = \lambda x_1 \neq 0$, and so $y_1 = Ux_1 \neq 0$. Clearly $\sup \|y_i\|^{1/i} < \infty$. Hence for $\lambda \neq 0$ we have $\lambda \in S_0(TU) \Leftrightarrow \lambda \in S_0(UT)$.

The same equivalence is true also for $\lambda = 0$. Let $0 \in S_0(TU)$. Let $x_i \in X$ satisfy $TUx_i = x_{i-1} \quad (i \geq 2)$, $TUx_1 = 0$, $x_1 \neq 0$ and $\sup \|x_i\|^{1/i} < \infty$. Set $y_i = Ux_i$. Then $UTy_i = UTUx_i = Ux_{i-1} = y_{i-1} \quad (i \geq 2)$ and $UTy_1 = UTUx_1 = 0$. We have $\sup \|y_i\|^{1/i} < \infty$ and $Ty_2 = TUx_2 = x_1 \neq 0$. Hence $y_2 \neq 0$ and $0 \in S_0(UT)$. $\square$

Definition 12. Let T be an operator on a Banach space X and let $x \in X$. The *local spectral radius* $r_x(T)$ is defined by $r_x(T) = \limsup_{n\to\infty} \|T^n x\|^{1/n}$.

The standard notation of the local spectral radius is $r(T, x)$; we choose the present notation in order to have an agreement with the notation for the local spectra.

The formula defining the local spectral radius is similar to the spectral radius formula $r(T) = \lim \|T^n\|^{1/n}$. It is easy to see that in general the limit $\lim \|T^n x\|^{1/n}$ does not exist. The meaning of the local spectral radius is the following: the series $f(z) = -\sum_{i=0}^{\infty} \frac{T^i x}{z^{i+1}}$ is convergent for $|z| > r_x(T)$ and $(T-z)f(z) = x$; for $|z| > r(T)$ we have $f(z) = (T-z)^{-1}x$.

The connection between the local spectral radius and the local spectra is the following:

Theorem 13. *Let $T \in \mathcal{B}(X)$ and $x \in X$. Then*

$$\max\{|\lambda| : \lambda \in \gamma_x(T)\} \le r_x(T) \le \max\{|\lambda| : \lambda \in \sigma_x(T)\}$$

(if $\gamma_x(T) = \emptyset$, then the left maximum is considered to be 0).

Proof. Set $f(z) = -\sum_{i=0}^{\infty} \frac{T^i x}{z^{i+1}}$. This series converges for $|z| > r_x(T)$ and $(T - z)f(z) = x$. Thus $|z| > r_x(T)$ implies $x \notin \gamma_x(T)$, which is the first inequality.

For $|z| > \max\{|\lambda| : \lambda \in \sigma_x(T)\}$ there is a unique analytic function $g(z)$ satisfying $(T-z)g(z) = x$. For $|z| > r(T)$ necessarily $g(z) = (T-z)^{-1}x = -\sum_{i=0}^{\infty} T^i x z^{-(i+1)}$, and so $\lim_{z\to\infty} g(z) = 0$.

Consider the function $h(z) = g(\frac{1}{z})$ $(0 \ne |z| < \max\{|\lambda| : \lambda \in \sigma_x(T)\}^{-1})$, $h(0) = 0$. Clearly, $h(z) = \sum_{i=0}^{\infty}(T^i x)z^{i+1}$ for $|z| < \max\{|\lambda| : \lambda \in \sigma_x(T)\}^{-1}$. Since the radius of convergence of the series $\sum (T^i x)z^{i+1}$ is equal to

$$\left(\limsup_{i\to\infty} \|T^i x\|^{1/i}\right)^{-1} = (r_x(T))^{-1},$$

we conclude that $r_x(T)^{-1} \ge \left(\max\{|\lambda| : \lambda \in \sigma_x(T)\}\right)^{-1}$. Hence $r_x(T) \le \max\{|\lambda| : \lambda \in \sigma_x(T)\}$. □

Example 14. In general, the inequalities in the previous theorem are strict. Consider the backward shift $S^* \in \mathcal{B}(H)$ and the vector $x = \sum_{i=0}^{\infty} 2^{-i} e_i$ which were studied in Example 5 (i). We have $S^* x = \frac{x}{2}$, and so

$$r_x(S^*) = \limsup \|S^{*i} x\|^{1/i} = \frac{1}{2}.$$

Further, $\gamma_x(S^*) = \emptyset$, and so $\max\{|\lambda| : \lambda \in \gamma_x(S^*)\} = 0$. Finally, the function $g(z) = \sum_{i=0}^{\infty} z^i e_i$ is convergent for $|z| < 1$ and $(S^* - z)g(z) = 0$. So $\sigma_x(S^*) = \{z : |z| \le 1\}$ and $\max\{|\lambda| : \lambda \in \sigma_x(S^*)\} = 1$.

The relation between the local spectral radius and the local spectra is closer for operators with SVEP. In this case $\gamma_x(T)$ and $\sigma_x(T)$ coincide, and so we have

Corollary 15. *Let $T \in \mathcal{B}(X)$ be an operator with SVEP and $x \in X$. Then*

$$\max\{|\lambda| : \lambda \in \gamma_x(T)\} = r_x(T) = \max\{|\lambda| : \lambda \in \sigma_x(T)\}.$$

In general, $\sigma_x(T) \subset \sigma(T)$ and $r_x(T) \le r(T)$. Our next goal is to show that there are always many points $x \in X$ with the local spectrum $\sigma_x(T)$ equal to the global spectrum $\sigma(T)$.

We start with the following observation:

Proposition 16. *Let $T \in \mathcal{B}(X)$ satisfy $TX = X$ and $0 \in \sigma(T)$. Then $0 \in S_0(T)$.*

Proof. Since $0 \in \sigma(T)$ and $TX = X$, there exists a non-zero vector $x_0 \in X$ with $Tx_0 = 0$. By the open mapping theorem, there is a constant $k > 0$ such that $TB_X \supset k \cdot B_X$. We can construct inductively vectors $x_1, x_2, \dots \in X$ such that $Tx_i = x_{i-1}$ and $\|x_i\| \le k^{-1}\|x_{i-1}\|$, so $\|x_i\| \le k^{-i}\|x_0\|$. Set $f(z) = \sum_{i=0}^{\infty} x_i z^i$. This series is convergent for $|z| < k$ and $(T - z)f(z) = 0$. Since $f(0) = x_0 \neq 0$, we have $0 \in S_0(T)$. □

Corollary 17. $\sigma(T) = S_0(T) \cup \{\lambda \in \mathbb{C} : (T - \lambda)X \neq X\}$.

Recall that the set $\{\lambda \in \mathbb{C} : (T-\lambda)X \neq X\}$ is the surjective spectrum $\sigma_\delta(T)$.

Theorem 18. *Let $T \in \mathcal{B}(X)$. Then the set $\{x \in X : \sigma_x(T) \neq \sigma(T)\}$ is of the first category.*

Proof. Let $\{\lambda_1, \lambda_2, \dots\}$ be a countable set dense in $\sigma(T)$. Set $M_i = \{x \in X : \lambda_i \notin \sigma_x(T)\}$. Since $\{x \in X : \sigma_x(T) \neq \sigma(T)\} = \bigcup_{i=1}^{\infty} M_i$, it is sufficient to show that each set M_i is of the first category. If $x \in M_i$, then $\lambda_i \notin S_0(T)$, and so $(T - \lambda_i)X \neq X$. Furthermore, $\lambda_i \notin \gamma_x(T)$, which implies $x \in (T - \lambda_i)X$. Thus $M_i \subset (T - \lambda_i)X \neq X$. By A.1.8, M_i is of the first category. □

Theorem 19. *Let $T \in \mathcal{B}(X)$. Then the set $\{x \in X : \sigma_\delta(T) \setminus \gamma_x(T) \neq \emptyset\}$ is of the first category.*

In particular, the set $\{x \in X : \partial\sigma(T) \setminus \gamma_x(T) \neq \emptyset\}$ is of the first category.

Proof. Choose a countable subset $\{\lambda_1, \lambda_2, \dots\}$ dense in $\sigma_\delta(T)$ and let $M_i = \{x \in X : \lambda_i \notin \gamma_x(T)\}$. Since

$$\{x \in X : \sigma_\delta(T) \setminus \gamma_x(T) \neq \emptyset\} = \bigcup_{i=1}^{\infty} M_i,$$

it is sufficient to show that each M_i is of the first category. Let $x \in M_i$. Since $\lambda_i \in \sigma_\delta(T)$, we have $(T - \lambda_i)X \neq X$. Further, $\lambda_i \notin \gamma_x(T)$, and so $x \in (T - \lambda_i)X$. Thus $M_i \subset (T - \lambda_i)X$, which is of the first category by A.1.8. □

Corollary 20. *Let $T \in \mathcal{B}(X)$. The set of all $x \in X$ for which $\sigma_x(T) = \sigma(T)$, $r_x(T) = r(T)$ and $\gamma_x(T) \supset \sigma_\delta(T)$ is residual (= complement of a set of the first category), and therefore dense in X.*

Proof. Follows from Theorems 13, 18 and 19. □

Theorem 21. *Let $T \in \mathcal{B}(X)$. Then the following statements are equivalent:*

(i) *T has SVEP;*

(ii) *$\gamma_x(T) \neq \emptyset$ for every non-zero $x \in X$.*

Proof. (i) $\Rightarrow$ (ii): If T has SVEP, then $\gamma_x(T) = \sigma_x(T)$ for all $x \in X$ and the statement follows from Proposition 8.

(ii) $\Rightarrow$ (i): Suppose that T has not SVEP. Then there exist a non-empty open subset $U \subset \mathbb{C}$ and a non-zero analytic function $f : U \to X$ such that $(T - z)f(z) = 0 \quad (z \in U)$.

Fix $\lambda \in U$ such that $f(\lambda) \neq 0$ and let $f(z) = \sum_{i=0}^{\infty} f_i(z-\lambda)^i$ be the Taylor expansion of f about λ. Then $(T - \lambda)f_0 = 0$ and $(T - \lambda)f_i = f_{i-1}$ for all $i \geq 1$. We show that $\gamma_{f_0}(T) = \emptyset$. We have

$$
\begin{aligned}
(T - z)\sum_{i=1}^{\infty} f_i(z-\lambda)^{i-1} &= \big((T-\lambda) + (\lambda - z)\big)\sum_{i=1}^{\infty} f_i(z-\lambda)^{i-1} \\
&= \sum_{i=1}^{\infty} f_{i-1}(z-\lambda)^{i-1} - \sum_{i=1}^{\infty} f_i(z-\lambda)^{i} = f_0
\end{aligned}
$$

in a neighbourhood of λ, and so $\lambda \notin \gamma_{f_0}(T)$. On the other hand, for $z \neq \lambda$ we have

$$
(T - z)\Big(\frac{f_0}{\lambda - z}\Big) = (T - \lambda)\Big(\frac{f_0}{\lambda - z}\Big) + (\lambda - z)\Big(\frac{f_0}{\lambda - z}\Big) = f_0,
$$

and so $\gamma_{f_0}(T) \subset \{\lambda\}$. Hence $\gamma_{f_0}(T) = \emptyset$. □

Comments on Chapter II

C.9.1. The basic results concerning the approximate point spectrum and the surjective spectrum of n-tuples of operators are due to Harte [Ha1], [Ha2], [Ha3].

Theorem 9.17 is a folklore (but, surprisingly, it seems that it has never been published in this form).

C.9.2. Let $T = (T_1, \dots, T_n)$ be a commuting n-tuple of operators on a Hilbert space. By the Dash lemma [Das],

$$
\sigma_l(T) = \sigma_\pi(T) = \Big\{\lambda \in \mathbb{C}^n : \sum_{j=1}^{n} (T_j - \lambda_j)^*(T_j - \lambda_j) \text{ is not invertible}\Big\}
$$

and

$$
\sigma_r(T) = \sigma_\delta(T) = \Big\{\lambda \in \mathbb{C}^n : \sum_{j=1}^{n} (T_j - \lambda_j)(T_j - \lambda_j)^* \text{ is not invertible}\Big\}.
$$

C.9.3. Theorem 9.23 was proved in [Mü8] and [Re5]; it gives a positive answer to a problem of Bollobás [Bo4].

By [Re4], if T is an operator on a Hilbert space H, then it is possible to find a Hilbert space $K \supset H$ and an extension $S \in \mathcal{B}(K)$ such that $\sigma(S) = \sigma_\pi(T)$.

C.9.4. The basic results concerning the spectra of the multiplication operators L_a and R_a (Theorem 9.26) are due to Harte [Ha3]. More generally, let $(a_1, \dots, a_n)$ and $(b_1, \dots, b_n)$ be commuting n-tuples of elements in a Banach algebra $\mathcal{A}$. Operators of the form $x \mapsto \sum_{i=1}^n a_i x b_i \quad (x \in \mathcal{A})$ are called *elementary operators*; they have been studied intensely, see, e.g., [LR], [DR], [Cu5], [Fi1], [Fi2], [CF].

C.9.5. Let T be an operator on a Banach space X such that $r(T) < 1$. Then there is an $n \in \mathbb{N}$ such that $\|T^n\| < 1$. Define a new norm on X by

$$|||x||| = \sup_{n \geq 0} \|T^n x\|. \tag{1}$$

Clearly, $|||\cdot|||$ is equivalent to the original norm and $|||Tx||| \leq |||x|||$ for each $x \in X$. Thus $T : (X, |||\cdot|||) \to (X, |||\cdot|||)$ is a contraction.

Hence for an operator $T \in \mathcal{B}(X)$ we have $r(T) = \inf\{|||T|||\}$ where the infimum is taken over all operator norms that arise from the norms on X equivalent to the original one. This is a stronger result than the corresponding statement for Banach algebras, see Corollary 1.33.

If X is a Hilbert space, then it is possible to consider only equivalent Hilbert space norms on X: we can replace (1) by

$$|||x||| = \Big(\sum_{i=0}^{n-1} \|T^i x\|^2 \Big)^{1/2}.$$

C.9.6. Complemented and uncomplemented subspaces play an important role in spectral theory. The uncomplemented subspaces are quite common. In fact, the following result is true, see [LT].

Theorem. All closed subspaces of a Banach space X are complemented if and only if X is isomorphic to a Hilbert space.

On the other hand, it is not easy to construct concrete examples of uncomplemented subspaces. The best-known example of an uncomplemented subspace is the space c_0 in ℓ^∞.

C.9.7. It is easy to see that if $T \in \mathcal{B}(X)$ is left (right) invertible, then $T^* \in \mathcal{B}(X^*)$ is right (left) invertible. The converse implications are true in reflexive Banach spaces but not in general, see [Pi1, pp. 366–367].

Thus in reflexive Banach spaces $\sigma_l(T) = \sigma_r(T^*)$ and $\sigma_r(T) = \sigma_l(T^*)$. In general, we have only $\sigma_l(T) \supset \sigma_r(T^*)$ and $\sigma_r(T) \supset \sigma_l(T^*)$.

C.9.8. If X is a finite-dimensional Banach space, then $\sigma(T) = \sigma_l(T) = \sigma_\pi(T)$ for each operator $T \in \mathcal{B}(X)$. By [GwM] there is an example of an infinite-dimensional Banach space with the same property (compare with C.8.4).

C.10.1. The notion of gap was introduced and the fundamental lemma 10.9 proved in [KKM], see also [GhK1], [Kat1].

For a survey of results concerning the reduced minimum modulus and the gap see [Kat2].

C.10.2. Instead of the gap $\widehat{\delta}(M, N)$ between two subspaces $M, N \subset X$ it is possible to consider the Hausdorff distance $\widehat{\Delta}(S_M, S_N)$ between their unit spheres, see Appendix A.4. These two quantities are closely related by

$$\widehat{\delta}(M, N) \leq \widehat{\Delta}(S_M, S_N) \leq 2\widehat{\delta}(M, N), \tag{2}$$

see [GhM]. The advantage of $\widehat{\Delta}$ is that it satisfies the triangular inequality. On the other hand, the gap $\widehat{\delta}$ is more convenient to work with.

By (2), the gap and the Hausdorff distance $\widehat{\Delta}$ define the same topology on the set of all closed subspaces of a Banach space. Moreover, this topology is complete.

C.10.3. Theorem 10.14 is due to Apostol [Ap3]. Theorem 10.17 was proved by Markus [Mar].

C.10.4. A subspace M of a Banach space is called *paraclosed* if there are a Banach space Y and $T \in \mathcal{B}(Y, X)$ with $\operatorname{Ran} T = M$. Paraclosed subspaces (sometimes also called paracomplete or operator ranges) were studied by a number of authors, see, e.g., [FW], [Em], [Cr].

Equivalently, a subspace M of $(X, \|\cdot\|)$ is paraclosed if and only if there is a complete norm $|||\cdot|||$ on M which is greater than the original norm $\|\cdot\|$.

Clearly, each closed subspace is paraclosed but the opposite is not true. Although there are many linear subspaces of X that are not paraclosed, practically all subspaces that appear in operator theory are paraclosed.

If M and L are paraclosed subspaces of X, then both $M \cap L$ and $M + L$ are paraclosed. Thus paraclosed subspaces form a lattice.

C.10.5. Let M, L be closed subspaces of a Hilbert space H. Then the gap between M and L can be expressed in a simpler way by $\widehat{\delta}(M, L) = \|P_M - P_L\|$ where P_M, P_L are orthogonal projections onto M and L, respectively.

Consequently, in Hilbert spaces the gap satisfies the triangular inequality.

By [Ap6], for $T \in \mathcal{B}(H)$ we have $\gamma(T) = \inf\bigl(\sigma(|T|) \setminus \{0\}\bigr)$ where $|T| = (T^*T)^{1/2}$.

C.11.1. The factorization of continuous vector-valued functions (Theorem 11.1) was proved by Taylor [Ta1] (formulated for exact sequences).

Lemma 11.3 is a folklore; the proof is, e.g., in [Fa2] or [Va5].

C.11.2. Factorization of analytic vector-valued functions was proved for exact sequences by Taylor [Ta1] and in general by Słodkowski [Sl4], see also [Jan]. Here we presented a different proof of the local result Theorem 11.9. The original proof of Słodkowski was done by induction and based on the following linearization result [GKL], [Boe]:

Theorem. Let $U \subset \mathbb{C}$ be an open connected set. Let $T : U \to \mathcal{B}(X, Y)$ be a function analytic on U. Then there exist a Banach space Z, operators $S, V \in \mathcal{B}(X \oplus Z, Y \oplus Z)$ and analytic functions $C_1 : U \to \mathcal{B}(X \oplus Z)$, $C_2 : U \to \mathcal{B}(Y \oplus Z)$ such that $C_1(z)$ and $C_2(z)$ are invertible operators and

$$C_2(z)(T(z) \oplus I_Z)C_1(z) = S - zV \qquad (z \in U).$$

The global result (Corollary 11.12) uses a result of Leiterer [Le]; its proof is based on the operator version of the classical Cartan lemma.

Corollary 11.14 was proved by Allan [All2], [All3], see also [Shu]. In this case an elementary proof is available.

C.11.3. Note that the statement of Theorem 11.9 is weaker than the corresponding one-dimensional result Theorem 11.5. In Theorem 11.5 we assumed only that $\inf \gamma(T(w_k)) > 0$ for a convergent sequence (w_k); in Theorem 11.9 we assumed in fact that $\inf_{z\in U} \gamma(T(z) > 0$ for some neighbourhood U of $w \in G$. This suggests the following

Conjecture. In Theorem 11.9 it is sufficient to assume that $\inf_{z\in M} \gamma(T(z)) > 0$ for some set $M \subset G$ with the property

$$f \text{ analytic in a neighbourhood } U \text{ of } w, f|(M \cap U) \equiv 0 \Longrightarrow f \equiv 0 \text{ on } U.$$

C.11.4. A factorization result analogous to Theorem 11.1 for C^∞-functions was proved by Mantlik [Man].

C.12.1. Kato and Saphar operators were studied under various names by a number of authors, see, e.g., [Kat1], [Ka1], [GlK], [Gr1], [Ap6], [Mb1], [MO1], [MO2], [Mü15], [Sm1], [Sm2], [Sap], [Ra4]. Many authors call Kato and Saphar operators semi-regular and regular, respectively. In the present monograph we prefer the names introduced by Schmoeger, since the words like "regular" are overused in mathematics (by a regular operator many mathematicians would understand an invertible operator).

C.12.2. For $T \in \mathcal{B}(X)$ it is possible to define a family of ranges $R_\alpha(T)$ indexed by ordinal numbers by $R_0(T) = X$, $R_{\alpha+1}(T) = TR_\alpha(T)$ and $R_\alpha(T) = \bigcap_{\beta<\alpha} R_\beta(T)$ for limit ordinals α. The transfinite ranges form a monotone family; its intersection $\operatorname{co}(T)$ is called the *couer* of T. Properties of the transfinite ranges were studied in [Sap]. It is easy to see that $T\operatorname{co}(T) = \operatorname{co}(T)$ and $\operatorname{co}(T)$ is the maximal subspace (not necessarily closed) with this property. Clearly, $\mathrm{co}(T) \subset R^\infty(T)$. For Kato operators we have $\mathrm{co}(T) = R^\infty(T)$ but this equality is not true in general.

C. 12.3. Propositions 12.28, 12.29 and 13.11 implying $\tilde{\sigma}(TS) \setminus \{0\} = \tilde{\sigma}(ST) \setminus \{0\}$ for many types of spectrum were proved by Barnes [Ba3].

C.13.1. For information about generalized inverses see Groetsch [Gro]. The generalized inverses originated in the Moore-Penrose inverses of matrices [Mo], [Pe].

C.13.2. Let $T \in \mathcal{B}(X)$ be Saphar and let $S : U \to \mathcal{B}(X)$ be the analytic generalized inverse $S(z) = \sum_{i=0}^\infty S^{i+1} z^i$ constructed in a neighbourhood U of 0, see Theorem 13.9. It is easy to verify that $\operatorname{Ker} S(z)$ and $\operatorname{Ran} T(z)$ are constant and S satisfies the resolvent identity

$$S(z) - S(w) = (z - w)S(z)S(w) \qquad (z, w \in U)$$

cf. C.8.5. It is an interesting open question, see [ApC1], [ApC2], whether there is a global analytic generalized inverse defined on $\mathbb{C} \setminus \sigma_{\mathrm{Sap}}(T)$ (or at least on a neighbourhood of a given compact set $K \subset \mathbb{C} \setminus \sigma_{\mathrm{Sap}}(T)$) satisfying this additional condition. For a positive answer in Hilbert spaces, see [BM].

C.13.3. The existence of a global analytic generalized inverse on the complement of the Saphar spectrum (Theorem 13.10) was proved in [Shu], see also [Mü15].

C.13.4. The Saphar spectrum in Banach algebras was studied in [Ko1] and [Mb2].

C.14.1. The local spectral theory was originated by Dunford [Du2]. It was motivated by the theory of scalar and decomposable operators that were studied intensely by a number of authors, see, e.g., [CF]. For a recent survey see [LN].

The spectral mapping property is due to Vasilescu [Va1], see also [Vr]; the present proof was given in [KM2]. The existence of many points with the local spectrum equal to the (global) spectrum was proved by Vrbová [Vr].

C.14.2. Let $T \in \mathcal{B}(X)$ and $x \in X$. In general, the limit $\lim_{n\to\infty} \|T^n x\|^{1/n}$ does not exist. In fact [Dan], the set of all accumulation points of the sequence $(\|T^n x\|^{1/n})$ is the whole interval $\langle \liminf \|T^n x\|^{1/n}, r_x(T)\rangle$.

C.14.3. Let $T \in \mathcal{B}(X)$ and $x \in X$. Then the local resolvent $g : \mathbb{C} \setminus \sigma_x(T) \to X$ satisfying $(T - z)g(z) = x \quad (z \in \mathbb{C} \setminus \sigma_x(T))$ is a uniquely determined analytic function.

For functions f analytic on a neighbourhood of $\sigma_x(T)$ define the local functional calculus by

$$f(T)x = \frac{1}{2\pi i} \int_\Gamma f(z)g(z)\,\mathrm{d}z,$$

where Γ is a contour surrounding $\sigma_x(T)$. In general, $f(T)$ is defined only for those $x \in X$ such that $\sigma_x(T)$ is contained in the domain of definition of f.

Chapter III

Essential Spectrum

In this chapter we study various types of essential spectra of operators on a Banach space X. They are closely connected with the Calkin algebra $\mathcal{B}(X)/\mathcal{K}(X)$, where $\mathcal{K}(X)$ denotes the ideal of compact operators.

We start with the classical theory of compact and Fredholm operators.

15 Compact operators

Definition 1. Let X, Y be Banach spaces. An operator $T \in \mathcal{B}(X, Y)$ is called *compact* if the set TB_X is totally bounded (i.e., if $\overline{TB_X}$ is compact, where B_X denotes the closed unit ball in X).

We say that T is of *finite rank* if $\dim \operatorname{Ran} T < \infty$.

The set of all compact (finite-rank) operators from X to Y will be denoted by $\mathcal{K}(X, Y)$ and $\mathcal{F}(X, Y)$, respectively. If $Y = X$, then we write $\mathcal{K}(X) = \mathcal{K}(X, X)$ and $\mathcal{F}(X) = \mathcal{F}(X, X)$ for short.

Clearly, $T \in \mathcal{B}(X, Y)$ is compact if and only if for every $\varepsilon > 0$ there exists a finite set $\{x_1, \dots, x_n\} \subset B_X$ such that $\min\{\|Tx - Tx_i\| : 1 \le i \le n\} \le \varepsilon$ for all $x \in B_X$.

The next theorem summarizes the basic properties of compact and finite-rank operators.

Theorem 2. *Let X and Y be Banach spaces. Then:*

(i) *$\mathcal{K}(X, Y)$ is a closed subspace of $\mathcal{B}(X, Y)$;*

(ii) *$\mathcal{F}(X, Y)$ is a subspace of $\mathcal{B}(X, Y)$ and $\overline{\mathcal{F}(X, Y)} \subset \mathcal{K}(X, Y)$;*

(iii) *if X_1 and Y_1 are Banach spaces, $U \in \mathcal{B}(X_1, X)$, $T \in \mathcal{K}(X, Y)$ and $S \in \mathcal{B}(Y, Y_1)$, then $STU \in \mathcal{K}(X_1, Y_1)$. If $T \in \mathcal{F}(X, Y)$, then $STU \in \mathcal{F}(X_1, Y_1)$;*

(iv) *in particular, if* $Y = X$, *then* $\mathcal{F}(X)$ *is a 2-sided ideal and* $\mathcal{K}(X)$ *is a closed 2-sided ideal in* $\mathcal{B}(X)$;

(v) *if* $T \in \mathcal{K}(X, Y)$, *then* T *is of finite rank if and only if* $\operatorname{Ran} T$ *is closed.*

Proof. (i) Let $T, S \in \mathcal{K}(X, Y)$ and let $\varepsilon > 0$. Then there exist finite subsets $\{x_1, \dots, x_n\}$ and $\{x'_1, \dots, x'_m\}$ of B_X such that $\min\{\|Tx - Tx_i\| : 1 \le i \le n\} \le \varepsilon/2$ and $\min\{\|Sx - Sx'_j\| : 1 \le j \le m\} \le \varepsilon/2$ for every $x \in B_X$. Consider the finite set $\{Tx_i + Sx'_j : 1 \le i \le n, 1 \le j \le m\}$. For $x \in B_X$ we have

$$\min\{\|(T+S)x - (Tx_i + Sx'_j)\| : 1 \le i \le n, 1 \le j \le m\} \le \varepsilon.$$

Since ε was arbitrary, the set $\overline{(T+S)B_X}$ is compact.

In a similar way it is possible to show that the set $\mathcal{K}(X, Y)$ is closed and that a scalar multiple of a compact operator is compact.

(ii) The inclusion $\mathcal{F}(X, Y) \subset \mathcal{K}(X, Y)$ follows from the fact that closed balls in finite-dimensional Banach spaces are compact.

(iii) Let $U \in \mathcal{B}(X_1, X)$, $T \in \mathcal{K}(X, Y)$ and $S \in \mathcal{B}(Y, Y_1)$. Then $\overline{TB_X}$ and $S(\overline{TB_X})$ are compact. Hence the set $\overline{(STU)B_X} \subset \overline{ST(\|U\| \cdot B_X)} = \|U\| \cdot S(\overline{TB_X})$ is also compact and $STU \in \mathcal{K}(X, Y)$.

The statement for finite-rank operators is clear.

This implies also (iv).

(v) Clearly, each finite-rank operator has closed range.

For the converse, let $T : X \to Y$ be compact and $\operatorname{Ran} T$ closed. By the open mapping theorem, there is a positive constant k with $TB_X \supset k \cdot B_{\operatorname{Ran} T}$. Since T is compact, we conclude that $k \cdot B_{\operatorname{Ran} T}$ is compact. Hence $\dim \operatorname{Ran} T < \infty$. □

Proposition 3. *If* $T \in \mathcal{K}(X, Y)$, *then* $\overline{\operatorname{Ran} T}$ *is separable.*

Proof. We have $\overline{\operatorname{Ran} T} = (\bigcup_{k=1}^{\infty} kTB_X)^-$ and kTB_X is totally bounded for every k. □

Theorem 4. *Let* $T \in \mathcal{B}(X, Y)$. *Then* T *is compact if and only if* T^* *is compact.*

Proof. Suppose that T is compact and let $\varepsilon > 0$. We must show that there exists a finite subset $\{y_1^*, \dots, y_p^*\} \subset B_{Y^*}$ such that for every $y^* \in B_{Y^*}$ there exists r, $1 \le r \le p$ with $\|T^*y^* - T^*y_r^*\| \le \varepsilon$.

Since T is compact, there exists a finite subset $\{x_1, \dots, x_n\} \subset B_X$ such that $\min\{\|Tx - Tx_j\| : 1 \le j \le n\} \le \varepsilon/3$ for every $x \in B_X$.

The set $\{(\langle Tx_1, y^* \rangle, \dots, \langle Tx_n, y^* \rangle) : y^* \in B_{Y^*}\}$ is a bounded subset of $\mathbb{C}^n$, therefore there exists a finite subset $\{y_1^*, \dots, y_p^*\}$ of B_{Y^*} such that for each $y^* \in B_{Y^*}$ there exists $r \in \{1, \dots, p\}$ with the property

$$|\langle Tx_j, y^* - y_r^* \rangle| \le \varepsilon/3 \qquad (1 \le j \le n). \tag{1}$$

We show that $\{y_1^*, \dots, y_p^*\}$ is the required subset of B_{Y^*}.

Let $y^* \in B_{Y^*}$. Find $r \in \{1, \ldots, p\}$ with (1). Let $x \in B_X$. Then there is a $j \in \{1, \ldots, n\}$ such that $\|Tx - Tx_j\| \le \varepsilon/3$ and we have

$$\begin{aligned}
&|\langle x, T^*y^* \rangle - \langle x, T^*y_r^* \rangle| \\
&\quad \le |\langle x - x_j, T^*y^* \rangle| + |\langle x_j, T^*(y^* - y_r^*) \rangle| + |\langle x_j - x, T^*y_r^* \rangle| \\
&\quad = |\langle Tx - Tx_j, y^* \rangle| + |\langle Tx_j, y^* - y_r^* \rangle| + |\langle Tx_j - Tx, y_r^* \rangle| \le \varepsilon.
\end{aligned}$$

Thus

$$\|T^*y^* - T^*y_r^*\| = \sup\{|\langle x, T^*y^* - T^*y_r^* \rangle| : x \in B_X\} \le \varepsilon$$

and T^* is compact.

Suppose now that T^* is compact. Then T^{**} is compact and so is $T = T^{**}|X$, since $\overline{TB_X} = \overline{T^{**}(B_{X^{**}} \cap X)} \subset \overline{T^{**}B_{X^{**}}}$, which is compact. □

Theorem 5. *Let $T \in \mathcal{B}(X, Y)$. The following conditions are equivalent:*

(i) *T is compact;*

(ii) *if $(x_\alpha)_\alpha$ is a net of elements of B_X and $x_\alpha \to 0$ weakly, then $\|Tx_\alpha\| \to 0$;*

(iii) *the restriction $T|B_X$ is a continuous mapping from B_X with the weak topology into Y with the norm topology;*

(iv) *for every $\varepsilon > 0$ there exists a closed subspace $M \subset X$ with $\operatorname{codim} M < \infty$ such that $\sup\{\|Tx\| : x \in M, \|x\| = 1\} \le \varepsilon$.*

Proof. (i) ⇒ (ii): Let (x_α) be a net of elements of B_X, $x_\alpha \to 0$ weakly. It is easy to show that also $Tx_a \to 0$ weakly. Suppose on the contrary that $\|Tx_\alpha\| \not\to 0$. Then there exists a positive constant c and a subnet $(x_{\alpha_\beta})_\beta$ such that $\|Tx_{\alpha_\beta}\| \ge c$ for all β. Since $\overline{TB_X}$ is compact, the net (Tx_{α_β}) has an accumulation point y. Clearly, $\|y\| \ge c$. On the other hand, $Tx_{\alpha_\beta} \to 0$ weakly and so $y = 0$, a contradiction.

(ii) ⇒ (iii): Let $(x_\alpha) \subset B_X$, $x_\alpha \to x$ weakly. Then $x_\alpha - x \to 0$ weakly, and so $\|Tx - Tx_\alpha\| \to 0$. Hence $Tx_\alpha \to Tx$ in the norm topology.

(iii) ⇒ (iv): Suppose that (iv) is not true, so there exists $c > 0$ such that, for every closed subspace $M \subset X$ of finite codimension, there exists $x_M \in M$ with $\|x_M\| = 1$ and $\|Tx_M\| \ge c$. Consider the net $(x_M)_M$ directed by the inclusion, $M \ge M' \Leftrightarrow M \subset M'$. It is easy to see that $x_M \to 0$ weakly and $\|Tx_M\| \not\to 0$.

(iv) ⇒ (i): Suppose that T is not compact, so there exists $c > 0$ such that TB_X can not be covered by a finite number of balls of radius c. Choose $x_1 \in B_X$ arbitrarily and construct inductively a sequence (x_i) of points in B_X such that $\|Tx_i - Tx_j\| \ge c$ for all $i, j \in \mathbb{N}, i \ne j$. Clearly, $\|x_i - x_j\| \ge \frac{c}{\|T\|}$.

Let $M \subset X$ be a closed subspace of finite codimension and let $P \in \mathcal{B}(X)$ be a projection onto M. Let $\varepsilon > 0$. Since $\dim \operatorname{Ran}(I - P) < \infty$, we can find $j, k \in \mathbb{N}$, $j \ne k$ such that $\|(I - P)x_j - (I - P)x_k\| < \varepsilon$. Then

$$\|P(x_j - x_k)\| \le \|x_j - x_k\| + \|(I - P)(x_j - x_k)\| \le 2 + \varepsilon$$

and

$$\|TP(x_j - x_k)\| \geq \|Tx_j - Tx_k\| - \|T(I-P)(x_j - x_k)\| \geq c - \varepsilon\|T\|.$$

Thus

$$\sup\{\|Tu\| : u \in M, \|u\| = 1\} \geq \frac{\|TP(x_j - x_k)\|}{\|P(x_j - x_k)\|} \geq \frac{c - \varepsilon\|T\|}{2+\varepsilon}.$$

Letting $\varepsilon \to 0$ gives $\sup\{\|Tu\| : u \in M, \|u\| = 1\} \geq c/2$ and (iv) is not true. □

Corollary 6. *Let $T \in \mathcal{B}(X, Y)$ be a compact operator and (x_n) a sequence of elements of X, $x_n \to 0$ weakly. Then $\|Tx_n\| \to 0$.*

Proof. A weakly converging sequence is bounded by the Banach-Steinhaus theorem. □

Theorem 7. *Let $T \in \mathcal{B}(X)$ be a compact operator and $\lambda \in \mathbb{C}, \lambda \neq 0$. Then* $\mathrm{Ran}(T - \lambda)$ *is closed.*

Proof. Let M be a subspace of X of finite codimension satisfying the condition $\sup\{\|Tx\| : x \in M, \|x\| = 1\} \leq \frac{|\lambda|}{2}$. Then

$$\|(T-\lambda)x\| \geq |\lambda| \cdot \|x\| - \|Tx\| \geq \frac{|\lambda|}{2}\|x\|$$

for all $x \in M$. Thus the restriction $(T-\lambda)|M : M \to X$ is bounded below and $(T-\lambda)M$ is closed. Let N be a finite-dimensional subspace of X such that $X = M \oplus N$. Then $(T-\lambda)X = (T-\lambda)M + (T-\lambda)N$, where $\dim(T-\lambda)N < \infty$. Hence $\mathrm{Ran}(T-\lambda)$ is closed. □

Lemma 8. *Let X be a Banach space, let $M \subset X$ be a closed subspace, $M \neq X$ and let $\varepsilon > 0$. Then there exists $x \in X$ such that $\|x\| \leq 1 + \varepsilon$ and $\mathrm{dist}\{x, M\} = 1$.*

If $\operatorname{codim} M = \infty$, then there exists a sequence $(x_i)_{i=1}^{\infty}$ of elements of X such that $\|x_k\| \leq 1 + \varepsilon$ and $\mathrm{dist}\{x_{k+1}, M \vee \bigvee_{i=1}^{k} x_i\} = 1$ for every k.

Proof. Choose any $x_0 \in X$ with $\|x_0 + M\|_{X/M} = 1$. Since $1 = \|x_0 + M\|_{X/M} = \inf\{\|x_0 + m\| : m \in M\}$, we can find $m \in M$ such that $x = x_0 + m$ satisfies $\mathrm{dist}\{x, M\} = \mathrm{dist}\{x_0, M\} = 1$ and $\|x\| \leq 1 + \varepsilon$.

Using the first statement repeatedly gives the second statement. □

Theorem 9. *Let $T \in \mathcal{B}(X)$ be a compact operator and let $\lambda \in \mathbb{C}$, $\lambda \neq 0$. Then* $\dim \mathrm{Ker}(T - \lambda) < \infty$.

Proof. Suppose on the contrary that $\dim \mathrm{Ker}(T-\lambda) = \infty$. By Lemma 8, there exists a sequence $(x_i)_{i=1}^{\infty}$ of elements of $\mathrm{Ker}(T-\lambda)$ such that $\|x_k\| = 1$ and $\mathrm{dist}\{x_{k+1}, \bigvee_{i=1}^{k} x_i\} \geq 1/2$ for all k. Then (x_i) is a bounded sequence and (Tx_i) contains no convergent subsequence, since $\|Tx_i - Tx_j\| = \|\lambda x_i - \lambda x_j\| = |\lambda| \cdot \|x_i - x_j\| \geq \frac{|\lambda|}{2}$ for all $i \neq j$. □

Theorem 10. *Let $K \in \mathcal{B}(X)$ be a compact operator. Then:*

(i) *there exists $k \in \mathbb{N}$ such that* $\operatorname{Ker}(I+K)^{k+1} = \operatorname{Ker}(I+K)^k$;

(ii) *there exists $j \in \mathbb{N}$ such that* $\operatorname{Ran}(I+K)^{j+1} = \operatorname{Ran}(I+K)^j$;

(iii) $\dim \operatorname{Ker}(I+K) \geq \operatorname{codim} \operatorname{Ran}(I+K)$.

Proof. (i) Write $T = I + K$. We have $\operatorname{Ker} T \subset \operatorname{Ker} T^2 \subset \cdots$. Suppose on the contrary that $\operatorname{Ker} T^{k+1} \neq \operatorname{Ker} T^k$ for all n. Using Lemma 8 inductively we find a sequence of elements $x_k \in \operatorname{Ker} T^k$ such that $\|x_k\| \leq 2$ and $\operatorname{dist}\{x_k, \operatorname{Ker} T^{k-1}\} = 1$ for each k. For $i > j$ we have

$$\|Kx_j - Kx_i\| = \|Tx_j - x_j - Tx_i + x_i\| \geq \operatorname{dist}\{x_i, \operatorname{Ker} T^{i-1}\} = 1,$$

since $Tx_j - x_j - Tx_i \in \operatorname{Ker} T^{i-1}$. Thus the sequence (Kx_i) contains no convergent subsequence, which is a contradiction with the compactness of K. Hence there is a $k \in \mathbb{N}$ such that $\operatorname{Ker} T^{k+1} = \operatorname{Ker} T^k$.

(ii) For each $k \geq 0$, the operator $(I+K)^k$ can be expressed as $I + K'$ for some compact operator $K' \in \mathcal{B}(X)$. Thus $\operatorname{Ran}(I+K)^k$ is closed for each k.

By (i), there exists $j \geq 0$ such that $\operatorname{Ker}(I_{X^*} + K^*)^{j+1} = \operatorname{Ker}(I_{X^*} + K^*)^j$. Thus $\operatorname{Ran}(I+K)^{j+1} = {}^\perp\operatorname{Ker}(I_{X^*} + K^*)^{j+1} = {}^\perp\operatorname{Ker}(I_{X^*} + K^*)^j = \operatorname{Ran}(I+K)^j$.

(iii) Let $T = I + K$ and let k satisfy $\operatorname{Ker} T^{k+1} = \operatorname{Ker} T^k$. By Theorem 9, we have $\dim \operatorname{Ker} T < \infty$. Since $T^k = (I+K)^k$, we can write T^k in the form $I + K'$ for some compact operator K', and so $\dim \operatorname{Ker} T^k < \infty$.

Let $a_1, \dots, a_m$ be a basis of $\operatorname{Ker} T$ and let $a_1, \dots, a_m, a_{m+1}, \dots, a_r$ be its completion to a basis of $\operatorname{Ker} T^k = \operatorname{Ker} T^{k+1}$. Since $T \operatorname{Ker} T^k \subset \operatorname{Ker} T^k$, we have $T \operatorname{Ker} T^k \subset \bigvee\{Ta_{m+1}, \dots, Ta_r\}$, and so $\dim T \operatorname{Ker} T^k \leq r - m$. Choose another basis $b_1, \dots, b_r$ in $\operatorname{Ker} T^k$ such that $b_{n+1}, \dots, b_r$ is a basis of $T \operatorname{Ker} T^k$. Thus $r - n = \dim T \operatorname{Ker} T^k \leq r - m$, and so $m \leq n$. We now prove that $\operatorname{codim} \operatorname{Ran} T \geq n$. To show this, it is sufficient to prove that $b_1, \dots, b_n$ are linearly independent modulo $\operatorname{Ran} T$.

Suppose on the contrary that $\sum_{i=1}^n \alpha_i b_i \in \operatorname{Ran} T$ for some $\alpha_i \in \mathbb{C}$. Let $b \in X$ satisfy $\sum_{i=1}^n \alpha_i b_i = Tb$. Since $\sum_{i=1}^n \alpha_i b_i \in \operatorname{Ker} T^k$, we have $b \in \operatorname{Ker} T^{k+1} = \operatorname{Ker} T^k$. Thus $Tb \in T \operatorname{Ker} T^k$ and Tb is a linear combination of $b_{n+1}, \dots, b_r$. Therefore $\alpha_1 = \cdots = \alpha_n = 0$. Hence $\operatorname{codim} \operatorname{Ran} T \geq n \geq m = \dim \operatorname{Ker} T$. □

Theorem 11. *Let $K \in \mathcal{B}(X)$ be a compact operator and let $\lambda \in \mathbb{C}$, $\lambda \neq 0$. Then*

$$\begin{aligned} \dim \operatorname{Ker}(K - \lambda) &= \dim \operatorname{Ker}(K^* - \lambda) \\ &= \operatorname{codim} \operatorname{Ran}(K - \lambda) = \operatorname{codim} \operatorname{Ran}(K^* - \lambda). \end{aligned}$$

Proof. By A.1.14 and Theorem 7, $\dim \operatorname{Ker}(K - \lambda) = \operatorname{codim} \operatorname{Ran}(K^* - \lambda)$ and $\dim \operatorname{Ker}(K^* - \lambda) = \operatorname{codim} \operatorname{Ran}(K - \lambda)$. By the preceding theorem, we have $\dim \operatorname{Ker}(K - \lambda) \geq \operatorname{codim} \operatorname{Ran}(K - \lambda) = \dim \operatorname{Ker}(K^* - \lambda)$. Since K^* is also compact, $\dim \operatorname{Ker}(K^* - \lambda) \geq \dim \operatorname{Ker}(K^{**} - \lambda)$. Further, $K - \lambda = (K^{**} - \lambda)|X$, and so $\operatorname{Ker}(K - \lambda) \subset \operatorname{Ker}(K^{**} - \lambda)$. Hence $\dim \operatorname{Ker}(K - \lambda) = \dim \operatorname{Ker}(K^* - \lambda)$. □

Lemma 12. *Let $T \in \mathcal{B}(X)$, let $\lambda_1, \lambda_2, \dots$ be distinct non-zero eigenvalues of T, and let $x_1, x_2, \dots$ be corresponding non-zero eigenvectors. Then the vectors $x_i \quad (i \in \mathbb{N})$ are linearly independent.*

Proof. We prove by induction on n that $x_1, \dots, x_n$ are linearly independent. This is clear for $n = 1$ since $x_1 \neq 0$.

Suppose that $x_1, \dots, x_n$ are linearly independent and $x_1, \dots, x_{n+1}$ are linearly dependent. So $x_{n+1} = \sum_{i=1}^n \alpha_i x_i$ for some $\alpha_i \in \mathbb{C}$. We have $x_{n+1} = \lambda_{n+1}^{-1} T x_{n+1} = \lambda_{n+1}^{-1} \sum_{i=1}^n \alpha_i T x_i = \sum_{i=1}^n \frac{\lambda_i \alpha_i}{\lambda_{n+1}} x_i$. Thus $\alpha_i = \frac{\lambda_i \alpha_i}{\lambda_{i+1}}$, which implies $\alpha_i = 0$ for $i = 1, \dots, n$. Hence $x_{n+1} = 0$, a contradiction. □

Theorem 13. *Let $K \in \mathcal{B}(X)$ be a compact operator, $\lambda \in \sigma(K)$ and $\lambda \neq 0$. Then λ is an isolated point of $\sigma(K)$.*

Proof. Suppose on the contrary that there exists a sequence $(\lambda_k) \subset \sigma(K)$ converging to λ. By Theorem 11, λ_k are eigenvalues of K. Let x_k be corresponding eigenvectors, i.e., $x_k \neq 0$, $Tx_k = \lambda_k x_k \quad (k \in \mathbb{N})$. For $k \in \mathbb{N}$ let $M_k = \bigvee\{x_1, \dots, x_k\}$. By Lemma 12, $M_1 \neq M_2 \neq \cdots$. Clearly, $KM_k \subset M_k$ and $(K - \lambda_k)M_k \subset M_{k-1} \quad (k \in \mathbb{N})$. By Lemma 8, we can find elements $y_k \in M_k$ such that $\|y_k\| \leq 2$ and $\operatorname{dist}\{y_k, M_{k-1}\} = 1 \quad (k \geq 2)$. For $i > j > 1$ we have

$$\|Ky_i - Ky_j\| = \|\lambda_i y_i + (K - \lambda_i)y_i - \lambda_j y_j - (K - \lambda_j)y_j\| \geq \operatorname{dist}\{\lambda_i y_i, M_{i-1}\} = |\lambda_i|.$$

Since $\lambda_k \to \lambda \neq 0$, the sequence (Ky_i) contains no convergent subsequence. This is a contradiction with the compactness of K. □

Corollary 14. *Let $K \in \mathcal{B}(X)$ be a compact operator. Then $\sigma(K)$ is at most countable. If $\lambda \in \sigma(K)$, $\lambda \neq 0$, then λ is an isolated point of $\sigma(K)$ and*

$$\begin{aligned} 0 \neq \dim \operatorname{Ker}(K - \lambda) &= \operatorname{codim} \operatorname{Ran}(K - \lambda) \\ &= \dim \operatorname{Ker}(K^* - \lambda) = \operatorname{codim} \operatorname{Ran}(T^* - \lambda) < \infty. \end{aligned}$$

Examples 15. (i) An important example of compact operators are integral operators.

Consider the Banach space $C\langle a, b\rangle$ of all continuous complex-valued functions on a bounded closed interval $\langle a, b\rangle$ with the sup-norm.

A continuous function $K(s,t)$ defined on $\langle a, b\rangle \times \langle a, b\rangle$ defines an operator T on $C\langle a, b\rangle$ by

$$(Tf)(s) = \int_a^b K(s,t) f(t) \mathrm{d}t.$$

It follows from classical results of analysis that T is a compact operator.

The classical Fredholm integral equation is

$$\lambda f(s) - \int_a^b K(s,t) f(t) \mathrm{d}t = g(s) \qquad (a \leq s \leq b),$$

where $g \in C\langle a, b\rangle$ is given, λ is a parameter and f is unknown. Clearly, we can write the equation as $(\lambda I - T)f = g$.

This was the original motivation that led to the study of operators of the form $\lambda I - T$ where T is compact, see Riesz [Ri2]. The theory of these operators is sometimes referred to as the Riesz-Schauder theory.

(ii) Let X, Y be Banach spaces, $x^* \in X^*$ and $y \in Y$. Denote by $y \otimes x^* : X \to Y$ the operator defined by $(y \otimes x^*)x = \langle x, x^* \rangle y \quad (x \in X)$. Obviously, $\|y \otimes x^*\| = \|y\| \cdot \|x^*\|$ and $\dim \operatorname{Ran}(y \otimes x^*) = 1$.

Finite-rank operators are precisely finite linear combinations of operators of this form.

Operators that can be expressed as $\sum_{i=1}^{\infty} y_i \otimes x_i^*$ for some $y_i \in Y$ and $x_i^* \in X^*$ with $\sum_i \|y_i\| \cdot \|x_i^*\| < \infty$ are called *nuclear*. It is easy to see that nuclear operators are norm-limits of finite-rank operators and therefore they are compact.

Nuclear operators acting on X form a non-closed two-sided ideal.

(iii) A diagonal operator $\operatorname{diag}(c_1, c_2, \dots)$ acting on a separable Hilbert space is compact if and only if $\lim c_i = 0$; it is nuclear if and only if $\sum |c_i| < \infty$. The same characterization is true for unilateral weighted shifts with weights c_i.

(iv) Let H be a Hilbert space with an orthonormal basis $(e_i)_{i \geq 1}$. Operators $T \in \mathcal{B}(H)$ defined by $Te_i = \sum_{j=1}^{\infty} \alpha_{i,j} e_j \quad (j \geq 1)$, where $\alpha_{i,j} \in \mathbb{C}$ satisfy $\sum_{i,j} |\alpha_{i,j}|^2 < \infty$, are called *Hilbert-Schmidt*. Clearly, $\sum_{i,j} |\alpha_{i,j}|^2 = \sum_j \|Te_j\|^2$; this number does not depend on the choice of an orthonormal basis (e_j).

Hilbert-Schmidt operators are an important example of compact operators.

(v) The Volterra operator $V : L^2(0,1) \to L^2(0,1)$ is defined by $(Vf)(x) = \int_0^x f(y) dy$. V is an example of a compact operator with $\sigma(V) = \{0\}$.

(vi) Let $1 \leq p < q < \infty$. By the Pitt theorem, every operator $T : \ell^q \to \ell^p$ is compact. Furthermore, if X is reflexive, then each operator $T \in \mathcal{B}(X, \ell^1)$ or $T \in \mathcal{B}(c_0, X)$ is compact.

16 Fredholm and semi-Fredholm operators

Definition 1. Let X, Y be Banach spaces, let $T \in \mathcal{B}(X, Y)$. We say that:

(i) T is *upper semi-Fredholm* if $\operatorname{Ran} T$ is closed and $\dim \operatorname{Ker} T < \infty$;

(ii) T is *lower semi-Fredholm* if $\operatorname{codim} \operatorname{Ran} T < \infty$;

(iii) T is *Fredholm* if $\dim \operatorname{Ker} T < \infty$ and $\operatorname{codim} \operatorname{Ran} T < \infty$.

The set of all upper semi-Fredholm, lower semi-Fredholm and Fredholm operators will be denoted by $\Phi_+(X, Y)$, $\Phi_-(X, Y)$ and $\Phi(X, Y)$, respectively. Obviously, $\Phi(X, Y) = \Phi_+(X, Y) \cap \Phi_-(X, Y)$. Operators in $\Phi_+(X, Y) \cup \Phi_-(X, Y)$ will be called shortly semi-Fredholm. If $Y = X$, then we write $\Phi(X) = \Phi(X, X)$ and similarly $\Phi_+(X)$ and $\Phi_-(X)$ for short.

The definition of semi-Fredholm operators is seemingly asymmetrical. However, the condition $\operatorname{codim} \operatorname{Ran} T < \infty$ implies that the range of T is closed.

Lemma 2. *Let $T \in \mathcal{B}(X,Y)$ and let $F \subset Y$ be a finite-dimensional subspace. Suppose that $\operatorname{Ran} T + F$ is closed. Then $\operatorname{Ran} T$ is closed.*

In particular, if $\operatorname{codim} \operatorname{Ran} T < \infty$, then $\operatorname{Ran} T$ is closed.

Proof. Let $F_0 = \operatorname{Ran} T \cap F$ and choose a subspace F_1 such that $F_0 \oplus F_1 = F$. Let $S : (X/\operatorname{Ker} T) \oplus F_1 \to Y$ be the operator defined by $S(x + \operatorname{Ker} T) \oplus f = Tx + f$. Then $\operatorname{Ran} S = \operatorname{Ran} T + F$ and S is one-to-one. Hence S is bounded below, and consequently, the space $\operatorname{Ran} T = S(X/\operatorname{Ker} T \oplus \{0\})$ is closed. □

Definition 3. Let $T \in \mathcal{B}(X,Y)$ be an operator with closed range. We write $\alpha(T) = \dim \operatorname{Ker} T$ and $\beta(T) = \operatorname{codim} \operatorname{Ran} T$. If T is semi-Fredholm (either upper or lower), then the *index* of T is defined by $\operatorname{ind}(T) = \alpha(T) - \beta(T)$.

Clearly, if T is Fredholm, then $\operatorname{ind}(T) < \infty$. If $T \in \Phi_+(X,Y) \setminus \Phi(X,Y)$, then $\operatorname{ind}(T) = -\infty$. If $T \in \Phi_-(X,Y) \setminus \Phi(X,Y)$, then $\operatorname{ind}(T) = +\infty$.

Theorem 4. *Let $T \in \mathcal{B}(X,Y)$ be an operator with closed range. Then $\alpha(T^*) = \beta(T)$ and $\beta(T^*) = \alpha(T)$. Thus:*

$$\begin{aligned} T \in \Phi(X,Y) &\quad\Leftrightarrow\quad T^* \in \Phi(Y^*,X^*); \\ T \in \Phi_+(X,Y) &\quad\Leftrightarrow\quad T^* \in \Phi_-(Y^*,X^*); \\ T \in \Phi_-(X,Y) &\quad\Leftrightarrow\quad T^* \in \Phi_+(Y^*,X^*). \end{aligned}$$

If T is semi-Fredholm, then $\operatorname{ind} T^* = -\operatorname{ind} T$.

Proof. We have

$$\beta(T) = \dim\, Y/\operatorname{Ran} T = \dim\,(Y/\operatorname{Ran} T)^* = \dim \operatorname{Ran} T^{\perp} = \dim \operatorname{Ker} T^* = \alpha(T^*).$$

Similarly,

$$\begin{aligned} \alpha(T) &= \dim \operatorname{Ker} T = \dim\,(\operatorname{Ker} T)^* \\ &= \dim\, X^*/(\operatorname{Ker} T)^{\perp} = \dim\, X^*/\operatorname{Ran} T^* = \beta(T^*). \end{aligned}$$

□

Theorem 5. *Let X, Y and Z be Banach spaces, $T \in \mathcal{B}(X,Y)$ and $S \in \mathcal{B}(Y,Z)$. Then:*

(i) *if T and S are lower semi-Fredholm, then ST is lower semi-Fredholm;*

(ii) *if T and S are upper semi-Fredholm, then ST is upper semi-Fredholm;*

(iii) *if T and S are Fredholm, then ST is Fredholm.*

Proof. (i) Let $Y_0 \subset Y$ and $Z_0 \subset Z$ be finite-dimensional subspaces such that $\operatorname{Ran} T + Y_0 = Y$ and $\operatorname{Ran} S + Z_0 = Z$. Then $Z = Z_0 + SY = Z_0 + S \operatorname{Ran} T + SY_0 = \operatorname{Ran}(ST) + (Z_0 + SY_0)$, where $\dim(Z_0 + SY_0) < \infty$. Thus $ST \in \Phi_-(X,Z)$.

(ii) If T, S are upper semi-Fredholm, then T^*, S^* are lower semi-Fredholm and, by (i), T^*S^* is lower semi-Fredholm. Thus ST is upper semi-Fredholm.

(iii) Follows from (i) and (ii). □

Theorem 6. *Let X, Y and Z be Banach spaces, $T \in \mathcal{B}(X,Y)$ and $S \in \mathcal{B}(Y,Z)$. Then:*

(i) *if ST is lower semi-Fredholm, then S is lower semi-Fredholm;*

(ii) *if ST is upper semi-Fredholm, then T is upper semi-Fredholm;*

(iii) *if ST is Fredholm, then S is lower semi-Fredholm and T is upper semi-Fredholm.*

Proof. (i) We have $\operatorname{Ran} S \supset \operatorname{Ran}(ST)$, so $\operatorname{codim} \operatorname{Ran} S \le \operatorname{codim} \operatorname{Ran}(ST) < \infty$.

(ii) If ST is upper semi-Fredholm, then its adjoint $(ST)^* = T^*S^*$ is lower semi-Fredholm, so T^* is lower semi-Fredholm and T is upper semi-Fredholm.

(iii) Follows from (i) and (ii). □

Corollary 7. *The sets $\Phi(X)$, $\Phi_+(X)$ and $\Phi_-(X)$ are regularities.*

Proof. If $T, S \in \mathcal{B}(X)$ and $TS = ST$, then the previous two theorems imply

$$TS \in \Phi(X) \Leftrightarrow T \in \Phi(X) \text{ and } S \in \Phi(X),$$

and analogous equivalences also for $\Phi_+(X)$ and $\Phi_-(X)$.

Thus the sets of all Fredholm, upper and lower semi-Fredholm operators satisfy property (P1) of Section 6. □

The corresponding spectra are called *essential, essential approximate point and essential surjective*, respectively:

$$\sigma_e(T) = \{\lambda \in \mathbb{C} : T - \lambda \notin \Phi(X)\};$$
$$\sigma_{\pi e}(T) = \{\lambda \in \mathbb{C} : T - \lambda \notin \Phi_+(X)\};$$
$$\sigma_{\delta e}(T) = \{\lambda \in \mathbb{C} : T - \lambda \notin \Phi_-(X)\}.$$

Properties of these important spectra will be studied later where we also show that they can be extended to commuting n-tuples of operators.

Clearly, $\lambda \in \sigma_{\delta e}(T)$ if and only if $\operatorname{codim} \operatorname{Ran}(T - \lambda) = \infty$. The name of the essential approximate point spectrum is justified by the following result:

Theorem 8. *Let $T \in \mathcal{B}(X,Y)$. Then T is upper semi-Fredholm if and only if there exists a closed subspace $M \subset X$ of finite codimension such that $\inf\{\|Tx\| : x \in M, \|x\| = 1\} > 0$. Consequently, $\lambda \in \sigma_{\pi e}(T)$ if and only if*

$$\inf\{\|(T - \lambda)x\| : x \in M, \|x\| = 1\} = 0$$

for each closed subspace M of finite codimension.

Proof. Let $T \in \Phi_+(X,Y)$. Then $\dim \operatorname{Ker} T < \infty$ and we can find a closed subspace $M \subset X$ such that $X = \operatorname{Ker} T \oplus M$. The restriction $T|M : M \to \operatorname{Ran} T$ is one-to-one and onto, and so it is bounded below.

Conversely, let $M \subset X$ be a closed subspace of finite codimension and $\inf\{\|Tx\| : x \in M, \|x\| = 1\} > 0$. Since $\operatorname{Ker} T \cap M = \{0\}$, we have $\dim \operatorname{Ker} T < \infty$. Let F be a finite-dimensional subspace of X such that $F \oplus M = X$. Then $\operatorname{Ran} T = TF + TM$, where TM is closed since $T|M$ is bounded below and $\dim TF < \infty$. Hence $\operatorname{Ran} T$ is closed. Therefore T is upper semi-Fredholm. □

Theorem 9. *Let $T \in \mathcal{B}(X,Y)$ and $K \in \mathcal{K}(X,Y)$. Then:*

(i) $T \in \Phi_+(X,Y) \Rightarrow T + K \in \Phi_+(X,Y)$;

(ii) $T \in \Phi_-(X,Y) \Rightarrow T + K \in \Phi_-(X,Y)$;

(iii) $T \in \Phi(X,Y) \Rightarrow T + K \in \Phi(X,Y)$.

Proof. (i) Let $T \in \Phi_+(X,Y)$ and let M_1 be a closed subspace of X such that $\operatorname{codim} M_1 < \infty$ and $\inf\{\|Tx\| : x \in M_1, \|x\| = 1\} = c > 0$. Since K is compact, there exists a closed subspace $M_2 \subset X$ with $\operatorname{codim} M_2 < \infty$ and $\sup\{\|Tx\| : x \in M_2, \|x\| = 1\} < c/2$. Set $M = M_1 \cap M_2$. Then $\operatorname{codim} M < \infty$ and $\inf\{\|(T + K)x\| : x \in M, \|x\| = 1\} \geq \inf\{\|Tx\| - \|Kx\| : x \in M, \|x\| = 1\} \geq c/2$. Hence $T + K \in \Phi_+(X,Y)$.

(ii) If $T \in \Phi_-(X,Y)$ and $K \in \mathcal{K}(X,Y)$, then T^* is upper semi-Fredholm and K^* is compact. By (i), $T^* + K^*$ is upper semi-Fredholm, and so $T + K$ is lower semi-Fredholm.

(iii) Follows from (i) and (ii). □

Theorem 10. *Let $T \in \Phi_+(X,Y)$ and let M be a closed subspace of X. Then TM is closed.*

Proof. Let $X = \operatorname{Ker} T \oplus M_1$; so $T|M_1$ be bounded below. Then $M = (M \cap M_1) + M_0$ for some finite-dimensional subspace M_0. Thus $TM = T(M \cap M_1) + TM_0$ where $T(M \cap M_1)$ is closed since $T|M_1$ is bounded below, and $\dim TM_0 < \infty$. Hence TM is closed. □

Theorem 11. *The sets $\Phi_+(X,Y)$, $\Phi_-(X,Y)$ and $\Phi(X,Y)$ are open.*

More precisely, if $T \in \Phi_+(X,Y)$ $\big(T \in \Phi_-(X,Y)\big)$, then there exists $\varepsilon > 0$ such that $\alpha(T + S) \leq \alpha(T)$ $\big(\beta(T + S) \leq \beta(T)\big)$ for every $S \in \mathcal{B}(X,Y)$ with $\|S\| < \varepsilon$. In particular, the functions $\alpha : \Phi_+(X,Y) \to \langle 0, \infty)$ and $\beta : \Phi_-(X,Y) \to \langle 0, \infty)$ are upper semicontinuous.

Proof. Let $T \in \Phi_+(X,Y)$. Let M be a closed subspace of X such that $X = \operatorname{Ker} T \oplus M$. Then $T|M$ is bounded below. If $S \in \mathcal{B}(X,Y)$ satisfies $\|S\| < j(T|M)$, then $(T+S)|M$ is bounded below, and so $T+S \in \Phi_+(X,Y)$. Since $\operatorname{Ker}(T+S) \cap M = \{0\}$, we have $\alpha(T + S) = \dim \operatorname{Ker}(T + S) \leq \operatorname{codim} M = \dim \operatorname{Ker} T = \alpha(T)$.

By taking adjoints we get the corresponding statements for lower semi-Fredholm operators. Finally, the set $\Phi(X,Y) = \Phi_+(X,Y) \cap \Phi_-(X,Y)$ is open. □

We show later that it is possible to take $\varepsilon = \gamma(T)$ in the previous theorem.

Theorem 12. *Let $S \in \mathcal{B}(X,Y)$ and $T \in \mathcal{B}(Y,Z)$. If both S and T are lower semi-Fredholm (or both are upper semi-Fredholm), then* $\operatorname{ind} TS = \operatorname{ind} T + \operatorname{ind} S$.

Proof. Let us first suppose that both T and S are Fredholm. Let $Y_0 = \operatorname{Ran} S \cap \operatorname{Ker} T$. Then $\dim Y_0 < \infty$. Choose subspaces Y_1 and Y_2 such that $\operatorname{Ran} S = Y_0 \oplus Y_1$ and $\operatorname{Ker} T = Y_0 \oplus Y_2$. We have $Y_2 \cap \operatorname{Ran} S \subset \operatorname{Ker} T \cap \operatorname{Ran} S = Y_0$, and so $Y_2 \cap \operatorname{Ran} S = \{0\}$. Choose a finite-dimensional subspace Y_3 such that $Y = \operatorname{Ran} S \oplus Y_2 \oplus Y_3 = Y_0 \oplus Y_1 \oplus Y_2 \oplus Y_3$. Then

$$TY = TY_1 \oplus TY_3 = T(Y_1 \oplus Y_0) \oplus TY_3 = T \operatorname{Ran} S \oplus TY_3,$$

and so $\operatorname{codim} \operatorname{Ran}(TS) = \operatorname{codim} \operatorname{Ran} T + \dim TY_3 = \operatorname{codim} \operatorname{Ran} T + \dim Y_3$.

Furthermore, $\dim \operatorname{Ker} T = \dim Y_0 + \dim Y_2$ and $\operatorname{codim} \operatorname{Ran} S = \dim Y_2 + \dim Y_3$. Consider the operator $\widehat{S} = S|\operatorname{Ker}(TS) : \operatorname{Ker}(TS) \to Y_0$. Clearly, $\widehat{S}$ is onto and $\operatorname{Ker} \widehat{S} = \operatorname{Ker} S$. Hence $\dim Y_0 = \dim \operatorname{Ker}(TS) - \dim \operatorname{Ker} S$. Thus

$$\begin{aligned}
\operatorname{ind}(TS) &= \dim \operatorname{Ker}(TS) - \operatorname{codim} \operatorname{Ran}(TS) \\
&= \dim Y_0 + \dim \operatorname{Ker} S - \operatorname{codim} \operatorname{Ran} T - \dim Y_3 \\
&= \dim \operatorname{Ker} T - \dim Y_2 + \dim \operatorname{Ker} S - \operatorname{codim} \operatorname{Ran} T - \operatorname{codim} \operatorname{Ran} S + \dim Y_2 \\
&= \operatorname{ind} T + \operatorname{ind} S.
\end{aligned}$$

If $S \in \Phi_-(X,Y) \setminus \Phi(X,Y)$, then $\dim \operatorname{Ker}(TS) \geq \dim \operatorname{Ker} S = \infty$, and so $\operatorname{ind}(TS) = \infty = \operatorname{ind} T + \operatorname{ind} S$.

Let S be Fredholm and $T \in \Phi_-(Y,Z) \setminus \Phi(Y,Z)$. Then $\dim \operatorname{Ker} T = \infty$. Since $\operatorname{codim} \operatorname{Ran} S < \infty$, we have $\dim(\operatorname{Ran} S \cap \operatorname{Ker} T) = \infty$. Since S maps $\operatorname{Ker}(TS)$ onto $\operatorname{Ran} S \cap \operatorname{Ker} T$, we conclude that $\dim \operatorname{Ker}(TS) = \infty$. Hence $\operatorname{ind} TS = -\infty = \operatorname{ind} T + \operatorname{ind} S$.

Thus $\operatorname{ind} TS = -\infty = \operatorname{ind} T + \operatorname{ind} S$ if both S and T are lower semi-Fredholm.

If S and T are both upper semi-Fredholm, then the statement follows by duality. □

Theorem 13. *Let $T \in \mathcal{B}(X,Y)$. The following statements are equivalent:*

(i) *T is Fredholm;*

(ii) *there exist $S \in \mathcal{B}(Y,X)$, $F_1 \in \mathcal{F}(X)$ and $F_2 \in \mathcal{F}(Y)$ such that $ST = I_X + F_1$ and $TS = I_Y + F_2$;*

(iii) *there exist $S \in \mathcal{B}(Y,X)$, $K_1 \in \mathcal{K}(X)$ and $K_2 \in \mathcal{K}(Y)$ such that $ST = I_X + K_1$ and $TS = I_Y + K_2$.*

Proof. (i) ⇒ (ii): Let $Q \in \mathcal{B}(Y)$ be a projection onto $\operatorname{Ran} T$, let M be a closed subspace of X such that $X = \operatorname{Ker} T \oplus M$. Then $T|M : M \to \operatorname{Ran} T$ is one-to-one and onto. Let $S_1 : \operatorname{Ran} T \to M$ be its inverse and set $S = S_1 Q$. Then $(ST - I_X)m = 0$ for all $m \in M$, and so $ST - I_X \in \mathcal{F}(X)$.

Further $(TS - I_Y)|\operatorname{Ran} T = 0$, and so $TS - I_Y \in \mathcal{F}(Y)$.

(ii) ⇒ (iii): Clear.

(iii) ⇒ (i): Since $ST = I_X + K_1 \in \Phi(X)$, we have $T \in \Phi_+(X,Y)$. Similarly, $TS = I_Y + K_2 \in \Phi(Y)$ implies that $T \in \Phi_-(X,Y)$. Hence T is Fredholm. □

Thus the Fredholm operators are precisely the operators invertible modulo the ideal of compact operators. Next, we characterize also the one-sided invertibility.

Theorem 14. *Let $T \in \mathcal{B}(X,Y)$. The following conditions are equivalent:*

(i) *$T \in \Phi_+(X,Y)$ and $\operatorname{Ran} T$ is complemented;*

(ii) *there exist $S \in \mathcal{B}(Y,X)$ and $F \in \mathcal{F}(X)$ such that $ST = I_X + F$;*

(iii) *there exist $S \in \mathcal{B}(Y,X)$ and $K \in \mathcal{K}(X)$ such that $ST = I_X + K$.*

Proof. (i) $\Rightarrow$ (ii): Let $Q \in \mathcal{B}(Y)$ be a projection onto $\operatorname{Ran} T$, let M be a closed subspace of X such that $X = \operatorname{Ker} T \oplus M$. Then $T|M : M \to \operatorname{Ran} T$ is one-to-one and onto. Let $S_1 : \operatorname{Ran} T \to M$ be its inverse and set $S = S_1 Q$. Then $(I_X - ST)|M = 0$, and so $I_X - ST$ is a finite-rank operator.

(ii) $\Rightarrow$ (iii): Clear.

(iii) $\Rightarrow$ (i): Let $ST = I_X + K$ for some $S \in \mathcal{B}(Y,X)$ and $K \in \mathcal{K}(X)$. Then ST is Fredholm and $T \in \Phi_+(X,Y)$ by Theorem 6.

By Theorem 15.10, there exists k such that $\operatorname{Ran}(ST)^k = \operatorname{Ran}(ST)^{k+1}$. Set $M = \operatorname{Ran}(ST)^k$. Since $(ST)^k \in \Phi(X)$, $\operatorname{codim} M < \infty$. Let $P \in \mathcal{B}(X)$ be a projection onto M. Write $T_0 = T|M : M \to Y$ and $S_0 = PS : Y \to M$. Then $S_0T_0 = (ST)|M = I_M + K|M$, $K|M$ is compact and $S_0T_0M = M$. By Theorem 15.11, $S_0T_0 \in \mathcal{B}(M)$ is invertible, so $(S_0T_0)^{-1}S_0T_0 = I_M$. By Theorem 9.16, $\operatorname{Ran} T_0 = TM$ is complemented in Y. Since $\operatorname{Ran} T = TM + T(I-P)X$ and $\dim T(I-P)X < \infty$, Lemma A.1.25 (iii) implies that $\operatorname{Ran} T$ is complemented. This finishes the proof. □

Operators satisfying any of the conditions of the previous theorem will be called left essentially invertible. The right essentially invertible operators can be characterized similarly.

Theorem 15. *Let $T \in \mathcal{B}(X,Y)$. The following conditions are equivalent:*

(i) *$T \in \Phi_-(X,Y)$ and $\operatorname{Ker} T$ is complemented;*

(ii) *there exist $S \in \mathcal{B}(Y,X)$ and $F \in \mathcal{F}(Y)$ such that $TS = I_X + F$;*

(iii) *there exist $S \in \mathcal{B}(Y,X)$ and $K \in \mathcal{K}(Y)$ such that $TS = I_X + K$.*

Proof. The proof is analogous to the previous theorem. □

It is clear that the sets of all left (right) essentially invertible operators from X to X satisfy (P1) and therefore they are regularities. We define the corresponding left (right) essential spectra of $T \in \mathcal{B}(X)$ by

$$\sigma_{le}(T) = \{\lambda \in \mathbb{C} : T - \lambda \text{ is not left essentially invertible}\},$$
$$\sigma_{re}(T) = \{\lambda \in \mathbb{C} : T - \lambda \text{ is not right essentially invertible}\}.$$

Again, we will show later that these spectra can be extended to commuting n-tuples of operators.

For operators on a Hilbert space the left (right) essential spectrum coincides with the essential approximate point spectrum and the essential surjective spectrum, respectively (from this reason, some authors call $\sigma_{\pi e}$ and $\sigma_{\delta e}$ the left and right essential spectrum even for Banach space operators; this terminology is rather confusing).

Theorem 16. *Let $T \in \mathcal{B}(X,Y)$ be a semi-Fredholm operator and let $K \in \mathcal{K}(X,Y)$. Then $T+K$ is semi-Fredholm and* $\operatorname{ind}(T+K) = \operatorname{ind} T$.

Proof. By Theorem 9, $\Phi(X,Y)$, $\Phi_+(X,Y)$ and $\Phi_-(X,Y)$ are invariant under compact perturbations.

Let $T \in \Phi(X,Y)$. By Theorem 13, there exist $S \in \mathcal{B}(Y,X)$ and $K_1 \in \mathcal{K}(X)$ such that $ST = I_X + K_1$. By Theorem 12 and Corollary 15.14, $\operatorname{ind} T + \operatorname{ind} S = \operatorname{ind}(I_X + K_1) = 0$, so $\operatorname{ind} T = -\operatorname{ind} S$. Further, $S(T+K) = I_X + (K_1 + SK)$, where $K_1 + SK \in \mathcal{K}(X)$, and so $\operatorname{ind} S + \operatorname{ind}(T+K) = 0$. Hence $\operatorname{ind}(T+K) = -\operatorname{ind} S = \operatorname{ind} T$.

If $T \in \Phi_+(X,Y) \setminus \Phi(X,Y)$, then $T+K \in \Phi_+(X,Y) \setminus \Phi(X,Y)$, and so $\operatorname{ind}(T+K) = \operatorname{ind} T = -\infty$.

The statement for $T \in \Phi_-(X,Y) \setminus \Phi(X,Y)$ can be proved similarly. □

Theorem 17. *Let $T \in \Phi(X,Y)$. Then there exists $\varepsilon > 0$ such that $T+R \in \Phi(X,Y)$ and* $\operatorname{ind}(T+R) = \operatorname{ind} T$ *for every $R \in \mathcal{B}(X,Y)$ with $\|R\| < \varepsilon$.*

Proof. By Theorem 13, there exist $S \in \mathcal{B}(Y,X)$ and $K \in \mathcal{K}(X)$ such that $ST = I_X + K$; we can assume that S is Fredholm. Thus $\operatorname{ind} T + \operatorname{ind} S = \operatorname{ind} ST = 0$. If $R \in \mathcal{B}(X,Y)$, $\|R\| < \|S\|^{-1}$, then $S(T+R) = (I_X + SR) + K$. Since $I_X + SR$ is invertible, we have $\operatorname{ind} S + \operatorname{ind}(T+R) = \operatorname{ind}(I_X + SR) + K = \operatorname{ind}(I_X + SR) = 0$. Thus $\operatorname{ind}(T+R) = -\operatorname{ind} S = \operatorname{ind} T$. □

The stability of the index for semi-Fredholm operators is true too; it will be proved later. Moreover, the precise estimate of ε will be given.

Theorem 18. *Let $T \in \mathcal{B}(X,Y)$. The following conditions are equivalent:*

(i) *$T \in \Phi_+(X,Y)$;*

(ii) $\dim \operatorname{Ker}(T-K) < \infty$ *for every $K \in \mathcal{K}(X,Y)$;*

(iii) *there exists $\varepsilon > 0$ such that* $\dim \operatorname{Ker}(T-K) < \infty$ *for every $K \in \mathcal{K}(X,Y)$ with $\|K\| \le \varepsilon$.*

Proof. (i) ⇒ (ii) and (ii) ⇒ (iii): Clear.

(iii) ⇒ (i): Suppose that $T \notin \Phi_+(X,Y)$, so $j(T|M) = 0$ for every subspace $M \subset X$ of finite codimension. Let $\varepsilon > 0$. We construct inductively points $x_k \in X$, $x_k^* \in X^*$ $(k \in \mathbb{N})$ such that $\langle x_k, x_j^* \rangle = \delta_{kj}$, $\|x_k\| = 1$, $\|x_j^*\| \le 2^j$ and $\|Tx_k\| \le \varepsilon \cdot 2^{-2k}$ for all $j,k \in \mathbb{N}$. The existence of x_1 with $\|x_1\| = 1$ and $\|Tx_1\| \le \varepsilon/4$ follows from the fact that $j(T) = 0$. By the Hahn-Banach theorem there exists $x_1^* \in X^*$ such that $\|x_1^*\| = 1 = \langle x_1, x_1^* \rangle$.

Suppose that we have constructed vectors $x_1, \dots, x_n \in X$, $x_1^*, \dots, x_n^* \in X^*$ satisfying all the conditions required. Consider the space $M = \bigcap_{i=1}^n \operatorname{Ker} x_i^*$. Since $\operatorname{codim} M < \infty$, we can find $x_{n+1} \in M$ such that $\|x_{n+1}\| = 1$ and $\|Tx_{n+1}\| < \varepsilon \cdot 2^{-2(n+1)}$. Let $g \in X^*$ satisfy $\|g\| = 1 = \langle x_{n+1}, g\rangle$. Set $x_{n+1}^* = g - \sum_{i=1}^n \langle x_i, g\rangle x_i^*$. Then $\langle x_i, x_{n+1}^*\rangle = \delta_{i,n+1}$ and $\|x_{n+1}^*\| \le 1 + \sum_{i=1}^n \|x_i^*\| \le 2^{n+1}$.

Let $x_k \in X$ and $x_k^* \in X^*$ be the vectors constructed above. For $n \in \mathbb{N}$ define $F_n \in \mathcal{B}(X)$ by $F_n x = \sum_{i=1}^n \langle x, x_i^*\rangle Tx_i$. For $m > n$ and $x \in X$ we have

$$\|(F_m - F_n)x\| = \left\| \sum_{i=n+1}^m \langle x, x_i^*\rangle Tx_i \right\| \le \sum_{i=n+1}^m \|x\| \cdot \|x_i^*\| \cdot \|Tx_i\|$$

$$\le \|x\| \cdot \sum_{i=n+1}^m \frac{\varepsilon 2^i}{2^{2i}} \le \|x\| \cdot \varepsilon \sum_{i=n+1}^m 2^{-i} < 2^{-n}\varepsilon \|x\|.$$

Thus the sequence (F_n) is convergent (in the norm) and its limit K is a compact operator. Further, $Kx = \sum_{i=1}^\infty \langle x, x_i^*\rangle Tx_i$, $\|K\| \le \varepsilon$ and $(T-K)x_i = 0$ for all i. Since the elements x_i are linearly independent, we have $\dim \operatorname{Ker}(T-K) = \infty$. □

Theorem 19. *Let $T \in \mathcal{B}(X,Y)$. The following conditions are equivalent:*

(i) $T \in \Phi_-(X,Y)$;

(ii) $\operatorname{codim} \overline{\operatorname{Ran}(T-K)} < \infty$ *for every* $K \in \mathcal{K}(X,Y)$;

(iii) *there exists* $\varepsilon > 0$ *such that* $\operatorname{codim} \overline{\operatorname{Ran}(T-K)} < \infty$ *for each compact operator* $K \in \mathcal{K}(X,Y)$ *with* $\|K\| \le \varepsilon$.

Proof. (i) ⇒ (ii) and (ii) ⇒ (iii): Clear.

(iii) ⇒ (i): The argument is similar to the previous proof. Suppose that T is not lower semi-Fredholm, so $T^* \notin \Phi_+(Y^*, X^*)$ and for every subspace $M \subset Y^*$ of finite codimension we have $j(T^*|M) = 0$. Let $\varepsilon > 0$. We construct inductively points $y_k \in Y$, $y_k^* \in Y^*$ $(k \in \mathbb{N})$ such that $\langle y_k, y_j^*\rangle = \delta_{kj}$, $\|y_k\| \le 4^k$, $\|y_j^*\| = 1$ and $\|T^* y_k^*\| < \varepsilon 8^{-k}$ for all $j, k \in \mathbb{N}$. The existence of y_1^* with $\|y_1^*\| = 1$ and $\|T^* y_1^*\| \le \varepsilon/8$ follows from the fact that T^* is not bounded below. We can find $y_1 \in T$ such that $\|y_1\| \le 4$ and $\langle y_1, y_1^*\rangle = 1$.

Suppose that we have constructed vectors $y_1, \dots, y_n \in Y$, $y_1^*, \dots, y_n^* \in Y^*$ satisfying all the conditions required. Consider the space $M = \{y_1, \dots, y_n\}^\perp$. Since $\operatorname{codim} M < \infty$ and $j(T^*|M) = 0$, we can find $y_{n+1}^* \in M$ such that $\|y_{n+1}^*\| = 1$ and $\|T^* y_{n+1}^*\| < \varepsilon 8^{-(n+1)}$. There exists $y \in Y$ such that $\|y\| \le 2$ and $\langle y, y_{n+1}^*\rangle = 1$. Set $y_{n+1} = y - \sum_{i=1}^n \langle y, y_i^*\rangle y_i$. Then we have $\|y_{n+1}\| \le 2 + \sum_{i=1}^n 2 \cdot 4^i \le 4^{n+1}$, $\langle y_{n+1}, y_{n+1}^*\rangle = \langle y, y_{n+1}^*\rangle = 1$ and $\langle y_{n+1}, y_j^*\rangle = \langle y, y_j^*\rangle - \sum_{i=1}^n \langle y, y_i^*\rangle \langle y_i, y_j^*\rangle = 0$ for $1 \le j \le n$.

Let $y_k \in Y$ and $y_k^* \in Y^*$ be the vectors constructed above. For $n \in \mathbb{N}$ define $F_n \in \mathcal{B}(X,Y)$ by $F_n x = \sum_{i=1}^n \langle x, T^* y_i^*\rangle y_i$. For $m > n$ and $x \in X$ we have

$$\|(F_m - F_n)x\| = \left\| \sum_{i=n+1}^m \langle x, T^* y_i^*\rangle y_i \right\| \le \sum_{i=n+1}^m \|x\| \varepsilon 8^{-i} \cdot 4^i \le \|x\| \varepsilon 2^{-n},$$

so the sequence (F_n) is convergent, its limit K is a compact operator and $\|K\| \le \varepsilon$. Further, $Kx = \sum_{i=1}^{\infty} \langle x, T^* y_i^* \rangle y_i$ and $\langle Kx, y_k^* \rangle = \langle x, T^* y_k^* \rangle = \langle Tx, y_k^* \rangle$ for all $x \in X$ and $k \in \mathbb{N}$. Thus $y_k^* \in \operatorname{Ran}(T-K)^\perp$ for all k. Since the elements y_k^* are linearly independent, we have $\dim \operatorname{Ran}(T-K)^\perp = \operatorname{codim} \overline{\operatorname{Ran}(T-K)} = \infty$. □

We finish the section with the Kato decomposition of semi-Fredholm operators. We start with the following lemma:

Lemma 20. *Let $T \in \mathcal{B}(X)$ be an operator with closed range. Suppose that for each $k \in \mathbb{N}$ there exists a finite-dimensional subspace $F_k \subset X$ such that $\operatorname{Ker} T \subset \overline{\operatorname{Ran} T^k} + F_k$. Then:*

(i) *$\operatorname{Ran} T^k$ is closed for each $k \in \mathbb{N}$;*

(ii) *either T is Kato, or there exist closed subspaces $Y_1, Y_2 \subset Y$ invariant with respect to T such that $X = Y_1 \oplus Y_2$, $1 \le \dim Y_1 < \infty$ and $T|Y_1$ is nilpotent.*

Proof. (i) We prove by induction on k that $\operatorname{Ran} T^k$ is closed for all k.

By assumption, this is true for $k=1$. Suppose that $k \ge 1$, $\operatorname{Ran} T^k$ is closed and let $\operatorname{Ker} T \subset \overline{\operatorname{Ran} T^{k+1}} + F_{k+1}$ for some finite-dimensional subspace F_{k+1}. We may assume that $F_{k+1} \subset \operatorname{Ker} T$.

Let $u \in \overline{\operatorname{Ran} T^{k+1}}$. By the induction assumption, $u \in \operatorname{Ran} T^k$, and so $u = T^k v$ for some $v \in X$. Further, there are vectors $v_j \in X \quad (j = 1, 2, \dots)$ such that $T^{k+1} v_j \to u \quad (j \to \infty)$. Thus $T(T^k v_j - T^{k-1} v) \to 0$. Consider the operator $\widehat{T} : X / \operatorname{Ker} T \to \operatorname{Ran} T$ induced by T. Clearly, $\widehat{T}$ is bounded below and $\widehat{T}(T^k v_j - T^{k-1} v + \operatorname{Ker} T) \to 0$, so $T^k v_j + T^{k-1} v + \operatorname{Ker} T \to 0 \quad (j \to \infty)$ in the quotient space $X / \operatorname{Ker} T$. Thus there exist vectors $w_j \in \operatorname{Ker} T$ such that $T^k v_j + w_j \to T^{k-1} v$. Since $w_j \in \operatorname{Ker} T \subset \operatorname{Ran} T^k + F_{k+1}$ and $\operatorname{Ran} T^k + F_{k+1}$ is closed, we have $T^{k-1} v = T^k x + f$ for some $x \in X$ and $f \in F_{k+1} \subset \operatorname{Ker} T$. Hence $u = T^k v = T^{k+1} x \in \operatorname{Ran} T^{k+1}$ and $\operatorname{Ran} T^{k+1}$ is closed.

(ii) Let $k \ge 1$ be the smallest integer such that $\operatorname{Ker} T \not\subset \operatorname{Ran} T^k$; so $\operatorname{Ker} T \subset \operatorname{Ran} T^{k-1}$. Since $\operatorname{Ker} T \subset \operatorname{Ran} T^k + F_k$, for all $u_1, \dots, u_n \in \operatorname{Ker} T$ with $n > \dim F_k$ there is a non-trivial linear combination $\sum_{i=1}^n \alpha_i u_i \in \operatorname{Ran} T^k \cap \operatorname{Ker} T$. Thus $\dim \operatorname{Ker} T / (\operatorname{Ker} T \cap \operatorname{Ran} T^k) \le \dim F_k$ and there is a finite-dimensional subspace L_{k-1} such that

$$\operatorname{Ker} T = L_{k-1} \oplus (\operatorname{Ker} T \cap \operatorname{Ran} T^k).$$

Let $r = \dim L_{k-1}$. Then $1 \le r < \infty$.

As $L_{k-1} \subset \operatorname{Ker} T \subset \operatorname{Ran} T^{k-1}$, we can find a subspace L_0 such that $\dim L_0 = r$ and $T^{k-1} L_0 = L_{k-1}$. For $i = 1, \dots, k-1$ set $L_i = T^i L_0$. Clearly, $L_i \subset \operatorname{Ran} T^i$ and $L_i \cap \operatorname{Ran} T^{i+1} = \{0\}$ for all i. Therefore the subspaces L_0, L_1, ..., L_{k-1} and $\operatorname{Ran} T^k$ are linearly independent in the following sense: if $l_i \in L_i$ $(0 \le i \le k-1)$, $x \in R(T^k)$ and $x + l_0 + \dots + l_{k-1} = 0$, then $x = l_0 = \dots = l_{k-1} = 0$.

Let $x_1, \dots, x_r$ be a basis in L_0. Since $T^{k-1} x_1, \dots, T^{k-1} x_r$ are linearly independent modulo $\operatorname{Ran} T^k + L_0 + \dots + L_{k-2}$, we can find linear functionals

$f_1, \dots, f_r \in \left(\operatorname{Ran} T^k + L_0 + \cdots + L_{k-2}\right)^\perp$ such that $\langle T^{k-1}x_i, f_j\rangle = \delta_{ij}$ for all $i, j = 1, \dots, r$. Set

$$Y_1 = \bigvee_{i=0}^{k-1} L_i \qquad \text{and} \qquad Y_2 = \bigcap_{i=0}^{k-1} \bigcap_{j=1}^{r} \operatorname{Ker}(T^{*i} f_j).$$

Clearly, $\dim Y_1 < \infty$, $TY_1 \subset Y_1$ and $(T|Y_1)^k = 0$. Further, $TY_2 \subset Y_2$. Indeed, if $x \in Y_2$, then

$$\langle Tx, T^{*i} f_j\rangle = \langle x, T^{*(i+1)} f_j\rangle = 0 \qquad \text{for} \qquad 0 \le i \le k-2$$

and $\langle Tx, T^{*(k-1)} f_j\rangle = \langle T^k x, f_j\rangle = 0$.

Since

$$\langle T^i x_j, T^{*i'} f_{j'}\rangle = \langle T^{i+i'} x_j, f_{j'}\rangle = \begin{cases} 1 & (j = j', i + i' = k-1), \\ 0 & \text{otherwise}, \end{cases}$$

the sets $\{T^i x_j : 0 \le i \le k-1, 1 \le j \le r\}$ and $\{T^{*i} f_j : 0 \le i \le k-1, 1 \le j \le r\}$ form a biorthogonal system. Thus it is easy to show that $X = Y_1 \oplus Y_2$. □

Theorem 21. (Kato decomposition) *Let $T \in \mathcal{B}(X)$ be semi-Fredholm. Then there exist closed subspaces $X_1, X_2 \subset X$ invariant with respect to T such that $X = X_1 \oplus X_2$, $\dim X_1 < \infty$, $T|X_1$ is nilpotent and $T|X_2$ is Kato.*

Proof. Suppose first that T is upper semi-Fredholm. If T is Kato, then we can take $X_1 = \{0\}$, $X_2 = X$.

If T is not Kato, then we can use the previous lemma to find a decomposition $X = Y_1 \oplus Y_2$ such that $TY_i \subset Y_i \quad (i = 1, 2)$, $1 \le \dim Y_1 < \infty$ and $T|Y_1$ nilpotent. Let $T_i = T|Y_i \quad (i = 1, 2)$. Clearly, $\operatorname{Ker} T = \operatorname{Ker} T_1 \oplus \operatorname{Ker} T_2$, and so $\dim \operatorname{Ker} T_2 < \dim \operatorname{Ker} T$. If T_2 is Kato, then the proof is finished, otherwise we can repeat the same construction for the operator T_2. After a finite number of steps we obtain the required decomposition.

If T is lower semi-Fredholm, then we can proceed similarly. In this case

$$\operatorname{codim} \operatorname{Ran} T_2 < \operatorname{codim} \operatorname{Ran} T_1 + \operatorname{codim} \operatorname{Ran} T_2 = \operatorname{codim} \operatorname{Ran} T.$$

Again, after a finite number of steps we obtain the required decomposition. □

17 Construction of Sadovskii/Buoni, Harte, Wickstead

In this section we introduce a construction which is very useful in the study of Fredholm operators and essential spectra.

Let X be a Banach space. Denote by $\ell^\infty(X)$ the set of all bounded sequences of elements of X. It is clear that $\ell^\infty(X)$ with the natural algebraic operations and with the norm $\|(x_i)_{i=1}^\infty\| = \sup_{i \in \mathbb{N}} \|x_i\|$ is a Banach space.

For $\tilde{x} = (x_i) \in \ell^\infty(X)$ let $q(\tilde{x})$ be the infimum of all $\varepsilon > 0$ such that the set $\{x_i : i \in \mathbb{N}\}$ is contained in the union of a finite number of open balls with radius ε.

It is easy to see that $q(\tilde{x}) = 0$ if and only if the set $\{x_i : i \in \mathbb{N}\}$ is totally bounded (i.e., its closure is compact). So q is a "measure of non-compactness". Let further $m(X) = \{\tilde{x} \in \ell^\infty(X) : q(\tilde{x}) = 0\}$.

The basic properties of q and $m(X)$ are given in the next lemma.

Lemma 1. *Let X be a Banach space. Then for all $\tilde{x}, \tilde{y} \in \ell^\infty(X)$ and $\lambda \in \mathbb{C}$ we have:*

(i) $q(\tilde{x}) \le \|\tilde{x}\|$;

(ii) $q(\tilde{x} + \tilde{y}) \le q(\tilde{x}) + q(\tilde{y})$, $q(\lambda\tilde{x}) = |\lambda| q(\tilde{x})$ *(so q is a seminorm on $\ell^\infty(X)$);*

(iii) *$m(X)$ is a closed subspace of $\ell^\infty(X)$;*

(iv) *$m(X)$ is the closure of the set of all sequences $(x_i) \in \ell^\infty(X)$ such that $\dim \bigvee\{x_i : i \in \mathbb{N}\} < \infty$;*

(v) $q(\tilde{x}) = \inf\{\|\tilde{x} + \tilde{m}\| : \tilde{m} \in m(X)\}$.

Proof. Straightforward. □

Lemma 2. *Let M, N be closed subspaces of a Banach space X and let $M \subset N$. Then:*

(i) *if $\dim N/M < \infty$, then $\ell^\infty(M) + m(X) = \ell^\infty(N) + m(X)$;*

(ii) *if $\dim N/M = \infty$, then $\dim\big(\ell^\infty(M) + m(X)\big)/\big(\ell^\infty(N) + m(X)\big) = \infty$.*

Proof. (i) Let $P : N \to M$ be a bounded projection onto M. If $\tilde{n} = (n_i) \in \ell^\infty(N)$, then $(n_i) = (Pn_i) + ((I - P)n_i)$, where $(Pn_i) \in \ell^\infty(M)$ and $((I - P)n_i) \in \ell^\infty(\operatorname{Ker} P) \subset m(X)$.

(ii) By Lemma 15.8, there exist vectors $x_1, x_2, \ldots$ in N such that $\|x_k\| \le 2$ and $\operatorname{dist}\{x_{k+1}, M \vee \{x_1, \ldots, x_k\}\} = 1$ for every k. Consider the sequences $\tilde{x}^{(1)}, \tilde{x}^{(2)}, \cdots \in \ell^\infty(N)$ defined by $\tilde{x}^{(k)} = (x_{k+1}, x_{k+2}, \ldots)$.

We show that these sequences are linearly independent modulo $\ell^\infty(M) + m(X)$. Suppose on the contrary that for some k there are $\tilde{m} = (m_i) \in \ell^\infty(M)$, $\tilde{l} = (l_i) \in m(X)$ and $\alpha_1, \ldots, \alpha_{k-1} \in \mathbb{C}$ such that $\tilde{x}^{(k)} = \tilde{m} + \tilde{l} + \sum_{i=1}^{k-1} \alpha_i \tilde{x}^{(i)}$. Equivalently,

$$l_j = x_{k+j} - m_j - \sum_{i=1}^{k-1} \alpha_i x_{i+j}$$

for all $j \ge 1$. Thus for $1 \le j < j'$ we have

$$\begin{aligned}\|l_{j'} - l_j\| &= \left\| x_{k+j'} - x_{k+j} - m_{j'} + m_j - \sum_{i=1}^{k-1} \alpha_i x_{i+j'} - \sum_{i=1}^{k-1} \alpha_i x_{i+j} \right\| \\ &\ge \operatorname{dist}\{x_{k+j'}, M \vee \{x_1, \ldots, x_{k+j'-1}\}\} \ge 1,\end{aligned}$$

and so $\tilde{l} \notin m(X)$, a contradiction. □

For a Banach space X write $\tilde{X} = \ell^\infty(X)/m(X)$. Thus the elements of $\tilde{X}$ can be viewed as bounded sequences of elements of X with the new norm $q(\tilde{x})$; we identify $\tilde{x}$ and $\tilde{y}$ if $q(\tilde{x} - \tilde{y}) = 0$.

Let X, Y be Banach spaces and let $T : X \to Y$ be an operator. Define $T^\infty : \ell^\infty(X) \to \ell^\infty(Y)$ by $T^\infty(x_i) = (Tx_i)$. It is easy to see that

$$q(T^\infty(\tilde{x})) \le \|T\| \cdot q(\tilde{x}) \qquad (\tilde{x} \in \ell^\infty(X)),$$

and so $T^\infty m(X) \subset m(Y)$.

Let $\tilde{T} : \tilde{X} \to \tilde{Y}$ be the operator defined by $\tilde{T}(\tilde{x} + m(X)) = T^\infty \tilde{x} + m(Y)$.

Lemma 3. *Let $T \in \mathcal{B}(X, Y)$, $S \in \mathcal{B}(Y, Z)$. Then:*

(i) $\|\tilde{T}\| \le \|T\|$;

(ii) $\widetilde{ST} = \tilde{S}\tilde{T}$;

(iii) *if I_X is the identity operator on X, then $\tilde{I}_X = I_{\tilde{X}}$;*

(iv) *$\tilde{T} = 0$ if and only if T is compact.*

Proof. Straightforward. □

Lemma 4. *Let $T : X \to Y$ be an operator with closed range. Then* $\operatorname{Ran} T^\infty = \ell^\infty(\operatorname{Ran} T)$.

Proof. It is clear that $\operatorname{Ran} T^\infty \subset \ell^\infty(\operatorname{Ran} T)$. The opposite inclusion is a consequence of the open mapping theorem. □

Theorem 5. *Let X, Y be infinite-dimensional Banach spaces and let $T : X \to Y$ be an operator. Then:*

(i) *if T is onto, then $\tilde{T}$ is onto. More precisely, $k(\tilde{T}) \ge k(T)$, where k is the surjectivity modulus;*

(ii) *if T is bounded below, then $\tilde{T}$ is bounded below. More precisely, $j(\tilde{T}) \ge \frac{1}{2} j(T)$, where j is the injectivity modulus;*

(iii) *if* $\operatorname{Ran} T$ *is closed, then* $\operatorname{Ran} \tilde{T}$ *is closed.*

Proof. (i) If T is onto, then $\tilde{T}$ is onto by the preceding lemma. Let $\tilde{y} = (y_i) \in \ell^\infty(Y)$ satisfy $q(\tilde{y}) < 1$. Then there are a finite number of elements $z_1, \dots, z_n \in Y$ such that $\operatorname{dist}\{y_i, \{z_1, \dots, z_n\}\} < 1 \quad (i \in \mathbb{N})$. For every $j = 1, \dots, n$ choose $u_j \in X$ such that $Tu_j = z_j$. For $i \in \mathbb{N}$ find $j(i) \in \{1, \dots, n\}$ such that $\|y_i - z_{j(i)}\| < 1$ and $v_i \in X$ such that $Tv_i = y_i - z_{j(i)}$ and $\|v_i\| \le k(T)^{-1}$. Set $\tilde{x} = (x_i)$, where $x_i = v_i + u_{j(i)}$. Then $Tx_i = y_i$ and $\operatorname{dist}\{x_i, \{u_1, \dots, u_n\}\} \le \|x_i - u_{j(i)}\| = \|v_i\| \le k(T)^{-1}$. Thus $q(\tilde{x}) \le k(T)^{-1}$ and $k(\tilde{T}) \ge k(T)$.

(ii) Let $(x_i) \in \ell^\infty(X)$ and let $q((Tx_i)) < 1$. Then the set $\{Tx_i : i \in \mathbb{N}\}$ can be covered by a finite number of balls $B_1, \dots, B_n$ of radius 1. We can assume that each ball B_j contains at least one element Tx_{i_j}. Then $\{Tx_i : i \in \mathbb{N}\} \subset \bigcup_{j=1}^n \{y \in Y : \|y - Tx_{i_j}\| \le 2\}$. We have $\{x_i : i \in \mathbb{N}\} \subset \bigcup_{j=1}^n \{x \in X : \|x - x_{i_j}\| \le 2j(T)^{-1}\}$, and so $q((x_i)) \le 2j(T)^{-1}$. Hence $j(\tilde{T}) \ge \frac{1}{2} j(T)$.

(iii) Let $\operatorname{Ran} T$ be closed. Then T can be expressed as the composition $T = T_2 T_1$, where $T_1 : X \to \operatorname{Ran} T$ is induced by T, and $T_2 : \operatorname{Ran} T \to Y$ is the natural embedding. By (i) and (ii), $\tilde{T}_1$ is onto and $\tilde{T}_2$ is bounded below. Thus $\tilde{T} = \tilde{T}_2 \tilde{T}_1$, and so $\operatorname{Ran} \tilde{T} = \tilde{T}_2 \operatorname{Ran} \tilde{T}_1$, which is closed. □

The factor $\frac{1}{2}$ in Theorem 5 (ii) is really necessary, see Example 24.6.

Theorem 6. *Let $T : X \to Y$. The following statements are equivalent:*

(i) *T is lower semi-Fredholm (i.e., $\operatorname{codim} \operatorname{Ran} T < \infty$);*

(ii) *$\tilde{T}$ is onto;*

(iii) *$\tilde{T}$ is lower semi-Fredholm;*

(iv) *$\operatorname{Ran} \tilde{T}$ is dense in $\tilde{Y}$.*

Proof. Suppose that T is lower semi-Fredholm. Since $\operatorname{Ran} T$ is closed and of finite codimension, we have $\ell^\infty(\operatorname{Ran} T) + m(Y) = \ell^\infty(Y)$ by Lemma 2. Thus $\tilde{T}$ is onto.

Conversely, suppose that T is not lower semi-Fredholm. By Theorem 16.19, there exists $T_1 \in \mathcal{B}(X, Y)$ such that $T - T_1$ is compact and $\operatorname{codim} \overline{\operatorname{Ran} T_1} = \infty$. Then $\tilde{T}_1 = \tilde{T}$, $\operatorname{Ran} T_1^\infty \subset \ell^\infty(\overline{\operatorname{Ran} T_1})$ and, by Lemma 2,

$$\dim \ell^\infty(Y) / \big(\ell^\infty(\overline{\operatorname{Ran} T_1}) + m(Y)\big) = \infty.$$

Thus $\operatorname{codim} \operatorname{Ran} \tilde{T}_1 = \infty$ and $\tilde{T}$ is not lower semi-Fredholm.

Also, there exists a sequence $\tilde{y} = (y_i) \in \ell^\infty(Y)$ such that $\|y_i\| \le 2$ and $\operatorname{dist}\{y_i, \overline{\operatorname{Ran} T_1} \vee \{y_1, \dots, y_{i-1}\}\} = 1$ for all $i \in \mathbb{N}$. Thus for all $\tilde{x} \in \ell^\infty(X)$ and $i \ne j$ we have $\|(y_i - T_1 x_i) - (y_j - T_1 x_j)\| \ge 1$, and so $q\big(\tilde{y} - (T_1^\infty \tilde{x})\big) \ge 1/2$. Hence $\tilde{y} + m(Y) \notin \overline{\operatorname{Ran} \tilde{T}_1} = \overline{\operatorname{Ran} \tilde{T}}$ and $\operatorname{Ran} \tilde{T}$ is not dense in $\tilde{Y}$. □

Lemma 7. *Let $(x_i) \in m(X)$. Then there exist numbers $1 \le n_1 < n_2 < \cdots$ with the property that*

$$\operatorname{dist}\{x_i, \{x_1, \dots, x_{n_k}\}\} \le 2^{-k}$$

for all $k, i \in \mathbb{N}$.

Proof. We construct the numbers n_k inductively. Let $k \in \mathbb{N}$. There exists a finite set $F \subset X$ such that $\operatorname{dist}\{x_i, F\} \le 2^{-(k+1)}$ for each i. We can assume that F is a minimal set with this property. For every $f \in F$ choose $i_f \in \mathbb{N}$ such that $\|x_{i_f} - f\| \le \frac{1}{2^{k+1}}$. Choose $n_k \ge \max\{i_f : f \in F\}$ such that $n_k > n_{k-1}$. Obviously, $\operatorname{dist}\{x_i, \{x_1, \dots, x_{n_k}\}\} \le 2^{-k}$ for each i. □

Lemma 8. *Let $T : X \to Y$ be an operator with closed range, let $(x_i) \in \ell^\infty(X)$ and $(Tx_i) \in m(Y)$. Then there exists a sequence $(x_i') \in m(X)$ such that $x_i - x_i' \in \operatorname{Ker} T \quad (i \in \mathbb{N})$. In particular, $\operatorname{Ker} \tilde{T} = \ell^\infty(\operatorname{Ker} T) + m(X)$.*

If T is onto, then $T^\infty m(X) = m(Y)$.

Proof. Let $(x_i) \in \ell^\infty(X)$ and $(Tx_i) \in m(Y)$. Let $0 < s < \gamma(T)$. By Lemma 7, there exist numbers $1 \le n_1 < n_2 < \cdots$ such that

$$\operatorname{dist}\{Tx_i, \{Tx_1, \dots, Tx_{n_k}\}\} \le 2^{-k} \qquad (k, i \in \mathbb{N}).$$

We construct inductively points $x_i' \in X$ such that $Tx_i' = Tx_i$ and

$$\operatorname{dist}\{x_i', \{x_1', \dots, x_{n_k}'\}\} \le \frac{1}{s \cdot 2^k} \qquad (k, i \in \mathbb{N}, n_k < i \le n_{k+1}). \tag{1}$$

For $i \le n_1$ set $x_i' = x_i$.

Suppose that $k \ge 1$ and the points $x_1', \dots, x_{n_k}'$ satisfying (1) have already been constructed. For each i, $n_k < i \le n_{k+1}$, there exists $j(i) \le n_k$ such that $\|Tx_i - Tx_{j(i)}\| \le 2^{-k}$. Find $u_i \in X$ such that $Tu_i = Tx_i - Tx_{j(i)}$ and $\|u_i\| \le \frac{1}{s \cdot 2^k}$. Set $x_i' = x_{j(i)}' + u_i$. Then $Tx_i' = Tx_{j(i)} + Tu_i = Tx_i$ and we have $\|x_i' - x_{j(i)}'\| = \|u_i\| \le \frac{1}{s \cdot 2^k}$. Thus x_i' satisfy (1) and we can continue the induction.

Note that the sequence (x_i') satisfying (1) also satisfies

$$\operatorname{dist}\{x_i', \{x_1', \dots, x_{n_k}'\}\} \le \frac{1}{s \cdot 2^{k-1}} \qquad (k, i \in \mathbb{N}). \tag{2}$$

This is clear for $i \le n_{k+1}$. Let $n_l < i \le n_{l+1}$ for some $l \ge k+1$. Then there exists $j_l \le n_l$ with $\|x_i' - x_{j_l}'\| \le \frac{1}{s \cdot 2^l}$ and we can construct inductively indices $j_{l-1} \le n_{l-1}, \dots, j_k \le n_k$ such that $\|x_{j_l}' - x_{j_{l-1}}'\| \le \frac{1}{s \cdot 2^{l-1}}, \dots, \|x_{j_{k+1}}' - x_{j_k}'\| \le \frac{1}{s \cdot 2^k}$. Thus

$$\|x_i' - x_{j_k}\| \le \frac{1}{s \cdot 2^{l-1}} + \frac{1}{s \cdot 2^{l-2}} + \dots + \frac{1}{s \cdot 2^k} \le \frac{1}{s \cdot 2^{k-1}}.$$

Clearly, (2) implies that $(x_i') \in m(X)$. □

Theorem 9. *Let $T \in \mathcal{B}(X, Y)$. The following conditions are equivalent:*

(i) *T is upper semi-Fredholm;*

(ii) *$\tilde{T}$ is one-to-one;*

(iii) *$\tilde{T}$ is bounded below;*

(iv) *$\tilde{T}$ is upper semi-Fredholm.*

Proof. Suppose that T is upper semi-Fredholm. Then $\operatorname{Ran} \tilde{T}$ is closed. By Lemma 8, $\operatorname{Ker} \tilde{T} = \ell^\infty(\operatorname{Ker} T) + m(X) = m(X)$. So $\tilde{T}$ is one-to-one and therefore bounded below.

Conversely, if T is not upper semi-Fredholm, then there exists a compact operator $K : X \to Y$ such that $T_1 = T + K$ satisfies $\dim \operatorname{Ker} T_1 = \infty$. Clearly, $\operatorname{Ker} T_1^\infty = \ell^\infty(\operatorname{Ker} T_1)$ and, by Lemma 2, $\dim\bigl(\operatorname{Ker} T_1^\infty + m(X)\bigr)/m(X) = \infty$. So $\dim \operatorname{Ker} \tilde{T}_1 = \infty$. Thus $\tilde{T} = \tilde{T}_1$ is neither one-to-one nor upper semi-Fredholm. □

Corollary 10. Let $T \in \mathcal{B}(X)$. Then $\sigma(\tilde{T}) = \sigma_e(T)$, $\sigma_\pi(\tilde{T}) = \sigma_{\pi e}(T)$ and $\sigma_\delta(\tilde{T}) = \sigma_{\delta e}(T)$.

Theorem 11. *Let X, Y, Z be Banach spaces, let $T : X \to Y$, $S : Y \to Z$ be operators such that $ST = 0$. The following conditions are equivalent:*

(i) *$\dim \operatorname{Ker} S / \operatorname{Ran} T < \infty$ and $\operatorname{Ran} S$ is closed;*

(ii) *$\operatorname{Ran} \tilde{T} = \operatorname{Ker} \tilde{S}$;*

(iii) *$\dim \operatorname{Ker} \tilde{S} / \operatorname{Ran} \tilde{T} < \infty$.*

Proof. Suppose that (i) is true. Then $\operatorname{Ran} T$ is closed and

$$\operatorname{Ran} T^\infty + m(Y) = \ell^\infty(\operatorname{Ran} T) + m(Y) = \ell^\infty(\operatorname{Ker} S) + m(Y) = \operatorname{Ker} S^\infty + m(Y).$$

Thus $\operatorname{Ran} \tilde{T} = \operatorname{Ker} \tilde{S}$ by Lemma 8.

Conversely, suppose that (i) is false. Suppose first that $\dim \operatorname{Ker} S / \operatorname{Ran} T = \infty$. By Theorem 16.19, there exists $T_1 : X \to Y$ such that $T - T_1$ is compact, $\operatorname{Ran} T_1 \subset \operatorname{Ker} S$ and $\dim \operatorname{Ker} S / \overline{\operatorname{Ran} T_1} = \infty$. Thus $\operatorname{Ran} T_1^\infty + m(Y) \subset \ell^\infty(\overline{\operatorname{Ran} T_1}) + m(Y)$ and

$$\dim\bigl(\ell^\infty(\operatorname{Ker} S) + m(Y))\bigr) / \bigl(\ell^\infty(\overline{\operatorname{Ran} T_1} + m(Y))\bigr) = \infty.$$

So $\dim \operatorname{Ker} \tilde{S} / \operatorname{Ran} \tilde{T} = \dim \operatorname{Ker} \tilde{S} / \operatorname{Ran} \tilde{T}_1 = \infty$.

Suppose now that $\dim \operatorname{Ker} S / \operatorname{Ran} T < \infty$ and $\operatorname{Ran} S$ is not closed. Then $\operatorname{Ran} T$ is closed. Consider the operator $S' : Y / \operatorname{Ran} T \to Z$ induced by S. Since $\operatorname{Ran} S' = \operatorname{Ran} S$, which is not closed, S' is not upper semi-Fredholm and there exists operator $S_1' : Y / \operatorname{Ran} T \to Z$ such that $S' - S_1'$ is compact and $\dim \operatorname{Ker} S_1' = \infty$. Let $P : Y \to Y / \operatorname{Ran} T$ be the canonical projection and $S_1 = S_1' P$. Then $S_1 T = 0$, $\dim \operatorname{Ker} S_1 / \operatorname{Ran} T = \infty$ and $S - S_1 = (S' - S_1')P$, which is compact. Hence $\dim \operatorname{Ker} \tilde{S} / \operatorname{Ran} \tilde{T} = \dim \operatorname{Ker} \tilde{S}_1 / \operatorname{Ran} \tilde{T} = \infty$. □

18 Perturbation properties of Fredholm and semi-Fredholm operators

In this section we give quantitative stability results for Fredholm and semi-Fredholm operators.

Lemma 1. *Let $T \in \partial\Phi(X, Y)$. Then $T \notin \Phi_+(X, Y) \cup \Phi_-(X, Y)$.*

Proof. Let $T_n \in \mathcal{B}(X, Y)$ be a sequence of Fredholm operators that converges in the norm topology to $T \notin \Phi(X, Y)$. Then the operators $\tilde{T}_n$ are invertible, $\tilde{T}_n \to \tilde{T}$ and $\tilde{T}$ is not invertible, and so $\tilde{T}$ is neither one-to-one nor onto. Thus T is not semi-Fredholm. □

Corollary 2. *Let $T \in \mathcal{B}(X, Y)$ be a semi-Fredholm operator. Then there exists $\varepsilon > 0$ such that $T + S$ is semi-Fredholm and $\operatorname{ind}(T + S) = \operatorname{ind} T$ for every $S \in \mathcal{B}(X, Y)$ with $\|S\| < \varepsilon$.*

Proof. If T is Fredholm, then the result was proved in Theorem 16.17. Suppose that $T \in \Phi_+(X, Y) \setminus \Phi(X, Y)$. Then $\operatorname{ind} T = -\infty$. Since the set of all upper semi-Fredholm operators is open and, by Lemma 1, T is not a limit of Fredholm operators, $\operatorname{ind}(T + S) = -\infty$ for every S with norm small enough.

The same argument can be used for lower semi-Fredholm operators. □

Corollary 3. *The index is constant on every component of connectivity of*

$$\Phi_+(X, Y) \cup \Phi_-(X, Y).$$

Proof. Let G be a component of connectivity of $\Phi_+(X,Y) \cup \Phi_-(X,Y)$. Fix $T_0 \in G$. By Corollary 2, function $T \mapsto \operatorname{ind} T$ is continuous on G. Therefore the set $\{T \in G : \operatorname{ind} T = \operatorname{ind} T_0\}$ is both open and closed, and thus it is equal to G. Hence the index is constant on G. □

Theorem 4. *If $T \in \mathcal{B}(X,Y)$ is semi-Fredholm and $S \in \mathcal{B}(X,Y)$ satisfies $\|S\| < \gamma(T)$, then $T+S$ is semi-Fredholm,* $\operatorname{ind}(T+S) = \operatorname{ind} T$, *$\alpha(T+S) \le \alpha(T)$ and $\beta(T+S) \le \beta(T)$.*

Proof. Suppose that T is upper semi-Fredholm and $\|S\| < \gamma(T)$.

By Lemma 10.12, $\delta\big(\operatorname{Ker}(T+S), \operatorname{Ker} T\big) \le \gamma(T)^{-1}\|S\| < 1$, and so, by Corollary 10.10, $\alpha(T+S) = \dim \operatorname{Ker}(T+S) \le \dim \operatorname{Ker} T = \alpha(T)$.

From the same reason $\dim \operatorname{Ker}(T+S+K) \le \dim \operatorname{Ker} T < \infty$ for every compact operator K with $\|K\| < \gamma(T) - \|S\|$. By Theorem 16.18 we conclude that $T+S$ is upper semi-Fredholm.

By Corollary 3, $\operatorname{ind}(T+S) = \operatorname{ind} T$.

For lower semi-Fredholm operators the statement follows by duality. □

Proposition 5. *Let $T \in \Phi_+(X,Y)$. Then*

$$\gamma(T) = \sup\big\{s > 0 : \alpha(T+S) \le \alpha(T) \text{ for every } S \text{ with } \|S\| < s\big\}.$$

If $T \in \Phi_-(X,Y)$, then

$$\gamma(T) = \sup\big\{s > 0 : \beta(T+S) \le \beta(T) \text{ for every } S \text{ with } \|S\| < s\big\}.$$

Proof. Let $T \in \Phi_+(X,Y) \cup \Phi_-(X,Y)$ and $\varepsilon > 0$. We show that there exists $S \in \mathcal{B}(X,Y)$ such that $\|S\| < \gamma(T) + \varepsilon$, $\operatorname{Ker}(T+S) \supset \operatorname{Ker} T$, $\operatorname{Ran}(T+S) \subset \operatorname{Ran} T$ and both inclusions are strict.

By the definition of the reduced minimum modulus there exists $x_0 \in X$ such that $\operatorname{dist}\{x_0, \operatorname{Ker} T\} = 1$ and $\|Tx_0\| < \gamma(T) + \varepsilon$. Let $x^* \in (\operatorname{Ker} T)^\perp$ satisfy $\langle x_0, x^* \rangle = 1$ and $\|x^*\| = \operatorname{dist}\{x_0, \operatorname{Ker} T\} = 1$. Let S be defined by $Sx = -\langle x, x^* \rangle \cdot Tx_0 \quad (x \in X)$. Then $\|S\| = \|x^*\| \cdot \|Tx_0\| < \gamma(T) + \varepsilon$ and $\operatorname{Ker}(T+S) \supset \operatorname{Ker} T \cup \{x_0\}$.

Furthermore, $\operatorname{Ran}(T+S) = (T+S) \operatorname{Ker} x^* \vee \{(T+S)x_0\} = T \operatorname{Ker} x^*$ and $\operatorname{Ran} T - T \operatorname{Ker} x^* \vee \{Tx_0\}$.

We show that $Tx_0 \notin T \operatorname{Ker} x^*$. Suppose on the contrary that there is an $x' \in \operatorname{Ker} x^*$ such that $Tx' = Tx_0$. Then $x_0 - x' \in \operatorname{Ker} T$ and

$$1 = \langle x_0, x^* \rangle = \langle x_0 - x', x^* \rangle = 0,$$

a contradiction. Thus $\operatorname{Ran}(T+S)$ is strictly smaller that $\operatorname{Ran} T$. □

Remark 6. In general, it is possible that $\gamma(T)$ is strictly smaller than the number $\sup\big\{s > 0 : T+S \in \Phi_+(X,Y) \text{ for every } S \text{ with } \|S\| < s\big\}$. Consider the diagonal operator $\operatorname{diag}(1,2,2,2,\dots)$ on a separable Hilbert space. Then $\gamma(T) = 1$ and $T+S$ is Fredholm for every S with $\|S\| < 2$.

By Theorem 4, semi-Fredholm operators are inner points of the set of all operators with closed range. Conversely, if $T : X \to Y$ is not semi-Fredholm and $\varepsilon > 0$, then using the technique of Theorem 16.18 it is possible to construct an operator $S : X \to Y$ with $\|S\| < \varepsilon$ such that the range of $T - S$ is not closed.

Theorem 7. (punctured neighbourhood theorem) *Let $T \in \Phi_+(X) \cup \Phi_-(X)$. Then there exists $\varepsilon > 0$ such that the functions $\alpha(T - z)$ and $\beta(T - z)$ are constant for $0 < |z| < \varepsilon$.*

Proof. Let $X = X_1 \oplus X_2$ be the Kato decomposition of T, i.e., $\dim X_1 < \infty$, both X_1 and X_2 are invariant with respect to T, $T|X_1$ is nilpotent and $T|X_2$ Kato. Let $T_1 = T|X_1$ and $T_2 = T|X_2$. By Corollary 12.4, there exists $\varepsilon > 0$ such that $\dim \operatorname{Ker}(T_2 - z)$ and $\operatorname{codim} \operatorname{Ran}(T_2 - z)$ are constant for $|z| < \varepsilon$. Also, for $z \neq 0$, $T_1 - z$ is invertible. Thus for $0 < |z| < \varepsilon$, $\alpha(T - z) = \dim \operatorname{Ker}(T - z) = \dim \operatorname{Ker}(T_2 - z) = \text{const}$ and $\beta(T - z) = \operatorname{codim} \operatorname{Ran}(T_2 - z) = \text{const}$. □

Theorem 8. *Let $T \in \mathcal{B}(X)$ be upper semi-Fredholm. Then $\lim_{n\to\infty} \gamma(T^n)^{1/n}$ exists and is equal to*

$$\sup\{s > 0 : T - z \in \Phi_+(X) \text{ and } \alpha(T - z) \text{ is constant for } 0 < |z| < s\}.$$

An analogous statement is true for lower semi-Fredholm operators (with α replaced by β).

Proof. Let $T \in \Phi_+(X)$. By Theorem 16.21, there exists a decomposition $X = X_1 \oplus X_2$ such that $\dim X_1 < \infty, T_1 = T|X_1$ is nilpotent and $T_2 = T|X_2$ is Kato and upper semi-Fredholm (the Kato decomposition). Let P be the projection with $\operatorname{Ran} P = X_2$ and $\operatorname{Ker} P = X_1$. Let $x_2 \in X_2$. For $n \geq \dim X_1$ we have

$$\begin{aligned}
\operatorname{dist}\{x_2, \operatorname{Ker} T_2^n\} &= \inf\{\|x_2 - y_2\| : y_2 \in X_2, T_2^n y_2 = 0\} \\
&\leq \|P\| \inf\{\|y_1 \oplus (x_2 - y_2)\| : y_1 \in X_1, y_2 \in X_2, T_2^n y_2 = 0\} \\
&= \|P\| \operatorname{dist}\{x_2, \operatorname{Ker} T^n\} \leq \|P\| \operatorname{dist}\{x_2, \operatorname{Ker} T_2^n\}.
\end{aligned}$$

Then

$$\begin{aligned}
\gamma(T_2^n) &= \inf\left\{\frac{\|T_2^n x_2\|}{\operatorname{dist}\{x_2, \operatorname{Ker} T_2^n\}} : x_2 \in X_2 \setminus \operatorname{Ker} T_2^n\right\} \\
&\leq \inf\left\{\frac{\|T^n x_2\|}{\operatorname{dist}\{x_2, \operatorname{Ker} T^n\}} : x_2 \in X_2 \setminus \operatorname{Ker} T^n\right\} \\
&= \inf\left\{\frac{\|T^n (x_1 \oplus x_2)\|}{\operatorname{dist}\{x_1 \oplus x_2, \operatorname{Ker} T^n\}} : x_1 \oplus x_2 \in X \setminus \operatorname{Ker} T^n\right\} = \gamma(T^n)
\end{aligned}$$

and

$$\begin{aligned}
\gamma(T^n) &\leq \inf\left\{\frac{\|T_2^n x_2\|}{\operatorname{dist}\{x_2, \operatorname{Ker} T^n\}} : x_2 \in X_2 \setminus \operatorname{Ker} T_2^n\right\} \\
&\leq \|P\| \inf\left\{\frac{\|T_2^n x_2\|}{\operatorname{dist}\{x_2, \operatorname{Ker} T_2^n\}} : x_2 \in X_2 \setminus \operatorname{Ker} T_2^n\right\} = \|P\| \gamma(T_2^n).
\end{aligned}$$

Hence $\lim_{n\to\infty} \gamma(T^n)^{1/n} = \lim_{n\to\infty} \gamma(T_2^n)^{1/n}$.

By Theorem 12.26, the limit $r = \lim(\gamma(T_2^n))^{1/n}$ is equal to

$$\sup\{s > 0 : T_2 - z \text{ is Kato for every } z \in \mathbb{C}, |z| < s\}.$$

For $0 < |z| < r$ we have $\alpha(T - z) = \dim \operatorname{Ker}(T - z) = \dim \operatorname{Ker}(T_2 - z)$, which is constant. Denote this constant by a.

On the other hand, there exists $w \in \mathbb{C}$ with $|w| = r$ such that $T_2 - w$ is not Kato. We prove that either $T - w \notin \Phi_+(X)$ or $\alpha(T - w) > a$. Suppose on the contrary that $T - w$ is upper semi-Fredholm. Then $T_2 - w = (T - w)|X_2$ is also upper semi-Fredholm. Since $T_2 - w$ is not Kato, the Kato decomposition of $T_2 - w$ is non-trivial, and so $\alpha(T_2 - w) > \lim_{z\to 0} \alpha(T_2 - w + z) = a$. Hence $\alpha(T - w) > a$.

The statement for lower semi-Fredholm operators follows by duality. □

19 Essential spectra

We assume in this section that X is an infinite-dimensional Banach space (for finite-dimensional spaces all results would be trivial).

In Section 16 we showed that the sets of all Fredholm, upper (lower) semi-Fredholm and left (right) essentially invertible operators in X form regularities. Recall that the corresponding spectra – the *essential spectrum, essential approximate point spectrum, essential surjective spectrum and left (right) essential spectrum* – were defined by

$$\begin{aligned}
\sigma_e(T) &= \{\lambda \in \mathbb{C} : T - \lambda \notin \Phi(X)\},\\
\sigma_{\pi e}(T) &= \{\lambda \in \mathbb{C} : T - \lambda \notin \Phi_+(X)\},\\
\sigma_{\delta e}(T) &= \{\lambda \in \mathbb{C} : T - \lambda \notin \Phi_-(X)\},\\
\sigma_{le}(T) &= \{\lambda \in \mathbb{C} : T - \lambda \text{ is not left essentially invertible}\},\\
\sigma_{re}(T) &= \{\lambda \in \mathbb{C} : T - \lambda \text{ is not right essentially invertible}\}.
\end{aligned}$$

The essential spectra satisfy the same relations as the one-sided and approximate point spectra:

Proposition 1. *Let $T \in \mathcal{B}(X)$. Then $\sigma_{\pi e}(T) \subset \sigma_{le}(T)$, $\sigma_{\delta e}(T) \subset \sigma_{re}(T)$, $\sigma_e(T) = \sigma_{le}(T) \cup \sigma_{re}(T) = \sigma_{\pi e}(T) \cup \sigma_{\delta e}(T)$, $\partial\sigma_e(T) \subset \sigma_{\pi e}(T) \cap \sigma_{\delta e}(T)$.*

Proof. All statements with the exception of the last one are trivial. The inclusion $\partial\sigma_e(T) \subset \sigma_{\pi e}(T) \cap \sigma_{\delta e}(T)$ follows from Lemma 18.1. □

Definition 2. Let $T \in \mathcal{B}(X)$. The *essential spectral radius* of T is defined by $r_e(T) = \max\{|\lambda| : \lambda \in \sigma_e(T)\}$ and the *essential norm* by $\|T\|_e = \inf\{\|T+K\| : K \in \mathcal{K}(X)\}$.

Clearly, $\|T\|_e$ is the norm of the class $T + \mathcal{K}(X)$ in the Calkin algebra $\mathcal{B}(X)/\mathcal{K}(X)$, and $\sigma_e(T)$ is the spectrum of the class $T + \mathcal{K}(X)$ in this algebra.

Corollary 3. *Let $T \in \mathcal{B}(X)$. Then $r_e(T) = \lim_{n\to\infty} \|T^n\|_e^{1/n} = \inf_n \|T^n\|_e^{1/n}$. Since $r_e(T) = r(\tilde{T})$ where $\tilde{T}$ is the operator considered in Section 17, we also have $r_e(T) = \lim_{n\to\infty} \|\tilde{T}^n\|^{1/n} = \inf_n \|\tilde{T}^n\|^{1/n}$.*

Theorem 4. *Let $T \in \mathcal{B}(X)$, let G be a component of $\mathbb{C} \setminus \sigma_e(T)$. Then either $G \subset \sigma(T)$ or $G \cap \sigma(T)$ consists of at most countably many isolated points.*

Proof. Let $\lambda \in G \cap \sigma(T)$. By the punctured neighbourhood theorem, there is an open neighbourhood U of λ such that either $U \cap \sigma(T) = \{\lambda\}$ or $U \subset \sigma(T)$.

Denote by G_0 the union of all open subsets of G which are contained in $\sigma(T)$. Obviously, G_0 is both open and relatively closed and therefore either $G_0 = G$ (in this case $G \subset \sigma(T)$) or $G_0 = \emptyset$, so $G \cap \sigma(T)$ has no limit point in G and therefore it is at most countable. □

Remark 5. The punctured neighbourhood theorem implies similarly that if G is a component of $\mathbb{C} \setminus (\sigma_{\pi e}(T) \cap \sigma_{\delta e}(T))$, then both $\alpha(T - z)$ and $\beta(T - z)$ are constant in G with the exception of at most countable many isolated points of G.

In particular, either $G \subset \sigma_\pi(T)$ or $G \cap \sigma_\pi(T)$ consists of at most countably many isolated points. An analogous statement is also true for σ_δ.

Theorem 6. *Let $T, S \in \mathcal{B}(X)$. Then $\tilde{\sigma}(TS) \setminus \{0\} = \tilde{\sigma}(ST) \setminus \{0\}$, where $\tilde{\sigma}$ stands for any of σ_e, $\sigma_{\pi e}$, $\sigma_{\delta e}$, σ_{le}, σ_{re}.*

Proof. For σ_e, $\sigma_{\pi e}$ and $\sigma_{\delta e}$ this follows from Propositions 12.28 and 12.29.

For σ_{le} and σ_{re} note that an operator is left (right) essentially invertible if and only if it is upper (lower) semi-Fredholm and has a generalized inverse, see Theorems 16.14 and 16.15. Thus the statement follows from Proposition 13.11. □

Clearly all spectra σ_e, $\sigma_{\pi e}$, $\sigma_{\delta e}$, σ_{le} and σ_{re} are invariant with respect to compact perturbations. Thus:

$$\sigma_e(T) \subset \bigcap\{\sigma(T + K) : K \in \mathcal{K}(X)\};$$

$$\sigma_{\pi e}(T) \subset \bigcap\{\sigma_\pi(T + K) : K \in \mathcal{K}(X)\};$$

$$\sigma_{\delta e}(T) \subset \bigcap\{\sigma_\delta(T + K) : K \in \mathcal{K}(X)\};$$

$$\sigma_{le}(T) \subset \bigcap\{\sigma_l(T + K) : K \in \mathcal{K}(X)\};$$

$$\sigma_{re}(T) \subset \bigcap\{\sigma_r(T + K) : K \in \mathcal{K}(X)\}.$$

In general, the opposite inclusions are not true. The next result characterizes the above intersections.

Theorem 7. *Let $T \in \mathcal{B}(X)$. Then T can be expressed as $T = S + K$, where $S, K \in \mathcal{B}(X)$, K is compact and S is invertible (bounded below, onto, left invertible, right invertible) if and only if T is Fredholm with $\operatorname{ind} T = 0$ (T upper semi-Fredholm with $\operatorname{ind} T \leq 0$, T lower semi-Fredholm with $\operatorname{ind} T \geq 0$, T is left* *essentially invertible with* $\operatorname{ind} T \leq 0$*, and* T *is right essentially invertible with* $\operatorname{ind} T \geq 0$*, respectively*). *Thus:*

$$\bigcap\{\sigma(T+K) : K \in \mathcal{K}(X)\} = \sigma_e(T) \cup \{\lambda \in \mathbb{C} : \operatorname{ind}(T-\lambda) \neq 0\};$$

$$\bigcap\{\sigma_\pi(T+K) : K \in \mathcal{K}(X)\} = \sigma_{\pi e}(T) \cup \{\lambda \in \mathbb{C} : \operatorname{ind}(T-\lambda) > 0\};$$

$$\bigcap\{\sigma_\delta(T+K) : K \in \mathcal{K}(X)\} = \sigma_{\delta e}(T) \cup \{\lambda \in \mathbb{C} : \operatorname{ind}(T-\lambda) < 0\};$$

$$\bigcap\{\sigma_l(T+K) : K \in \mathcal{K}(X)\} = \sigma_{le}(T) \cup \{\lambda \in \mathbb{C} : \operatorname{ind}(T-\lambda) > 0\};$$

$$\bigcap\{\sigma_r(T+K) : K \in \mathcal{K}(X)\} = \sigma_{re}(T) \cup \{\lambda \in \mathbb{C} : \operatorname{ind}(T-\lambda) < 0\}.$$

The compact operators in all statements can be replaced by finite-rank operators.

Proof. If $T = S + K$ where S is invertible and K compact, then T is Fredholm and $\operatorname{ind} T = \operatorname{ind} S = 0$.

Conversely, suppose that T is Fredholm and $\operatorname{ind} T = 0$. Then $\dim \operatorname{Ker} T = \operatorname{codim} \operatorname{Ran} T < \infty$. Let $P \in \mathcal{B}(X)$ be a projection onto $\operatorname{Ker} T$. Let $x_1, \dots, x_n$ be a basis in $\operatorname{Ker} T$ and let $y_1, \dots, y_n$ be linearly independent vectors in X such that $\operatorname{Ran} T \vee \{y_1, \dots, y_n\} = X$. Let $U : \operatorname{Ker} T \to X$ be the operator defined by $Ux_i = y_i \quad (i = 1, \dots, n)$ and let $F = UP$. Then F is a finite-rank operator and $T + F$ is invertible.

The remaining statements can be proved similarly. □

The sets described in the previous theorem do not satisfy in general the spectral mapping property; for their properties see Section 23.

Recall that the essential spectrum of an operator $T \in \mathcal{B}(X)$ is equal to the ordinary spectrum of the class $T + \mathcal{K}(X)$ in the Calkin algebra $\mathcal{B}(X)/\mathcal{K}(X)$. This suggests the following definition:

Definition 8. Let $T_1, \dots, T_n$ be a commuting n-tuple of operators on X. Denote by $Q : \mathcal{B}(X) \to \mathcal{B}(X)/\mathcal{K}(X)$ the canonical projection. We define the *left* (*right*) *essential spectrum* of $(T_1, \dots, T_n)$ by

$$\sigma_{le}(T_1, \dots, T_n) = \sigma_l\bigl(Q(T_1), \dots, Q(T_n)\bigr),$$
$$\sigma_{re}(T_1, \dots, T_n) = \sigma_r\bigl(Q(T_1), \dots, Q(T_n)\bigr).$$

The *essential Harte spectrum* of $T_1, \dots, T_n$ is defined by

$$\sigma_{He}(T_1, \dots, T_n) = \sigma_H\bigl(Q(T_1), \dots, Q(T_n)\bigr) = \sigma_{le}(T_1, \dots, T_n) \cup \sigma_{re}(T_1, \dots, T_n).$$

We say that $(T_1, \dots, T_n)$ is *left essentially invertible* if and only if there are $K \in \mathcal{K}(X)$ and operators $S_1, \dots, S_n \in \mathcal{B}(X)$ such that $S_1T_1 + \dots + S_nT_n = I + K$. The *right essentially invertible* n-tuples are defined analogously.

The following properties of the essential spectra follow easily from the definition.

Theorem 9. *σ_{le}, σ_{re} and σ_{He} are upper semicontinuous spectral systems.*

For a commuting n-tuple $T = (T_1, \dots, T_n) \in \mathcal{B}(X)^n$ the sets $\sigma_{le}(T)$, $\sigma_{re}(T)$ and $\sigma_{He}(T)$ are non-empty compact subsets of $\mathbb{C}^n$, $\sigma_{le}(T) \subset \sigma_l(T)$, $\sigma_{re}(T) \subset \sigma_r(T)$ and $\sigma_{He}(T) \subset \sigma_H(T)$.

Although in general $\overline{\mathcal{F}(X)} \neq \mathcal{K}(X)$, Theorem 16.13 gives for $T_1 \in \mathcal{B}(X)$ that

$$\sigma^{\mathcal{B}(X)/\mathcal{K}(X)}(T_1) = \sigma^{\mathcal{B}(X)/\overline{\mathcal{F}(X)}}(T_1).$$

An analogous equality also holds for n-tuples of operators.

Theorem 10. *Let $T = (T_1, \dots, T_n)$ be a commuting n-tuple of operators on a Banach space X. Then*

$$\sigma_l^{\mathcal{B}(X)/\mathcal{K}(X)}(T_1, \dots, T_n) = \sigma_l^{\mathcal{B}(X)/\overline{\mathcal{F}(X)}}(T_1, \dots, T_n).$$

The same equalities are also true for σ_r and σ_H.

Proof. It is sufficient to show that $(0, \dots, 0) \notin \sigma_{le}(T_1, \dots, T_n)$ if and only if $(0, \dots, 0) \notin \sigma_l^{\mathcal{B}(X)/\overline{\mathcal{F}(X)}}(T_1, \dots, T_n)$.

If $(0, \dots, 0) \notin \sigma_{le}(T_1, \dots, T_n)$, then there exist operators $S_1, \dots, S_n \in \mathcal{B}(X)$ and $K \in \mathcal{K}(X)$ such that $\sum_{i=1}^n S_i T_i = I + K$. By Theorem 16.13, there exist operators $A \in \mathcal{B}(X)$ and $F \in \mathcal{F}(X)$ such that $A(I + K) = I + F$. Thus $\sum_{i=1}^n (AS_i)T_i = A(I + K) = I + F$ and $(0, \dots, 0) \notin \sigma_l^{\mathcal{B}(X)/\overline{\mathcal{F}(X)}}(T_1, \dots, T_n)$.

The opposite implication is clear. □

Definition 11. Let $T = (T_1, \dots, T_n)$ be a commuting n-tuple of operators in a Banach space X. We say that the n-tuple T is *upper semi-Fredholm* if there exists a subspace $M \subset X$ of finite codimension such that

$$\inf \left\{ \sum_{i=1}^n \|T_i x\| : x \in M, \|x\| = 1 \right\} > 0.$$

T is called *lower semi-Fredholm* if $\operatorname{codim}(T_1 X + \cdots + T_n X) < \infty$.

Clearly, $T = (T_1, \dots, T_n)$ is upper semi-Fredholm if and only if the operator $\delta_T : X \to X^n$ defined by $\delta_T x = (T_1 x, \dots, T_n x)$ is upper semi-Fredholm. Similarly, the n-tuple T is lower semi-Fredholm if and only if the operator $\eta_T : X^n \to X$ defined by $\eta_T(x_1, \dots, x_n) = \sum_{i=1}^n T_i x_i$ is lower semi-Fredholm. As in the proof of Corollary 9.14 we get:

Corollary 12. *Let $T = (T_1, \dots, T_n)$ be a commuting n-tuple of operators on a Banach space X. Then:*

$(T_1, \dots, T_n)$ is upper semi-Fredholm $\Leftrightarrow$ $(T_1^*, \dots, T_n^*)$ is lower semi-Fredholm;

$(T_1, \dots, T_n)$ is lower semi-Fredholm $\Leftrightarrow$ $(T_1^*, \dots, T_n^*)$ is upper semi-Fredholm.

In the following we use the construction $T \in \mathcal{B}(X) \mapsto \tilde{T} \in \mathcal{B}(\tilde{X})$, which was defined in Section 17; for this purpose it is more convenient to consider the space X^n in the definition of operators δ_T and η_T with the ℓ^∞ norm.

With this convention $\ell^\infty(X^n)$ can be identified with $(\ell^\infty(X))^n$ and $\widetilde{X^n}$ with $\tilde{X}^n$. Thus $\widetilde{\delta_T}\tilde{x} = (\tilde{T}_1\tilde{x}, \ldots, \tilde{T}_n\tilde{x})$ and $\widetilde{\eta_T}(\tilde{x}_1, \ldots, \tilde{x}_n) = \sum_{i=1}^n \tilde{T}_i\tilde{x}_i$.

Using the corresponding results for the operators δ_T and η_T (Theorems 17.6 and 17.9) we get:

Corollary 13. *An n-tuple $T = (T_1, \ldots, T_n)$ is upper semi-Fredholm if and only if $\inf\{\sum_{i=1}^n \|\tilde{T}_i\tilde{x}\| : \tilde{x} \in \tilde{X}, \|\tilde{x}\| = 1\} > 0$. T is lower semi-Fredholm if and only if $\tilde{T}_1\tilde{X} + \cdots + \tilde{T}_n\tilde{X} = \tilde{X}$.*

Definition 14. The *essential approximate point spectrum and the essential surjective spectrum* of $T = (T_1, \ldots, T_n)$ are defined by

$$\sigma_{\pi e}(T) = \{\lambda \in \mathbb{C}^n : T - \lambda \text{ is not upper semi-Fredholm}\},$$
$$\sigma_{\delta e}(T) = \{\lambda \in \mathbb{C}^n : T - \lambda \text{ is not lower semi-Fredholm}\}.$$

Theorem 15. *Let $T = (T_1, \ldots, T_n)$ be an n-tuple of commuting operators acting on a Banach space X. Then $\sigma_{\pi e}(T_1, \ldots, T_n) = \sigma_\pi(\tilde{T}_1, \ldots, \tilde{T}_n)$ and $\sigma_{\delta e}(T_1, \ldots, T_n) = \sigma_\delta(\tilde{T}_1, \ldots, \tilde{T}_n)$.*

In particular, $\sigma_{\pi e}$ and $\sigma_{\delta e}$ are upper semicontinuous spectral systems.

Proof. Follows from Theorems 17.6 and 17.9. □

Corollary 16. *Let $T = (T_1, \ldots, T_n)$ be an n-tuple of commuting operators on a Banach space X. Denote by $\mathcal{P}(n)$ the algebra of all polynomials in n variables. Then $\Gamma(\sigma_{He}(T), \mathcal{P}(n)) \subset \sigma_{\pi e}(T) \cap \sigma_{\delta e}(T)$. In particular, the polynomially convex hulls of $\sigma_{He}(T)$, $\sigma_{\pi e}(T)$ and $\sigma_{\delta e}(T)$ coincide.*

Proof. Follows from Theorem 8.8 and Corollary 9.13. □

Theorem 17. *An n-tuple $T = (T_1, \ldots, T_n)$ is left essentially invertible if and only if T is upper semi-Fredholm and the operator δ_T has complemented range. T is right essentially invertible if and only if T is lower semi-Fredholm and the operator η_T has complemented kernel.*

Proof. Let $K \in \mathcal{K}(X)$ and $S_1, \ldots, S_n \in \mathcal{B}(X)$. Then $S_1T_1 + \cdots + S_nT_n = I + K$ if and only if $\eta_S\delta_T = I + K$, where $\eta_S : X^n \to X$ is defined by $\eta_S(x_1, \ldots, x_n) = \sum_{i=1}^n S_ix_i$. By Theorem 16.14, δ_T is essentially left invertible if and only if δ_T is upper semi-Fredholm and $\operatorname{Ran}\delta_T$ is complemented in X^n.

The second statement can be proved similarly. □

For a single operator T_1 the set $\sigma(T_1) \setminus \sigma_e(T_1)$ can be easily described, see Theorem 4. For n-tuples of operators the situation is more complicated.

Theorem 18. *Let $T = (T_1, \dots, T_n)$ be an n-tuple of commuting operators on an infinite-dimensional Banach space X. Let $\lambda = (\lambda_1, \dots, \lambda_n) \in \sigma_H(T) \setminus \widehat{\sigma}_{He}(T)$. Then:*

(i) *λ is an isolated point of $\sigma_H(T)$;*

(ii) *λ is a joint eigenvalue of operators $T_1, \dots, T_n$ of finite multiplicity, i.e.,*

$$1 \le \dim \bigcap_{i=1}^{n} \operatorname{Ker}(T_i - \lambda_i) < \infty.$$

In particular, $\sigma_H(T) \setminus \widehat{\sigma}_{He}(T)$ is at most countable.

Proof. (i) Let $\lambda \in \sigma_H(T) \setminus \widehat{\sigma}_{He}(T)$. Then there exists a polynomial p of n variables such that $|p(\lambda)| > \max\{|p(\mu)| : \mu \in \sigma_{He}(T)\}$. By the spectral mapping theorem for the Harte spectrum we have $p(\lambda) \in \sigma(p(T))$ and

$$\max\{|\mu| : \mu \in \sigma_e(p(T))\} = \max\{|p(\mu)| : \mu \in \sigma_{He}(T)\} < |p(\lambda)|.$$

Thus $p(\lambda)$ lies in the unbounded component of $\mathbf{C} \setminus \sigma_e(p(T))$, and so it is an isolated point of $\sigma(p(T))$. This means that there is a neighbourhood U_0 of λ such that $p(\sigma_H(T) \cap U_0) = \{p(\lambda)\}$.

Let ε be a sufficiently small positive number, so that

$$|q_i(\lambda)| > \max\{|q_i(\mu)| : \mu \in \sigma_{He}(T)\} \qquad (i = 1, \dots, n),$$

where q_i are polynomials of n variables defined by

$$q_i(z_1, \dots, z_n) = p(z_1, \dots, z_n) + \varepsilon z_i.$$

Repeating the same considerations for q_i instead of p, we get that there are neighbourhoods U_i of λ such that $q_i(\sigma_H(T) \cap U_i) = \{q_i(\lambda)\} \quad (i = 1, \dots, n)$. Let $W = \bigcap_{i=0}^{n} U_i$ and let $\mu = (\mu_1, \dots, \mu_n) \in \sigma_H(T) \cap W$. Then $p(\mu) = p(\lambda)$ and

$$p(\mu) + \varepsilon\mu_i = q_i(\mu) = q_i(\lambda) = p(\lambda) + \varepsilon\lambda_i.$$

So $\mu_i = \lambda_i \quad (i = 1, \dots, n)$ and $\mu = \lambda$. Hence λ is an isolated point of $\sigma_H(T)$.

This implies that the set $\sigma_H(T) \setminus \widehat{\sigma}_{He}(T)$ is at most countable.

(ii) Let p be the polynomial constructed in part (i) such that

$$|p(\lambda)| > \max\{|p(\mu)| : \mu \in \sigma_{He}(T)\}.$$

So there is an open neighbourhood V of $\widehat{\sigma}_{He}(T)$ such that $|p(\lambda)| > \sup\{|p(\mu)| : \mu \in V\}$. By (i), $\sigma_H(T) \setminus V$ is a finite set, $\sigma_H(T) \setminus (V \cup \{\lambda\}) = \{\lambda^{(1)}, \dots, \lambda^{(k)}\}$. Find polynomials $q_1, \dots, q_n$ such that $q_i(\lambda^{(i)}) = 0 \neq q_i(\lambda) \quad (i = 1, \dots, k)$. Then, for a positive integer s large enough, the polynomial $h = p^s q_1 \cdots q_k$ satisfies

$$|h(\lambda)| > \sup\{|h(\mu)| : \mu \in V\} = \sup\{|h(\mu)| : \mu \in \sigma_H(T) \setminus \{\lambda\}\}.$$

Thus $\lambda \in \Gamma(\sigma_H(T), \mathcal{P}(n))$. By Corollary 9.13 and Theorem 8.8, $\lambda \in \sigma_\pi(T)$, and so the operator $\delta_{T-\lambda} : X \to X^n$ defined by $\delta_{T-\lambda}x = \big((T_1 - \lambda_1)x, \dots, (T_n - \lambda_n)x\big)$ is not bounded below. Furthermore, $\lambda \notin \sigma_{He}(T)$, and so $\delta_{T-\lambda}$ is upper semi-Fredholm. In particular, $\operatorname{Ker} \delta_{T-\lambda} = \bigcap_{i=1}^n \operatorname{Ker}(T_i - \lambda_i)$ is finite dimensional and $\operatorname{Ran} \delta_{T-\lambda}$ is closed. Thus $\operatorname{Ker} \delta_{T-\lambda} \neq \{0\}$.

This completes the proof. □

20 Ascent, descent and Browder operators

Let T be an operator on a Banach space X. It is easy to see that $\operatorname{Ker} T \subset \operatorname{Ker} T^2 \subset \operatorname{Ker} T^3 \subset \cdots$ and $\operatorname{Ran} T \supset \operatorname{Ran} T^2 \supset \cdots$.

Lemma 1. *If $k \geq 0$ and $\operatorname{Ker} T^{k+1} = \operatorname{Ker} T^k$, then $\operatorname{Ker} T^p = \operatorname{Ker} T^k$ for every $p \geq k$. Similarly, if $\operatorname{Ran} T^{k+1} = \operatorname{Ran} T^k$, then $\operatorname{Ran} T^p = \operatorname{Ran} T^k \quad (p \geq k)$.*

Proof. We show the implications $\operatorname{Ker} T^{k+1} = \operatorname{Ker} T^k \Rightarrow \operatorname{Ker} T^{k+2} = \operatorname{Ker} T^{k+1}$ and $\operatorname{Ran} T^{k+1} = \operatorname{Ran} T^k \Rightarrow \operatorname{Ran} T^{k+2} = \operatorname{Ran} T^{k+1}$; the rest follows easily by induction.

Suppose that $\operatorname{Ker} T^{k+1} = \operatorname{Ker} T^k$ and let $x \in \operatorname{Ker} T^{k+2}$. Then

$$Tx \in \operatorname{Ker} T^{k+1} = \operatorname{Ker} T^k,$$

and so $x \in \operatorname{Ker} T^{k+1}$.

Similarly, if $\operatorname{Ran} T^{k+1} = \operatorname{Ran} T^k$ and $x \in \operatorname{Ran} T^{k+1}$, then $x = T^{k+1}y$ for some $y \in X$ and $T^k y \in \operatorname{Ran} T^k = \operatorname{Ran} T^{k+1}$. So $x = T(T^k y) \in \operatorname{Ran} T^{k+2}$. □

Definition 2. Let $T \in \mathcal{B}(X)$. The *ascent* of T is defined by

$$a(T) = \min\{n : \operatorname{Ker} T^n = \operatorname{Ker} T^{n+1}\}$$

(if no such n exists, then we set $a(T) = \infty$). Similarly, the *descent* of T is defined by

$$d(T) = \min\{n : \operatorname{Ran} T^n = \operatorname{Ran} T^{n+1}\}.$$

The following simple lemma is useful in many situations.

Lemma 3. *Let $T \in \mathcal{B}(X, Y)$, let M be a closed subspace of Y such that both $M + \operatorname{Ran} T$ and $M \cap \operatorname{Ran} T$ are closed. Then $\operatorname{Ran} T$ is closed.*

Proof. Write for short $N = M \cap \operatorname{Ran} T$ and $L = T^{-1}N$. Clearly, L is a closed subspace of X. Let $\Phi : (X/L) \oplus (M/N) \to (\operatorname{Ran} T + M)/N$ be the operator defined by

$$\Phi\big((x + L) \oplus (m + N)\big) = (Tx + m) + N.$$

It is easy to check that the definition of Φ is correct, Φ is onto and one-to-one. Thus Φ is bounded below, and so $\Phi(X/L)$ is closed.

Let $Q : \operatorname{Ran} T + M \to (\operatorname{Ran} T + M)/N$ be the canonical projection. Then $\operatorname{Ran} T = Q^{-1}\big(\operatorname{Ran} T/N\big) = Q^{-1}\Phi(X/L)$, and so $\operatorname{Ran} T$ is closed. □

Theorem 4. *Let $T \in \mathcal{B}(X)$, $k \geq 0$ and suppose that* $\operatorname{Ker} T^{k+1} = \operatorname{Ker} T^k$ *and* $\operatorname{Ran} T^{k+1} = \operatorname{Ran} T^k$. *Then* $\operatorname{Ran} T^k$ *is closed and* $X = \operatorname{Ker} T^k \oplus \operatorname{Ran} T^k$.

Proof. We must show that $\operatorname{Ran} T^k \cap \operatorname{Ker} T^k = \{0\}$ and $\operatorname{Ran} T^k + \operatorname{Ker} T^k = X$.

Let $x \in \operatorname{Ran} T^k \cap \operatorname{Ker} T^k$. Then $x = T^k y$ for some $y \in X$ and $0 = T^k x = T^{2k} y$. Thus $y \in \operatorname{Ker} T^{2k} = \operatorname{Ker} T^k$ and $x = T^k y = 0$. Hence $\operatorname{Ran} T^k \cap \operatorname{Ker} T^k = \{0\}$.

Let $x \in X$. Then $T^k x \in \operatorname{Ran} T^k = \operatorname{Ran} T^{2k}$, and so $T^k x = T^{2k} y$ for some $y \in X$. Consequently, $x = T^k y + (x - T^k y)$ where $T^k y \in \operatorname{Ran} T^k$ and $x - T^k y \in \operatorname{Ker} T^k$.

Lemma 3 for the operator T^k now implies that $\operatorname{Ran} T^k$ is closed. □

Corollary 5. *Let $T \in \mathcal{B}(X)$ and let $a(T) < \infty$ and $d(T) < \infty$. Then $a(T) = d(T)$. If $k = a(T) = d(T)$, then* $X = \operatorname{Ker} T^k \oplus \operatorname{Ran} T^k$.

Proof. Let $k = \max\{a(T), d(T)\}$. By Theorem 4, $X = \operatorname{Ker} T^k \oplus \operatorname{Ran} T^k$ and both $\operatorname{Ran} T^k$ and $\operatorname{Ker} T^k$ are invariant with respect to T. Write $T_1 = T|\operatorname{Ker} T^k : \operatorname{Ker} T^k \to \operatorname{Ker} T^k$ and $T_2 = T|\operatorname{Ran} T^k : \operatorname{Ran} T^k \to \operatorname{Ran} T^k$. Then $\operatorname{Ran} T_2 = T \operatorname{Ran} T^k = \operatorname{Ran} T^{k+1} = \operatorname{Ran} T^k$. Further, if $x \in \operatorname{Ker} T_2$, then $x = T^k y$ for some $y \in X$, so $y \in \operatorname{Ker} T^{k+1} = \operatorname{Ker} T^k$ and $x = T^k y = 0$. Thus T_2 is invertible. Furthermore, $T_1^k = 0$. Hence

$$a(T) = a(T_1) = \min\{n : T_1^n = 0\} = d(T_1) = d(T).$$ □

Definition 6. We say that an operator $T \in \mathcal{B}(X)$ is *upper semi-Browder* if it is upper semi-Fredholm and has finite ascent.

Similarly, T is *lower semi-Browder* if it is lower semi-Fredholm and has finite descent. An operator T is *Browder* if it is both lower and upper semi-Browder. Equivalently, this means that T is Fredholm and has finite both ascent and descent.

By Section 15, any operator of the form $K + \lambda I$ where $K : X \to X$ is compact and $\lambda \neq 0$ is Browder.

Theorem 7. *Let $T \in \mathcal{B}(X)$. Then:*

(i) *T is upper semi-Browder $\Leftrightarrow$ T^* is lower semi-Browder;*
(ii) *T is lower semi-Browder $\Leftrightarrow$ T^* is upper semi-Browder;*
(iii) *T is Browder $\Leftrightarrow$ T^* is Browder.*

Proof. Follows from the corresponding results for semi-Fredholm operators and Theorem A.1.14.

Recall that $R^\infty(T) = \bigcap_n \operatorname{Ran} T^n$ and $N^\infty(T) = \bigcup_n \operatorname{Ker} T^n$.

Proposition 8. *Let $T \in \mathcal{B}(X)$. Then:*

(i) *T is upper semi-Browder $\Leftrightarrow$ $\operatorname{Ran} T$ is closed and $\dim N^\infty(T) < \infty$;*
(ii) *T is lower semi-Browder $\Leftrightarrow$ $\operatorname{codim} R^\infty(T) < \infty$;*
(iii) *T is Browder $\Leftrightarrow$ $\dim N^\infty(T) < \infty$ and $\operatorname{codim} R^\infty(T) < \infty$.*

Proof. If T is upper semi-Browder, then $\operatorname{Ran} T$ is closed and $k = a(T) < \infty$. Since T^k is upper semi-Fredholm, we have $\dim N^\infty(T) = \dim \operatorname{Ker} T^k < \infty$.

Conversely, if $\operatorname{Ran} T$ is closed and $\dim N^\infty(T) < \infty$, then T is upper semi-Fredholm. Further, $\operatorname{Ker} T \subset \operatorname{Ker} T^2 \subset \cdots \subset N^\infty(T)$, and so there exists k with $\operatorname{Ker} T^{k+1} = \operatorname{Ker} T^k$. Hence $a(T) < \infty$.

The remaining statements can be proved similarly. □

Lemma 9. *Let $T \in \mathcal{B}(X)$ be upper semi-Browder and Kato. Then T is bounded below. If T is lower semi-Browder and Kato, then T is onto.*

Proof. Suppose that there exists a non-zero vector $x_0 \in \operatorname{Ker} T$. Since $\operatorname{Ker} T \subset \operatorname{Ran} T$, there exists $x_1 \in X$ such that $Tx_1 = x_0$. Further, $x_1 \in \operatorname{Ker} T^2 \subset \operatorname{Ran} T$ and we can construct inductively vectors $x_i \in X$ satisfying $Tx_i = x_{i-1} \quad (i \geq 1)$. It is easy to show that the vectors x_i are linearly independent and $x_i \in N^\infty(T)$, a contradiction with Proposition 8.

The second statement can be proved by duality. □

Theorem 10. *An operator $T \in \mathcal{B}(X)$ is upper semi-Browder (lower semi-Browder, Browder) if and only if there exists a decomposition $X = X_1 \oplus X_2$ such that $\dim X_1 < \infty$, $TX_i \subset X_i \quad (i = 1, 2)$, $T|X_1$ is nilpotent and $T|X_2$ is bounded below (onto, invertible, respectively).*

If T is upper semi-Browder, then the space X_1 is uniquely determined and $X_1 = N^\infty(T)$. If T is lower semi-Browder, then $X_2 = R^\infty(T)$. If T is Browder, then the decomposition is unique: $N^\infty(T) \oplus R^\infty(T)$.

Proof. Suppose that $T \in \mathcal{B}(X)$ is upper semi-Browder and let $X = X_1 \oplus X_2$ be the Kato decomposition, i.e., $\dim X_1 < \infty$, $T|X_1$ is nilpotent and $T_2 = T|X_2$ is Kato. Evidently, T_2 is upper semi-Browder, and so T_2 is bounded below by Lemma 9. Clearly, $X_1 \subset N^\infty(T)$. Furthermore, if $T^n(x_1 \oplus x_2) = 0$ for some n, then $T^n x_2 = 0$, and so $x_2 = 0$. Thus $\operatorname{Ker} T^n \subset X_1$ for all n and hence $X_1 = N^\infty(T)$.

In the opposite direction, it is easy to see that an operator that can be written as a direct sum of a finite-dimensional nilpotent and an operator bounded below is upper semi-Browder.

The statements for lower semi-Browder and Browder operators can be obtained similarly. □

Corollary 11. *If T is upper semi-Browder, then $\operatorname{ind} T \leq 0$. If T is lower semi-Browder, then $\operatorname{ind} T \geq 0$. If T is Browder, then $\operatorname{ind} T = 0$.*

Proof. Let $T \in \mathcal{B}(X)$ be upper semi-Browder, and let $X = X_1 \oplus X_2$ be the decomposition from the preceding theorem: $\dim X_1 < \infty$, $TX_i \subset X_i \quad (i = 1, 2)$, $T|X_1$ nilpotent and $T|X_2$ bounded below. Then $\operatorname{ind} T = \operatorname{ind}(T|X_2) = -\dim X_2 / \operatorname{Ran} T_2 \leq 0$.

The statements for lower semi-Browder and Browder operators can be proved analogously. □

It is not difficult to show that the Browder and upper (lower) semi-Browder operators form regularities. Our aim is to prove a stronger result and to extend these notions to commuting n-tuples of operators.

We discuss the lower semi-Browder case; the upper case will be dual.

Let $T = (T_1, \dots, T_n)$ be an n-tuple of mutually commuting operators on a Banach space X.

For $k = 0, 1, 2, \dots$ set $M_k(T) = \operatorname{Ran} T_1^k + \dots + \operatorname{Ran} T_n^k$ and let $M'_k(T)$ be the smallest subspace of X containing the set $\bigcup\{\operatorname{Ran} T^\alpha : \alpha \in \mathbb{Z}_+^n, |\alpha| = k\}$. Clearly, $X = M_0(T) \supset M_1(T) \supset M_2(T) \supset \cdots$ and $X = M'_0(T) \supset M'_1(T) \supset M'_2(T) \supset \cdots$. Furthermore,

$$M'_{n(k-1)+1}(T) \subset M_k(T) \subset M'_k(T). \tag{1}$$

Indeed, if $\alpha = (\alpha_1, \dots, \alpha_n) \in \mathbb{Z}_+^n$ and $|\alpha| = n(k-1)+1$, then there exists i, $1 \le i \le n$ such that $\alpha_i \ge k$. So $\operatorname{Ran} T^\alpha \subset \operatorname{Ran} T_i^k \subset M_k(T)$. This proves the first inclusion of (1) and the second inclusion is clear.

Let $R^\infty(T) = \bigcap_{k=0}^\infty M_k(T) = \bigcap_{k=0}^\infty M'_k(T)$.

If $M'_k(T) = M'_{k+1}(T)$ for some k, then it is easy to see by induction that $M'_m(T) = M'_k(T)$ for every $m \ge k$, and so $R^\infty(T) = M'_k(T)$.

We say that $T = (T_1, \dots, T_n)$ is *lower semi-Browder* if $\operatorname{codim} R^\infty(T) < \infty$. It is clear that the lower semi-Browder n-tuples are contained in $\Phi_-^{(n)}(X)$, where $\Phi_-^{(n)}(X)$ denotes the set of all lower semi-Fredholm n-tuples of commuting operators on X.

Define the corresponding *lower semi-Browder spectrum* by

$$\sigma_{B_-}(T) = \{\lambda \in \mathbb{C}^n : T - \lambda \text{ is not lower semi-Browder}\}.$$

By Theorem 19.15, $(T_1, \dots, T_n) \in \Phi_-^{(n)}(X)$ if and only if $(T_1^k, \dots, T_n^k) \in \Phi_-^{(n)}(X)$. Thus $\operatorname{codim} M_1(T) < \infty$ implies $\operatorname{codim} M_k(T) < \infty$ for all k.

Theorem 12. *Let $T = (T_1, \dots, T_n)$ be an n-tuple of mutually commuting operators on a Banach space X. The following statements are equivalent:*

(i) *T is lower semi-Browder;*

(ii) *$T \in \Phi_-^{(n)}(X)$ and there exists k such that $M'_k(T) = M'_{k+1}(T)$;*

(iii) *$T \in \Phi_-^{(n)}(X)$ and there exists k such that $M_k(T) = M_{k+1}(T)$;*

(iv) *there exists a closed subspace $Y \subset X$ invariant with respect to all T_i ($i = 1, \dots, n$) such that $\operatorname{codim} Y < \infty$ and $T_1 Y + \dots + T_n Y = Y$. It is possible to take $Y = R^\infty(T)$.*

Proof. (iii) $\Rightarrow$ (ii): Let $M_k(T) = M_{k+1}(T)$ for some k. Using the same argument as in the proof of (1) it is possible to show that $M'_{n(k-1)+1}(T) = M'_{n(k-1)+2}(T)$.

(ii) $\Rightarrow$ (i): Let $M'_k(T) = M'_{k+1}(T)$ for some k. Then $M_k(T) \subset M'_k(T) = R^\infty(T)$. Moreover, $T \in \Phi_-^{(n)}(X)$, which implies that $\operatorname{codim} M_k(T) < \infty$, and so T is lower semi-Browder.

(i) ⇒ (iv): Set $Y = R^\infty(T)$. Clearly Y is invariant with respect to T_i ($i = 1, \dots, n$), $\operatorname{codim} Y < \infty$ and $Y = M_k(T) = M_{k+1}(T)$ for some k. If $y \in Y$, then, for some $x_1, \dots, x_n \in X$, we have

$$y = \sum_{i=1}^{n} T_i^{k+1} x_i = \sum_{i=1}^{n} T_i(T_i^k x_i) \in T_1 Y + \cdots + T_n Y.$$

(iv) ⇒ (iii): Since $M_1(T) \supset M_1(T|Y) = Y$, we have $\operatorname{codim} M_1(T) < \infty$, and so $T \in \Phi_-^{(n)}(X)$. Further, $R^\infty(T) \supset Y$, and so $\operatorname{codim} R^\infty(T) < \infty$. □

Theorem 13. *Let $T = (T_1, \dots, T_n)$ be a lower semi-Browder n-tuple of operators on a Banach space X. Then there exists $\epsilon > 0$ such that S is lower semi-Browder for every commuting n-tuple $S = (S_1, \dots, S_n) \in \mathcal{B}(X)^n$ with $\sum_{i=1}^n \|S_i - T_i\| < \epsilon$.*

Proof. Choose k such that $M_k(T) = R^\infty(T)$ and $\operatorname{codim} R^\infty(T) \le k$. Then $(T_1^{k+1}, \dots, T_n^{k+1}) \in \Phi_-^{(n)}(X)$. Consider the operator $\eta : X^n \to X$ defined by

$$\eta(x_1, \dots, x_n) = \sum_{i=1}^{n} T_i^{k+1} x_i.$$

Clearly,

$$\operatorname{codim} \operatorname{Ran} \eta = \operatorname{codim} M_{k+1}(T) = \operatorname{codim} R^\infty(T) \le k.$$

By Theorem 16.11, there exists $\nu > 0$ such that $\operatorname{codim} \operatorname{Ran} \eta' \le k$ for each operator $\eta' : X \to X^n$ with $\|\eta' - \eta\| < \nu$. If $S = (S_1, \dots, S_n)$ is a commuting n-tuple of operators in X close enough to T, then $(S_1^{k+1}, \dots, S_n^{k+1}) \in \Phi_-^{(n)}(X)$ and

$$\operatorname{codim} M_1(S_1^{k+1}, \dots, S_n^{k+1}) \le \operatorname{codim} \eta \le k.$$

Since $M_1(S) \supset M_2(S) \supset \cdots \supset M_{k+1}(S)$ and $\operatorname{codim} M_{k+1}(S) \le k$, there exists $j \le k$ such that $M_j(S) = M_{j+1}(S)$. Consequently, S is lower semi-Browder. □

Proposition 14. *Let $T_1, \dots, T_n, S_1, \dots, S_n$ be mutually commuting operators on X such that $\sum_{i=1}^n T_i S_i = I$. Then the n-tuple $(T_1, \dots, T_n)$ is lower semi-Browder.*

Proof. Clearly $M_1(T_1, \dots, T_n) = X = M_0(T_1, \dots, T_n)$, and so $(T_1, \dots, T_n)$ is lower semi-Browder. □

Lemma 15. *Let $T_0, T_1, \dots, T_n$ be mutually commuting operators on a Banach space X. Suppose that $\operatorname{codim} R^\infty(T_1, \dots, T_n) = \infty$ and let $k \in \mathbb{N}$. Then there exists a complex number λ such that*

$$\operatorname{codim}\big(\operatorname{Ran}(T_0 - \lambda)^k + \operatorname{Ran} T_1^k + \cdots + \operatorname{Ran} T_n^k\big) \ge k. \tag{2}$$

Proof. Using condition (iii) of Theorem 12, we can distinguish two cases:

(a) Let $(T_1,\dots,T_n) \notin \Phi_-^{(n)}(X)$. By the spectral mapping property for the essential surjective spectrum (Theorem 19.14), there exists $\lambda \in \mathbb{C}$ such that $(T_0 - \lambda, T_1,\dots,T_n) \notin \Phi_-^{(n+1)}(X)$. Thus

$$\begin{aligned}&\operatorname{codim}\bigl(\operatorname{Ran}(T_0-\lambda)^k + \operatorname{Ran} T_1^k + \cdots + \operatorname{Ran} T_n^k\bigr)\\ &\geq \operatorname{codim}\bigl(\operatorname{Ran}(T_0-\lambda) + \operatorname{Ran} T_1 + \cdots + \operatorname{Ran} T_n\bigr) = \infty.\end{aligned}$$

(b) Suppose that $T = (T_1,\dots,T_n) \in \Phi_-^{(n)}(X)$ and $\operatorname{codim} R^\infty(T) = \infty$. Then $\operatorname{codim} M_j(T) < \infty$ for all j. Since $\operatorname{codim} R^\infty(T) = \infty$, we have $M_j(T) \neq M_{j+1}(T)$ for all $j \geq 1$.

Fix $k \in \mathbb{N}$. Then there exists $i, 1 \leq i \leq n$ such that $\operatorname{Ran} T_i^{k-1} \not\subset M_k(T)$ (otherwise $M_{k-1}(T) = M_k(T)$). Let $Y = X/M_k(T)$, and let $S : Y \mapsto Y$ be defined by $S(x + M_k(T)) = T_i x + M_k(T)$. Clearly, $\dim Y < \infty$, $S^k = 0$ and $S^{k-1} \neq 0$.

Consider the operator $U : Y \mapsto Y$ defined by $U(x + M_k(T)) = T_0 x + M_k(T)$. Obviously, $US = SU$. Let Z be a subspace of Y satisfying $Z \oplus \operatorname{Ker} S^{k-1} = Y$. In this decomposition U can be written as

$$U = \begin{pmatrix} U_{11} & 0 \\ U_{12} & U_{22} \end{pmatrix}.$$

Choose an eigenvalue λ of $U_{11} - \lambda$ and let $z \in Z$ be a corresponding eigenvector; so $z \neq 0$ and $(U-\lambda)z \in \operatorname{Ker} S^{k-1}$. Since $z \in \operatorname{Ker} S^k \setminus \operatorname{Ker} S^{k-1}$, we have

$$S^{k-j} z \in \operatorname{Ker} S^j \setminus \operatorname{Ker} S^{j-1} \qquad (j = 1,\dots,k).$$

Furthermore,

$$(U-\lambda)S^{k-j}z = S^{k-j}(U-\lambda)z \in S^{k-j}\operatorname{Ker} S^{k-1} \subset \operatorname{Ker} S^{j-1}.$$

Let $1 \leq j \leq k$ and write $M = \operatorname{Ker} S^{j-1} \vee \{S^{k-j}z\}$. Then $M \subset \operatorname{Ker} S^j$ and $(U-\lambda)^j M \subset (U-\lambda)^{j-1}\operatorname{Ker} S^{j-1} \subset M$. We have

$$\begin{aligned}&\dim\bigl(\operatorname{Ker} S^j/(U-\lambda)^j \operatorname{Ker} S^j\bigr) = \dim \operatorname{Ker}\bigl((U-\lambda)^j|\operatorname{Ker} S^j\bigr)\\ &\geq \dim\operatorname{Ker}\bigl((U-\lambda)^j|M\bigr) = \dim\bigl(M/(U-\lambda)^j M\bigr)\\ &\geq \dim\bigl(M/(U-\lambda)^{j-1}\operatorname{Ker} S^{j-1}\bigr) = \dim\bigl(\operatorname{Ker} S^{j-1}/(U-\lambda)^{j-1}\operatorname{Ker} S^{j-1}\bigr) + 1,\end{aligned}$$

since $S^{k-j}z \notin \operatorname{Ker} S^{j-1}$. Thus, by induction,

$$\dim\bigl(\operatorname{Ker} S^j/(U-\lambda)^j \operatorname{Ker} S^j\bigr) \geq j \qquad (j = 1,\dots,k).$$

In particular, $\dim(Y/(U-\lambda)^k Y) \geq k$. Consequently,

$$\operatorname{codim}\bigl(\operatorname{Ran}(T_0-\lambda)^k\bigr) + \operatorname{Ran} T_1^k + \cdots + \operatorname{Ran} T_n^k\bigr) \geq k. \qquad \square$$

Corollary 16. *Let $T_0, T_1, \dots, T_n$ be mutually commuting operators on a Banach space X. Suppose that* $\operatorname{codim} R^\infty(T_1, \dots, T_n) = \infty$. *Then there exists $\lambda \in \mathbb{C}$ such that*

$$\operatorname{codim} R^\infty(T_0 - \lambda, T_1, \dots, T_n) = \infty.$$

Proof. For each $k \geq 1$ we can find $\lambda_k \in \mathbb{C}$ such that

$$\begin{aligned}&\operatorname{codim} R^\infty(T_0 - \lambda_k, T_1, \dots, T_n)\\ &\geq \operatorname{codim}\bigl(\operatorname{Ran}(T_0 - \lambda_k)^k + \operatorname{Ran} T_1^k + \dots + \operatorname{Ran} T_n^k\bigr) \geq k.\end{aligned}$$

It is clear that $\lambda_k \in \sigma(T_0)$ for every k. Thus we may assume (by passing to a subsequence if necessary) that the sequence (λ_k) is convergent, $\lambda_k \to \lambda \in \sigma(T_0)$. We have

$$\lim_{k\to\infty} \operatorname{codim} R^\infty(T_0 - \lambda_k, T_1, \dots, T_n) = \infty.$$

By Theorem 13, this implies that $\operatorname{codim} R^\infty(T_0 - \lambda, T_1, \dots, T_n) = \infty$. □

Corollary 17. *σ_{B_-} is an upper semicontinuous spectral system.*

Upper semi-Browder n-tuples can be defined similarly. Let $T = (T_1, \dots, T_n)$ be an n-tuple of commuting operators on a Banach space X. Recall that T is upper semi-Fredholm if the mapping $\delta_T : X \mapsto X^n$ defined by $\delta_T x = (T_1 x, \dots, T_n x)$ is upper semi-Fredholm. We say that T is *upper semi-Browder* if T is upper semi-Fredholm and $\dim N^\infty(T) < \infty$, where

$$N^\infty(T) = \bigcup_{k=1}^{\infty} \bigl(\operatorname{Ker} T_1^k \cap \dots \cap \operatorname{Ker} T_n^k\bigr).$$

We say that T is *Browder* if it is both upper and lower semi-Browder.

Write $T^* = (T_1^*, \dots, T_n^*) \in \mathcal{B}(X^*)^n$.

Theorem 18. *Let $T = (T_1, \dots, T_n)$ be an n-tuple of mutually commuting operators in a Banach space X. Then:*

(i) *T is lower semi-Browder $\iff$ T^* is upper semi-Browder;*
(ii) *T is upper semi-Browder $\iff$ T^* is lower semi-Browder;*
(iii) *T is Browder $\iff$ T^* is Browder.*

Proof. The corresponding equivalences for semi-Fredholm n-tuples were proved in Corollary 19.12. Furthermore,

$$\operatorname{Ker} T_1^k \cap \dots \cap \operatorname{Ker} T_n^k = {}^\perp\bigl(\operatorname{Ran} T_1^{*k} + \dots + \operatorname{Ran} T_n^{*k}\bigr).$$

and

$$\bigl(\operatorname{Ran} T_1^k + \dots + \operatorname{Ran} T_n^k\bigr)^\perp = \operatorname{Ker} T_1^{*k} \cap \dots \cap \operatorname{Ker} T_n^{*k}.$$

The statement of Theorem 18 is now an easy consequence of these identities. □

For a commuting n-tuple $T = (T_1, \ldots, T_n) \in \mathcal{B}(X)^n$ we define the *upper semi-Browder spectrum* of T by

$$\sigma_{B_+}(T) = \{\lambda \in \mathbb{C}^n : T - \lambda \text{ is not upper semi-Browder}\}$$

and the *Browder spectrum*

$$\sigma_B(T) = \{\lambda \in \mathbb{C}^n : T - \lambda \text{ is not Browder}\} = \sigma_{B_-}(T) \cup \sigma_{B_+}(T).$$

By the previous theorem, it is easy to see that σ_{B_+} and σ_B satisfy the same properties as σ_{B_-}.

From the general theory of spectral systems it is easy to deduce the following properties:

Theorem 19.

(i) *σ_{B_-}, σ_{B_+} and σ_B are upper semicontinuous spectral systems.*

(ii) *(spectral mapping property) If $(T_1, \ldots, T_n)$ is a commuting n-tuple of operators on a Banach space X and $p = (p_1, \ldots, p_m)$ is an m-tuple of polynomials, then:*

$$\begin{aligned} \sigma_{B_+}(p(T_1, \ldots, T_n)) &= p(\sigma_{B_+}(T_1, \ldots, T_n)); \\ \sigma_{B_-}(p(T_1, \ldots, T_n)) &= p(\sigma_{B_-}(T_1, \ldots, T_n)); \\ \sigma_B(p(T_1, \ldots, T_n)) &= p(\sigma_B(T_1, \ldots, T_n)). \end{aligned}$$

(iii) *(continuity on commuting operators) If $\{T_k\}_{k=1}^\infty \subset \mathcal{B}(X)$, $T \in \mathcal{B}(X)$, $\lim T_k = T$ and $T_k T = T T_k$, $k = 1, 2, \ldots$, then:*

$$\begin{aligned} \lambda \in \sigma_{B_-}(T) &\iff \text{there exist } \lambda_k \in \sigma_{B_-}(T_k) \text{ such that } \lambda_k \to \lambda; \\ \lambda \in \sigma_{B_+}(T) &\iff \text{there exist } \lambda_k \in \sigma_{B_+}(T_k) \text{ such that } \lambda_k \to \lambda; \\ \lambda \in \sigma_B(T) &\iff \text{there exist } \lambda_k \in \sigma_B(T_k) \text{ such that } \lambda_k \to \lambda. \end{aligned}$$

(iv) *(property (P1) of Section 6) Let $T, S \in \mathcal{L}(X)$, $TS = ST$. Then TS is lower semi-Browder (upper semi-Browder, Browder) if and only if both T and S have the same property.*

(v) *$\partial\sigma_e(T) \subset \sigma_{B_+}(T) \cap \sigma_{B_-}(T)$. In particular, $\max\{|z| : z \in \sigma_{B_+}(T)\} = \max\{|z| : z \in \sigma_{B_-}(T)\} = \max\{|z| : z \in \sigma_B(T)\} = r_e(T)$ for all $T \in \mathcal{B}(X)$.*

(vi) *Let $T, S \in \mathcal{L}(X)$, $TS = ST$. Then:*

$$\begin{aligned} \widehat{\Delta}\big(\sigma_{B_-}(T), \sigma_{B_-}(S)\big) &\le r_e(T - S); \\ \widehat{\Delta}\big(\sigma_{B_+}(T), \sigma_{B_+}(S)\big) &\le r_e(T - S); \\ \widehat{\Delta}\big(\sigma_B(T), \sigma_B(S)\big) &\le r_e(T - S), \end{aligned}$$

where $\widehat{\Delta}$ denotes the Hausdorff distance and r_e the essential spectral radius.

(vii) *If $T \in \mathcal{B}(X)$ is upper semi-Browder (lower semi-Browder, Browder), $U \in \mathcal{B}(X)$, $UT = TU$ and $r_e(U) = 0$, then $T + U$ is upper semi-Browder (lower semi-Browder, Browder, respectively).*

In particular, this is true if U is either quasinilpotent or compact.

For single operators we have another characterization of the Browder and semi-Browder spectrum.

Denote by $\operatorname{acc} L$ the set of all accumulation points of a set $L \subset \mathbb{C}$.

Corollary 20. *Let $T \in \mathcal{B}(X)$. Then:*

$$\sigma_B(T) = \sigma_e(T) \cup \operatorname{acc} \sigma(T);$$
$$\sigma_{B_+}(T) = \sigma_{\pi e}(T) \cup \operatorname{acc} \sigma_\pi(T);$$
$$\sigma_{B_-}(T) = \sigma_{\delta e}(T) \cup \operatorname{acc} \sigma_\delta(T).$$

Proof. We prove the statement for the Browder spectrum; the remaining statements can be proved similarly.

Suppose that $\lambda \notin \sigma_B(T)$, so $T - \lambda$ is Browder. Let $X = X_1 \oplus X_2$ be the Kato decomposition of $T - \lambda$, so $(T - \lambda)|X_1$ is a finite-dimensional nilpotent, and $(T - \lambda)|X_2$ is invertible. Clearly, $T - \mu$ is invertible for all $\mu \neq \lambda$ close enough to λ. Thus λ is not an accumulation point of $\sigma(T)$. Also, $\lambda \notin \sigma_e(T)$ since $T - \lambda$ is Fredholm.

Conversely, let $\lambda \notin \sigma_e(T) \cup \operatorname{acc} \sigma(T)$. Then $T - \lambda$ is Fredholm. Let $X = X_1 \oplus X_2$ be the Kato decomposition of $T - \lambda$, so $(T - \lambda)|X_1$ is a finite-dimensional nilpotent and $(T - \lambda)|X_2$ is Kato. By assumption, $T - \mu$ is invertible for all μ sufficiently close to λ, $\mu \neq \lambda$, and so $(T - \mu)|X_2$ is invertible. Since $(T - \lambda)|X_2$ is Kato, $(T - \lambda)|X_2$ is also invertible. Thus $T - \lambda$ is Browder. □

Theorem 21. *Let $T \in \mathcal{B}(X)$. Then:*

$$\sigma_{B_-}(T) = \bigcap \{\sigma_\delta(T + K) : K \in \mathcal{K}(X), TK = KT\};$$
$$\sigma_{B_+}(T) = \bigcap \{\sigma_\pi(T + K) : K \in \mathcal{K}(X), TK = KT\};$$
$$\sigma_B(T) = \bigcap \{\sigma(T + K) : K \in \mathcal{K}(X), TK = KT\}.$$

Proof. By Theorem 19 (vii), $\sigma_{B_-}(T) = \sigma_{B_-}(T + K) \subset \sigma_\delta(T + K)$ if K is a compact operator commuting with T. Thus $\sigma_{B_-}(T) \subset \bigcap \{\sigma_\delta(T+K) : K \in \mathcal{K}(X), TK = KT\}$.

Conversely, let $\lambda \notin \sigma_{B_-}(T)$; so $T - \lambda$ is lower semi-Browder. Let $X = X_1 \oplus X_2$ be the Kato decomposition of $T - \lambda$ satisfying $\dim X_1 < \infty, TX_i \subset X_i \quad (i = 1, 2)$, $(T - \lambda)|X_1$ is nilpotent and $(T - \lambda)|X_2$ onto. Set $K = I \oplus 0$. Clearly, K is a compact (even finite rank) operator commuting with T and $T - \lambda + K$ is onto. Thus $\lambda \notin \sigma_\delta(T + K)$.

The statements for the upper semi-Browder and Browder spectrum can be proved similarly. □

Clearly, the compact operators in the last theorem can be replaced by finite-rank operators.

Theorem 22. *Let $T, S \in \mathcal{B}(X)$. Then $\tilde{\sigma}(TS) \setminus \{0\} = \tilde{\sigma}(ST) \setminus \{0\}$, where $\tilde{\sigma}$ stands for any of σ_B, σ_{B_-}, σ_{B_+}.*

Proof. Follows from Propositions 12.28 and 12.29.

21 Essentially Kato operators

Recall that an operator $T \in \mathcal{B}(X)$ is Kato if $\operatorname{Ran} T$ is closed and $\operatorname{Ker} T \subset R^\infty(T)$. For equivalent conditions see Theorem 12.2.

In this section we study an essential version of this class of operators.

For subspaces M, N of X we write $M \overset{e}{\subset} N$ (M is *essentially contained* in N) if there exists a finite-dimensional subspace $F \subset X$ such that $M \subset N + F$.

Similarly, we write $M \overset{e}{=} N$ if both $M \overset{e}{\subset} N$ and $N \overset{e}{\subset} M$.

We recall one simple algebraic result.

Lemma 1. *Let M, N be subspaces of a vector space X. Then $\dim M/(M \cap N) = \dim(M+N)/N$.*

Proof. The identity operator induces an isomorphism from $M/(M \cap N)$ onto $(M+N)/N$. □

We summarize the basic properties of the relation $\overset{e}{\subset}$.

Proposition 2. *Let M, N, L be subspaces of a Banach space X. Then:*

(i) $M \overset{e}{\subset} N \Leftrightarrow \dim M/(M \cap N) < \infty \Leftrightarrow \dim(M+N)/N < \infty \Leftrightarrow$ *there is a finite-dimensional subspace $F \subset M$ such that $M \subset N + F$;*

(ii) $M \overset{e}{\subset} N,\ N \overset{e}{\subset} L \Rightarrow M \overset{e}{\subset} L$;

(iii) $M \overset{e}{\subset} N,\ L \overset{e}{\subset} N \Rightarrow M + L \overset{e}{\subset} N$;

(iv) $M \overset{e}{\subset} N,\ M \overset{e}{\subset} L \Rightarrow M \overset{e}{\subset} N \cap L$.

Proof. A simple verification. □

Theorem 3. *Let $T \in B(X)$ be an operator with closed range. Then the following conditions are equivalent:*

(i) $\operatorname{Ker} T \overset{e}{\subset} R^\infty(T)$;

(ii) $\operatorname{Ker} T \overset{e}{\subset} \overline{R^\infty(T)}$;

(iii) $N^\infty(T) \overset{e}{\subset} \operatorname{Ran} T$;

(iv) $N^\infty(T) \overset{e}{\subset} R^\infty(T)$;

(v) *(Kato decomposition) there exists a decomposition $X = X_1 \oplus X_2$ with the properties that $TX_1 \subset X_1$, $TX_2 \subset X_2$, $\dim X_1 < \infty$, $T|X_1$ is nilpotent and $T|X_2$ is Kato;*

(vi) $\operatorname{Ker} T \overset{e}{\subset} \bigvee_{z \neq 0} \operatorname{Ker}(T - z)$;

(vii) $\operatorname{Ran} T \overset{e}{\supset} \bigcap_{z \neq 0} \overline{\operatorname{Ran}(T - z)}$;

Proof. Clearly (v) implies any of the remaining conditions (see the corresponding properties of Kato operators, Theorems 12.2 and 12.20).

(i) $\Rightarrow$ (v): By Lemma 16.20, either T is Kato or there exists a decomposition $X = Y_1 \oplus Y_2$ such that $1 \leq \dim Y_1 < \infty$, $TY_i \subset Y_i$ $(i = 1, 2)$ and $T|Y_1$ is nilpotent. Write $T_i = T|Y_i \quad (i = 1, 2)$. We have $\dim\bigl(\operatorname{Ker} T/(\operatorname{Ker} T \cap R^\infty(T))\bigr) < \infty$, $R^\infty(T) = R^\infty(T_2)$ and $\operatorname{Ker} T = \operatorname{Ker} T_1 \oplus \operatorname{Ker} T_2$; so

$$\dim \operatorname{Ker} T/(\operatorname{Ker} T \cap R^\infty(T)) = \dim \operatorname{Ker} T_1 + \dim\ \operatorname{Ker} T_2/(\operatorname{Ker} T_2 \cap R^\infty(T_2)).$$

Thus $\dim \operatorname{Ker} T_2/\bigl(\operatorname{Ker} T_2 \cap R^\infty(T_2)\bigr) < \dim \operatorname{Ker} T/\bigl(\operatorname{Ker} T \cap R^\infty(T)\bigr)$.

Using the construction of Lemma 16.20 for T_2 repeatedly, after a finite number of steps we obtain a decomposition $X = X_1 \oplus X_2$ such that $\dim X_1 < \infty$, $TX_1 \subset X_1$, $TX_2 \subset X_2$, $T|X_1$ is nilpotent and $\operatorname{Ker}(T|X_2) \subset R^\infty(T|X_2)$, i.e., $T|X_2$ is Kato.

(ii) $\Rightarrow$ (i): By Lemma 16.20 (i), $R(T^k)$ is closed for all k, and so is $R^\infty(T)$.

(iii) $\Rightarrow$ (v): We prove by induction on k that $\operatorname{Ker} T \overset{e}{\subset} \operatorname{Ran} T^k$. This is clear for $k = 1$. Let $k \geq 1$ and suppose that the statement is true for all $l \leq k$. Thus there are finite-dimensional subspaces $F_l \subset \operatorname{Ker} T$ such that $\operatorname{Ker} T \subset \operatorname{Ran} T^l + F_l \quad (l = 1, \ldots, k)$. We prove the statement for $k + 1$. We have $\operatorname{Ker} T^{k+1} \overset{e}{\subset} \operatorname{Ran} T$, so there is a finite-dimensional subspace G such that $\operatorname{Ker} T^{k+1} \subset \operatorname{Ran} T + G$. So

$$\operatorname{Ker} T \subset (\operatorname{Ker} T \cap \operatorname{Ran} T^k) + F_k = T^k \operatorname{Ker} T^{k+1} + F_k \subset \operatorname{Ran} T^{k+1} + T^k G + F_k.$$

Hence $\operatorname{Ker} T \overset{e}{\subset} \operatorname{Ran} T^{k+1}$, which completes the induction step.

By Lemma 16.20, either T is Kato or there is a decomposition $X = Y_1 \oplus Y_2$ such that $1 \leq \dim Y_1 < \infty$, $TY_i \subset Y_i$ and $T_1 = T|Y_1$ is nilpotent. Write $T_i = T|Y_i \quad (i = 1, 2)$. Since $\dim N^\infty(T)/(N^\infty(T) \cap \operatorname{Ran} T) < \infty$ and $N^\infty(T) = Y_1 \oplus N^\infty(T_2)$, we have

$$\begin{aligned}\dim N^\infty(T)/(N^\infty(T) \cap \operatorname{Ran} T)\\ = \dim Y_1/\operatorname{Ran} T_1 + \dim N^\infty(T_2)/(N^\infty(T_2) \cap \operatorname{Ran} T_2),\end{aligned}$$

so $\dim N^\infty(T_2)/(N^\infty(T_2) \cap \operatorname{Ran} T_2) < \dim N^\infty(T)/(N^\infty(T) \cap \operatorname{Ran} T)$.

Thus, using the same construction for T_2, after a finite number of steps we obtain the required decomposition $X = X_1 \oplus X_2$.

The implication (iv) $\Rightarrow$ (i) is clear.

(vi) $\Rightarrow$ (ii): It is easy to see that $\operatorname{Ker}(T - z) \subset R^\infty(T)$ for $z \neq 0$. Thus

$$\operatorname{Ker} T \overset{e}{\subset} \bigvee_{z \neq 0} \operatorname{Ker}(T - z) \subset \overline{R^\infty(T)}.$$

(vii) $\Rightarrow$ (iii): Let $x \in \operatorname{Ker} T^n$ and $z \neq 0$. Then

$$(T - z)(T^{n-1} + zT^{n-2} + \cdots + z^{n-1})x = (T^n - z^n)x = -z^n x,$$

and so $x \in \operatorname{Ran}(T - z)$. Thus $N^\infty(T) \subset \bigcap_{z \neq 0} \operatorname{Ran}(T - z) \overset{e}{\subset} \operatorname{Ran} T$

This completes the proof. □

Definition 4. We say that an operator $T \in B(X)$ is *essentially Kato* if $\operatorname{Ran} T$ is closed and T satisfies any of the equivalent conditions of Theorem 3.

Clearly, any semi-Fredholm operator is essentially Kato.

Theorem 5. *Let $T \in B(X)$. Then:*

(i) *if T is essentially Kato, then T^n is essentially Kato for every n;*

(ii) *T is essentially Kato if and only if $T^* \in B(X^*)$ is essentially Kato.*

Proof. (i) Let $X = X_1 \oplus X_2$ be the Kato decomposition of T (see condition (v) of Theorem 3). The same decomposition is clearly also the Kato decomposition of T^n.

(ii) Suppose that T is essentially Kato, so $N^\infty(T) \subset \operatorname{Ran} T + F$ for some finite-dimensional subspace $F \subset X$. Then $\operatorname{Ran} T^n$ and $\operatorname{Ran} T^{*n}$ are closed for all n and

$$\begin{aligned} R^\infty(T^*) &= \bigcap_{n=1}^{\infty} \operatorname{Ran} T^{*n} = \bigcap_{n=1}^{\infty} (\operatorname{Ker} T^n)^\perp = \Big(\bigcup_{n=1}^{\infty} \operatorname{Ker} T^n\Big)^\perp \\ &= N^\infty(T)^\perp \supset \big(\operatorname{Ran} T + F\big)^\perp = (\operatorname{Ran} T)^\perp \cap F^\perp = \operatorname{Ker} T^* \cap F^\perp. \end{aligned}$$

Since $\operatorname{codim} F^\perp < \infty$, we have $\operatorname{Ker} T^* \overset{e}{\subset} R^\infty(T^*)$, and T^* is essentially Kato.

Conversely, if T^* is essentially Kato, then $\operatorname{Ran} T^n$ and $\operatorname{Ran} T^{*n}$ are closed for all n and $T^{**} \in \mathcal{B}(X^{**})$ is essentially Kato; so $\operatorname{Ker} T^{**} \overset{e}{\subset} R^\infty(T^{**})$. Further, $\operatorname{Ker} T = \operatorname{Ker} T^{**} \cap X$ and $\operatorname{Ran} T^n = \operatorname{Ran} T^{**n} \cap X$ for all n, so $R^\infty(T) = R^\infty(T^{**}) \cap X$ and $\operatorname{Ker} T \overset{e}{\subset} R^\infty(T)$. □

Theorem 6. *Let $A, B \in B(X)$, $AB = BA$. If AB is essentially Kato, then A and B are essentially Kato.*

Proof. We have $\operatorname{Ker} A \subset \operatorname{Ker}(AB) \overset{e}{\subset} R^\infty(AB) \subset R^\infty(A)$, so it is sufficient to prove that $\operatorname{Ran} A$ is closed.

There exists a finite-dimensional subspace $F \subset X$ such that $\operatorname{Ker}(AB) \subset \operatorname{Ran}(AB) + F$. We prove that $\operatorname{Ran} A + F$ is closed. Let $v_j \in X$, $f_j \in F$ and $Av_j + f_j \to u$. Then $BAv_j + Bf_j \to Bu$ and $Bu \in \operatorname{Ran}(AB) + BF$ since $\operatorname{Ran}(AB) + BF$ is closed. Thus $Bu = ABv + Bf$ for some $v \in X$ and $f \in F$; so

$$Av + f - u \in \operatorname{Ker} B \subset \operatorname{Ker}(AB) \subset \operatorname{Ran}(AB) + F \subset \operatorname{Ran} A + F.$$

Hence $u \in \operatorname{Ran} A + F$ and $\operatorname{Ran} A + F$ is closed.

The closeness of $\operatorname{Ran} A$ follows from Lemma 16.2. □

The following lemma is an analogue of Theorem 12.21 for essentially Kato operators:

Theorem 7. *Let $T \in \mathcal{B}(X)$. The following conditions are equivalent:*

(i) *T is essentially Kato;*

(ii) *there exists a closed subspace $M \subset X$ such that $TM = M$ and the operator $\widehat{T} : X/M \to X/M$ induced by T is upper semi-Fredholm;*

(iii) *there exists a closed subspace $M \subset X$ such that $TM \subset M$, $T|M$ is lower semi-Fredholm and the operator $\widehat{T} : X/M \to X/M$ induced by T is upper semi-Fredholm.*

If T is essentially Kato, then it is possible to take $M = R^\infty(T)$ in (ii) *or* (iii).

Proof. (i) $\Rightarrow$ (ii): Let T be essentially Kato. Set $M = R^\infty(T)$. If $X = X_1 \oplus X_2$ is the Kato decomposition of T ($\dim X_1 < \infty, TX_1 \subset X_1, TX_2 \subset X_2, T_1 = T|X$ is nilpotent and $T_2 = T|X_2$ Kato), then $M = R^\infty(T_2) \subset X_2$ and $TM = T_2M = M$. If $x = x_1 \oplus x_2$ satisfies $Tx \in M$, then $T_2x_2 \in M$, and so $x_2 \in M$. Thus $x \in X_1 + M$ and $\operatorname{Ker} \widehat{T} \subset X_1 + M$. Hence $\dim \operatorname{Ker} \widehat{T} \le \dim X_1 < \infty$.

Let $Q : X \to X/M$ be the canonical projection. Since $M \subset \operatorname{Ran} T$ and $\operatorname{Ran} \widehat{T} = \{Tx + M : x \in X\} = Q \operatorname{Ran} T$, the range of $\widehat{T}$ is closed. Thus $\widehat{T}$ is upper semi-Fredholm.

(ii) $\Rightarrow$ (iii): Clear.

(iii) $\Rightarrow$ (i): We first prove that $\operatorname{Ran} T$ is closed. Let $Q : X \to X/M$ be the canonical projection and let F be a finite-dimensional subspace of M satisfying $M = TM + F$. Clearly, $\operatorname{Ran} T \subset Q^{-1} \operatorname{Ran} \widehat{T}$. If $y \in X$ and $Qy \in \operatorname{Ran} \widehat{T}$, then $y \in \operatorname{Ran} T + M \subset \operatorname{Ran} T + F$. Thus $\operatorname{Ran} T$ is a subspace of finite codimension of the closed space $Q^{-1} \operatorname{Ran} \widehat{T}$. By Lemma 16.2, $\operatorname{Ran} T$ is also closed.

Since $Q \operatorname{Ker} T \subset \operatorname{Ker} \widehat{T}$ and $\dim \operatorname{Ker} \widehat{T} < \infty$, we have $\operatorname{Ker} T \overset{e}{\subset} M$. Thus there is a finite-dimensional subspace $F_1 \subset \operatorname{Ker} T$ such that $\operatorname{Ker} T \subset F_1 + (\operatorname{Ker} T \cap M)$. Thus

$$\operatorname{Ker} T \subset F_1 + \operatorname{Ker} T|M \overset{e}{\subset} R^\infty(T|M) \subset R^\infty(T).$$

Hence T is essentially Kato. □

Theorem 8. *Let $T \in \mathcal{B}(X)$ be essentially Kato. Then $\lim_{n\to\infty} \gamma(T^n)^{1/n}$ exists and*

$$\lim_{n\to\infty} \gamma(T^n)^{1/n} = \max\{r : T - \lambda \text{ is Kato for } 0 < |\lambda| < r\} = \operatorname{dist}\{0, \sigma_K(T) \setminus \{0\}\}.$$

Moreover, if $S \in \mathcal{B}(X)$, $ST = TS$ and $\|S\| < \lim_{n\to\infty} \gamma(T^n)^{1/n}$, then $T + S$ is essentially Kato and $(T + S)R^\infty(T) = R^\infty(T)$.

Proof. Let $X = X_1 \oplus X_2$ be the Kato decomposition of T (i.e., $\dim X_1 < \infty$, $TX_1 \subset X_1$, $TX_2 \subset X_2$, $T_1 = T|X_1$ is nilpotent and $T_2 = T|X_2$ is Kato). As in the proof of Theorem 18.8 we can show that

$$\lim_{n\to\infty} \gamma(T^n)^{1/n} = \lim_{n\to\infty} \gamma(T_2^n)^{1/n}.$$

By Theorem 12.26, this limit is equal to $\operatorname{dist}\{0, \sigma_K(T_2)\}$.

If $\lambda \neq 0$, then $T-\lambda$ is Kato if and only if $T_2-\lambda$ is Kato. Then

$$\max\{r : T-\lambda \text{ is Kato for } 0<|\lambda|<r\} = \operatorname{dist}\{0, \sigma_K(T_2)\} = \lim_{n\to\infty} \gamma(T^n)^{1/n}.$$

Let $S \in \mathcal{B}(X)$, $ST = TS$ and $\|S\| < \lim_{n\to\infty} \gamma(T^n)^{1/n}$. Write $M = R^\infty(T)$. Clearly $TM = M$ and $SM \subset M$. Denote by $T_M, S_M \in \mathcal{B}(M)$ the restrictions of T and S to M, respectively. Similarly, let $\widehat{T}, \widehat{S} \in \mathcal{B}(X/M)$ be the operators induced by T and S, respectively.

If $M = X$, then T is onto and $(T+S)M = M = X$ by Theorems 9.25 and 7.14 (ii).

If $M = \{0\}$, then T is upper semi-Fredholm and so is $T+S$ by Theorems 18.8 and 17.14.

Suppose that $\{0\} \neq M \neq X$. By Theorem 7, T_M is onto and $\widehat{T}$ is upper semi-Fredholm. Using lemma 12.25 for T_2 and the proof of Theorem 18.8 we have

$$\lim_{k\to\infty} \gamma(T^k)^{1/k} = \min\Big\{\lim_{k\to\infty} \gamma((T_M)^k)^{1/k}, \lim_{k\to\infty} \gamma(\widehat{T}^k)^{1/k}\Big\},$$

Clearly $\|S_M\| \le \|S\| < \lim_{k\to\infty} \gamma((T_M)^k)^{1/k}$. By Theorem 9.25, $T_M + S_M$ is onto, i.e., $(T+S)R^\infty(T) = R^\infty(T)$.

Similarly, $\|\widehat{S}\| \le \|S\| < \lim_{k\to\infty} \gamma(\widehat{T}^k)^{1/k}$. By Theorems 18.8 and 7.14, $\widehat{T}+\widehat{S}$ is upper semi-Fredholm. By Theorem 7, $T+S$ is essentially Kato. □

Theorem 9. *Let $T, A \in \mathcal{B}(X), TA = AT$, let T be essentially Kato and let A be either compact or a quasinilpotent. Then $T+A$ is essentially Kato.*

Proof. Let T be essentially Kato and let A be an operator commuting with T. Let $M = R^\infty(T), T_1 = T|M$ and let $\widehat{T} : X/M \to X/M$ be the operator induced by T. Since $AM \subset M$, we can define the operators $A_1 = A|M$ and $\widehat{A} : X/M \to X/M$ induced by A. Suppose that A is either compact or quasinilpotent. Then both A_1 and $\widehat{A}$ have the same property, and consequently, $T_1 + A_1$ is lower semi-Fredholm and $\widehat{T} + \widehat{A}$ is upper semi-Fredholm. Thus $T+A$ is essentially Kato by Theorem 7. □

Theorem 10. *The set of all essentially Kato operators in X is a regularity satisfying* (P3) (*upper semicontinuity on commuting elements*).

Proof. By Theorems 5, 6 and 6.12, it is sufficient to show that if $A, B, C, D \in \mathcal{B}(X)$ are mutually commuting operators satisfying $AC + BD = I$ and A, B are essentially Kato, then AB is essentially Kato. By Lemma 12.8, we have $\operatorname{Ran}(AB) = \operatorname{Ran} A \cap \operatorname{Ran} B$, and so $\operatorname{Ran}(AB)$ is closed. Further, $\operatorname{Ker} A \overset{e}{\subset} R^\infty(A)$ and $\operatorname{Ker} A \subset R^\infty(B)$. So $\operatorname{Ker} A \overset{e}{\subset} R^\infty(A) \cap R^\infty(B) = R^\infty(AB)$. By symmetry, we also have $\operatorname{Ker} B \overset{e}{\subset} R^\infty(AB)$, and thus $\operatorname{Ker}(AB) = \operatorname{Ker} A + \operatorname{Ker} B \overset{e}{\subset} R^\infty(AB)$. Hence AB is essentially Kato. □

Denote by

$$\sigma_{Ke}(T) = \{\lambda \in \mathbf{C} : T - \lambda \text{ is not essentially Kato}\}$$

the corresponding spectrum.

Theorem 11. *Let* $\dim X = \infty$ *and* $T \in \mathcal{B}(X)$. *Then:*

(i) $\sigma_{Ke}(T) \subset \sigma_K(T)$ *and* $\sigma_K(T) \setminus \sigma_{Ke}(T)$ *consists of at most countably many isolated points;*

(ii) $\sigma_{Ke}(T)$ *is a non-empty compact set;*

(iii) $\partial\sigma_e(T) \subset \sigma_{Ke}(T) \subset \sigma_{\pi e}(T) \cap \sigma_{\delta e}(T) \subset \sigma_e(T)$;

(iv) $\sigma_{Ke}(f(T)) = f(\sigma_{Ke}(T))$ *for every function* f *analytic on a neighbourhood of* $\sigma(T)$.

Proof. (i) Let $\lambda \in \sigma_K(T) \setminus \sigma_{Ke}(T)$. Then $T-\lambda$ is essentially Kato, so there exists a decomposition $X = X_1 \oplus X_2$ with $TX_1 \subset X_1$, $TX_2 \subset X_2$, $\dim X_1 < \infty$, $(T-\lambda)|X_1$ nilpotent and $(T-\lambda)|X_2$ Kato. Then $(T-z)|X_2$ is Kato for all z in a certain neighbourhood U of λ and $(T-z)|X_1$ is Kato (even invertible) for every $z \neq \lambda$. Thus $T - z$ is Kato for $z \in U - \{\lambda\}$ and λ is an isolated point of $\sigma_K(T)$.

Clearly, $\sigma_K(T) \setminus \sigma_{Ke}(T)$ is at most countable.

(ii) By Theorem 8, $\sigma_{Ke}(T)$ is closed.

The non-emptiness of $\sigma_{Ke}(T)$ follows from the inclusion $\partial\sigma_e(T) \subset \sigma_{Ke}(T)$, which will be proved next.

(iii) Suppose $\lambda \in \partial\sigma_e(T)$ and $\lambda \notin \sigma_{Ke}(T)$. Then $T - \lambda$ is essentially Kato, so $\operatorname{Ran}(T-\lambda)$ is closed and there exists a decomposition $X = X_1 \oplus X_2$ such that $\dim X_1 < \infty$, $TX_1 \subset X_1$, $TX_2 \subset X_2$, $(T-\lambda)|X_1$ is nilpotent and $(T-\lambda)|X_2$ is Kato. Choose a sequence $\lambda_n \to \lambda$ such that $\lambda_n \notin \sigma_e(T)$, i.e., $T - \lambda_n$ is Fredholm. We have

$$\dim \operatorname{Ker}(T-\lambda_n)|X_2 \le \dim \operatorname{Ker}(T-\lambda_n) < \infty$$

and, since $T|X_2$ is Kato, we conclude that

$$\dim \operatorname{Ker}(T-\lambda)|X_2 < \infty$$

and $\dim \operatorname{Ker}(T-\lambda) < \infty$.

Similarly, we can prove that $\operatorname{codim} \operatorname{Ran}(T-\lambda) < \infty$, so $T-\lambda$ is a Fredholm operator and $\lambda \notin \sigma_e(T)$, a contradiction.

Thus $\partial\sigma_e(T) \subset \sigma_{Ke}(T)$.

Since semi-Fredholm operators are Kato, we have $\sigma_{Ke}(T) \subset \sigma_{\pi e}(T) \cap \sigma_{\delta e}(T)$.

(iv) If $X = X_1 \oplus X_2$ and $T_i \in \mathcal{B}(X_i)$ $(i = 1, 2)$, then $\sigma_{Ke}(T_1 \oplus T_2) = \sigma_{Ke}(T_1) \cup \sigma_{Ke}(T_2)$. The non-emptiness of σ_{Ke} and Theorem 6.8 imply (iv). □

Theorem 12. *Let* $T \in \mathcal{B}(X)$ *be essentially Kato and let* $F \in \mathcal{B}(X)$ *be a finite-rank operator. Then* $T + F$ *is also essentially Kato.*

Proof. Clearly, it is sufficient to consider only the case of $\dim \operatorname{Ran} F = 1$. Thus F is of the form $Tx = f(x)v$ for some $v \in X$ and $f \in X^*$.

Since $\operatorname{Ran}(T+F) \overset{e}{=} \operatorname{Ran} T$, $\operatorname{Ran}(T+F)$ is closed and it is sufficient to show only the algebraic condition in the definition of essentially Kato operators for $T+F$.

Since T is essentially Kato, $\operatorname{Ran} T^k$ is closed for all k. The existence of the Kato decomposition implies that there exists $d \in \mathbb{N}$ such that $\operatorname{Ker} T \cap \operatorname{Ran} T^d \subset R^\infty(T)$. Let $M = \operatorname{Ran} T^d$ and $T_1 = T|M$. Then $\operatorname{Ker} T_1 = \operatorname{Ker} T \cap \operatorname{Ran} T^d \subset R^\infty(T) = R^\infty(T_1)$, and so T_1 is Kato.

It is sufficient to show that

$$\operatorname{Ker} T_1 \overset{e}{\subset} R^\infty(T+F). \tag{1}$$

Indeed, since

$$\operatorname{Ker} T_1 = \operatorname{Ker} T \cap \operatorname{Ran} T^d \overset{e}{=} \operatorname{Ker}(T+F) \cap \operatorname{Ran}(T+F)^d,$$

(1) implies that

$$\operatorname{Ker}(T+F) \cap \operatorname{Ran}(T+F)^d \overset{e}{\subset} R^\infty(T+F).$$

Consequently,

$$\operatorname{Ker}(T+F) \overset{e}{=} \operatorname{Ker} T \overset{e}{\subset} \operatorname{Ker} T \cap \operatorname{Ran} T^d \overset{e}{=} \operatorname{Ker}(T+F) \cap \operatorname{Ran}(T+F)^d \overset{e}{\subset} R^\infty(T+F)$$

and $T+F$ is essentially Kato.

To prove (1), we distinguish two cases:

(a) Let $N^\infty(T_1) \subset \operatorname{Ker} f$. Let $x_0 \in \operatorname{Ker} T_1$. Since T_1 is Kato, there exist vectors $x_1, x_2, \dots \in R^\infty(T_1)$ such that $Tx_i = x_{i-1}$ for all i. By assumption, $f(x_i) = 0$, and so $Fx_i = 0$ for all i. For $n \in \mathbb{N}$ we have

$$(T+F)^n x_n = (T+F)^{n-1} x_{n-1} = \cdots = (T+F)x_1 = x_0,$$

and so $x_0 \in \operatorname{Ran}(T+F)^n$. Since n was arbitrary, $\operatorname{Ker} T_1 \subset R^\infty(T+F)$.

(b) There exists $k \geq 1$ such that $\operatorname{Ker} T_1^k \not\subset \operatorname{Ker} f$. Choose the minimal k with this property, so $\operatorname{Ker} T_1^{k-1} \subset \operatorname{Ker} f$ and there exists $u \in \operatorname{Ker} T_1^k$ with $f(u) = 1$.

Denote by Y the set of all vectors $x_0 \in \operatorname{Ker} T_1 \cap \operatorname{Ker} f$ for which there exist $x_1, \dots, x_{k-1} \in M \cap \operatorname{Ker} f$ satisfying $Tx_i = x_{i-1} \quad (i = 1, \dots, k-1)$.

Clearly, $Y \overset{e}{=} \operatorname{Ker} T_1$, and so it is sufficient to show $Y \subset R^\infty(T+F)$.

Let $x_0 \in Y$. We prove by induction on n the following statement:

there exist $x_n \in M$ such that

$$T^n x_n = x_0 \qquad \text{and} \qquad T^i x_n \in \operatorname{Ker} f \quad (i = 0, \dots, n-1). \tag{2}$$

If (2) is proved, then of course

$$(T+F)^n x_n = (T+F)^{n-1} T x_n = \cdots = (T+F) T^{n-1} x_n = x_0,$$

and so $x_0 \in \operatorname{Ran}(T+F)^n$ for all n. Thus $Y \subset R^\infty(T+F)$ and the theorem is proved.

Statement (2) is clear for $n \leq k-1$. Suppose (2) is true for some $n \geq k-1$, so there is an $x_n \in M$ such that $T^n x_n = x_0$ and $T^i x_n \in \operatorname{Ker} f \quad (i = 0, \dots, n-1)$. Since T_1 is Kato, we can find $x'_{n+1} \in M$ such that $T x'_{n+1} = x_n$. Set $x_{n+1} = x'_{n+1} - f(x'_{n+1}) u$. Clearly, $T^{n+1} x_{n+1} = T^n x_n - f(x'_{n+1}) T^{n+1} u = x_0$ and $f(x_{n+1}) = 0$. For $i \geq 1$ we have $f(T^i x_{n+1}) = f(T^{i-1} x_n) - f(x'_{n+1}) f(T^i u) = 0$ since $T^i u \in \operatorname{Ker} T_1^{k-1} \subset \operatorname{Ker} f$.

This finishes the proof of (2) and also of the theorem. □

Theorem 13. *Let $T, S \in \mathcal{B}(X)$. Then $\sigma_{Ke}(TS) \setminus \{0\} = \sigma_{Ke}(ST) \setminus \{0\}$.*

Proof. Let $\lambda \in \mathbb{C}$, $\lambda \neq 0$. By Proposition 12.29, $\operatorname{Ran}(TS - \lambda)$ is closed if and only if $\operatorname{Ran}(ST - \lambda)$ is closed.

Suppose that $TS - \lambda$ is essentially Kato, i.e., $N^\infty(TS - \lambda) \subset \operatorname{Ran}(TS - \lambda)$. By Proposition 12.28,

$$\begin{aligned} N^\infty(ST - \lambda) &= \bigcup_n \operatorname{Ker}(ST - \lambda)^n = S\Big(\bigcup_n \operatorname{Ker}(TS - \lambda)^n\Big) \overset{e}{\subset} S \operatorname{Ran}(TS - \lambda) \\ &= ST \operatorname{Ran}(ST - \lambda) \subset \operatorname{Ran}(TS - \lambda). \end{aligned}$$

□

Let T be essentially Kato. We can write $N^\infty(T) = F + \big(R^\infty(T) \cap N^\infty(T)\big)$, where F is a finite-dimensional subspace and $F \cap R^\infty(T) = \{0\}$. Then $\overline{N^\infty(T)} = F + \overline{R^\infty(T) \cap N^\infty(T)}$ and

$$R^\infty(T) \cap \overline{N^\infty(T)} = \overline{R^\infty(T) \cap N^\infty(T)}. \tag{3}$$

Similarly, by Appendix 1.22, one can show that

$$R^\infty(T^*) \cap \overline{N^\infty(T^*)}^{w^*} = \overline{R^\infty(T^*) \cap N^\infty(T^*)}^{w^*}. \tag{4}$$

Theorem 14. Let $T \in B(X)$ be essentially Kato, let $S \in B(X)$, $ST = TS$ and $\|S\| < \lim \gamma(T^k)^{1/k}$. Then $T + S$ is essentially Kato,

$$R^\infty(T+S) \cap \overline{N^\infty(T+S)} = R^\infty(T) \cap \overline{N^\infty(T)}$$

and

$$R^\infty(T+S) + N^\infty(T+S) = R^\infty(T) + N^\infty(T).$$

Proof. By Theorem 8, $T + S$ is essentially Kato and $(T+S) R^\infty(T) = R^\infty(T)$. Thus $R^\infty(T+S) \supset R^\infty(T)$.

(a) $\overline{N^\infty(T+S)} \subset \overline{N^\infty(T)}$.

Proof. We have

$$R^\infty(T) = \bigcap_{k=0}^{\infty} \operatorname{Ran} T^k = \bigcap_{k=0}^{\infty} {}^\perp \operatorname{Ker} T^{*k} = {}^\perp \bigcup_{k=0}^{\infty} \operatorname{Ker} T^{*k} = {}^\perp N^\infty(T^*)$$

and

$$\overline{N^\infty(T)} = {}^\perp\left(N^\infty(T)^\perp\right) = {}^\perp\Big(\bigcap_{k=0}^{\infty} (\operatorname{Ker} T^k)^\perp\Big) = {}^\perp\Big(\bigcap_{k=0}^{\infty} \operatorname{Ran} T^{*k}\Big) = {}^\perp R^\infty(T^*).$$

The analogous equalities are true also for the operator $T+S$. By duality argument, we have

$$\overline{N^\infty(T)} = {}^\perp R^\infty(T^*) \supset {}^\perp R^\infty(T^*+S^*) = \overline{N^\infty(T+S)}.$$

(b) $R^\infty(T+S) \cap \overline{N^\infty(T+S)} \subset R^\infty(T)$.

Proof. Using (3) for $T+S$ it is sufficient to show that $R^\infty(T+S) \cap \operatorname{Ker}(T+S)^k \subset R^\infty(T)$ for $k = 1, 2, \dots$. We will do this by induction on k. The statement is clear for $k = 0$. Let $k \geq 1$ and assume that the inclusion holds for $k-1$.

Let $x_0 \in R^\infty(T+S) \cap \operatorname{Ker}(T+S)^k$. Since $T+S$ maps $R^\infty(T+S)$ onto itself, we can find an infinite sequence $x_0, x_1, \dots$ in $R^\infty(T+S)$ such that $(T+S)x_j = x_{j-1}$ $(j = 1, \dots)$. This sequence is contained in $\overline{N^\infty(T)}$ by (a). We have $\overline{N^\infty(T)} \overset{e}{\subset} R^\infty(T)$, i.e., $m := \dim \overline{N^\infty(T)}/(R^\infty(T) \cap \overline{N^\infty(T)})$ is finite. Thus $x_0, \dots, x_m$ are linearly dependent, i.e., there exists a non-trivial linear combination $x := \sum_{i=0}^m \alpha_i x_i \in R^\infty(T)$. Let l be such that $\alpha_l \neq 0$ and $\alpha_j = 0$ for $j = l+1, \dots, m$. We obtain

$$(T+S)^l x = \alpha_l x_0 + \sum_{j=0}^{l-1} \alpha_j (T+S)^l x_j \in \alpha_l x_0 + \big(\operatorname{Ker}(T+S)^{k-1} \cap R^\infty(T+S)\big).$$

Since $R^\infty(T)$ is invariant for T and S, we have $(T+S)^l x \in R^\infty(T)$. Therefore $x_0 \in R^\infty(T)$.

Let $M = R^\infty(T)$. Let T_M and $S_M \in \mathcal{B}(M)$ be the restrictions of T and S to M, respectively.

(c) Let c be a positive number such that $S' = cS$ satisfies $\|S'\| < \frac{1}{2}\gamma(T_M)$. Then $R^\infty(T) \cap \overline{N^\infty(T)} \subset \overline{N^\infty(T+S')}$.

Proof. By (3), it is sufficient to show that $R^\infty(T) \cap \operatorname{Ker} T^n \subset \overline{N^\infty(T+S')}$ for all n.

Let $n \geq 1$ and $x_0 \in \operatorname{Ker} T^n \cap M$. Since $TM = M$, $SM \subset M$ and $\|S'\| < \gamma(T_M)$, we have $(T+S')M = M$ and

$$\gamma((T+S')|M) \geq \gamma(T_M) - \|S'\| > \frac{1}{2}\gamma(T_M).$$

Therefore we can find inductively vectors $x_1, x_2, \dots \in M$ such that $(T+S')x_k = x_{k-1}$ and $\|x_k\| < 2\gamma(T_M)^{-1}\|x_{k-1}\|$ for all $k \geq 1$.

For $k \geq n$ set $y_k = x_0 - \sum_{j=0}^{n-1} \binom{k}{j} T^j S'^{k-j} x_k$. Then $y_k \in M$ and we have

$$(T+S')^k y_k = (T+S')^k x_0 - \sum_{j=0}^{n-1} \binom{k}{j} T^j S'^{k-j} x_0 = 0.$$

Thus $y_k \in N^\infty(T+S')$ for all k. Moreover,

$$\|y_k - x_0\| = \Big\| \sum_{j=0}^{n-1} \binom{k}{j} T^j S'^{k-j} x_k \Big\| \leq \sum_{j=0}^{n-1} \binom{k}{j} \|T^j\| \cdot \|S'\|^{k-j} \cdot \|x_k\|$$
$$\leq \Big(\frac{2\|S'\|}{\gamma(T_M)} \Big)^k \cdot \sum_{j=0}^{n-1} \binom{k}{j} \frac{\|T^j\| \cdot \|x_0\|}{\|S'\|^j} \to 0$$

as $k \to \infty$. Thus $x_0 \in \overline{N^\infty(T+S')}$, and so $\overline{N^\infty(T)} \cap R^\infty(T) \subset \overline{N^\infty(T+S')}$.

Proof of Theorem 14. By statements (a)–(c), the spaces $R^\infty(T+zS) \cap \overline{N^\infty(T+zS)}$ are constant for all complex numbers z with $|z|$ small enough $\big(|z| < \frac{\gamma(T_M)}{2\|S\|}\big)$. By a standard argument, these spaces are constant on each connected set for which $T+zS$ is essentially Kato. In particular,

$$R^\infty(T+S) \cap \overline{N^\infty(T+S)} = R^\infty(T) \cap \overline{N^\infty(T)}.$$

The second statement can be obtained by duality argument. As in (a), we have $N^\infty(T)^\perp = R^\infty(T^*)$ and $R^\infty(T)^\perp = \big({}^\perp N^\infty(T^*)\big)^\perp = \overline{N^\infty(T^*)}^{w^*}$.

By (4), we have

$$N^\infty(T) + R^\infty(T) = {}^\perp\Big(\big(N^\infty(T) + R^\infty(T)\big)^\perp\Big) = {}^\perp\Big(N^\infty(T)^\perp \cap R^\infty(T)^\perp\Big)$$
$$= {}^\perp\Big(R^\infty(T^*) \cap \overline{N^\infty(T^*)}^{w^*}\Big) = {}^\perp\Big(R^\infty(T^*) \cap \overline{N^\infty(T^*)}\Big)^{-w^*}.$$

Similarly,

$$N^\infty(T+S) + R^\infty(T+S) = {}^\perp\Big(R^\infty(T^*+S^*) \cap \overline{N^\infty(T^*+S^*)}\Big)^{-w^*},$$

and so

$$N^\infty(T+S) + R^\infty(T+S) = N^\infty(T) + R^\infty(T).$$ □

Corollary 15. *Let $T \in \mathcal{B}(X)$ be Kato, $S \in \mathcal{B}(X)$, $ST = TS$ and*

$$\|S\| < \lim_{n\to\infty} \gamma(T^n)^{1/n}.$$

Then $T+S$ is Kato, $R^\infty(T+S) = R^\infty(T)$ and $\overline{N^\infty(T+S)} = \overline{N^\infty(T)}$.

22 Classes of operators defined by means of kernels and ranges

In this section we give a systematic survey of various classes of operators that are defined by means of kernels and ranges of powers of an operator. Some of the classes were already introduced; we mention them again for the sake of completeness of the survey.

For an operator $T \in \mathcal{B}(X)$ we have already defined $\alpha(T) = \dim \operatorname{Ker} T$ and $\beta(T) = \operatorname{codim} \operatorname{Ran} T$. More generally, for every $n \geq 0$ we define numbers $\alpha_n(T) = \dim \operatorname{Ker} T^{n+1} / \operatorname{Ker} T^n$ and $\beta_n(T) = \dim \operatorname{Ran} T^n / \operatorname{Ran} T^{n+1}$. In this notation $\alpha_0(T) = \alpha(T)$ and $\beta_0(T) = \beta(T)$.

Write further $k_n(T) = \dim\bigl(\operatorname{Ran} T^n \cap \operatorname{Ker} T\bigr)/\bigl(\operatorname{Ran} T^{n+1} \cap \operatorname{Ker} T\bigr)$.

Lemma 1. *Let $T \in \mathcal{B}(X)$ and $n \geq 0$. Then $\alpha_n(T) = \dim\bigl(\operatorname{Ran} T^n \cap \operatorname{Ker} T\bigr)$ and $\beta_n(T) = \operatorname{codim}\bigl(\operatorname{Ran} T + \operatorname{Ker} T^n\bigr)$.*

Proof. Clearly, T^n induces an isomorphism from $\operatorname{Ker} T^{n+1} / \operatorname{Ker} T^n$ onto the space $\operatorname{Ran} T^n \cap \operatorname{Ker} T$. Thus $\alpha_n(T) = \dim\bigl(\operatorname{Ran} T^n \cap \operatorname{Ker} T\bigr)$.

In the same way, T^n induces an isomorphism from $X/(\operatorname{Ran} T + \operatorname{Ker} T^n)$ onto $\operatorname{Ran} T^n / \operatorname{Ran} T^{n+1}$, and so $\beta_n(T) = \operatorname{codim}\bigl(\operatorname{Ran} T + \operatorname{Ker} T^n\bigr)$. □

To characterize the numbers $k_n(T)$, we need the following elementary lemma.

Lemma 2. *Let U, V and W be subspaces of a Banach space X and let $U \subset W$. Then $(U + V) \cap W = U + (V \cap W)$.*

Proof. A simple verification. □

Lemma 3. *Let T be an operator on a Banach space X and $n \geq 0$. Then $k_n(T)$ defined by $k_n(T) = \dim\bigl(\operatorname{Ran} T^n \cap \operatorname{Ker} T\bigr)/\bigl(\operatorname{Ran} T^{n+1} \cap \operatorname{Ker} T\bigr)$ is equal to any of the following quantities:*

(i) *the dimension of the kernel of the operator*

$$\widehat{T} : \operatorname{Ran} T^n / \operatorname{Ran} T^{n+1} \longrightarrow \operatorname{Ran} T^{n+1} / \operatorname{Ran} T^{n+2}$$

induced by T; this operator is onto;

(ii) *the codimension of the range of the operator*

$$T' : \operatorname{Ker} T^{n+2} / \operatorname{Ker} T^{n+1} \longrightarrow \operatorname{Ker} T^{n+1} / \operatorname{Ker} T^n$$

induced by T; this operator is one-to-one;

(iii) $\dim\bigl(\operatorname{Ran} T + \operatorname{Ker} T^{n+1}\bigr)/\bigl(\operatorname{Ran} T + \operatorname{Ker} T^n\bigr)$.

Proof. Clearly, $\widehat{T}$ is onto and T' is one-to-one. By Lemmas 2 and 21.1, we have

$$\begin{aligned}\dim \operatorname{Ker} \widehat{T} &= \dim\big(T^{-1} \operatorname{Ran} T^{n+2} \cap \operatorname{Ran} T^n\big) / \operatorname{Ran} T^{n+1} \\ &= \dim\big((\operatorname{Ker} T + \operatorname{Ran} T^{n+1}) \cap \operatorname{Ran} T^n\big) / \operatorname{Ran} T^{n+1} \\ &= \dim\big((\operatorname{Ker} T \cap \operatorname{Ran} T^n) + \operatorname{Ran} T^{n+1}\big) / \operatorname{Ran} T^{n+1} \\ &= \dim\big(\operatorname{Ran} T^n \cap \operatorname{Ker} T\big)/\big(\operatorname{Ran} T^{n+1} \cap \operatorname{Ker} T\big) = k_n(T).\end{aligned}$$

Similarly,

$$\begin{aligned}\operatorname{codim} \operatorname{Ran} T' &= \dim \operatorname{Ker} T^{n+1}/\big(T \operatorname{Ker} T^{n+2} + \operatorname{Ker} T^n\big) \\ &= \dim \operatorname{Ker} T^{n+1}/\big((\operatorname{Ran} T \cap \operatorname{Ker} T^{n+1}) + \operatorname{Ker} T^n\big) \\ &= \dim \operatorname{Ker} T^{n+1}/\big((\operatorname{Ran} T + \operatorname{Ker} T^n) \cap \operatorname{Ker} T^{n+1}\big) \\ &= \dim\big(\operatorname{Ran} T + \operatorname{Ker} T^{n+1}\big)/\big(\operatorname{Ran} T + \operatorname{Ker} T^n\big).\end{aligned}$$

Finally, T^n induces an isomorphism from

$$\operatorname{Ker} T^{n+1}/\big((\operatorname{Ran} T \cap \operatorname{Ker} T^{n+1}) + \operatorname{Ker} T^n\big)$$

onto the space $\big(\operatorname{Ran} T^n \cap \operatorname{Ker} T\big)/\big(\operatorname{Ran} T^{n+1} \cap \operatorname{Ker} T\big)$. As we have proved above, the former space is isomorphic to $\big(\operatorname{Ran} T + \operatorname{Ker} T^{n+1}\big)/\big(\operatorname{Ran} T + \operatorname{Ker} T^n\big)$. □

Corollary 4. *Let $T \in \mathcal{B}(X)$. Then $\alpha_0(T) \geq \alpha_1(T) \geq \alpha_2(T) \geq \cdots$ and $\beta_0(T) \geq \beta_1(T) \geq \cdots$. If $\alpha_{n+1}(T) < \infty$, then $k_n(T) = \alpha_n(T) - \alpha_{n+1}(T)$. Similarly, if $\beta_{n+1}(T) < \infty$ for some $n \geq 0$, then $k_n(T) = \beta_n(T) - \beta_{n+1}(T)$.*

Proof. The inequality $\beta_{n+1}(T) \leq \beta_n(T)$ follows from the surjectivity of the operator $\widehat{T}$. If $\beta_{n+1}(T) < \infty$, then $k_n(T) = \beta_n(T) - \beta_{n+1}(T)$ by Lemma 3 (i). The analogous statements for α_n follow from Lemma 3 (ii). □

Note that if $\alpha_n(T) = \alpha_{n+1}(T) = \infty$, then $k_n(T)$ can be arbitrary. For an example, let $0 \leq m \leq \infty$ and $T = \bigoplus_{i=1}^{\infty} S \oplus \bigoplus_{j=1}^{m} S_{n+1}$ where S is the backward shift on a separable Hilbert space and S_{n+1} is a shift on an $(n+1)$-dimensional space. Then $\alpha_n(T) = \alpha_{n+1}(T) = \infty$ and $k_n(T) = m$.

In general, direct sums of various shift operators can serve as model examples for all classes considered in this section.

Sequences $\alpha_i(T), \beta_i(T)$ and $k_i(T)$ give rise to three families of reasonable classes of operators.

A. Descent

We start with the numbers $\beta_i(T)$. Recall that the descent of T is defined by $d(T) = \min\{n : \beta_n(T) = 0\}$. Similarly we define the *essential descent* $d_e(T) = \min\{n : \beta_n(T) < \infty\} = \min\{n : \operatorname{Ran} T^{n+1} \overset{e}{=} \operatorname{Ran} T^n\}$. If $d = d_e(T) < \infty$, then $\operatorname{Ran} T^d \overset{e}{=} \operatorname{Ran} T^n$ for all $n \geq d$ (of course $\operatorname{Ran} T^d \overset{e}{=} R^{\infty}(T)$ is not true in general; an example is the unilateral shift).

The following two lemmas enable us an easy verification of axioms of regularity:

Lemma 5. *Let $T \in \mathcal{B}(X)$, $m \geq 1$, $n \geq 0$. Then $\beta_n(T^m) = \sum_{i=0}^{m-1} \beta_{mn+i}(T)$. In particular,*

$$\beta_{mn}(T) \leq \beta_n(T^m) \leq m \cdot \beta_{mn}(T).$$

Proof. We have

$$\begin{aligned}\beta_n(T^m) &= \dim\bigl(\operatorname{Ran} T^{mn} / \operatorname{Ran} T^{mn+m}\bigr)\\ &= \sum_{i=0}^{m-1} \dim\bigl(\operatorname{Ran} T^{mn+i} / \operatorname{Ran} T^{mn+i+1}\bigr) = \sum_{i=0}^{m-1} \beta_{mn+i}(T)\,. \qquad \square\end{aligned}$$

Lemma 6. *Let A, B, C, D be mutually commuting operators on a Banach space X satisfying $AC + BD = I$ and let $n \geq 0$. Then*

$$\max\{\beta_n(A), \beta_n(B)\} \leq \beta_n(AB) \leq \beta_n(A) + \beta_n(B)\,.$$

Proof. We first prove $\beta_n(A) \leq \beta_n(AB)$. This is clear if $\beta_n(AB) = \infty$. Suppose that $\beta_n(AB) < \infty$. Set $m = \beta_n(AB) + 1$ and let $x_1, \dots, x_m$ be arbitrary elements of $\operatorname{Ran} A^n$. Then $B^n x_i \in \operatorname{Ran} A^n B^n \quad (i = 1, \dots, m)$, and so there exists a non-trivial linear combination

$$\sum_{i=1}^{m} c_i B^n x_i \in \operatorname{Ran}(A^{n+1} B^{n+1}).$$

By Lemma 12.8, we have

$$\sum_{i=1}^{m} c_i x_i \in \operatorname{Ran}(A^{n+1} B) + \operatorname{Ker} B^n \subset \operatorname{Ran} A^{n+1}.$$

Since the vectors $x_1, \dots, x_m$ were arbitrary, we conclude that

$$\beta_n(A) = \dim\bigl(\operatorname{Ran} A^n / \operatorname{Ran} A^{n+1}\bigr) \leq \beta_n(AB).$$

This implies the first inequality.

The second inequality is clear if $\beta_n(A) + \beta_n(B) = \infty$. Let $\beta_n(A) + \beta_n(B)$ be finite. If $m > \beta_n(A) + \beta_n(B)$ and $x_1, \dots, x_m$ are arbitrary vectors in $\operatorname{Ran}(A^n B^n) = \operatorname{Ran} A^n \cap \operatorname{Ran} B^n$, then there exists a non-trivial linear combination such that $\sum_{i=1}^{m} c_i x_i \in \operatorname{Ran} A^{n+1}$ and $\sum_{i=1}^{m} c_i x_i \in \operatorname{Ran} B^{n+1}$. By Lemma 12.8, $\sum_{i=1}^{m} c_i x_i \in \operatorname{Ran}(A^{n+1} B^{n+1})$. Hence $\beta_n(AB) \leq \beta_n(A) + \beta_n(B)$. $\square$

Let us consider the following classes of operators:

(1) $R_1^a = \{T \in \mathcal{B}(X) : d(T) = 0\}$.
Equivalently, $T \in R_1^a \Leftrightarrow \beta_0(T) = 0 \Leftrightarrow \beta_n(T) = 0$ for all $n \Leftrightarrow T$ is onto.

(2) $R_2^a = \{T \in \mathcal{B}(X) : d(T) < \infty \text{ and } d_e(T) = 0\}$.
Equivalently, $\sum_{i=0}^{\infty} \beta_i(T) < \infty \Leftrightarrow \beta_0(T) < \infty$ and there exists $d \in \mathbb{N}$ such that $\beta_d(T) = 0 \Leftrightarrow T$ is lower semi-Fredholm and T has finite descent $\Leftrightarrow T$ is lower semi-Browder.

(3) $R_3^a = \{T \in \mathcal{B}(X) : d_e(T) = 0\}$.

Equivalently, $\beta_0(T) < \infty \Leftrightarrow \beta_n(T) < \infty$ for every $n \Leftrightarrow T$ is lower semi-Fredholm.

(4) $R_4^a = \{T \in \mathcal{B}(X) : d(T) < \infty\}$.

Equivalently, there exists $d \in \mathbb{N}$ such that $\beta_n(T) = 0 \quad (n \geq d) \Leftrightarrow T$ has finite descent.

(5) $R_5^a = \{T \in \mathcal{B}(X) : d_e(T) < \infty\}$.

Equivalently, there exists $d \in \mathbb{N}$ such that $\beta_n(T) < \infty \quad (n \geq d) \Leftrightarrow T$ has finite essential descent.

In case of ambiguity we write $R_i^a(X)$ instead of $R_i^a \quad (i = 1, \dots, 5)$.

It is easy to see, by Lemmas 5 and 6, that the sets $R_1^a, \dots, R_5^a$ are regularities. So the corresponding spectra satisfy the spectral mapping theorem (for locally non-constant analytic functions).

The conditions defining the sets $R_1^a, \dots, R_5^a$ are purely algebraic (therefore we use the upper index a). We could define these classes for linear mappings in an arbitrary vector space. The spectral mapping theorem would remain true (of course, for non-constant polynomials only).

An operator $T \in \mathcal{B}(X)$ with codim $\operatorname{Ran} T < \infty$ has automatically closed range (and in this case $\operatorname{Ran} T^n$ is also closed for every n). This is not the case for operators with finite descent.

Example 7. Let H be a separable Hilbert space and let $K \in \mathcal{B}(H)$ be an operator with non-closed range. Consider the operator $T : \bigoplus_{i=0}^{\infty} H \to \bigoplus_{i=0}^{\infty} H$ defined by $T(h_0, h_1, h_2, \dots) = (Kh_1, h_2, h_3, \dots)$. Clearly, $\operatorname{Ran} T^2 = \operatorname{Ran} T$ and $\operatorname{Ran} T$ is not closed.

From the point of view of operator theory it is more interesting to combine the algebraic conditions defining regularities R_4^a and R_5^a with a topological condition – closeness of $\operatorname{Ran} T^d$. It is easy to see that if $\beta_d(T) = \dim(\operatorname{Ran} T^d / \operatorname{Ran} T^{d+1}) < \infty$, then $\operatorname{Ran} T^d$ is closed if and only if $\operatorname{Ran} T^{d+1}$ is closed. Thus, by induction, if $\beta_d(T) < \infty$ and $\operatorname{Ran} T^n$ is closed for some $n \geq d$, then $\operatorname{Ran} T^i$ is closed for every $i \geq d$.

The classes of operators which we are really interested in are the following ones (the first three sets remain unchanged since the topological condition is already implicitly contained in the definition; we repeat them only in order to preserve the symmetry with the subsequent situations):

$$
\begin{aligned}
R_1 &= \{T \in \mathcal{B}(X) : T \text{ is onto}\}; \\
R_2 &= \{T \in \mathcal{B}(X) : T \text{ is lower semi-Browder}\}; \\
R_3 &= \phi_-(X); \\
R_4 &= \{T \in \mathcal{B}(X) : d(T) < \infty \text{ and } \operatorname{Ran} T^{d(T)} \text{ is closed}\}; \\
R_5 &= \{T \in \mathcal{B}(X) : d_e(T) < \infty \text{ and } \operatorname{Ran} T^{d_e(T)} \text{ is closed}\}.
\end{aligned}
$$

Obviously, $R_1 \subset R_2 = R_3 \cap R_4 \subset R_3 \cup R_4 \subset R_5$.

It is easy to see that the sets $R_1,\dots,R_5$ are regularities, cf. Lemma 12.8. The first three of them were already studied and it was shown that they can be extended to commuting n-tuples of operators. The classes R_4 and R_5 are new.

Denote by $\sigma_i \quad (i=1,\dots,5)$ the corresponding spectra.

Corollary 8. *Let $T\in\mathcal{B}(X)$ and let f be a function analytic on a neighbourhood of $\sigma(T)$. Then:*

(i) $\sigma_i(f(T))=f(\sigma_i(T)) \qquad (i=1,2,3)$;

(ii) *if f is non-constant on each component of its domain of definition, then*

$$\sigma_i(f(T))=f(\sigma_i(T)) \qquad (i=4,5).$$

B. Ascent

Similar considerations apply to the dual situation.

Recall that the ascent of T is defined by

$$a(T)=\inf\{n:\alpha_n(T)=0\}=\inf\{n:\operatorname{Ker}T^{n+1}=\operatorname{Ker}T^n\}.$$

The *essential ascent* is defined similarly by

$$a_e(T)=\inf\{n:\alpha_n(T)<\infty\}=\inf\{n:\operatorname{Ker}T^{n+1}\overset{e}{=}\operatorname{Ker}T^n\}.$$

As in Lemmas 5 and 6 it is possible to show that

$$\alpha_{nm}(T)\le\alpha_n(T^m)\le m\cdot\alpha_{nm}(T) \qquad (m\ge 1, n\ge 0)$$

and, for commuting A,B,C,D satisfying $AC+BD=I$,

$$\max\{\alpha_n(A),\alpha_n(B)\}\le\alpha_n(AB)\le\alpha_n(A)+\alpha_n(B)\ .$$

The dual versions of the regularities $R_1^a,\dots,R_5^a$ are the following classes:

$$\begin{aligned}
R_6^a &= \{T\in\mathcal{B}(X): T \text{ is one-to-one}\};\\
R_7^a &= \{T\in\mathcal{B}(X): \dim\operatorname{Ker}T<\infty \text{ and } a(T)<\infty\};\\
R_8^a &= \{T\in\mathcal{B}(X): \dim\operatorname{Ker}T<\infty\};\\
R_9^a &= \{T\in\mathcal{B}(X): a(T)<\infty\};\\
R_{10}^a &= \{T\in\mathcal{B}(X): a_e(T)<\infty\}\ .
\end{aligned}$$

It is easy to see that the sets $R_6^a,\dots,R_{10}^a$ are regularities. So the corresponding spectra satisfy the spectral mapping theorem (for locally non-constant analytic functions). Note that R_6^a defines the point spectrum (= the set of all eigenvalues).

If we consider the topological versions of these regularities, there is a small difference from the dual case, since the ranges of operators in R_6^a, R_7^a and R_8^a need not be closed.

The dual versions of $R_1, \dots R_5$ are:

$$\begin{aligned} R_6 &= \{T \in \mathcal{B}(X) : T \text{ is bounded below}\}; \\ R_7 &= \{T \in \mathcal{B}(X) : T \text{ is upper semi-Browder}\}; \\ R_8 &= \phi_+(X); \\ R_9 &= \{T \in \mathcal{B}(X) : a(T) < \infty \text{ and } \operatorname{Ran} T^{a(T)+1} \text{ is closed}\}; \\ R_{10} &= \{T \in \mathcal{B}(X) : a_e(T) < \infty \text{ and } \operatorname{Ran} T^{a_e(T)+1} \text{ is closed}\}. \end{aligned}$$

Obviously, $R_6 \subset R_7 = R_8 \cap R_9 \subset R_8 \cup R_9 \subset R_{10}$.

To explain the exponents in the definitions of R_9 and R_{10} we need the following lemma:

Lemma 9. *Let T be an operator on a Banach space X with $a_e(T) < \infty$. Then the following two statements are equivalent:*

(i) *there exists $n \geq a_e(T) + 1$ such that* $\operatorname{Ran} T^n$ *is closed;*

(ii) $\operatorname{Ran} T^n$ *is closed for all $n \geq a_e(T)$.*

Proof. The implication (ii) ⇒ (i) is trivial.

(i) ⇒ (ii): Let $n \geq a_e(T) + 1$ and let $\operatorname{Ran} T^n$ be closed. We first prove that $\operatorname{Ran} T^{n-1}$ is also closed. To see this, note that $\operatorname{Ran} T^n \cap \operatorname{Ker} T$ is closed and $k_{n-1}(T) = \alpha_{n-1}(T) - \alpha_n(T) < \infty$.

Thus $\operatorname{Ran} T^n \cap \operatorname{Ker} T$ is of finite codimension in $\operatorname{Ran} T^{n-1} \cap \operatorname{Ker} T$ by Lemma 3, and so $\operatorname{Ran} T^{n-1} \cap \operatorname{Ker} T$ is closed. Further, $\operatorname{Ran} T^{n-1} + \operatorname{Ker} T = T^{-1}(\operatorname{Ran} T^n)$ is closed. By Lemma 20.3, we conclude that $\operatorname{Ran} T^{n-1}$ is closed.

Repeating these considerations, we get that $\operatorname{Ran} T^i$ is closed for all i with

$$a_e(T) \leq i \leq n.$$

Furthermore, $T|\operatorname{Ran} T^{n-1}$ is an upper semi-Fredholm operator; so

$$\operatorname{Ran} T^i = \operatorname{Ran}\big((T|\operatorname{Ran} T^{n-1})^{i-n+1}\big)$$

is closed for all $i \geq n$. □

It is easy to see that the sets R_i $(i = 6, \dots, 10)$ are regularities; so the corresponding spectra $\sigma_i(T) = \{\lambda : T - \lambda \notin R_i\}$ satisfy the spectral mapping theorem (in the case of $i = 6, 7, 8$ for all analytic functions; in the case of $i = 9, 10$ for analytic functions which are locally non-constant).

Further, $T \in \mathcal{B}(X)$ belongs to $R_i(X)$ $(i = 1, \dots, 5)$ if and only if $T^* \in R_{i+5}(X^*)$. Similarly, $T \in R_i(X)$ $(i = 6, \dots, 10)$ if and only if $T^* \in R_{i-5}(X^*)$.

Moreover, since the intersection of two (or more) regularities is again a regularity, we can obtain the spectral mapping theorem for a large number of combinations of $R_1, \dots, R_{10}$. Of particular interest are the symmetrical combinations $R_i \cap R_{i+5}$ $(i = 1, \dots, 5)$.

Clearly, $R_1 \cap R_6 = R_1^a \cap R_6^a$ is the set of all invertible operators, $R_2 \cap R_7 = R_2^a \cap R_7^a$ is the set of all Browder operators, and $R_3 \cap R_8 = R_3^a \cap R_8^a$ is the set of all Fredholm operators. We characterize the remaining combinations $R_4 \cap R_9$ and $R_5 \cap R_{10}$.

Theorem 10. *Let T be an operator on a Banach space X. The following statements are equivalent:*

(i) $T \in R_4 \cap R_9$;

(ii) $T \in R_4^a \cap R_9^a$;

(iii) $a(T) < \infty$ *and* $d(T) < \infty$;

(iv) *T can be written as $T_1 \oplus T_2$ where T_1 is nilpotent and T_2 invertible;*

(v) *either T is invertible or 0 is an isolated point of $\sigma(T)$ and $T|X_1$ is nilpotent, where X_1 is the spectral subspace corresponding to $\{0\}$;*

(vi) *either T is invertible or the resolvent $z \mapsto (T-z)^{-1}$ has a pole at 0;*

(vii) *there are $S \in \mathcal{B}(X)$ and $n \in \mathbb{N}$ such that $ST = TS$, $STS = S$ and $T^n ST = T^n$ (such an operator T is sometimes called Drazin invertible; the operator S with properties described here is called the Drazin inverse of T);*

(viii) *there are $S \in \mathcal{B}(X)$ and $n \in \mathbb{N}$ such that $ST = TS$ and $T^{n+1}S = T^n$.*

Proof. The implication (i) $\Rightarrow$ (ii) and the equivalence (ii) $\Leftrightarrow$ (iii) are clear.

(iii) $\Rightarrow$ (i): Follows from Theorem 20.4.

(iii) $\Rightarrow$ (iv): By Theorem 20.4, we have $X = \operatorname{Ker} T^n \oplus \operatorname{Ran} T^n$ where $n = a(T) = d(T)$, and $\operatorname{Ran} T^n$ is closed. Let $T_1 = T|\operatorname{Ker} T^n$ and $T_2 = T|\operatorname{Ran} T^n$. Then $T_1^n = 0$. Since $\operatorname{Ker} T^{n+1} = \operatorname{Ker} T^n$ and $\operatorname{Ran} T^{n+1} = \operatorname{Ran} T^n$, it is easy to see that T_2 is invertible.

(iv) $\Rightarrow$ (v): If $X = X_1 \oplus X_2$ with $T|X_1$ nilpotent and $T|X_2$ invertible, then either T is invertible or 0 is an isolated point of $\sigma(T)$. It is easy to see that the set $X_1 = \{x \in X : r_x(T) = 0\}$ is the spectral subspace corresponding to $\{0\}$.

The implication (v) $\Rightarrow$ (iv) is clear.

(iv) $\Rightarrow$ (vi): In a punctured neighbourhood U of 0 we have

$$(T-z)^{-1} = (T_1 - z)^{-1} \oplus (T_2 - z)^{-1} = \sum_{i=0}^{\infty} T_1^i z^{-(i+1)} \oplus (T_2 - z)^{-1},$$

where $(T_2 - z)^{-1}$ is analytic in $U \cup \{0\}$. Thus the resolvent $z \mapsto (T-z)^{-1}$ has a pole at 0 of order $n = \min\{j : T_1^j = 0\}$.

The implication (vi) $\Rightarrow$ (v) can be proved similarly.

(iv) $\Rightarrow$ (vii): Set $S = 0 \oplus T_2^{-1}$ and let n satisfy $T_1^n = 0$. Then S satisfies (vii).

(vii) $\Rightarrow$ (viii): Obvious.

(viii) $\Rightarrow$ (iii): We have $\operatorname{Ran} T^n = \operatorname{Ran}(T^{n+1}S) \subset \operatorname{Ran} T^{n+1}$. So $d(T) \le n$. Let $x \in \operatorname{Ker} T^{n+1}$. Then $T^n x = ST^{n+1}x = 0$, so $x \in \operatorname{Ker} T^n$ and $a(T) \le n$. □

Lemma 11. *Let $T \in \mathcal{B}(X)$ satisfy $a_e(T) < \infty$ and $d_e(T) < \infty$. Then $a_e(T) = d_e(T)$ and $\operatorname{Ran} T^{a_e(T)}$ is a closed and complemented subspace of X.*

Proof. Let $n = \max\{a_e(T), d_e(T)\}$. The operator from $\operatorname{Ker} T^{2n}/\operatorname{Ker} T^n$ to $\operatorname{Ker} T^n \cap \operatorname{Ran} T^n$ induced by T^n is onto, so

$$\begin{aligned}\dim(\operatorname{Ker} T^n \cap \operatorname{Ran} T^n) &\le \dim \operatorname{Ker} T^{2n}/\operatorname{Ker} T^n \\ &= \sum_{i=n}^{2n-1} \dim \operatorname{Ker} T^{i+1}/\operatorname{Ker} T^i = \sum_{i=n}^{2n-1} \alpha_i(T) < \infty.\end{aligned}$$

Similarly, the operator from $X/(\operatorname{Ran} T^n + \operatorname{Ker} T^n)$ to $\operatorname{Ran} T^n/\operatorname{Ran} T^{2n}$ induced by T^n is one-to-one, and so $\operatorname{codim}(\operatorname{Ran} T^n + \operatorname{Ker} T^n) \le \dim \operatorname{Ran} T^n/\operatorname{Ran} T^{2n} < \infty$.

Let F be a finite-dimensional subspace satisfying $F \oplus (\operatorname{Ran} T^n + \operatorname{Ker} T^n) = X$. Lemma 20.3 for the spaces $\operatorname{Ran} T^n$ and $\operatorname{Ker} T^n + F$ implies that $\operatorname{Ran} T^n$ is closed.

Let L be a closed subspace satisfying $L \oplus (\operatorname{Ran} T^n \cap \operatorname{Ker} T^n) = \operatorname{Ker} T^n$. It is easy to see that $\operatorname{Ran} T^n \oplus L = \operatorname{Ran} T^n + \operatorname{Ker} T^n$ and $X = \operatorname{Ran} T^n \oplus (F + L)$. Hence $\operatorname{Ran} T^n$ is a complemented subspace.

It remains to show that $a_e(T) = d_e(T) = n$. This is clear if $n = 0$. Suppose that $n \ge 1$. We have $\alpha_n(T) < \infty$, $\beta_n(T) < \infty$, and $\alpha_{n-1}(T) - \alpha_n(T) = k_{n-1}(T) = \beta_{n-1}(T) - \beta_n(T)$. Thus $\alpha_{n-1}(T) < \infty \Leftrightarrow \beta_{n-1}(T) < \infty$. Since $n = \max\{a_e(T), d_e(T)\}$, we conclude that $a_e(T) = n = d_e(T)$. □

Next we characterize the intersection $R_5 \cap R_{10}$. Operators in this class are sometimes called *B-Fredholm.*

Theorem 12. *Let T be an operator on a Banach space X. The following statements are equivalent:*

(i) $T \in R_5 \cap R_{10}$;

(ii) $T \in R_5^a \cap R_{10}^a$;

(iii) $a_e(T) < \infty$ *and* $d_e(T) < \infty$;

(iv) *there exists n such that $\operatorname{Ran} T^n$ is closed and $T|\operatorname{Ran} T^n$ is Fredholm;*

(v) *(Kato decomposition) there are closed subspaces X_1, X_2 such that $X = X_1 \oplus X_2$, $TX_i \subset X_i \quad (i = 1, 2)$, $T|X_1$ is nilpotent and $T|X_2$ Fredholm.*

Proof. The implication (i) ⇒ (ii) and the equivalence (ii) ⇔ (iii) are clear. The implication (iii) ⇒ (i) follows from Lemmas 9 and 11.

(v) ⇒ (iv): Let $X = X_1 \oplus X_2$, $TX_i \subset X_i \quad (i = 1, 2)$, $T^n|X_1 = 0$ and let $T|X_2$ be Fredholm. Then $\operatorname{Ran} T^n = \operatorname{Ran} T^n|X_2$, which is of finite codimension in X_2. Therefore $\operatorname{Ran} T^n$ is closed. It is easy to see that $T|\operatorname{Ran} T^n$ is Fredholm.

(iv) ⇒ (iii): Let $\operatorname{Ran} T^n$ be closed and $T_2 = T|\operatorname{Ran} T^n$ Fredholm. Then

$$\alpha_n(T) = \dim \operatorname{Ker} T \cap \operatorname{Ran} T^n = \alpha(T_2) < \infty$$

and

$$\beta_n(T) = \dim \operatorname{Ran} T^n / \operatorname{Ran} T^{n+1} = \beta(T_2) < \infty.$$

Hence $a_e(T) < \infty$ and $d_e(T) < \infty$.

(iii) ⇒ (v): Since $\alpha_j(T)$ and $\beta_j(T)$ are finite for all j sufficiently large, and these sequences are non-increasing, there exists $n \in \mathbb{N}$ such that $\alpha_j(T) = \alpha_n(T) < \infty$ and $\beta_j(T) = \beta_n(T) < \infty$ for all $j \geq n$. Therefore $k_j(T) = 0$ for $j \geq n$ and $\operatorname{Ker} T \cap \operatorname{Ran} T^n \subset R^\infty(T)$. By Lemma 11, $\operatorname{Ran} T^n$ is closed. By Lemma 12.1 for the restriction $T|\operatorname{Ran} T^n$, we also have $N^\infty(T|\operatorname{Ran} T^n) = N^\infty(T) \cap \operatorname{Ran} T^n \subset R^\infty(T)$.

If $n = 0$, then T is Fredholm and the decomposition is trivial. In the following we assume that $n \geq 1$.

Since $\dim(\operatorname{Ran} T^n \cap \operatorname{Ker} T) = \alpha_n(T) < \infty$, there exists a closed subspace L such that $X = L \oplus (\operatorname{Ran} T^n \cap \operatorname{Ker} T)$.

We define closed subspaces $N_j \quad (j = 0, \dots, n)$ inductively by $N_0 = \{0\}$ and $N_{j+1} = T^{-1} N_j \cap L \quad (j < n)$.

Clearly, $T N_{j+1} \subset N_j \cap \operatorname{Ran} T$. Conversely, let $x \in N_j \cap \operatorname{Ran} T$. Then $x = Tu$ for some $u \in X$. Express $u = l + v$ with $l \in L$ and $v \in \operatorname{Ker} T \cap \operatorname{Ran} T^n$. Then $u - v = l \in L$ and $T(u - v) = Tu = x$. Thus $u - v \in N_{j+1}$ and $x \in T N_{j+1}$.

Hence

$$T N_{j+1} = N_j \cap \operatorname{Ran} T \quad (j < n).$$

We prove by induction on j that $N_j \subset N_{j+1}$. The statement is clear for $j = 0$. Suppose that $j \geq 0$, $N_j \subset N_{j+1}$ and let $x \in N_{j+1}$. Then $Tx \in N_j \subset N_{j+1}$, and so $x \in T^{-1} N_{j+1}$. Since $x \in N_{j+1} \subset L$, we conclude that $x \in N_{j+2}$.

Hence

$$N_j \subset N_{j+1} \quad (j = 0, 1, \dots, n-1).$$

One can see easily that $N_j \subset \operatorname{Ker} T^j$ for all j.

We now prove by induction on j that

$$\operatorname{Ker} T^j \subset N_j + (\operatorname{Ker} T^j \cap \operatorname{Ran} T^n). \tag{1}$$

The inclusion is clear for $j = 0$. For $j = 1$ we have $\operatorname{Ker} T = (\operatorname{Ker} T \cap L) + (\operatorname{Ker} T \cap \operatorname{Ran} T^n) = N_1 + (\operatorname{Ker} T \cap \operatorname{Ran} T^n)$. Let $j \geq 1$, $\operatorname{Ker} T^j \subset N_j + (\operatorname{Ker} T^j \cap \operatorname{Ran} T^n)$ and let $x \in \operatorname{Ker} T^{j+1}$. Then $Tx \in \operatorname{Ker} T^j$, and so $Tx = v_1 + v_2$ for some $v_1 \in N_j$ and $v_2 \in \operatorname{Ker} T^j \cap \operatorname{Ran} T^n = \operatorname{Ker} T^j \cap \operatorname{Ran} T^{n+1} = T(\operatorname{Ker} T^{j+1} \cap \operatorname{Ran} T^n)$. Thus $v_1 \in N_j \cap \operatorname{Ran} T = T N_{j+1}$ and

$$\begin{aligned} x &\in N_{j+1} + (\operatorname{Ker} T^{j+1} \cap \operatorname{Ran} T^n) + \operatorname{Ker} T \\ &= N_{j+1} + (\operatorname{Ker} T^{j+1} \cap \operatorname{Ran} T^n) + (\operatorname{Ker} T \cap L) + (\operatorname{Ker} T \cap \operatorname{Ran} T^n) \\ &= N_{j+1} + (\operatorname{Ker} T^{j+1} \cap \operatorname{Ran} T^n). \end{aligned}$$

Hence (1).

Finally, we prove by induction that $N_j \cap \operatorname{Ran} T^n = \{0\}$. This is clear for $j = 0$. Let $j \geq 0$, $N_j \cap \operatorname{Ran} T^n = \{0\}$ and let $x \in N_{j+1} \cap \operatorname{Ran} T^n$. Then $Tx \in N_j \cap \operatorname{Ran} T^n$ and so, by the induction assumption, $Tx = 0$. Thus $x \in \operatorname{Ker} T \cap \operatorname{Ran} T^n$ and $x \in N_{j+1} \subset L$, and so $x = 0$. Hence

$$N_j \cap \operatorname{Ran} T^n = \{0\} \qquad (j \leq n).$$

Set $N = N_n$. Then $TN \subset N$ and $N \subset \operatorname{Ker} T^n$. Further, $N + \operatorname{Ran} T^n \supset \operatorname{Ker} T^n$ by (1), and $N \cap \operatorname{Ran} T^n = \{0\}$. Note also that $N + \operatorname{Ran} T^n = \operatorname{Ker} T^n + \operatorname{Ran} T^n = T^{-n} \operatorname{Ran}(T^{2n})$, which is closed, since $\operatorname{Ran} T^{2n}$ is closed by Lemma 9.

Consider the dual operator $T^* \in \mathcal{B}(X^*)$. It is easy to see that $\operatorname{Ran}(T^{*j})$ is closed, $\alpha_j(T^*) = \beta_j(T)$ and $\beta_j(T^*) = \alpha_j(T)$ for all $j \geq n$. Thus we can use the same construction for T^*.

Since $\dim(\operatorname{Ker} T^* \cap \operatorname{Ran} T^{*n}) = \alpha_n(T^*) = \beta_n(T) < \infty$, there exists a finite-dimensional subspace $G \subset X$ such that ${}^\perp(\operatorname{Ker} T^* \cap \operatorname{Ran} T^{*n}) \oplus G = X$. Set $L' = G^\perp$. Then L' is a w^*-closed subspace and $L' \oplus (\operatorname{Ker} T^* \cap \operatorname{Ran} T^{*n}) = X^*$.

Define subspaces $M_0' \subset M_1' \subset \cdots \subset M_n' \subset X^*$ by $M_0' = \{0\}$ and $M_{j+1}' = T^{*-1} M_j \cap L'$. By induction, M_j' is w^*-closed for all j.

Set $M' = M_n'$. As above we have $T^* M' \subset M' \subset \operatorname{Ker} T^{*n}$, $M' \cap \operatorname{Ran} T^{*n} = \{0\}$ and $\operatorname{Ker} T^{*n} \subset M' + \operatorname{Ran} T^{*n}$. Moreover, $M' + \operatorname{Ran} T^{*n}$ is a closed subspace.

Set $M = {}^\perp M'$. Then $TM \subset M$ and $M = {}^\perp M' \supset {}^\perp \operatorname{Ker} T^{*n} = \operatorname{Ran} T^n$.

Further,

$$\operatorname{Ran} T^n = {}^\perp \operatorname{Ker} T^{*n} \supset {}^\perp\big(M' + \operatorname{Ran} T^{*n}\big) = {}^\perp M' \cap {}^\perp \operatorname{Ran} T^{*n} = M \cap \operatorname{Ker} T^n$$

and $M + \operatorname{Ker} T^n = {}^\perp M' + {}^\perp \operatorname{Ran} T^{*n} = {}^\perp\big(M' \cap \operatorname{Ran} T^{*n}\big) = X$ (the middle equality follows from A.1.13).

Thus

$$M + N \supset M + \operatorname{Ran} T^n + N \supset M + \operatorname{Ran} T^n + \operatorname{Ker} T^n = X$$

and

$$M \cap N \subset M \cap \operatorname{Ker} T^n \cap N \subset \operatorname{Ran} T^n \cap N = \{0\}.$$

Hence $X = N \oplus M$, $TN \subset N$, $TM \subset M$ and $(T|N)^n = 0$.

Let $T_2 = T|M$. We have

$$\begin{aligned}\operatorname{Ker} T_2 = \operatorname{Ker} T \cap M \subset \operatorname{Ker} T^n \cap M &= {}^\perp(\operatorname{Ran} T^{*n} + M') \\ &= {}^\perp(\operatorname{Ran} T^{*n} + \operatorname{Ker} T^{*n}) = \operatorname{Ker} T^n \cap \operatorname{Ran} T^n \subset R^\infty(T).\end{aligned}$$

Thus $k_j(T_2) = 0$ for all $j \geq 0$. Hence the sequences $\alpha_j(T_2)$ and $\beta_j(T_2)$ are constant. Since $\alpha_n(T_2) = \alpha_n(T) < \infty$ and $\beta_n(T_2) = \beta_n(T) < \infty$, we conclude that $\alpha(T_2) < \infty$ and $\beta(T_2) < \infty$. So T_2 is Fredholm. □

Proposition 13. Let $T, S \in \mathcal{B}(X)$. Then $\sigma_i(TS) \setminus \{0\} = \sigma_i(ST) \setminus \{0\}$ for $i = 1, 2, \ldots, 10$. The same relation is true also for all unions of these spectra, in particular for $\sigma_4 \cup \sigma_9$ and $\sigma_5 \cup \sigma_{10}$.

Proof. Follows from Propositions 12.28 and 12.29.

C. Kato, essentially Kato and quasi-Fredholm operators

In this section we replace the numbers $\beta_n(T) = \dim \operatorname{Ran} T^n / \operatorname{Ran} T^{n+1}$ and $\alpha_n(T) = \dim \operatorname{Ker} T^{n+1} / \operatorname{Ker} T^n$ by the numbers

$$\begin{aligned} k_n(T) &= \dim(\operatorname{Ran} T + \operatorname{Ker} T^{n+1})/(\operatorname{Ran} T + \operatorname{Ker} T^n) \\ &= \dim(\operatorname{Ker} T \cap \operatorname{Ran} T^n)/(\operatorname{Ker} T \cap \operatorname{Ran} T^{n+1}). \end{aligned}$$

Note that $k_n(T) = 0$ if and only if $\operatorname{Ker} T \cap \operatorname{Ran} T^n \subset \operatorname{Ran} T^{n+1}$. Similarly, $k_n(T) < \infty$ if and only if $\operatorname{Ker} T \cap \operatorname{Ran} T^n \overset{e}{\subset} \operatorname{Ran} T^{n+1}$.

We start with an analogue of Lemmas 5 and 6.

Lemma 14. *Let A, B, C, D be mutually commuting operators on a Banach space X satisfying $AC + BD = I$ and let $n \geq 0$. Then:*

(i) $\operatorname{Ran}(A^n B^n) \cap \operatorname{Ker}(AB) = \big(\operatorname{Ran} A^n \cap \operatorname{Ker} A\big) + \big(\operatorname{Ran} B^n \cap \operatorname{Ker} B\big)$;

(ii) $\max\{k_n(A), k_n(B)\} \leq k_n(AB) \leq k_n(A) + k_n(B)$.

Proof. (i) By Lemma 12.8, we have

$$\begin{aligned} \operatorname{Ran}(A^n B^n) \cap \operatorname{Ker}(AB) &= \operatorname{Ran} A^n \cap \operatorname{Ran} B^n \cap (\operatorname{Ker} A + \operatorname{Ker} B) \\ &\supset \big(\operatorname{Ran} A^n \cap \operatorname{Ran} B^n \cap \operatorname{Ker} A\big) + \big(\operatorname{Ran} A^n \cap \operatorname{Ran} B^n \cap \operatorname{Ker} B\big) \qquad (2) \\ &= \big(\operatorname{Ran} A^n \cap \operatorname{Ker} A\big) + \big(\operatorname{Ran} B^n \cap \operatorname{Ker} B\big). \end{aligned}$$

On the other hand, if $x \in \operatorname{Ran} A^n \cap \operatorname{Ran} B^n \cap \big(\operatorname{Ker} A + \operatorname{Ker} B\big)$, then $x = y + z$ for some $y \in \operatorname{Ker} A \subset \operatorname{Ran} B^n$ and $z \in \operatorname{Ker} B \subset \operatorname{Ran} A^n$. Thus we also have $y = x - z \in \operatorname{Ran} A^n$ and $z = x - y \in \operatorname{Ran} B^n$. So

$$x \in \big(\operatorname{Ran} A^n \cap \operatorname{Ran} B^n \cap \operatorname{Ker} A\big) + \big(\operatorname{Ran} A^n \cap \operatorname{Ran} B^n \cap \operatorname{Ker} B\big)$$

and we have equality in (2).

(ii.a) We prove $k_n(A) \leq k_n(AB)$. If $x_1, \dots, x_m \in \operatorname{Ran} A^n \cap \operatorname{Ker} A$ where $m > k_n(AB)$, then $B^n x_i \in \operatorname{Ran}(A^n B^n) \cap \operatorname{Ker} A \subset \operatorname{Ran}(A^n B^n) \cap \operatorname{Ker}(AB)$ for all $i = 1, \dots, m$. Thus there exists a non-trivial linear combination

$$\sum_{i=1}^m c_i B^n x_i \in \operatorname{Ran}(A^{n+1} B^{n+1}) \subset B^n \operatorname{Ran} A^{n+1}.$$

So

$$\sum_{i=1}^m c_i x_i \in \operatorname{Ran} A^{n+1} + \operatorname{Ker} B^n \subset \operatorname{Ran} A^{n+1}.$$

Hence $k_n(A) = \dim\big(\operatorname{Ran} A^n \cap \operatorname{Ker} T\big)/\big(\operatorname{Ran} A^{n+1} \cap \operatorname{Ker} T\big) \leq k_n(AB)$.

(ii.b) To prove the second inequality, let $x_1, \dots, x_m \in \operatorname{Ran}(A^n B^n) \cap \operatorname{Ker}(AB)$ where $m > k_n(A) + k_n(B)$. By (i), we can write $x_i = y_i + z_i \quad (i = 1, \dots, m)$

for some $y_i \in \operatorname{Ran} A^n \cap \operatorname{Ker} A$ and $z_i \in \operatorname{Ran} B^n \cap \operatorname{Ker} B$. Thus there exists a non-trivial linear combination such that $\sum_{i=1}^m c_i y_i \in \operatorname{Ran} A^{n+1} \cap \operatorname{Ker} A$ and $\sum_{i=1}^m c_i z_i \in \operatorname{Ran} B^{n+1} \cap \operatorname{Ker} B$. Hence $\sum_{i=1}^m c_i x_i \in \operatorname{Ran}(A^{n+1}B^{n+1}) \cap \operatorname{Ker}(AB)$ and $k_n(AB) < m$. This proves the second inequality. □

Lemma 15. *Let $T \in \mathcal{B}(X)$, $n \geq 0$ and $m \geq 1$. Then*

$$\begin{aligned} k_n(T^m) = k_{mn}(T) + 2k_{mn+1}(T) + 3k_{mn+2}(T) + \cdots + mk_{mn+m-1}(T) \\ + (m-1)k_{mn+m}(T) + \cdots + k_{mn+2m-2}(T). \end{aligned}$$

In particular,

$$k_{mn}(T) \leq k_n(T^m) \leq m^2 \max_{0 \leq i \leq 2m-2} k_{mn+i}(T).$$

Proof. Consider the mapping

$$\widehat{T}_j : \operatorname{Ran} T^j / \operatorname{Ran} T^{j+m} \to \operatorname{Ran} T^{j+1} / \operatorname{Ran} T^{j+m+1}$$

induced by T. By Lemmas 2 and 21.1, we have

$$\begin{aligned} \dim \operatorname{Ker} \widehat{T}_j &= \dim\big(T^{-1} \operatorname{Ran} T^{j+m+1} \cap \operatorname{Ran} T^j\big) / \operatorname{Ran} T^{j+m} \\ &= \dim\big((\operatorname{Ker} T + \operatorname{Ran} T^{j+m}) \cap \operatorname{Ran} T^j\big) / \operatorname{Ran} T^{j+m} \\ &= \dim\big((\operatorname{Ker} T \cap \operatorname{Ran} T^j) + \operatorname{Ran} T^{j+m}\big) / \operatorname{Ran} T^{j+m} \\ &= \dim\big(\operatorname{Ker} T \cap \operatorname{Ran} T^j\big) / \big(\operatorname{Ker} T \cap \operatorname{Ran} T^{j+m}\big) = \sum_{i=0}^{m-1} k_{i+j}(T). \end{aligned}$$

Since the mapping $\operatorname{Ran} T^{mn} / \operatorname{Ran} T^{mn+m} \to \operatorname{Ran} T^{mn+m} / \operatorname{Ran} T^{mn+2m}$ induced by T^m is equal to the composition $\widehat{T}_{mn+m-1} \widehat{T}_{mn+m-2} \cdots \widehat{T}_{mn}$ and all these mappings are onto, we have

$$k_n(T^m) = \sum_{j=mn}^{mn+m-1} \dim \operatorname{Ker} \widehat{T}_j = \sum_{j=mn}^{mn+m-1} \sum_{i=0}^{m-1} k_{j+i}(T),$$

which gives the statement of the lemma. □

We now define the classes of operators analogous to $R_1^a, \ldots, R_5^a$:

$R_{11}^a = \{T \in \mathcal{B}(X) : k_n(T) = 0 \text{ for all } n \geq 0\}$;

$R_{12}^a = \{T \in \mathcal{B}(X) : \sum_{i=0}^{\infty} k_i(T) < \infty\}$;

$R_{13}^a = \{T \in \mathcal{B}(X) : k_n(T) < \infty \text{ for all } n \geq 0\}$;

$R_{14}^a = \{T \in \mathcal{B}(X) : \text{ there exists } d \in \mathbb{N} \text{ such that } k_n(T) = 0 \text{ for all } n \geq d\}$;

$R_{15}^a = \{T \in \mathcal{B}(X) : \text{ there exists } d \in \mathbb{N} \text{ such that } k_n(T) < \infty \text{ for all } n \geq d\}$.

The condition in R_{11}^a means that

$$\operatorname{Ker} T = \operatorname{Ker} T \cap \operatorname{Ran} T = \operatorname{Ker} T \cap \operatorname{Ran} T^2 = \cdots = \operatorname{Ker} T \cap R^{\infty}(T),$$

and so $R_{11}^a = \{T : \operatorname{Ker} T \subset R^{\infty}(T)\}$.

Similarly, $\sum_{i=0}^{\infty} k_i(T) < \infty$ means that there is a $d \in \mathbb{N}$ such that

$$\operatorname{Ker} T \stackrel{e}{=} \operatorname{Ker} T \cap \operatorname{Ran} T \stackrel{e}{=} \operatorname{Ker} T \cap \operatorname{Ran} T^2 \stackrel{e}{=} \cdots \stackrel{e}{=} \operatorname{Ker} T \cap \operatorname{Ran} T^d = \operatorname{Ker} T \cap R^\infty(T),$$

and so $R_{12_e}^a = \{T : \operatorname{Ker} T \stackrel{e}{\subset} R^\infty(T)\}$. The condition defining R_{13}^a can be rewritten as $\operatorname{Ker} T^m \stackrel{e}{\subset} \operatorname{Ran} T^n$ for all $m, n \in \mathbb{N}$. The condition in R_{14}^a is equivalent to $\operatorname{Ker} T \cap \operatorname{Ran} T^d \subset R^\infty(T)$.

It follows from Lemmas 14 and 15 that the sets $R_{11}^a \cdots R_{15}^a$ are regularities; so the corresponding spectra satisfy the spectral mapping theorem (for locally non-constant analytic functions).

Before we introduce the topological version of $R_{11}^a, \dots, R_{15}^a$ we state several simple lemmas.

Lemma 16. *Let $T \in \mathcal{B}(X)$, $m \geq 0$ and $n \geq i \geq 1$. If $\operatorname{Ran} T^n + \operatorname{Ker} T^m$ is closed, then $\operatorname{Ran} T^{n-i} + \operatorname{Ker} T^{m+i}$ is closed.*

Proof. It is sufficient to show that

$$\operatorname{Ran} T^{n-i} + \operatorname{Ker} T^{m+i} = T^{-i}\big(\operatorname{Ran} T^n + \operatorname{Ker} T^m\big). \tag{3}$$

The inclusion $\subset$ is clear. Conversely, suppose that $T^i z \in \operatorname{Ran} T^n + \operatorname{Ker} T^m$, so $T^i z = T^n x + u$ for some $x \in X$ and $u \in \operatorname{Ker} T^m$. Then $u \in \operatorname{Ran} T^i$, and so $u = T^i v$ for some $v \in \operatorname{Ker} T^{m+i}$. Consequently, $z - T^{n-i}x - v \in \operatorname{Ker} T^i$. Thus $z \in \operatorname{Ran} T^{n-i} + \operatorname{Ker} T^{m+i} + \operatorname{Ker} T^i = \operatorname{Ran} T^{n-i} + \operatorname{Ker} T^{m+i}$ and we have equality in (3). □

Lemma 17. *Let $T \in \mathcal{B}(X)$ and let $n \geq 0$. If $\operatorname{Ran} T^n$ is closed and $\operatorname{Ran} T + \operatorname{Ker} T^n$ is closed, then $\operatorname{Ran} T^{n+1}$ is closed.*

Proof. Let $u_j \in X$ $(j = 1, 2, \dots)$ and let $T^{n+1}u_j \to z$ as $j \to \infty$. Then $z \in \operatorname{Ran} T^n$, $z = T^n u$ for some $u \in X$ and $T^n(u - Tu_j) \to 0$.

Consider the operator $\widehat{T^n} : X/\operatorname{Ker} T^n \to X$ induced by T^n.

Clearly, $\widehat{T^n}$ is one-to-one and has closed range, therefore it is bounded below, and $\widehat{T^n}(u - Tu_j + \operatorname{Ker} T^n) \to 0$ $(j \to \infty)$ implies $u - Tu_j + \operatorname{Ker} T^n \to 0$ in $X/\operatorname{Ker} T^n$. Thus there are elements $v_j \in \operatorname{Ker} T^n$ such that $Tu_j + v_j \to u$, and so $u \in \operatorname{Ran} T + \operatorname{Ker} T^n$. Hence $z = T^n u \in \operatorname{Ran} T^{n+1}$. □

Lemma 18. *Let $T \in \mathcal{B}(X)$, $d \in \mathbb{N}$ and let $k_i(T) < \infty$ for every $i \geq d$. Then the following statements are equivalent:*

(i) *there exists $n \geq d+1$ such that $\operatorname{Ran} T^n$ is closed;*

(ii) *$\operatorname{Ran} T^n$ is closed for all $n \geq d$;*

(iii) *$\operatorname{Ran} T^n + \operatorname{Ker} T^m$ is closed for all m, n with $m + n \geq d$.*

Proof. Clearly, (iii) $\Rightarrow$ (ii) $\Rightarrow$ (i). The implication (ii) $\Rightarrow$ (iii) follows from Lemma 16.

(i) $\Rightarrow$ (ii) : If $\operatorname{Ran} T^n$ is closed, then, by Lemma 16, $\operatorname{Ran} T + \operatorname{Ker} T^{n-1}$ is closed. Since $\operatorname{Ran} T+\operatorname{Ker} T^{n-1} \overset{e}{\subset} \operatorname{Ran} T+\operatorname{Ker} T^n \overset{e}{\subset} \cdots$, we get that $\operatorname{Ran} T+\operatorname{Ker} T^i$ is closed for every $i \geq n$. Thus, by Lemma 17, we get inductively that $\operatorname{Ran} T^i$ is closed for every $i \geq n$.

To show that $\operatorname{Ran} T^i$ is closed for all i, $d \leq i \leq n$, we can proceed exactly as in the proof of Lemma 9. □

We use the following notation:

$$\begin{aligned}
R_{11} &= \{T \in \mathcal{B}(X) : \operatorname{Ker} T \subset R^\infty(T) \text{ and } \operatorname{Ran} T \text{ is closed}\};\\
R_{12} &= \{T \in \mathcal{B}(X) : \operatorname{Ker} T \overset{e}{\subset} R^\infty(T) \text{ and } \operatorname{Ran} T \text{ is closed}\};\\
R_{13} &= \{T \in \mathcal{B}(X) : k_n(T) < \infty \text{ for every } n \in \mathbb{N} \text{ and } \operatorname{Ran} T \text{ is closed}\};\\
R_{14} &= \{T \in \mathcal{B}(X) : \text{there exists } d \in \mathbb{N} \text{ such that}\\
&\qquad k_n(T) = 0 \quad (n \geq d) \text{ and } \operatorname{Ran} T^{d+1} \text{ is closed}\};\\
R_{15} &= \{T \in \mathcal{B}(X) : \text{there exists } d \in \mathbb{N} \text{ such that}\\
&\qquad k_n(T) < \infty \ (n \geq d) \text{ and } \operatorname{Ran} T^{d+1} \text{ is closed}\}.
\end{aligned}$$

The sets R_{11} and R_{12} are the classes of all Kato and essentially Kato operators, respectively. The operators in R_{14} are called *quasi-Fredholm*.

Clearly, $R_{11} \subset R_{12} = R_{13} \cap R_{14} \subset R_{13} \cup R_{14} \subset R_{15}$, $R_1 \cup R_6 \subset R_{11}$, $R_2 \cup R_7 \subset R_3 \cup R_8 \subset R_{12}$, $R_4 \cup R_9 \subset R_{14}$ and $R_5 \cup R_{10} \subset R_{15}$.

It is easy to see that the sets $R_{11} \cdots R_{15}$ are regularities.

Let σ_i $(i = 11, \dots, 15)$ be the corresponding spectra defined by $\sigma_i(T) = \{\lambda : T - \lambda \notin R_i\}$. If $X = X_1 \oplus X_2$ is a decomposition of X with closed X_1, X_2 and if $T_1 \in \mathcal{B}(X_1)$, $T_2 \in \mathcal{B}(X_2)$, then

$$\sigma_i(T_1 \oplus T_2) = \sigma_i(T_1) \cup \sigma_i(T_2) \qquad (i = 11, \dots, 15).$$

Since $\sigma_{11}(T_1) \neq \emptyset \Leftrightarrow X_1 \neq \{0\}$, and $\sigma_i(T_1) \neq \emptyset \Leftrightarrow \dim X_1 = \infty$ for $i = 12, 13$, we have the following spectral mapping theorems:

Theorem 19. *Let $T \in \mathcal{B}(X)$ and let f be a function analytic on a neighbourhood of $\sigma(T)$. Then*

$$\sigma_i(f(T)) = f(\sigma_i(T)) \qquad (i = 11, 12, 13).$$

If f is non-constant on each component of its domain of definition, then

$$\sigma_i(f(T)) = f(\sigma_i(T)) \qquad (i = 14, 15).$$

Examples 20. Let S_n be the shift in an n-dimensional Hilbert space. A typical example of an operator in the class R_{13} is $T = \bigoplus_{n=1}^\infty S_n$; then $k_n(T) = 1$ for all $n \geq 0$.

An example of a quasi-Fredholm operator (class R_{14}) is $T' = \bigoplus_{j=1}^\infty S_n$, where $n \in \mathbb{N}$ is fixed. Then $k_{n-1}(T') = \infty$ and $k_i(T') = 0 \quad (i \neq n-1)$.

The operator $T'' = T \oplus T'$ is an example of an operator in the class R_{15}.

Lemma 21. If $T \in \mathcal{B}(X)$, $n \in \mathbb{N}$ and $\operatorname{Ran} T^n$, $\operatorname{Ran} T^{n+1}$ and $\operatorname{Ran} T^{n+2}$ are closed, then $k_n(T^*) = k_n(T)$.

Proof. The space $\operatorname{Ran} T + \operatorname{Ker} T^n = T^{-n} \operatorname{Ran} T^{n+1}$ is closed, and similarly, $\operatorname{Ran} T + \operatorname{Ker} T^{n+1}$ is closed. Thus

$$\begin{aligned} k_n(T) &= \dim(\operatorname{Ran} T + \operatorname{Ker} T^{n+1})/(\operatorname{Ran} T + \operatorname{Ker} T^n) \\ &= \dim(\operatorname{Ran} T + \operatorname{Ker} T^n)^\perp/(\operatorname{Ran} T + \operatorname{Ker} T^{n+1})^\perp \\ &= \dim(\operatorname{Ker} T^* \cap \operatorname{Ran} T^{*n})/(\operatorname{Ker} T^* \cap \operatorname{Ran} T^{*n+1}) = k_n(T^*). \end{aligned}$$ □

Corollary 22. $T^* \in R_i \Leftrightarrow T \in R_i$ for $i = 11, \dots, 15$.

23 Semiregularities and miscellaneous spectra

Some spectra studied in literature satisfy the conditions required in Section 6 only partially. We met some examples of this kind in the previous sections.

Recall that a non-empty subset R of a Banach algebra $\mathcal{A}$ is called a regularity if it satisfies the following two conditions:

(i) if $a \in \mathcal{A}$ and $n \in \mathbb{N}$, then $a \in R \Leftrightarrow a^n \in R$,

(ii) if a, b, c, d are mutually commuting elements of $\mathcal{A}$ satisfying $ac + bd = 1_{\mathcal{A}}$, then $ab \in R \Leftrightarrow a, b \in R$.

Axioms (i) and (ii) can be divided into two parts, each of them implying a one-way spectral mapping theorem.

There are many natural examples of classes of operators or Banach algebras elements that satisfy only one half of the axioms of regularities. This motivates the definition of semiregularities.

Lower semiregularities

Definition 1. Let R be a non-empty subset of a Banach algebra $\mathcal{A}$. Then R is called a *lower semiregularity* if

(i) $a \in \mathcal{A}, n \in \mathbb{N}, a^n \in R \Rightarrow a \in R$,

(ii) if a, b, c, d are mutually commuting elements of $\mathcal{A}$ satisfying $ac + bd = 1_{\mathcal{A}}$ and $ab \in R$, then $a, b \in R$.

For a lower semiregularity R let σ_R be the corresponding spectrum defined by σ_R by $\sigma_R(a) = \{\lambda \in \mathbb{C} : a - \lambda \notin R\}$.

Clearly the intersection $R = \bigcap_\alpha R_\alpha$ of any system of lower semiregularities is again a lower semiregularity. The corresponding spectra satisfy $\sigma_R(a) = \bigcup_\alpha \sigma_{R_\alpha}(a)$ for all $a \in \mathcal{A}$.

Note that also the union $R = \bigcup_\alpha R_\alpha$ of any system of lower semiregularities is again a lower semiregularity. The corresponding spectrum then satisfies $\sigma_R(a) = \bigcap_\alpha \sigma_{R_\alpha}(a)$.

Lemma 2. *Let $R \subset \mathcal{A}$ be a lower semiregularity. Then:*

(i) $1_\mathcal{A} \in R$;

(ii) $\mathrm{Inv}(\mathcal{A}) \subset R$;

(iii) *if $a \in R$, $b \in \mathrm{Inv}(\mathcal{A})$ and $ab = ba$, then $ab \in R$;*

(iv) $\sigma_R(a) \subset \sigma(a)$;

(v) *(translation property) $\sigma_R(a + \lambda) = \lambda + \sigma_R(a)$.*

Proof. (i) Let $b \in R$. We have $1 \cdot 1 + b \cdot 0 = 1$ and $1 \cdot b = b \in R$. Thus $1 \in R$.

(ii) Let $a \in \mathrm{Inv}(\mathcal{A})$. Then $a \cdot a^{-1} + a^{-1} \cdot 0 = 1$ and $a \cdot a^{-1} = 1 \in R$. Hence $a \in R$.

(iii) We have $(ab) \cdot 0 + b^{-1} \cdot b = 1$ and $(ab) \cdot b^{-1} = a \in R$, so $ab \in R$.

The remaining statements are clear. □

Remark 3. Suppose that $R \subset \mathcal{A}$ is a non-empty subset satisfying

$$a, b \in \mathcal{A}, ab = ba, ab \in R \Rightarrow a, b \in R. \tag{1}$$

Then clearly R is a lower semiregularity.

Theorem 4. *Let $R \subset \mathcal{A}$ be a lower semiregularity and $a \in \mathcal{A}$. Then*

$$f(\sigma_R(a)) \subset \sigma_R(f(a))$$

for each locally non-constant function f analytic on a neighbourhood of $\sigma(a)$.

Proof. Suppose on the contrary that $\lambda \in f(\sigma_R(a)) \setminus \sigma_R(f(a))$. Since the function $f(z) - \lambda$ has only a finite number of zeros $\alpha_1, \dots, \alpha_n$ in $\sigma(a)$, we can write

$$f(z) - \lambda = (z - \alpha_1)^{k_1} \dots (z - \alpha_n)^{k_n} g(z),$$

for some $k_i \geq 1$ and a function g analytic on a neighbourhood of $\sigma(a)$ such that $g(z) \neq 0 \quad (z \in \sigma(a))$. Thus

$$f(a) - \lambda = (a - \alpha_1)^{k_1} \dots (a - \alpha_n)^{k_n} g(a),$$

where $f(a) - \lambda \in R$ and $g(a) \in \mathrm{Inv}(\mathcal{A})$. By Lemma 2 (iii),

$$(a - \alpha_1)^{k_1} \cdots (a - \alpha_n)^{k_n} \in R.$$

Let $i \in \{1, \dots, n\}$. For certain polynomials p, q we have

$$(z - \alpha_i)^{k_i} \cdot p(z) + \Big(\prod_{j \neq i} (z - \alpha_j)^{k_j}\Big) \cdot q(z) = 1.$$

The corresponding identity for z replaced by a gives $(a - \alpha_i)^{k_i} \in R$. Thus $a - \alpha_i \in R$ and $\alpha_i \notin \sigma_R(a) \quad (i = 1, \dots, n)$. Hence $\lambda \notin f(\sigma_R(a))$, a contradiction. □

Corollary 5. *Let $R \subset \mathcal{A}$ be a lower semiregularity and $0 \notin R$. Then $p(\sigma_R(a)) \subset \sigma_R(p(a))$ for all polynomials p.*

Proof. It is sufficient to verify the inclusion for the constant polynomials $p(z) \equiv \lambda$. In this case we have $p(\sigma_R(a)) \subset \{\lambda\}$ and $\sigma_R(p(a)) = \sigma_R(\lambda \cdot 1_{\mathcal{A}}) = \{\lambda\}$. □

The assumption in Theorem 4 that the function f is locally non-constant can be frequently omitted.

Theorem 6. *Let $R \subset \mathcal{A}$ be a lower semiregularity satisfying the following condition: if $c = c^2 \in R$, $a \in \mathcal{A}$ and $ac = ca$, then $c+(1-c)a \in R$. Then $f(\sigma_R(a)) \subset \sigma_R(f(a))$ for all $a \in \mathcal{A}$ and f analytic on a neighbourhood of $\sigma(a)$.*

Proof. Let U be the domain of definition of f. Suppose on the contrary that $\lambda \in f(\sigma_R(a)) \setminus \sigma_R(f(a))$. Let U_1 be the union of all components of U where f is identically equal to λ, and $U_2 = U \setminus U_1$. Let h be defined by

$$h(z) = \begin{cases} 0 & (z \in U_1), \\ 1 & (z \in U_2). \end{cases}$$

Then we can write

$$f(z) - \lambda = h(z)(z-\alpha_1)^{k_1} \cdots (z-\alpha_n)^{k_n} \cdot g(z)$$

where $\alpha_1, \dots, \alpha_n \in \sigma(a) \cap U_2$, g is analytic on U and $g(z) \neq 0 \quad (z \in \sigma(a))$. Set $p(z) = (z-\alpha_1)^{k_1} \cdots (z-\alpha_n)^{k_n}$. Thus $f(a) - \lambda = h(a)p(a)g(a) = h(a)p(a)\big(1 - h(a) + g(a)h(a)\big)$, where $1 - h(a) + g(a)h(a) \in \operatorname{Inv}(\mathcal{A})$. We have $f(a) - \lambda \in R$ and so, by Lemma 2 (iii), $h(a)p(a) \in R$.

Consider the function r defined by

$$r(z) = \begin{cases} p(z)^{-1} & (z \in U_1), \\ 0 & (z \in U_2). \end{cases}$$

Then $p(a)(1-h(a)) \cdot r(a) + h(a) \cdot 1 = 1$ and $p(a)h(a) \in R$, and so $p(a) \in R, h(a) \in R$. As in Theorem 4, $p(a) \in R$ implies $a - \alpha_i \in R \quad (i = 1, \dots, n)$ and so $\alpha_i \notin \sigma_R(a)$.

Since $\lambda \in f(\sigma_R(a))$, there is a $\beta \in U_1 \cap \sigma_R(a)$. Further, $h(a)$ is an idempotent in R and, by assumption, we have $(a-\beta)(1-h(a)) + h(a) \in R$. Since $(1-h(a)) + (a-\beta)h(a) \in \operatorname{Inv}(\mathcal{A})$, we have

$$a - \beta = \big((a-\beta)(1-h(a)) + h(a)\big) \cdot \big((1-h(a)) + (a-\beta)h(a)\big) \in R.$$

This contradicts to the fact that $\beta \in \sigma_R(a)$. □

Remark 7. In particular the condition of the previous theorem is satisfied if the unit element is the unique idempotent in R.

Another typical application is when $\mathcal{A}$ is the algebra of all bounded operators on a Banach space, all idempotents in R are projections onto subspaces of finite codimension and R is invariant under finite rank perturbations (for example Fredholm operators, upper (lower) semi-Fredholm operators etc.).

Theorem 8. *Let $R \subset \mathcal{A}$ be a lower semiregularity. The following conditions are equivalent:*

(i) *R is open;*

(ii) *$\sigma_R(a)$ is closed for each $a \in \mathcal{A}$ and the set-valued function $a \mapsto \sigma_R(a)$ is upper semicontinuous.*

Proof. Straightforward. □

Remark 9. Let $R \subset \mathcal{A}$ be a lower semiregularity. Then the spectrum σ_R can be extended to n-tuples of commuting elements of $\mathcal{A}$ in such a way that

$$p(\sigma_R(a_1, \dots, a_n)) \subset \sigma_R(p(a_1, \dots, a_n))$$

for all commuting n-tuples $a_1, \dots, a_n \in \mathcal{A}$ and all non-constant polynomials p in n variables. Indeed, define

$$\sigma_R(a_1, \dots, a_n) = \{(\lambda_1, \dots, \lambda_n) : p(a_1, \dots, a_n) - p(\lambda_1, \dots, \lambda_n) \notin R \text{ for all } p\}.$$

The extension is not unique; other (trivial) extension is $\sigma_R(a_1, \dots, a_n) = \emptyset$ whenever $n \geq 2$.

The first extension is maximal among all extensions satisfying the one-way spectral mapping property (clearly the trivial extension is minimal).

We show now some examples of lower semiregularities. Of course every regularity is also a lower semiregularity. Therefore we restrict here only to examples of lower semiregularities that are not regularities.

Let X be a Banach space. The set $\Phi_+(X) \cup \Phi_-(X)$ of all semi-Fredholm operators is a lower semiregularity (since it is a union of two regularities). The corresponding *semi-Fredholm spectrum* is defined by $\sigma_{sF}(T) = \{\lambda \in \mathbb{C} : T - \lambda \text{ is not semi-Fredholm}\}$.

By Remark 7, the one-way spectral mapping $f(\sigma_{sF}(T)) \subset \sigma_{sF}(f(T))$ is satisfied for all functions analytic on a neighbourhood of $\sigma(T)$.

Another example of a lower semiregularity in a Banach algebra $\mathcal{A}$ is the set $\operatorname{Inv}_l(\mathcal{A}) \cup \operatorname{Inv}_r(\mathcal{A})$ of all one-side invertible elements.

Let $T \in \mathcal{B}(X)$ and $n \geq 0$. Recall the definitions from the previous section:

$$\begin{aligned}
\alpha_n(T) &= \dim \operatorname{Ker} T^{n+1} / \operatorname{Ker} T^n, \\
\beta_n(T) &= \dim \operatorname{Ran} T^n / \operatorname{Ran} T^{n+1}, \\
k_n(T) &= \dim(\operatorname{Ran} T^n \cap \operatorname{Ker} T)/(\operatorname{Ran} T^{n+1} \cap \operatorname{Ker} T).
\end{aligned}$$

Fix $m \geq 0$. It follows from the results in the previous sections that the following subsets of $\mathcal{B}(X)$ are lower semiregularities:

(i) $\{T \in \mathcal{B}(X) : \dim \operatorname{Ker} T \le m\} = \{T : \sup \alpha_i(T) \le m\}$,
(ii) $\{T \in \mathcal{B}(X) : \dim N^\infty(T) \le m\} = \{T : \sum \alpha_i(T) \le m\}$,
(iii) $\{T \in \mathcal{B}(X) : \lim \alpha_i(T) \le m\}$,
(iv) $\{T \in \mathcal{B}(X) : \operatorname{codim} \operatorname{Ran} T \le m\} = \{T : \sup \beta_i(T) \le m\}$,
(v) $\{T \in \mathcal{B}(X) : \operatorname{codim} R^\infty(T) \le m\} = \{T : \sum \beta_i(T) \le m\}$,
(vi) $\{T \in \mathcal{B}(X) : \lim \beta_i(T) \le m\}$,
(vii) $\{T \in \mathcal{B}(X) : \sup k_i(T) \le m\}$,
(viii) $\{T \in \mathcal{B}(X) : \sum k_i(T) \le m\}$,
(ix) $\{T \in \mathcal{B}(X) : \limsup k_i(T) \le m\}$.
(x) $\{T \in \mathcal{B}(X) : \dim \operatorname{Ker} T \le m$ and $\operatorname{Ran} T$ is closed$\}$,
(xi) $\{T \in \mathcal{B}(X) : \dim N^\infty(T) \le m$ and $\operatorname{Ran} T$ is closed$\}$,
(xii) $\{T \in \mathcal{B}(X) :$ there is a j such that $\alpha_j(T) \le m$ and $\operatorname{Ran} T^{j+1}$ is closed$\}$,
(xiii) $\{T \in \mathcal{B}(X) : \sup k_i(T) \le m$ and $\operatorname{Ran} T$ is closed$\}$,
(xiv) $\{T \in \mathcal{B}(X) : \sum k_i(T) \le m$ and $\operatorname{Ran} T$ is closed$\}$,
(xv) $\{T \in \mathcal{B}(X) : k_n(T) \le m$ and $\operatorname{Ran} T^n$ is closed for every $n \ge n_0\}$.

Note that the ranges in classes (iv)–(vi) are closed automatically.

This shows that it is rather easy to find examples of lower semiregularities.

Upper semiregularities

Definition 10. A subset R of a Banach algebra $\mathcal{A}$ is called an *upper semiregularity* if

(i) $a \in R, n \in \mathbb{N} \Rightarrow a^n \in R$,
(ii) if a, b, c, d are mutually commuting elements of $\mathcal{A}$ satisfying $ac + bd = 1_\mathcal{A}$ and $a, b \in R$, then $ab \in R$,
(iii) R contains a neighbourhood of the unit element $1_\mathcal{A}$.

The definitions of upper and lower semiregularities are only seemingly asymmetric. In fact, condition (iii) for lower semiregularities was satisfied automatically.

Clearly R is a regularity if and only if it is both a lower and upper semiregularity.

Define again $\sigma_R(a) = \{\lambda \in \mathbb{C} : a - \lambda \notin R\}$. Clearly the intersection of any family of upper semiregularities is again an upper semiregularity. Also the mapping $a \mapsto \sigma_R(a)$ is upper semicontinuous if and only if R is open.

Remark 11. If $R \subset \mathcal{A}$ is a semigroup, then conditions (i) and (ii) of Definition 10 are satisfied. Thus a semigroup containing a neighbourhood of the unit element is an upper semiregularity.

Lemma 12. *Let $R \subset \mathcal{A}$ be an upper semiregularity, let $a \in R \cap \operatorname{Inv}(\mathcal{A})$. Then there exists $\varepsilon > 0$ such that $\{b \in \mathcal{A} : ab = ba, \|b - a\| < \varepsilon\} \subset R$.*

Proof. Let $\delta > 0$ satisfy $\{c \in \mathcal{A} : \|c - 1_{\mathcal{A}}\| < \delta\} \subset R$. Let $a \in R \cap \operatorname{Inv}(\mathcal{A})$. Set $\varepsilon = \frac{\delta}{\|a^{-1}\|}$. Suppose that $b \in \mathcal{A}$, $ab = ba$ and $\|b - a\| < \varepsilon$. Then $\|a^{-1}b - 1\| = \|a^{-1}(b-a)\| \le \|a^{-1}\| \cdot \|b-a\| < \delta$, and so $a^{-1}b \in R$. Further, $a \cdot a^{-1} + (a^{-1}b) \cdot 0 = 1$, hence $b = a \cdot (a^{-1}b) \in R$. □

Lemma 13. *Let $R \subset \mathcal{A}$ be an upper semiregularity, $a_n \in R \cap \operatorname{Inv}(\mathcal{A})$ $(n = 1, 2, \dots)$, $a \in \operatorname{Inv}(\mathcal{A})$, $a_n \to a$ and $a_n a = a a_n$. Then $a \in R$.*

Proof. For each n we have $a_n \cdot a_n^{-1} + (a_n^{-1}a) \cdot 0 = 1$. Further, $a_n^{-1}a \to 1$, and so $a_n^{-1}a \in R$ for n large enough. Thus $a = a_n \cdot (a_n^{-1}a) \in R$. □

Theorem 14. *Let $R \subset \mathcal{A}$ be an upper semiregularity, let $a \in \mathcal{A}$. Let M be a component of $\mathbb{C} \setminus \sigma(a)$. Then either $M \subset \sigma_R(a)$ or $M \cap \sigma_R(a) = \emptyset$.*

Proof. Let $L = \{a - \lambda : \lambda \in M, a - \lambda \in R\}$. By Lemma 12, L is open and, by Lemma 13, it is relatively closed in M. Thus either $L = \emptyset$ or $L = M$. □

Corollary 15. *Let $R \subset \mathcal{A}$ be an upper semiregularity. Then $\lambda \cdot 1_{\mathcal{A}} \in R$ for each non-zero complex number λ.*

Proof. Consider the element $a = 0$. The set $M = \{\lambda \in \mathbb{C} : \lambda \neq 0\}$ is a component of $\mathbb{C} \setminus \sigma(0)$. Further, $1 \in R$, so $\lambda \in R$ for all $\lambda \in M$. □

Lemma 16. *Let $R \subset \mathcal{A}$ be an upper semiregularity, let $a \in R$, $b \in R \cap \operatorname{Inv}(\mathcal{A})$ and $ab = ba$. Then $ab \in R$.*

Proof. We have $a \cdot 0 + b \cdot b^{-1} = 1$, so $ab \in R$. □

Theorem 17. *Let $R \subset \mathcal{A}$ be an upper semiregularity, let $a \in \mathcal{A}$. Then $\sigma_R(a) \subset \widehat{\sigma}(a)$. Further, $\sigma_R(a) \setminus \sigma(a)$ is a union of some bounded components of $\mathbb{C} \setminus \sigma(a)$.*

Proof. For $|\lambda|$ large enough we have $1 - \frac{a}{\lambda} \in R$, so $a - \lambda = -\lambda(1 - \frac{a}{\lambda}) \in R$. By Theorem 14 the unbounded component of $\mathbb{C} \setminus \sigma(a)$ is disjoint with $\sigma_R(a)$ and thus $\sigma_R(a) \subset \widehat{\sigma}(a)$. □

Theorem 18. *Let $R \subset \mathcal{A}$ be an upper semiregularity, let $a \in \mathcal{A}$. Then $\sigma_R(p(a)) \subset p(\sigma_R(a))$ for all non-constant polynomials p.*

Moreover, if $\sigma_R(b) \neq \emptyset$ for all $b \in \mathcal{A}$, then $\sigma_R(p(a)) \subset p(\sigma_R(a))$ for all polynomials p.

Proof. Let p be a non-constant polynomial. Let $\lambda \notin p(\sigma_R(a))$. Write $p(z) - \lambda = \beta \cdot (z - \alpha_1)^{k_1} \cdots (z - \alpha_n)^{k_n}$ where $n \ge 1$ and $\alpha_1, \dots, \alpha_n, \beta \in \mathbb{C}$, $\beta \neq 0$. Thus

$$p(a) - \lambda = \beta \cdot (a - \alpha_1)^{k_1} \cdots (a - \alpha_n)^{k_n}.$$

By assumption, $\alpha_i \notin \sigma_R(a)$ $\quad (i = 1, \dots, n)$. Thus $a - \alpha_i \in R$ and $(a - \alpha_i)^{k_i} \in R$. As in Theorem 4 we have $(z - \alpha_1)^{k_1} \cdots (z - \alpha_n)^{k_n} \in R$ and $p(a) - \lambda \in R$, i.e., $\lambda \notin \sigma_R(p(a))$. Thus $\sigma_R(p(a)) \subset p(\sigma_R(a))$ for all non-constant polynomials.

Suppose that $\sigma_R(b) \neq \emptyset$ for all $b \in \mathcal{A}$. Let $p(z) = \lambda$ be a constant polynomial. Then

$$\sigma_R(p(a)) = \sigma_R(\lambda \cdot 1_{\mathcal{A}}) = \{\lambda\} = p(\sigma_R(a)).$$ □

Theorem 19. *Let $R \subset \mathcal{A}$ be an upper semiregularity. Suppose that R satisfies the condition*

$$b \in R \cap \mathrm{Inv}(\mathcal{A}) \Rightarrow b^{-1} \in R. \tag{2}$$

Then $\sigma_R(f(a)) \subset f(\sigma_R(a))$ for all $a \in \mathcal{A}$ and all locally non-constant functions f analytic on a neighbourhood of $\sigma(a) \cup \sigma_R(a)$.

Further, $\sigma_R(f(a)) \subset f(\sigma_R(a) \cup \sigma(a))$ for all functions f analytic on a neighbourhood of $\sigma_R(a) \cup \sigma(a)$.

Proof. Suppose first that f is locally non-constant and suppose on the contrary that there is a $\lambda \in \sigma_R(f(a)) \setminus f(\sigma_R(a))$. Then $f(a)-\lambda = q(a)g(a)$ where $q(a) = (z-\alpha_1)^{k_1} \cdots (a-\alpha_n)^{k_n}$ and g is a function analytic and non-zero on a neighbourhood of $\sigma(a) \cup \sigma_R(a)$. By assumption, $f(a)-\lambda \notin R$ and $\alpha_i \notin \sigma_R(a)$, i.e., $a-\alpha_i \in R \quad (i = 1,\dots,n)$. As in Theorem 4 we obtain that $q(a) \in R$. Further, there are a compact neighbourhood V of $\sigma(a) \cup \sigma_R(a)$ and rational functions $\frac{p_n(z)}{q_n(z)}$ with poles outside V such that $\frac{p_n(z)}{q_n(z)} \to g(z)$ uniformly on V. We can assume that the polynomials p_n, q_n are non-constant and $p_n(z) \neq 0$ on $\sigma(a) \cup \sigma_R(a)$.

By Theorem 18, this means that $p_n(a) \in R$, $q_n(a) \in R$. By assumption, $q_n(a)^{-1} \in R$. Thus we have $p_n(a)q_n(a)^{-1} \in R$ and, by Lemma 13, $g(a) = \lim p_n(a)q_n(a)^{-1} \in R$. Since $q(a) \in R$ and $g(a) \in R \cap \mathrm{Inv}(\mathcal{A})$, we have $f(a)-\lambda \in R$, a contradiction.

Suppose now that f is analytic on a neighbourhood of $\sigma(a) \cup \sigma_R(a)$ and $\lambda \in \sigma_R(f(a)) \setminus \bigl(f(\sigma(a) \cup \sigma_R(a))\bigr)$. Let U be the domain of definition of f, $U = U_1 \cup U_2$ where U_1, U_2 are disjoint open sets, $f|U_1 \equiv \lambda$ and f is not identically equal to λ on any non-empty open subset of U_2. By assumption, $(\sigma_R(a) \cup \sigma(a)) \cap U_1 = \emptyset$, so U_2 is an open neighbourhood of $\sigma(a) \cup \sigma_R(a)$. The proof proceeds as in the first part. □

In many cases the inclusion $\sigma_R(f(a)) \subset f(\sigma_R(a))$ is true for all analytic functions. By Theorem 19, this is true if R satisfies (2) and $R \subset \mathrm{Inv}(\mathcal{A})$, i.e., $\sigma_R(a) \supset \sigma(a)$ for all a.

Another typical situation is described in the following theorem.

Theorem 20. *Let $R \subset \mathcal{B}(X)$ be an upper semiregularity satisfying (2) such that*

(i) *if $T \in R$ and $F \in \mathcal{B}(X)$ is a finite rank operator commuting with T, then $T + F \in R$,*

(ii) *if $T \in \mathcal{B}(X)$, U_1, U_2 are disjoint open sets, $\sigma(T) \subset U_1 \cup U_2$ and $\sigma_R(T) \subset U_2$, then the spectral projection of T corresponding to U_1 is of finite rank.*

Then $\sigma_R(f(T) \subset f(\sigma_R(T))$ for all $T \in \mathcal{B}(X)$ and f analytic on a neighbourhood of $\sigma(T) \cup \sigma_R(T)$.

Proof. Let f be analytic on a neighbourhood of $\sigma(T) \cup \sigma_R(T)$ and suppose that there is $\lambda \in \sigma_R(f(T)) \setminus f(\sigma_R(T))$. Let U_1, U_2 be disjoint open sets, $f|U_1 \equiv \lambda$ and f is not identically equal to λ on any non-empty open subset of U_2. By assumption, $\sigma_R(T) \cap U_1 = \emptyset$, so $\sigma_R(T) \subset U_2$. Let h be defined by

$$h(z) = \begin{cases} 0 & (z \in U_1), \\ 1 & (z \in U_2). \end{cases}$$

By (ii), $I - h(T)$ is a finite rank projection. We can write

$$f(z) - \lambda = h(z)(z - \alpha_1)^{k_1} \cdots (z - \alpha_n)^{k_n} g(z)$$

for some $\alpha_1, \ldots, \alpha_n \in \sigma(a) \cap U_2$, g analytic on $U_1 \cup U_2$ and $g(z) \neq 0 \quad (z \in \sigma(a))$. Set $q(z) = (z - \alpha_1)^{k_1} \cdots (z - \alpha_n)^{k_n}$.

We have $\alpha_i \notin \sigma_R(T)$, so $T - \alpha_i \in R$ and, as in Theorem 4, $q(T) \in R$. As in Theorem 19 we get $g(T) \in R \cap \operatorname{Inv}(\mathcal{B}(X))$, and so $q(T)g(T) \in R$. By (i), $f(T) - \lambda = h(T)q(T)g(T) \in R$, a contradiction. □

Examples 21.

(i) Let R be the principal component of $\operatorname{Inv}(\mathcal{A})$, i.e., the component of $\operatorname{Inv}(\mathcal{A})$ containing the unit. Then R is an open semigroup and so an upper semiregularity. The corresponding spectrum is the *exponential spectrum* σ_{exp} (the name is justified by the fact that $R = \{\exp(a_1) \cdots \exp(a_n) : n \in \mathbb{N}, a_1, \ldots, a_n \in \mathcal{A}\}$).

By Theorems 17 and 19, $\sigma(a) \subset \sigma_{exp}(a) \subset \widehat{\sigma}(a)$ and $\sigma_{exp}(f(a)) \subset f\sigma_{exp}(a)$ for each function f analytic on a neighbourhood of $\sigma_{exp}(a)$.

(ii) Let $R = \{T \in \Phi(X) : \operatorname{ind} T = 0\}$. Then R is an open semigroup and thus an upper semiregularity. The corresponding spectrum is the *Weyl spectrum* (sometimes also called the *Schechter spectrum*) $\sigma_W(T) = \{\lambda \in \mathbb{C} : T - \lambda \notin \Phi(X)$ or $\operatorname{ind} T \neq 0\}$. By Theorem 19.7, $\sigma_W(T) = \bigcap \sigma(T + K)$ where the intersection is taken over the set of all compact operators K. By Theorem 20, we have $\sigma_W(f(T)) \subset f(\sigma_W(T))$ for each function f analytic on a neighbourhood of $\sigma(T)$.

In fact, the Weyl spectrum is closely related to the exponential spectrum. It is easy to see that T is a Fredholm operator with $\operatorname{ind} T = 0$ if and only if T lies in the principal component of $\Phi(X)$.

(iii) More generally, let J be a closed two-sided ideal in a Banach algebra $\mathcal{A}$ and $R = \{a + b : a \in \operatorname{Inv}(\mathcal{A}), b \in J\}$. It is easy to check that R is a semigroup containing $\operatorname{Inv}(\mathcal{A})$, and so an upper semiregularity.

(iv) Let $\mathcal{A} = \mathcal{B}(X)$. Then the sets

$$\Phi_+^-(X) = \{T \in \Phi_+(X) : \operatorname{ind} T \leq 0\} \qquad \text{and}$$
$$\Phi_-^+(X) = \{T \in \Phi_-(X) : \operatorname{ind} T \geq 0\}.$$

are upper semiregularities. The corresponding spectra $\sigma_{\Phi_+^-}$ and $\sigma_{\Phi_-^+}$ satisfy

$$\sigma_{\Phi_+^-}(T) = \bigcap\{\sigma_\pi(T+K) : K \text{ compact}\},$$
$$\sigma_{\Phi_-^+}(T) = \bigcap\{\sigma_\delta(T+K) : K \text{ compact}\}$$

and the one-way spectral mapping theorem for all analytic functions, cf. Theorem 20.

Closed-range spectrum

Most of the classes of "nice" operators require that the operators have closed ranges. Thus it is natural to consider the *closed-range spectrum* of an operator $T \in \mathcal{B}(X)$ defined by

$$\sigma_{cr}(T) = \{\lambda \in \mathbb{C} : \operatorname{Ran}(T-\lambda) \text{ is not closed}\}.$$

This spectrum is sometimes called the *Goldberg spectrum*. However, the closed-range spectrum has not good properties. For example, it is possible that $\operatorname{Ran} T$ is closed but $\operatorname{Ran} T^2$ is not. Conversely, it is also possible that $\operatorname{Ran} T^2$ is closed but $\operatorname{Ran} T$ is not. In particular, operators with closed range are not a semiregularity.

In fact it is possible to construct the following extreme example:

Example 22. Let $M \subset \mathbb{N}$ be any subset. Then there exists an operator T acting on a separable Hilbert space H such that $\operatorname{Ran} T^n$ is closed if and only if $n \in M$.

Construction: If $M = \mathbb{N}$, then the statement is clear (take for example $T = I$). So we may assume that $M \neq \mathbb{N}$.

Let K be a separable infinite-dimensional Hilbert space and fix an operator $V \in \mathcal{B}(K)$ with $\|V\| = 1$ and $\operatorname{Ran} V$ non-closed (for example, let $V = \operatorname{diag}(1, 1/2, 1/3, \dots)$).

Let $m \in \mathbb{N}$. We construct an operator T_m such that $\|T_m\| \le 2$, $\gamma(T_m^j) = 1$ for $1 \le j \le m-1$, $\operatorname{Ran} T_m^m$ is not closed and $T_m^{m+1} = 0$.

Set $H_m = \bigoplus_{i=-m+1}^m K$. The operator $T_m \in \mathcal{B}(H_m)$ will be defined by

$$\begin{aligned} &T_m\big(x_{-m+1}, x_{-m+2}, \dots, x_{-1}, x_0, x_1, \dots, x_m\big) \\ &\quad = \big(0, x_{-m+1}, x_{-m+2}, \dots, x_{-2}, x_{-1} + Vx_1, x_2, \dots, x_m, 0\big). \end{aligned}$$

Clearly $\|T_m\| \le 2$ and $T_m^{m+1} = 0$.

For $1 \le j \le m-1$ we have

$$\operatorname{Ran} T_m^j = \underbrace{0 \oplus \cdots \oplus 0}_{j} \oplus \underbrace{K \oplus \cdots \oplus K}_{2m-2j} \oplus \underbrace{0 \oplus \cdots \oplus 0}_{j}.$$

Thus $\operatorname{Ran} T_m^j$ is closed for $j = 1, \dots, m-1$. It is easy to see that $\gamma(T_m^j) = 1$.

Further,

$$\operatorname{Ran} T_m^m = \underbrace{0 \oplus \cdots \oplus 0}_{m-1} \oplus \operatorname{Ran} V \oplus \underbrace{0 \oplus \cdots \oplus 0}_{m},$$

and so $\operatorname{Ran} T_m^m$ is not closed.

Set now $H = \bigoplus_{m \in \mathbb{N} \setminus M} H_m$ and $T = \bigoplus_{m \in \mathbb{N} \setminus M} T_m$. Then

$$\operatorname{Ran} T^j = \bigoplus_{m \in \mathbb{N} \setminus M} \operatorname{Ran} T_m^j$$

for each j. Consequently, $\operatorname{Ran} T^j$ is closed if and only if $m \in M$.

Note that the Kato spectrum, which has very nice spectral properties, is not too far from the closed-range spectrum. Clearly $\sigma_{cr}(T) \subset \sigma_K(T)$ and, by Theorem 12.13, $\sigma_K(T) \setminus \sigma_{cr}(T)$ is at most countable. Thus the Kato spectrum can be considered as a nice completion of the closed-range spectrum.

Generalized spectrum

Another type of spectrum considered in literature is the *generalized spectrum*, $\sigma_g(T) = \{\lambda \in \mathbb{C} : T - \lambda \text{ has not a generalized inverse}\}$.

For operators on Hilbert spaces, the generalized spectrum coincides with the closed range spectrum. Therefore the last example is also valid for the generalized spectrum.

24 Measures of non-compactness and other operator quantities

The basic operator quantities connected with an operator $T \in \mathcal{B}(X)$ are the norm, the injectivity modulus $j(T)$ and the surjectivity modulus $k(T)$.

In this section we study various essential versions of these quantities.

We start with the Hausdorff measure of non-compactness:

Definition 1. Let Ω be a bounded subset of a Banach space X. The *Hausdorff measure of non-compactness* of Ω is defined by

$$q_X(\Omega) = \inf\{\varepsilon > 0 : \text{ there exists a finite set } F \subset X \text{ such that } \Omega \subset F + \varepsilon B_X\}.$$

If no confusion can arise, then we write simply $q(\Omega)$ instead of $q_X(\Omega)$.

For sequences this definition coincides with the norm in the space $\tilde{X}$, see Section 17.

Proposition 2. *Let Ω, Ψ be bounded subsets of X, let $c > 0$. Then:*

(i) $q(\overline{\Omega}) = q(\Omega)$;

(ii) $q(\Omega) = 0 \Leftrightarrow \overline{\Omega}$ *is compact;*

(iii) *if* $\Omega \subset \Psi$, *then* $q(\Omega) \leq q(\Psi)$;

(iv) $q(c\Omega) = cq(\Omega)$;

(v) $q(\Omega + \Psi) \leq q(\Omega) + q(\Psi)$.

Proof. (v): Let $\delta > 0$. There are finite subsets $F, F' \subset X$ such that $\Omega \subset F + (q(\Omega) + \delta)B_X$ and $\Psi \subset F' + (q(\Psi) + \delta)B_X$. Thus $\Omega + \Psi \subset (F + F') + (q(\Omega) + q(\Psi) + 2\delta)B_X$ and $q(\Omega + \Psi) \leq q(\Omega) + q(\Psi) + 2\delta$. Letting $\delta \to 0$ yields (v).

The remaining statements are straightforward. □

Proposition 3. *Let X be a closed subspace of a Banach space Y and let $\Omega \subset X$ be a bounded set. Then $q_Y(\Omega) \leq q_X(\Omega) \leq 2q_Y(\Omega)$.*

Proof. The first inequality is clear.

Let $s > q_Y(\Omega)$ and let $F \subset Y$ be a finite set such that $\Omega \subset F + sB_Y$. We can assume that F is a minimal set with this property. So for each $f \in F$ there is an $x_f \in \Omega$ with $\|f - x_f\| \leq s$. Set $F' = \{x_f : f \in F\}$. Clearly, F' is a finite subset of X. Let $x \in \Omega$ and let $f \in F$ satisfy $\|x - f\| \leq s$. Then

$$\text{dist}\{x, F'\} \leq \|x - x_f\| \leq \|x - f\| + \|f - x_f\| \leq 2s.$$

Thus $q_X(\Omega) \leq 2s$. Letting $s \to q_Y(\Omega)$ yields $q_X(\Omega) \leq 2q_Y(\Omega)$. □

Later we give an example that the estimates given in Proposition 3 are the best possible.

Proposition 4. *If X is an infinite-dimensional Banach space, then $q(B_X) = 1$.*

Proof. Clearly, $q(B_X) \leq 1$. Suppose on the contrary that $q(B_X) < s < 1$. Then there exists a finite set $F \subset X$ such that $B_X \subset F + sB_X$. Thus $B_X \subset F + s(F + sB_X) \subset (F + sF) + s^2 B_X$, and so $q(B_X) \leq s^2$. Since $s > q(B_X)$ was arbitrary, we have $q(B_X) \leq (q(B_X))^2$. Hence $q(B_X) = 0$, and so B_X is compact, a contradiction with the assumption that $\dim X = \infty$. □

Proposition 5. *Let $\Omega \subset X$ be a bounded set. Then*

$$\sup\{q(C) : C \subset \Omega, C \text{ countable}\} \geq \frac{1}{2} q(\Omega).$$

Proof. Let $r < q(\Omega)$. Choose $x_1 \in \Omega$ arbitrarily. Since $\{x_1\} + rB_X \not\supset \Omega$, there exists $x_2 \in \Omega$ such that $\|x_2 - x_1\| > r$. We can construct inductively a sequence $x_1, x_2, \ldots$ of points of Ω such that $\|x_i - x_j\| > r$ for all $i, j \in \mathbb{N}, i \neq j$. Set $C = \{x_1, x_2, \ldots\}$. Since every closed ball of radius $r/2$ contains at most one point of C, C can not be covered by a finite number of closed balls of radius $r/2$. Hence $q(C) \geq r/2$. □

Example 6. Let I be an uncountable set and let

$$X = \{f : I \to \mathbb{C} : \operatorname{supp} f \text{ countable}, \ \sup |f(i)| < \infty\}.$$

Clearly, X with the sup-norm is a Banach space. Let $\Omega = \{e_i : i \in I\}$. Then $q(\Omega) = 1$ and $q(C) = 1/2$ for each countable subset $C \subset \Omega$. Thus the estimate in Proposition 5 is the best possible.

Let Y be the Banach space of all bounded functions $f : I \to \mathbb{C}$ with the sup-norm. Let $y \in Y$ be the constant function $y(i) = 1/2 \quad (i \in I)$. Then $\|\omega - y\| = 1/2$ for all $\omega \in \Omega$; so $q_Y(\Omega) = 1/2 = 1/2 q_X(\Omega)$. Thus the estimate in Proposition 3 is also the best possible.

Let M be a closed subspace of X. Denote by J_M the natural embedding $J_M : M \to X$ and by $Q_M : X \to X/M$ the canonical projection.

Definition 7. Let $T \in \mathcal{B}(X, Y)$. We consider the following quantities:

$$\begin{aligned}
&\|T\|_e = \inf\{\|T + K\| : K \in \mathcal{K}(X, Y)\};\\
&\|\tilde{T}\|;\\
&\|T\|_\mu = \inf\{\|TJ_M\| : M \subset X, \operatorname{codim} M < \infty\};\\
&\|T\|_q = \inf\{\|Q_N T\| : N \subset Y, \dim N < \infty\}
\end{aligned}$$

(where $\tilde{T}$ is the operator acting in $\ell^\infty/m(X)$ which was studied in Section 17).

It is easy to see that all these quantities are seminorms.

The first quantity $\|T\|_e = \operatorname{dist}\{T, \mathcal{K}(X, Y)\}$ is the essential norm of T. Clearly, $\|T\|_e = 0$ if and only if T is compact. Also, $\|\tilde{T}\| = 0 \Leftrightarrow T$ is compact $\Leftrightarrow \|T\|_\mu = 0$ by Lemma 17.3 and Theorem 15.5. It will be shown later that the latter three seminorms are equivalent. Thus, in particular, $\|T\|_q = 0 \Leftrightarrow T$ is compact.

From this reason the quantities defined in Definition 7 are usually called measures of non-compactness.

The first result gives the connection with the Hausdorff measure of non-compactness:

Proposition 8. *Let $T \in \mathcal{B}(X, Y)$. Then $\|T\|_q = q(TB_X)$.*

Proof. Let $\varepsilon > 0$ and let $F \subset Y$ be a finite set such that $\operatorname{dist}\{Tx, F\} \le q(TB_X) + \varepsilon \quad (x \in B_X)$. Let N be the subspace of Y spanned by F. Then

$$\|Q_N T\| = \sup_{\substack{x \in X\\ \|x\|=1}} \inf_{u \in N} \|Tx + u\| \le \sup_{\substack{x \in X\\ \|x\|=1}} \operatorname{dist}\{Tx, F\} \le q(TB_X) + \varepsilon.$$

Thus $\|T\|_q \le q(TB_X) + \varepsilon$, and letting $\varepsilon \to 0$ yields $\|T\|_q \le q(TB_X)$.

Conversely, let $\varepsilon > 0$ and let N be a finite-dimensional subspace of Y. Let F be a finite ε-net in the ball in N with radius $\|Q_N T\| + \varepsilon + \|T\|$. Let $x \in B_X$ and $y = Tx$. Then $\|Q_N y\| = \|Q_N Tx\| \le \|Q_N T\|$, and so there exists $u \in N$ with $\|y - u\| \le \|Q_N T\| + \varepsilon$. Clearly, $\|u\| \le \|y - u\| + \|y\| \le \|Q_N T\| + \varepsilon + \|T\|$, and so there exists $y' \in F$ with $\|y' - u\| \le \varepsilon$. Thus $\|y - y'\| \le \|y - u\| + \|u - y'\| \le \|Q_N T\| + 2\varepsilon$. Hence $q(TB_X) \le \|Q_N T\| + 2\varepsilon$. Since ε and N were arbitrary, we conclude that $q(TB_X) \le \|T\|_q$. □

Proposition 9. *Let $T \in \mathcal{B}(X,Y)$. Then*

$$\|\tilde{T}\| = \sup\{q(C) : C \subset TB_X, C \text{ countable}\}.$$

Consequently, $\|\tilde{T}\| \le \|T\|_q \le 2\|\tilde{T}\|$.

Proof. By definition, $\|\tilde{T}\| = \sup\{q(TC) : C \subset X, C \text{ countable}, q(C) < 1\} \ge \sup\{q(TC) : C \subset B_X, C \text{ countable}\}$.

On the other hand, if $C = \{x_1, x_2, \dots\}$ is a countable subset of X with $q(C) < 1$, then there are a sequence $C' = \{x_i'\} \subset B_X$ and a finite set $F \subset X$ such that $x_i - x_j' \in F$ for all i. Thus $q(TC) \subset q(TC' + TF) = q(TC')$ and we have the equality $\|\tilde{T}\| = \sup\{q(C) : C \subset TB_X, C \text{ countable}\}$.

The second statement follows from Proposition 5. □

Theorem 10. *Let $T \in \mathcal{B}(X,Y)$. Then:*

(i) $\|T^*\|_q = \|T\|_\mu$;

(ii) $\|T^*\|_\mu \le \|T\|_q \le 2\|T^*\|_\mu$.

Proof. We have

$$\begin{aligned}\|T\|_\mu &= \inf\{\|TJ_M\| : M \subset X, \operatorname{codim} M < \infty\}\\ &= \inf\{\|J_M^* T^*\| : M \subset X, \operatorname{codim} M < \infty\}\\ &= \inf\{\|Q_{M^\perp} T^*\| : M \subset X, \operatorname{codim} M < \infty\}\\ &\ge \inf\{\|Q_N T^*\| : N \subset X^*, \dim N < \infty\} = \|T^*\|_q.\end{aligned}$$

Similarly,

$$\begin{aligned}\|T\|_q &= \inf\{\|Q_N T\| : N \subset Y, \dim N < \infty\}\\ &= \inf\{\|T^* J_{N^\perp}\| : N \subset Y, \dim N < \infty\}\\ &\ge \inf\{\|T^* J_L\| : L \subset Y^*, \operatorname{codim} L < \infty\} = \|T^*\|_\mu.\end{aligned}$$

Further, $\|T\|_\mu \ge \|T^*\|_q \ge \|T^{**}\|_\mu$. If $M'' \subset X^{**}$ and $\operatorname{codim} M'' < \infty$, then $M'' \cap X$ is a subspace of finite codimension in X and

$$\begin{aligned}\|T^{**}\|_\mu &= \inf\{\|T^{**} J_{M''}\| : M'' \subset X^{**}, \operatorname{codim} M'' < \infty\}\\ &\ge \inf\{\|TJ_{M''\cap X}\| : M'' \subset X^{**}, \operatorname{codim} M'' < \infty\} \ge \|T\|_\mu.\end{aligned}$$

Thus $\|T^*\|_q = \|T\|_\mu$.

Finally,

$$\|T^*\|_\mu = \|T^{**}\|_q = q_{Y^{**}}(T^{**}B_{X^{**}}) \ge q_{Y^{**}}(TB_X) \ge \frac{1}{2} q_Y(TB_X) = \frac{1}{2}\|T\|_q. \qquad \square$$

Theorem 11. *Let $T \in \mathcal{B}(X,Y)$ and $S \in \mathcal{B}(Y,Z)$. Then $\|ST\|_q \le \|S\|_q \cdot \|T\|_q$ and $\|ST\|_\mu \le \|S\|_\mu \cdot \|T\|_\mu$.*

Proof. Let $\varepsilon > 0$ and let $F_Y \subset Y, F_Z \subset Z$ be finite sets satisfying $TB_X \subset F_Y + (\|T\|_q + \varepsilon)B_Y$, $SB_Y \subset F_Z + (\|S\|_q + \varepsilon)B_Z$. Then

$$STB_X \subset S\big(F_Y + (\|T\|_q + \varepsilon)B_Y\big) \subset SF_Z + (\|T\|_q + \varepsilon)\big(F_Z + (\|S\|_q + \varepsilon)B_Z\big).$$

Thus $\|ST\|_q \le (\|T\|_q + \varepsilon)(\|S\|_q + \varepsilon)$. Letting $\varepsilon \to 0$ gives $\|ST\|_q \le \|S\|_q \|T\|_q$.

Further,

$$\|ST\|_\mu = \|T^*S^*\|_q \le \|T^*\|_q \|S^*\|_q = \|S\|_\mu \|T\|_\mu. \qquad \square$$

Theorem 12. *Let $T \in \mathcal{B}(X, Y)$. Then $\|T\|_q \le 2\|T\|_\mu \le 4\|T\|_q$.*

Proof. $\|T\|_q \le 2\|T\|_\mu$: Let $\varepsilon > 0$ and let $M \subset X$ be a closed subspace of finite codimension such that $\|TJ_M\| \le \|T\|_\mu + \varepsilon$. Let P be a projection onto M.

Since $I - P$ is compact, there exists a finite set $F \subset B_X$ such that

$$\min\big\{\|(I - P)(x - x_0)\| : x_0 \in F\big\} \le \varepsilon \tag{1}$$

for every $x \in B_X$. Let $x \in B_X$ and find $x_0 \in F$ satisfying (1). We have

$$\|P(x - x_0)\| \le \|x - x_0\| + \|(I - P)(x - x_0)\| \le \|x - x_0\| + \varepsilon \le 2 + \varepsilon$$

and

$$\begin{aligned} \|T(x - x_0)\| &\le \|TP(x - x_0)\| + \|T(I - P)(x - x_0)\| \\ &\le (\|T\|_\mu + \varepsilon) \cdot \|P(x - x_0)\| + \|T\| \cdot \varepsilon \\ &\le \|T\| \cdot \varepsilon + (\|T\|_\mu + \varepsilon)(2 + \varepsilon). \end{aligned}$$

Thus $\operatorname{dist}\{TB_X, TF\} \le 2\|T\|_\mu + \varepsilon(\|T\| + \|T\|_\mu + 2 + \varepsilon)$. Letting $\varepsilon \to 0$ gives $\|T\|_q = q(TB_X) \le 2\|T\|_\mu$.

$\|T\|_\mu \le 2\|T\|_q$: By Theorem 10 and the preceding inequality, we have

$$\|T\|_\mu = \|T^*\|_q \le 2\|T^*\|_\mu \le 2\|T\|_q. \qquad \square$$

The essential versions of the injectivity modulus of an operator $T \in \mathcal{B}(X, Y)$ are:

$$\begin{aligned} & j_e(T) = \sup\{j(T + K) : K \in \mathcal{K}(X, Y)\}; \\ & j(\tilde{T}); \\ & j_\mu(T) = \sup\{j(TJ_M) : M \subset X, \operatorname{codim} M < \infty\}. \end{aligned}$$

The first two quantities do not depend on compact perturbations. The same is true for j_μ:

Lemma 13. *Let $T, K \in \mathcal{B}(X, Y)$ and let K be compact. Then $j_\mu(T + K) = j_\mu(T)$.*

Proof. Let $\varepsilon > 0$ and let $M \subset X$ be a closed subspace of finite codimension such that $j(TJ_M) > j_\mu(T) - \varepsilon$. By Theorem 15.5, there exists a closed subspace $M' \subset X$ of finite codimension such that $\|KJ_{M'}\| \leq \varepsilon$. Thus

$$j\big((T+K)J_{M\cap M'}\big) \geq j(TJ_{M\cap M'}) - \|KJ_{M\cap M'}\| \geq j_\mu(T) - 2\varepsilon.$$

Letting $\varepsilon \to 0$ yields $j_\mu(T+K) \geq j_\mu(T)$ and the opposite inequality follows by symmetry. □

Theorem 14. *Let $T \in \mathcal{B}(X,Y)$. Then*

$$j(T) \leq j_e(T) \leq j_\mu(T) \leq 2j(\tilde{T}) \leq 4j_\mu(T).$$

Proof. $j(T) \leq j_e(T)$: Clear.

$j_e(T) \leq j_\mu(T)$: For every $K \in \mathcal{K}(X,Y)$ we have $j(T+K) \leq j_\mu(T+K) = j_\mu(T)$, and so $j_e(T) \leq j_\mu(T)$.

$j_\mu(T) \leq 2j(\tilde{T})$: Let $M \subset X$ be a closed subspace of finite codimension. Then $\widetilde{J_M} : \tilde{M} \to \tilde{X}$ is bounded below and onto, and so, by Theorem 17.5, $j(TJ_M) \leq 2j(\tilde{T}\widetilde{J_M}) \leq 2j(\tilde{T})\|\widetilde{J_M}\| \leq 2j(\tilde{T})$.

$j(\tilde{T}) \leq 2j_\mu(T)$: Let $s > j_\mu(T)$. Then for every closed subspace $M \subset X$ with $\operatorname{codim} M < \infty$ there exists $x \in M$ with $\|x\| = 1$ and $\|Tx\| < s$. Choose $x_1 \in X$ with $\|x_1\| = 1$ and $\|Tx_1\| < s$. Let $x_1^* \in X^*$ be a functional satisfying $\|x_1^*\| = 1 = \langle x_1, x_1^*\rangle$ and set $M_1 = \operatorname{Ker} x_1^*$. Clearly, $\operatorname{codim} M_1 = 1 < \infty$, so there exists $x_2 \in M_1$ with $\|x_2\| = 1$ and $\|Tx_2\| < s$. Further, $\|x_2 - x_1\| \geq |\langle x_1 - x_2, x_1^*\rangle| = 1$. We can construct inductively a sequence $\{x_1, x_2, \dots\} \subset X$ such that $\|x_i\| = 1$, $\|Tx_i\| < s \quad (i \in \mathbb{N})$ and $\|x_i - x_j\| \geq 1 \quad (i \neq j)$. Let $\tilde{x} = (x_i) \in \tilde{X}$. It is easy to see that $q(\tilde{x}) \geq 1/2$ and $q(\tilde{T}\tilde{x}) = q((Tx_i)) \leq s$. Thus $j(\tilde{T}) \leq 2s$. Since $s > j_\mu(T)$ was arbitrary, we have $j(\tilde{T}) \leq 2j_\mu(T)$. □

Theorem 15. *Let $T \in \mathcal{B}(X,Y)$. Then:*

(i) $j_\mu(T) > 0 \Leftrightarrow j(\tilde{T}) > 0 \Leftrightarrow T$ *is upper semi-Fredholm;*
(ii) $j_e(T) > 0 \Leftrightarrow T$ *is upper semi-Fredholm and* $\operatorname{ind} T \leq 0$.

Proof. (i) The first equivalence follows from Theorem 14. The second one was proved in Theorem 17.9.

(ii) Clearly, $j_e(T) > 0$ if and only if T can be written as $T = S + K$ where S is bounded below and K is compact. By Theorem 19.6, this is equivalent to the condition that T is upper semi-Fredholm and $\operatorname{ind} T \leq 0$. □

Theorem 16. *The quantities j_e, $j(\tilde{\cdot})$ and j_μ are supermultiplicative, i.e., $j_e(ST) \geq j_e(S)j_e(T)$, $j(\widetilde{ST}) \geq j(\tilde{S})j(\tilde{T})$ and $j_\mu(ST) \geq j_\mu(S) \cdot j_\mu(T)$ for all $T \in \mathcal{B}(X,Y)$ and $S \in \mathcal{B}(Y,Z)$.*

Proof. The first two statements follow immediately from the supermultiplicativity of the injectivity modulus j.

To prove the third inequality, let $M \subset X$ and $L \subset Y$ be closed subspaces of finite codimension. Set $M' = M \cap T^{-1}L$. Then $\operatorname{codim} M' < \infty$ and

$$j(STJ_{M'}) \geq j(TJ_M) \cdot j(SJ_L).$$

Since M and L were arbitrary, we have $j_\mu(ST) \geq j_\mu(S) \cdot j_\mu(T)$. □

The essential versions of the surjectivity modulus of an operator $T \in \mathcal{B}(X,Y)$ are:

$$\begin{aligned}
&k_e(T) = \sup\{k(T+K) : T \in \mathcal{K}(X,Y)\};\\
&k(\tilde{T});\\
&k_q(T) = \sup\{k(Q_N T) : N \subset Y, \dim N < \infty\}.
\end{aligned}$$

It is easy to see that the first two quantities are supermultiplicative and do not depend on compact perturbations. The same is true for k_q.

Proposition 17.

(i) *If* $T \in \mathcal{B}(X,Y)$ *and* $K \in \mathcal{K}(X,Y)$, *then* $k_q(T+K) = k_q(T)$.

(ii) *If* $T \in \mathcal{B}(X,Y)$ *and* $S \in \mathcal{B}(Y,Z)$, *then* $k_q(ST) \geq k_q(S)k_q(T)$.

Proof. (i) Let $\varepsilon > 0$. Let $F \subset Y$ be a finite set such that $\operatorname{dist}\{Kx, F\} \leq \varepsilon$ for all $x \in B_X$. Let N be the subspace generated by F. Then $\|Q_n K\| \leq \varepsilon$ and

$$\begin{aligned}
k_q(T+K) &= \sup\{k(Q_{N'}(T+K)) : N' \subset N \subset Y, \dim N' < \infty\}\\
&\geq \sup\{k(Q_{N'}T) - \|Q_{N'}K\| : N' \subset N \subset Y, \dim N' < \infty\} \geq k_q(T) - \varepsilon.
\end{aligned}$$

Letting $\varepsilon \to 0$ gives $k_q(T+K) \leq k_q(T)$ and the second inequality follows by symmetry.

(ii) Let $N \subset Y$ and $N' \subset Z$ be finite-dimensional subspaces. Then

$$\begin{aligned}
k_q(ST) &= \sup\{k(Q_F ST) : F \subset Z, \dim F < \infty\}\\
&= \sup\{j(T^*S^*J_M) : M \subset N'^{\perp} \cap S^{*-1}(N^\perp), \operatorname{codim} M < \infty\}\\
&\geq j(T^*J_{N^\perp}S^*J_{N'^\perp}) \geq j(T^*J_{N^\perp})j(S^*J_{N'^\perp}) = k(Q_N T)k(Q_{N'}S).
\end{aligned}$$

Taking the supremum over all N and N', we obtain the required inequality. □

Theorem 18. *Let* $T \in \mathcal{B}(X,Y)$. *Then*

$$k(T) \leq k_e(T) \leq k_q(T) \leq k(\tilde{T}) \leq 2k_q(T).$$

Proof. $k(T) \leq k_e(T)$: Clear.

$k_e(T) \leq k_q(T)$: For every compact operator $K : X \to Y$ we have $k(T+K) \leq k_q(T+K) = k_q(T)$, and so $k_e(T) \leq k_q(T)$.

$k_q(T) \le k(\tilde{T})$: Let $N \subset Y$, $\dim N < \infty$. Then $\widetilde{Q_N} : \tilde{Y} \to \widetilde{Y/N}$ is an isometry onto $\widetilde{Y/N}$, by Lemma 17.5, and so $k(Q_N T) \le k(\widetilde{Q_N}\tilde{T}) = k(\tilde{T})$.

$k(\tilde{T}) \le 2k_q(T)$: Let $\varepsilon > 0$ and let s be a number satisfying $k_q(T) < s$. Let $N \subset Y$ be a finite-dimensional subspace. Then $k(Q_N T) < s$ or, equivalently, $j(T^* J_{N^\perp}) < s$.

Find $y_1^* \in Y^*$ such that $\|y_1^*\| = 1$ and $\|T^* y_1^*\| < s$. Choose $y_1 \in Y$ such that $\|y_1\| = 1$ and $\langle y_1, y_1^*\rangle > 1 - \varepsilon$. We construct inductively vectors $y_i^* \in \{y_1, \dots, y_{i-1}\}^\perp$ and $y_i \in Y$ such that $\|y_i^*\| = 1 = \|y_i\|$, $\|T^* y_i^*\| < s$ and $\langle y_i, y_i^*\rangle > 1 - \varepsilon$ for all $i \in \mathbb{N}$. Set $\tilde{y} = (y_i)$. Clearly, $q(\tilde{y}) \le 1$. Let $\tilde{x} = (x_i) \in \tilde{X}$ satisfy $\tilde{T}\tilde{x} = \tilde{y}$. Thus there exists a totally bounded sequence $(u_i) \in \ell^\infty(Y)$ such that $Tx_i - u_i = y_i \quad (i \in \mathbb{N})$. Passing to a subsequence if necessary, we can assume that the sequence (u_i) is convergent, and so $\|u_i - u_j\| < \varepsilon$ for all i, j sufficiently large. For $i > j$ we have

$$\|x_i - x_j\| \ge \frac{|\langle x_i - x_j, T^* y_i^*\rangle|}{\|T^* y_i^*\|} \ge s^{-1} \cdot \Big| \langle y_i - y_j, y_i^*\rangle + \langle u_i - u_j, y_i^*\rangle \Big| \ge \frac{1 - 2\varepsilon}{s}$$

and $q(\tilde{x}) \ge \frac{1-2\varepsilon}{2s}$. Consequently, $k(\tilde{T}) \le \frac{2s}{1-2\varepsilon}$. Letting $\varepsilon \to 0$ and $s \to k_q(T)$ yields $k(\tilde{T}) \le 2k_q(T)$. □

Theorem 19. *Let $T \in \mathcal{B}(X, Y)$. Then:*

(i) $k_q(T) > 0 \Leftrightarrow k(\tilde{T}) > 0 \Leftrightarrow T$ *is lower semi-Fredholm;*

(ii) $k_e(T) > 0 \Leftrightarrow T$ *is lower semi-Fredholm and* $\operatorname{ind} T \ge 0$.

Proof. (i) The first equivalence follows from the preceding theorem. The second one was proved in Theorem 17.6.

(ii) Clearly, $k_e(T) > 0$ if and only if T can be written as $T = S + K$ where S is onto and K is compact. By Theorem 19.6, this is equivalent to the condition that T is lower semi-Fredholm and $\operatorname{ind} T = \operatorname{ind} S \ge 0$. □

Theorem 20. *Let $T \in \mathcal{B}(X, Y)$. Then $k_q(T^*) = j_\mu(T)$ and $k_q(T) \le j_\mu(T^*) \le 16k_q(T)$.*

Proof. We have

$$\begin{aligned} j_\mu(T) &= \sup\{j(TJ_M) : M \subset X, \operatorname{codim} M < \infty\} \\ &= \sup\{k(Q_{M^\perp} T^*) : M \subset X, \operatorname{codim} M < \infty\} \\ &\le \sup\{k(Q_N T^*) : N \subset X^*, \dim N < \infty\} = k_q(T^*). \end{aligned}$$

Similarly,

$$\begin{aligned} k_q(T) &= \sup\{k(Q_N T) : N \subset Y, \dim N < \infty\} \\ &= \sup\{j(T^* J_{N^\perp}) : N \subset Y, \dim N < \infty\} \\ &\le \sup\{j(T^* J_{M'}) : M' \subset Y^*, \operatorname{codim} M' < \infty\} = j_\mu(T^*). \end{aligned}$$

Further, $j_\mu(T) \le k_q(T^*) \le j_\mu(T^{**})$. If M'' is a subspace of X^{**} of finite codimension, then $M'' \cap X$ is of finite codimension in X. Thus

$$\begin{aligned} j_\mu(T^{**}) &= \sup\{j(T^{**}J_{M''}) : M'' \subset X^{**}, \operatorname{codim} M'' < \infty\} \\ &\le \sup\{j(TJ_{M''\cap X}) : M'' \subset X^{**}, \operatorname{codim} M'' < \infty\} \\ &\le \sup\{j(TJ_M) : M \subset X, \operatorname{codim} M < \infty\} = j_\mu(T). \end{aligned}$$

Hence $j_\mu(T) = k_q(T^*)$.

The remaining inequality $j_\mu(T^*) \le 16k_q(T)$ is clear if T is not lower semi-Fredholm, see Theorems 15 and 19.

Suppose that T is lower semi-Fredholm, let $T_0 : X/\operatorname{Ker} T \to Y$ be the operator induced by T and let $Q : X \to X/\operatorname{Ker} T$ be the canonical projection. Then $\widetilde{T_0}$ is invertible, and so $j_\mu(T_0) \le 2j(\widetilde{T_0}) = 2k(\widetilde{T_0}) \le 4k_q(T_0)$. Similarly, $j_\mu(T_0^*) \le 4k(T_0^*)$. We have $T = T_0Q$ and $T^* = Q^*T_0^*$, where Q^* is the embedding of $(X/\operatorname{Ker} T)^* = (\operatorname{Ker} T)^\perp = \operatorname{Ran} T^*$ into X^*. It is easy to see that $j_\mu(T^*) = j_\mu(T_0^*)$ and, by Theorem 9.6,

$$\begin{aligned} k_q(T) &= \sup\{k(Q_NT_0Q) : N \subset Y, \dim N < \infty\} \\ &= \sup\{k(Q_NT_0) : N \subset Y, \dim N < \infty\} = k_q(T_0). \end{aligned}$$

Hence

$$j_\mu(T^*) = j_\mu(T_0^*) \le 4k_q(T_0^*) = 4j_\mu(T_0) \le 16k_q(T_0) = 16k_q(T). \qquad \square$$

Comments on Chapter III

C.15.1. The results in section 15 are classical. The definition and basic properties are due to Hilbert [Hi] and Riesz [Ri2]. Theorem 15.4 (T^* is compact $\Leftrightarrow$ T is compact) was proved by Schauder [Sd].

C.15.2. A Banach space X is reflexive if and only if B_X is compact in the weak topology.

If X is a reflexive Banach space, Y a Banach space and $T \in \mathcal{B}(X,Y)$, then TB_X is a convex weakly compact subset of Y and therefore it is closed in the norm topology.

Consequently, for reflexive Banach spaces it is not necessary to take the closure $\overline{TB_X}$ in the definition of compact operators.

C.15.3. For a long time it was an open question whether each compact operator $T \in \mathcal{B}(X)$ is a norm-limit of finite-rank operators. This is true for Hilbert spaces and for most of naturally defined Banach spaces (including all Banach spaces having a Schauder basis). The problem was solved by Enflo [En1] who constructed an example of a separable reflexive Banach space X where $\mathcal{F}(X) \ne \mathcal{K}(X)$.

C.15.4. Let X, Y be Banach spaces, let $T : X \to Y$ be a linear mapping. Denote by w the weak topology. The following statements are equivalent, cf. [Hal1, Problem 130]:

(i) $T : (X, \|\cdot\|) \to (Y, \|\cdot\|)$ is continuous;

(ii) $T : (X, w) \to (Y, w)$ is continuous;

(iii) $T : (X, \|\cdot\|) \to (Y, w)$ is continuous.

Further, $T : (X, w) \to (Y, \|\cdot\|)$ is continuous if and only if T is of finite rank.

C.15.5. If X is reflexive, then $T \in \mathcal{B}(X, Y)$ is compact if and only if T maps weakly converging sequences to convergent sequences, cf. Theorem 15.6 (this was the original definition of Hilbert of compact operators in ℓ^2; operators satisfying this condition are sometimes called *completely continuous*).

C.15.6. The ideal $\mathcal{F}(X)$ is the smallest ideal in $\mathcal{B}(X)$. Every left, right or two-sided ideal in $\mathcal{B}(X)$ contains the ideal of finite-rank operators.

C.15.7. If H is a separable Hilbert space, then $\mathcal{K}(X)$ is the only non-trivial closed two-sided ideal in $\mathcal{B}(X)$, see Calkin [Ca]. The same is true for the spaces $\ell^p \quad (1 \le p \le \infty)$ and c_0, see Gohberg, Markus, Fel'dman [GMF].

On the other hand, there are Banach spaces with a rich structure of closed two-sided ideals in $\mathcal{B}(X)$, see [CPY, p. 80].

C.15.8. In general, the Calkin algebra $\mathcal{B}(X)/\mathcal{K}(X)$ is not semisimple. An example is the Banach space $L^1(m)$ of all Lebesgue integrable functions f on $\langle 0, 1\rangle$ such that $\int_0^1 |f| \mathrm{d}m < \infty$, see [CPY, p. 33].

Operators T with the property that $T + \mathcal{K}(X) \in \operatorname{rad} \mathcal{B}(X)/\mathcal{K}(X)$ are called *inessential*. Clearly, the set $I(X)$ of all inessential operators is a closed two sided ideal.

The ideal $I(X)$ can be characterized by $T \in I(X) \Leftrightarrow T + \Phi(X) \subset \Phi(X)$.

C.15.9. An operator $T : X \to Y$ is called strictly singular if $T|M$ is not bounded below for each closed infinite-dimensional subspace $M \subset X$.

Denote by $S(X)$ the set of all strictly singular operators acting on X. Then $S(X)$ is a closed two-sided ideals in $\mathcal{B}(X)$ and $\mathcal{K}(X) \subset S(X) \subset I(X)$.

If X is a Hilbert space, then $\mathcal{K}(X) = S(X) = I(X)$. In Banach spaces the inclusions can be strict, see [CPY, p.101].

C.15.10. There is an example of a Banach space X with the property that each operator $T \in \mathcal{B}(X)$ can be written as $\lambda I + K$ for some $\lambda \in \mathbb{C}$ and a strictly singular operator K, see [GwM].

It is an open problem whether there is a Banach space X with the property that each operator on X can be expressed as a sum of a scalar multiple of the identity and a compact operator.

C.15.11. A particular case of Theorem 15.11 is called the Fredholm alternative: if $T \in \mathcal{B}(X)$ is compact and $\lambda \neq 0$, then either $T - \lambda$ is both one-to-one and onto, or it is neither one-to-one nor onto (in the notation of Section 15, $\alpha(T-\lambda) = 0 \Leftrightarrow \beta(T-\lambda) = 0$).

C.15.12. An operator $T \in \mathcal{B}(X,Y)$ is called *weakly compact* if the weak closure of TB_X is compact in the weak topology of Y. By [DS, VI-4], an operator $T \in \mathcal{B}(X,Y)$ is weakly compact if and only if $T^{**}X^{**} \subset Y$ (where we identify Y with a subspace of Y^{**}). If either X or Y is reflexive, then each operator in $\mathcal{B}(X,Y)$ is weakly compact. Furthermore, $T \in \mathcal{B}(X,Y)$ is weakly compact if and only if T^* is weakly compact.

For any Banach space X, the weakly compact operators form a closed two-sided ideal in $\mathcal{B}(X)$.

C.16.1. The basic ideas concerning Fredholm and semi-Fredholm operators appeared in [At], [Goh1], [Goh2] and [Yo1]. For a survey of results see [GhK1], [Kat2], and [CPY].

Further results including detailed historical comments can be found in [RN].

C.16.2. Many results from the Fredholm theory can be extended to unbounded closed operators. For a survey of results in this direction see [GhK1], [Kat2] and [Sch2].

C.16.3. A generalization of the Fredholm theory to the Banach algebras setting was done by Barnes [Ba1], [Ba2]. For a survey of results see [BMSW].

C.16.4. The characterizations of semi-Fredholm operators given in Theorems 16.18 and 16.19 are due to Lebow and Schechter [LS].

Theorem 16.21 (Kato decomposition) was proved in [Kat1].

C.17.1. The construction of Section 17 was given in [Sa] and independently in [BHW], [HW].

A similar construction of the space $\ell^\infty(X)/c_0(X)$ where $c_0(X)$ denotes the set of all sequences of elements of X converging to zero was used by Berberien and Quigly (we used this construction for Banach algebras, see C.3.1).

The latter construction can be used for reducing the approximate point spectrum to the point spectrum.

For properties of these constructions see also [FL].

C.17.2. If X is a Hilbert space, then it is possible to modify both constructions of the previous comment in order to obtain again a Hilbert space.

Fix a Banach limit and denote it by LIM. Then $|||(x_n)||| = \left(LIM\|x_n\|^2\right)^{1/2}$ is a seminorm on $\ell^\infty(X)$. Set $N = \{(x_n) : |||(x_n)||| = 0\}$. Clearly, $N \supset c_0(X)$ and the completion of $(\ell^\infty(X)/N, |||\cdot|||)$ is a Hilbert space (since it satisfies the parallelogram law) containing X as constant sequences.

As for the construction of Section 17, set $|||(x_n)|||' = \lim_M |||(P_M x_n)|||$, where $M \subset X$ is a subspace of finite codimension and P_M is the orthogonal projection onto M. Clearly, $N' := \{(x_n) : |||(x_n)|||' = 0\} \supset m(X)$ and the completion of $(\ell^\infty(X)/N', |||\cdot|||')$ is a Hilbert space.

C.17.3. Let X be the ℓ^1 space over the set $\{e_i, f_i : i = 1, 2, \dots\}$. Let $M = \bigvee\{e_i : i \in \mathbb{N}\}$ and $L = \bigvee\{e_i + \frac{1}{i} f_i : i \in \mathbb{N}\}$. Then $\ell^\infty(M) + m(X) = \ell^\infty(N) + m(X)$ and $M \cap L = \{0\}$. So M is not essentially equal to L.

Thus the condition $M \subset L$ in Lemma 17.2 is necessary.

C.17.4. It is possible that $T \in \mathcal{B}(X)$ has not closed range but $\operatorname{Ran} \tilde{T}$ is closed: consider any compact operator with non-closed range.

We do not know whether each operator T with $\operatorname{Ran} \tilde{T}$ closed can be written as $T = S + K$ where $\operatorname{Ran} S$ is closed and K is compact.

C.18.1. The punctured neighbourhood theorem 18.7 was proved by Gohberg [Goh2].

C.18.2. Using the linearization technique (cf. C.11.2) it is possible to prove the following generalization of the punctured neighbourhood theorem [Mü21]:

Theorem. Let X, Y, Z be Banach spaces, let U be an open subset of $\mathbb{C}$ and $w \in U$. Suppose that $S : U \to \mathcal{B}(X, Y)$, $T : U \to \mathcal{B}(Y, Z)$ are analytic functions satisfying $T(z)S(z) = 0 \quad (z \in U)$. Write $\alpha(z) = \dim \operatorname{Ker} T(z) / \operatorname{Ran} S(z)$. Suppose that $\alpha(w) < \infty$ and $\operatorname{Ran} T(w)$ is closed. Then there exist $\varepsilon > 0$ and a constant $c \le \alpha(w)$ such that $\alpha(z) = c$ for all z, $0 < |z - w| < \varepsilon$.

Clearly, the classical punctured neighbourhood theorem follows easily from this generalization for the sequences $0 \longrightarrow X \xrightarrow{T-z} Y$ and $X \xrightarrow{T-z} Y \longrightarrow 0$, respectively.

C.18.3. Theorem 18.8 was proved in [FK] and [Ze6]. For further results concerning the stability of semi-Fredholm operators see [SW], [RZ].

C.19.1. It is clear that $\|T^*\|_e \le \|T\|_e$ for each operator $T \in \mathcal{B}(X)$. For reflexive Banach spaces the equality $\|T^*\|_e = \|T\|_e$ holds. This is no longer true for non-reflexive Banach spaces. By [Ty2], these two quantities are even not equivalent.

C.19.2. The Calkin algebras $\mathcal{B}(X)/\mathcal{K}(X)$ were first studied by Calkin [Ca] in the most important case of a separable Hilbert space. The Calkin algebras over Banach spaces were first studied by Yood [Yo2].

C.19.3. An operator $T \in \mathcal{B}(X)$ is called *Riesz* if $\sigma_e(T) = \{0\}$. Riesz operators are a generalization of compact operators and exhibit many of their properties.

C.19.4. Let T be a Riesz operator on a Hilbert space H. Then there exists a compact operator $K \in \mathcal{K}(H)$ and a quasinilpotent operator $Q \in \mathcal{B}(X)$ such that $T = Q + K$ (so-called West decomposition [Wes1]). The same statement is true for the spaces ℓ^p and L^p; it is not known for general Banach spaces.

In general, it is not possible to require that Q and K in the West decomposition commute.

C.19.5. The West decomposition is a particular case of the following more general result [St], see also [Ap2]:

Theorem. Let T be an operator on a Hilbert space H. Then there exists a compact operator $K \in \mathcal{K}(H)$ such that

$$\sigma(T+K) = \bigcap_{K' \in \mathcal{K}(H)} \sigma(T+K') \qquad (= \sigma_W(T), \text{ see Section 23}).$$

C.19.6. By Theorem 16.13, an operator $T \in \mathcal{B}(X)$ is Fredholm if and only if it is invertible modulo the ideal of compact operators. A characterization of one-sided inverses in the Calkin algebra was given in Theorems 16.14 and 16.15.

If X is a Hilbert space, then $T \in \Phi_+(X)$ $(T \in \Phi_-(X))$ if and only if the class $T + \mathcal{K}(X)$ is not a left (right) topological divisor of zero in the Calkin algebra $\mathcal{B}(X)/\mathcal{K}(X)$. In general, for Banach space operators there is no relation between these notions, see [AT] and [Ty1].

C.19.7. Let $T \in \Phi_+^-(X) = \{T \in \Phi_+(X) : \operatorname{ind} T \le 0\}$. Then there is a finite-rank operator F such that $T+F$ is bounded below. In general, it is not possible to choose F commuting with T, cf. Theorem 20.21. It is always possible to find F such that $(TF - FT)^2 = 0$, see [LW], [Se].

Similar statements are true for $T \in \Phi_-^+(X)$ and for the intersection $\Phi_+^-(X) \cap \Phi_-^+(X) = \{T \in \Phi(X) : \operatorname{ind} T = 0\}$.

C.20.1. The ascent and descent were introduced and studied first in [Tay], [La] and [Kat1].

Heuser [He] proved the following relations between the ascent, descent and the defect numbers $\alpha(T) = \dim \operatorname{Ker} T$, $\beta(T) = \operatorname{codim} \operatorname{Ran} T$ of an operator $T \in \mathcal{B}(X)$:

Let $\min\{\alpha(T), \beta(T)\} < \infty$. Then $a(T) < \infty \Rightarrow \beta(T) \ge \alpha(T)$ and $d(T) < \infty \Rightarrow \beta(T) \le \alpha(T)$. Moreover, if $\alpha(T) = \beta(T) < \infty$, then $a(T) = d(T)$.

C.20.2. Semi-Browder operators were studied by a number of authors, see [Gr1], [KV], [Ra2], [Ra5], [Ra6], [Wes2]. The name was introduced in [Ha8].

The Browder operators are sometimes also called Riesz-Schauder, cf. [CPY].

C.20.3. The extension of the Browder and semi-Browder spectra to commuting n-tuples presented in Section 20 appeared in [KMR].

For a single operator $A \in \mathcal{B}(X)$ we have $\sigma_B(A) = \sigma_e(A) \cup \operatorname{acc} \sigma(A)$ where $\operatorname{acc} \sigma(A)$ denotes the set of all accumulation points of $\sigma(A)$. This equality was used by Curto and Dash [CD] for other extension of the Browder spectrum to commuting n-tuples:

$$\sigma_b(A_1, \dots, A_n) = \sigma_{Te}(A_1, \dots, A_n) \cup \operatorname{acc} \sigma_T(A_1, \dots, A_n),$$

where σ_T and σ_{Te} denote the Taylor and essential Taylor spectrum, see Chapter IV.

Then σ_b defined in this way is also a spectral system, which in general differs from σ_B defined in Section 20, see [KMR].

Thus the extension of the Browder spectrum for single operators to a spectral system is not unique.

C.20.4. Lemma 20.3 is due to Neubauer, see [Lab], Proposition 2.1.1. In fact, a more general formulation is also true: if M_1, M_2 are paraclosed subspaces of a Banach space X (see C.10.4) and both $M_1 \cap M_2$, $M_1 + M_2$ are closed, then M_1 and M_2 are closed.

C.21.1. Essentially Kato operators or similar classes of operators were studied by a number of authors, see, e.g., [Ka1], [GlK], [Gr2], [Ra4], [Mü15], [Ko2], [KM3], [BO].

C.21.2. Theorem 21.12 is due to Kordula [Ko2]. Theorem 21.15 was proved by Livčak [Liv], implicitly it is also contained in papers of M.A. Gol'dman and S.N. Kračkovskiĭ.

C.21.3. An operator $T \in \mathcal{B}(X)$ is called *essentially Saphar* if $\operatorname{Ker} T \overset{e}{\subset} R^\infty(T)$ and T has a generalized inverse. Clearly, the essentially Saphar operators form a regularity since they are the intersection of the classes of essentially Kato and Saphar operators. Consequently, the corresponding spectrum satisfies the spectral mapping property.

C.22.1. The numbers $k_n(T)$ were introduced and studied by Grabiner [Gr2].

Most of the results in Section 22 are taken from [MM].

C.22.2. Let $R_1, \dots, R_5$ be the regularities introduced in Section 22. The following properties of them and the corresponding spectra were studied in [MM] (to avoid trivialities we consider infinite-dimensional Banach spaces X only):

(A) $\sigma_i(T) \neq \emptyset$ for every $T \in \mathcal{B}(X)$;

(B) $\sigma_i(T)$ is closed for every $T \in \mathcal{B}(X)$;

(C) if $T \in R_i$, then there exists $\varepsilon > 0$ such that $T + U \in R_i$ whenever $TU = UT$ and $\|U\| < \varepsilon$ (this means property (P3), the upper semicontinuity of σ_i on commuting elements);

(D) if $T \in R_i$ and $F \in \mathcal{B}(X)$ is a finite-rank operator, then $T + F \in R_i$;

(E) if $T \in R_i$ and K is a compact operator commuting with T, then $T + K \in R_i$;

(F) if $T \in R_i$ and $Q \in \mathcal{B}(X)$ is a quasinilpotent operator commuting with T, then $T + Q \in R_i$.

These properties for R_i $(i = 1, \dots, 5)$ are summarized in Table 1. For details we refer to [MM].

	(A) $\sigma_i \neq \emptyset$	(B) σ_i closed	(C) small comm. perturb.	(D) finite-rank perturb.	(E) commut. comp. perturb	(F) commut. quasinilp. perturb.
R_1 onto	yes	yes	yes	no	no	yes
R_2 $\phi_-(X)$ and $d(T) < \infty$	yes	yes	yes	no	yes	yes
R_3 $\phi_-(X)$	yes	yes	yes	yes	yes	yes
R_4 $d(T) < \infty$	no	yes	no	no	no	no
R_5 $d_e(T) < \infty$	no	yes	no	yes	no	no

Table 1.

Since the properties (A)–(F) considered above are preserved by taking adjoints, the regularities $R_6 \dots R_{10}$ satisfy exactly those properties as $R_1, \dots, R_5$. So Table 1 remains valid for $R_1, \dots, R_5$ replaced by $R_6, \dots, R_{10}$.

C.22.3. Drazin invertible operators (class $R_4 \cap R_9$) were studied, e.g., in [RoS] or [Kol]. B-Fredholm operators (class $R_5 \cap R_{10}$) were introduced and studied in [Be1] and [Be2].

C.22.4. The properties (A)–(F) of C.22.2 for regularities $R_{11}, \dots, R_{15}$ are summarized in Table 2. For details see [MM] and [KMMP].

C.22.5. Quasi-Fredholm operators (class R_{14}) in Hilbert spaces were introduced and studied by Labrousse [Lab]. Equivalently, an operator T on a Hilbert space H is quasi-Fredholm if and only if there is a Kato decomposition $H = H_1 \oplus H_2$ with $TH_i \subset H_i$, $T|H_1$ nilpotent and $T|H_2$ Kato.

The same decomposition exists also for quasi-Fredholm operators on Banach spaces under the additional assumption that the subspaces $\operatorname{Ker} T \cap \operatorname{Ran} T^d$ and $\operatorname{Ker} T^d + \operatorname{Ran} T$ are complemented, see Remark after Theorem 3.2.2 in [Lab]. In fact, the proof of Theorem 22.12, (iii) $\Rightarrow$ (v) is a simplified version of the proof of Labrousse; without any change it works also for quasi-Fredholm operators.

C.23.1. The notions of upper and lower semiregularity were introduced and basic properties proved in [Mü21].

C.23.2. Joint spectra satisfying the one-way spectral mapping property (see Remark 23.9) were studied in [MW].

	(A) $\sigma_i \neq \emptyset$	(B) σ_i closed	(C) small comm. perturb.	(D) finite-rank perturb.	(E) commut. comp. perturb	(F) commut. quasinilp. perturb.
R_{11} Kato	yes	yes	yes	no	no	yes
R_{12} ess. Kato	yes	yes	yes	yes	yes	yes
R_{13}	yes	no	no	yes	no	no
R_{14} $q\phi$	no	yes	no	yes	no	no
R_{15}	no	no	no	yes	no	no

Table 2.

C.23.3. The semi-Fredholm spectrum

$$\sigma_{sF}(T) = \{\lambda \in \mathbb{C} : T - \lambda \text{ is not semi-Fredholm}\}$$

was studied by Kato [Kat2], Oberai [Ob2], Gramsch and Lay [GL], and others. It is sometimes called the Kato essential spectrum; we used this name for something else.

C.23.4. The exponential spectrum was introduced by Harte [Ha4].

For the Weyl/Schechter spectrum

$$\sigma_W(T) = \{\lambda : T - \lambda \text{ is not Fredholm or } \operatorname{ind}(T-\lambda) \neq 0\} = \bigcap_{K \text{ compact}} \sigma(T+K)$$

see [Ob1], [Sch1]).

The spectrum corresponding to the regularity $R = \{a+b : a \in \operatorname{Inv}(\mathcal{A}), b \in J\}$, where J is a closed two-sided ideal in a Banach algebra $\mathcal{A}$ was studied in [Ha7] under the name of *T-Weyl spectrum.*

The spectra

$$\sigma_{\Phi_+^-}(T) = \bigcap \{\sigma_\pi(T+K) : K \in \mathcal{K}(X)\}$$

and

$$\sigma_{\Phi_-^+}(T) = \bigcap \{\sigma_\delta(T+K) : K \in \mathcal{K}(X)\}$$

were studied in [Ra2], [Ra3] and [Ze7] under the names of *essential approximate point spectrum* and *essential defect spectrum.*

C.24.1. The operational quantities defined in Section 24 were considered by a number of authors. The measures of non-compactness $\|T\|_\mu$ and $\|T\|_q$ were studied in [GGM], [GlM] and [LS]. For remaining operational quantities and related topics see [Sch4], [Ra1], [Fa4], [GnM1], [GnM2], [Mb3].

C.24.2. Further operational quantities closely related to those studied in Section 24 are (see [Pi2]): the *approximation numbers*

$$a_n(T) = \inf\{\|T - L\| : L \in \mathcal{B}(X,Y), \dim \operatorname{Ran} L < n\};$$

Gelfand numbers

$$c_n(T) = \inf\{\|TJ_M\| : \operatorname{codim} M < n\};$$

Kolmogorov numbers

$$k_n(T) = \inf\{\|Q_F T\| : \dim F < n\};$$

Berstein numbers

$$u_n(T) = \sup\{j(TJ_M) : \dim M \geq n\};$$

Mytiagin numbers

$$v_n(T) = \sup\{k(Q_M T) : \operatorname{codim} M \geq n\}.$$

All these numbers have lower analogues, see [RZ] and [RZi].

C.24.3. The essential version of the reduced minimal modulus

$$\gamma_e(T) = \sup\{\gamma(T+K) : K \text{ compact}\}$$

for operators on Hilbert spaces was studied in [MP].

It was shown that the supremum is attained for some compact operator K and $\gamma_e(T) = \inf \sigma_e(T) \setminus \{0\}$, cf. C.10.5.

Chapter IV

Taylor Spectrum

In this chapter we introduce and study another important spectral system for commuting operators – the Taylor spectrum. Although the definition of the Taylor spectrum is rather complicated, the Taylor spectrum has a distinguished property among other spectral systems, namely the existence of the functional calculus for functions analytic on a neighbourhood of the Taylor spectrum. From this reason many experts consider the Taylor spectrum to be the proper generalization of the ordinary spectrum for single operators.

25 Basic properties

Let $s = (s_1, \dots, s_n)$ be a system of indeterminates. Denote by $\Lambda[s]$ the exterior algebra generated by $s_1, \dots, s_n$, i.e., $\Lambda[s]$ is the free complex algebra generated by $s_1, \dots, s_n$, where the multiplication operation $\wedge$ in $\Lambda[s]$ satisfies the anticommutative relations $s_i \wedge s_j = -s_j \wedge s_i \quad (i, j = 1, \dots, n)$.

In particular, $s_i \wedge s_i = 0$ for all i.

For $F \subset \{1, \dots, n\}$, $F = \{i_1, \dots, i_p\}$ with $1 \le i_1 < i_2 < \cdots < i_p \le n$ write $s_F = s_{i_1} \wedge \cdots \wedge s_{i_p}$. Every element of $\Lambda[s]$ can be written uniquely in the form

$$\sum_{F \subset \{1, \dots, n\}} \alpha_F s_F$$

with complex coefficients α_F. Clearly, $s_\emptyset$ is the unit in $\Lambda[s]$.

For $p = 0, 1, \dots, n$ let $\Lambda^p[s]$ be the set of all elements of $\Lambda[s]$ of degree p, i.e., $\Lambda^p[s, X]$ is the subspace generated by the elements s_F with $\operatorname{card} F = p$. Thus $\Lambda[s] = \bigoplus_{p=0}^n \Lambda^p[s]$, $\dim \Lambda^p[s] = \binom{n}{p}$ and $\dim \Lambda[s] = 2^n$.

Let X be a vector space. Write $\Lambda[s, X] = X \otimes \Lambda[s]$. So

$$\Lambda[s, X] = \left\{ \sum_{F \subset \{1, \dots, n\}} x_F s_F : x_F \in X \right\};$$

to simplify the notation, we omit the symbol "$\otimes$". Similarly, for $p = 0, \dots, n$ write $\Lambda^p[s, X] = X \otimes \Lambda^p$; so

$$\Lambda^p[s, X] = \Big\{ \sum_{\substack{F \subset \{1,\dots,n\} \\ \operatorname{card} F = p}} x_F s_F : x_F \in X \Big\}.$$

Thus $\Lambda^p[s, X]$ is a direct sum of $\binom{n}{p}$ copies of X and $\Lambda[s, X]$ is a direct sum of 2^n copies of X.

In the following X will be a fixed Banach space. Then $\Lambda[s, X]$ can be considered to be also a Banach space. For the following considerations it is not essential which norm we take on $\Lambda[s, X]$; we can assume it to be $\|\sum x_F s_F\| = \left(\sum \|x_F\|^2\right)^{1/2}$.

For $j = 1, \dots, n$ let $S_j : \Lambda[s, X] \to \Lambda[s, X]$ be the operators of left multiplication by s_j,

$$S_j\Big(\sum_F x_F s_F\Big) = \sum_F x_F s_j \wedge s_F = \sum_{\substack{F \subset \{1,\dots,n\} \\ j \notin F}} (-1)^{\operatorname{card}\{i \in F : i < j\}} x_F s_{F \cup \{j\}}. \tag{1}$$

Clearly, $S_j S_i = -S_i S_j \quad (i, j = 1, \dots, n)$. In particular, $S_i^2 = 0$ for all i.

For an operator $T \in \mathcal{B}(X)$ we denote by the same symbol the operator $T : \Lambda[s, X] \to \Lambda[s, X]$ defined by

$$T\Big(\sum_F x_F s_F\Big) = \sum_F (T x_F) s_F.$$

Obviously, $TSj = S_j T$ for all j.

Let $A = (A_1, \dots, A_n)$ be an n-tuple of mutually commuting operators on X. Denote by δ_A the operator $\delta_A : \Lambda[s, X] \to \Lambda[s, X]$ defined by

$$\delta_A = \sum_{i=1}^n A_i S_i.$$

Clearly,

$$(\delta_A)^2 = \sum_{i=1}^n \sum_{j=1}^n A_i S_i A_j S_j = \sum_{1 \le i < j \le n} A_i A_j (S_i S_j + S_j S_i) = 0,$$

and so $\operatorname{Ran} \delta_A \subset \operatorname{Ker} \delta_A$ (note that we have used the commutativity of the operators A_i).

Definition 1. An n-tuple $A = (A_1, \dots, A_n)$ of mutually commuting operators on a Banach space X is called *Taylor regular* if $\operatorname{Ker} \delta_A = \operatorname{Ran} \delta_A$.

The *Taylor spectrum* $\sigma_T(A)$ is the set of all $\lambda \in \mathbb{C}^n$ such that the n-tuple $(A_1 - \lambda_1, \dots, A_n - \lambda_n)$ is not Taylor regular.

Since $\delta_A \Lambda^p[s,X] \subset \Lambda^{p+1}[s,X] \quad (p = 0,1,\dots,n-1)$, we can define operators $\delta_A^p : \Lambda^p[s,X] \to \Lambda^{p+1}[s,X]$ as the restrictions of δ_A to $\Lambda^p[s,X]$. Thus δ_A defines the following sequence of operators

$$0 \to \Lambda^0[s,X] \xrightarrow{\delta_A^0} \Lambda^1[s,X] \xrightarrow{\delta_A^1} \cdots \xrightarrow{\delta_A^{n-1}} \Lambda^n[s,X] \to 0, \tag{2}$$

where $\delta^{p+1}\delta^p = 0$ for each p. A sequence of this type is called a complex; we are going to study complexes in more details in one of the subsequent sections.

Complex (2) is called the *Koszul complex* of A. It is easy to see that A is Taylor regular if and only if the Koszul complex is exact, i.e., if $\operatorname{Ran} \delta_A^i = \operatorname{Ker} \delta_A^{i+1}$ for all i, where we set formally δ_A^p to be the zero operators for $p < 0$ or $p \geq n$.

Remark 2. (i) Let $n = 1$. We can identify $\Lambda^0[s,X]$ and $\Lambda^1[s,X]$ with X, and so the Koszul complex of a single operator $A_1 \in \mathcal{B}(X)$ becomes

$$0 \to X \xrightarrow{A_1} X \to 0.$$

This complex is exact if and only if A_1 is invertible. Thus for single operators the Taylor spectrum coincides with the ordinary spectrum.

(ii) Let $n = 2$ and let $A = (A_1, A_2)$ be a commuting pair of operators on X. Then the Koszul complex of A becomes

$$0 \to X \xrightarrow{\delta_A^0} X \oplus X \xrightarrow{\delta_A^1} X \to 0,$$

where δ_A^0 and δ_A^1 are defined by $\delta_A^0 x = A_1 x \oplus A_2 x \quad (x \in X)$ and $\delta_A^1(x \oplus y) = -A_2 x + A_1 y \quad (x, y \in X)$.

(iii) The most important parts of the Koszul complex of an n-tuple $A = (A_1, \dots, A_n)$ are its ends. The first mapping δ_A^0 can be interpreted as $\delta_A^0 : X \to X^n$ defined by $\delta_A^0 x = \bigoplus_{i=1}^n A_i x \quad (x \in X)$, cf. Section 9. Thus the Koszul complex of A is exact at $\Lambda^0[s,X]$ if and only if $0 \notin \sigma_\pi(A)$. Similarly, $\delta_A^{n-1} : X^n \to X$ is defined by $\delta_A^{n-1}(x_1 \oplus \cdots \oplus x_n) = \sum_{i=1}^n (-1)^{i-1} A_i x_i$, and so the exactness at $\Lambda^n[s,X]$ means that $0 \notin \sigma_\delta(A)$.

The main result of this section will be that the Taylor regular n-tuples form a joint regularity, and so the Taylor spectrum is a spectral system.

Proposition 3. *Let $A_1, \dots, A_n, B_1, \dots, B_n$ be mutually commuting operators on a Banach space X satisfying $\sum_{i=1}^n A_i B_i = I$. Then the n-tuple $A = (A_1, \dots, A_n)$ is Taylor regular.*

Consequently, $\sigma_T(A) \subset \sigma^{\langle A \rangle}(A)$ for all commuting n-tuples A.

Proof. For $j = 1, \dots, n$ let $H_j : \Lambda[s,X] \to \Lambda[s,X]$ be the operators defined by

$$H_j\Big(\sum_{F \subset \{1,\dots,n\}} x_F s_F\Big) = \sum_{\substack{F \subset \{1,\dots,n\} \\ j \in F}} (-1)^{\operatorname{card}\{i \in F : i < j\}} x_F s_{F \setminus \{j\}}. \tag{3}$$

We have

$$(H_jS_j + S_jH_j)x_F s_F = \begin{cases} (-1)^{\operatorname{card}\{i\in F: i<j\}} S_j x_F s_{F\setminus\{j\}} = x_F s_F & (j \in F), \\ (-1)^{\operatorname{card}\{i\in F: i<j\}} H_j x_F s_{F\cup\{j\}} = x_F s_F & (j \notin F), \end{cases}$$

and so $H_jS_j + S_jH_j = I \quad (j = 1, \dots, n)$.

Further, for $i \neq j$, we have $H_iS_j + S_jH_i = 0$. Indeed, if either $j \in F$ or $i \notin F$, then $(H_iS_j + S_jH_i)xs_F = 0$. Suppose that $j \notin F$ and $i \in F$. Then

$$xs_F = (-1)^{\operatorname{card}\{k\in F: k<i\}} xs_i \wedge s_{F\setminus\{i\}}$$

and

$$\begin{aligned}(H_iS_j + S_jH_i)xs_i \wedge s_{F\setminus\{i\}} &= H_i xs_j \wedge s_i \wedge s_{F\setminus\{i\}} + S_j xs_{F\setminus\{i\}} \\ &= -xs_j \wedge s_{F\setminus\{i\}} + xs_j \wedge s_{F\setminus\{i\}} = 0.\end{aligned}$$

Suppose that $\sum_{i=1}^n A_iB_i = I$. As above, denote by the same symbols B_i the operators acting on $\Lambda[s, X]$. Let $\varepsilon_B : \Lambda[s, X] \to \Lambda[s, X]$ be the operator defined by $\varepsilon_B = \sum_{j=1}^n H_jB_j$. Then

$$\begin{aligned}\varepsilon_B\delta_A + \delta_A\varepsilon_B &= \sum_{i=1}^n\sum_{j=1}^n (H_jB_jS_iA_i + S_iA_iH_jB_j) \\ &= \sum_{i=1}^n (H_iS_iB_iA_i + S_iH_iA_iB_i) + \sum_{i\neq j}(H_jS_i + S_iH_j)B_jA_i \\ &= \sum_{i=1}^n B_iA_i = I_{\Lambda[s,X]}.\end{aligned}$$

Let $\psi \in \operatorname{Ker}\delta_A$. Then $\psi = (\varepsilon_B\delta_A + \delta_A\varepsilon_B)\psi = \delta_A\varepsilon_B\psi$, and so $\psi \in \operatorname{Ran}\delta_A$. Hence $\operatorname{Ker}\delta_A = \operatorname{Ran}\delta_A$ and the n-tuple A is Taylor regular. □

Theorem 4. *The set of all commuting Taylor regular n-tuples is relatively open in the set of all commuting n-tuples.*

Consequently, $\sigma_T(A)$ is a closed subset of $\mathbb{C}^n$.

Proof. Consider the sequence

$$\Lambda[s, X] \xrightarrow{\delta_A} \Lambda[s, X] \xrightarrow{\delta_A} \Lambda[s, X]$$

and apply Lemma 11.3. □

Lemma 5. *Let Z_1, Z_2 be Banach spaces, let $B : Z_1 \to Z_1$, $D : Z_1 \to Z_2$ and $C : Z_2 \to Z_2$ be operators satisfying $DB = CD$, see the following diagram:*

$$\begin{array}{ccc} Z_1 & \xrightarrow{D} & Z_2 \\ \downarrow B & & \downarrow C \\ Z_1 & \xrightarrow{D} & Z_2 \end{array}$$

Suppose that $DZ_1 \neq Z_2$. Then there exists a complex number λ such that $DZ_1 + (C-\lambda)Z_2 \neq Z_2$.

Proof. We reduce the statement of Lemma 5 to the projection property of the surjective spectrum. Consider the Banach space $Z = Z_2 \oplus Z_1 \oplus Z_1 \oplus \cdots$ (for example with the ℓ^1 norm) and operators $U, V \in \mathcal{B}(Z)$ given in the matrix form by

$$U = \begin{pmatrix} 0 & D & 0 & 0 & \cdots \\ 0 & 0 & I & 0 & \\ 0 & 0 & 0 & I & \\ \vdots & & & & \ddots \end{pmatrix} \quad \text{and} \quad V = \begin{pmatrix} C & 0 & 0 & \cdots \\ 0 & B & 0 & \\ 0 & 0 & B & \\ \vdots & & & \ddots \end{pmatrix}.$$

It is easy to check that $UV = VU$. Furthermore, $UZ \neq Z$ since $DZ_1 \neq Z_2$. By the projection property for the surjective spectrum, there exists $\lambda \in \mathbb{C}$ such that $UZ + (V-\lambda)Z \neq Z$. Since $UZ \supset 0 \oplus Z_1 \oplus Z_1 \oplus \cdots$, this is equivalent to the condition $DZ_1 + (C-\lambda)Z_2 \neq Z_2$. □

To prove the projection property for the Taylor spectrum we are going to investigate the exactness of the Koszul complex in more details.

For $k = 0, \dots, n$ denote by $\Gamma_k^{(n)}$ the set of all commuting n-tuples of operators $A = (A_1, \dots, A_n) \in \mathcal{B}(X)^n$ such that the Koszul complex of A is exact at $\Lambda^k(X)$, i.e., $\operatorname{Ran} \delta_A^{k-1} = \operatorname{Ker} \delta_A^k$. In agreement with our convention we set formally $\Gamma_{-1}^{(n)}$ to be the set of all commuting n-tuples of operators.

Proposition 6. *Let $A_0, A_1, \dots, A_n$ be commuting operators on a Banach space X, let $0 \le k \le n$ and $(A_1, \dots, A_n) \notin \Gamma_k^{(n)}$. Then there exists $\lambda \in \mathbb{C}$ such that $(A_0 - \lambda, A_1, \dots, A_n) \notin \Gamma_{k+1}^{(n+1)}$.*

Proof. Write for short $A = (A_1, \dots, A_n)$, $A_\lambda = (A_0 - \lambda, A_1, \dots, A_n)$ $(\lambda \in \mathbb{C})$ and $s = (s_1, \dots, s_n)$. Suppose that $A \notin \Gamma_k^{(n)}$. We show that there is a $\lambda \in \mathbb{C}$ such that $(A_0 - \lambda, A) \notin \Gamma_{k+1}^{(n+1)}$. Clearly, $A_0 \operatorname{Ker} \delta_A^k \subset \operatorname{Ker} \delta_A^k$. Consider the following diagram

$$\begin{array}{ccc} \Lambda^{k-1}[s,X] & \xrightarrow{\delta_A^{k-1}} & \operatorname{Ker} \delta_A^k \\ \big\downarrow A_0 & & \big\downarrow A_0 \\ \Lambda^{k-1}[s,X] & \xrightarrow{\delta_A^{k-1}} & \operatorname{Ker} \delta_A^k. \end{array}$$

By Lemma 5, there is a $\lambda \in \mathbb{C}$ such that $\operatorname{Ran} \delta_A^{k-1} + (A_0 - \lambda) \operatorname{Ker} \delta_A^k \neq \operatorname{Ker} \delta_A^k$. We prove that $\operatorname{Ran} \delta_{A_\lambda}^k \neq \operatorname{Ran} \operatorname{Ker}_{A_\lambda}^{k+1}$.

Let $\psi \in \operatorname{Ker} \delta_A^k \setminus \left(\operatorname{Ran} \delta_A^{k-1} + (A_0 - \lambda) \operatorname{Ker} \delta_A^k\right)$. Then

$$\delta_{A_\lambda}^{k+1} S_0 \psi = \Big((A_0 - \lambda) S_0 + \sum_{i=1}^n A_i S_i\Big) S_0 \psi = \sum_{i=1}^n A_i S_i S_0 \psi = -S_0 \delta_A^k \psi = 0.$$

We show that $S_0 \psi \notin \operatorname{Ran} \delta_{A_\lambda}^k$.

Suppose on the contrary that there is a $\varphi \in \Lambda^k[s_0, s, X]$ with $\delta^k_{A_\lambda}\varphi = S_0\psi$. Write $\varphi = S_0\varphi_{k-1}+\varphi_k$ for $\varphi_{k-1} \in \Lambda^{k-1}[s, X]$, $\varphi_k \in \Lambda^k[s, X]$. Then $S_0\psi = \delta^k_{A_\lambda}\varphi = S_0(-\delta^{k-1}_A\varphi_{k-1} + (A_0 - \lambda)\varphi_k) + \delta^k_A\varphi_k$. Thus $\varphi_k \in \operatorname{Ker}\delta^k_A$ and $\psi = -\delta_A\varphi_{k-1} + (A_0 - \lambda)\varphi_k \in \operatorname{Ran}\delta^{k-1}_A + (A_0 - \lambda)\operatorname{Ker}\delta^k_A$, which is a contradiction. □

Proposition 7. *Let $A_1, \dots, A_n, A_{n+1}$ be commuting operators on a Banach space X, let $0 \le k \le n$. Suppose that $(A_1, \dots, A_n) \in \Gamma^{(n)}_{k-1} \cap \Gamma^{(n)}_k$. Then*

$$(A_1, \dots, A_n, A_{n+1}) \in \Gamma^{(n+1)}_k.$$

Proof. Write $A = (A_1, \dots, A_n)$, $s = (s_1, \dots, s_n)$, and $A' = (A_1, \dots, A_n, A_{n+1})$. Suppose that $\operatorname{Ker}\delta^{k-1}_A = \operatorname{Ran}\delta^{k-2}_A$ and $\operatorname{Ker}\delta^k_A = \operatorname{Ran}\delta^{k-1}_A$.

We prove that $\operatorname{Ker}\delta^k_{A'} \subset \operatorname{Ran}\delta^{k-1}_{A'}$ (the opposite inclusion is always true).

Let $\psi \in \operatorname{Ker}\delta^k_{A'}$. Express $\psi = \eta_k + S_{n+1}\eta_{k-1}$ for some $\eta_k \in \Lambda^k[s, X]$ and $\eta_{k-1} \in \Lambda^{k-1}[s, X]$. Then

$$0 = \delta^k_{A'}\psi = \sum_{i=1}^{n+1} A_iS_i\eta_k + \sum_{i=1}^{n} A_iS_iS_{n+1}\eta_{k-1} = \delta^k_A\eta_k + S_{n+1}(A_{n+1}\eta_k - \delta^{k-1}_A\eta_{k-1}).$$

Thus $\delta^k_A\eta_k = 0$, and so $\eta_k = \delta^{k-1}_A\xi_{k-1}$ for some $\xi_{k-1} \in \Lambda^{k-1}[s_1, \dots, s_n, X]$. Further,

$$0 = A_{n+1}\eta_k - \delta^{k-1}_A\eta_{k-1} = \delta^{k-1}_A(A_{n+1}\xi_{k-1} - \eta_{k-1}),$$

and so $A_{n+1}\xi_{k-1} - \eta_{k-1} = \delta^{k-2}_A\xi_{k-2}$ for some $\xi_{k-2} \in \Lambda^{k-2}[s, X]$. Hence

$$\begin{aligned}\psi = \eta_k + S_{n+1}\eta_{k-1} &= \delta^{k-1}_A\xi_{k-1} + S_{n+1}A_{n+1}\xi_{k-1} - S_{n+1}\delta^{k-2}_A\xi_{k-2}\\ &= \delta^{k-1}_{A'}\xi_{k-1} + \delta^{k-1}_{A'}S_{n+1}\xi_{k-2} \in \operatorname{Ran}\delta^{k-1}_{A'}.\end{aligned}$$ □

Corollary 8. *The Taylor spectrum is an upper semicontinuous spectral system.*

Proof. By Propositions 6 and 7, the Taylor regular tuples of operators form a joint regularity. Thus the Taylor spectrum is a spectral system. The upper semicontinuity follows from Theorem 4. □

In fact, we have proved more. For a fixed $k \ge 0$ and a commuting n-tuple $A = (A_1, \dots, A_n)$ of operators in X define

$$\sigma_{\delta,k}(A) = \left\{\lambda \notin \mathbb{C}^n : A - \lambda \notin \bigcap_{j=0}^{k} \Gamma^{(n)}_{n-j}\right\}$$

where $A - \lambda = (A_1 - \lambda_1, \dots, A_n - \lambda_n)$. Clearly, the condition in the definition of $\sigma_{\delta,k}$ means the non-exactness of the Koszul complex at some of the last k positions. By Remark 2 (iii), we have $\sigma_\delta(A) = \sigma_{\delta,0}(A) \subset \sigma_{\delta,1}(A) \subset \cdots \subset \sigma_{\delta,n}(A) = \sigma_T(A)$.

Dually we define

$$\sigma_{\pi,k}(A) = \Big\{\lambda \notin \mathbb{C}^n : A - \lambda \notin \bigcap_{j=0}^{k} \Gamma_j^{(n)} \text{ or } \operatorname{Ran} \delta_A^k \text{ is not closed}\Big\}.$$

Evidently, $\sigma_\pi(A) = \sigma_{\pi,0}(A) \subset \sigma_{\pi,1}(A) \subset \cdots \subset \sigma_{\pi,n}(A) = \sigma_T(A)$.

Theorem 9. *$\sigma_{\delta,k}$ and $\sigma_{\pi,k}$ are upper semicontinuous spectral systems for each $k \geq 0$.*

Proof. By Propositions 6 and 7, $\sigma_{\delta,k}$ is a spectral system. The upper semicontinuity follows from Lemma 11.3. The statements for $\sigma_{\pi,k}$ follow from the following duality result. □

Theorem 10. *Let $A = (A_1, \dots, A_n)$ be a commuting n-tuple of operators on a Banach space X and let $0 \leq k \leq n$. Then:*

(i) $\sigma_{\delta,k}(A_1^*, \dots, A_n^*) = \sigma_{\pi,k}(A_1, \dots, A_n)$;

(ii) $\sigma_{\pi,k}(A_1^*, \dots, A_n^*) = \sigma_{\delta,k}(A_1, \dots, A_n)$.

Proof. We may identify $\Lambda[s, X^*]$ with the dual of $\Lambda[s, X]$; the duality is given by the formula

$$\langle x s_F, x^* s_G\rangle = \begin{cases} (-1)^{c(F)}\langle x, x^*\rangle & \text{if } G = \{1, \dots, n\} \setminus F, \\ 0 & \text{otherwise,} \end{cases}$$

where $c(F) = \sum_{i \in F}(i-1)$. For $1 \leq j \leq n$ denote by $S_{j,X}$ (S_{j,X^*}) the left multiplication by s_j in $\Lambda[s, X]$ and in $\Lambda[s, X^*]$, respectively, see (1). Let $x \in X$, $x^* \in X^*$ and $F, G \subset \{1, \dots, n\}$. If $j \in F$ or $G \neq \{1, \dots, j-1, j+1, \dots, n\} \setminus F$, then

$$\langle S_{j,X} x s_F, x^* s_G\rangle = 0 = \langle x s_F, S_{j,X^*} x^* s_G\rangle.$$

Suppose that $j \notin F$ and $G = \{1, \dots, j-1, j+1, \dots, n\} \setminus F$. Then

$$\begin{aligned} \langle S_{j,X} x s_F, x^* s_G\rangle &= (-1)^{\operatorname{card}\{i \in F : i < j\}}\langle x s_{F \cup \{j\}}, x^* s_G\rangle \\ &= (-1)^{\operatorname{card}\{i \in F : i < j\}}(-1)^{c(F)+j-1}\langle x, x^*\rangle = (-1)^{\operatorname{card}\{i \in G : i < j\}}(-1)^{c(F)}\langle x, x^*\rangle \\ &= (-1)^{\operatorname{card}\{i \in G : i < j\}}\langle x s_F, x^* s_{G \cup \{j\}}\rangle = \langle x s_F, S_{j,X^*} x^* s_G\rangle. \end{aligned}$$

Thus $(S_{j,X})^* = S_{j,X^*}$ and $(\delta_A)^* = \delta_{A^*}$. More precisely, we have $\big(\Lambda^p[s, X]\big)^* = \Lambda^{n-p}[s, X^*]$ and $(\delta_A^p)^* = \delta_{A^*}^{n-p}$.

Let $0 \leq k \leq n$. By Lemma 11.2, the following statements are equivalent:

- A is $\sigma_{\pi,k}$ regular;
- the complex

$$0 \to \Lambda^0[s, X] \xrightarrow{\delta_A^0} \Lambda^1[s, X] \xrightarrow{\delta_A^1} \cdots \xrightarrow{\delta_A^{n-1}} \Lambda^n[s, X] \to 0$$

  is exact at $\Lambda^j[s, X]$ $(j \leq k)$ and $\operatorname{Ran} \delta_A^k$ is closed;

– the complex

$$0 \leftarrow \Lambda^n[s,X^*] \overset{\delta^{n-1}_{A^*}}{\longleftarrow} \Lambda^{n-1}[s,X^*] \overset{\delta^{n-2}_{A^*}}{\longleftarrow} \cdots \overset{\delta^0_{A^*}}{\longleftarrow} \Lambda^0[s,X^*] \leftarrow 0$$

is exact at $\Lambda^j[s,X^*]$ $(j \geq n-k)$;

– A^* is $\sigma_{\delta,k}$-regular.

This proves (i).

The second statement can be proved similarly. □

Proposition 11. Let $A = (A_1, \ldots, A_n)$ be a commuting n-tuple of operators acting on a Banach space X. Let $j \in \{1, \ldots, n\}$. Then $A_j \operatorname{Ker} \delta_A \subset \operatorname{Ran} \delta_A$.

Proof. Let $\psi \in \operatorname{Ker} \delta_A$. Write $\psi = s_j \wedge \psi_1 + \psi_2$, where ψ_2 does not contain s_j. We have

$$0 = \delta_A \psi = s_j \wedge A_j \psi_2 + \sum_{i \neq j} s_i \wedge s_j \wedge A_i \psi_1 + \sum_{i \neq j} s_i \wedge A_i \psi_2.$$

In particular, $A_j \psi_2 - \sum_{i \neq j} s_i \wedge A_i \psi_1 = 0$. Thus

$$\delta_A \psi_1 = s_j \wedge A_j \psi_1 + \sum_{i \neq j} s_i \wedge A_i \psi_1 = s_j \wedge A_j \psi_1 + A_j \psi_2 = A_j \psi.$$ □

Remark 12. The precise name of complex (2) is the *cochain Koszul complex* of A. It is possible to assign to a commuting n-tuple $A = (A_1, \ldots, A_n) \in \mathcal{B}(X)^n$ also another "dual" complex (called the *chain Koszul complex* of A). As in the proof of Proposition 3, for $j = 1, \ldots, n$ define operators $H_j : \Lambda[s,X] \to \Lambda[s,X]$ by (3) and set $\varepsilon_A = \sum_{j=1}^n A_j H_j : \Lambda[s,X] \to \Lambda[s,X]$. Equivalently, for $1 \leq i_1 < i_2 < \cdots < i_p \leq n$ we have

$$\varepsilon_A x s_{i_1} \wedge \cdots \wedge s_{i_p} = \sum_{k=1}^{p} (-1)^{k-1} A_{i_k} x s_{i_1} \wedge \cdots \wedge \widehat{s_{i_k}} \wedge \cdots \wedge s_{i_p},$$

where the hat denotes the omitted term.

It is easy to verify that $H_j B = B H_j$ for all $B \in \mathcal{B}(X)$ and $H_i H_j = -H_j H_i$ $(1 \leq i,j \leq n)$. Thus $(\varepsilon_A)^2 = 0$. Clearly, $\varepsilon_A \Lambda^p[s,X] \subset \Lambda^{p-1}[s,X]$ for all p, and so ε_A defines a complex

$$0 \leftarrow \Lambda^0[s,X] \overset{\varepsilon^0_A}{\longleftarrow} \Lambda^1[s,X] \overset{\varepsilon^1_A}{\longleftarrow} \cdots \overset{\varepsilon^{n-1}_A}{\longleftarrow} \Lambda^n[s,X] \leftarrow 0, \tag{4}$$

where ε^p_A is the restriction of ε_A to $\Lambda^{p+1}[s,X]$ $(p = 0, \ldots, n)$. Complex (4) is called the chain Koszul complex of A.

The chain complex can also be used for the definition of the Taylor spectrum of A (in fact this was the original definition of Taylor). Fortunately, these two definitions coincide since the chain Koszul complex of A is exact if and only if the cochain Koszul complex is exact.

To see this, denote by $J : \Lambda[s,X] \to \Lambda[s,X]$ the operator defined by

$$Jxs_F = (-1)^{c(F)} xs_{\bar F} \qquad (x \in X, F \subset \{1,\dots,n\}),$$

where $c(F) = \sum_{i\in F}(i-1)$ and $\bar F = \{1,\dots,n\} \setminus F$ denotes the complement of F. It is easy to see that J is an invertible operator.

Let $x \in X$ and $F \subset \{1,\dots,n\}$. If $j \notin F$, then $JH_j xs_F = 0 = S_j Jxs_F$; if $j \in F$, then

$$\begin{aligned} S_jJxs_F &= (-1)^{c(F)} xs_j \wedge s_{\bar F} = (-1)^{c(F)}(-1)^{\operatorname{card}\{k\in\bar F: k<j\}} xs_{\bar F\cup\{j\}} \\ &= (-1)^{c(F\setminus\{j\})}(-1)^{\operatorname{card}\{k\in F:k<j\}} xs_{\bar F\cup\{j\}} = JH_jxs_F. \end{aligned}$$

Thus $S_jJ = JH_j$ $(j = 1,\dots,n)$ and $\delta_A J = J\varepsilon_A$. Consequently, $\operatorname{Ran}\delta_A = \operatorname{Ran}(\delta_A J) = \operatorname{Ran}(J\varepsilon_A) = J\operatorname{Ran}\varepsilon_A$, $\operatorname{Ker}\varepsilon_A = \operatorname{Ker}(\delta_A J) = J^{-1}\operatorname{Ker}\delta_A$ and $\operatorname{Ker}\delta_A = J\operatorname{Ker}\varepsilon_A$. Thus $\operatorname{Ker}\delta_A = \operatorname{Ran}\delta_A$ if and only if $\operatorname{Ker}\varepsilon_A = \operatorname{Ran}\varepsilon_A$ and the exactness of both Koszul complexes is equivalent.

Note also that for $0 \le p \le n$ the exactness of one of the Koszul complexes at $\Lambda^p[s,X]$ is equivalent to the exactness of the other Koszul complex at $\Lambda^{n-p}[s,X]$.

26 Split spectrum

In this section we study a variant of the Taylor spectrum. The relation between the split spectrum and the Taylor spectrum is analogous to the relation between the left (right) spectrum and the approximate point (surjective) spectrum.

Definition 1. Let $A = (A_1,\dots,A_n)$ be an n-tuple of commuting operators on a Banach space X. We say that A is *split regular* if it is Taylor regular and the mapping $\delta_A : \Lambda[s,X] \to \Lambda[s,X]$ has a generalized inverse.

The *split spectrum* $\sigma_S(A)$ is the set of all $\lambda = (\lambda_1,\dots,\lambda_n) \in \mathbb{C}$ such that the n-tuple $(A_1-\lambda_1,\dots,A_n-\lambda_n)$ is not split regular.

The following result characterizes the split regular n-tuples of operators.

Proposition 2. *Let $A = (A_1,\dots,A_n)$ be an n-tuple of mutually commuting operators on a Banach space X. The following conditions are equivalent:*

(i) *A is split regular;*

(ii) *A is Taylor regular and $\operatorname{Ker}\delta_A^p$ is a complemented subspace of $\Lambda^p[s,X]$ for each $p = 0,\dots,n-1$;*

(iii) *there exist operators $W_1, W_2 : \Lambda[s,X] \to \Lambda[s,X]$ such that $W_1\delta_A + \delta_A W_2 = I_{\Lambda[s,X]}$;*

(iv) *there exists an operator $V : \Lambda[s,X] \to \Lambda[s,X]$ such that $V\delta_A + \delta_A V = I$, $V^2 = 0$ and $V\Lambda^p[s,X] \subset \Lambda^{p-1}[s,X]$ $(p = 0,\dots,n)$. Equivalently, there are operators $V_p : \Lambda^{p+1}[s,X] \to \Lambda^p[s,X]$ (see the diagram below) such that*

$V_{p-1}V_p = 0$ and $V_p\delta_A^p + \delta_A^{p-1}V_{p-1} = I_{\Lambda^p[s,X]}$ for every p (for $p = 0$ and $p = n$ this reduces to $V_0\delta_A^0 = I_{\Lambda^0[s,X]}$ and $\delta_A^{n-1}V_{n-1} = I_{\Lambda^n[s,X]}$, respectively).

$$0 \to \Lambda^0[s,X] \underset{V_0}{\overset{\delta_A^0}{\rightleftarrows}} \Lambda^1[s,X] \underset{V_1}{\overset{\delta_A^1}{\rightleftarrows}} \cdots \underset{V_{n-1}}{\overset{\delta_A^{n-1}}{\rightleftarrows}} \Lambda^n[s,X] \to 0$$

Proof. (iv) $\Rightarrow$ (iii) is clear.

(iii) $\Rightarrow$ (i): If $W_1\delta_A + \delta_A W_2 = I$, then $\delta_A W_1 \delta_A = \delta_A$, so δ_A has generalized inverse and $\operatorname{Ker}\delta_A$ is complemented. Furthermore, if $x \in \operatorname{Ker}\delta_A$, then $x = \delta_A W_2 x \in \operatorname{Ran}\delta_A$, and so $\operatorname{Ker}\delta_A = \operatorname{Ran}\delta_A$. Hence A is Taylor regular.

(i) $\Rightarrow$ (ii): Denote by $J_p : \Lambda^p[s,X] \to \Lambda[s,X]$ the natural embedding, let $Q_p : \Lambda[s,X] \to \Lambda^p[s,X]$ be the natural projection and let $P : \Lambda[s,X] \to \operatorname{Ker}\delta_A$ be a bounded projection onto $\operatorname{Ker}\delta_A$.

Clearly, $Q_p(\operatorname{Ker}\delta_A) = \operatorname{Ker}\delta_A^p$. Then Q_pPJ_p is a bounded projection from $\Lambda^p[s,X]$ onto $\operatorname{Ker}\delta_A^p$.

(ii) $\Rightarrow$ (iv): Let M_p be a closed subspace of $\Lambda^p[s,X]$ such that $\operatorname{Ker}\delta_A^p \oplus M_p = \Lambda^p[s,X]$. The operator $\delta_A^p|M_p : M_p \to \operatorname{Ran}\delta_A^p = \operatorname{Ker}\delta_A^{p+1}$ is a bijection. In the decompositions $\Lambda^p[s,X] = \operatorname{Ker}\delta_A^p \oplus M_p$, $\Lambda^{p+1}[s,X] = \operatorname{Ker}\delta_A^{p+1} \oplus M_{p+1}$ we have

$$\delta_A^p = \begin{array}{c} \\ \operatorname{Ran}\delta_A^p \\ M_{p+1} \end{array} \begin{array}{c} \begin{array}{cc} \operatorname{Ker}\delta_A^p & M_p \end{array} \\ \begin{pmatrix} 0 & \delta_A^p|M_p \\ 0 & 0 \end{pmatrix} \end{array}.$$

Set

$$V_p = \begin{array}{c} \\ \operatorname{Ker}\delta_A^p \\ M_p \end{array} \begin{array}{c} \begin{array}{cc} \operatorname{Ran}\delta_A^p & M_{p+1} \end{array} \\ \begin{pmatrix} 0 & 0 \\ (\delta_A^p|M_p)^{-1} & 0 \end{pmatrix} \end{array}.$$

Then $V_{p-1}V_p = 0$ since $\operatorname{Ran}V_p \subset M_p \subset \operatorname{Ker}V_{p-1}$. For $x \in M_p$ we have

$$(V_p\delta_A^p + \delta_A^{p-1}V_{p-1})x = V_p\delta_A^p x = x.$$

For $x \in \operatorname{Ker}\delta_A^p$ we have

$$(V_p\delta_A^p + \delta_A^{p-1}V_{p-1})x = \delta_A^{p-1}V_{p-1}x = x.$$

Thus $V_p\delta_A^p + \delta_A^{p-1}V_{p-1} = I_{\Lambda^p[s,X]}$ for each p (for $p = 0$ and $p = n$ we set formally $V_{-1} = 0$ and $V_n = 0$). □

Remark 3. For single operators on a Banach space the split spectrum coincides with the Taylor spectrum (and with the ordinary spectrum).

By Proposition 2 (ii), the split spectrum coincides with the Taylor spectrum also for n-tuples of commuting operators on a Hilbert space. For general Banach spaces the split spectrum differs from the Taylor spectrum, see C.26.1.

Theorem 4. *Let $A = (A_1, \dots, A_n)$ be an n-tuple of commuting operators on a Banach space X. Then $\sigma_S(A) \subset \sigma^{\langle A\rangle}(A)$.*

Proof. The result was essentially proved already in Proposition 25.3.

It is sufficient to show that if $(0, \dots, 0) \notin \sigma^{\langle A\rangle}(A)$, then A is split regular.

Suppose that there are operators $B_1, \dots, B_n$ in the algebra generated by A satisfying $\sum_{i=1}^n A_i B_i = I$. The operator $\varepsilon_B : \Lambda[s, X] \to \Lambda[s, X]$, see the proof of Proposition 25.3, satisfies $\varepsilon_B \delta_A + \delta_A \varepsilon_B = I$. By the previous theorem, this implies that A is split regular. $\square$

More generally, it is possible to define analogues of the partial Taylor spectra $\sigma_{\pi,k}$ and $\sigma_{\delta,k}$.

Definition 5. Let $A = (A_1, \dots, A_n)$ be an n-tuple of commuting operators on a Banach space X, let $0 \le k \le n$. We say that A is $\sigma_{l,k}$-regular if $\operatorname{Ker} \delta_A^j = \operatorname{Ran} \delta_A^{j-1}$ $(j = 0, \dots, k)$ and the operators $\delta_A^0, \delta_A^1, \dots, \delta_A^k$ have generalized inverses.

Dually, A is called $\sigma_{r,k}$-regular if $\operatorname{Ker} \delta_A^j = \operatorname{Ran} \delta_A^{j-1}$ $(j = n-k, \dots, n)$ and the operators $\delta_A^{n-k-1}, \delta_A^{n-k}, \dots, \delta_A^{n-1}$ have generalized inverses.

Write

$$\sigma_{l,k}(A) = \{\lambda \in \mathbb{C}^n : (A_1 - \lambda_1, \dots, A_n - \lambda_n) \text{ is not } \sigma_{l,k} \text{ regular}\}$$

and

$$\sigma_{r,k}(A) = \{\lambda \in \mathbb{C}^n : (A_1 - \lambda_1, \dots, A_n - \lambda_n) \text{ is not } \sigma_{r,k} \text{ regular}\}.$$

Then $\sigma_{l,k}(A) \supset \sigma_{\pi,k}(A)$ and $\sigma_{r,k}(A) \supset \sigma_{\delta,k}(A)$, and so all the sets $\sigma_{l,k}(A)$ and $\sigma_{r,k}$ are non-empty. Furthermore,

$$\sigma_l(A) = \sigma_{l,0}(A) \subset \sigma_{l,1}(A) \subset \cdots \subset \sigma_{l,n}(A) = \sigma_S(A)$$

and

$$\sigma_r(A) = \sigma_{r,0}(A) \subset \sigma_{r,1}(A) \subset \cdots \subset \sigma_{r,n}(A) = \sigma_S(A).$$

To characterize the spectra $\sigma_{l,k}$ and $\sigma_{r,k}$, we need the following modification of Proposition 2:

Lemma 6. *Let X, Y, Z be Banach spaces, and let $A_1 : X \to Y$, $A_2 : Y \to Z$ be operators satisfying $A_2 A_1 = 0$. The following statements are equivalent:*

(i) *A_1 and A_2 have generalized inverses and $\operatorname{Ran} A_1 = \operatorname{Ker} A_2$;*

(ii) *there exist operators $V_1 : Y \to X$ and $V_2 : Z \to Y$ such that $A_1 V_1 + V_2 A_2 = I_Y$;*

(iii) *there exist operators $V_1 : Y \to X$ and $V_2 : Z \to Y$ satisfying $V_1 V_2 = 0$ and $A_1 V_1 + V_2 A_2 = I_Y$.*

Proof. (iii) $\Rightarrow$ (ii): Clear.

(ii) $\Rightarrow$ (i): If we multiply the relation $A_1V_1 + V_2A_2 = I_Y$ by A_2 from the left-hand side (by A_1 from the right-hand side), then we get $A_2V_2A_2 = A_2$ and $A_1V_1A_1 = A_1$. Thus A_1 and A_2 have generalized inverses.

Furthermore, if $y \in \operatorname{Ker} A_2$, then $y = A_1V_1y \in \operatorname{Ran} A_1$, and so $\operatorname{Ran} A_1 = \operatorname{Ker} A_2$.

(i) $\Rightarrow$ (iii): As in the proof of Proposition 2. □

The next result shows the relations between the partial Taylor and split spectra of an n-tuple A and left (right) multiplication operators L_A and R_A. It is a generalization of Theorem 9.26.

Theorem 7. *Let $A = (A_1, \dots, A_n)$ be an n-tuple of commuting operators on a Banach space X. Then, for all k, $0 \le k \le n$:*

(i) $\sigma_{r,k}(A) = \sigma_{\delta,k}(L_A) = \sigma_{r,k}(L_A)$;

(ii) $\sigma_{l,k}(A) = \sigma_{\delta,k}(R_A) = \sigma_{r,k}(R_A)$.

In particular, $\sigma_S(A) = \sigma_T(L_A) = \sigma_S(L_A) = \sigma_T(R_A) = \sigma_S(R_A)$.

Proof. (i) For $\psi = \sum_{F\subset\{1,\dots,n\}} T_F s_F \in \Lambda[s, \mathcal{B}(X)]$ and $x \in X$ it is natural to define $\psi x \in \Lambda[s, X]$ by $\psi x = \sum_{F\subset\{1,\dots,n\}} (T_F x) s_F$. With this notation we have $(\delta_{L_A}\psi)x = \delta_A(\psi x)$.

(i.1) Suppose that A is $\sigma_{r,k}$-regular and let $n - k \le p \le n$. By Lemma 6, there are operators $V_{p-1} : \Lambda^p[s, X] \to \Lambda^{p-1}[s, X]$ and $V_p : \Lambda^{p+1}[s, X] \to \Lambda^p[s, X]$ satisfying $\delta_A^{p-1}V_{p-1} + V_p\delta_A^p = I_{\Lambda^p[s,X]}$, see the diagram below.

$$\Lambda^{p-1}[s, X] \underset{V_{p-1}}{\overset{\delta_A^{p-1}}{\rightleftarrows}} \Lambda^p[s, X] \underset{V_p}{\overset{\delta_A^p}{\rightleftarrows}} \Lambda^{p+1}[s, X]. \tag{1}$$

For $q = p - 1, p$ express V_q in the matrix form as

$$V_q\Big(\sum_{\operatorname{card} G=q+1} x_G s_G\Big) = \sum_{\operatorname{card} G=q+1}\ \sum_{\operatorname{card} F=q} V_{F,G} x_G s_F, \tag{2}$$

where $V_{G,F} \in \mathcal{B}(X)$ are uniquely determined operators.

Define operators $W_q : \Lambda^{q+1}[s, \mathcal{B}(X)] \to \Lambda^q[s, \mathcal{B}(X)]$ as the multiplication by the matrices defining V_q. More precisely,

$$W_q\Big(\sum_{\operatorname{card} G=q+1} T_G s_G\Big) = \sum_{\operatorname{card} G=q+1}\ \sum_{\operatorname{card} F=q} V_{F,G} T_G s_F$$

for $q = p-1, p, T_G \in \mathcal{B}(X)$. So $(W_q\psi)x = V_q(\psi x)$ for all $x \in X$. It is easy to verify that $\delta_{L_A}^{p-1}W_{p-1} + W_p\delta_{L_A}^p = I_{\Lambda^p[s,\mathcal{B}(X)]}$.

Consequently, if A is $\sigma_{r,k}$-regular, then L_A is $\sigma_{r,k}$-regular. So

$$\sigma_{r,k}(A) \supset \sigma_{r,k}(L_A) \supset \sigma_{\delta,k}(L_A).$$

(i.2) We prove by induction on k that if L_A is $\sigma_{\delta,k}$-regular, then A is $\sigma_{r,k}$-regular. For $k=0$ this was proved in Theorem 9.26.

Let $k \geq 1$ and suppose that L_A is $\sigma_{\delta,k}$-regular. By the induction hypothesis, A is $\sigma_{r,k-1}$-regular. In particular, δ_A^{n-k} has generalized inverse, and so there exists an operator $V_{n-k} : \Lambda^{n-k+1}[s,X] \to \Lambda^{n-k}[s,X]$ such that $\delta_A^{n-k} V_{n-k} \delta_A^{n-k} = \delta_A^{n-k}$. It is sufficient to show that there is an operator $V_{n-k-1} : \Lambda^{n-k}[s,X] \to \Lambda^{n-k-1}[s,X]$ (see the diagram below) such that $\delta_A^{n-k-1} V_{n-k-1} + V_{n-k}\delta_A^{n-k} = I_{\Lambda^{n-k}[s,X]}$.

$$\Lambda^{n-k-1}[s,X] \underset{V_{n-k-1}}{\overset{\delta_A^{n-k-1}}{\rightleftarrows}} \Lambda^{n-k}[s,X] \underset{V_{n-k}}{\overset{\delta_A^{n-k}}{\rightleftarrows}} \Lambda^{n-k+1}[s,X]$$

Set $U = I - V_{n-k}\delta_A^{n-k} : \Lambda^{n-k}[s,X] \to \Lambda^{n-k}[s,X]$. Express U in the matrix form as

$$Uxs_G = \sum_{\operatorname{card} G' = n-k} U_{G',G}\, x s_{G'} \qquad (x \in X, \operatorname{card} G = n-k), \tag{3}$$

where $U_{G',G} \in \mathcal{B}(X)$.

For each G with $\operatorname{card} G = n-k$ set $\psi_G = \sum_{\operatorname{card} G' = n-k} U_{G',G} s_{G'} \in \Lambda^k[s, \mathcal{B}(X)]$. Then $\psi_G x = U(xs_G)$ for all $x \in X$ and $(\delta_{L_A}\psi_G)x = \delta_A(\psi_G x) = \delta_A U x s_G = 0$. So $\psi_G \in \operatorname{Ker} \delta_{L_A}^{n-k} = \operatorname{Ran} \delta_{L_A}^{n-k-1}$. Therefore there is an $\eta_G \in \Lambda^{n-k-1}[s, \mathcal{B}(X)]$ such that $\delta_{L_A}^{n-k-1}\eta_G = \psi_G$. Let $V_{n-k-1}^A : \Lambda^{n-k}[s,X] \to \Lambda^{n-k-1}[s,X]$ be the operator defined by $V_{n-k-1} x s_G = \eta_G x$. We have

$$\delta_A^{n-k-1} V_{n-k-1} x s_G = \delta_A^{n-k-1}(\eta_G x) = (\delta_{L_A}^{n-k-1}\eta_G)x = \psi_G x = U x s_G.$$

Thus $\delta_A^{n-k-1} V_{n-k-1} = U = I - V_{n-k}\delta_A^{n-k}$ and A is $\sigma_{r,k}$-regular.

By Lemma 6, this implies that $\sigma_{r,k}(A) \subset \sigma_{\delta,k}(L_A)$, which finishes the proof of (i).

(ii) The proof of the second statement is similar to part (i). Since the left spectra of A correspond the the right spectra of R_A, we use instead of the cochain Koszul complex of R_A rather the chain Koszul complex, see Remark 25.12. Recall the operators $\varepsilon_{R_A}^q : \Lambda^{q+1}[s, \mathcal{B}(X)] \to \Lambda^q[s, \mathcal{B}(X)]$ defined in Proposition 25.3,

$$\varepsilon_{R_A}^q T s_F = \sum_{i \in F} T A_i (-1)^{\operatorname{card}\{j \in F : j < i\}} s_{F \setminus \{i\}}$$

for $T \in \mathcal{B}(X), F \subset \{1, \dots, n\}, \operatorname{card} F = q+1$.

(ii.1) Let $0 \leq p \leq n$ and suppose that there are operators $V_{p-1} : \Lambda^p[s,X] \to \Lambda^{p-1}[s,X]$ and $V_p : \Lambda^{p+1}[s,X] \to \Lambda^p[s,X]$ satisfying $\delta_A^{p-1} V_{p-1} + V_p \delta_A^p = I_{\Lambda^p[s,X]}$, see (1). Express V_{p-1} and V_p in the matrix form (2). For all $x \in X$ and $G \subset \{1,\dots,n\}$, $\operatorname{card} G = p$ we have

$$\begin{aligned} xs_G &= (\delta_A^{p-1}V_{p-1} + V_p\delta_A^p)xs_G = \delta_A^{p-1} \sum_{\operatorname{card} F=p-1} V_{F,G}xs_F + V_p \sum_{i\notin G} A_i xs_i \wedge s_G \\ &= \sum_{i\notin F} \sum_{\operatorname{card} F=p-1} A_i V_{F,G} xs_i \wedge s_F + \sum_{\operatorname{card} G'=p} \sum_{i\notin G} V_{G',G\cup\{i\}} A_i (-1)^{\operatorname{card}\{j\in G: j<i\}} xs_{G'} \\ &= \sum_{\operatorname{card} G'=p} \Bigg(\sum_{i\in G'} A_i V_{G'\setminus\{i\},G}(-1)^{\operatorname{card}\{j\in G': j<i\}} \\ &\qquad + \sum_{i\notin G} V_{G',G\cup\{i\}} A_i (-1)^{\operatorname{card}\{j\in G: j<i\}} \Bigg) xs_{G'}, \end{aligned}$$

so

$$\begin{aligned} &\sum_{i\in G'} A_i V_{G'\setminus\{i\},G}(-1)^{\operatorname{card}\{j\in G': j<i\}} \\ &\qquad + \sum_{i\notin G} V_{G',G\cup\{i\}} A_i (-1)^{\operatorname{card}\{j\in G: j<i\}} = \begin{cases} I & G = G', \\ 0 & G \neq G'. \end{cases} \end{aligned} \tag{4}$$

Define operators W_{p-1} and W_p (see the diagram below) by

$$W_q T s_G = \sum_{\operatorname{card} F=q+1} T V_{G,F} s_F \qquad (q = p-1, p, T \in \mathcal{B}(X), \operatorname{card} G = q).$$

$$\Lambda^{p-1}[s,\mathcal{B}(X)] \underset{\varepsilon_{R_A}^{p-1}}{\overset{W_{p-1}}{\rightleftarrows}} \Lambda^p[s,\mathcal{B}(X)] \underset{\varepsilon_{R_A}^{p}}{\overset{W_{p}}{\rightleftarrows}} \Lambda^{p+1}[s,\mathcal{B}(X)]$$

By (4), we have

$$\begin{aligned} &(W_{p-1}\varepsilon_{R_A}^{p-1} + \varepsilon_{R_A}^p W_p) T s_G \\ &= W_{p-1} \sum_{i\in G} (-1)^{\operatorname{card}\{j\in G: j<i\}} T A_i s_{G\setminus\{i\}} + \varepsilon_{R_A}^p \sum_{\operatorname{card} F=p+1} T V_{G,F} s_F \\ &= \sum_{\operatorname{card} G'=p} \sum_{i\in G} (-1)^{\operatorname{card}\{j\in G: j<i\}} T A_i V_{G\setminus\{i\},G'} s_{G'} \\ &\qquad + \sum_{i\in F} \sum_{\operatorname{card} F=p+1} T V_{G,F} A_i s_{F\setminus\{i\}} (-1)^{\operatorname{card}\{j\in F: j<i\}} \\ &= \sum_{\operatorname{card} G'=p} T \Bigg(\sum_{i\in G} A_i V_{G\setminus\{i\},G'}(-1)^{\operatorname{card}\{j\in G: j<i\}} \\ &\qquad + \sum_{i\notin G'} V_{G,G'\cup\{i\}} A_i (-1)^{\operatorname{card}\{j\in G': j<i\}} \Bigg) s_{G'}. \end{aligned}$$

By (4), the last expression is equal to Ts_G, and so $W_{p-1}\varepsilon_{R_A}^{p-1} + \varepsilon_{R_A}^p W_p = I_{\Lambda^p[s,\mathcal{B}(X)]}$. This implies that $\sigma_{r,k}(R_A) \subset \sigma_{l,k}(A)$.

(ii.2) We prove by induction on k that the $\sigma_{\delta,k}$-regularity of R_A implies the $\sigma_{l,k}$-regularity of A. For $k=0$ this was proved in Theorem 9.26.

Let $k \geq 1$ and let R_A be $\sigma_{\delta,k}$-regular. By the induction hypothesis, A is $\sigma_{l,k-1}$-regular. In particular, δ_A^{k-1} has a generalized inverse, and so there exists an operator $V_{k-1} : \Lambda^k[s,X] \to \Lambda^{k-1}[s,X]$ such that $\delta_A^{k-1} V_{k-1} \delta_A^{k-1} = \delta_A^{k-1}$. It is sufficient to show that there is an operator $V_k : \Lambda^{k+1}[s,X] \to \Lambda^k[s,X]$ such that $V_k \delta_A^k + \delta_A^{k-1} V_{k-1} = I_{\Lambda^k[s,X]}$.

Set $U = I - \delta_A^{k-1} V_{k-1} : \Lambda^k[s,X] \to \Lambda^k[s,X]$. Express U in the matrix form by

$$U x s_G = \sum_{\operatorname{card} G' = k} U_{G',G} x s_{G'} \qquad (x \in X, \operatorname{card} G = k). \tag{5}$$

For $x \in X$ and $\operatorname{card} F = k-1$ we have

$$\begin{aligned} 0 = U \delta_A^{k-1} x s_F &= U \sum_{i \notin F} A_i x s_i \wedge s_F \\ &= \sum_{\operatorname{card} G' = k} \sum_{i \notin F} (-1)^{\operatorname{card}\{j \in F : j < i\}} U_{G', F \cup \{i\}} A_i x s_{G'}. \end{aligned}$$

Hence

$$\sum_{i \notin F} (-1)^{\operatorname{card}\{j \in F : j < i\}} U_{G', F \cup \{i\}} A_i = 0 \tag{6}$$

for all $G', F \subset \{1, \dots, n\}$ with $\operatorname{card} G' = k$ and $\operatorname{card} F = k-1$. For each G with $\operatorname{card} G = k$ define $\psi_G \in \Lambda^k[s, \mathcal{B}(X)]$ by $\psi_G = \sum_{\operatorname{card} G' = k} U_{G,G'} s_{G'}$. Then, by (6),

$$\begin{aligned} \varepsilon_{R_A} \psi_G &= \sum_{\operatorname{card} G' = k} \sum_{i \in G'} U_{G,G'} A_i s_{G' \setminus \{i\}} (-1)^{\operatorname{card}\{j \in G' : j < i\}} \\ &= \sum_{\operatorname{card} F = k-1} \Big(\sum_{i \notin F} U_{G, F \cup \{i\}} A_i (-1)^{\operatorname{card}\{j \in F : j < i\}} \Big) s_F = 0. \end{aligned}$$

Thus $\psi_G \in \operatorname{Ker} \varepsilon_{R_A}^{k-1} = \operatorname{Ran} \varepsilon_{R_A}^k$ and therefore $\varepsilon_{R_A}^k \eta_G = \psi_G$ for some $\eta_G = \sum_{\operatorname{card} K = k+1} T_{G,K} s_K \in \Lambda^{k+1}[s, \mathcal{B}(X)]$. For each G with $\operatorname{card} G = k$ we have

$$\begin{aligned} \sum_{\operatorname{card} G' = k} U_{G,G'} s_{G'} = \psi_G = \varepsilon_{R_A}^k \eta_G &= \varepsilon_{R_A}^k \sum_{\operatorname{card} K = k+1} T_{G,K} s_K \\ &= \sum_{\operatorname{card} K = k+1} \sum_{i \in K} T_{G,K} A_i s_{K \setminus \{i\}} (-1)^{\operatorname{card}\{j \in K : j < i\}}. \end{aligned}$$

Hence, for all G, G' with $\operatorname{card} G = \operatorname{card} G' = k$, we have

$$U_{G,G'} = \sum_{i \notin G'} T_{G, G' \cup \{i\}} A_i (-1)^{\operatorname{card}\{j \in G' : j < i\}}. \tag{7}$$

Define $V_k : \Lambda^{k+1}[s,X] \to \Lambda^k[s,X]$ by $V_k x s_K = \sum_{\operatorname{card} G=k} T_{G,K} x s_G$. For $x \in X$ and $\operatorname{card} G' = k$ we have, using (6) and (5),

$$\begin{aligned} V_k \delta_A^k x s_{G'} &= V_k \sum_{i \notin G'} A_i x s_i \wedge s_{G'} \\ &= \sum_{\operatorname{card} G=k} \sum_{i \notin G'} T_{G,G' \cup \{i\}} A_i (-1)^{\operatorname{card}\{j \in G' : j<i\}} x s_G \\ &= \sum_{\operatorname{card} G=k} U_{G,G'} x s_G = U x s_{G'}. \end{aligned}$$

Hence $V_k \delta_A^k = U$ and $\delta_A^{k-1} V_{k-1} + V_k \delta_A^k = I_{\Lambda^k[s,X]}$. Thus A is $\sigma_{l,k}$-regular. Hence $\sigma_{l,k}(A) \subset \sigma_{\delta,k}(R_A) \subset \sigma_{r,k}(R_A)$, which finishes the proof. □

Corollary 8. *$\sigma_{r,k}$, $\sigma_{l,k}$ and σ_S are upper semicontinuous spectral systems for all k, $0 \le k \le n$.*

Proof. The statement follows from Theorem 7 and from the corresponding statements for $\sigma_{\delta,k}$ and $\sigma_{\pi,k}$. □

Proposition 9. *Let $\mathcal{A}$ be a commutative Banach algebra and $a = (a_1, \dots, a_n) \in \mathcal{A}^n$. Let $L_a = (L_{a_1}, \dots, L_{a_n}) \in \mathcal{B}(\mathcal{A})^n$, where $L_{a_j} : \mathcal{A} \to \mathcal{A}$ is defined by $L_{a_j} b = a_j b \quad (b \in \mathcal{A}, j = 1, \dots, n)$. Then $\sigma_T(L_a) = \sigma_S(L_a) = \sigma(a)$.*

Proof. If $0 \notin \sigma(a)$, then $\sum a_i b_i = 1_{\mathcal{A}}$ for some $b_i \in \mathcal{A}$. Then $\sum L_{a_i} L_{b_i} = I_{\mathcal{A}}$, and so L_a is split regular by the proof of Theorem 4. So $\sigma_S(L_a) \subset \sigma(a)$.

Conversely, by Theorem 9.26, $\sigma(a) = \sigma_r(a) = \sigma_\delta(L_a) \subset \sigma_T(L_a) \subset \sigma_S(L_a)$. □

27 Some non-linear results

A great advantages of Hilbert spaces is the existence of projections onto all closed subspaces. In general Banach spaces it is sometimes possible to use non-linear techniques instead of it. We have already used the Borsuk antipodal theorem; in this section we give several other results which are based essentially on the Michael selection theorem, see Appendix A.4. Some applications in operator theory will be given in this and the subsequent sections.

Definition 1. Let X, Y be Banach spaces. Denote by $\mathcal{H}(X,Y)$ the set of all continuous mappings $f : X \to Y$ that are homogeneous (i.e., $f(\alpha x) = \alpha f(x)$ for all $\alpha \in \mathbb{C}$ and $x \in X$). Write for short $\mathcal{H}(X)$ instead of $\mathcal{H}(X,X)$.

Let $f \in \mathcal{H}(X,Y)$. The continuity of f at 0 implies that f is bounded, i.e., $\sup\{\|f(x)\| : x \in X, \|x\| \le 1\} < \infty$. Clearly, $\mathcal{H}(X,Y)$ with this norm is a Banach space and $\mathcal{B}(X,Y) \subset \mathcal{H}(X,Y)$.

Lemma 2. *Let X, Y be Banach spaces. For each $x \in X$ let $G(x) \subset Y$ be a non-empty closed convex set such that the mapping $x \mapsto G(x)$ is lower semicontinuous and $\bigcup\{G(x) : x \in X, \|x\| = 1\}$ is a bounded subset of Y.*

Suppose further that $G(\alpha x) = \alpha G(x)$ for all $x \in X$ and $\alpha \in \mathbb{C}$. Then there exists a mapping $g \in \mathcal{H}(X, Y)$ such that $g(x) \in G(x)$ for all $x \in X$.

Proof. Let S_X be the unit sphere in X. By Micheal's theorem (see Theorem A.4.5) there exists a continuous selection $f : S_X \to Y$ satisfying $f(x) \in G(x)$ for all $x \in S_X$. For $x \in S_X$ set

$$g(x) = \frac{1}{2\pi} \int_0^{2\pi} e^{-it} f(e^{it}x) \mathrm{d}t.$$

We have $e^{-it} f(e^{it}x) \in e^{-it} G(e^{it}x) = G(x)$. Since $G(x)$ is closed and convex, $g(x)$ (as a limit of convex combinations) also belongs to $G(x)$. Evidently, $g : S_X \to Y$ is continuous by the Lebesgue dominated convergence theorem. It is easy to see that $g(\lambda x) = \lambda g(x)$ for all $x \in S_X$ and $\lambda \in \mathbb{C}$, $|\lambda| = 1$.

Extend g to X by $g(0) = 0$ and $g(x) = \|x\| \cdot g(\frac{x}{\|x\|})$ $(x \neq 0)$. It is easy to see that g is continuous and satisfies all the conditions required. □

Theorem 3. (Bartle-Graves) *Let M be a closed subspace of a Banach space X and let $\varepsilon > 0$. Then there exists $h \in \mathcal{H}(X/M, X)$ such that $\|h\| < 1 + \varepsilon$ and $h(x + M) \in x + M$ for each class $x + M \in X/M$.*

Proof. For $x + M \in X/M$, $x + M \neq M$ set

$$G(x + M) = \{u \in X : u \in x + M, \|u\| < (1 + \varepsilon)\|x + M\|\}^{-}.$$

Let $G(M) = \{0\}$. Clearly, $G(\alpha x + M) = \alpha G(x + M)$ and $G(x + M)$ is a non-empty closed convex subset of X for all $x \in X$ and $\alpha \in \mathbb{C}$.

We show that the mapping G is lower semicontinuous. Clearly, G is lower semicontinuous at M. Let $x \in X \setminus M$ and let U be an open subset of X with $U \cap G(x + M) \neq \emptyset$. Find $u \in U$ such that $u \in x + M$ and $\|u\| < (1 + \varepsilon)\|x + M\|$.

Let δ be a positive number satisfying $\{v \in X : \|v - u\| < \delta\} \subset U$ and $\|u\| < (1 + \varepsilon)(\|x + M\| - 2\delta)$. If $\xi \in X/M$ satisfies $\|\xi - (x + M)\| < \delta$, then there exists $x' \in \xi$ such that $\|x' - x\| < \delta$. Then $x' + u - x \in \xi$ and $\|(x' + u - x) - u\| = \|x' - x\| < \delta$, and so $x' + u - x \in U$. Further,

$$\begin{aligned} \|x' + u - x\| &\leq \|u\| + \|x' - x\| < (1 + \varepsilon)(\|x + M\| - 2\delta) + \delta \\ &\leq (1 + \varepsilon)(\|x + M\| - \delta\|) \leq (1 + \varepsilon)\|\xi\|. \end{aligned}$$

Thus $x' + u - x \in U \cap G(\xi)$ and G is lower semicontinuous.

By the previous lemma, there exists a homogeneous continuous selection $h : X/M \to X$. Clearly, h satisfies all the conditions required. □

Corollary 4. *Let M be a closed subspace of a Banach space X and let $\varepsilon > 0$. Then there exists a continuous homogeneous mapping $p : X \to M$ such that $\|x - px\| \le (1+\varepsilon)\operatorname{dist}\{x, M\}$ and $p(x+m) = p(x) + m$ for all $x \in X$ and $m \in M$.*

In particular, $pm = m$ for all $m \in M$, and so p is a (non-linear) projection onto M.

Proof. Let $h : X/M \to X$ be the mapping constructed in Theorem 3. So $h(x + M) \in x + M$ and $\|h(x+M)\| \le (1+\varepsilon)\|x + M\|$ for all $x + M \in X/M$. Set $px = x - hQx$ where $Q : X \to X/M$ is the canonical projection. Then $\|x - px\| = \|hQx\| \le (1+\varepsilon)\|Qx\| = (1+\varepsilon)\operatorname{dist}\{x, M\}$. For $x \in X$ and $m \in M$ we have $p(x+m) = x + m - hQ(x+m) = x + m - hQx = px + m$. □

Proposition 5. *Let X, Y be Banach spaces and let $T : X \to Y$ be a bounded linear operator with closed range. Then:*

(i) if $f \in \mathcal{H}(Y)$ satisfies $f(Y) \subset \operatorname{Ran} T$, then there exists $g \in \mathcal{H}(Y, X)$ such that $f = Tg$. In particular, if T is onto, then there exists $g \in \mathcal{H}(Y, X)$ such that $Tg = I_Y$;

(ii) if $T \in \mathcal{B}(X, Y)$ is bounded below, then there exists $g \in \mathcal{H}(Y, X)$ such that $gT = I_X$.

Proof. (i) Let $h : X/\operatorname{Ker} T \to X$ be the selection given by the Bartle-Graves theorem. Let $T_0 : X/\operatorname{Ker} T \to \operatorname{Ran} T$ be the operator induced by T. Set $g = hT_0^{-1}f$. For $y \in Y$ we have $Tgy = ThT_0^{-1}fy = fy$, and so $Tg = f$.

(ii) Let $T \in \mathcal{B}(X, Y)$ be bounded below. Let $p \in \mathcal{H}(Y)$ be the non-linear projection onto $\operatorname{Ran} T$ constructed in Corollary 4. Then $T^{-1}p \in \mathcal{H}(Y, X)$ is the required non-linear left inversion of T. □

By the preceding proposition, the operators that are bounded below or onto become one-sided invertible if we admit non-linear mappings. In the same way, the Taylor regular n-tuples of operators get the split property.

Proposition 6. *Let $A = (A_1, \dots, A_n)$ be an n-tuple of mutually commuting operators on a Banach space X. Then A is Taylor regular if and only if there are mappings $V_i \in \mathcal{H}(\Lambda^{i+1}[s, X], \Lambda^i[s, X])$ $(i = 0, \dots, n-1)$ such that*

$$\begin{aligned} &V_0\delta_A^0 = I_{\Lambda^0[s,X]}, \\ &V_j\delta_A^j + \delta_A^{j-1}V_{j-1} = I_{\Lambda^j[s,X]} \qquad (j = 1, \dots, n-1), \\ &\delta_A^{n-1}V_{n-1} = I_{\Lambda^n[s,X]}. \end{aligned}$$

Proof. Suppose that the mappings V_i satisfy the conditions of the theorem. Let $\psi \in \operatorname{Ker}\delta_A^j$. Then $\psi = (V_j\delta_A^j + \delta_A^{j-1}V_{j-1})\psi = \delta_A^{j-1}V_{j-1}\psi \in \operatorname{Ran}\delta_A^{j-1}$ (note that the same relation also holds for $j = 0, n$). Hence A is Taylor regular.

Conversely, suppose that A is Taylor regular. Since δ_A^{n-1} is onto, by Proposition 5 there exists $V_{n-1} \in \mathcal{H}(\Lambda^n[s, X], \Lambda^{n-1}[s, X])$ satisfying $\delta_A^{n-1}V_{n-1} = I$.

We construct the mappings V_j inductively. Suppose that $1 \le j \le n-1$ and the mappings V_j and V_{j+1} satisfy $V_{j+1}\delta_A^{j+1} + \delta_A^j V_j = I$ (for $j = n-1$ set formally $V_n = 0$). We have

$$\delta_A^j(I - V_j\delta_A^j) = \delta_A^j - \delta_A^j V_j \delta_A^j = \delta_A^j - (I - V_{j+1}\delta_A^{j+1})\delta_A^j = 0,$$

and so, by Proposition 5, there exists $V_{j-1} \in \mathcal{H}(\Lambda^j[s,X], \Lambda^{j-1}[s,X])$ such that $\delta_A^{j-1}V_{j-1} = I - V_j\delta_A^j$, i.e., $V_j\delta_A^j + \delta_A^{j-1}V_{j-1} = I$.

At the end, suppose that $V_0 \in \mathcal{H}(\Lambda^1[s,X], \Lambda^0[s,X])$ satisfies $V_1\delta_A^1 + \delta_A^0 V_0 = I$. Then $\delta_A^0 = \delta_A^0 V_0 \delta_A^0$. Since δ_A^0 is injective, we have $V_0\delta_A^0 = I$. □

Let $A = (A_1, \dots, A_n)$ be an n-tuple of commuting operators on a Banach space X. By Theorem 26.7, $\sigma_T(L_A) = \sigma_S(A)$, where $L_A = (L_{A_1}, \dots, L_{A_n})$ and L_{A_j} are the operators of left multiplication by A_j on $\mathcal{B}(X)$.

The situation changes if we consider the operators of left multiplication acting on $\mathcal{H}(X)$.

Let $L'_A = (L'_{A_1}, \dots, L'_{A_n})$, where $L'_{A_i} f = A_i f \quad (f \in \mathcal{H}(X), i = 1, \dots, n)$. Then L'_A is an n-tuple of commuting operators acting on $\mathcal{H}(X)$.

Corollary 7. $\sigma_T(L'_A) = \sigma_T(A)$.

Proof. It is sufficient to show that A is Taylor regular if and only if L'_A is Taylor regular. We can consider an element $\varphi \in \Lambda[s, \mathcal{H}(X)]$ as a continuous homogeneous mapping $\varphi : X \to \Lambda[s, X]$. With this convention we have $(\delta_{L'_A}\varphi)x = \delta_A(\varphi(x))$ for all $x \in X$.

Let A be Taylor regular. Let $V_j \;\; (0 \le j \le n-1)$ be the mappings constructed in Proposition 6. Define $V \in \mathcal{H}(\Lambda[s,X])$ by $V\left(\bigoplus_{i=0}^n \psi_i\right) = \bigoplus_{i=0}^n V_{i-1}\psi_i$. Then $V\delta_A + \delta_A V = I_{\Lambda[s,X]}$. If we lift V naturally to a mapping acting on $\Lambda[s, \mathcal{H}(X)]$, we have $V\delta_{L'_A} + \delta_{L'_A}V = I_{\Lambda[s,\mathcal{H}(X)]}$. Let $\psi \in \operatorname{Ker}\delta_{L'_A}$. Then $\psi = (\delta_{L'_A}V + V\delta_{L'_A})\psi = \delta_{L'_A}V\psi \in \operatorname{Ran}\delta_{L'_A}$.

Conversely, suppose that A is not Taylor regular. Then there exists $\psi \in \operatorname{Ker}\delta_A \setminus \operatorname{Ran}\delta_A$. Let $x^* \in X^*$ be any non-zero functional and let $\varphi \in \Lambda[s, \mathcal{H}(X)]$ be defined by $\varphi(x) = x^*(x) \cdot \psi \quad (x \in X)$. For each $x \in X$ we have $(\delta_{L'_A}\varphi)(x) = \delta_A\varphi(x) = x^*(x) \cdot \delta_A\psi = 0$. Hence $\varphi \in \operatorname{Ker}\delta_{L'_A}$ and similarly one can show that $\varphi \notin \operatorname{Ran}\delta_{L'_A}$. □

The next result is a generalization of Corollary 10.10.

Theorem 8. *Let R, R_1, N, N_1 be closed subspaces of a Banach space X and let $R \subset N$. Suppose that $\delta(R,R_1) + \delta(N_1,N) + \delta(R,R_1)\delta(N_1,N) < 1$. Then*

$$\dim N_1/(R_1 \cap N_1) \le \dim N/R + \dim R_1/(R_1 \cap N_1).$$

In particular, if also $R_1 \subset N_1$, then $\dim N_1/R_1 \le \dim N/R$. Consequently, if $R \subset N$, $R_1 \subset N_1$, $\widehat{\delta}(R,R_1) < 1/3$ and $\widehat{\delta}(N_1,N) < 1/3$, then $\dim N_1/R_1 = \dim N/R$.

Proof. Suppose on the contrary that $\delta(R,R_1)+\delta(N_1,N)+\delta(R,R_1)\delta(N_1,N)<1$ and

$$\dim N_1/(R_1\cap N_1)>\dim N/R+\dim R_1/(R_1\cap N_1).$$

Let ε be a positive number small enough (the exact condition on ε will be clear from the proof). Let $Q_R : N\to N/R$ and $Q_{N_1\cap R_1} : R_1\to R_1/(N_1\cap R_1)$ be the canonical projections, let $h : N_1/(N_1\cap R_1)\to N_1$ be the mapping constructed in Theorem 3 (in particular, $h(n_1+(N_1\cap R_1))\in n_1+(N_1\cap R_1)$ and $\bigl\|h(n_1+(N_1\cap R_1))\bigr\|\le(1+\varepsilon)\|n_1+(N_1\cap R_1)\|$ for all $n_1\in N_1$) and let $p_N : X\to N$, $p_{R_1} : X\to R_1$ be the continuous non-linear projections constructed in Corollary 4 (in particular, $\|x-p_Nx\|\le(1+\varepsilon)\operatorname{dist}\{x,N\}$, $\|x-p_{R_1}x\|\le(1+\varepsilon)\operatorname{dist}\{x,R_1\}$ for all $x\in X$).

Let $\Phi : N_1/(N_1\cap R_1)\to N/R\oplus R_1/(N_1\cap R_1)$ be the mapping defined by

$$\Phi(\xi)=Q_Rp_Nh(\xi)\oplus Q_{N_1\cap R_1}p_{R_1}h(\xi)\qquad(\xi\in N_1/(N_1\cap R_1)).$$

Clearly, Φ is a continuous homogeneous mapping. By the Borsuk antipodal theorem, there exists $\xi\in N_1/(N_1\cap R_1)$ such that $\|\xi\|=1$ and $\Phi(\xi)=0$. Set $x=h(\xi)\in N_1$. Then $\|x\|\le1+\varepsilon$, $Q_Rp_Nx=0$ and $Q_{N_1\cap R_1}p_{R_1}x=0$; so $p_Nx\in R$ and $p_{R_1}x\in N_1\cap R_1$.

We have

$$\|x-p_Nx\|\le(1+\varepsilon)\operatorname{dist}\{x,N\}\le(1+\varepsilon)\|x\|\delta(N_1,N)\le(1+\varepsilon)^2\delta(N_1,N)$$

and $\|p_Nx\|\le\|x\|+\|x-p_Nx\|\le1+\varepsilon+(1+\varepsilon)^2\delta(N_1,N)$. Thus

$$\begin{aligned}1=\|\xi\|=\operatorname{dist}\{x,N_1\cap R_1\}&\le\|x-p_{R_1}x\|\le(1+\varepsilon)\operatorname{dist}\{x,R_1\}\\&\le(1+\varepsilon)\bigl(\|x-p_Nx\|+\operatorname{dist}\{p_Nx,R_1\}\bigr)\\&\le(1+\varepsilon)^3\delta(N_1,N)+(1+\varepsilon)\|p_Nx\|\delta(R,R_1)\\&\le(1+\varepsilon)^3\bigl(\delta(N_1,N)+\delta(R,R_1)+\delta(N_1,N)\delta(R,R_1)\bigr)<1\end{aligned}$$

for ε small enough. This gives a contradiction. □

The assumption $R\subset N$ in the preceding theorem is not necessary. First, we replace this condition by the assumption $R\overset{e}{\subset}N$.

Theorem 9. *Let R,N be closed subspaces of a Banach space X, let $R\overset{e}{\subset}N$. Then there exists $\varepsilon>0$ such that, for all closed subspaces R_1 and N_1 of X with $\delta(R,R_1)<\varepsilon$ and $\delta(N_1,N)<\varepsilon$, we have*

$$\dim N_1/(R_1\cap N_1)+\dim R/(R\cap N)\le\dim N/(R\cap N)+\dim R_1/(R_1\cap N_1).$$

Proof. For $R\subset N$ this was proved in the previous theorem. We reduce the general situation to this case.

Choose a finite-dimensional subspace $F\subset R$ such that $(R\cap N)\oplus F=R$. Let $\dim F=k<\infty$ and let $f_1,\dots,f_k$ be a basis in F with $\|f_1\|=\dots=\|f_k\|=1$. Clearly, $F\cap N=\{0\}$.

For $f = \sum_{i=1}^k \alpha_i f_i \in F$ where $\alpha_i \in \mathbb{C}$ consider three norms: $\|f\|$, $\operatorname{dist}\{f, N\}$ and $\sum_{i=1}^k |\alpha_i|$. Since these three norms are equivalent, there exists $c > 0$ such that

$$c \cdot \sum_{i=1}^k |\alpha_i| \le \operatorname{dist}\Big\{\sum_{i=1}^k \alpha_i f_i, N\Big\} \le \Big\|\sum_{i=1}^k \alpha_i f_i\Big\| \le \sum_{i=1}^k |\alpha_i|$$

for all $\alpha_1, \dots, \alpha_k \in \mathbb{C}$. Clearly, $c \le 1$.

Set $\varepsilon = \frac{c}{8}$. Let R_1 and N_1 be closed subspaces of X such that $\delta(R, R_1) < \varepsilon$ and $\delta(N_1, N) < \varepsilon$.

For $i = 1, \dots, k$ find elements $g_i \in R_1$ such that $\|f_i - g_i\| < \varepsilon$. Then $\|g_i\| < 1 + \varepsilon \quad (i = 1, \dots, k)$.

Denote by G the subspace of R_1 generated by $g_1, \dots, g_k$.

We prove that the elements $g_1, \dots, g_k$ are linearly independent modulo N_1. Suppose that $\sum_{i=1}^k \alpha_i g_i \in N_1$ for some $\alpha_i \in \mathbb{C}$. Then

$$\begin{aligned}\sum_{i=1}^k |\alpha_i| &\le c^{-1} \operatorname{dist}\Big\{\sum_{i=1}^k \alpha_i f_i, N\Big\} \le c^{-1}\Big(\sum_{i=1}^k |\alpha_i| \|f_i - g_i\| + \operatorname{dist}\Big\{\sum_{i=1}^k \alpha_i g_i, N\Big\}\Big) \\ &\le c^{-1}\varepsilon \sum_{i=1}^k |\alpha_i| + c^{-1}\Big\|\sum_{i=1}^k \alpha_i g_i\Big\| \cdot \delta(N_1, N) \\ &\le \Big(\frac{\varepsilon}{c} + \frac{\varepsilon(1+\varepsilon)}{c}\Big) \cdot \sum_{i=1}^k |\alpha_i| \le \frac{1}{2}\sum_{i=1}^k |\alpha_i|,\end{aligned}$$

and so $\alpha_1 = \dots = \alpha_k = 0$. In particular, $\dim G = k$ and $G \cap N_1 = \{0\}$.

Write $N' = N + F$ and $N_1' = N_1 + G$. Then $N' = N + R \supset R$.

We prove that $\delta(N_1', N') < 3/4$. Let $n_1 + \sum_{i=1}^k \alpha_i g_i \in N_1'$, where $n_1 \in N_1$, $\alpha_i \in \mathbb{C} \quad (i = 1, \dots, k)$, and $\Big\|n_1 + \sum_{i=1}^k \alpha_i g_i\Big\| = 1$. Then

$$\|n_1\| \le 1 + (1+\varepsilon)\sum_{i=1}^k |\alpha_i|.$$

There exists $n \in N$ such that $\|n_1 - n\| \le \varepsilon \|n_1\| \le \varepsilon + \varepsilon(1+\varepsilon)\sum_{i=1}^k |\alpha_i|$. We have

$$\begin{aligned}c\sum_{i=1}^k |\alpha_i| &\le \operatorname{dist}\Big\{\sum_{i=1}^k \alpha_i f_i, N\Big\} \le \Big\|\sum_{i=1}^k \alpha_i f_i + n\Big\| \\ &\le \Big\|\sum_{i=1}^k \alpha_i (f_i - g_i)\Big\| + \Big\|\sum_{i=1}^k \alpha_i g_i + n_1\Big\| + \|n - n_1\| \\ &\le \varepsilon \sum_{i=1}^k |\alpha_i| + 1 + \varepsilon + \varepsilon(1+\varepsilon)\sum_{i=1}^k |\alpha_i| \le 1 + \varepsilon + 3\varepsilon\sum_{i=1}^k |\alpha_i|.\end{aligned}$$

Thus

$$\sum_{i=1}^{k} |\alpha_i| \le \frac{1+\varepsilon}{c-3\varepsilon} \le \frac{1}{4\varepsilon}$$

and

$$\begin{aligned}\operatorname{dist}\Big\{n_1 + \sum_{i=1}^{k} \alpha_i g_i, N'\Big\} &\le \|n_1 - n\| + \Big\|\sum_{i=1}^{k} \alpha_i (f_i - g_i)\Big\| \\ &\le \varepsilon + \varepsilon(1+\varepsilon)\sum_{i=1}^{k}|\alpha_i| + \varepsilon \sum_{i=1}^{k} |\alpha_i| < \frac{11}{16}.\end{aligned}$$

Hence $\delta(N_1', N') < 3/4$ and

$$\delta(N_1', N') + \delta(R, R_1) + \delta(N_1', N')\delta(R, R_1) < 1.$$

By Theorem 8, we have

$$\dim N_1'/(R_1 \cap N_1') \le \dim N'/R + \dim R_1/(R_1 \cap N_1'). \tag{1}$$

By Lemma 21.1,

$$\begin{aligned}\dim N_1/(R_1 \cap N_1) &= \dim(N_1 + R_1)/R_1 \\ &= \dim(N_1' + R_1)/R_1 = \dim N_1'/(R_1 \cap N_1')\end{aligned} \tag{2}$$

and

$$\dim N/(R \cap N) = \dim(N + R)/R = \dim N'/R. \tag{3}$$

Furthermore,

$$\dim R/(R \cap N) = k \tag{4}$$

and

$$\begin{aligned}\dim R_1/(R_1 \cap N_1) &= \dim(N_1 + R_1)/N_1 \\ &= \dim(N_1 + R_1)/(N_1 + G) + \dim(N_1 + G)/N_1 \\ &= \dim(N_1' + R_1)/N_1' + k = \dim R_1/(R_1 \cap N_1') + k.\end{aligned} \tag{5}$$

Thus, by (1)–(5), we have

$$\begin{aligned}&\dim N_1/(R_1 \cap N_1) + \dim R/(R \cap N) \\ &= \dim N_1'/(R_1 \cap N_1') + k \\ &\le \dim N'/R + \dim R_1/(R_1 \cap N_1') + k \\ &= \dim N/(R \cap N) + \dim R_1/(R_1 \cap N_1).\end{aligned}$$

□

The assumption $R \overset{e}{\subset} N$ in the previous theorem can be omitted. To show this, recall the construction of Section 17. Let $m(X) \subset \ell^\infty(X)$ be the subspace of all totally bounded sequences.

We need the following lemma.

Lemma 10. *Let M, L be closed subspaces of a Banach space X. Let $\widehat{M} = \ell^\infty(M) + m(X)$ and $\widehat{L} = \ell^\infty(L) + m(X)$ be the corresponding subspaces of $\ell^\infty(X)$. Then $\delta(\widehat{M}, \widehat{L}) \le 2\delta(M, L)$.*

Proof. Let $\tilde{x} = (x_i) \in \widehat{M}$, $\|\tilde{x}\| = \sup \|x_i\| = 1$. We can write $x_i = m_i + g_i$, where $m_i \in M$ and $(g_i) \in m(X)$. Let $\varepsilon > 0$. Then there is a finite set $F = \{f_1, \dots, f_n\}$ and indices $j_i \in \{1, \dots, n\}$ such that $\|g_i - f_{j_i}\| < \varepsilon$. Replacing f_j by some $f'_j \in f_j + M$ if necessary, we can assume that $\|f_j\| < (1+\varepsilon) \operatorname{dist}\{f_j, M\} \quad (j = 1, \dots, n)$. Thus we have

$$\begin{aligned}\|f_{j_i}\| &< (1+\varepsilon) \operatorname{dist}\{f_{j_i}, M\} \le (1+\varepsilon)\|f_{j_i} + m_i\| \\ &\le (1+\varepsilon)\big(\|f_{j_i} - g_i\| + \|x_i\|\big) \le (1+\varepsilon)^2,\end{aligned}$$

and so $\|m_i\| = \|x_i - g_i\| \le \|x_i\| + \|g_i - f_{j_i}\| + \|f_{j_i}\| \le 1 + \varepsilon + (1+\varepsilon)^2$. Thus there are $l_i \in L$ with $\|m_i - l_i\| \le 2(1+\varepsilon)^2 \delta(M, L)$. Let $\tilde{y} = (l_i + g_i) \in \ell^\infty(L) + m(X)$. Then $\|\tilde{x} - \tilde{y}\| = \sup \|m_i - l_i\| \le 2(1+\varepsilon)^2 \delta(M, L)$ and so $\delta(\widehat{M}, \widehat{L}) \le 2(1+\varepsilon)^2 \delta(M, L)$. Letting $\varepsilon \to 0$ gives $\delta(\widehat{M}, \widehat{L}) \le 2\delta(M, L)$. □

Theorem 11. *Let R, N be closed subspaces of a Banach space X. Then there exists $\varepsilon > 0$ such that*

$$\dim N_1/(R_1 \cap N_1) + \dim R/(N \cap R) \le \dim N/(N \cap R) + \dim R_1/(R_1 \cap N_1)$$

for all closed subspaces $R_1, N_1 \subset X$ with $\delta(R, R_1) < \varepsilon$ and $\delta(N_1, N) < \varepsilon$.

Proof. If $R \overset{e}{\subset} N$, then the statement (i) was proved in Theorem 9.

Suppose that $\dim R/(N \cap R) = \infty$. Let $N_1, R_1 \subset X$ be closed subspaces satisfying $\delta(R, R_1) < 1/6$ and $\delta(N_1, N) < 1/6$.

It is sufficient to show that $\dim N/(N \cap R) + \dim R_1/(N_1 \cap R_1) = \infty$. Suppose on the contrary that $R_1 \overset{e}{\subset} N_1$ and $N \overset{e}{\subset} R$.

Set

$$\begin{aligned}\widehat{N} &= \ell^\infty(N) + m(X), \\ \widehat{R} &= \ell^\infty(R) + m(X), \\ \widehat{N_1} &= \ell^\infty(N_1) + m(X), \\ \widehat{R_1} &= \ell^\infty(R_1) + m(X).\end{aligned}$$

By Lemma 17.2, $\widehat{R_1} \subset \widehat{N_1}$ and $\widehat{N} \subset \widehat{R}$. By Lemma 10, $\delta(\widehat{R}, \widehat{R_1}) < 1/3$ and $\delta(\widehat{N_1}, \widehat{N}) < 1/3$. Thus

$$\begin{aligned}\delta(\widehat{R}, \widehat{N}) &\le \delta(\widehat{R}, \widehat{R_1}) + \delta(\widehat{R_1}, \widehat{N}) + \delta(\widehat{R}, \widehat{R_1})\delta(\widehat{R_1}, \widehat{N}) \\ &< 1/3 + 4/3\ \delta(\widehat{R_1}, \widehat{N}) \le 1/3 + 4/3\ \delta(\widehat{N_1}, \widehat{N}) \le 1/3 + 4/9 < 1.\end{aligned}$$

Therefore $\widehat{N} = \widehat{R}$. Since $N \overset{e}{\subset} R$, Lemma 17.2 implies that $R \overset{e}{=} N$, a contradiction. □

28 Taylor functional calculus for the split spectrum

The most important property of the Taylor spectrum is the existence of the functional calculus for functions analytic on a neighbourhood of the Taylor spectrum.

As the construction of the Taylor functional calculus is rather technical, in this section we introduce a simpler version for functions analytic on a neighbourhood of the split spectrum. Since the split spectrum contains the Taylor spectrum, this split Taylor functional calculus is less rich. However, the construction of the calculus is much simpler.

Note that for Hilbert space operators the split spectrum coincides with the Taylor spectrum and so the corresponding functional calculi also coincide. The split functional calculus is also sufficient for the construction of the functional calculus in commutative Banach algebras, cf. Section 2.

Theorem 1. *Let $A = (A_1, \dots, A_n)$ be an n-tuple of mutually commuting operators in a Banach space X. Suppose that A is split regular, i.e., $\operatorname{Ker} \delta_A = \operatorname{Ran} \delta_A$ and δ_A has a generalized inverse. Then there exists a neighbourhood U of 0 in $\mathbb{C}^n$ and an analytic function $V : U \to \mathcal{B}(\Lambda[s, X])$ such that $V(\lambda)\delta_{A-\lambda} + \delta_{A-\lambda}V(\lambda) = I_{\Lambda[s,X]}$ for every $\lambda \in U$.*

Moreover, we may assume that $V(\lambda)^2 = 0 \quad (\lambda \in U)$ and

$$V(\lambda)\Lambda^p[s, X] \subset \Lambda^{p-1}[s, X] \qquad (\lambda \in U, p = 0, \dots, n).$$

Proof. By Proposition 26.2, there exists an operator $V : \Lambda[s, X] \to \Lambda[s, X]$ such that $V^2 = 0$, $\delta_A V + V\delta_A = I_{\Lambda[s,X]}$, and $V\Lambda^p[s, X] \subset \Lambda^{p-1}[s, X]$ for every p.

For $\lambda \in \mathbb{C}^n$ write $H_\lambda = \delta_{A-\lambda} - \delta_A$. Let U be the set of all $\lambda \in \mathbb{C}^n$ such that $\|H_\lambda\| < \|V\|^{-1}$. Clearly, U is a neighbourhood of 0 in $\mathbb{C}^n$ and, for $\lambda \in U$, the operators $I+H_\lambda V$ and $I+VH_\lambda$ are invertible. We have $V(I+H_\lambda V) = (I+VH_\lambda)V$, and so $(I + VH_\lambda)^{-1}V = V(I + H_\lambda V)^{-1}$. For $\lambda \in U$ set $V(\lambda) = (I + VH_\lambda)^{-1}V$. Then

$$\begin{aligned}
&\delta_{A-\lambda}V(\lambda) + V(\lambda)\delta_{A-\lambda}\\
&= (\delta_A + H_\lambda)V(I + H_\lambda V)^{-1} + (I + VH_\lambda)^{-1}V(\delta_A + H_\lambda)\\
&= (I + VH_\lambda)^{-1}\big((I + VH_\lambda)(\delta_A + H_\lambda)V\\
&\qquad + V(\delta_A + H_\lambda)(I + H_\lambda V)\big)(I + H_\lambda V)^{-1}.
\end{aligned}$$

The expression in the middle is equal to

$$\begin{aligned}
&\delta_A V + H_\lambda V + VH_\lambda\delta_A V + VH_\lambda^2 V + V\delta_A + VH_\lambda + V\delta_A H_\lambda V + VH_\lambda^2 V\\
&= (I + VH_\lambda)(I + H_\lambda V) + V(H_\lambda\delta_A + \delta_A H_\lambda + H_\lambda^2)V\\
&= (I + VH_\lambda)(I + H_\lambda V) + V\big((\delta_A + H_\lambda)^2 - (\delta_A)^2\big)V = (I + VH_\lambda)(I + H_\lambda V)
\end{aligned}$$

since $(\delta_A)^2 = 0$ and $(\delta_A + H_\lambda)^2 = (\delta_{A-\lambda})^2 = 0$. Thus

$$\delta_{A-\lambda}V(\lambda) + V(\lambda)\delta_{A-\lambda} = I_{\Lambda[s,X]} \qquad (\lambda \in U).$$

Further,

$$V(\lambda)^2 = (I + VH_\lambda)^{-1}V \cdot V(I + H_\lambda V)^{-1} = 0.$$

Finally, $V(\lambda) = \sum_{i=0}^{\infty}(-1)^i(VH_\lambda)^iV$ where

$$(VH_\lambda)\Lambda^p[s,X] \subset \Lambda^p[s,X] \qquad (p = 0,\dots,n),$$

and so

$$V(\lambda)\Lambda^p[s,X] \subset \Lambda^{p-1}[s,X] \qquad (\lambda \in U, p = 0,\dots,n).$$

□

Corollary 2. *Let $A = (A_1,\dots,A_n)$ be an n-tuple of mutually commuting operators on a Banach space X. Let $G = \mathbb{C}^n \setminus \sigma_S(A)$. Then there exists an operator-valued C^∞-function $V : G \to \mathcal{B}(\Lambda[s,X])$ such that $\delta_{A-\lambda}V(\lambda) + V(\lambda)\delta_{A-\lambda} = I_{\Lambda[s,X]}$ and*

$$V(\lambda)\Lambda^p[s,X] \subset \Lambda^{p-1}[s,X] \qquad (\lambda \in G, p = 0,\dots,n).$$

Proof. For every $\mu \in G$ there exists a neighbourhood U_μ of μ and an analytic operator-valued function $V_\mu : U_\mu \to \mathcal{B}(\Lambda[s,X])$ such that $V_\mu(\lambda)\delta_{A-\lambda} + \delta_{A-\lambda}V_\mu(\lambda) = I_{\Lambda[s,X]}$ and

$$V_\mu(\lambda)\Lambda^p[s,X] \subset \Lambda^{p-1}[s,X] \qquad (\lambda \in U_\mu, p = 0,\dots,n).$$

Let $\{\psi_i\}_{i=1}^\infty$ be a C^∞-partition of unity subordinated to the cover $\{U_\mu, \mu \in G\}$ of G, i.e., ψ_i are C^∞-functions, $0 \le \psi_i \le 1$, $\operatorname{supp}\psi_i \subset U_{\mu_i}$ for some $\mu_i \in G$, for each $\mu \in G$ there exists a neighbourhood U of μ such that all but finitely many of the functions ψ_i are 0 on U and $\sum_{i=1}^\infty \psi_i(\mu) = 1$ for each $\mu \in G$.

For $\lambda \in G$ set $V(\lambda) = \sum_{i=1}^\infty \psi_i(\lambda)V_{\mu_i}(\lambda)$. Then

$$\delta_{A-\lambda}V(\lambda) + V(\lambda)\delta_{A-\lambda} = \sum_{i=1}^{\infty}\big(\delta_{A-\lambda}V_{\mu_i}(\lambda) + V_{\mu_i}(\lambda)\delta_{A-\lambda}\big)\psi_i(\lambda) = I_{\Lambda[s,X]}$$

and

$$V(\lambda)\Lambda^p[s,X] \subset \Lambda^{p-1}[s,X]$$

for all $\lambda \in G$ and $p = 0,1,\dots,n$. □

Remark 3. It is possible to require also that $V(z)^2 = 0$ and $V(z)\delta_{A-z}V(z) = V(z)$ for all $z \in G$. In particular, in this case $V(z)$ is a generalized inverse of δ_{A-z}.

Indeed, let $V : G \to \mathcal{B}(\Lambda[s,X])$ be the C^∞-function constructed in Corollary 2, i.e., $\delta_{A-z}V(z) + V(z)\delta_{A-z} = I$ and $V(z)\Lambda^p[s,x] \subset \Lambda^{p-1}[s,X]$.

Clearly, $\delta_{A-z}V(z)\delta_{A-z} = \delta_{A-z}$. Set $V'(z) = V(z)\delta_{A-z}V(z)$. Then

$$\delta_{A-z}V'(z)\delta_{A-z} = \delta_{A-z}V(z)\delta_{A-z}V(z)\delta_{A-z} = \delta_{A-z}$$

and

$$V'(z)\delta_{A-z}V'(z) = V(z)\delta_{A-z}V(z)\delta_{A-z}V(z)\delta_{A-z}V(z) = V(z)\delta_{A-z}V(z) = V'(z).$$

Furthermore,

$$\begin{aligned}\delta_{A-z}V'(z)+V'(z)\delta_{A-z}&=\delta_{A-z}V(z)\delta_{A-z}V(z)+V(z)\delta_{A-z}V(z)\delta_{A-z}\\&=\delta_{A-z}V(z)+V(z)\delta_{A-z}=I.\end{aligned}$$

Finally, we have

$$V'(z)=\big(V'(z)\delta_{A-z}+\delta_{A-z}V'(z)\big)V'(z)=V'(z)+\delta_{A-z}V'(z)^2,$$

and so $\delta_{A-z}V'(z)^2=0$. Thus $V'(z)^2=\big(V'(z)\delta_{A-z}+\delta_{A-z}V'(z)\big)V'(z)^2=0$.

These additional properties of the generalized inverse V, however, are not essential for our purpose and we are not going to use them in the sequel.

In the following we fix a commuting n-tuple $A=(A_1,\dots,A_n)$ of bounded linear operators on a Banach space X, the set $G=\mathbb{C}^n\setminus\sigma_S(A)$ and a C^∞-function $V:G\to\mathcal{B}(\Lambda[s,X])$ with the properties of Corollary 2.

Consider the space $C^\infty(G,\Lambda[s,X])$. Clearly, this space can be identified with the set $\Lambda[s,C^\infty(G,X)]$.

The function $V:G\to\mathcal{B}(\Lambda[s,X])$ induces naturally the operator (denoted by the same symbol) $V:C^\infty(G,\Lambda[s,X])\to C^\infty(G,\Lambda[s,X])$ by

$$(Vy)(z)=V(z)y(z)\qquad\big(z\in G,y\in C^\infty(G,\Lambda[s,X])\big).$$

Similarly, we define the operator δ_{A-z} (or δ for short if no ambiguity can arise) acting in $C^\infty(G,\Lambda[s,X])$ by

$$(\delta y)(z)=\delta_{A-z}\,y(z)\qquad\big(z\in G,y\in C^\infty(G,\Lambda[s,X])\big).$$

Clearly, $\delta^2=0$, $V\delta+\delta V=I_{\Lambda[s,C^\infty(G,X)]}$ and both V and δ are "graded", i.e.,

$$\begin{aligned}&V\Lambda^p[s,C^\infty(G,X)]\subset\Lambda^{p-1}[s,C^\infty(G,X)]\qquad\text{and}\\&\delta\Lambda^p[s,C^\infty(G,X)]\subset\Lambda^{p+1}[s,C^\infty(G,X)].\end{aligned}$$

Consider now another indeterminates $\mathrm{d}\bar z=(\mathrm{d}\bar z_1,\dots,\mathrm{d}\bar z_n)$ and the space $\Lambda[s,\mathrm{d}\bar z,C^\infty(G,X)]$. Let $\bar\partial:\Lambda[s,\mathrm{d}\bar z,C^\infty(G,X)]\to\Lambda[s,\mathrm{d}\bar z,C^\infty(G,X)]$ be the linear mapping defined by

$$\bar\partial f s_{i_1}\wedge\cdots\wedge s_{i_p}\wedge\mathrm{d}\bar z_{j_1}\wedge\cdots\wedge\mathrm{d}\bar z_{j_q}=\sum_{k=1}^n\frac{\partial f}{\partial\bar z_k}\mathrm{d}\bar z_k\wedge s_{i_1}\wedge\cdots\wedge s_{i_p}\wedge\mathrm{d}\bar z_{j_1}\wedge\cdots\wedge\mathrm{d}\bar z_{j_q},$$

see Appendix A.3. Obviously, $\bar\partial^2=0$.

The operators V and δ can be lifted to $\Lambda[s,\mathrm{d}\bar z,C^\infty(G,X)]$ in a natural way. Clearly, the properties of V and δ are preserved: $\delta^2=0$, $V\delta+\delta V=I$ and both V and δ are graded. Note also that $\delta\bar\partial=-\bar\partial\delta$ and $(\bar\partial+\delta)^2=0$.

Let $W : \Lambda[s, \mathrm{d}\bar{z}, C^\infty(G, X)] \to \Lambda[s, \mathrm{d}\bar{z}, C^\infty(G, X)]$ be the mapping defined in the following way: if $\psi \in \Lambda[s, \mathrm{d}\bar{z}, C^\infty(G, X)]$, $\psi = \psi_0 + \cdots + \psi_n$, where ψ_j is the part of ψ of degree j in $\mathrm{d}\bar{z}$, then set $W\psi = \eta_0 + \cdots + \eta_n$, where

$$\begin{aligned} \eta_0 &= V\psi_0, \\ \eta_1 &= V(\psi_1 - \bar{\partial}\eta_0), \\ &\vdots \\ \eta_n &= V(\psi_n - \bar{\partial}\eta_{n-1}). \end{aligned} \tag{1}$$

Note that η_j is the part of $W\psi$ of degree j in $\mathrm{d}\bar{z}$.

Lemma 4. *Let* $W : \Lambda[s, \mathrm{d}\bar{z}, C^\infty(G, X)] \to \Lambda[s, \mathrm{d}\bar{z}, C^\infty(G, X)]$ *be the mapping defined by (1). Then:*

(i) $\operatorname{supp} W\psi \subset \operatorname{supp} \psi$ *for all* ψ;

(ii) *if* G' *is an open subset of* G *and* $\psi \in \Lambda[s, \mathrm{d}\bar{z}, C^\infty(G, X)]$ *satisfies* $(\bar{\partial} + \delta)\psi = 0$ *on* G', *then* $(\bar{\partial} + \delta)W\psi = \psi$ *on* G';

(iii) $(\bar{\partial} + \delta)W(\bar{\partial} + \delta) = \bar{\partial} + \delta$.

Proof. (i) Clear.

(ii) Let $\psi = \psi_0 + \cdots + \psi_n$, where ψ_j is the part of ψ of degree j in $\mathrm{d}\bar{z}$. The condition $(\bar{\partial} + \delta)\psi = 0$ on G' can be rewritten as

$$\begin{aligned} \delta\psi_0 &= 0, \\ \bar{\partial}\psi_0 + \delta\psi_1 &= 0, \\ &\vdots \\ \bar{\partial}\psi_{n-1} + \delta\psi_n &= 0 \end{aligned} \tag{2}$$

(the condition $\bar{\partial}\psi_n = 0$ is satisfied automatically).

Let $W\psi = \eta_0 + \cdots + \eta_n$, where η_j are defined by (1). The required condition $(\bar{\partial} + \delta)W\psi = \psi$ becomes

$$\begin{aligned} \delta\eta_0 &= \psi_0, \\ \bar{\partial}\eta_0 + \delta\eta_1 &= \psi_1, \\ &\vdots \\ \bar{\partial}\eta_{n-1} + \delta\eta_n &= \psi_n \end{aligned} \tag{3}$$

on G' (again, $\bar{\partial}\eta_n = 0$ automatically).

By (1) and (2), we have $\delta\eta_0 = \delta V\psi_0 = (\delta V + V\delta)\psi_0 = \psi_0$ and $\bar{\partial}\eta_0 + \delta\eta_1 = \bar{\partial}\eta_0 + \delta V(\psi_1 - \bar{\partial}\eta_0) = \bar{\partial}\eta_0 + (I - V\delta)(\psi_1 - \bar{\partial}\eta_0) = \psi_1 - V\delta(\psi_1 - \bar{\partial}\eta_0) = \psi_1$, since $\delta(\psi_1 - \bar{\partial}\eta_0) = \delta\psi_1 + \bar{\partial}\delta\eta_0 = \delta\psi_1 + \bar{\partial}\psi_0 = 0$.

We prove (3) by induction. Suppose that $\bar{\partial}\eta_{j-1} + \delta\eta_j = \psi_j$ for some $j \geq 1$. Then $\delta(\psi_{j+1} - \bar{\partial}\eta_j) = \delta\psi_{j+1} + \bar{\partial}\delta\eta_j = \delta\psi_{j+1} + \bar{\partial}\psi_j = 0$ and, by the induction assumption, $\bar{\partial}\eta_j + \delta\eta_{j+1} = \bar{\partial}\eta_j + \delta V(\psi_{j+1} - \bar{\partial}\eta_j) = \bar{\partial}\eta_j + (I - V\delta)(\psi_{j+1} - \bar{\partial}\eta_j) = \psi_{j+1}$.

(iii) Since $(\bar{\partial} + \delta)^2 = 0$, the statement follows from (ii). □

Let P be the natural projection $P : \Lambda[s, \mathrm{d}\bar{z}, C^\infty(G,X)] \to \Lambda[\mathrm{d}\bar{z}, C^\infty(G,X)]$ that annihilates all terms containing at least one of the indeterminates $s_1, \ldots, s_n$ and leaves invariant all the remaining terms.

Let U be a neighbourhood of $\sigma_S(A)$. Let f be a function analytic in U. It is possible to find a compact neighbourhood Δ of $\sigma_S(A)$ such that $\Delta \subset U$ and the boundary $\partial\Delta$ is a smooth surface. Define $f(A) : X \to X$ by

$$f(A)x = \frac{-1}{(2\pi i)^n} \int_{\partial\Delta} Pf(z)Wxs \wedge \mathrm{d}z \qquad (x \in X), \tag{4}$$

where $\mathrm{d}z$ stands for $\mathrm{d}z_1 \wedge \cdots \wedge \mathrm{d}z_n$ and $s = s_1 \wedge \cdots \wedge s_n$. By the Stokes formula,

$$f(A)x = \frac{-1}{(2\pi i)^n} \int_{\Delta} \bar{\partial}\varphi Pf(z)Wxs \wedge \mathrm{d}z,$$

where φ is a C^∞-function equal to 0 on a neighbourhood of $\sigma_S(A)$ and to 1 on $\mathbb{C}^n \setminus \Delta$ (consequently, $\varphi = 1$ also on $\partial\Delta$).

On $\mathbb{C}^n \setminus \Delta$ we have

$$\bar{\partial}\varphi PfWxs = Pf(\bar{\partial} + \delta)Wxs = Pfxs = 0.$$

Thus we can write

$$f(A)x = \frac{-1}{(2\pi i)^n} \int_{\mathbb{C}^n} \bar{\partial}\varphi Pf(z)Wxs \wedge \mathrm{d}z. \tag{5}$$

It is clear from the Stokes theorem that the definition of $f(A)x$ does not depend on the choice of the function φ and, by (5), it is independent of Δ.

We show that $f(A)$ does not depend on the choice of the generalized inverse V which determines W.

Suppose that W_1, W_2 are two operators satisfying

$$(\bar{\partial} + \delta)W_i xs = xs \qquad (i = 1, 2).$$

For those z where $\varphi \equiv 1$ we have

$$(\bar{\partial} + \delta)\varphi f(z)(W_1 - W_2)xs = 0,$$

and so the form $\eta = (\bar{\partial} + \delta)\varphi f(z)(W_1 - W_2)xs$ has a compact support. We have

$$\begin{aligned}
&\int_{\mathbb{C}^n} \bar{\partial}\varphi Pf(z)W_1xs \wedge \mathrm{d}z - \int_{\mathbb{C}^n} \bar{\partial}\varphi Pf(z)W_2xs \wedge \mathrm{d}z \\
&= \int_{\mathbb{C}^n} P\bar{\partial}\varphi f(z)(W_1 - W_2)xs \wedge \mathrm{d}z = \int_{\mathbb{C}^n} P(\bar{\partial} + \delta)\varphi f(z)(W_1 - W_2)xs \wedge \mathrm{d}z \\
&= \int_{\mathbb{C}^n} P\eta \wedge \mathrm{d}z = \int_{\mathbb{C}^n} P(\bar{\partial} + \delta)W_1\eta \wedge \mathrm{d}z = \int_{\mathbb{C}^n} \bar{\partial}PW_1\eta \wedge \mathrm{d}z = 0
\end{aligned}$$

by the Stokes theorem.

In fact, in the same way it is possible to show that

$$f(A)x = \frac{-1}{(2\pi i)^n} \int_{\mathbb{C}^n} \bar{\partial}\varphi f P\psi \wedge \mathrm{d}z \tag{6}$$

for any form ψ satisfying $(\bar{\partial} + \delta)\psi = fxs$ on $\mathbb{C}^n \setminus \sigma_S(A)$.

It is possible to express the mapping PW that appears in the definition of the functional calculus more explicitly. By the definition of W, we have

$$PWxs = (-1)^{n-1}V(\bar{\partial}V)^{n-1}xs = (-1)^{n-1}V_0\bar{\partial}V_1\bar{\partial}\cdots\bar{\partial}V_{n-1}xs.$$

Note that we can write formulas (4) and (5) also globally:

$$\begin{aligned} f(A) &= \frac{-1}{(2\pi i)^n} \int_{\partial\Delta} Pf(z)WIs \wedge \mathrm{d}z = \frac{-1}{(2\pi i)^n} \int_{\mathbb{C}^n} \bar{\partial}\varphi Pf(z)WIs \wedge \mathrm{d}z \\ &= \frac{(-1)^n}{(2\pi i)^n} \int_{\mathbb{C}^n} \bar{\partial}\varphi fV(\bar{\partial}V)^{n-1}Is \wedge \mathrm{d}z, \end{aligned} \tag{7}$$

where $I = I_X$ is the identity operator on X. The coefficients of forms in (7) are $\mathcal{B}(X)$-valued C^∞-functions. Therefore $f(A) \in \mathcal{B}(X)$.

Proposition 7. *For $n = 1$, the functional calculus defined by (7) coincides with the classical functional calculus given by the Cauchy formula.*

Proof. Let $A \in \mathcal{B}(X)$ and let f be a function analytic on a neighbourhood of $\sigma(A)$. Then $Wxs = Vxs = (A-z)^{-1}x$. Thus, for a suitable contour Σ surrounding $\sigma(A)$, we have

$$f(A) = \frac{-1}{2\pi i}\int_\Sigma PfWIs \wedge \mathrm{d}z = \frac{-1}{2\pi i}\int_\Sigma f(z)(A-z)^{-1}I\mathrm{d}z = \frac{1}{2\pi i}\int_\Sigma f(z)(z-A)^{-1}\mathrm{d}z,$$

which is the Cauchy formula. □

We postpone the proof of basic properties of this functional calculus to Section 30 where we prove it more generally, for functions analytic on a neighbourhood of the Taylor spectrum.

It is worth to note that this simpler split functional calculus is sufficient for introducing the functional calculus for n-tuples of elements in commutative Banach algebras. Indeed, let $\mathcal{A}$ be a commutative Banach algebra and $a = (a_1, \dots, a_n) \in \mathcal{A}^n$. Consider the n-tuple $L_A = (L_{A_1}, \dots, L_{A_n}) \in \mathcal{B}(\mathcal{A})^n$. Then $\sigma_S(L_A) = \sigma^{\mathcal{A}}(a)$ and for any function f analytic on a neighbourhood of $\sigma^{\mathcal{A}}(a)$ we may define $f(L_A) \in \mathcal{B}(\mathcal{A})$. Then the functional calculus for a may be defined by $f(a) = f(L_A)(1_{\mathcal{A}})$. We postpone the details to Section 30.

29 Local spectrum for n-tuples of operators

Definition 1. Let $A = (A_1, \dots, A_n)$ be a commuting n-tuple of operators on a Banach space X and let $x \in X$. The *local spectrum* of A at the point x is the subset $\gamma_x(A) \subset \mathbb{C}^n$ defined by: $\lambda = (\lambda_1, \dots, \lambda_n) \notin \gamma_x(A)$ if and only if there exist a neighbourhood U of λ and analytic functions $f_1, \dots, f_n : U \to X$ such that $\sum_{i=1}^n (A_i - z_i) f_i(z) = x \quad (z \in U)$.

Clearly, $\gamma_x(A)$ is a closed subset of $\mathbb{C}^n$. For single operators this definition coincides with the definition given in Section 14.

Theorem 2. *Let $A = (A_1, \dots, A_n)$ be a commuting n-tuple of operators on a Banach space X, let $x \in X$. Then $\lambda \notin \gamma_x(A)$ if and only if there exist a neighbourhood U of λ and $\psi \in \Lambda^{n-1}\big[s, \mathrm{d}\bar{z}, C^\infty(U, X)\big]$ such that $(\bar\partial + \delta_{A-z})\psi = xs \quad (z \in U)$, where $s = s_1 \wedge \dots \wedge s_n$.*

Proof. As in the previous section, write for short δ instead of δ_{A-z}.

Suppose that $\sum_{i=1}^n (A_i - z_i) f_i(z) = x \quad (z \in U)$ for some analytic functions $f_i : U \to X$. Set $\psi = \sum_{i=1}^n (-1)^{i-1} f_i(z) s_1 \wedge \dots \wedge s_{i-1} \wedge s_{i+1} \wedge \dots \wedge s_n$. Then $(\bar\partial + \delta)\psi = \delta\psi = \sum_{i=1}^n (A_i - z_i) f_i(z) s_1 \wedge \dots \wedge s_n = xs$.

Let now U be a polydisc centered at λ, and let $\psi \in \Lambda^{n-1}\big[s, \mathrm{d}\bar{z}, C^\infty(U, X)\big]$ be a form satisfying $(\bar\partial + \delta)\psi = xs$. Write $\psi = \psi_0 + \dots + \psi_{n-1}$, where ψ_j is of degree j in $\mathrm{d}\bar{z}$. Then

$$
\begin{aligned}
\delta\psi_0 &= xs, \\
\bar\partial\psi_0 + \delta\psi_1 &= 0, \\
&\;\;\vdots \\
\bar\partial\psi_{n-2} + \delta\psi_{n-1} &= 0, \\
\bar\partial\psi_{n-1} &= 0.
\end{aligned}
$$

By A.3.5, the sequence

$$
\begin{aligned}
0 \to H(U, X) \xrightarrow{j} C^\infty(U, X) \xrightarrow{\bar\partial} \Lambda^1[s, \mathrm{d}\bar{z}, C^\infty(U, X)] \xrightarrow{\bar\partial} \quad \cdots \\
\cdots \quad \xrightarrow{\bar\partial} \Lambda^n[s, \mathrm{d}\bar{z}, C^\infty(U, X)] \to 0
\end{aligned}
$$

is exact, where j is the natural embedding. Thus there exists φ_{n-2} of degree $n-2$ in $\mathrm{d}\bar{z}$ such that $\bar\partial\varphi_{n-2} = \psi_{n-1}$. Then $0 = \bar\partial\psi_{n-2} + \delta\bar\partial\varphi_{n-2} = \bar\partial(\psi_{n-2} - \delta\varphi_{n-2})$, and so there exists φ_{n-3} of degree $n-3$ in $\mathrm{d}\bar{z}$ such that $\bar\partial\varphi_{n-3} = \psi_{n-2} - \delta\varphi_{n-2}$.

If we continue in this way we can construct forms $\varphi_i \quad (i = n-2, n-3, \dots, 0)$ of degree i in $\mathrm{d}\bar{z}$ such that $\bar\partial\varphi_i = \psi_{i+1} - \delta\varphi_{i+1}$. Set $\xi = \psi_0 - \delta\varphi_0$. Then $\bar\partial\xi = \bar\partial\psi_0 + \delta\bar\partial\varphi_0 = \bar\partial\psi_0 + \delta\psi_1 = 0$. Thus $\xi \in \Lambda^{n-1}\big[s, C^\infty(U, X)\big]$ has analytic coefficients, $\xi = \sum_{i=1}^n f_i(z) s_1 \wedge \dots \wedge s_{i-1} \wedge s_{i+1} \wedge \dots \wedge s_n$ for some analytic functions $f_i : U \to X$. Further, $\delta\xi = \delta\psi_0 = xs$, and so $\sum_{i=1}^n (-1)^{i-1}(A_i - z_i) f_i(z) x = xs \quad (z \in U)$. □

Recall that $\sigma_\delta(A)$ denotes the surjective spectrum of A.

Theorem 3. *Let $A = (A_1, \ldots, A_n)$ be a commuting n-tuple of operators on a Banach space X and let $x \in X$. Then:*

(i) *$\gamma_x(A) \subset \sigma_\delta(A)$ for all $x \in X$;*

(ii) *the set of all $x \in X$ with $\gamma_x(A) = \sigma_\delta(A)$ is residual.*

Proof. (i) Let $x \in X$ and $\lambda \notin \sigma_\delta(A)$. Then there is an open neighbourhood U of λ disjoint with $\sigma_\delta(A)$. Thus the operator $\delta_{A-z}^{n-1} : \Lambda^{n-1}[s, X] \to \Lambda^n[s, X]$ is onto for $z \in U$. By Corollary 11.10, there is a neighbourhood V of λ and an analytic function $f : V \to \Lambda^{n-1}[s, X]$ such that $\delta_{A-z}^{n-1} f(z) = xs \quad (z \in V)$. This means exactly that $\lambda \notin \gamma_x(A)$.

(ii) Let $\{w^{(j)}\}$ be a countable dense subset of $\sigma_\delta(A)$. For each j let $M_j = (A_1 - w_1^{(j)})X + \cdots + (A_n - w_n^{(j)})X$. Since $w^{(j)} \in \sigma_\delta(A)$, we have $M_j \neq X$, and so it is a set of the first category by A.1.8. We have

$$\{x \in X : \gamma_x(A) \neq \sigma_\delta(A)\} = \bigcup_j \{x \in X : w^{(j)} \notin \gamma_x(A)\} \subset \bigcup_j M_j,$$

which is a set of the first category. □

Next we define an analogue of the analytic residuum and the local spectrum σ_x for n-tuples of operators.

Definition 4. Let $A = (A_1, \ldots, A_n)$ be a commuting n-tuple of operators on a Banach space X. Denote by $\rho(A)$ the union of all open sets $U \subset \mathbb{C}^n$ with the following property: the sequence

$$\begin{aligned} 0 \to \Lambda^0\big[s, \mathrm{d}\bar{z}, C^\infty(G, X)\big] &\xrightarrow{\bar{\partial}+\delta} \Lambda^1\big[s, \mathrm{d}\bar{z}, C^\infty(G, X)\big] \xrightarrow{\bar{\partial}+\delta} \quad \cdots \\ \cdots \quad &\xrightarrow{\bar{\partial}+\delta} \Lambda^n\big[s, \mathrm{d}\bar{z}, C^\infty(G, X)\big] \end{aligned} \tag{1}$$

is exact for each open subset $G \subset U$.

Let $S(A) = \mathbb{C}^n \setminus \rho(A)$. For $x \in X$ define $\sigma_x(A) = \gamma_x(A) \cup S(A)$.

For $n = 1$, this definition coincides with that in Section 14. Indeed, if $A_1 \in \mathcal{B}(X)$, $G \subset \mathbb{C}$ is open and the sequence

$$0 \to \Lambda^0[s_1, \mathrm{d}\bar{z}_1, C^\infty(G, X)] \xrightarrow{\bar{\partial}+\delta} \Lambda^1[s_1, \mathrm{d}\bar{z}_1, C^\infty(G, X)]$$

is not exact, then there exists a non-zero C^∞-function $f : G \to X$ such that $(\bar{\partial} + \delta)f = 0$. This means that f is analytic and $(A_1 - z_1)f(z) = 0 \quad (z \in G)$.

We show that $S(A)$ is contained in the Taylor spectrum $\sigma_T(A)$, i.e., that sequence (1) is exact for each open subset $G \subset \mathbb{C}^n \setminus \sigma_T(A)$.

The exactness of (1) for $p = 0$ is clear. If $f \in C^\infty(G, X)$ and $(\bar{\partial} + \delta)f = 0$, then f is analytic on G and $\delta f = 0$. Since δ_{A-z} is injective for $z \in G \subset \mathbb{C}^n \setminus \sigma_T(A)$, we have $f = 0$ on G.

To prove the exactness of (1) for $p \geq 1$ we need several lemmas.

Proposition 5. Let $w \in \mathbb{C}^n \setminus \sigma_T(A)$, $1 \leq p \leq 2n$ and let $G \subset \mathbb{C}^n \setminus \sigma_T(A)$ be an open neighbourhood of w. Let $\eta \in \Lambda^p[s, \mathrm{d}\bar{z}, C^\infty(G, X)]$ and $(\bar{\partial} + \delta)\eta = 0$. Then there exists an open set G', $w \in G' \subset G$ and $\psi \in \Lambda[s, \mathrm{d}\bar{z}, C^\infty(G', X)]$ such that $(\bar{\partial} + \delta)\psi = \eta$ on G'.

Proof. Let $\eta \in \Lambda^p[s, \mathrm{d}\bar{z}, C^\infty(G, X)]$ and $(\bar{\partial} + \delta)\eta = 0$. Let $\eta = \eta_0 + \cdots + \eta_p$ where η_j is the part of η of degree j in $\mathrm{d}\bar{z}$. The condition $(\bar{\partial} + \delta)\eta = 0$ means

$$\begin{aligned} \bar{\partial}\eta_p &= 0, \\ \bar{\partial}\eta_{p-1} + \delta\eta_p &= 0, \\ &\vdots \\ \bar{\partial}\eta_0 + \delta\eta_1 &= 0, \\ \delta\eta_0 &= 0. \end{aligned}$$

Let D be an open polydisc, $w \in D \subset G$.

Since $\bar{\partial}\eta_p = 0$, there exists $\psi_{p-1} \in \Lambda^{p-1}[s, \mathrm{d}\bar{z}, C^\infty(D, X)]$ such that $\bar{\partial}\psi_{p-1} = \eta_p$ by A.3.5. We have $\bar{\partial}(\eta_{p-1} - \delta\psi_{p-1}) = \bar{\partial}\eta_{p-1} + \delta\bar{\partial}\psi_{p-1} = \bar{\partial}\eta_{p-1} + \delta\eta_p = 0$, and so there exists $\psi_{p-2} \in \Lambda^{p-2}[s, \mathrm{d}\bar{z}, C^\infty(D, X)]$ with $\bar{\partial}\psi_{p-2} = \eta_{p-1} - \delta\psi_{p-1}$. In the same way we construct inductively $\psi_j \in \Lambda^j[s, \mathrm{d}\bar{z}, C^\infty(D, X)]$ $(j = p-1, \dots, 0)$ such that $\bar{\partial}\psi_j = \eta_{j+1} - \delta\psi_{j+1}$ $(j = p-1, \dots, 0)$.

At the end we have $\delta\bar{\partial}\psi_0 = \delta\eta_1 = -\bar{\partial}\eta_0$, and so $\bar{\partial}(\delta\psi_0 - \eta_0) = 0$. Since the degree of $\delta\psi_0 - \eta_0$ in $\mathrm{d}\bar{z}$ is zero, the coefficients of $\delta\psi_0 - \eta_0$ are analytic functions and $\delta(\delta\psi_0 - \eta_0) = 0$. By Theorem 11.9, there are an open set G', $w \in G' \subset D \subset G$ and ψ_0' on G' whose coefficients are analytic functions such that $\delta\psi_0' = \delta\psi_0 - \eta_0$.

Set $\psi = \psi_{p-1} + \cdots + \psi_1 + \psi_0 - \psi_0'$. Then $(\bar{\partial} + \delta)\psi = \bar{\partial}\psi_{p-1} + (\bar{\partial}\psi_{p-2} + \delta\psi_{p-1}) + \cdots + (\bar{\partial}\psi_0 + \delta\psi_1) + \delta\psi_0 - \delta\psi_0' = \eta_p + \cdots + \eta_1 + \eta_0 = \eta$ on G'. □

Lemma 6. *Let U_1, U_2 be open subsets of $\mathbb{C}^n$ and let $f \in C^\infty(U_1 \cap U_2)$. Then $f = f_1 - f_2$ on $U_1 \cap U_2$ for some functions $f_1 \in C^\infty(U_1)$ and $f_2 \in C^\infty(U_2)$.*

Proof. Let $\{\varphi_j\}$ be a C^∞-partition of unity subordinate to the cover $\{U_1, U_2\}$ of $U_1 \cup U_2$. This means that $0 \leq \varphi_j \leq 1$, $\sum_j \varphi_j = 1$, each $z \in U_1 \cup U_2$ has a neighbourhood intersecting only finitely many supports of φ_j, and either $\operatorname{supp}\varphi_j \subset U_1$ or $\operatorname{supp}\varphi_j \subset U_2$.

Define $f_1 \in C^\infty(U_1)$ and $f_2 \in C^\infty(U_2)$ by

$$f_1(z) = \begin{cases} \sum_{\operatorname{supp}\varphi_j \subset U_2} f(z)\varphi_j(z) & (z \in U_1 \cap U_2), \\ 0 & (z \in U_1 \setminus U_2) \end{cases}$$

and

$$f_2(z) = \begin{cases} -\sum_{\operatorname{supp}\varphi_j \not\subset U_2} f(z)\varphi_j(z) & (z \in U_1 \cap U_2), \\ 0 & (z \in U_2 \setminus U_1). \end{cases}$$

Obviously, $f_1 - f_2 = f$ on $U_1 \cap U_2$ and f_1 is smooth on $U_1 \setminus \partial U_2$. Let $\lambda \in U_1 \cap \partial U_2$ Then there exists a neighbourhood V of λ intersecting only finitely many supports of φ_i. Thus there exists a neighbourhood $V' \subset V$ of λ such that $V' \cap \operatorname{supp} \varphi_i = \emptyset$ whenever $\operatorname{supp} \varphi_i \subset U_2$. So f_1 vanishes on V'. Hence $f_1 \in C^\infty(U_1)$.

Since $\operatorname{supp} \varphi_j \not\subset U_2$ implies $\operatorname{supp} \varphi_j \subset U_1$, the same considerations can be done for f_2. □

Lemma 7. *Let $2 \le p \le 2n$, let U_1, U_2 be open subsets of $\mathbb{C}^n$ and suppose that the sequence*

$$\Lambda^{p-2}[s, \mathrm{d}\bar{z}, C^\infty(U_1 \cap U_2, X)] \xrightarrow{\bar\partial + \delta} \Lambda^{p-1}[s, \mathrm{d}\bar{z}, C^\infty(U_1 \cap U_2, X)] \xrightarrow{\bar\partial + \delta} \Lambda^{p}[s, \mathrm{d}\bar{z}, C^\infty(U_1 \cap U_2, X)]$$

is exact. Let $\eta \in \Lambda^p[s, \mathrm{d}\bar{z}, C^\infty(U_1 \cup U_2, X)]$ and $(\bar\partial + \delta)\psi_i = \eta$ on U_i for some $\psi_i \in \Lambda^{p-1}[s, \mathrm{d}\bar{z}, C^\infty(U_i, X)]$ $(i = 1, 2)$.

Then there exists a form ψ in $\Lambda^{p-1}[s, \mathrm{d}\bar{z}, C^\infty(U_1 \cup U_2, X)]$ such that $(\bar\partial + \delta)\psi = \eta$ on $U_1 \cup U_2$.

Proof. We have $(\bar\partial + \delta)(\psi_1 - \psi_2) = 0$ on $U_1 \cap U_2$, and so $\psi_1 - \psi_2 = (\bar\partial + \delta)\xi$ for some $\xi \in \Lambda^{p-2}[s, \mathrm{d}\bar{z}, C^\infty(U_1 \cap U_2, X)]$. By the previous lemma, we can write $\xi = \xi_1 - \xi_2$ on $U_1 \cap U_2$ for some $\xi_i \in \Lambda^{p-2}[s, \mathrm{d}\bar{z}, C^\infty(U_i, X)]$ $(i = 1, 2)$. Define

$$\psi(z) = \begin{cases} \psi_1(z) - (\bar\partial + \delta)\xi_1(z) & (z \in U_1), \\ \psi_2(z) - (\bar\partial + \delta)\xi_2(z) & (z \in U_2). \end{cases}$$

The definition is correct, since for $z \in U_1 \cap U_2$ we have $\psi_1(z) - (\bar\partial + \delta)\xi_1(z) = \psi_1(z) - (\bar\partial + \delta)(\xi + \xi_2(z)) = \psi_2(z) - (\bar\partial + \delta)\xi_2(z)$. Clearly, ψ is the required solution of $(\bar\partial + \delta)\psi = \eta$ on $U_1 \cup U_2$. □

Lemma 8. *Let $2 \le p \le 2n$ and let G be an open subset of $\mathbb{C}^n$. Suppose that the sequence*

$$\Lambda^{p-2}[s, \mathrm{d}\bar{z}, C^\infty(G', X)] \xrightarrow{\bar\partial + \delta} \Lambda^{p-1}[s, \mathrm{d}\bar{z}, C^\infty(G', X)] \xrightarrow{\bar\partial + \delta} \Lambda^{p}[s, \mathrm{d}\bar{z}, C^\infty(G', X)]$$

is exact for each open subset $G' \subset G$. Let $\eta \in \Lambda^p[s, \mathrm{d}\bar{z}, C^\infty(G, X)]$ and suppose that, for every $w \in G$, there are a neighbourhood $U_w \subset G$ of w and a form $\psi_w \in \Lambda^{p-1}[s, \mathrm{d}\bar{z}, C^\infty(U_w, X)]$ satisfying $(\bar\partial + \delta)\psi_w = \eta$ on U_w. Then there exists a global solution $\psi \in \Lambda^{p-1}[s, \mathrm{d}\bar{z}, C^\infty(G, X)]$ satisfying $(\bar\partial + \delta)\psi = \eta$ on G.

Proof. Since every compact subset $K \subset G$ can be covered by a finite number of neighbourhoods U_w, the repetitive use of Lemma 7 gives that there exist a neighbourhood U_K of K and a form $\psi_K \in \Lambda^{p-1}[C^\infty(U_K, X), s, \mathrm{d}\bar{z}]$ with $(\bar\partial + \delta)\psi_K = \eta$ on U_K.

Let $(K_i)_{i=1}^\infty$ be an increasing sequence of compact subsets of G such that $\bigcup_{i=1}^\infty K_i = G$. We will construct inductively $(p-1)$-forms ψ_i defined on a neighbourhood of K_i such that $(\bar\partial + \delta)\psi_i = \eta$ and $\psi_{i+1} = \psi_i$ on a neighbourhood of K_i.

Suppose that $\psi_1, \dots, \psi_k$ have already been constructed and let ψ'_{k+1} be an arbitrary solution of $(\bar{\partial} + \delta)\psi'_{k+1} = \eta$ on a neighbourhood of K_{k+1}. Then $(\bar{\partial} + \delta)(\psi'_{k+1} - \psi_k) = 0$ on a neighbourhood of K_k, and so $\psi'_{k+1} - \psi_k = (\bar{\partial}+\delta)\xi$ for some $(p-2)$-form ξ defined on a neighbourhood V of K_k. Let $\varphi \in C^\infty(G)$ satisfy $\varphi = 1$ on a neighbourhood of K_k and $\varphi = 0$ outside V. Set $\psi_{k+1} = \psi'_{k+1} - (\bar{\partial}+\delta)\varphi\xi$. Then $(\bar{\partial} + \delta)\psi_{k+1} = \eta$ on a neighbourhood of K_{k+1} and $\psi_{k+1} = \psi_k$ on a neighbourhood of K_k. Define $\psi(z) = \lim_{k\to\infty} \psi_k(z)$. Clearly, ψ is the required global solution. □

Using Lemmas 5 and 8 inductively we get the following theorem.

Theorem 9. *Let $G \subset \mathbb{C}^n \setminus \sigma_T(A)$ be an open subset. Then the sequence*

$$0 \to \Lambda^0[s, \mathrm{d}\bar{z}, C^\infty(G, X)] \xrightarrow{\bar{\partial}+\delta} \cdots \xrightarrow{\bar{\partial}+\delta} \Lambda^{2n}[s, \mathrm{d}\bar{z}, C^\infty(G, X)] \to 0$$

is exact. In particular, $S(A) \subset \sigma_T(A)$ and $\sigma_x(A) \subset \sigma_T(A)$ for all $x \in X$.

Corollary 10. For each $x \in X$ there is a solution $\psi \in \Lambda^{n-1}[s, \mathrm{d}\bar{z}, C^\infty(\mathbb{C}^n \setminus \sigma_x(A), X)]$ satisfying $(\bar{\partial} + \delta)\psi = xs$.

Proof. Use Lemma 8 for $G = \mathbb{C}^n \setminus \sigma_x(A)$ and $p = n$. The assumptions of Lemma 8 are satisfied by the definition of $\sigma_x(A) = S(A) \cup \gamma_x(A)$. □

Corollary 11. *Let $G \subset \mathbb{C}^n \setminus \sigma_T(A)$, $\eta \in \Lambda[s, \mathrm{d}\bar{z}, C^\infty(G, X)]$ and $(\bar{\partial} + \delta)\eta = 0$. Then it is possible to find a form $\psi \in \Lambda[s, \mathrm{d}\bar{z}, C^\infty(G, X)]$ such that $(\bar{\partial} + \delta)\psi = \eta$ and the support of ψ is contained in any given neighbourhood of* $\operatorname{supp} \eta$.

Proof. Let V be a neighbourhood of $\operatorname{supp} \eta$. By Theorem 9, there exists $\xi \in \Lambda[s, \mathrm{d}\bar{z}, C^\infty(G, X)]$ such that $(\bar{\partial} + \delta)\xi = \eta$ on G. Then $(\bar{\partial} + \delta)\xi = 0$ on $G \setminus \operatorname{supp} \eta$. By Theorem 9, there exists $\xi' \in \Lambda[s, \mathrm{d}\bar{z}, C^\infty(G\setminus \operatorname{supp} \eta, X)]$ such that $(\bar{\partial}+\delta)\xi' = \xi$.

Let φ be a C^∞-function such that $\varphi = 0$ on a neighbourhood of $\operatorname{supp} \eta$ and $\varphi = 1$ on a neighbourhood of $\mathbb{C}^n \setminus V$. Set $\psi = \xi - (\bar{\partial} + \delta)\varphi\xi'$. Then $(\bar{\partial} + \delta)\psi = \eta$ on G and $\operatorname{supp} \psi \subset V$. □

Remark 12. Without any change it is possible to prove the preceding theorem in a more general form. Let $z \mapsto A(z)$ be an analytic function defined on an open subset $G \subset \mathbb{C}^n$ such that the values $A(z)$ are Taylor regular n-tuples of operators on X for all $z \in G$. Let $\psi \in \Lambda[s, \mathrm{d}\bar{z}, C^\infty(G, X)]$ satisfy $(\bar{\partial} + \delta_{A(z)})\psi = 0$. Then there exists a form $\theta \in \Lambda[s, \mathrm{d}\bar{z}, C^\infty(G, X)]$ such that $\psi = (\bar{\partial} + \delta_{A(z)})\theta$. Moreover, θ can be chosen in such a way that $\operatorname{supp} \theta$ is contained in any given neighbourhood of $\operatorname{supp} \psi$.

As for single operators, $\sigma_x(A) \neq \emptyset$ for each $x \neq 0$. We postpone the proof of this fact till the next section.

Theorem 13. *Let $w = (w_1, \dots, w_n) \notin S(A)$ and $(A_1 - w_1)X + \cdots + (A_n - w_n)X = X$. Then $A - w$ is Taylor regular.*

Proof. We must prove the exactness of the Koszul complex

$$0 \to \Lambda^0[s, X] \xrightarrow{\delta^0_{A-w}} \Lambda^1[s, X] \xrightarrow{\delta^1_{A-w}} \cdots \xrightarrow{\delta^{n-2}_{A-w}} \Lambda^{n-1}[s, X] \xrightarrow{\delta^{n-1}_{A-w}} \Lambda^n[s, X] \to 0. \qquad (2)$$

The condition $(A_1 - w_1)X + \cdots + (A_n - w_n)X = X$ means that δ^{n-1}_{A-w} is onto and the Koszul complex (2) is exact at $\Lambda^n[s, X]$.

We prove the exactness of the complex (2) by "downward" induction.

Suppose that $1 \le p \le n$ and $\operatorname{Ker} \delta^p_{A-w} = \operatorname{Ran} \delta^{p-1}_{A-w}$. Let $\psi \in \operatorname{Ker} \delta^{p-1}_{A-w}$. By Lemma 11.3, there exists a neighbourhood U of w such that $\operatorname{Ker} \delta^p_{A-z} = \operatorname{Ran} \delta^{p-1}_{A-z}$ for all $z \in U$. By Example 10.24 (iv), the function $z \mapsto \delta^{p-1}_{A-z}$ is regular and analytic in U. Thus there exist a neighbourhood U_0 of w and an analytic function $f : U_0 \to \Lambda^{p-1}[s, X]$ such that $f(w) = \psi$ and $\delta^{p-1}_{A-z} f(z) = 0 \quad (z \in U_0)$. Then $(\bar\partial + \delta_{A-z})f = 0$ on U_0. We can assume that $U_0 \cap S(A) = \emptyset$.

Since $w \notin S(A)$, there exists $g \in C^\infty(U_0, \Lambda^{p-2}[s, \mathrm{d}\bar z, X])$ such that $(\bar\partial + \delta_{A-z})g = f$. Let $g_0 \in C^\infty(U_0, \Lambda^{p-2}[s, X])$ be the part of g of degree 0 in $\mathrm{d}\bar z$. Then $\delta_{A-z} g_0 = f$. In particular, $\psi = f(w) = \delta_{A-w} g_0(w) \in \operatorname{Ran} \delta^{p-2}_{A-w}$.

Continuation of the induction argument gives that $T - w$ is Taylor regular. □

Theorem 14. *The set* $\{x \in X : \sigma_x(A) \ne \sigma_T(A)\}$ *is of the first category.*

Proof. Let $w \in \sigma_T(A)$. We prove first that $\{x \in X : w \notin \sigma_x(A)\}$ is of the first category. Indeed, if $w \notin \sigma_x(A)$, then $x = \sum_{j=1}^n (A_j - z_j) f_j(z)$ for some analytic X-valued functions defined on a neighbourhood of w. In particular, $x \in (A_1 - w_1)X + \cdots + (A_n - z_n)X = \operatorname{Ran} \delta^{n-1}_{A-w}$. The previous theorem implies that $(A_1 - w_1)X + \cdots + (A_n - w_n)X \ne X$, and so $\{x \in X : w \notin \sigma_x(A)\} \subset \operatorname{Ran} \delta^{n-1}_{A-w}$, which is a set of the first category.

Let $\{w^{(j)}\}$ be a countable dense subset of $\sigma_T(A)$. Then

$$\{x \in X : \sigma_x(A) \ne \sigma_T(A)\} = \bigcup_j \{x \in X : w^{(j)} \notin \sigma_x(A)\},$$

which is of the first category. □

30 Taylor functional calculus

The most important property of the Taylor spectrum is the existence of the analytic functional calculus.

Let $A = (A_1, \dots, A_n)$ be an n-tuple of commuting operators on a Banach space X. Let $G = \mathbb{C}^n \setminus \sigma_T(A)$.

Let $x \in X$. By Corollary 29.10, there exists $\psi \in \Lambda[s, \mathrm{d}\bar z, C^\infty(G, X)]$ such that $(\bar\partial + \delta)\psi = xs$. As it was noted in Section 28, this form can be used for the definition of the Taylor functional calculus. However, it is possible to consider such a form also globally, on the whole space X.

For $i = 1, \dots, n$ let $L'_{A_i} : \mathcal{H}(X) \to \mathcal{H}(X)$ be defined by $L'_{A_i} f = A_i f \quad (f \in \mathcal{H}(X))$. Let $L'_A = (L'_{A_1}, \dots, L'_{A_n})$. Clearly L'_A is a commuting n-tuple of bounded linear operators acting on the Banach space $\mathcal{H}(X)$.

By Corollary 27.7, $\sigma_T(L'_A) = \sigma_T(A)$. By Corollary 29.10, there is a form $W_A \in \Lambda^{n-1}[s, \mathrm{d}\bar{z}, C^\infty(G, \mathcal{H}(X))]$ such that $(\bar{\partial} + \delta_{L'_{A-\lambda}})W_A(\lambda) = Is$, where I is the identity operator on X. The form W_A can be also considered to be a mapping $W_A : X \to \Lambda^{n-1}[s, \mathrm{d}\bar{z}, C^\infty(G, X)]$. Then $(\bar{\partial} + \delta_{A-\lambda})W_A(\lambda)x = xs$ for all $x \in X$.

The definition of the Taylor functional calculus is analogous to the definition of the split functional calculus.

Recall that we interpret the differential form

$$(2i)^{-n}\mathrm{d}\bar{z}_1 \wedge \cdots \wedge \mathrm{d}\bar{z}_n \wedge \mathrm{d}z_1 \wedge \cdots \wedge \mathrm{d}z_n \tag{1}$$

as the Lebesgue measure in $\mathbb{C}^n = \mathbb{R}^{2n}$.

Let P be the natural projection $P : \Lambda[s, \mathrm{d}\bar{z}, C^\infty(G, X)] \to \Lambda[\mathrm{d}\bar{z}, C^\infty(G, X)]$ that annihilates all terms containing at least one of the indeterminates $s_1, \ldots, s_n$ and leaves invariant all the remaining terms.

Let U be a neighbourhood of $\sigma_T(A)$ and let f be a function analytic on U. It is possible to find a compact neighbourhood Δ of $\sigma_T(A)$ such that $\Delta \subset U$ and the boundary $\partial\Delta$ is a smooth surface. Define $f(A) : X \to X$ by

$$f(A) = \frac{-1}{(2\pi i)^n} \int_{\partial\Delta} PfW_A \wedge \mathrm{d}z. \tag{2}$$

By the Stokes formula,

$$f(A) = \frac{-1}{(2\pi i)^n} \int_{\Delta} \bar{\partial}\varphi PfW_A \wedge \mathrm{d}z,$$

where φ is a C^∞-function equal to 0 on a neighbourhood of $\sigma_T(A)$ and to 1 on $\mathbb{C}^n \setminus \Delta$ (consequently, $\varphi = 1$ also on $\partial\Delta$).

On $\mathbb{C}^n \setminus \Delta$ we have

$$\bar{\partial}\varphi PfW_A = Pf(\bar{\partial} + \delta)W_A = PfIs = 0.$$

Thus we can write

$$f(A) = \frac{-1}{(2\pi i)^n} \int_{\mathbb{C}^n} \bar{\partial}\varphi PfW_A \wedge \mathrm{d}z. \tag{3}$$

It is clear from the Stokes theorem that the definition of $f(A)$ does not depend on the choice of the function φ and, by (3), it is independent of Δ.

We show that $f(A)$ does not depend on the choice of the form W_A.

The following simple lemma will be used frequently.

Proposition 1. *Let $\eta \in \Lambda^n[s, d\bar{z}, C^\infty(G, X)]$ be a differential form with compact support disjoint with $\sigma_T(A)$ such that $(\bar{\partial} + \delta)\eta = 0$. Then*

$$\int_{\mathbb{C}^n} P\eta \wedge \mathrm{d}z = 0.$$

Proof. By Corollary 29.11, there exists $\psi \in \Lambda[s, \mathrm{d}\bar{z}, C^\infty(G, X)]$ with a compact support disjoint with $\sigma_T(A)$ such that $(\delta + \bar{\partial})\psi = \eta$. We have

$$P\eta = P(\bar{\partial} + \delta)\psi = P\bar{\partial}\psi.$$

By the Stokes theorem, we have

$$\int_{\mathbb{C}^n} P\eta \wedge \mathrm{d}z = \int_{\mathbb{C}^n} \bar{\partial} P\psi \wedge \mathrm{d}z = 0. \qquad \square$$

Let $x \in X$ and let $\psi_1, \psi_2 \in \Lambda[s, \mathrm{d}\bar{z}, C^\infty(G, X)]$ satisfy $(\delta+\bar{\partial})\psi_1 = (\delta+\bar{\partial})\psi_2 = xs$. Let φ be a C^∞-function equal to 0 on a neighbourhood of $\sigma_T(A)$ and to 1 on $\mathbb{C}^n \setminus U$. Then

$$\int \bar{\partial}\varphi P f \psi_1 \wedge \mathrm{d}z - \int \bar{\partial}\varphi P f \psi_2 \wedge \mathrm{d}z = \int P(\delta + \bar{\partial})\varphi f(\psi_1 - \psi_2) \wedge \mathrm{d}z.$$

On $\mathbb{C}^n \setminus \Delta$ we have $\varphi \equiv 1$, and so $(\delta + \bar{\partial})\varphi f(\psi_1 - \psi_2) = f(\delta + \bar{\partial})(\psi_1 - \psi_2) = 0$. Thus the form $(\delta + \bar{\partial})\varphi f(\psi_1 - \psi_2)$ has a compact support disjoint with $\sigma_T(A)$. By Proposition 1, $\int P(\delta + \bar{\partial})\varphi f(\psi_1 - \psi_2) \wedge \mathrm{d}z = 0$. In particular, the definition of $f(A)$ does not depend on the choice of W_A.

Note that for the definition of $f(A)x$ we can use any form ψ satisfying $(\bar{\partial} + \delta_{A-z})\psi = xs$. This implies that for functions analytic on a neighbourhood of $\sigma_S(A)$ the Taylor functional calculus coincides with the split functional calculus introduced in Section 28. By Proposition 28.7, for $n = 1$ the Taylor functional calculus coincides with the standard definition used in Section 1.

Lemma 2. $f(A) \in \mathcal{B}(X)$.

Proof. Clearly $f(A) \in \mathcal{H}(X)$, so it is sufficient to show the additivity of $f(A)$.

Let $x, y \in X$. Then $(\delta + \bar{\partial})(W_A x + W_A y) = (x + y)s$, and so $f(A)(x + y) = \int_{\mathbb{C}^n} \bar{\partial}\varphi P f(W_A x + W_A y) \wedge \mathrm{d}z = f(A)x + f(A)y$. $\square$

Proposition 3. *Let f be a function analytic on a neighbourhood of $\sigma_T(A)$, $1 \le j \le n$ and $g(z) = z_j f(z)$. Then $g(A) = A_j f(A)$.*

Proof. The statement is well known for $n = 1$. Suppose that $n \ge 2$. Then

$$\begin{aligned} -(2\pi i)^n \big(A_j f(A) - g(A)\big) &= A_j \int_{\mathbb{C}^n} \bar{\partial}\varphi P f W_A \wedge \mathrm{d}z - \int_{\mathbb{C}^n} \bar{\partial}\varphi P f z_j W_A \wedge \mathrm{d}z \\ &= \int_{\mathbb{C}^n} \bar{\partial}\varphi f \cdot (A_j - z_j) P W_A \wedge \mathrm{d}z. \end{aligned}$$

For $F \subset \{1, \dots, n\}$, $F = \{i_1, \dots, i_p\}$ with $i_1 < i_2 < \dots < i_p$ write $s_F = s_{i_1} \wedge \dots \wedge s_{i_p}$. Express $W_A \in \Lambda^{n-1}\big[s, \mathrm{d}\bar{z}, C^\infty(G, \mathcal{H}(X))\big]$ as

$$W_A = \sum_{F \subset \{1, \dots, n\}} s_F \wedge \xi_F,$$

where ξ_F contains no variable from $s_1, \dots, s_n$. Since $(\bar\partial + \delta_{A-z})W_A = Is$, for each $F \neq \{1, \dots, n\}$ we have

$$\bar\partial \xi_F + \sum_{k \in F} (-1)^{\operatorname{card}\{k' \in F : k' < k\}} (A_k - z_k) \xi_{F \setminus \{k\}} = 0.$$

In particular, for $F = \{j\}$ we have

$$\bar\partial \xi_{\{j\}} = -(A_j - z_j)\xi_\emptyset = -(A_j - z_j) P W_A.$$

Thus

$$\int_{\mathbb{C}^n} \bar\partial \varphi f \cdot (A_j - z_j) P W_A \wedge \mathrm{d}z = -\int_{\mathbb{C}^n} \bar\partial \varphi f \bar\partial \xi_{\{j\}} \wedge \mathrm{d}z$$
$$= -\int_{\mathbb{C}^n} \bar\partial \big(\varphi \bar\partial f \xi_{\{j\}} - \bar\partial \varphi f \xi_{\{j\}}\big) \wedge \mathrm{d}z = 0$$

by the Stokes theorem. Hence $g(A) = A_j f(A)$. □

Proposition 3 implies that the definition of the Taylor functional calculus for polynomials coincides with the usual definition. This also implies that the local spectrum $\sigma_x(A)$ of any non-zero vector x is non-empty.

Theorem 4. $\sigma_x(A) \neq \emptyset$ for every $x \neq 0$.

Proof. Let $x \in X$ and suppose that $\sigma_x(A) = \emptyset$. By Corollary 29.10, there exists $\psi \in \Lambda[s, \mathrm{d}\bar z, C^\infty(\mathbb{C}^n, X)]$ such that $(\bar\partial + \delta_{A-z})\psi = xs$. We have

$$-(2\pi i)^n x = -(2\pi i)^n Ix = \int_{\mathbb{C}^n} \bar\partial P \psi \wedge \mathrm{d}z = \int_{\mathbb{C}^n} P(\bar\partial + \delta)\psi \wedge \mathrm{d}z = \int_{\mathbb{C}^n} Pxs \wedge \mathrm{d}z = 0,$$

and so $x = 0$. □

Proposition 5. *Let $A = (A_1, \dots, A_n) \in \mathcal{B}(X)^n$, $B = (B_1, \dots, B_m) \in \mathcal{B}(X)^m$. Suppose that $(A, B) = (A_1, \dots, A_n, B_1, \dots, B_m)$ is a commuting $(n+m)$-tuple and let f and g be functions analytic on a neighbourhood of $\sigma_T(A)$ and $\sigma_T(B)$, respectively. Let h be defined by $h(z, w) = f(z) \cdot g(w)$. Then $h(A, B) = g(B) f(A)$.*

Proof. Write $z = (z_1, \dots, z_n)$ and $w = (w_1, \dots, w_m)$. Denote by $\bar\partial_z$, $\bar\partial_w$ and $\bar\partial_{z,w}$ the $\bar\partial$ mapping corresponding to z, w and (z, w), respectively. We associate with B another system $t = (t_1, \dots, t_m)$ of exterior indeterminates when defining the operator δ_{B-w}.

Choose forms W_A, W_B and $W_{A,B}$ corresponding to the tuples A, B and (A, B). Let Δ' and Δ'' be compact neighbourhoods of $\sigma_T(A)$ and $\sigma_T(B)$ contained in the domains of definition of f and g, respectively. Let φ, ψ and χ be C^∞-functions equal to 0 on a neighbourhood of $\sigma_T(A)$ $\big(\sigma_T(B)$ and $\sigma_T(A, B)\big)$, and to 1 on a neighbourhood of $\mathbb{C}^n \setminus \Delta'$ $\big(\mathbb{C}^m \setminus \Delta''$ and $\mathbb{C}^{n+m} \setminus \Delta' \times \Delta''$, respectively$\big)$.

Denote by P_s and P_t the projections which annihilate all terms containing at least one of the variables $s_1, \dots, s_n$ ($t_1, \dots, t_m$, respectively) and leave invariant the remaining terms. Set $P = P_s P_t$.

Let $x \in X$. We have

$$f(A)x = \frac{-1}{(2\pi i)^n} \int_{\mathbb{C}^n} \bar{\partial}_z \varphi P_s f W_A x \wedge \mathrm{d}z = \frac{-1}{(2\pi i)^n} \int_{\mathbb{C}^n} P_s \xi \wedge \mathrm{d}z,$$

where $\xi = (\bar{\partial}_z + \delta_{A-z})\varphi f W_A x - f x s$. If $\varphi \equiv 1$, then $\xi \equiv 0$. Thus $\operatorname{supp} \xi$ is compact, $\operatorname{supp} \xi \subset \operatorname{Int} \Delta'$. Further,

$$g(B)f(A)x = \frac{1}{(2\pi i)^{n+m}} \int_{\mathbb{C}^m} P_t(\bar{\partial}_w + \delta_{B-w})\psi g W_B \Big(\int_{\mathbb{C}^n} P_s \xi \wedge \mathrm{d}z \Big) \wedge \mathrm{d}w. \qquad (4)$$

On the other hand, $-(2\pi i)^{m+n} h(A,B)x = \int P\eta_1 \wedge \mathrm{d}z \wedge \mathrm{d}w$, where

$$\eta_1 = (\bar{\partial}_{z,w} + \delta_{A-z,B-w})\chi h W_{A,B} x - h x s \wedge t.$$

Clearly, $\operatorname{supp} \eta_1 \subset \Delta' \times \Delta''$.

We have $(\bar{\partial}_{z,w} + \delta_{A-z,B-w})\xi \wedge t = (\bar{\partial}_z + \delta_{A-z})\xi \wedge t = 0$. By Corollary 29.11, there exists $\alpha \in \Lambda[s, t, \mathrm{d}\bar{z}, \mathrm{d}\bar{w}, C^\infty(\mathbb{C}^{n+m} \setminus \sigma_T(A,B), X)]$ such that $(\bar{\partial}_{z,w} + \delta_{A-z,B-w})\alpha = \xi \wedge t$. Moreover, we can assume that $\operatorname{supp} \alpha \subset \Delta' \times \mathbb{C}^m$. Let

$$\eta_2 = (\bar{\partial}_{z,w} + \delta_{A-z,B-w})\psi g \alpha - g\xi \wedge t.$$

We have $(\bar{\partial}_{z,w} - \delta_{A-z,B-w})(\eta_1 - \eta_2) = 0$. Clearly, $\operatorname{supp} \eta_2 \subset \Delta' \times \mathbb{C}^m$. Moreover, if $\psi \equiv 1$, then $\eta_2 \equiv 0$, and so $\operatorname{supp} \eta_2$ is compact. On a neighbourhood of $\sigma_T(A,B)$ we have $\eta_2 = -g\xi \wedge t = fgxs \wedge t = -\eta_1$. By Proposition 1, we have $\int P(\eta_1 + \eta_2) \wedge \mathrm{d}z \wedge \mathrm{d}w = 0$, and so

$$\begin{aligned}(2\pi i)^{m+n} h(A,B)x &= \int_{\mathbb{C}^{n+m}} P\eta_2 \wedge \mathrm{d}z \wedge \mathrm{d}w \\ &= (-1)^{mn} \int_{\mathbb{C}^m} \Big(\int_{\mathbb{C}^n} P_t(\bar{\partial}_{z,w} + \delta_{B-w})\psi g P_s \alpha \wedge \mathrm{d}z \Big) \wedge \mathrm{d}w\end{aligned}$$

by the Fubini theorem (the factor $(-1)^{mn}$ is caused by convention (1) defining the Lebesgue measures in $\mathbb{C}^n$, $\mathbb{C}^m$ and $\mathbb{C}^{m+n}$, respectively). By the Stokes theorem, we have

$$(2\pi i)^{m+n} h(A,B)x = (-1)^{mn} \int_{\mathbb{C}^m} P_t(\bar{\partial}_w + \delta_{B-w}) g \Big(\int_{\mathbb{C}^n} \psi P_s \alpha \wedge \mathrm{d}z \Big) \wedge \mathrm{d}w.$$

Consider the form

$$\eta_3 = (-1)^{mn}(\bar{\partial}_w + \delta_{B-w}) g \int_{\mathbb{C}^n} \psi P_s \alpha \wedge \mathrm{d}z - (\bar{\partial}_w + \delta_{B-w})\psi g W_B \int_{\mathbb{C}^n} P_s \xi \wedge \mathrm{d}z \wedge t.$$

Clearly, $(\bar\partial_w+\delta_{B-w})\eta_3=0$. If $\psi\equiv 1$, then, by the Stokes theorem,

$$\eta_3=(-1)^{mn}g\int_{\mathbb{C}^n}P_s(\bar\partial_{z,w}+\delta_{A-z,B-w})\alpha\wedge\mathrm{d}z-(-1)^{mn}g\int_{\mathbb{C}^n}\bar\partial_z P_s\alpha\wedge\mathrm{d}z$$
$$-g\int_{\mathbb{C}^n}P_s\xi\wedge\mathrm{d}z\wedge t=(-1)^{mn}g\int_{\mathbb{C}^n}P_s\xi\wedge t\wedge\mathrm{d}z-g\int_{\mathbb{C}^n}P_s\xi\wedge\mathrm{d}z\wedge t=0.$$

Thus $\operatorname{supp}\eta_3$ is compact and disjoint with $\sigma_T(B)$. Hence $\int P_t\eta_3\wedge\mathrm{d}w=0$ and

$$(2\pi i)^{n+m}h(A,B)x=\int_{\mathbb{C}^m}P_t(\bar\partial_w+\delta_{B-w})\psi g W_B\int_{\mathbb{C}^n}P_s\xi\wedge\mathrm{d}z\wedge\mathrm{d}w$$
$$=(2\pi i)^{m+n}g(B)f(A)x$$

by (4). Hence $h(A,B)=g(B)f(A)$. □

We will use the following simple lemma:

Lemma 6. *Let K be a compact subset of $\mathbb{C}^n$ and let f be a function analytic on an open neighbourhood of K. Then there are functions h_j $(j=1,\dots,n)$ analytic on a neighbourhood of the set $D=\{(z,z):z\in K\}$ such that*

$$f(z)-f(w)=\sum_{j=1}^n(z_j-w_j)\cdot h_j(z,w).$$

Proof. For $j=1,\dots,n$ define g_j by

$$g_j(z_1,\dots,z_n,w_1,\dots,w_n)$$
$$=f(z_1,\dots,z_j,w_{j+1},\dots,w_n)-f(z_1,\dots,z_{j-1},w_j,\dots,w_n).$$

It is easy to see that g_j is analytic on a neighbourhood of D.

Let $h_j(z,w)=\frac{g_j(z,w)}{z_j-w_j}$. Clearly, h_j is analytic at each point (z,w) with $z_j\neq w_j$. By the Weierstrass division theorem (see [GR], p. 70), h_j can be defined and is analytic also on a neighbourhood of each point (z,w) with $z_j=w_j$. Thus h_j is analytic on a neighbourhood of D. Hence

$$\sum_{j=1}^n(z_j-w_j)\cdot h_j(z,w)=\sum_{j=1}^n g_j(z,w)=f(z)-f(w).$$ □

Recall that H_K denotes the algebra of all functions analytic on a neighbourhood of a compact set $K\subset\mathbb{C}^n$ (more precisely, the algebra of all germs of functions analytic on a neighbourhood of K).

Theorem 7. *Let $A=(A_1,\dots,A_n)$ be an n-tuple of mutually commuting operators on X. Then:*

(i) *the mapping $f\mapsto f(A)$ is linear and multiplicative, i.e., the Taylor functional calculus is a homomorphism from $H_{\sigma_T(A)}$ to $\mathcal{B}(X)$;*

(ii) *if p is a polynomial, $p(z) = \sum_{\alpha\in\mathbb{Z}_+^n} c_\alpha z^\alpha$, then $p(A) = \sum_{\alpha\in\mathbb{Z}_+^n} c_\alpha A^\alpha$;*

(iii) *if $f_n \to f$ uniformly on a compact neighbourhood of $\sigma_T(A)$, then $f_n(A) \to f(A)$ in the norm topology;*

(iv) *$f(A) \in (A)''$ for each $f \in H_{\sigma_T(A)}$.*

Proof. (i) The linearity of the mapping $f \mapsto f(A)$ is clear. Let f and g be functions analytic on a neighbourhood of $\sigma_T(A)$. Consider the $(2n)$-tuple (A, A). It is easy to see that $\sigma_T(A, A) = \{(z, z) : z \in \sigma_T(A)\}$. Define functions $h_1(z, w) = f(z)g(w)$ and $h_2(z, w) = f(z)g(z)$. By Lemma 6, we can write $g(z) - g(w) = \sum_{i=1}^n (z_i - w_i)q_i(z, w)$ for some functions $q_1, \dots, q_n$ analytic on a neighbourhood of $\sigma_T(A, A)$. By Proposition 5, we have $h_1(A, A) = f(A)g(A)$ and $h_2(A, A) = (fg)(A)$. Thus, by Proposition 3,

$$(fg)(A) - f(A)g(A) = h_2(A, A) - h_1(A, A) = \sum_{i=1}^n (A_i - A_i)(fq_i)(A, A) = 0.$$

Hence $(fg)(A) = f(A)g(A)$.

(ii) The statement follows from Proposition 5.

(iii) Follows from the definition.

(iv) Let $S \in \mathcal{B}(X)$ be an operator commuting with $A_1, \dots, A_n$. By Proposition 5, it is possible to consider $f(A)$ to be a function of the $(n + 1)$-tuple $(A_1, \dots, A_n, S)$. Therefore $f(A)$ commutes with its argument S. Hence $f(A) \in (A)''$. □

It follows from the general theory that the Taylor spectrum satisfies the spectral mapping property for all polynomials (and consequently, for all functions that can be approximated by polynomials uniformly on a neighbourhood of the Taylor spectrum). In fact, the spectral mapping property is true for all analytic functions.

By Proposition 25.11, each A_j behaves as the zero operator on the quotient $\operatorname{Ker}\delta_A/\operatorname{Ran}\delta_A$. It is natural to expect that $f(A)$ behaves as $f(0)$ on this quotient space. However, there is a technical difficulty because in general $\operatorname{Ran}\delta_A$ is not closed, and so the quotient $\operatorname{Ker}\delta_A/\operatorname{Ran}\delta_A$ is not a Banach space. Therefore the proof is a little bit more complicated.

Lemma 8. *Let $A = (A_1, \dots, A_n)$ be a commuting n-tuple of operators on X, let $c = (c_1, \dots, c_n) \in \sigma_T(A)$ and let f be a function analytic on a neighbourhood of $\sigma_T(A)$. Consider exterior indeterminates $t = (t_1, \dots, t_n)$ and the operator $\delta_{A-c,t} : \Lambda[t, X] \to \Lambda[t, X]$ defined by $\delta_{A-c,t}\psi = \sum_{j=1}^n (A_j - c_j)t_j \wedge \psi$ for all $\psi \in \Lambda[t, X]$. Let $\eta \in \operatorname{Ker}\delta_{A-c,t}$. Then $(f(A) - f(c))\eta \in \delta_{A-c,t}\Lambda[t, X]$.*

Proof. To define $f(A)$, consider exterior indeterminates $s = (s_1, \dots, s_n)$, the mapping δ_{A-z} acting on $\Lambda[s, d\bar{z}, C^\infty(\mathbb{C}^n \setminus \sigma_T(A), X)]$ defined by the formula $\delta_{A-z}\psi = \sum_{j=1}^n (A_j - z_j)s_j \wedge \psi$, and the mapping W_A corresponding to A. Note that δ_{A-z} and W_A are connected with variables s; the mapping $\delta_{A-c,t}$ is related to variables t.

Without loss of generality we can assume that η is homogeneous of degree p, $0 \le p \le n$.

Since $\eta \in \Lambda^p[t,X]$ and $\Lambda^p[t,X]$ is a direct sum of $\binom{n}{p}$ copies of X, it is possible to define the form $\xi_0 := W_A\eta \in \Lambda[s,t,\mathrm{d}\bar{z},C^\infty(G,X)]$. We have $(\bar{\partial}+\delta_{A-z})\xi_0 = s\wedge\eta$ and $(\bar{\partial}+\delta_{A-z})\delta_{A-c,t}\xi_0 = -\delta_{A-c,t}(\bar{\partial}+\delta_{A-z})\xi_0 = 0$. Thus there exists $\xi_1 \in \Lambda[s,t,\mathrm{d}\bar{z},C^\infty(G,X)]$ such that $(\bar{\partial}+\delta_{A-z})\xi_1 = \delta_{A-c,t}\xi_0$.

In the same way we can construct forms $\xi_1,\dots,\xi_{n-p} \in \Lambda[s,t,\mathrm{d}\bar{z},C^\infty(G,X)]$ such that $(\bar{\partial}+\delta_{A-z})\xi_{k+1} = \delta_{A-c,t}\xi_k$. Clearly the degree of ξ_k in t is $p+k$.

Set $\xi = \sum_{k=0}^{n-p}(-1)^k\xi_k$ Then

$$(\bar{\partial}+\delta_{A-z}+\delta_{A-c,t})\xi = \sum_{k=0}^{n-p}(-1)^k(\bar{\partial}+\delta_{A-z})\xi_k + \sum_{k=0}^{n-p}(-1)^k\delta_{A-c,t}\xi_k = s\wedge\eta,$$

since $\delta_{A-c,t}\xi_{n-p} = 0$.

Let Δ be a compact neighbourhood of $\sigma_T(A)$ contained in the domain of definition of f. Let φ be a C^∞-function equal to 0 on a neighbourhood of $\sigma_T(A)$ and to 1 on a neighbourhood of $\mathbb{C}^n \setminus \Delta$. Let P_s be the projection annihilating all terms that contain at least one of the variables $s_1,\dots,s_n$ and leaving invariant all other terms.

Consider the integral

$$\int(\bar{\partial}+\delta_{A-c,t})P_s\varphi\xi\wedge\mathrm{d}z = \int(\bar{\partial}+\delta_{A-c,t})P_s\varphi\sum_{k=0}^{n-p}(-1)^k\xi_k\wedge\mathrm{d}z.$$

Since ξ_k has degree $p+k$ in t and $n-k-1$ in $(s,\mathrm{d}\bar{z})$, the only relevant term in the integral above is ξ_0. Thus

$$\begin{aligned}\int(\bar{\partial}+\delta_{A-c,t})P_s\varphi\xi\wedge\mathrm{d}z &= \int(\bar{\partial}+\delta_{A-c,t})P_s\varphi\xi_0\wedge\mathrm{d}z\\ &= \int\bar{\partial}P_s\varphi W_A\eta\wedge\mathrm{d}z\\ &= -(2\pi i)^n f(A)\eta.\end{aligned}$$

Consider now the n-tuple $B = (c_1I,\dots,c_nI) \in \mathcal{B}(X)^n$. Since f can be approximated by polynomials uniformly on a neighbourhood of c, we note that $f(B) = f(c)\cdot I$.

As above, consider the mappings δ_{B-z} and W_B connected with variables s. Let $\xi_0' = W_B\eta$ and inductively define $\xi_k' \in \Lambda[s,t,\mathrm{d}\bar{z},C^\infty(G,X)]$ satisfying $(\bar{\partial}+\delta_{B-z})\xi_{k+1}' = \delta_{A-c,t}\xi_k'$.

Let $\xi' = \sum_{k=0}^{n-p}(-1)^k \xi'_k$. As above, we have $(\bar\partial + \delta_{B-z} + \delta_{A-c,t})\xi' = s \wedge \eta$ and

$$\int (\bar\partial + \delta_{A-c,t}) P_s \varphi \xi' \wedge \mathrm{d}z = \int (\bar\partial + \delta_{A-c,t}) P_s \varphi W_B \eta \wedge \mathrm{d}z$$
$$= \int \bar\partial P_s \varphi W_B \eta \wedge \mathrm{d}z = -(2\pi i)^n f(B)\eta = -(2\pi i)^n f(c)\eta.$$

To show that $(f(A)-f(c))\eta_0 \in \delta_{A-c,t}\Lambda[t, X]$, consider the linear mapping U acting on $\Lambda[s, t, \mathrm{d}\bar z, C^\infty(\mathbb{C}^n \setminus \sigma_T(A), X)]$ defined by

$$U\Big(t_{i_1} \wedge \cdots \wedge t_{i_m} \wedge \psi\Big) = (t_{i_1} - s_{i_1}) \wedge \cdots \wedge (t_{i_m} - s_{i_m}) \wedge \psi$$

for all $i_1, \dots, i_m$ and $\psi \in \Lambda[s, \mathrm{d}\bar z, C^\infty(\mathbb{C}^n \setminus \sigma_T(A), X)]$. We have $P_s U = P_s$ and, for each $\psi \in \Lambda[s, t, \mathrm{d}\bar z, C^\infty(\mathbb{C}^n \setminus \sigma_T(A), X)]$,

$$\begin{aligned}
&U(\bar\partial + \delta_{A-z} + \delta_{A-c,t})\psi \\
&\quad = \bar\partial U\psi + \sum (A_j - z_j)s_j \wedge U\psi + \sum (A_j - c_j)(t_j - s_j) \wedge U\psi \\
&\quad = (\bar\partial + \delta_{B-z} + \delta_{A-c,t})U\psi.
\end{aligned}$$

We have

$$-(2\pi i)^n f(A)\eta = \int (\bar\partial + \delta_{A-c,t}) P_s \varphi \xi \wedge \mathrm{d}z = \int P_s(\bar\partial + \delta_{A-z} + \delta_{A-c,t})\varphi\xi \wedge \mathrm{d}z$$
$$= \int P_s U(\bar\partial + \delta_{A-z} + \delta_{A-c,t})\varphi\xi \wedge \mathrm{d}z = \int P_s(\bar\partial + \delta_{B-z} + \delta_{A-c,t})\varphi U\xi \wedge \mathrm{d}z.$$

Thus

$$-(2\pi i)^n (f(A) - f(c))\eta = \int P_s(\bar\partial + \delta_{B-z} + \delta_{A-c,t})\varphi(U\xi - \xi') \wedge \mathrm{d}z = \int P_s\theta \wedge \mathrm{d}z,$$

where $\theta = (\bar\partial + \delta_{B-z} + \delta_{A-c,t})\varphi(U\xi - \xi')$. If $\varphi \equiv 1$, then

$$\theta = (\bar\partial + \delta_{B-z} + \delta_{A-c,t})U\xi - s\wedge\eta = U(\bar\partial + \delta_{A-z} + \delta_{A-c,t})\xi - \eta = U(s\wedge\eta) - s\wedge\eta = 0;$$

so $\operatorname{supp}\theta \subset \operatorname{Int}\Delta$. Furthermore, θ can be written as $\theta = (\bar\partial + \delta_{B-z} + \delta_{A-c,t})\psi$ for some form $\psi \in \Lambda[s, t, \mathrm{d}\bar z, C^\infty(\mathbb{C}^n, X)]$ with compact support. Indeed, by Remark 29.12, there exists a form $\vartheta \in \Lambda[s, t, \mathrm{d}\bar z, \mathrm{d}\bar w, C^\infty(\mathbb{C}^{2n}, X)]$ with $\operatorname{supp}\vartheta \subset \Delta \times \mathbb{C}^n$ such that $(\bar\partial_{z,w} + \delta_{B-z} + \delta_{A-c,t})\vartheta = \theta$.

Set $\psi(z) = \vartheta_0(z, c)$, where ϑ_0 is the part of ϑ containing none of the variables $\mathrm{d}\bar w_j$. Then $\operatorname{supp}\psi \subset \Delta$ and $(\bar\partial_z + \delta_{B-z} + \delta_{A-c,t})\psi = \theta$. By the Stokes theorem,

$$\begin{aligned}
\int P_s\theta \wedge \mathrm{d}z &= \int P_s(\bar\partial_z + \delta_{B-z} + \delta_{A-c,t})\psi \wedge \mathrm{d}z \\
&= \int \bar\partial_z P_s\psi \wedge \mathrm{d}z + \int P_s\delta_{A-c,t}\psi \wedge \mathrm{d}z \\
&= \delta_{A-c,t}\int P_s\psi \wedge \mathrm{d}z \in \delta_{A-c,t}\Lambda[t, X].
\end{aligned}$$

□

Proposition 9. *Let $A = (A_1, \dots, A_n)$ be a commuting n-tuple of operators on X, $c = (c_1, \dots, c_n) \in \sigma_T(A)$ and let f be a function analytic on a neighbourhood of $\sigma_T(A)$. Then the $(n+1)$-tuple $\big(A_1 - c_1, \dots, A_n - c_n, f(A)\big)$ is Taylor regular if and only if $f(c) \neq 0$.*

Proof. To the $(n+1)$-tuple $(A - c, f(A))$ we relate exterior variables $s_1, \dots, s_{n+1}$. Write for short $s = (s_1, \dots, s_n)$. Let $\delta_{A-c} : \Lambda[s, X] \to \Lambda[s, X]$ be defined by $\delta_{A-c}\psi = \sum (A_j - c_j) s_j \wedge \psi \quad (\psi \in \Lambda[s, X])$. We have $\Lambda[s, s_{n+1}, X] = \Lambda[s, X] \oplus s_{n+1} \wedge \Lambda[s, X]$. The operator $\delta_{A-c,f(A)}$ corresponding to the $(n+1)$-tuple $(A - c, f(A))$ can be written in this decomposition in the matrix form

$$\delta_{A-c,f(A)} = \begin{pmatrix} \delta_{A-c} & 0 \\ f(A) & -\delta_{A-c} \end{pmatrix}.$$

We distinguish two cases:

(a) $f(c) = 0$.

Since $c \in \sigma_T(A)$, there is a $\psi \in \Lambda[s, X]$ such that $\delta_{A-c}\psi = 0$ and $\psi \notin \delta_{A-c}\Lambda[s, X]$. By the preceding lemma, there is an $\eta \in \Lambda[s, X]$ such that $f(A)\psi = \delta_{A-c}\eta$. Then $\delta_{A-c,f(A)}(\psi + s_{n+1} \wedge \eta) = 0$ and $(\psi + s_{n+1} \wedge \eta) \notin \delta_{A-c,f(A)}\Lambda[s, s_{n+1}, X]$ since $\psi \notin \delta_{A-c}\Lambda[s, X]$.

Thus the $(n+1)$-tuple $(A - c, f(A))$ is Taylor singular.

(b) $f(c) \neq 0$. Without loss of generality we can assume that $f(c) = 1$.

Let $\psi, \xi \in \Lambda[s, X]$, $\delta_{A-c,f(A)}(\psi + s_{n+1} \wedge \xi) = 0$. Then $\delta_{A-c}\psi = 0$ and $f(A)\psi - \delta_{A-c}\xi = 0$. By the preceding lemma, $f(A)\psi - \psi \in \delta_{A-c}\Lambda[s, X]$. Since $f(A)\psi \in \delta_{A-c}\Lambda[s, X]$, we have $\psi = \delta_{A-c}\eta$ for some $\eta \in \Lambda[s, X]$.

Further, $\delta_{A-c}(f(A)\eta - \xi) = f(A)\psi - \delta_{A-c}\xi = 0$. Thus there is a $\theta \in \Lambda[s, X]$ with $f(A)(f(A)\eta - \xi) - (f(A)\eta - \xi) = \delta_{A-c}\theta$. Set $\eta' = \eta - (f(A)\eta - \xi)$. Then $\delta_{A-c}\eta' = \delta_{A-c}\eta = \psi$ and $f(A)\eta' - \delta_{A-c}\theta = f(A)\eta - f(A)(f(A)\eta - \xi) + \delta_{A-c}\theta = f(A)\eta - (f(A)\eta - \xi) = \xi$. Hence $\delta_{A-c,f(A)}(\eta' - s_{n+1} \wedge \theta) = (\psi + s_{n+1} \wedge \xi)$ and the $(n+1)$-tuple $(A - c, f(A))$ is Taylor regular. □

Lemma 10. *Let $A = (A_1, \dots, A_n)$ be a commuting n-tuple of operators on X, let f be a function analytic on a neighbourhood of $\sigma_T(A)$. Denote by $\mathcal{A}$ the commutative Banach algebra generated by $A_1, \dots, A_n$ and $f(A)$. Let φ be a multiplicative functional on $\mathcal{A}$ such that $\varphi(B) \in \sigma_T(B)$ for all tuples $B = (B_1, \dots, B_m)$ of operators in $\mathcal{A}$. Then $\varphi(f(A)) = f(\varphi(A))$.*

Proof. Consider the $(n+1)$-tuple $\big(A_1 - \varphi(A_1), \dots, A_n - \varphi(A_n), f(A) - \varphi(f(A))\big)$. By assumption, this $(n+1)$-tuple is Taylor singular. By Proposition 9, we have $f(\varphi(A)) - \varphi(f(A)) = 0$. □

Corollary 11. (spectral mapping property) *Let $\tilde{\sigma}$ be a compact-valued spectral system on $\mathcal{B}(X)$ which is contained in the Taylor spectrum. Let $A = (A_1, \dots, A_n)$ be a commuting n-tuple of operators on X and let $f = (f_1, \dots, f_m)$ be an m-tuple of functions analytic on a neighbourhood of $\sigma_T(A)$. Then $\tilde{\sigma}(f(A)) = f(\tilde{\sigma}(A))$.*

In particular, $\sigma_T(f(A)) = f(\sigma_T(A))$*. Similarly,* $\sigma_{\pi k}(f(A)) = f(\sigma_{\pi k}(A))$ *and* $\sigma_{\delta k}(f(A)) = f(\sigma_{\delta k}(A))$ *for all* $k = 0, \dots, n$.

Proof. Consider the commutative Banach algebra $\mathcal{A}$ generated by $A_1, \dots, A_n$ and $f_1(A), \dots, f_m(A)$. Since the restriction of $\tilde{\sigma}$ to $\mathcal{A}$ is again a compact-valued spectral system, there is a compact subset $K \subset \mathcal{M}(\mathcal{A})$ such that $\tilde{\sigma}(B) = \{\varphi(B) : \varphi \in K\}$ for each tuple $B = (B_1, \dots, B_k) \subset \mathcal{A}$.

Then

$$\begin{aligned}\tilde{\sigma}(f(A)) &= \{(\varphi(f_1(A), \dots, \varphi(f_m(A))) : \varphi \in K\} \\ &= \{(f_1(\varphi(A)), \dots, f_m(\varphi(A))) : \varphi \in K\} \\ &= \{f(c) : c \in \tilde{\sigma}(A)\} = f(\tilde{\sigma}(A)).\end{aligned}$$

□

Theorem 12. (superposition principle) *Let* $A = (A_1, \dots, A_n)$ *be a commuting* n*-tuple of operators on* X*, let* $f = (f_1, \dots, f_m)$ *be an* m*-tuple of function analytic on a neighbourhood of* $\sigma_T(A)$*, let* $B = f(A)$*, let* g *be a function analytic on a neighbourhood of* $\sigma_T(B)$ *and let* $h(z) = g(f_1(z), \dots, f_m(z))$*. Then* $h(A) = g(B)$.

Proof. By Lemma 5, $g(v) - g(w) = \sum_{j=1}^{m}(v_j - w_j)r_j(v, w)$ for some functions $r_1, \dots, r_m$ analytic on a neighbourhood of the set $\{(v, v) : v \in \sigma_T(B)\}$. Thus $g(f(z)) - g(w) = \sum_{j=1}^{m}(f_j(z) - w_j)r'_j(z, w)$, where $r'_j(z, w) = r_j(f(z), w))$ and the functions r'_j are analytic on a neighbourhood of the set $\sigma_T(A, f(A)) = \{(z, f(z)) : z \in \sigma_T(A)\}$. Thus $h(A) - g(B) = \sum_{j=1}^{m}(f_j(A) - B_j)r'_j(A, B) = 0$. Hence $h(A) = g(B)$. □

As a corollary of the Taylor functional calculus we obtain the properties of the functional calculus in commutative Banach algebras which were formulated without proof in Section 2. For convenience, we state them here once more in an extended form.

Theorem 13. *Let* $\mathcal{A}$ *be a commutative Banach algebra. To each finite family* $a = (a_1, \dots, a_n)$ *of elements of* $\mathcal{A}$ *and each function* $f \in H_{\sigma(a)}$ *it is possible to assign an element* $f(a) \in \mathcal{A}$ *such that the following conditions are satisfied:*

(i) *if* $f(z_1, \dots, z_n) = \sum_{\alpha \in \mathbb{Z}_+^n} c_\alpha z_1^{\alpha_1} \cdots z_n^{\alpha_n}$ *is a polynomial in* n *indeterminates, then* $f(a_1, \dots, a_n) = \sum_{\alpha \in \mathbb{Z}_+^n} c_\alpha a_1^{\alpha_1} \cdots a_n^{\alpha_n}$*;*

(ii) *the mapping* $f \mapsto f(a_1, \dots, a_n)$ *is an algebra homomorphism from the algebra* $H_{\sigma(a_1, \dots, a_n)}$ *to* $\mathcal{A}$*;*

(iii) *if* U *is a neighbourhood of* $\sigma(x_1, \dots, x_n)$*,* f*,* f_k $(k \in \mathbb{N})$ *are analytic in* U *and* f_k *converge to* f *uniformly on* U*, then*

$$f_k(a_1, \dots, a_n) \to f(a_1, \dots, a_n);$$

(iv) *if* $\varphi \in \mathcal{M}(\mathcal{A})$ *and* $f \in H_{\sigma(a_1, \dots, a_n)}$*, then*

$$\varphi(f(a_1, \dots, a_n)) = f(\varphi(a_1), \dots, \varphi(a_n));$$

(v) $\tilde{\sigma}\big(f(a_1,\ldots,a_n)\big) = f\big(\tilde{\sigma}(a_1,\ldots,a_n)\big)$ *for each compact-valued spectral system in* $\mathcal{A}$;

(vi) *if* $a_1,\ldots,a_m \in \mathcal{A}$, $n < m$, $f \in H_{\sigma(a_1,\ldots,a_n)}$ and $\tilde{f} \in H_{\sigma(a_1,\ldots,a_m)}$ *satisfy* $\tilde{f}(z_1,\ldots,z_m) = f(z_1,\ldots,z_n)$ *for all* $z_1,\ldots,z_m$ *in a neighbourhood of* $\sigma(a_1,\ldots,a_m)$, *then*

$$\tilde{f}(a_1,\ldots,a_m) = f(a_1,\ldots,a_n);$$

(vii) *if* $f_1,\ldots,f_m \in H_{\sigma(a)}$, $b_i = f_i(a)$, $g \in H_{\sigma(b_1,\ldots,b_m)}$ and $h \in H_{\sigma(a)}$ *is defined by* $h(z) = g(f_1(z),\ldots,f_m(z))$, *then* $h(a) = g(b)$;

(viii) *properties* (i), (ii), (iii) *and* (vi) *determine the functional calculus* $(a,f) \mapsto f(a)$ *uniquely.*

Proof. For an n-tuple $a = (a_1,\ldots,a_n) \in \mathcal{A}^n$ consider the left multiplication operators $L_{a_i} \in \mathcal{B}(\mathcal{A})$ defined by $L_{a_i}x = a_ix \quad (x \in \mathcal{A}, i = 1,\ldots,n)$. Then $L_a = (L_{a_1},\ldots,L_{a_n})$ is a commuting n-tuple of operators. Further, $\sigma(a) = \sigma_T(L_a)$ by Proposition 26.9.

For a function f analytic on a neighbourhood of $\sigma(a)$ set $f(a) = f(L_a)1_{\mathcal{A}}$.

Since $f(L_a) \in (L_a)''$, for each $b \in \mathcal{A}$ we have $f(L_a)(b) = f(L_a)L_b(1_{\mathcal{A}}) = L_bf(L_a)(1_{\mathcal{A}}) = b\cdot f(a) = L_{f(a)}(b)$. Thus $f(L_a) = L_{f(a)}$.

Properties (i), (ii), (iii), (vi) and (vii) follow from the corresponding properties of the Taylor functional calculus; the multiplicativity follows from the observation that

$$(fg)(a) = (fg)(L_a)(1_{\mathcal{A}}) = f(L_a)g(L_a)(1_{\mathcal{A}}) = L_{f(a)}g(a) = f(a)g(a).$$

Property (iv) follows from Lemma 10; this implies also (v).

It remains to show the uniqueness of the functional calculus (viii). Let f be a function analytic in an open neighbourhood U of $\sigma(a_1,\ldots,a_n)$.

(a) We first show that there are elements

$$a_{n+1},\ldots,a_m \in \mathcal{A} \quad \text{such that} \quad \sigma^{\langle a_1,\ldots,a_m\rangle}(a_1,\ldots,a_n) \subset U.$$

Set

$$Z = \big\{(z_1,\ldots,z_n) \in \mathbb{C}^n : |z_j| \le \|a_j\| \quad (j = 1,\ldots,n)\big\}.$$

If $\lambda = (\lambda_1,\ldots,\lambda_n) \in Z \setminus U$, then $\lambda \notin \sigma^{\mathcal{A}}(a_1,\ldots,a_n)$, and so there exist

$$y_{\lambda,1},\ldots,y_{\lambda,n} \quad \text{such that} \quad \sum_{j=1}^{n}(a_j - \lambda_j)y_{\lambda,j} = 1.$$

Thus

$$\lambda \notin \sigma^{\langle a_1,\ldots,a_n,y_{\lambda,1},\ldots,y_{\lambda,n}\rangle}(a_1,\ldots,a_n)$$

and there exists an open neighbourhood U_λ of λ such that

$$U_\lambda \cap \sigma^{\langle a_1,\dots,a_n,y_{\lambda,1},\dots,y_{\lambda,n}\rangle}(a_1,\dots,a_n) = \emptyset.$$

Since $Z\setminus U$ is compact, there are points $\lambda^{(1)},\dots,\lambda^{(m)} \in Z\setminus U$ such that $\bigcup_{i=1}^m U_{\lambda^{(i)}} \supset Z\setminus U$. If $\mathcal{A}_0$ is the algebra generated by $a_1,\dots,a_m$ where

$$\{a_{n+1},\dots,a_m\} = \{y_{\lambda^{(1)},1},\dots,y_{\lambda^{(1)},n},\dots,y_{\lambda^{(m)},1},\dots,y_{\lambda^{(m)},n}\},$$

then $\sigma^{\mathcal{A}_0}(a_1,\dots,a_n) \subset U$.

Extend f to $U\times\mathbb{C}^{m-n}$ by $\tilde{f}(z_1,\dots,z_m) = f(z_1,\dots,z_n)$. Clearly, $\tilde{f}$ is analytic on a neighbourhood of $\sigma^{\mathcal{A}_0}(a_1,\dots,a_m)$.

(b) Write $K = \sigma^{\mathcal{A}_0}(a_1,\dots,a_m)$. By Theorem 2.18, K is a polynomially convex set. We show that there is a polynomially convex neighbourhood V of K such that $V \subset U \times \mathbb{C}^{m-n}$.

Choose $r > 0$ such that $\sigma^{\mathcal{A}_0}(a_1,\dots,a_m) \subset \overline{\Delta}(0,r) = \{(z_1,\dots,z_m) \in \mathbb{C}^m : |z_i| \le r, i = 1,\dots,m\}$. For every $\lambda \in \overline{\Delta}(0,r)\setminus(U\times\mathbb{C}^{m-n})$ there exists a polynomial p such that $|p(\lambda)| > \max\{|p(z)| : z \in K\}$. Using the compactness of the set $\overline{\Delta}(0,r) \setminus (U \times \mathbb{C}^{m-n})$, we get that there exist a finite number of polynomials $p_1,\dots,p_s$ and positive numbers $\varepsilon_1,\dots,\varepsilon_s$ such that

$$V = \{z \in \overline{\Delta}(0,r) : |p_i(z)| \le \|p_i\|_K + \varepsilon_i \ (i = 1,\dots,s)\} \subset U \times \mathbb{C}^{m-n}.$$

Clearly, V is a compact polynomially convex neighbourhood of $\sigma^{\mathcal{A}_0}(a_1,\dots,a_m)$.

Thus $\tilde{f}$ can be approximated by polynomials uniformly on V. Consequently, $f(a_1,\dots,a_n) = \tilde{f}(a_1,\dots,a_m)$ is determined uniquely. □

31 Taylor functional calculus in Banach algebras

A natural idea how to define the Taylor spectrum and Taylor functional calculus for a commuting n-tuple $a = (a_1,\dots,a_n)$ in a (non-commutative) Banach algebra $\mathcal{A}$ is to consider the n-tuple $L_A = (L_{A_1},\dots,L_{A_n}) \in \mathcal{B}(\mathcal{A})^n$. However, if $\mathcal{A} = \mathcal{B}(X)$ for some Banach space X and $A = (A_1,\dots,A_n) \in \mathcal{B}(X)^n$ is a commuting n-tuple of operators, then $\sigma_T(L_A) = \sigma_S(A)$, which is in general bigger than the Taylor spectrum of A. So this is not a proper way how to define the Taylor spectrum and Taylor functional calculus for commuting elements in Banach algebras.

In this section we suggest a way how to overcome this difficulty. We introduce the concept of semidistributive Banach algebras.

Definition 1. By a *semidistributive Banach algebra* we mean a Banach space $(\mathcal{A}, \|\cdot\|)$ together with a multiplication in $\mathcal{A}$ satisfying the following conditions (for all $x, y, z \in \mathcal{A}$, $\alpha \in \mathbb{C}$):

(i) $(xy)z = x(yz)$;

(ii) $(x + y)z = xz + yz$;

(iii) $(\alpha x)y = \alpha(xy) = x(\alpha y)$;
(iv) $\|xy\| \le \|x\| \cdot \|y\|$;
(v) $\mathcal{A}$ has a unit $1_{\mathcal{A}}$ satisfying $\|1_{\mathcal{A}}\| = 1$.

In other words, semidistributive Banach algebras satisfy all axioms of Banach algebras except of one of the distributive laws.

Let $\mathcal{A}$ be a semidistributive Banach algebra. Denote by $\mathcal{D}(\mathcal{A})$ the *distributive center* of $\mathcal{A}$, i.e., the set of all elements $x \in \mathcal{A}$ such that $x(y+z) = xy + xz$ for all $y, z \in \mathcal{A}$.

Example 2. Let X be a Banach space and $\mathcal{H}(X)$ the set of all continuous homogeneous mappings $\varphi : X \to X$. It is easy to check that $\mathcal{H}(X)$ with the norm $\|\varphi\| = \sup\{\|\varphi x\| : x \in X, \|x \le 1\}$ is a semidistributive Banach algebra.

Let $A \in \mathcal{B}(X)$. It is easy to see that $A(\varphi_1 + \varphi_2) = A\varphi_1 + A\varphi_2$ for all $\varphi_1, \varphi_2 \in \mathcal{H}(X)$. So $A \in \mathcal{D}(\mathcal{H}(X))$. In fact, it is easy to check that $\mathcal{D}(\mathcal{H}(X)) = \mathcal{B}(X)$.

Let $\mathcal{A}$ be a semidistributive Banach algebra and $a = (a_1, \dots, a_n)$ an n-tuple of commuting elements of $\mathcal{D}(\mathcal{A})$. Since $a_j \in \mathcal{D}(\mathcal{A})$, the mappings L_{a_j} defined by $L_{a_j} b = a_j b \quad (j = 1, \dots, n, b \in \mathcal{A})$ are commuting bounded linear operators acting on $\mathcal{A}$. Let $L_a = (L_{a_1}, \dots, L_{a_n})$.

Definition 3. Let $\mathcal{A}$ be a semidistributive Banach algebra and $a = (a_1, \dots, a_n)$ an n-tuple of commuting elements of $\mathcal{D}(\mathcal{A})$. We say that a is Taylor regular if L_a is Taylor regular (in the sense of Section 25).

The Taylor spectrum $\sigma_T^{\mathcal{A}}(a)$ of a is defined as the set of all $\lambda \in \mathbb{C}^n$ such that the n-tuple $a - \lambda = (a_1 - \lambda_1, \dots, a_n - \lambda_n)$ is not Taylor regular, i.e., $\sigma_T^{\mathcal{A}}(a) = \sigma_T(L_a)$.

Clearly, the Taylor regularity and the Taylor spectrum depend on the choice of the semidistributive algebra $\mathcal{A}$.

Let X be a Banach space, $A = (A_1, \dots, A_n)$ commuting linear operators on X. Then $\{A_1, \dots, A_n\} \subset \mathcal{D}(\mathcal{H}(X))$ and, by Corollary 27.7, $\sigma_T^{\mathcal{H}(X)}(A) = \sigma_T(A)$ where $\sigma_T(A)$ is the Taylor spectrum of the n-tuple A of operators in the sense of Section 25.

The basic property of the Taylor spectrum in semidistributive algebras is the existence of the functional calculus.

Theorem 4. *Let $\mathcal{A}$ be a semidistributive Banach algebra, let $a = (a_1, \dots, a_n)$ be a commuting n-tuple of elements of $\mathcal{D}(\mathcal{A})$. Let U be an open neighbourhood of $\sigma_T^{\mathcal{A}}(a)$. Then there exists an algebraic homomorphism $\Phi : H(U) \to \mathcal{A}$, $f \mapsto f(a)$ such that*

(i) *if $f \equiv 1$, then $f(a) = 1_{\mathcal{A}}$;*
if $f \equiv z_j$, then $f(a) = a_j \quad (j = 1, \dots, n)$;
(ii) *if $f_n \to f$ uniformly on U, then $f_n(a) \to f(a)$;*
(iii) *$f(a) \in \mathcal{D}(\mathcal{A})$ for all $f \in H(U)$.*

Proof. We have $\sigma_T(L_a) = \sigma_T^{\mathcal{A}}(a) \subset U$. Thus for $f \in H(U)$ we can define $f(L_a) \in \mathcal{B}(\mathcal{A})$ by the Taylor functional calculus constructed in Section 30.

Define $f(a) \in \mathcal{A}$ by $f(a) = f(L_a)(1_{\mathcal{A}})$. Clearly, the mapping $f \mapsto f(a)$ is linear and satisfies (i) and (ii).

Let $f \in H(U)$ and $b \in \mathcal{A}$. Then the mapping $R_b : \mathcal{A} \to \mathcal{A}$ defined by $R_b c = cb$ is linear and $L_{a_j} R_b = R_b L_{a_j} \quad (j = 1, \dots, n)$. Since $f(L_a) \in (L_a)''$, we have $f(L_a)R_b = R_b f(L_a)$. Hence

$$f(L_a)b = f(L_a)R_b(1_{\mathcal{A}}) = R_b f(L_a)(1_{\mathcal{A}}) = f(a)b$$

for every $b \in \mathcal{A}$. Consequently, $f(L_a) = L_{f(a)}$.

Let $f, g \in H(U)$. We have

$$(fg)(a) = (fg)(L_a)(1_{\mathcal{A}}) = f(L_a)g(L_a)(1_{\mathcal{A}}) = L_{f(a)}g(a) = f(a)g(a).$$

Hence the functional calculus $f \mapsto f(a)$ is multiplicative.

(iii) Let $x, y \in \mathcal{A}$. Since $f(L_a)$ is a linear operator, we have

$$f(a)(x+y) = L_{f(a)}(x+y) = f(L_a)(x+y) = f(L_a)x + f(L_a)y = f(a)x + f(a)y.$$

Hence $f(a) \in \mathcal{D}(\mathcal{A})$. □

32 k-regular functions

In Sections 10 and 11 we studied operator-valued functions with continuously changing ranges and kernels. In this section we generalize this notion to operator-valued functions with finite-dimensional "jumps" in the range (kernel).

Definition 1. Let L, N be subspaces of a Banach space X. Let $k \geq 0$. Set

$$\delta_k(L,N) = \inf\{\delta(L,N') : N' \supset N, \dim N'/N \leq k\},$$
$$\widehat{\delta}_k(L,N) = \inf\{\widehat{\delta}(L,N') : N' \supset N, \dim N'/N \leq k\} \qquad \text{and}$$
$$\vartheta_k(L,N) = \inf\{\delta(L',N) : L' \subset L, \dim L/L' \leq k\}.$$

Let M be a closed subspace of a Banach space X with $\operatorname{codim} M = k < \infty$. Recall that there exists a projection $P \in \mathcal{B}(X)$ with $\|P\| < k+1$ and $\operatorname{Ker} P = M$, see A.1.25.

The quantities δ_k and ϑ_k are closely connected.

Lemma 2. *Let L, N be closed subspaces of X. Then:*

(i) $\delta_k(L,N) \leq (k+2)\vartheta_k(L,N)$;

(ii) $\vartheta_k(L,N) \leq (k+1)^2 \delta_k(L,N)$.

Proof. (i) Let $L' \subset L$, $\dim L/L' \le k$. Let $P \in \mathcal{B}(L)$ be a projection satisfying $\operatorname{Ker} P = L'$ and $\|P\| < k+1$. Let $F = \operatorname{Ran} P$. Then $L = L' \oplus F$ and $\dim F \le k$.

We show that $\delta(L, N+F) \le (k+2) \cdot \delta(L', N)$. Let $x \in L$, $\|x\| = 1$. Then $x = l' + f$ for some $l' \in L'$ and $f \in F$, and $\|l'\| = \|(I-P)x\| < k+2$. There is an $n \in N$ with $\|l' - n\| \le (k+2)\delta(L', N)$. Thus

$$\operatorname{dist}\{x, N+F\} \le \|x - n - f\| = \|l' - n\| \le (k+2)\delta(L', N).$$

Hence $\delta(L, N+F) \le (k+2)\delta(L', N)$. This proves (i).

(ii) Let $N' \supset N$, $\dim N'/N \le k$. Let $P \in \mathcal{B}(N')$ be a projection such that $\operatorname{Ker} P = N$ and $\|P\| < k+1$. Let $F = \operatorname{Ran} P$. Then $\dim F \le k$ and $N' = N \oplus F$.

By the Auerbach Lemma, see A.1.23, there is a biorthogonal system $f_j \in F$, $f_j^* \in F^*$ $(j = 1, 2, \dots, \dim F)$ such that $\|f_i\| = 1 = \|f_j^*\|$ and $\langle f_i, f_j^* \rangle = \delta_{ij}$ for all i, j.

Extend f_j^* to functionals on y_j^* on N' by setting $y_j^* | N = 0$. For $x \in N'$ we have

$$|\langle x, y_j^* \rangle| = |\langle Px, f_j^* \rangle| \le \|P\| \cdot \|x\|.$$

By the Hahn-Banach theorem we can extend y_j^* to a functional on X (denoted by the same symbol y_j^*) with the same norm $\|y_j^*\| \le \|P\| < k+1$. Let $L' = L \cap \bigcap_{j=1}^{\dim F} \operatorname{Ker} y_j^*$. Clearly, $\dim L/L' \le k$.

We prove that $\delta(L', N) \le (k+1)^2 \delta(L, N')$. Let $x \in L' \subset L$, $\|x\| < 1$. Then there exists $y \in N'$ with $\|x - y\| \le \delta(L, N')$. Let $Q \in \mathcal{B}(X)$ be defined by $Qx = \sum_j \langle x, y_j^* \rangle f_j$. Then Q is a projection onto F, $\operatorname{Ker} Q \supset L'$ and $\|Q\| \le k \cdot \max\{\|y_j^*\|\} \le k(k+1)$. We have

$$\begin{aligned} \operatorname{dist}\{x, N\} &\le \|x - (I-Q)y\| \le \|I - Q\| \cdot \|x - y\| \\ &\le \big(1 + k(k+1)\big)\delta(L, N') \le (k+1)^2 \delta(L, N'). \end{aligned}$$

Hence $\delta(L', N) \le (k+1)^2 \delta(L, N')$. This proves (ii). □

Theorem 3. *Let $k \ge 0$ and $\varepsilon > 0$. Then there exists a positive number $\eta = \eta(k, \varepsilon)$ with the following property: if M, N, M' are closed subspaces of a Banach space X, $M \subset M'$, $\dim M'/M \le k$, $\delta(M, N) < \eta$ and $\delta(N, M') < \eta$, then there exists a subspace $G \subset X$ with $\dim G \le k$ and $\widehat{\delta}(N, M+G) \le \varepsilon$.*

Proof. We prove the statement by induction on k. For $k = 0$ the statement is clear with $G = \{0\}$ and $\eta(0, \varepsilon) = \varepsilon$.

Suppose that the statement is true for $k - 1 \ge 0$. Let $\varepsilon' = \eta(k-1, \varepsilon)$ be the number given by the induction assumption; we may assume that $0 < \varepsilon' \le \varepsilon \le 1$. Choose $\eta < \frac{\varepsilon\varepsilon'}{4(k+2)^2}$. Let M, N, M' satisfy the conditions of the theorem, so $M \subset M'$, $\dim M'/M \le k$, $\delta(M, N) < \eta$ and $\delta(N, M') < \eta$. Let $P \in \mathcal{B}(M')$ be a projection such that $\operatorname{Ker} P = M$ and $\|P\| < k+1$. Set $F = \operatorname{Ran} P$. So $\dim F \le k$ and $M' = M \oplus F$.

The statement is clear if $\delta(N,M) \le \varepsilon$; in this case it is sufficient to take $G = \{0\}$. Suppose on the contrary that $\delta(N,M) > \varepsilon$, so there is an $x \in N$ with $\|x\| = 1$ and $\operatorname{dist}\{x, M\} \ge \varepsilon$. By assumption, there are $m_0 \in M$ and $f \in F$ such that $\|x - (m_0 + f)\| < \eta$. We have $\|m_0 + f\| \le \|x - (m_0+f)\| + \|x\| < 1 + \eta$ and $\|m_0\| = \|(I-P)(m_0+f)\| < (k+2)(1+\eta)$. Furthermore,

$$\varepsilon \le \operatorname{dist}\{x, M\} \le \|x - m_0\| \le \|x - (m_0+f)\| + \|f\| < \eta + \|f\|,$$

and so $\|f\| > \varepsilon - \eta$.

Set $M_1 = M \vee \{f\}$. Since $\operatorname{dist}\{x, M\} \ge \varepsilon$ and $\operatorname{dist}\{x, M_1\} < \eta < \varepsilon$, we have $M_1 \ne M$ and $\dim M'/M_1 \le k-1$. It is sufficient to show that $\delta(M_1, N) \le \varepsilon'$. Indeed, the induction assumption then gives the existence of G' with $\dim G' \le k-1$ and $\widehat{\delta}(M_1 + G, N) \le \varepsilon$, and so we can take $G = G' \vee \{f\}$.

To show that $\delta(M_1, N) \le \varepsilon'$, let $u = m + \alpha f \in M_1$ with $\|u\| = 1$, $m \in M$ and $\alpha \in \mathbb{C}$. Then $\|m\| = \|(I-P)u\| < k+2$ and $\|\alpha f\| = \|Pu\| < k+1$. Thus $|\alpha| = \frac{\|\alpha f\|}{\|f\|} < \frac{k+1}{\varepsilon - \eta}$. Then there exists $n \in N$ with

$$\|m - n\| \le (k+2)\delta(M,N) < (k+2)\eta.$$

Similarly, there exists $n_0 \in N$ with

$$\|m_0 - n_0\| \le (k+2)(1+\eta)\delta(M,N) < (k+2)(1+\eta)\eta.$$

Thus

$$\begin{aligned}\operatorname{dist}\{u, N\} &\le \|m + \alpha f - n - \alpha(x - n_0)\| \le \|m - n\| + |\alpha| \cdot \|f - x + n_0\| \\ &\le \|m-n\| + |\alpha|\big(\|f - x + m_0\| + \|n_0 - m_0\|\big) \\ &< (k+2)\eta + \frac{k+1}{\varepsilon - \eta}\big(\eta + (k+2)(1+\eta)\eta\big) \le \frac{4\eta(k+2)^2}{\varepsilon} < \varepsilon'. \qquad \square\end{aligned}$$

Lemma 4. *Let $k \ge 0$ and $\varepsilon > 0$. Then there exists $\eta > 0$ with the following property: if M, N, M' are closed subspaces of a Banach space X, $M \subset M'$, $\dim M'/M = k$ and $\widehat{\delta}(M', N) < \eta$, then there exists a subspace $G \subset N$ such that $\dim G = k$, $\widehat{\delta}(M + G, N) \le \varepsilon$ and*

$$\|m + g\| \ge \frac{1}{4(k+1)} \max\{\|m\|, \|g\|\} \qquad (m \in M, g \in G).$$

Proof. We can assume that $\varepsilon \le 1$. Let M, M', N satisfy the conditions of the lemma and let η be a positive number satisfying $\eta < \frac{\varepsilon}{4(k+2)^2}$.

Let $P \in \mathcal{B}(M')$ be a projection satisfying $\operatorname{Ker} P = M$ and $\|P\| < k+1$. Set $F = \operatorname{Ran} P$. Then $\dim F = k$ and $M' = M \oplus F$. Choose a biorthogonal system $f_1, \dots, f_k \in F$, $f_1^*, \dots, f_k^* \in F^*$ such that $\|f_i\| = 1 = \|f_j^*\|$ and $\langle f_i, f_j^* \rangle = \delta_{ij}$ for all i, j. If $\alpha_i \in \mathbb{C}$ $\quad (i = 1, \dots, k)$, then

$$\Big\|\sum_{i=1}^k \alpha_i f_i\Big\| \ge \Big|\Big\langle \sum_{i=1}^k \alpha_i f_i, f_j^*\Big\rangle\Big| = |\alpha_j|$$

for all j. Hence

$$\max\{|\alpha_j| : j = 1, \ldots, k\} \leq \Big\| \sum_{i=1}^{k} \alpha_i f_i \Big\| \leq \sum_{j=1}^{k} |\alpha_j|.$$

For each $j=1,\ldots,k$ there exists $n_j \in N$ with $\|n_j - f_j\| < \eta$. Set $G = \bigvee\{n_1,\ldots,n_k\}$. Then $G \subset N$.

Let $\Phi : M + F \to M + G$ be the operator defined by

$$\Phi\Big(m + \sum_{i=1}^{k} \alpha_i f_i\Big) = m + \sum_{i=1}^{k} \alpha_i n_i \qquad (m \in M, \alpha_i \in \mathbb{C}).$$

Let $m \in M$, $f = \sum \alpha_i f_i \in F$ and let $x = m + f$ be an element of $M + F$ of norm one. Then $\|m\| = \|(I - P)x\| < k + 2$ and $|\alpha_i| \leq \|f\| = \|Px\| < k + 1$ for all $i = 1, \ldots, k$. Hence

$$\|x - \Phi x\| = \Big\| \sum_{i=1}^{k} \alpha_i (f_i - n_i) \Big\| \leq k(k+1)\eta.$$

Thus $\|\Phi\| \leq 1 + k(k+1)\eta \leq 2$. Further, Φ is onto and $\|\Phi x\| \geq 1 - k(k+1)\eta \geq 1/2$. Hence Φ is invertible and $\|\Phi^{-1}\| \leq 2$.

Let $x = m + g \in M + G$, $\|x\| = 1$. Let $f = \Phi^{-1} g$. Then

$$\|m\| = \|(I - P)(m + f)\| \leq (k+2)\|m + f\| = (k+2)\|\Phi^{-1}x\| \leq 2(k+2)$$

and

$$\|g\| = \|\Phi f\| \leq \|\Phi\| \cdot \|P\| \cdot \|m + f\| \leq 2(k+1)\|\Phi^{-1}x\| \leq 4(k+1).$$

Hence

$$\|m + g\| \geq \frac{1}{4(k+1)} \max\{\|m\|, \|g\|\}$$

for all $m \in M$ and $g \in G$.

We show that $\delta(N, M + G) < \varepsilon$. Let $n \in N$, $\|n\| = 1$. Then there exist $m \in M$ and $f = \sum \alpha_i f_i \in F$ such that $\|n - m - f\| < \eta$. Thus $\|m + f\| < 1 + \eta$, $\|m\| = \|(I - P)(m + f)\| < (k+2)(1+\eta)$ and $\|f\| = \|P(m+f)\| < (k+1)(1+\eta)$. Hence $|\alpha_i| \leq (k+1)(1+\eta)$ for all i. We have

$$\begin{aligned} \operatorname{dist}\{n, M + G\} &\leq \Big\| n - \Big(m + \sum_{i=1}^{k} \alpha_i n_i\Big) \Big\| \\ &\leq \|n - (m + f)\| + \sum_{i=1}^{k} |\alpha_i| \cdot \|f_i - n_i\| < \eta + k(k+1)(1+\eta)\eta < \varepsilon. \end{aligned}$$

Hence $\delta(N, M + G) \leq \varepsilon$.

Finally, we show that $\delta(M+G,N)\le\varepsilon$. Let $m\in M$, $g=\sum\alpha_i n_i\in G$ and $\|m+g\|=1$. Then $\|m\|=\|(I-P)\Phi^{-1}(m+g)\|<2(k+2)$ and there exists $n\in N$ with $\|m-n\|<\|m\|\cdot\eta\le 2(k+2)\eta$. Further $\|g\|\le\|m\|+\|m+g\|<2k+5$ and $|\alpha_i|\le\left\|\sum_{i=1}^k\alpha_i f_i\right\|=\|\Phi^{-1}g\|\le 2\|g\|\le 4k+10$. Hence

$$\begin{aligned}\operatorname{dist}\{m+g,N\}&\le\|m-n\|+\sum_{i=1}^k|\alpha_i|\cdot\|n_i-f_i\|\\&\le 2(k+2)\eta+k(4k+10)\eta<\varepsilon.\end{aligned}$$

□

Theorem 5. *Let X,Y be Banach spaces and U a metric space. Let $T:U\to\mathcal{B}(X,Y)$ be a norm-continuous function, $w\in U$, $k\ge 0$ and suppose that* $\operatorname{Ran}T(w)$ *is closed. The following statements are equivalent:*

(i) $\lim_{z\to w}\delta_k\big(\operatorname{Ran}T(z),\operatorname{Ran}T(w)\big)=0$;

(ii) $\lim_{z\to w}\vartheta_k\big(\operatorname{Ran}T(z),\operatorname{Ran}T(w)\big)=0$;

(iii) $\lim_{z\to w}\widehat{\delta}_k\big(\operatorname{Ran}T(z),\operatorname{Ran}T(w)\big)=0$;

(iv) $\lim_{z\to w}\delta_k\big(\operatorname{Ker}T(w),\operatorname{Ker}T(z)\big)=0$;

(v) $\lim_{z\to w}\vartheta_k\big(\operatorname{Ker}T(w),\operatorname{Ker}T(z)\big)=0$;

(vi) $\lim_{z\to w}\widehat{\delta}_k\big(\operatorname{Ker}T(w),\operatorname{Ker}T(z)\big)=0$;

(vii) $\lim_{z\to w}\vartheta_k\big(\operatorname{Ran}T(z)^*,\operatorname{Ran}T(w)^*\big)=0$.

Moreover, if any of condition (i)–(vii) *is satisfied, then* $\operatorname{Ran}T(z)$ *is closed for all z in a neighbourhood of w.*

Proof. By Lemma 2, (i) ⇔ (ii) and (iv) ⇔ (v).

Since $\delta\big(\operatorname{Ran}T(w),\operatorname{Ran}T(z)\big)\to 0$ and $\delta\big(\operatorname{Ker}T(z),\operatorname{Ker}T(w)\big)\to 0$ by Lemma 10.12, the equivalences (i) ⇔ (iii) and (iv) ⇔ (vi) follow from Theorem 3.

(iii) ⇒ (iv): Let ε be a positive number, $\varepsilon\le 1$. By Lemma 4, there exists a neighbourhood U_0 of w with the following property: if $z\in U_0$, then there exists a subspace $F\subset\operatorname{Ran}T(z)$ with $\dim F\le k$, $\widehat{\delta}\big(\operatorname{Ran}T(w)+F,\operatorname{Ran}T(z)\big)<\varepsilon/6$, and

$$\|T(w)x+f\|\ge\frac{1}{4(k+1)}\max\{\|T(w)x\|,\|f\|\}\tag{1}$$

for all $x\in X$ and $f\in F$. We may also assume that $\|T(z)-T(w)\|<\frac{\varepsilon\cdot\gamma(T(w))}{48(k+1)}$ for all $z\in U_0$.

Fix $z\in U_0$ and F with the above-described property. Let $S_w,S_z:X\oplus F\to Y$ be defined by

$$\begin{aligned}S_w(x\oplus f)&=T(w)x+\gamma(T(w))\cdot f,\\S_z(x\oplus f)&=T(z)x+\gamma(T(w))\cdot f\end{aligned}$$

for all $x\in X$, $f\in F$; here $X\oplus F$ denotes the ℓ^1 direct sum of X and F.

Clearly, $\operatorname{Ran} S_w = \operatorname{Ran} T(w) + F$ and $\operatorname{Ran} S_z = \operatorname{Ran} T(z) + F = \operatorname{Ran} T(z)$. Thus $\widehat{\delta}(\operatorname{Ran} S_w, \operatorname{Ran} S_z) < \varepsilon/6$.

We show that $\gamma(S_w) \geq \frac{\gamma(T(w))}{8(k+1)}$. Let $0 < s < \gamma(T(w))$ and let $y \in \operatorname{Ran} S_w = \operatorname{Ran} T(w) + F$ be a vector of norm one. Express $y = T(w)x + f$ for some $x \in X$ and $f \in F$. Then there exists $x' \in X$ with $T(w)x' = T(w)x$ and $\|x'\| \leq s^{-1}\|T(w)x\|$. We have $S_w\big(x' \oplus \gamma(T(w))^{-1} f\big) = T(w)x + f = y$ and, by (1),

$$\|x' \oplus \gamma(T(w))^{-1} f\| \leq s^{-1}(\|T(w)x\| + \|f\|) \leq 8(k+1)s^{-1}\|T(w)x + f\| = 8(k+1)s^{-1}.$$

Hence $\gamma(S_w) \geq \frac{s}{8(k+1)}$. Letting $s \to \gamma(T(w))$ gives $\gamma(S_w) \geq \frac{\gamma(T(w))}{8(k+1)}$.

By Lemma 10.13, we have

$$\begin{aligned}
\gamma(S_z) = \gamma(S_z^*) &\geq \gamma(S_w^*)\Big(1 - 2\delta(\operatorname{Ker} S_w^*, \operatorname{Ker} S_z^*)\Big) - \|S_w^* - S_z^*\| \\
&= \gamma(S_w)\Big(1 - 2\delta(\operatorname{Ran} S_z, \operatorname{Ran} S_w)\Big) - \|S_w - S_z\| \\
&\geq \frac{\gamma(T(w))}{8(k+1)}\Big(1 - \frac{\varepsilon}{3}\Big) - \frac{\varepsilon\gamma(T(w))}{48(k+1)} \geq \frac{\gamma(T(w))}{16(k+1)}.
\end{aligned}$$

In particular, $\operatorname{Ran} T(z) = \operatorname{Ran} S_z$ is closed for each $z \in U_0$.

Furthermore,

$$\delta(\operatorname{Ker} S_w, \operatorname{Ker} S_z) \leq \gamma(S_z)^{-1}\|S_w - S_z\| \leq \frac{16(k+1)}{\gamma(T(w))} \cdot \frac{\varepsilon\gamma(T(w))}{48(k+1)} < \varepsilon.$$

Find a subspace $G \subset X$ with $T(z)G = F$ and $\dim G = \dim F \leq k$. We have $\operatorname{Ker} S_w = \operatorname{Ker} T(w)$ and $\operatorname{Ker} S_z = \operatorname{Ker} T(z) + \big\{g \oplus -\gamma(T(w))^{-1}T(z)g : g \in G\big\}$. Consequently, $\delta_k\big(\operatorname{Ker} T(w), \operatorname{Ker} T(z)\big) \leq \delta\big(\operatorname{Ker} T(w), \operatorname{Ker} T(z) + G\big) \leq \delta\big(\operatorname{Ker} S_w, \operatorname{Ker} S_z\big) < \varepsilon$. Thus

$$\lim_{z \to w} \delta_k\big(\operatorname{Ker} T(w), \operatorname{Ker} T(z)\big) = 0.$$

(iv) $\Rightarrow$ (vii): Let $\varepsilon > 0$. Since $\lim_{z\to w} \delta_k\big(\operatorname{Ker} T(w), \operatorname{Ker} T(z)\big) = 0$, there exists a neighbourhood U_0 of w with the following property: if $z \in U_0$, then $\operatorname{Ran} T(z)$ is closed, and there is an $N \supset \operatorname{Ker} T(z)$ with $\dim N/\operatorname{Ker} T(z) \leq k$ and $\delta\big(\operatorname{Ker} T(w), N\big) < \varepsilon$. Thus $\delta(N^\perp, \operatorname{Ran} T(w)^*) < \varepsilon$ where $N^\perp \subset \operatorname{Ran} T(z)^*$ and $\dim \operatorname{Ran} T(z)^*/N^\perp = \dim N/\operatorname{Ker} T(z) \leq k$. Hence $\vartheta_k\big(\operatorname{Ran} T(z)^*, \operatorname{Ran} T(w)^*\big) < \varepsilon$ and

$$\lim_{z\to w} \vartheta_k\big(\operatorname{Ran} T(z)^*, \operatorname{Ran} T(w)^*\big) = 0.$$

(vii) $\Rightarrow$ (ii): Let $\varepsilon > 0$. Then there exists a neighbourhood U_1 of w with the following property: if $z \in U_1$, then $\operatorname{Ran} T(z)^*$ is closed and there is a subspace $F' \subset X^*$ with $\dim F' \leq k$ and $\delta\big(\operatorname{Ran} T(z)^*, \operatorname{Ran} T(w)^* + F'\big) < \varepsilon$. Hence $\delta\big(\operatorname{Ker} T(w) \cap {}^\perp F', \operatorname{Ker} T(z)\big) < \varepsilon$. Consequently, $\vartheta_k\big(\operatorname{Ker} T(w), \operatorname{Ker} T(z)\big) \to 0$. $\square$

Definition 6. Let X, Y be Banach spaces and U a metric space. Let $T : U \to \mathcal{B}(X,Y)$ be a norm-continuous function. Let $w \in U$ and $k \geq 0$. We say that T is *k-regular* at w if $\operatorname{Ran} T(w)$ is closed and T satisfies any of the equivalent conditions of Theorem 5.

Corollary 7. *A function $T : U \to \mathcal{B}(X,Y)$ is k-regular at w if and only if the function $z \mapsto T(z)^*$ is k-regular at w.*

As we have seen in Section 10, regular functions are closely related to exact sequences. Similarly, k-regular functions are connected with "Fredholm sequences".

Theorem 8. *Let X, Y, Z be Banach spaces and U a metric space. Let $T : U \to \mathcal{B}(X,Y)$ and $S : U \to \mathcal{B}(Y,Z)$ be norm-continuous functions satisfying $S(z)T(z) = 0$ for all $z \in U$. Let $w \in U$, let $\operatorname{Ran} S(w)$ be closed and $\dim \operatorname{Ker} S(w)/\operatorname{Ran} T(w) = k < \infty$. Then both T and S are k-regular at w.*

Proof. By Lemma 16.2, $\operatorname{Ran} T(w)$ is closed as a subspace of finite codimension in $\operatorname{Ker} S(w)$. Let F be a subspace satisfying $\operatorname{Ker} S(w) = \operatorname{Ran} T(w) \oplus F$ and $\dim F = k$. We have

$$\begin{aligned}\delta_k\big(\operatorname{Ran} T(z), \operatorname{Ran} T(w)\big) &\leq \delta\big(\operatorname{Ran} T(z), \operatorname{Ran} T(w) + F\big)\\ &= \delta\big(\operatorname{Ran} T(z), \operatorname{Ker} S(w)\big) \leq \delta\big(\operatorname{Ker} S(z), \operatorname{Ker} S(w)\big) \to 0.\end{aligned}$$

Thus T is k-regular at w.

Similarly, let $M \subset X$ be a subspace of codimension k such that $\operatorname{Ran} T(w) = \operatorname{Ker} S(w) \cap M$. Then

$$\begin{aligned}\vartheta_k\big(\operatorname{Ker} S(w), \operatorname{Ker} S(z)\big) &\leq \delta\big(\operatorname{Ker} S(w) \cap M, \operatorname{Ker} S(z)\big) = \delta\big(\operatorname{Ran} T(w), \operatorname{Ker} S(z)\big)\\ &\leq \delta\big(\operatorname{Ran} T(w), \operatorname{Ran} T(z)\big) \to 0.\end{aligned}$$

Hence S is k-regular at w.

Definition 9. Let M, N be closed subspaces of a Banach space X. Suppose that there exists a subspace $M' \supset M$ such that $\dim M'/M < \infty$ and $\widehat{\delta}(M', N) < \sqrt{2}-1$. Then we define $\operatorname{jump}(M, N) = \dim M'/M$.

The definition is correct: if $M'' \supset M$ is another subspace with $\dim M''/M < \infty$ and $\widehat{\delta}(M'', N) < \sqrt{2} - 1$, then

$$\widehat{\delta}(M', M'') \leq \widehat{\delta}(M', N) + \widehat{\delta}(N, M'') + \widehat{\delta}(M', N) \cdot \widehat{\delta}(N, M'') < 1.$$

By Theorem 27.8, this means that $\dim M'/M = \dim M''/M$.

Theorem 5 gives immediately an important result that for k-regular functions the jump in the kernel is always equal to the jump in the range. This is well known for operators in finite-dimensional spaces. Another classical result of this type is for the function $z \mapsto T - z$ where T is a compact operator, see Theorem 15.11. The Kato decomposition gives the same result also for the function $z \mapsto T - z$ where T is a semi-Fredholm operator, or more generally, an essentially Kato operator.

Theorem 10. *Let $T : U \to \mathcal{B}(X,Y)$ be k-regular at a point $w \in U$. Then there exists a neighbourhood U_0 of w such that*

$$\operatorname{jump}\bigl(\operatorname{Ker} T(z), \operatorname{Ker} T(w)\bigr) = \operatorname{jump}\bigl(\operatorname{Ran} T(w), \operatorname{Ran} T(z)\bigr)$$

for each $z \in U_0$.

Proof. There is a neighbourhood U_1 of w such that $\widehat{\delta}_k\bigl(\operatorname{Ran} T(z), \operatorname{Ran} T(w)\bigr) < \sqrt{2}-1$ and $\widehat{\delta}_k\bigl(\operatorname{Ker} T(w), \operatorname{Ker} T(z)\bigr) < \sqrt{2}-1$ for all $z \in U_1$. For each $z \in U_1$ fix subspaces F_z, G_z such that $\dim F_z \le k$, $\widehat{\delta}\bigl(\operatorname{Ran} T(z), \operatorname{Ran} T(w) + F_z\bigr) < \sqrt{2}-1$, $F_z \cap \operatorname{Ran} T(w) = \{0\}$, $\dim G_z \le k$, $\widehat{\delta}\bigl(\operatorname{Ker} T(w), \operatorname{Ker} T(z) + G_z\bigr) < \sqrt{2}-1$ and $G_z \cap \operatorname{Ker} T(z) = \{0\}$. Moreover, we can assume that $\lim_{z\to w} \widehat{\delta}\bigl(\operatorname{Ran} T(z), \operatorname{Ran} T(w) + F_z\bigr) = 0$ and $\lim_{z\to w} \widehat{\delta}\bigl(\operatorname{Ker} T(w), \operatorname{Ker} T(z) + G_z\bigr) = 0$.

By definition, we have $\dim F_z = \operatorname{jump}\bigl(\operatorname{Ran} T(w), \operatorname{Ran} T(z)\bigr)$ and $\dim G_z = \operatorname{jump}\bigl(\operatorname{Ker} T(z), \operatorname{Ker} T(w)\bigr)$.

We show that $\dim F_z = \dim G_z$ for all z in a certain neighbourhood of w. Suppose on the contrary that there is a sequence (z_n) converging to w with $\dim F_{z_n} \ne \dim G_{z_n}$. By passing to a subsequence if necessary we may assume that $\dim F_{z_n}$ and $\dim G_{z_n}$ are constant; denote this constants by a and b. By assumption, $a \ne b$. We have $\lim_{n\to\infty} \widehat{\delta}_a\bigl(\operatorname{Ran} T(z_n), \operatorname{Ran} T(w)\bigr) = 0$, and so, by Theorem 5, $\lim_{n\to\infty} \widehat{\delta}_a\bigl(\operatorname{Ker} T(w), \operatorname{Ker} T(z_n)\bigr) = 0$. Thus for all n sufficiently large we have $b = \operatorname{jump}\bigl(\operatorname{Ker} T(z_n), \operatorname{Ker} T(w)\bigr) \le a$. Similarly, it is possible to show that $a \le b$, which is a contradiction. Hence

$$\operatorname{jump}\bigl(\operatorname{Ker} T(z), \operatorname{Ker} T(w)\bigr) = \operatorname{jump}\bigl(\operatorname{Ran} T(w), \operatorname{Ran} T(z)\bigr)$$

for all z in a certain neighbourhood of w. □

Theorem 11. *Let M_0, M_1 and M_2 be closed subspaces of a Banach space X, let $k, n \ge 0$. Then*

$$\widehat{\delta}_{k+n}(M_2, M_0) \le 4(n+2)\max\bigl\{\widehat{\delta}_k(M_1, M_0), \widehat{\delta}_n(M_2, M_1)\bigr\}.$$

In particular, if $\max\bigl\{\widehat{\delta}_k(M_1, M_0), \widehat{\delta}_n(M_2, M_1)\bigr\} < \frac{\sqrt{2}-1}{4(n+2)}$, then

$$\operatorname{jump}(M_0, M_2) = \operatorname{jump}(M_0, M_1) + \operatorname{jump}(M_1, M_2).$$

Proof. We may assume that $\max\bigl\{\widehat{\delta}_k(M_1, M_0), \widehat{\delta}_n(M_2, M_1)\bigr\} < 1/8$, since the statement is trivial otherwise. Let $d > \max\bigl\{\widehat{\delta}_k(M_1, M_0), \widehat{\delta}_n(M_2, M_1)\bigr\}$. Then there exist subspaces $L_0 \supset M_0$ and $L_1 \supset M_1$ such that $\dim L_0/M_0 \le k$, $\dim L_1/M_1 \le n$, $\widehat{\delta}(L_0, M_1) < d$ and $\widehat{\delta}(L_1, M_2) < d$. Let $P \in \mathcal{B}(L_1)$ be a projection with $\|P\| < n+1$ and $\operatorname{Ker} P = M_1$. Set $F = \operatorname{Ran} P$.

We show that $\widehat{\delta}(L_0 + F, M_2) \le 4(n+2)d$. We may assume that $d < \frac{1}{4(n+2)}$.

(a) Let $x \in M_2$, $\|x\| = 1$. Then there exists $l_1 \in L_1$ with $\|x - l_1\| < d$. Express $l_1 = m_1 + f$ with $m_1 \in M_1$ and $f \in F$. Then $\|m_1\| = \|(I-P)l_1\| < (n+2)\|l_1\| \le (n+2)(1+d)$, and so there exists $l_0 \in L_0$ with $\|m_1 - l_0\| < (n+2)(1+d)d$. We have

$$\begin{aligned} \operatorname{dist}\{x, L_0 + F\} &\le \|x - (l_0 + f)\| \le \|x - l_1\| + \|l_1 - (l_0 + f)\| \\ &\le d + \|m_1 - l_0\| \le d + (n+2)(1+d)d < 2(n+2)d \le 4(n+2)d. \end{aligned}$$

Thus $\delta(M_2, L_0 + F) \le 4(n+2)d$.

(b) Let $x \in L_0 + F$ and $\|x\| = 1$. Express $x = l_0 + f$ with $l_0 \in L_0$ and $f \in F$. Then there exists $m_1 \in M_1$ with $\|l_0 - m_1\| < \|l_0\| \cdot d$, and so $\|m_1\| > \|l_0\| - \|l_0 - m_1\| > \|l_0\|(1-d)$. Since $m_1 = (I-P)(m_1+f)$, we have $\|m_1 + f\| \ge \frac{\|m_1\|}{\|I-P\|} > \frac{\|l_0\|(1-d)}{n+2}$. Hence

$$1 = \|l_0 + f\| \ge \|m_1 + f\| - \|m_1 - l_0\| > \frac{\|l_0\|(1-d)}{n+2} - \|l_0\|d \ge \|l_0\|\frac{1}{2(n+2)}.$$

Hence $\|l_0\| \le 2(n+2)$ and there exists $m_1 \in M_1$ with $\|l_0 - m_1\| < 2(n+2)d$. Therefore $\|m_1 + f\| \le \|l_0 + f\| + \|m_1 - l_0\| < 1 + 2(n+2)d$ and there exists $m_2 \in M_2$ with $\|m_2 - (m_1 + f)\| < (2(n+2)d + 1)d$. Hence

$$\begin{aligned} \operatorname{dist}\{x, M_2\} &\le \|l_0 + f - m_2\| \le \|l_0 - m_1\| + \|m_1 + f - m_2\| \\ &< 2(n+2)d + 2(n+2)d^2 + d \le 4(n+2)d. \end{aligned}$$

Thus $\delta(L_0 + F, M_2) \le 4(n+2)d$ and so $\widehat{\delta}_{k+n}(M_2, M_0) \le 4(n+2)d$.

Suppose now that $\max\{\widehat{\delta}_k(M_1, M_0), \widehat{\delta}_n(M_2, M_1)\} < \frac{\sqrt{2}-1}{4(n+2)}$. Using the previous construction we have $\operatorname{jump}(M_0, M_1) = \dim L_0/M_0$ and $\operatorname{jump}(M_1, M_2) = \dim L_1/M_1 = \dim F$. It is sufficient to show that $F \cap L_0 = \{0\}$. Indeed, this will imply that $\operatorname{jump}(M_0, M_2) = \dim(L_0 + F)/M_0 = \dim L_0/M_0 + \dim F = \operatorname{jump}(M_0, M_1) + \operatorname{jump}(M_1, M_2)$.

Suppose on the contrary that $F \cap L_0 \ne \{0\}$. Let $f \in F \cap L_0$, $\|f\| = 1$. Then there exists $m_1 \in M_1$ with $\|m_1 - f\| < \frac{\sqrt{2}-1}{4(n+2)}$. Hence $\|P(m_1 - f)\| = \|-f\| = 1$ and so $\|P\| \ge \frac{4(n+2)}{\sqrt{2}-1} > n+2$, a contradiction with the assumption that $\|P\| < n+1$. □

33 Stability of index of complexes

Recall the most important stability results concerning the index $\operatorname{ind} T = \alpha(T) - \beta(T) = \dim \operatorname{Ker} T - \operatorname{codim} \operatorname{Ran} T$ of a semi-Fredholm operator $T : X \to Y$:

(a) there exists $\varepsilon > 0$ such that $\alpha(T') \le \alpha(T)$ and $\beta(T') \le \beta(T')$ for all $T' : X \to Y$ with $\|T' - T\| < \varepsilon$;

(b) if $K : X \to Y$ is a compact operator, then $\operatorname{ind}(T + K) = \operatorname{ind} T$;

(c) there exists $\varepsilon > 0$ such that $\operatorname{ind} T' = \operatorname{ind} T$ for all $T' : X \to Y$ with $\|T' - T\| < \varepsilon$;

(d) there exists $\varepsilon > 0$ such that both $\alpha(T - \lambda)$ and $\beta(T - \lambda)$ are constant in the punctured neighbourhood $\{\lambda \in \mathbb{C} : 0 < |\lambda| < \varepsilon\}$.

In this and the next sections we discuss generalizations of these properties to complexes of Banach spaces.

Definition 1. By a *complex* $\mathcal{K} = (X_i, \psi_i)_{i=0}^n$ we mean an object of the following type:

$$0 \to X_0 \xrightarrow{\psi_0} X_1 \xrightarrow{\psi_1} \cdots \xrightarrow{\psi_{n-2}} X_{n-1} \xrightarrow{\psi_{n-1}} X_n \to 0 \tag{1}$$

where X_i are Banach spaces and $\psi_i : X_i \to X_{i+1}$ operators satisfying $\psi_{i+1}\psi_i = 0$ for all i. Formally we set $X_i = \{0\}$ for either $i < 0$ or $i > n$. Similarly, ψ_i is the zero operator if either $i < 0$ or $i \geq n$. Write $\alpha_i(\mathcal{K}) = \dim \operatorname{Ker} \psi_i / \operatorname{Ran} \psi_{i-1}$. In particular, $\alpha_0(\mathcal{K}) = \dim \operatorname{Ker} \psi_0$ and $\alpha_n(\mathcal{K}) = \operatorname{codim} \operatorname{Ran} \psi_{n-1}$.

Complex (1) is called *Fredholm* if $\alpha_i(\mathcal{K}) < \infty$ for all i. Obviously, in this case the operators ψ_i have closed ranges.

The *index* of a Fredholm complex $\mathcal{K}$ is defined by

$$\operatorname{ind} \mathcal{K} = \sum_{i=0}^{n} (-1)^i \alpha_i(\mathcal{K}). \tag{2}$$

A complex $\mathcal{K}$ is called *semi-Fredholm* if the operators ψ_i have closed ranges and the index $\operatorname{ind} \mathcal{K}$ is well defined by (2) (either $+\infty$ and $-\infty$ does not appear in the formula).

Note that this definition generalizes the classical definition of the index of a semi-Fredholm operator $T : X \to Y$. It is easy to see that $\operatorname{ind} T$ is equal to the index of the semi-Fredholm complex $0 \to X \xrightarrow{T} Y \to 0$.

For two complexes $\mathcal{K} = (X_i, \psi_i)_{i=0}^n$ and $\mathcal{K}' = (X_i, \psi_i')_{i=0}^n$ we set $\operatorname{dist}\{\mathcal{K}, \mathcal{K}'\} = \max\{\|\psi_j - \psi_j'\| : j = 0, \dots, n-1\}$.

Property (a) for semi-Fredholm complexes was essentially proved in Section 27 even in a more general form.

Theorem 2. *Let X, Y, Z be Banach spaces, let $S : X \to Y$ and $T : Y \to Z$ be operators with closed ranges. Then there exists $\eta > 0$ such that*

$$\begin{aligned} &\dim \operatorname{Ran} S / (\operatorname{Ran} S \cap \operatorname{Ker} T) + \dim \operatorname{Ker} T_1 / (\operatorname{Ran} S_1 \cap \operatorname{Ker} T_1) \\ &\quad \leq \dim \operatorname{Ran} S_1 / (\operatorname{Ran} S_1 \cap \operatorname{Ker} T_1) + \dim \operatorname{Ker} T / (\operatorname{Ran} S \cap \operatorname{Ker} T) \end{aligned} \tag{3}$$

for all operators $S_1 : X \to Y, T_1 : Y \to Z$ with closed ranges such that $\|T_1 - T\| < \eta$ and $\|S_1 - S\| < \eta$.

Proof. Set $N = \operatorname{Ker} T$ and $R = \operatorname{Ran} S$. Let ε be the number constructed in Theorem 27.11. Set $\eta = \frac{\varepsilon}{\max\{\gamma(S), \gamma(T)\}}$.

If $S_1 : X \to Y$ and $T_1 : Y \to Z$ are operators with closed ranges satisfying $\|S - S_1\| < \eta$ and $\|T - T_1\| < \eta$, then, by Lemma 10.12, $\delta(\operatorname{Ran} S, \operatorname{Ran} S_1) \le \gamma(S)^{-1}\|S - S_1\| < \varepsilon$ and $\delta(\operatorname{Ker} T_1, \operatorname{Ker} T) \le \gamma(T)^{-1}\|T - T_1\| < \varepsilon$. Thus the inequality follows from Theorem 27.11. □

Corollary 3. *Let $\mathcal{K} = (X_i, \psi_i)_{i=0}^n$ be a semi-Fredholm complex. Then there exists $\varepsilon > 0$ such that $\alpha_i(\mathcal{K}') \le \alpha_i(\mathcal{K}) \quad (i = 0, \dots, n)$ for each semi-Fredholm complex $\mathcal{K}' = (X_i, \psi_i')_{i=0}^n$ satisfying $\operatorname{dist}\{\mathcal{K}', \mathcal{K}\} < \varepsilon$.*

In particular, the functions $\mathcal{K} \mapsto \alpha_i(\mathcal{K})$ are upper semicontinuous.

Next, we show the stability of index under finite-rank perturbations.

Lemma 4. *Let X_1 and X_2 be Banach spaces, let $T, T' \in \mathcal{B}(X_1, X_2)$ be operators such that $T - T'$ is a finite-rank operator and let $Z_1 \subset X_1$, $Z_2 \subset X_2$ be closed subspaces such that $Z_1 \overset{e}{=} \operatorname{Ker} T$, $Z_2 \overset{e}{=} \operatorname{Ran} T$. Then*

$$\begin{aligned}
&\dim \operatorname{Ker} T/(Z_1 \cap \operatorname{Ker} T) - \dim Z_1/(Z_1 \cap \operatorname{Ker} T)\\
&\qquad - \dim Z_2/(Z_2 \cap \operatorname{Ran} T) + \dim \operatorname{Ran} T/(Z_2 \cap \operatorname{Ran} T)\\
&= \dim \operatorname{Ker} T'/(Z_1 \cap \operatorname{Ker} T') - \dim Z_1/(Z_1 \cap \operatorname{Ker} T')\\
&\qquad - \dim Z_2/(Z_2 \cap \operatorname{Ran} T') + \dim \operatorname{Ran} T'/(Z_2 \cap \operatorname{Ran} T').
\end{aligned} \tag{4}$$

Proof. Set

$$\begin{aligned}
M_1 &= \operatorname{Ker} T \cap \operatorname{Ker} T' \cap Z_1,\\
M_2 &= Z_2 \cap \operatorname{Ran} T \cap \operatorname{Ran} T',\\
Y_1 &= \operatorname{Ker} T + \operatorname{Ker} T' + Z_1 \qquad \text{and}\\
Y_2 &= Z_2 + \operatorname{Ran} T + \operatorname{Ran} T'.
\end{aligned}$$

Then $\dim Y_1/M_1 < \infty$ and $\dim Y_2/M_2 < \infty$. Note that $\operatorname{Ran} T$ and $\operatorname{Ran} T'$ are closed. Denote by L and R the expressions at the left- and right-hand sides of (4), respectively. Then

$$\begin{aligned}
L &= \dim \operatorname{Ker} T/M_1 - \dim Z_1/M_1 - \dim Z_2/M_2 + \dim \operatorname{Ran} T/M_2,\\
R &= \dim \operatorname{Ker} T'/M_1 - \dim Z_1/M_1 - \dim Z_2/M_2 + \dim \operatorname{Ran} T'/M_2
\end{aligned}$$

and

$$L - R = \dim \operatorname{Ker} T/M_1 - \dim \operatorname{Ker} T'/M_1 + \dim \operatorname{Ran} T/M_2 - \dim \operatorname{Ran} T'/M_2.$$

Let $\widehat{T}$, $\widehat{T'} : X_1/M_1 \to Y_2$ be the operators defined by $\widehat{T}(x_1 + M_1) = Tx_1$ and $\widehat{T'}(x_1 + M_1) = T'x_1$ for all $x_1 + M_1 \in X_1/M_1$. Clearly, $\widehat{T}, \widehat{T}' \in \Phi_+(X_1/M_1, Y_2)$, $\operatorname{Ran} \widehat{T} = \operatorname{Ran} T$, $\operatorname{Ran} \widehat{T}' = \operatorname{Ran} T'$, $\widehat{T} - \widehat{T}'$ is a finite-rank operator and

$$\dim Y_2/M_2 = \dim Y_2/\operatorname{Ran} T + \dim \operatorname{Ran} T/M_2 = \dim Y_2/\operatorname{Ran} T' + \dim \operatorname{Ran} T'/M_2.$$

Thus

$$\begin{aligned}L-R&=\dim\operatorname{Ker}\widehat{T}-\dim\operatorname{Ker}\widehat{T}'+\dim\operatorname{Ran}\widehat{T}/M_2-\dim\operatorname{Ran}\widehat{T}'/M_2\\&=\dim\operatorname{Ker}\widehat{T}-\dim\operatorname{Ker}\widehat{T}'-\operatorname{codim}\operatorname{Ran}\widehat{T}+\operatorname{codim}\operatorname{Ran}\widehat{T}'\\&=\operatorname{ind}\widehat{T}-\operatorname{ind}\widehat{T}'=0\end{aligned}$$

by the stability of index of semi-Fredholm operators, see Theorem 16.16. □

Theorem 5. *Let $\mathcal{K}=\{X_i,\psi_i\}_{i=0}^n$, $\mathcal{K}'=\{X_i,\psi_i'\}_{i=0}^n$ be semi-Fredholm complexes such that $\psi_i-\psi_i'$ are finite-rank operators for all $i=0,\dots,n-1$. Then $\operatorname{ind}\mathcal{K}=\operatorname{ind}\mathcal{K}'$.*

Proof. If $\operatorname{ind}\mathcal{K}=\pm\infty$, then $\pm\infty$ appears in the expression defining the index of $\mathcal{K}$. Clearly, the corresponding term for $\mathcal{K}'$ is also equal to $\pm\infty$, and so $\operatorname{ind}\mathcal{K}=\operatorname{ind}\mathcal{K}'$. Therefore we can assume that both complexes are Fredholm.

For $j=0,\dots,n$ define

$$\begin{aligned}c_j=&\sum_{i=0}^{j-1}(-1)^i\alpha_i(\mathcal{K}')+(-1)^j\dim\operatorname{Ker}\psi_j/(\operatorname{Ker}\psi_j\cap\operatorname{Ran}\psi_{j-1}')\\&-(-1)^j\dim\operatorname{Ran}\psi_{j-1}'/(\operatorname{Ker}\psi_j\cap\operatorname{Ran}\psi_{j-1}')+\sum_{i=j+1}^{j-1}(-1)^i\alpha_i(\mathcal{K}).\end{aligned}$$

We have

$$c_0=\dim\operatorname{Ker}\psi_0+\sum_{i=1}^{n}(-1)^i\alpha_i(\mathcal{K})=\operatorname{ind}\mathcal{K},$$

and similarly,

$$c_n=\sum_{i=0}^{n-1}(-1)^i\alpha_i(\mathcal{K}')+(-1)^n\dim X_n/\operatorname{Ran}\psi_{n-1}'=\operatorname{ind}\mathcal{K}'.$$

Thus it is sufficient to show that $c_{j+1}=c_j$ for $j=0,\dots,n-1$. Fix j, $0\le j\le n-1$. We have

$$\begin{aligned}&(-1)^j(c_j-c_{j+1})\\&=\dim\operatorname{Ker}\psi_j/(\operatorname{Ker}\psi_j\cap\operatorname{Ran}\psi_{j-1}')-\dim\operatorname{Ran}\psi_{j-1}'/(\operatorname{Ker}\psi_j\cap\operatorname{Ran}\psi_{j-1}')\\&\quad-\dim\operatorname{Ker}\psi_{j+1}/\operatorname{Ran}\psi_j-\dim\operatorname{Ker}\psi_j'/\operatorname{Ran}\psi_{j-1}'\\&\quad+\dim\operatorname{Ker}\psi_{j+1}/(\operatorname{Ker}\psi_{j+1}\cap\operatorname{Ran}\psi_j')-\dim\operatorname{Ran}\psi_j'/(\operatorname{Ker}\psi_{j+1}\cap\operatorname{Ran}\psi_j')=0\end{aligned}$$

by Lemma 4 for $Z_1=\operatorname{Ran}\psi_{j-1}'$, $Z_2=\operatorname{Ker}\psi_{j+1}$, $T=\psi_j$ and $T'=\psi_j'$. This completes the proof. □

The stability of index under compact perturbations is also true but the proof is much more complicated. We state the result here without proof; for an outline of the basic ideas see C.33.4.

Theorem 6. *Let* $\mathcal{K} = (X_i, \psi_i)_{i=0}^n$*,* $\mathcal{K}' = (X_i, \psi_i')_{i=0}^n$ *be semi-Fredholm complexes such that* $\psi_i - \psi_i'$ *are compact operators for all* $i = 0, \dots, n-1$*. Then* $\operatorname{ind} \mathcal{K} = \operatorname{ind} \mathcal{K}'$*.*

The stability of index of Fredholm complexes under small perturbations follows from the results of the previous section.

Theorem 7. *Let* $\mathcal{K} = (X_i, \psi_i)_{i=0}^n$ *be be a Fredholm complex. Then there exists* $\varepsilon > 0$ *such that* $\operatorname{ind} \mathcal{K}' = \operatorname{ind} \mathcal{K}$ *for each Fredholm complex* $\mathcal{K}' = (X_i, \psi_i')_{i=0}^n$ *satisfying* $\operatorname{dist}\{\mathcal{K}, \mathcal{K}'\} < \varepsilon$*.*

Proof. Let $k = \max\{\alpha_j(\mathcal{K}) : j = 0, \dots, n\}$. Let U be the metric space of all Fredholm complexes $(X_i, \psi_i)_{i=0}^n$ with the distance defined above.

By Theorem 32.8, the functions $(X_i, \xi_i)_{i=0}^n \mapsto \xi_j$ are k-regular at $\mathcal{K}$ for all $j = 0, \dots, n-1$. By Theorem 32.10, there exists $\varepsilon > 0$ such that, for all j and all complexes $\mathcal{K}' = (X_i, \psi_i')$ with $\operatorname{dist}\{\mathcal{K}', \mathcal{K}\} < \varepsilon$, we have $\widehat{\delta}_k(\operatorname{Ker} \psi_j', \operatorname{Ker} \psi_j) < \frac{\sqrt{2}-1}{16(k+2)^2}$, $\widehat{\delta}_k(\operatorname{Ran} \psi_j', \operatorname{Ran} \psi_j) < \frac{\sqrt{2}-1}{16(k+2)^2}$ and

$$\operatorname{jump}(\operatorname{Ker} \psi_j', \operatorname{Ker} \psi_j) = \operatorname{jump}(\operatorname{Ran} \psi_j, \operatorname{Ran} \psi_j').$$

Fix a complex $\mathcal{K}'$ with $\operatorname{dist}\{\mathcal{K}', \mathcal{K}\} < \varepsilon$. Let $c_j = \operatorname{jump}(\operatorname{Ker} \psi_j', \operatorname{Ker} \psi_j) = \operatorname{jump}(\operatorname{Ran} \psi_j, \operatorname{Ran} \psi_j')$. Then $c_j \leq k$. Formally set $c_{-1} = 0 = c_n$.

Using Theorem 32.12 twice we get

$$\begin{aligned}
&\alpha_j(\mathcal{K}) = \operatorname{jump}(\operatorname{Ran} \psi_{j-1}, \operatorname{Ker} \psi_j) \\
&= \operatorname{jump}(\operatorname{Ran} \psi_{j-1}, \operatorname{Ran} \psi_{j-1}') + \operatorname{jump}(\operatorname{Ran} \psi_{j-1}', \operatorname{Ker} \psi_j') + \operatorname{jump}(\operatorname{Ker} \psi_j', \operatorname{Ker} \psi_j) \\
&= c_{j-1} + \alpha_j(\mathcal{K}') + c_j.
\end{aligned}$$

Hence

$$\operatorname{ind} \mathcal{K} = \sum_{j=0}^n (-1)^j \alpha_j = \sum_{j=0}^n (-1)^j (c_{j-1} + \alpha_j' + c_j) = \sum_{j=0}^n (-1)^j \alpha_j' = \operatorname{ind} \mathcal{K}'. \qquad \square$$

Corollary 8. *Let* $\mathcal{K}$ *be a Fredholm complex and* $0 \leq m \leq n$*. Then there exists* $\varepsilon > 0$ *such that, for each Fredholm complex* $\mathcal{K}'$ *with* $\operatorname{dist}\{\mathcal{K}', \mathcal{K}\} < \varepsilon$*, we have:*

(i) *if* m *is odd, then* $\sum_{j=0}^m (-1)^j \alpha_j(\mathcal{K}') \geq \sum_{j=0}^m (-1)^j \alpha_j(\mathcal{K})$ *and* $\sum_{j=0}^m (-1)^j \alpha_{n-j}(\mathcal{K}') \geq \sum_{j=0}^m (-1)^j \alpha_{n-j}(\mathcal{K})$*;*

(ii) *if* m *is even, then* $\sum_{j=0}^m (-1)^j \alpha_j(\mathcal{K}') \leq \sum_{j=0}^m (-1)^j \alpha_j(\mathcal{K})$ *and* $\sum_{j=0}^m (-1)^j \alpha_{n-j}(\mathcal{K}') \geq \sum_{j=0}^m (-1)^j \alpha_{n-j}(\mathcal{K})$*.*

Proof. Let $k = \max\{\alpha_j(\mathcal{K}) : j = 0, \dots, n\}$ and let ε be the number constructed in the previous theorem. Let $\mathcal{K}'$ be a Fredholm complex satisfying $\operatorname{dist}\{\mathcal{K}', \mathcal{K}\} < \varepsilon$. As in the previous proof we set

$$c_j = \operatorname{jump}(\operatorname{Ran} \psi_j, \operatorname{Ran} \psi_j') = \operatorname{jump}(\operatorname{Ker} \psi_j', \operatorname{Ker} \psi_j).$$

We have

$$\sum_{j=0}^{m}(-1)^j\alpha_j(\mathcal{K}) = \sum_{j=0}^{m}(-1)^j(c_{j-1}+\alpha_j(\mathcal{K}')+c_j) = \sum_{j=0}^{m}(-1)^j\alpha_j(\mathcal{K}')+(-1)^m c_m.$$

This proves the first inequalities both in (i) and (ii).

The remaining statements can be proved similarly. □

The previous stability result can be extended to semi-Fredholm complexes. We need the following lemma.

Lemma 9. *Let X, Y be Banach spaces, let $S : X \to Y$ and $T : Y \to X$ be operators with closed ranges such that* $\operatorname{Ran} S = \operatorname{Ker} T$ *and* $\operatorname{Ran} T \subset \operatorname{Ker} S$. *Then there exists $\varepsilon > 0$ such that*

$$\dim \operatorname{Ker} S/\operatorname{Ran} T = \dim \operatorname{Ker} S_1/\operatorname{Ran} T_1$$

for all operators $S_1 : X \to Y$ and $T_1 : Y \to X$ such that $\|S_1-S\| < \varepsilon$, $\|T_1-T\| < \varepsilon$, $\operatorname{Ran} S_1 \subset \operatorname{Ker} T_1$ *and* $\operatorname{Ran} T_1 \subset \operatorname{Ker} S_1$.

Proof. The sequence $X\xrightarrow{S}Y\xrightarrow{T}X$ is exact in the middle. By Lemma 11.3, there exist positive constants $\varepsilon_1 > 0$ and c such that $\operatorname{Ran} S_1 = \operatorname{Ker} T_1$, $\gamma(S_1) \geq c$ and $\gamma(T_1) \geq c$ for all operators $S_1 : X \to Y$, $T_1 : Y \to X$ satisfying $\|S_1 - S\| < \varepsilon_1$, $\|T_1 - T\| < \varepsilon_1$ and $\operatorname{Ran} S_1 \subset \operatorname{Ker} T_1$.

Set $\varepsilon = \min\{\varepsilon_1, \frac{c}{9}\}$. Let S_1 and T_1 be operators satisfying $\|S_1 - S\| < \varepsilon$, $\|T_1 - T\| < \varepsilon$, $\operatorname{Ran} S_1 \subset \operatorname{Ker} T_1$ and $\operatorname{Ran} T_1 \subset \operatorname{Ker} S_1$. Then, by Lemma 10.12, we have $\widehat{\delta}(\operatorname{Ker} S, \operatorname{Ker} S_1) \leq c^{-1}\|S_1-S\| < 1/9$ and $\widehat{\delta}(\operatorname{Ran} T, \operatorname{Ran} T_1) \leq c^{-1}\|T_1-T\| < 1/9$. By Theorem 27.8, we have the required equality. □

Theorem 10. *Let $\mathcal{K} = (X_i, \psi_i)_{i=0}^n$ be a semi-Fredholm complex. Then there exists $\varepsilon > 0$ such that* $\operatorname{ind}\mathcal{K}' = \operatorname{ind}\mathcal{K}$ *for every semi-Fredholm complex $\mathcal{K}'$ satisfying* $\operatorname{dist}\{\mathcal{K}', \mathcal{K}\} < \varepsilon$.

Proof. To simplify the statement, set

$$X = \bigoplus_{i \text{ even}} X_i, \quad Y = \bigoplus_{i \text{ odd}} X_i, \quad T = \bigoplus_{i \text{ even}} \psi_i \quad \text{and} \quad S = \bigoplus_{i \text{ odd}} \psi_i.$$

Then

$$TS = 0, \quad ST = 0 \quad \text{and} \quad \operatorname{ind}\mathcal{K} = \dim \operatorname{Ker} T/\operatorname{Ran} S - \dim \operatorname{Ker} S/\operatorname{Ran} T.$$

Consider the operators $\tilde{T} : \tilde{X} \to \tilde{Y}$ and $\tilde{S} : \tilde{Y} \to \tilde{X}$ defined in Section 17. So $\tilde{S}\tilde{T} = 0$ and $\tilde{T}\tilde{S} = 0$.

Let $\mathcal{K}' = (X_i, \psi_i')_{i=0}^n$ be a semi-Fredholm complex close to $\mathcal{K}$ and define similarly the operators $T' : X \to Y$ and $S' : Y \to X$ corresponding to $\mathcal{K}'$. So $T'S' = 0$, $S'T' = 0$ and $\max\{\|T' - T\|, \|S' - S\|\} = \operatorname{dist}\{\mathcal{K}', \mathcal{K}\}$.

We distinguish two cases:

(a) Let $\dim \operatorname{Ker} S/\operatorname{Ran} T = \infty$. Since the complex $\mathcal{K}$ is semi-Fredholm, we have $\dim \operatorname{Ker} T/\operatorname{Ran} S < \infty$, and so $\operatorname{Ker} \tilde{T} = \operatorname{Ran} \tilde{S}$. By Theorem 17.10, $\dim \operatorname{Ker} \tilde{S}/\operatorname{Ran} \tilde{T} = \infty$. By the previous lemma, $\dim \operatorname{Ker} \tilde{S}'/\operatorname{Ran} \tilde{T}' = \infty$ for each semi-Fredholm complex $\mathcal{K}'$ sufficiently close to $\mathcal{K}$. So $\dim \operatorname{Ker} S'/\operatorname{Ran} T' = \infty$. Hence $\operatorname{ind} \mathcal{K}' = \infty = \operatorname{ind} \mathcal{K}$.

Similar considerations can be done if $\dim \operatorname{Ker} T/\operatorname{Ran} S = \infty$.

(b) It remains the case $\dim \operatorname{Ker} T/\operatorname{Ran} S < \infty$ and $\dim \operatorname{Ker} S/\operatorname{Ran} T < \infty$. Then $\mathcal{K}$ is a Fredholm complex, and so $\operatorname{Ran} \widetilde{S} = \operatorname{Ker} \widetilde{T}$ and $\operatorname{Ran} \widetilde{T} = \operatorname{Ker} \widetilde{S}$.

By Lemma 9, $\operatorname{Ran} \widetilde{S}' = \operatorname{Ker} \widetilde{T}'$ and $\operatorname{Ran} \widetilde{T}' = \operatorname{Ker} \widetilde{S}'$ whenever T', S' are operators induced by a semi-Fredholm complex $\mathcal{K}'$ which is sufficiently close to $\mathcal{K}$. Thus $\mathcal{K}'$ is also a Fredholm complex.

The equality $\operatorname{ind} \mathcal{K}' = \operatorname{ind} \mathcal{K}$ for Fredholm complexes $\mathcal{K}'$ close enough to $\mathcal{K}$ was proved in Theorem 7. □

Corollary 11. *Let $A = (A_1, A_2)$ be a pair of commuting operators on a Banach space X. Then $\partial \sigma_T(A) \subset \sigma_\pi(A) \cup \sigma_\delta(A)$.*

Proof. Let $\lambda \in \partial \sigma_T(A) \setminus \big(\sigma_\pi(A) \cup \sigma_\delta(A)\big)$. Then the Koszul complex of the pair $A - \lambda$ is semi-Fredholm. Furthermore, there is a sequence $\lambda^{(j)} \in \mathbb{C}^2 \setminus \sigma_T(A)$ converging to λ. Thus the Koszul complex $\mathcal{K}(A - \lambda^{(j)})$ of $A - \lambda^{(j)}$ is exact, and so $\operatorname{ind} \mathcal{K}(A - \lambda^{(j)}) = 0$ for all j. By the continuity of the index we have $\operatorname{ind} \mathcal{K}(A - \lambda) = 0$. Since $\lambda \notin \sigma_\pi(A) \cup \sigma_\delta(A)$, the Koszul complex $\mathcal{K}(A - \lambda)$ is exact, and so $\lambda \notin \sigma_T(A)$, which is a contradiction. □

34 Essential Taylor spectrum

Definition 1. Let $A = (A_1, \dots, A_n)$ be a commuting n-tuple of operators on a Banach space X. The *essential Taylor spectrum* $\sigma_{Te}(A)$ is the set of all $\lambda = (\lambda_1, \dots, \lambda_n) \in \mathbb{C}^n$ such that the Koszul complex of $A - \lambda = (A_1 - \lambda_1, \dots, A_n - \lambda_n)$ is not Fredholm.

Clearly, $\sigma_{Te}(A) \subset \sigma_T(A)$. For single operators we have $\sigma_{Te}(A_1) = \sigma_e(A_1)$.

Let $\tilde{X} = \ell^\infty(X)/m(X)$ and $\tilde{A} = (\tilde{A}_1, \dots, \tilde{A}_n) \in \mathcal{B}(\tilde{X})^n$ be the construction studied in Section 17. By Theorem 17.10, the Koszul complex of A is Fredholm if and only if the Koszul complex of $\tilde{A}$ is exact. Thus $\sigma_{Te}(A) = \sigma_T(\tilde{A})$.

Corollary 2. *The essential Taylor spectrum is an upper semicontinuous spectral system.*

Another corollary of the equality $\sigma_{Te}(A) = \sigma_T(\tilde{A})$ is:

Theorem 3. *Let $A = (A_1, \dots, A_n)$ and $B = (B_1, \dots, B_n)$ be commuting n-tuples of operators on a Banach space X such that operators $A_i - B_i$ are compact for $i = 1, \dots, n$. Then $\sigma_{Te}(A) = \sigma_{Te}(B)$.*

For one operator the difference $\sigma(A_1)\setminus\sigma_e(A_1)$ can be characterized easily. It consists of whole components of $\mathbb{C}\setminus\sigma_e(A_1)$ and at most countably many isolated points, see Theorem 19.4.

The difference $\sigma_T(A)\setminus\sigma_{Te}(A)$ for n-tuples of operators can be more complicated. As in Theorem 19.17 one can show that the set $\sigma_T(A)\setminus\widehat{\sigma}_{Te}(A)$ consists of countable many isolated points.

We are going to show that $\sigma_T(A)\setminus\sigma_{Te}(A)$ is always an analytic set.

Let $U\subset\mathbb{C}^n$ be an open set. Recall that a subset $M\subset U$ is called analytic in U if for every $w\in U$ there exists a neighbourhood U_0 of w and a family $\{f_\alpha:\alpha\in\Lambda\}\subset H(U_0)$ such that $M\cap U_0=\{z\in U_0: f_\alpha(z)=0 \text{ for all } \alpha\in\Lambda\}$.

The set $\{f_\alpha\}$ can be always chosen to be finite, see [GR, p. 86].

We start with the following lemma:

Lemma 4. *Let $U\subset\mathbb{C}^n$ be an open subset, let $T:U\to\mathcal{B}(X,Y)$ be an analytic function, let $k\in\mathbb{N}$. Then the set $\{z\in U:\dim\operatorname{Ran}T(z))<k\}$ is analytic.*

Proof. If $x_1,\dots,x_k\in X$, $y_1^*,\dots,y_k^*\in Y^*, z\in U$ and $\dim\operatorname{Ran}T(z)<k$, then the vectors $T(z)x_1,\dots,T(z)x_k$ are linearly dependent and $\det\big(\langle T(z)x_i,y_j^*\rangle\big)=0$.

On the other hand, if $\dim\operatorname{Ran}T(z)\ge k$, then there are vectors $x_1,\dots,x_k\in X$ and $y_1^*,\dots,y_k^*\in Y^*$ such that $\langle T(z)x_i,y_j^*\rangle=\delta_{ij}$, and so $\det\big(\langle T(z)x_i,y_j^*\rangle\big)\neq 0$. Thus

$$\begin{aligned}&\big\{z\in U:\dim\operatorname{Ran}T(z))<k\big\}\\&\quad=\big\{z\in U:\det\big(\langle T(z)x_i,y_j^*\rangle\big)=0 \text{ for all } x_1,\dots,x_k\in X, y_1^*,\dots,y_k^*\in Y^*\big\},\end{aligned}$$

which is an analytic set. □

Corollary 5. *Let X,Y,Z be Banach spaces and let U be an open subset of $\mathbb{C}^n$. Let $S:U\to\mathcal{B}(X,Y)$ and $T:U\to\mathcal{B}(Y,Z)$ be analytic functions and $k\in\mathbb{N}$. Then the set $\big\{z\in U:\dim\operatorname{Ran}S(z)/\big(\operatorname{Ran}S(z)\cap\operatorname{Ker}T(z)\big)<k\big\}$ is analytic.*

Proof. We have $\dim\operatorname{Ran}S(z)/\big(\operatorname{Ran}S(z)\cap\operatorname{Ker}T(z)\big)=\dim\operatorname{Ran}\big(T(z)S(z)\big)$, so the corollary follows from the preceding lemma. □

Lemma 6. *Let U be an open subset of $\mathbb{C}^n$, let $S:U\to\mathcal{B}(X,Y)$ and $T:U\to\mathcal{B}(Y,Z)$ be functions regular in U. Suppose that $\operatorname{Ker}T(z)\overset{e}{=}\operatorname{Ran}S(z)$ for all $z\in U$. Then the function α defined by*

$$\begin{aligned}\alpha(z)=\dim\operatorname{Ker}T(z)/(\operatorname{Ker}T(z)\cap\operatorname{Ran}S(z))\\ -\dim\operatorname{Ran}S(z)/(\operatorname{Ker}T(z)\cap\operatorname{Ran}S(z))\end{aligned}$$

is constant on each component of U.

Proof. Let U_0 be a component of U. If there is no $w \in U_0$ with $\alpha(w) < \infty$, then clearly $\alpha(z) = \infty$ on U_0.

Let $w \in U$ satisfy $\alpha(w) < \infty$. Then $\lim_{z\to w} \hat{\delta}(\operatorname{Ker} T(w), \operatorname{Ker} T(z)) = 0$ and $\lim_{z\to w} \hat{\delta}(\operatorname{Ran} S(w), \operatorname{Ran} S(z)) = 0$ by Theorem 10.21 (iv) and (vi). Thus, by Theorem 27.8, $\alpha(z) = \alpha(w)$ for all z sufficiently close to w.

A standard argument gives that $\alpha(z)$ is constant on the whole component of connectivity U_0. □

Lemma 7. *Let U be an open subset of $\mathbb{C}^n$, let $S : U \to \mathcal{B}(X,Y)$ and $T : U \to \mathcal{B}(Y,Z)$ be analytic functions satisfying $T(z)S(z) = 0 \quad (z \in U)$. Suppose that there are Banach spaces X_1 and Z_1 and regular analytic functions $S_1 : U \to \mathcal{B}(X_1,Y)$, $T_1 : U \to \mathcal{B}(Y,Z_1)$ satisfying $\operatorname{Ker} T_1(z) \subset \operatorname{Ran} S(z) \subset \operatorname{Ker} T(z) \subset \operatorname{Ran} S_1(z)$, see the diagram below. Suppose that $\dim \operatorname{Ran} S_1(z)/\operatorname{Ker} T_1(z) < \infty$ for all $z \in U$. Then the set $\{z \in G : \dim \operatorname{Ker} T(z)/\operatorname{Ran} S(z) \geq k\}$ is analytic in U for each $k \geq 0$.*

$$\begin{array}{ccccc} & & X_1 & & \\ & & \downarrow{\scriptstyle S_1(z)} & & \\ X & \xrightarrow{S(z)} & Y & \xrightarrow{T(z)} & Z \\ & & \downarrow{\scriptstyle T_1(z)} & & \\ & & Z_1 & & \end{array}$$

Proof. We can assume that U is connected. For each j set

$$A_j = \{z \in U : \dim \operatorname{Ran} S(z)/\operatorname{Ker} T_1(z)) \leq j\}$$

and

$$B_j = \{z \in U : \dim \operatorname{Ran} S_1(z)/\operatorname{Ker} T(z) \leq j\}.$$

By Corollary 5, A_j and B_j are analytic sets. By Lemma 6, there is a constant c such that $\dim \operatorname{Ran} S_1(z)/\operatorname{Ker} T_1(z) = c$ in U. Thus

$$\begin{aligned} &\{z \in G : \dim \operatorname{Ker} T(z)/\operatorname{Ran} S(z) \geq k\} \\ &= \{z \in G : \dim \operatorname{Ran} S_1(z)/\operatorname{Ker} T(z) + \dim \operatorname{Ran} S(z)/\operatorname{Ker} T_1(z) \leq c - k\} \\ &= \bigcup_{i=0}^{c-k} A_i \cap B_{c-k-i}. \end{aligned}$$

The last set is clearly analytic. □

Theorem 8. *Let U be an open subset of $\mathbb{C}^n$, let $S : U \to \mathcal{B}(X,Y)$ and $T : U \to \mathcal{B}(Y,Z)$ be analytic operator-valued functions. Suppose that $T(z)S(z) = 0$, $\dim \operatorname{Ker} T(z)/\operatorname{Ran} S(z) < \infty$ and the operators $S(z)$ and $T(z)$ have generalized inverses for all $z \in U$. Let $k \in \mathbb{N}$. Then the set $\{z \in G : \alpha(z) \geq k\}$ is analytic.*

Proof. Let $w \in U$. Let V be a generalized inverse of $S(w)$, so $VS(w)V = V$ and $S(w)VS(w) = S(w)$. Set $P = I - S(w)V$. Then P is a projection and $\operatorname{Ker} P = \operatorname{Ran} S(w)$.

The operator $I+(S(z)-S(w))V$ is invertible for all z close to w. Set $P(z) = P(I + (S(z) - S(w))V)^{-1} \in \mathcal{B}(Y)$. Since $\operatorname{Ran} S(z) = \operatorname{Ran} P$ is constant, the function $z \mapsto P(z)$ is regular at w. We prove that $\operatorname{Ker} P(z) \subset \operatorname{Ran} S(z)$. Let $y \in \operatorname{Ker} P(z)$. Then

$$\Big(I + \big(S(z) - S(w)\big)V\Big)^{-1} y \in \operatorname{Ker} P = \operatorname{Ran} S(w),$$

and so $y \in \Big(I + \big(S(z) - S(w)\big)V\Big)S(w)X = S(z)VS(w)X \subset \operatorname{Ran} S(z)$.

Similarly, let W be a generalized inverse of $T(w)$. Set $Q = I - WT(w)$. Then Q is a projection with $\operatorname{Ran} Q = \operatorname{Ker} T(w)$. For z close to w define $Q(z) \in \mathcal{B}(Y)$ by $Q(z) = \Big(I + W\big(T(z) - T(w)\big)\Big)^{-1} Q$. Since $\operatorname{Ker} Q(z) = \operatorname{Ker} Q$ is constant, the function $z \mapsto Q(z)$ is regular. We have

$$WT(z) = WT(w) + W\big(T(z) - T(w)\big) = I - Q + W\big(T(z) - T(w)\big)$$

and

$$\Big(I + W\big(T(z) - T(w)\big)\Big)^{-1} WT(z) = I - \Big(I + W\big(T(z) - T(w)\big)\Big)^{-1} Q = I - Q(z).$$

Consequently, $\operatorname{Ker} T(z) \subset \operatorname{Ran} Q(z)$. Thus we have

$$\operatorname{Ker} P(z) \subset \operatorname{Ran} S(z) \subset \operatorname{Ker} T(z) \subset \operatorname{Ran} Q(z)$$

and

$$\dim \operatorname{Ran} Q(w)/\operatorname{Ker} P(w) = \dim \operatorname{Ran} Q/\operatorname{Ker} P = \dim \operatorname{Ker} T(w)/\operatorname{Ran} S(w) < \infty.$$

The rest follows from Lemma 7. □

In particular, if $A = (A_1, \dots, A_n)$ is an n-tuple of commuting Hilbert space operators, then the difference $\sigma_T(A) \setminus \sigma_{Te}(A)$ is an analytic set.

To show the same statement for Banach space operators is a little bit more complicated. We need the following two lemmas.

Lemma 9. *Let U be an open subset of $\mathbb{C}^n$, let $S : U \to \mathcal{B}(X, Y)$ and $T : U \to \mathcal{B}(Y, Z)$ be analytic functions satisfying $T(z)S(z) = 0 \quad (z \in U)$. Suppose that there are Banach spaces X_1, Z_1, finite-dimensional Banach spaces F, G and regular analytic functions $S_1 : U \to \mathcal{B}(X_1, Y \oplus F)$ and $T_1 : U \to \mathcal{B}(Y \oplus G, Z_1)$ with the property that $\operatorname{Ran} S_1(z) \supset \operatorname{Ker} T(z)$, $\operatorname{Ran} S(z) + G \supset \operatorname{Ker} T_1(z)$ and $\dim\big(\operatorname{Ran} S_1(z) + G\big)/\operatorname{Ker} T_1(z) < \infty \quad (z \in U)$, see the diagram below. Let $k \in \mathbb{N}$.*

Then the set $\{z \in U : \dim \operatorname{Ker} T(z)/\operatorname{Ran} S(z) \geq k\}$ *is analytic in* U.

$$\begin{array}{ccccc} X_1 & \xrightarrow{S_1(z)} & \left\{\begin{array}{c} F \\ \oplus \\ Y \end{array}\right. & & \\ X & \xrightarrow[S(z)]{} & \left.\begin{array}{c} Y \\ \oplus \\ G \end{array}\right\} & \begin{array}{c} \xrightarrow{T(z)} \\ \xrightarrow[T_1(z)]{} \end{array} & \begin{array}{c} Z \\ Z_1 \end{array} \end{array}$$

Proof. Set $Y' = Y \oplus F \oplus G$. For $z \in U$ define operators $S'(z) : X \oplus G \to Y'$, $T'(z) : Y' \to Z \oplus F$, $S_1'(z) : X_1 \oplus G \to Y'$ and $T_1'(z) : Y' \to Z_1 \oplus F$ by

$$\begin{aligned} S'(z)(x \oplus g) &= S(z)x + g, \\ T'(z)(y \oplus f \oplus g) &= T(z)y + f, \\ S_1'(z)(x_1 \oplus g) &= S_1(z)x_1 + g \qquad \text{and} \\ T_1'(z)(y \oplus f \oplus g) &= T_1(z)(y \oplus g) + f \end{aligned}$$

for all $x \in X$, $f \in F$, $g \in G$ and $x_1 \in X_1$. Thus $\operatorname{Ran} S'(z) = \operatorname{Ran} S(z) + G$, $\operatorname{Ker} T'(z) = \operatorname{Ker} T(z) + G$, $\operatorname{Ran} S_1'(z) = \operatorname{Ran} S_1(z) + G$ and $\operatorname{Ker} T_1'(z) = \operatorname{Ker} T_1(z)$. We have $\operatorname{Ran} S_1'(z) \supset \operatorname{Ker} T'(z) \supset \operatorname{Ran} S'(z) \supset \operatorname{Ker} T_1'(z)$. Clearly, S_1' and T_1' are regular functions.

By Lemma 7, the set $\{z \in U : \dim \operatorname{Ker} T'(z)/\operatorname{Ran} S'(z) \geq k\}$ is analytic in U. This set, however, is equal to $\{z \in U : \dim \operatorname{Ker} T(z)/\operatorname{Ran} S(z) \geq k\}$. □

Lemma 10. *Let* U *be an open subset of* $\mathbb{C}^n$, *let* $S : U \to \mathcal{B}(X, Y)$ *and* $T : U \to \mathcal{B}(Y, Z)$ *be analytic operator-valued functions satisfying* $T(z)S(z) = 0$ *and* $\dim \operatorname{Ker} T(z)/\operatorname{Ran} S(z) < \infty$ $(z \in U)$. *Let* $w \in U$. *Suppose that there are finite-dimensional spaces* G, H, *a neighbourhood* U_1 *of* w *and a regular analytic function* $T_1 : U_1 \to \mathcal{B}(Y \oplus G, Z \oplus H)$ *such that* $T_1(z)|Y = T(z)$. *Then there exist a finite-dimensional space* F, *a neighbourhood* U_2 *of* w *and a regular analytic function* $S_1 : U_2 \to \mathcal{B}(X \oplus F, Y \oplus G)$ *such that* $S_1(z)|X = S(z)$ *and* $\operatorname{Ran} S_1(z) = \operatorname{Ker} T_1(z) \supset \operatorname{Ker} T(z)$.

$$\left.\begin{array}{c} X \\ \oplus \\ F \end{array}\right\} \begin{array}{c} \xrightarrow{S(z)} \\ \xrightarrow[S_1(z)]{} \end{array} \left.\begin{array}{c} Y \\ \oplus \\ G \end{array}\right\} \begin{array}{c} \xrightarrow{T(z)} \\ \xrightarrow[T_1(z)]{} \end{array} \left.\begin{array}{c} Z \\ \oplus \\ H \end{array}\right\}$$

Proof. We have

$$\begin{aligned} \dim \operatorname{Ker} T_1(z)/\operatorname{Ran} S(z) &= \dim \operatorname{Ker} T_1(z)/\operatorname{Ker} T(z) \\ &\quad + \dim \operatorname{Ker} T(z)/\operatorname{Ran} S(z) < \infty. \end{aligned}$$

Let $y_1, \dots, y_r$ be linearly independent vectors in $\operatorname{Ker} T_1(w)$ such that

$$\operatorname{Ran} S(w) \vee \{y_1, \dots, y_r\} = \operatorname{Ker} T_1(w).$$

Since T_1 is regular, for $i = 1, \dots, r$ there exists a $(Y \oplus G)$-valued analytic function φ_i defined in a neighbourhood of w such that $T_1(z)\varphi_i(z) = 0$ and $\varphi_i(w) = y_i$. Let

F be an r-dimensional space with a basis $f_1, \dots, f_r$ and define $S_1(z) : X \oplus F \to Y \oplus G$ by

$$S_1(z)\Big(x \oplus \sum_{i=1}^{r} \beta_i f_i\Big) = S(z)x + \sum_{i=1}^{r} \beta_i \varphi_i(z) \qquad (x \in X, \beta_i \in \mathbb{C}).$$

We have $T_1(z)S_1(z) = 0$ and $\operatorname{Ran} S_1(w) = \operatorname{Ker} T_1(w)$, so there is a neighbourhood of w where $\operatorname{Ker} T_1(z) = \operatorname{Ran} S_1(z)$. Thus S_1 is regular in a neighbourhood of w and satisfies all the conditions required. □

Theorem 11. *Let $X_0, X_1, \dots, X_n$ be Banach spaces, U an open subset of $\mathbb{C}^n$. Let*

$$0 \to X_0 \xrightarrow{\delta_0(z)} X_1 \xrightarrow{\delta_1(z)} \cdots \xrightarrow{\delta_{n-1}(z)} X_n \to 0$$

be a Fredholm complex analytically dependent on $z \in U$.

Let $0 \le j \le n$ and $k \in \mathbb{N}$. Then the set $\big\{z \in U : \dim \operatorname{Ker} \delta_j / \operatorname{Ran} \delta_{j-1} \ge k\big\}$ is analytic in U.

Proof. Let $w \in U$. Using Lemma 10 repeatedly it is easy to see by downward induction that there are finite-dimensional spaces F_{j-1}, F_j and a regular analytic function $S(z) : X_{j-1} \oplus F_{j-1} \to X_j \oplus F_j$ defined on a neighbourhood of w such that $S(z)|X_{j-1} = \delta_{j-1}(z)$ and $\operatorname{Ran} S(z) \supset \operatorname{Ker} \delta_j(z)$. In particular, $\dim \operatorname{Ran} S(z) / \operatorname{Ker} \delta_{j-1}(z) < \infty$.

Consider the "adjoint" complex

$$0 \to X_0^* \xleftarrow{\delta_0^*(z)} X_1^* \xleftarrow{\delta_1^*(z)} \cdots \xleftarrow{\delta_{n-1}^*(z)} X_n^* \to 0$$

where we write for short $\delta_j^*(z)$ instead of $(\delta_j(z))^*$. Since this complex is also Fredholm, similarly as above there exist finite-dimensional spaces G_j and G_{j+1} and a regular analytic function $T(z) : X_{j+1}^* \oplus G_{j+1} \to X_j^* \oplus G_j$ defined in a neighbourhood of w such that $\operatorname{Ran} T(z) \supset \operatorname{Ker} \delta_{j-1}^*(z)$ and

$$\dim \operatorname{Ran} T(z) / \operatorname{Ker} \delta_{j-1}^*(z) < \infty.$$

Further, the operator $S(z)^* : X_j^* \oplus F_j^* \to X_{j-1}^* \oplus F_{j-1}^*$ satisfies

$$\operatorname{Ker} S(z)^* = (\operatorname{Ran} S(z))^\perp \subset (\operatorname{Ker} \delta_j(z))^\perp + F_j^* = \operatorname{Ran} \delta_j^*(z) + F_j^*.$$

By Lemma 9, the set $\big\{z : \dim \operatorname{Ker} \delta_{j-1}^*(z) / \operatorname{Ran} \delta_j^*(z) \ge k\big\}$ is analytic. Since

$$\dim \operatorname{Ker} \delta_{j-1}^*(z) / \operatorname{Ran} \delta_j^*(z) = \dim \operatorname{Ker} \delta_j(z) / \operatorname{Ran} \delta_{j-1}(z),$$

this finishes the proof. □

Corollary 12. *Let $A = (A_1, \dots, A_n)$ be an n-tuple of commuting operators on a Banach space X. Then the set $\sigma_T(A) \setminus \sigma_{Te}(A)$ is analytic in $\mathbb{C}^n \setminus \sigma_{Te}(A)$.*

Comments on Chapter IV

C.25.1. The Taylor spectrum was introduced by J.L. Taylor in [Ta1]. The "partial Taylor spectra" $\sigma_{\pi,k}$ and $\sigma_{\delta,k}$ were defined by Słodkowski [Sl2].

The projection property for the Taylor spectrum was proved in [Ta1]; we followed a simpler argument from [Sl2].

C.25.2. Let $\mathcal{A}$, $\mathcal{B}$ be C^*-algebras, $\mathcal{A} \subset \mathcal{B}$ and let $a = (a_1, \dots, a_n)$ be commuting elements of $\mathcal{A}$. Consider the n-tuples $L_a^{\mathcal{A}} = (L_{a_1}^{\mathcal{A}}, \dots, L_{a_n}^{\mathcal{A}})$ and $L_a^{\mathcal{B}} = (L_{a_1}^{\mathcal{B}}, \dots, L_{a_n}^{\mathcal{B}})$ of multiplication operators by a in the algebras $\mathcal{A}$ and $\mathcal{B}$, respectively. By [Cu2], $\sigma_T(L_a^{\mathcal{A}}) = \sigma_T(L_a^{\mathcal{B}})$. This phenomenon, which is well known for single elements, is called the spectral permanence.

C.25.3. Let $A = (A_1, \dots, A_n)$ be commuting operators on a Banach space X. Let $\mathcal{M} \subset \mathcal{B}(X)$ be a commutative algebra containing $A_1, \dots, A_n$. Then $\mathcal{M}$ is contained in the commutant $(A)'$ of A. As in Proposition 25.3 one can show that

$$\sigma_T(A) \subset \sigma_l^{(A)'}(A) \subset \sigma^{\mathcal{M}}(A).$$

There is an example in [Ta1] (for $n = 5$) that the first inclusion is strict. Thus in general the Taylor spectrum is smaller than the spectrum in any commutative subalgebra of $\mathcal{B}(X)$ containing $A_1, \dots, A_n$.

The example was improved by Albrecht [Al2] who constructed a commuting n-tuple $A = (A_1, \dots, A_n)$ (for $n \geq 2$) such that the Taylor functional calculus is richer (it contains more analytic functions) than the calculus in any commutative algebra containing A. Moreover, there are two maximal commutative subalgebras $\mathcal{M}_1$, $\mathcal{M}_2$ containing A such that the spectra $\sigma^{\mathcal{M}_1}(A)$ and $\sigma^{\mathcal{M}_2}(A)$ are different.

C.26.1. The split spectrum is a natural modification of the Taylor spectrum. It was considered, e.g., in [Es3] and [Ha6]; some ideas appeared implicitly already in the original paper of Taylor [Ta1].

In Hilbert spaces the Taylor spectrum and the split spectrum coincide. The same is true for commuting tuples of operators in ℓ^∞ and ℓ^1, see [Ha6].

In general, these two spectra are different. An example of a commuting pair of operators $A = (A_1, A_2)$ with $\sigma_S(A) \neq \sigma_T(A)$ was given in [Mü17]. The example also shows that in general there is no inclusion between the Taylor and Harte spectrum, cf. [BS].

C.26.2. Theorem 26.7 was proved in [Cu5], see also [EP].

C.26.3. Equivalently, it is possible to use for the definition of $\sigma_{r,k}$ and $\sigma_{l,k}$ the chain Koszul complex, see Remark 25.12. From the considerations there it is easy to see that δ_A^p has generalized inverse if and only if ε_A^{n-p} has generalized inverse.

C.27.1. Theorem 27.3 is usually attributed to Bartle-Graves [BG], see also [ZKKP]. Theorem 27.7 is due to Fainshtein [Fa3]; the partial case of $R_1 \subset N_1$ was proved in [FS2], see also [Va7]. Theorem 10 was proved in [Mü18].

C. 28.1 The split functional calculus for Hilbert space operators was constructed by Vasilescu see [Va3], [Va6]. For Banach space operators it was generalized in [KM1].

For Hilbert space operators it is possible to choose $V = (\delta_{A-z} + \delta^*_{A-z})^{-1}$ in Corollary 28.2. Formula (7) defining the Taylor functional calculus is then quite explicit.

C.29.1. Definition 29.1 is due to Albrecht [Al1]. An alternative definition of the local spectrum was given by Frunza [Fr] as the complement of the union of all open sets $U \subset \mathbb{C}^n$ on which there is a solution of the equation $xs = (\bar\partial + \delta_{A-z})\psi$.

The equivalence of these two definitions (Theorem 29.2) was shown by Eschmeier [Es2].

C.29.2. An n-tuple A of commuting operators on X is said to have property SVEP (single value extension property) if $S(A) = \emptyset$. By Definition 29.4, A has SVEP if and only if the sequence

$$0 \to \Lambda^0\left[s, \mathrm{d}\bar z, C^\infty(G,X)\right] \xrightarrow{\bar\partial+\delta} \Lambda^0\left[s, \mathrm{d}\bar z, C^\infty(G,X)\right] \xrightarrow{\bar\partial+\delta} \cdots$$
$$\cdots \xrightarrow{\bar\partial+\delta} \Lambda^n\left[s, \mathrm{d}\bar z, C^\infty(G,X)\right]$$

is exact for each open set $G \subset \mathbb{C}^n$.

By [Es1], see also [Fr], this is equivalent to the condition that the sequence

$$0 \longrightarrow \Lambda^0\left[s, H(G,X)\right] \xrightarrow{\delta} \Lambda^0\left[s, H(G,X)\right] \xrightarrow{\delta} \cdots \xrightarrow{\delta} \Lambda^n\left[s, H(G,X)\right]$$

is exact for each open polydisc $G \subset \mathbb{C}^n$.

C.29.3. Let $A = (A_1, \dots, A_n)$ be a commuting n-tuple of operators on a Banach space X and let $x \in X$. As in the case of one operator (cf. C.14.3) it is possible to define the local functional calculus, see [Es1]. If f is analytic on a neighbourhood of $\sigma_x(A)$, then define

$$f(A)x = \frac{-1}{(2\pi i)^n} \int_\Gamma Pf\psi \wedge \mathrm{d}z,$$

where $(\bar\partial + \delta)\psi = xs$ and Γ is a smooth surface surrounding $\sigma_x(A)$. As in the case of $n = 1$, $f(A)$ is defined only for those x for which f is analytic on a neighbourhood of $\sigma_x(A)$.

C.29.4. The local spectrum $\gamma_x(A)$ does not satisfy the projection property. As an example, take $A_1 \in \mathcal{B}(X)$ and $x \in X$ such that $\gamma_x(A_1) = \emptyset$ (see Example 14.5 (i)), and let $A_2 = I_X$. Then $\gamma_x(A_2) \neq \emptyset$, which contradicts to Proposition 7.6.

A similar argument shows that neither the analytic residuum $S(A)$ nor the local spectrum $\sigma_x(A) = \gamma_x(A) \cup S_A$ satisfy the projection property. For example, let A_1 be the backward shift and $A_2 = I_X$; then $S_{A_1} \neq \emptyset$ and $S_{A_2} = \emptyset$. For σ_x, see Example 14.10 (i).

Trivially, $\gamma_x(A_1,\dots,A_n) \supset Q\gamma_x(A_1,\dots,A_n,A_{n+1})$ where $Q : \mathbb{C}^{n+1} \to \mathbb{C}^n$ is the natural projection. On the other hand, as in [Es1] one can show that $S_{(A_1,\dots,A_n)} \subset QS_{(A_1,\dots,A_n,A_{n+1})}$ and $\sigma_x(A_1,\dots,A_n) \subset Q\sigma_x(A_1,\dots,A_n,A_{n+1})$.

A better situation is for the n-tuples A with SVEP. Then $\sigma_x(f(A)) = f(\sigma_x(A))$ for all $x \in X$ and all m-tuples $f = (f_1,\dots,f_m)$ of functions analytic on a neighbourhood of $\sigma_x(A)$, see [Es1].

C.29.5. Theorem 29.12 was proved in [Va4].

C.30.1. The Taylor functional calculus was constructed in [Ta2]. For simplified versions of the calculus see [Lev1], [Va4], [Hel1], [Hel2], [Alb], [EP] and [An].

C.30.2. There are many variants of formulas (2), (3) in Section 30 defining the Taylor functional calculus that differ from each other in the sign in front of the integral. There are several sources of differences:

(i) Instead of the n-tuple $A - z = (A_1 - z_1,\dots,A_n - z_n)$ it is possible to consider the n-tuple $z - A$ (which appears naturally in the Cauchy formula). In this approach the additional factor $(-1)^n$ in front of the integrals (2), (3) appears.

(ii) Instead of (1) it is possible to use the convention that the Lebesgue measure in $\mathbb{C}^n$ is $(2i)^{-n} \mathrm{d}\bar{z}_1 \wedge \mathrm{d}z_1 \wedge \cdots \wedge \mathrm{d}\bar{z}_n \wedge \mathrm{d}z_n$. With this convention the Fubini theorem becomes more natural. In formula (2), however, the additional factor $(-1)^{\binom{n}{2}}$ would appear.

(iii) It is also possible to modify the definition of the mappings δ_A^p in the Koszul complex as in [Lev1]: $\delta_A^p x s_{i_1} \wedge \cdots \wedge s_{i_p} = \sum_j A_j x s_{i_1} \wedge \cdots \wedge s_{i_p} \wedge s_j$. This convention results also in the additional factor $(-1)^{\binom{n}{2}}$ in formula (2).

(iv) In order to obtain more symmetrical formulas, it is possible to replace the variables $s_1,\dots,s_n$ by $\mathrm{d}z_1,\dots,\mathrm{d}z_n$. In this case δ and $\bar{\partial}$ act in the space $\Lambda[\mathrm{d}z, \mathrm{d}\bar{z}, C^\infty(G,X)]$ and the projection P in formulas (2), (3) can be omitted.

C.30.3. The commutativity of the Gelfand transform with the Taylor functional calculus (Lemma 30.10 and Corollary 30.11) was proved by Putinar [Pu3], see also [AS].

The superposition principle (Theorem 30.12) is also due to Putinar [Pu1].

C.30.4. Let $A \in \mathcal{B}(X)^n$ be a commuting n-tuple and U an open neighbourhood of $\sigma_T(A)$. By [Pu2], there is exactly one continuous functional calculus from $H(U)$ to $\mathcal{B}(X)$, which satisfies the spectral mapping property. Thus the Taylor functional calculus is uniquely determined (in the above sense).

C.32.1. By Theorem 11.4, an operator-valued function $T : U \to \mathcal{B}(X,Y)$ is regular if and only if it can be completed to an exact sequence $\xrightarrow{T(z)} \xrightarrow{S(z)}$.

Let $T : U \to \mathcal{B}(X,Y)$ be k-regular at a point $w \in U$. In general, it is not possible to complete it to a Fredholm sequence $\xrightarrow{T(z)} \xrightarrow{S(z)}$ (the opposite implication was proved in Theorem 32.8). The reason is that, for all z in a certain neighbourhood of w, $\operatorname{Ran} T(z)$ is close to $\operatorname{Ran} T(w) + F_z$ for some finite-dimensional subspace F_z but the space F_z depends on z. If T can be completed to a Fredholm sequence $\xrightarrow{T(z)} \xrightarrow{S(z)}$, then it is possible to take F_z to be constantly equal to a complement of $\operatorname{Ran} T(w)$ in $\operatorname{Ker} S(w)$.

Example: Let H be a Hilbert space with an orthonormal basis $\{e_i : i \in \mathbb{N}\}$, let $U = \{1/n : n \in \mathbb{N}\} \cup \{0\}$, $T(0) = 0$ and let $T(1/n) : H \to H$ be defined by $T(1/n)e_n = n^{-1}e_n$ and $T(1/n)e_j = 0 \quad (j \neq n)$. Then we have $\widehat{\delta}_1(\operatorname{Ran} T(0), \operatorname{Ran} T(1/n)) = 0$ for all n and T is a 1-regular function. It is easy to see that it can not be completed to a Fredholm sequence. Indeed, if $S(z)T(z) = 0$ for all z, $\dim \operatorname{Ker} S(0) < \infty$ and $S(0)$ has closed range, then $\operatorname{Ker} S(1/n) \supset \{e_n\}$ and

$$1 = \limsup_{n\to\infty} \delta\big(\operatorname{Ker} S(1/n), \operatorname{Ker} S(0)\big) \leq \limsup_{n\to\infty} \gamma(S(0))^{-1}\|S(1/n) - S(0)\| = 0,$$

which is a contradiction.

C.33.1. The concepts of a Fredholm complex and its index (sometimes called the Euler characteristic) appeared in [Cu1], [Va5] and [FS1]; it is a particular case of a more general concept of Fredholm complexes of vector bundles [Seg]. An important motivation for the study of Fredholm complexes was the Taylor spectrum.

The index of Fredholm complexes of Hilbert spaces can be reduced to the index of a certain operator, see [AV]. In the context of Banach spaces the situation is much more complicated.

C.33.2. The basic stability results for index of Fredholm complexes of Banach spaces (Corollary 33.3 and Theorem 33.7) were proved by Vasilescu [Va5], see also [AV]. Here we present a new proof based on k-regular functions. The stability of index under compact perturbations (Theorem 33.6) was a long standing problem, which was answered by Ambrozie [Am2]. His proof is quite technical and he was forced to consider more general stability results for Fredholm chains, see below.

C.33.3 By a *Fredholm chain* we mean a sequence $\mathcal{K} = (X_i, \psi_i)_{i=0}^n$ of the following type:

$$0 \to X_0 \xrightarrow{\psi_0} X_1 \xrightarrow{\psi_1} \cdots \xrightarrow{\psi_{n-2}} X_{n-1} \xrightarrow{\psi_{n-1}} X_n \to 0, \tag{1}$$

where X_i are Banach spaces and ψ_i operators satisfying $\operatorname{Ker} \psi_{i+1} \stackrel{e}{=} \operatorname{Ran} \psi_i$ for all i.

Clearly, in this case the operators ψ_i have closed ranges.

The *index* of a Fredholm chain $\mathcal{K}$ is defined by

$$\operatorname{ind} \mathcal{K} = \sum_{i=0}^{n} (-1)^i \alpha_i(\mathcal{K}), \tag{2}$$

where

$$\alpha_j(\mathcal{K}) = \dim \operatorname{Ker} \psi_j / (\operatorname{Ker} \psi_j \cap \operatorname{Ran} \psi_{j-1}) - \dim \operatorname{Ran} \psi_{j-1} / (\operatorname{Ker} \psi_j \cap \operatorname{Ran} \psi_{j-1}).$$

A sequence $\mathcal{K}$ of the form (1) is called a *semi-Fredholm chain* if operators ψ_i have closed ranges and the index $\operatorname{ind}\mathcal{K}$ is well defined by (2) (either $+\infty$ and $-\infty$ does not appear in the formula).

For semi-Fredholm complexes the definition coincides with Definition 33.1.

An advantage of semi-Fredholm chains is that they are closed under finite-rank perturbations.

The stability results for index can be generalized to Fredholm and semi-Fredholm chains.

Some of the results were essentially proved already in Section 33. Let $\mathcal{K}$ be a semi-Fredholm chain. Then:

(i) $\alpha_j(\mathcal{K}') \le \alpha_j(\mathcal{K})$ for all semi-Fredholm chains sufficiently close to $\mathcal{K}$ (see Theorem 33.2).

(ii) $\operatorname{ind}\mathcal{K}' = \operatorname{ind}\mathcal{K}$ if $\mathcal{K} = (X_i, \psi_i)_{i=0}^n$ and $\mathcal{K}' = (X_i, \psi_i')_{i=0}^n$ are semi-Fredholm chains such that $\psi_i - \psi_i'$ are finite-rank operators for all i, see Lemma 33.4.

The stability of index of Fredholm chains under small perturbations was proved by Ambrozie [Am1].

C.33.4. The stability results for the index of Fredholm chains under small and finite-rank perturbations give immediately the stability of the index of Fredholm chains under perturbations that are limits of finite-rank operators. The stability of the index of Fredholm chains under compact perturbations is more difficult. It was proved by Ambrozie [Am2].

Let $\mathcal{K} = (X_i, \psi_i)_{i=0}^n$ and $\mathcal{K}' = (X_i, \psi_i')_{i=0}^n$ be a Fredholm chains and suppose that $\psi_i - \psi_i'$ is compact for all i. As in the proof of Theorem 33.10, set $X = \bigoplus_{i \text{ even}} X_i$, $Y = \bigoplus_{i \text{ odd}} X_i$, $T = \bigoplus_{i \text{ even}} \psi_i$ and $S = \bigoplus_{i \text{ odd}} \psi_i$; similarly define T' and S'.

The main idea is to consider X and Y as subspaces of greater spaces $E \supset X$ and $F \supset Y$ such that $T - T'$ and $S - S'$ are already limits of finite-rank operators $X \to E$ ($Y \to F$, respectively). As E one can take the space $\big(X \times C\langle 0,1\rangle\big)/\{(x, Jx) : x \in \overline{\operatorname{Ran}(T' - T)}\}$ where $J : \overline{\operatorname{Ran}(T' - T)} \to C\langle 0,1\rangle$ is an isometrical embedding. The space F can be constructed similarly.

C.33.5. All stability results can be extended to semi-Fredholm chains in a way similar to Theorem 33.10. It was done in [AV] and [Mü23].

C.33.6. The inclusion $\partial\sigma_T(A_1, A_2) \subset \sigma_\pi(A_1, A_2) \cup \sigma_\delta(A_1, A_2)$ (Corollary 33.11) for Hilbert spaces was proved in [ChT], for Banach spaces see [Wr] and [Cu3].

For more than two commuting operators the inclusion is not true, see [Cu3]. For an n-tuple A of commuting operators we have

$$\partial\sigma_T(A) \subset \sigma_{\pi,k}(A) \cup \sigma_{\delta,n-k-2}(A) \qquad (k = 0, \dots, n-2)$$

and $\partial\sigma_T(A) \subset \sigma_{\pi,n-1}(A) \cap \sigma_{\delta,n-1}(A)$, see [Mü14].

More generally, $\Gamma_k(\sigma_T(A), \mathcal{P}(n)) \subset \sigma_{\pi k}(A) \cap \sigma_{\delta k}(A)$ and $\Gamma_k(\sigma_T(A), \mathcal{P}(n)) \subset \sigma_{\pi,k} \cup \sigma_{\delta,k-j-1}(A) \quad (0 \le j < k \le n-1)$, where $\Gamma_k(\sigma_T(A), \mathcal{P}(n))$ are the higher Shilov boundaries of the set $\sigma_T(A)$; for details see [Mü14].

C.34.1. Let $S : U \to \mathcal{B}(X, Y)$ and $T : U \to \mathcal{B}(Y, Z)$ be analytic operator-valued functions defined in an open set $U \subset \mathbb{C}^n$. Suppose that $T(z)S(z) = 0$ and $\dim \operatorname{Ker} T(z)/\operatorname{Ran} S(z) < \infty \quad (z \in U)$. Let $k \ge 0$. The following problem is open:

Problem. Is the set $\{z \in U : \dim \operatorname{Ker} T(z)/\operatorname{Ran} S(z) \ge k\}$ analytic?

By Theorem 34.8, the answer is positive if $\operatorname{Ran} S(z)$ is complemented for all $z \in U$. By [Kab1], [Kab2], the answer is positive if either $S(z) \equiv 0$ or $T(z) \equiv 0$; so the remaining function is semi-Fredholm-valued.

For a discussion about this problem see [Mü23].

C.34.2. The essential Taylor spectrum was studied in [Lev1] and [Fa1]. The fact that the difference $\sigma_T(A) \setminus \sigma_{Te}(A)$ is an analytic set was announced without proof in [Lev1], [Lev2]. The present proof is taken from [Mü23], see also [Pu4].

Chapter V

Orbits and Capacity

Let T be an operator on a Banach space X. By an orbit of T we mean a sequence $(T^n x)_{n=0}^{\infty}$, where $x \in X$ is a fixed vector. This notion, which originated in the theory of dynamical systems, is closely related to the concepts of local spectral radius and capacity of an operator. Further motivations come from stability problems of semigroups of operators and the invariant subspace problem.

In this chapter we give a survey of results concerning orbits and related concepts. The main tool for most of the results will be the properties of the essential approximate point spectrum.

35 Joint spectral radius

One of the most important concepts of the spectral theory is that of the spectral radius of a Banach algebra element. In this section we generalize this notion to commuting n-tuples.

Two difficulties arise for $n \geq 2$: there are many reasonable spectra for commuting n-tuples and there is not a unique norm in $\mathbb{C}^n$.

The first difficulty is not a serious problem. We will see later that all reasonable spectral systems give the same joint spectral radius.

As for the norm in $\mathbb{C}^n$, we consider the ℓ^p-norms: for $z = (z_1, \dots, z_n) \in \mathbb{C}^n$ define $|z|_\infty = \max\{|z_i| : i = 1, \dots, n\}$ and $|z|_p = \left(\sum_{i=1}^n |z_i|^p\right)^{1/p}$ for $1 \leq p < \infty$.

Definition 1. Let $a = (a_1, \dots, a_n)$ be a commuting n-tuple of elements in a Banach algebra $\mathcal{A}$. For $1 \leq p \leq \infty$ define the *spectral radius* by

$$r_p(a) = \max\{|\lambda|_p : \lambda \in \sigma_H(a)\},$$

where $\sigma_H(a)$ is the Harte spectrum of a, see Definition 8.1.

Since the Harte spectrum is a compact set, the maximum really exists. For a single Banach algebra element $a_1 \in \mathcal{A}$ the just defined spectral radius $r_p(a)$ does not depend on p and coincides with the ordinary spectral radius $r(a_1) = \max\{|\lambda_1| : \lambda_1 \in \sigma(a_1)\}$.

The Harte spectrum can be replaced by any other spectral-radius-preserving spectral system (i.e., any compact-valued spectral system $\tilde{\sigma}$ satisfying $\max\{|\lambda_1| : \lambda_1 \in \tilde{\sigma}(a_1)\} = r(a_1)$ for each single element a_1).

Proposition 2. *Let $\tilde{\sigma}$ be a spectral-radius-preserving spectral system in a Banach algebra $\mathcal{A}$. Let $|||\cdot|||$ be any norm in $\mathbb{C}^n$. Then*

$$\max\{|||\lambda||| : \lambda \in \tilde{\sigma}(a_1, \dots, a_n)\} = \max\{|||\lambda||| : \lambda \in \sigma_H(a_1, \dots, a_n)\}$$

for all commuting n-tuples $a = (a_1, \dots, a_n) \in \mathcal{A}^n$.

In particular, $\max\{|\lambda|_p : \lambda \in \tilde{\sigma}(a_1, \dots, a_n)\} = r_p(a_1, \dots, a_n)$ for $1 \le p \le \infty$.

Proof. By Theorem 7.22, we have $\operatorname{conv} \tilde{\sigma}(a_1, \dots, a_n) = \operatorname{conv} \sigma_H(a_1, \dots, a_n)$. Thus

$$\begin{aligned}\max\{|||\lambda||| : \lambda \in \tilde{\sigma}(a_1, \dots, a_n)\} &= \max\{|||\lambda||| : \lambda \in \operatorname{conv} \tilde{\sigma}(a_1, \dots, a_n)\} \\ &= \max\{|||\lambda||| : \lambda \in \operatorname{conv} \sigma_H(a_1, \dots, a_n)\} \\ &= \max\{|||\lambda||| : \lambda \in \sigma_H(a_1, \dots, a_n)\}. \qquad \square\end{aligned}$$

The most important result for a single element a_1 of a Banach algebra $\mathcal{A}$ is the spectral radius formula,

$$r(a_1) = \lim_{k \to \infty} \|a_1^k\|^{1/k} = \inf_k \|a_1^k\|^{1/k}.$$

Our goal is to generalize this formula for commuting n-tuples.

Let $a = (a_1, \dots, a_n)$ be a commuting n-tuple of elements of a Banach algebra A. Instead of powers of a single element it is natural to consider all possible products of $a_1, \dots, a_n$.

Denote by $F(k, n)$ the set of all functions from $\{1, \dots, k\}$ to $\{1, \dots, n\}$. Let

$$s'_{k,\infty}(a) = \max_{f \in F(k,n)} r(a_{f(1)} \cdots a_{f(k)})$$

and

$$s'_{k,p}(a) = \Big(\sum_{f \in F(k,n)} r^p(a_{f(1)} \cdots a_{f(k)}) \Big)^{1/p} \qquad (1 \le p < \infty)$$

(for short we write $r^p(x)$ instead of $(r(x))^p$). Similarly, let

$$s''_{k,\infty}(a) = \max_{f \in F(k,n)} \|a_{f(1)} \cdots a_{f(k)}\|$$

and

$$s''_{k,p}(a) = \Big(\sum_{f \in F(k,n)} \|a_{f(1)} \cdots a_{f(k)}\|^p \Big)^{1/p} \qquad (1 \le p < \infty).$$

Using the standard multi-index notation, we can simplify the definitions of $s'_{k,p}$ and $s''_{k,p}$. For $\alpha = (\alpha_1, \dots, \alpha_n) \in \mathbb{Z}_+^n$ write $|\alpha| = \alpha_1 + \cdots + \alpha_n$, $\alpha! = \alpha_1! \cdots \alpha_n!$ and $a^\alpha = a_1^{\alpha_1} \cdots a_n^{\alpha_n}$. If k is an integer, $k \geq |\alpha|$, then write

$$\binom{k}{\alpha} = \frac{k!}{\alpha!(k-|\alpha|)!}$$

(for $n = 1$ this definition coincides with the classical binomial coefficients).

Using this notations, the definitions of s' and s'' for a commuting n-tuple $a = (a_1, \dots, a_n) \in \mathcal{A}^n$ assume a simpler form:

$$s'_{k,\infty}(a) = \max\{r(a^\alpha) : \alpha \in \mathbb{Z}_+^n, |\alpha| = k\},$$
$$s''_{k,\infty}(a) = \max\{\|a^\alpha\| : \alpha \in \mathbb{Z}_+^n, |\alpha| = k\},$$

and, for $1 \leq p < \infty$,

$$s'_{k,p}(a) = \Bigl(\sum_{|\alpha|=k} \binom{k}{\alpha} r^p(a^\alpha)\Bigr)^{1/p} \qquad \text{and} \qquad s''_{k,p}(a) = \Bigl(\sum_{|\alpha|=k} \binom{k}{\alpha} \|a^\alpha\|^p\Bigr)^{1/p}.$$

We will use frequently the following formula (for commuting variables x_i):

$$(x_1 + \cdots + x_n)^k = \sum_{|\alpha|=k} \binom{k}{\alpha} x^\alpha.$$

In particular, for $x_1 = \cdots = x_n = 1$ we have $\sum_{|\alpha|=k} \binom{k}{\alpha} = n^k$.

Lemma 3. *Let $a = (a_1, \dots, a_n)$ be a commuting n-tuple of elements of a Banach algebra $\mathcal{A}$, let $1 \leq p \leq \infty$. Then $s'_{k+l,p}(a) \leq s'_{k,p}(a) \cdot s'_{l,p}(a)$ and $s''_{k+l,p}(a) \leq s''_{k,p}(a) \cdot s''_{l,p}(a)$ for all $k, l \in \mathbb{N}$.*

Proof. The second statement is obvious for $p = \infty$. For $p < \infty$ we have

$$\begin{aligned}\bigl(s''_{k,p}(a) \cdot s''_{l,p}(a)\bigr)^p &= \sum_{f \in F(k,n)} \|a_{f(1)} \cdots a_{f(k)}\|^p \cdot \sum_{g \in F(l,n)} \|a_{g(1)} \cdots a_{g(l)}\|^p \\ &\geq \sum_{f,g} \|a_{f(1)} \cdots a_{f(k)} a_{g(1)} \cdots a_{g(l)}\|^p = \bigl(s''_{k+l,p}(a)\bigr)^p.\end{aligned}$$

The statements for s'_k can be proved similarly using the submultiplicativity of the spectral radius for commuting elements. □

By Lemma 1.21, the previous lemma implies that $\lim_{k\to\infty} \bigl(s'_{k,p}(a)\bigr)^{1/k}$ exists and is equal to $\inf_k \bigl(s'_{k,p}(a)\bigr)^{1/k}$. Similarly,

$$\lim_{k\to\infty} \bigl(s''_{k,p}(a)\bigr)^{1/k} = \inf_k \bigl(s''_{k,p}(a)\bigr)^{1/k}.$$

This leads to the following definition:

Definition 4. Let $a = (a_1, \dots, a_n)$ be a commuting n-tuple of elements in a Banach algebra $\mathcal{A}$. For $1 \le p \le \infty$ we write

$$r'_p(a) = \lim_{k\to\infty} \left(s'_{k,p}(a)\right)^{1/k}$$

and

$$r''_p(a) = \lim_{k\to\infty} \left(s''_{k,p}(a)\right)^{1/k}.$$

Thus

$$r'_\infty(a) = \lim_{k\to\infty} \left(\max\{r(a^\alpha) : |\alpha| = k\}\right)^{1/k}$$

and

$$r''_\infty(a) = \lim_{k\to\infty} \left(\max\{\|a^\alpha\| : |\alpha| = k\}\right)^{1/k}.$$

For $1 \le p < \infty$ we have

$$r'_p(a) = \lim_{k\to\infty} \Bigg(\sum_{f\in F(k,n)} r^p(a_{f(1)} \cdots a_{f(k)}) \Bigg)^{1/pk} = \lim_{k\to\infty} \Bigg(\sum_{|\alpha|=k} \binom{k}{\alpha} r^p(a^\alpha) \Bigg)^{1/pk}$$

and

$$r''_p(a) = \lim_{k\to\infty} \Bigg(\sum_{f\in F(k,n)} \|a_{f(1)} \cdots a_{f(k)}\|^p \Bigg)^{1/pk} = \lim_{k\to\infty} \Bigg(\sum_{|\alpha|=k} \binom{k}{\alpha} \|a^\alpha\|^p \Bigg)^{1/pk}.$$

We start with the spectral radius formula in the simplest case of $p = \infty$.

Theorem 5. *Let $a = (a_1, \dots, a_n)$ be an n-tuple of mutually commuting elements of a Banach algebra $\mathcal{A}$. Then*

$$r_\infty(a) = \max\{r(a_i) : i = 1, \dots, n\} = r'_\infty(a) = r''_\infty(a).$$

Proof. (i) The inequality $r'_\infty(a) \le r''_\infty(a)$ is clear.

(ii) $r''_\infty(a) \le \max\{r(a_i) : i = 1, \dots, n\}$: Write $R = \max\{r(a_i) : i = 1, \dots, n\}$. Let $\varepsilon > 0$. There exists $k \in \mathbb{N}$ such that $\|a_i^j\| \le (R+\varepsilon)^j$ for all $j > k$ and $1 \le i \le n$. Let

$$M = \max\big\{\|a_i^j\|(R+\varepsilon)^{-j} : 1 \le i \le n, 0 \le j \le k\big\}.$$

Then $M \ge 1$ and $\|a_i^j\| \le M(R+\varepsilon)^j$ for all $j \in \mathbb{N}, 1 \le i \le n$.

Let $\alpha \in \mathbb{Z}_+^n$. Then

$$\|a^\alpha\| \le \prod_{i=1}^n \|a_i^{\alpha_i}\| \le \prod_{i=1}^n M(R+\varepsilon)^{\alpha_i} = M^n(R+\varepsilon)^{|\alpha|}.$$

Thus $s''_{j,\infty}(a) \le M^n(R+\varepsilon)^j$ and $r''_\infty(a) = \lim_{j\to\infty}\left(s''_{j,\infty}(a)\right)^{1/j} \le R+\varepsilon$. Letting $\varepsilon \to 0$ gives $r''_\infty(a) \le R$.

(iii) $\max\{r(a_i) : i = 1, \ldots, n\} \le r_\infty(a)$: Let $i \in \{1, \ldots, n\}$ and let $\mu \in \sigma(a_i)$ satisfy $|\mu| = r(a_i)$. By the projection property, there exists $\lambda = (\lambda_1, \ldots, \lambda_n) \in \sigma_H(a)$ such that $\lambda_i = \mu$. Thus $r_\infty(a) \ge |\lambda|_\infty \ge |\lambda_i| = r(a_i)$.

(iv) $r_\infty(a) \le r'_\infty(a)$: Let $\lambda = (\lambda_1, \ldots, \lambda_n) \in \sigma_H(a)$ and $k \in \mathbb{N}$. Then

$$\begin{aligned} s'_{k,\infty} &= \max\{r(a^\alpha) : \alpha \in \mathbb{Z}^n_+, |\alpha| = k\} \ge \max\{r(a_i^k) : i = 1, \ldots, n\} \\ &\ge \max\{|\lambda_i|^k : i = 1, \ldots, n\} = |\lambda|^k_\infty. \end{aligned}$$

Thus $r'_\infty(a) = \lim_{k\to\infty} s'^{1/k}_{k,\infty} \ge |\lambda|_\infty$, and so $r'_\infty(a) \ge r_\infty(a)$. □

Similar formulas are also true for $p < \infty$; the proofs, however, are more complicated.

Theorem 6. *Let $a = (a_1, \ldots, a_n)$ be an n-tuple of mutually commuting elements of a Banach algebra $\mathcal{A}$. Let $1 \le p < \infty$. Then*

$$r_p(a) = r'_p(a) = r''_p(a).$$

Proof. The inequality $r'_p(a) \le r''_p(a)$ is clear.

We show that $r_p(a) \le r'_p(a)$.

Let $\lambda = (\lambda_1, \ldots, \lambda_n) \in \sigma_H(a)$. Denote by $\mathcal{A}_0$ the closed subalgebra of $\mathcal{A}$ generated by the unit 1 and the elements $a_1, \ldots, a_n$. Let $h : \mathcal{A}_0 \to \mathbb{C}$ be a multiplicative functional such that $h(a_j) = \lambda_j \quad (j = 1, \ldots, n)$. Then

$$\begin{aligned} (s'_{k,p}(a))^p &= \sum_{|\alpha|=k} \binom{k}{\alpha} r^p(a^\alpha) \ge \sum_{|\alpha|=k} \binom{k}{\alpha} |h(a^\alpha)|^p \\ &= \sum_{|\alpha|=k} \binom{k}{\alpha} |\lambda_1|^{p\alpha_1} \cdots |\lambda_n|^{p\alpha_n} = \left(|\lambda_1|^p + \cdots + |\lambda_n|^p\right)^k = |\lambda|_p^{pk}. \end{aligned}$$

So $(s'_{k,p}(a))^{1/k} \ge |\lambda|_p$ and

$$r'_p(a) = \lim_{k\to\infty} s'_{k,p}(a)^{1/k} \ge \max\{|\lambda|_p : \lambda \in \sigma_H(a)\} = r_p(a).$$

For the proof of the remaining inequalities it is convenient to reformulate the definitions of $r'_p(a)$ and $r''_p(a)$.

Recall that the number of all partitions of the set $\{1, \ldots, k\}$ into n parts is equal to $\binom{k+n-1}{n-1} \le (k + n - 1)^{n-1}$.

We have

$$\max_{|\alpha|=k} \binom{k}{\alpha} \|a^\alpha\|^p \le \sum_{|\alpha|=k} \binom{k}{\alpha} \|a^\alpha\|^p \le \binom{k+n-1}{n-1} \max_{|\alpha|=k} \binom{k}{\alpha} \|a^\alpha\|^p.$$

Note that

$$\lim_{k\to\infty} \binom{k+n-1}{n-1}^{1/k} = 1.$$

Thus

$$r'_p(a) = \lim_{k\to\infty} \max_{|\alpha|=k} \left(\binom{k}{\alpha} r(a^\alpha)^p \right)^{1/kp}.$$

Similarly,

$$r''_p(a) = \lim_{k\to\infty} \max_{|\alpha|=k} \left(\binom{k}{\alpha} \|a^\alpha\|^p \right)^{1/kp}.$$

We now prove the inequality $r'_p(a) \le r_p(a)$:

Choose k and $\alpha \in \mathbb{Z}_+^n$, $|\alpha| = k$. Let $\mu \in \sigma(a^\alpha)$ satisfy $|\mu| = r(a^\alpha)$. By the spectral mapping property, there exists $\lambda = (\lambda_1, \dots, \lambda_n) \in \sigma_H(a)$ such that $\mu = \lambda_1^{\alpha_1} \cdots \lambda_n^{\alpha_n}$. Then

$$\begin{aligned}\binom{k}{\alpha} r^p(a^\alpha) &= \binom{k}{\alpha} |\mu|^p = \binom{k}{\alpha} |\lambda_1|^{\alpha_1 p} \cdots |\lambda_n|^{\alpha_n p} \\ &\le \sum_{|\beta|=k} \binom{k}{\beta} |\lambda_1|^{\beta_1 p} \cdots |\lambda_n|^{\beta_n p} = \left(|\lambda_1|^p + \cdots |\lambda_n|^p\right)^k = |\lambda|_p^{pk} \le r_p^{pk}(a).\end{aligned}$$

Thus

$$r'_p(a) = \lim_{k\to\infty} \max_{|\alpha|=k} \left(\binom{k}{\alpha} r^p(a^\alpha) \right)^{1/kp} \le r_p(a).$$

The remaining inequality $r''_p(a) \le r'_p(a)$ will be proved by induction on n.

For $n = 1$, Theorem 6 reduces to the well-known spectral radius formula for a single element.

Let $n \ge 2$ and suppose that the inequality $r''_p \le r'_p$ is true for all commuting $(n-1)$-tuples. Let $a = (a_1, \dots, a_n)$ be a commuting n-tuple of elements of $\mathcal{A}$.

For each k there is an $\alpha \in \mathbb{Z}_+^n$, $|\alpha| = k$ such that

$$\binom{k}{\alpha} \|a^\alpha\|^p = \max_{|\beta|=k} \binom{k}{\beta} \|a^\beta\|^p.$$

Using the compactness of $\langle 0, 1\rangle^n$, we can choose a sequence

$$\{\alpha(i)\}_{i=1}^\infty = \left\{(\alpha_1(i), \dots, \alpha_n(i)\right\}_{i=1}^\infty \subset \mathbb{Z}_+^n$$

such that $\lim_{i\to\infty} |\alpha(i)| = \infty$,

$$\binom{|\alpha(i)|}{\alpha(i)} \|a^{\alpha(i)}\|^p = \max_{|\beta|=|\alpha(i)|} \binom{|\alpha(i)|}{\beta} \|a^\beta\|^p \qquad (i = 1, 2, \dots) \tag{1}$$

and the sequences $\left\{\frac{\alpha_j(i)}{|\alpha(i)|}\right\}_{i=1}^{\infty}$ are convergent for $j=1,\dots,n$. Write $k(i)=|\alpha(i)|$ and

$$t_j=\lim_{i\to\infty}\frac{\alpha_j(i)}{k(i)}\in\langle 0,1\rangle \qquad (j=1,\dots,n).$$

By (1), we have

$$r_p''^p(a)=\lim_{i\to\infty}\left(\binom{k(i)}{\alpha(i)}\|a^{\alpha(i)}\|^p\right)^{1/k(i)}.$$

We distinguish two cases:

(a) There exists $j\in\{1,\dots,n\}$ such that $t_j=0$. Without loss of generality we can assume that $t_n=0$.

Write $a'=(a_1,\dots,a_{n-1})$, $\alpha'(i)=\big(\alpha_1(i),\dots,\alpha_{n-1}(i)\big)\in\mathbb{Z}_+^{n-1}$ and

$$k'(i)=|\alpha'(i)|=k(i)-\alpha_n(i).$$

We have $\lim_{i\to\infty}\frac{k'(i)}{k(i)}=1$ and $\|a^{\alpha(i)}\|\le\|a'^{\alpha'(i)}\|\cdot\|a_n\|^{\alpha_n(i)}$. Then

$$r_p''^p(a')\ge\limsup_{i\to\infty}\left(\binom{k'(i)}{\alpha'(i)}\|a'^{\alpha'(i)}\|^p\right)^{1/k'(i)}\ge L_1\cdot L_2\cdot L_3,$$

where

$$L_1=\limsup_{i\to\infty}\left(\frac{\binom{k'(i)}{\alpha'(i)}}{\binom{k(i)}{\alpha(i)}}\right)^{1/k'(i)}, \qquad L_2=\lim_{i\to\infty}\left(\binom{k(i)}{\alpha(i)}\|a^{\alpha(i)}\|^p\right)^{1/k'(i)}$$

and

$$L_3=\lim_{i\to\infty}\|a_n\|^{-\alpha_n(i)p/k'(i)}.$$

Since $\lim_{i\to\infty}\frac{\alpha_n(i)}{k'(i)}=0$, we have $L_3=1$.

Furthermore,

$$L_2=\lim_{i\to\infty}\left(\left(\binom{k(i)}{\alpha(i)}\|a^{\alpha(i)}\|^p\right)^{1/k(i)}\right)^{k(i)/k'(i)}=r_p''^p(a).$$

Finally,

$$L_1=\limsup_{i\to\infty}\left(\frac{k'(i)!\cdot\alpha_n(i)!}{k(i)!}\right)^{1/k'(i)}\ge\limsup_{i\to\infty}\left(\frac{\left(\frac{\alpha_n(i)}{3}\right)^{\alpha_n(i)}}{k(i)^{\alpha_n(i)}}\right)^{1/k'(i)}$$

$$=\limsup_{i\to\infty}\left(\frac{\alpha_n(i)}{3k(i)}\right)^{\frac{\alpha_n(i)}{k(i)}\cdot\frac{k(i)}{k'(i)}}=1,$$

since $\lim_{i\to\infty}\frac{k(i)}{k'(i)}=1$ and

$$\lim_{i\to\infty}\left(\frac{\alpha_n(i)}{3k(i)}\right)^{\frac{\alpha_n(i)}{k(i)}}=\lim_{x\to 0_+}\left(\frac{x}{3}\right)^x=\lim_{x\to 0_+}e^{x(\ln x-\ln 3)}=1.$$

Thus $r''_p(a')\geq r''_p(a)$.

By the induction assumption, we have $r''_p(a')=r'_p(a')=r_p(a')$ and, by definition, $r_p(a')\leq r_p(a)=r'_p(a)$. Hence $r''_p(a)\leq r'_p(a)$.

(b) It remains the case $t_j>0$ $(j=1,\dots,n)$, where $t_j=\lim_{i\to\infty}\frac{\alpha_j(i)}{k(i)}$. Choose $\varepsilon>0$, $\varepsilon<\min_{1\leq j\leq n}\frac{t_j}{n}$. For i sufficiently large and $j=1,\dots,n$ we have

$$t_j-\frac{\varepsilon}{4}\leq\frac{\alpha_j(i)}{k(i)}\leq t_j+\frac{\varepsilon}{4}. \tag{2}$$

We approximate $t_1,\dots,t_n$ by rational numbers. Fix positive integers d and $c_1,\dots,c_n$ such that

$$t_j-\frac{\varepsilon}{2}\leq\frac{c_j}{d}\leq t_j-\frac{\varepsilon}{4}\qquad(j=1,\dots,n).$$

Let $\gamma=(c_1,\dots,c_n)$ and $u=a^\gamma=a_1^{c_1}\cdots a_n^{c_n}$. For each i we can write $k(i)=m(i)d+z(i)$, where $0\leq z(i)\leq d-1$. For i sufficiently large we have

$$\frac{c_j}{d}\leq\frac{\alpha_j(i)}{k(i)},\qquad\frac{\alpha_j(i)}{k(i)}-\frac{c_j}{d}\leq\frac{3\varepsilon}{4}$$

and

$$\alpha_j(i)-m(i)c_j=\alpha_j(i)-\frac{k(i)-z(i)}{d}\cdot c_j=k(i)\left(\frac{\alpha_j(i)}{k(i)}-\frac{c_j}{d}\right)+\frac{z(i)c_j}{d}.$$

Thus, for $1\leq j\leq n$ and for i large enough,

$$0\leq\alpha_j(i)-m(i)c_j\leq k(i)\frac{3\varepsilon}{4}+c_j\leq\varepsilon k(i), \tag{3}$$

and so

$$0\leq k(i)-m(i)|\gamma|=\sum_{j=1}^{n}(\alpha_j(i)-m(i)c_j)\leq\varepsilon nk(i). \tag{4}$$

We have

$$\begin{aligned}\|a^{\alpha(i)}\|&\leq\|a_1^{m(i)c_1}\cdots a_n^{m(i)c_n}\|\cdot\|a_1\|^{\alpha_1(i)-m(i)c_1}\cdots\|a_n\|^{\alpha_n(i)-m(i)c_n}\\&\leq\|u^{m(i)}\|\cdot K^{n\varepsilon k(i)},\end{aligned}$$

where $K = \max\{1, \|a_1\|, \dots, \|a_n\|\}$. Then

$$\begin{aligned} r_p'^p(a) &\ge \limsup_{i\to\infty}\left(\binom{m(i)|\gamma|}{m(i)\gamma} r^p(a^{m(i)\gamma})\right)^{1/m(i)|\gamma|} \\ &= \limsup_{i\to\infty}\binom{m(i)|\gamma|}{m(i)\gamma}^{1/m(i)|\gamma|} \cdot r(u)^{p/|\gamma|} \\ &= \limsup_{i\to\infty}\left(\binom{m(i)|\gamma|}{m(i)\gamma}\|u^{m(i)}\|^p\right)^{1/m(i)|\gamma|} \ge L_1 \cdot L_2 \cdot L_3, \end{aligned}$$

where

$$L_1 = \liminf_{i\to\infty}\left(\frac{\binom{m(i)|\gamma|}{m(i)\gamma}}{\binom{k(i)}{\alpha(i)}}\right)^{1/m(i)|\gamma|},$$

$$L_2 = \liminf_{i\to\infty}\left(\binom{k(i)}{\alpha(i)}\|a^{\alpha(i)}\|^p\right)^{1/m(i)|\gamma|}$$

and

$$L_3 = \liminf_{i\to\infty} K^{-n\varepsilon pk(i)/m(i)|\gamma|}.$$

By (4), we have

$$1 \le \frac{k(i)}{m(i)|\gamma|} \le \frac{1}{1-n\varepsilon} \tag{5}$$

for all i sufficiently large. Thus $L_3 \ge K^{-\frac{n\varepsilon p}{1-n\varepsilon}}$.

Since

$$\lim_{i\to\infty}\left(\binom{k(i)}{\alpha(i)}\|a^{\alpha(i)}\|^p\right)^{1/k(i)} = r_p''^p(a),$$

we have $L_2 \ge \min\{r_p''^p(a), (r_p''^p(a))^{1/1-n\varepsilon}\}$.

To estimate L_1, we use the well-known Stirling formula

$$l! = l^l e^{-l}\sqrt{2\pi l}(1+o(l)).$$

Consequently,

$$(1-\varepsilon)\Big(\frac{\alpha_j(i)}{e}\Big)^{\alpha_j(i)/m(i)|\gamma|} \le \big(\alpha_j(i)!\big)^{1/m(i)|\gamma|} \le (1+\varepsilon)\Big(\frac{\alpha_j(i)}{e}\Big)^{\alpha_j(i)/m(i)|\gamma|}$$

for $j = 1, \dots, n$ and for all i sufficiently large. Similar estimates we can use for $(m(i)c_j)!$, $(m(i)|\gamma|)!$ and $|\alpha(i)|!$. Thus, for i sufficiently large, we have (to simplify the expressions we write m, k and α instead of $m(i)$, $k(i)$ and $\alpha(i)$, respectively)

$$\begin{aligned}
\left(\frac{\binom{m|\gamma|}{m\gamma}}{\binom{k}{\alpha}}\right)^{1/m|\gamma|} &= \left(\frac{(m|\gamma|)!\alpha_1!\cdots\alpha_n!}{k!(mc_1)!\cdots(mc_n)!}\right)^{1/m|\gamma|}\\
&\geq \left(\frac{1-\varepsilon}{1+\varepsilon}\right)^{n+1} \frac{m|\gamma|\cdot\alpha_1^{\alpha_1/m|\gamma|}\cdots\alpha_n^{\alpha_n/m|\gamma|}\cdot e^{k/m|\gamma|}\cdot e^{c_1/|\gamma|}\cdots e^{c_n/|\gamma|}}{e\cdot e^{\alpha_1/m|\gamma|}\cdots e^{\alpha_n/m|\gamma|}\cdot k^{k/m|\gamma|}\cdot(mc_1)^{c_1/|\gamma|}\cdots(mc_n)^{c_n/|\gamma|}}\\
&= \left(\frac{1-\varepsilon}{1+\varepsilon}\right)^{n+1}\left(\frac{\alpha_1}{mc_1}\right)^{c_1/|\gamma|}\cdots\left(\frac{\alpha_n}{mc_n}\right)^{c_n/|\gamma|}\alpha_1^{(\alpha_1-mc_1)/m|\gamma|}\cdots\\
&\qquad\qquad \alpha_n^{(\alpha_n-mc_n)/m|\gamma|}\cdot\frac{m|\gamma|}{k^{k/m|\gamma|}}\\
&\geq \left(\frac{1-\varepsilon}{1+\varepsilon}\right)^{n+1}\cdot\left(\frac{\alpha_1}{k}\right)^{(\alpha_1-mc_1)/m|\gamma|}\cdots\left(\frac{\alpha_n}{k}\right)^{(\alpha_n-mc_n)/m|\gamma|}\cdot\frac{m|\gamma|}{k}.
\end{aligned}$$

By (3) and (5), we have

$$\frac{\alpha_j-mc_j}{m|\gamma|} = \frac{\alpha_j-mc_j}{k}\cdot\frac{k}{m|\gamma|} \leq \frac{\varepsilon}{1-n\varepsilon},$$

and so, by (2),

$$L_1 \geq \left(\frac{1-\varepsilon}{1+\varepsilon}\right)^{n+1}(1-n\varepsilon)\big((t_1-\varepsilon/4)\cdots(t_n-\varepsilon/4)\big)^{\frac{\varepsilon}{1-n\varepsilon}}.$$

Hence

$$\begin{aligned}
r_p'^p(a) \geq \left(\frac{1-\varepsilon}{1+\varepsilon}\right)^{n+1}&(1-n\varepsilon)\big((t_1-\varepsilon/4)\cdots(t_n-\varepsilon/4)\big)^{\frac{\varepsilon}{1-n\varepsilon}}\\
&\cdot K^{-\frac{n\varepsilon p}{1-n\varepsilon}}\cdot\min\big\{r_p''^p(a),(r_p''^p(a))^{1/1-n\varepsilon}\big\}.
\end{aligned}$$

Letting $\varepsilon\to 0$ yields $r_p'(a)\geq r_p''(a)$. □

We now apply the previous result to the case of n-tuples of operators.

Let $T=(T_1,\ldots,T_n)$ be an n-tuple of commuting operators on a Banach space X. Set

$$\|T\|_\infty = \sup\{\|T_jx\| : j=1,\ldots,n, x\in X, \|x\|=1\}$$

and, for $1\leq p<\infty$,

$$\|T\|_p = \sup\left\{\left(\sum_{j=1}^n\|T_jx\|^p\right)^{1/p} : x\in X, \|x\|=1\right\}.$$

Equivalently, $\|T\|_p$ is the norm of the operator $\delta_T : X\to X_p^n$, where X_p^n is the direct sum of n copies of X endowed with the ℓ^p-norm, and $\delta_T x = T_1x\oplus\cdots\oplus T_nx$.

Let $T=(T_1,\dots,T_n)\in\mathcal{B}(X)^n$ and $S=(S_1,\dots,S_m)\in\mathcal{B}(X)^m$. Denote by TS the mn-tuple

$$TS=(T_1S_1,\dots,T_1S_m,T_2S_1,\dots,T_2S_m,\dots,T_nS_1,\dots,T_nS_m).$$

Further, let $T^2=TT$ and $T^{k+1}=T\cdot T^k \quad (k\in\mathbb{N})$. With this notation we can state the spectral radius formula in a familiar way:

Theorem 7. *Let $T=(T_1,\dots,T_n)$ be an n-tuple of mutually commuting operators on a Banach space X, let $1\le p\le\infty$. Then $r_p(T)=\lim_{k\to\infty}\|T^k\|_p^{1/k}$.*

Proof. For $p=\infty$ the statement follows directly from Theorem 5. Suppose that $1\le p<\infty$. We have

$$\|T^k\|_p=\sup_{\|x\|=1}\Bigl(\sum_{|\alpha|=k}\binom{k}{\alpha}\|T^\alpha x\|^p\Bigr)^{1/p}$$

and

$$r_p(T)=\lim_{k\to\infty}\Bigl(\sum_{|\alpha|=k}\binom{k}{\alpha}\|T^\alpha\|^p\Bigr)^{1/kp}=\lim_{k\to\infty}\max_{|\alpha|=k}\Bigl(\binom{k}{\alpha}\|T^\alpha\|^p\Bigr)^{1/kp}$$

$$=\lim_{k\to\infty}\max_{|\alpha|=k}\sup_{\|x\|=1}\Bigl(\binom{k}{\alpha}\|T^\alpha x\|^p\Bigr)^{1/kp}$$

$$=\lim_{k\to\infty}\sup_{\|x\|=1}\max_{|\alpha|=k}\Bigl(\binom{k}{\alpha}\|T^\alpha x\|^p\Bigr)^{1/kp}$$

$$=\lim_{k\to\infty}\sup_{\|x\|=1}\Bigl(\sum_{|\alpha|=k}\binom{k}{\alpha}\|T^\alpha x\|^p\Bigr)^{1/kp}=\lim_{k\to\infty}\|T^k\|_p^{1/k}.$$

□

36 Capacity

In this section we study the notion of capacity in Banach algebras. This notion was inspired by the classical capacity of compact subsets of $\mathbb{C}$ that is studied in potential theory.

We start with the capacity of single Banach algebra elements.

For $k\in\mathbb{N}$ denote by $\mathcal{P}_k^1$ the set of all complex polynomials in one variable of degree k with leading coefficient equal to 1 (these polynomials are called monic).

Let a be an element of a Banach algebra $\mathcal{A}$. For $k\in\mathbb{N}$ set $\operatorname{cap}_k a=\inf\{\|p(a)\|:p\in\mathcal{P}_k^1\}$. Since the product of two monic polynomials is again monic, we have

$$\operatorname{cap}_{k+m} a\le\operatorname{cap}_k a\cdot\operatorname{cap}_m a$$

for all $k,m\in\mathbb{N}$. By Lemma 1.21, this implies that the limit $\lim_{k\to\infty}(\operatorname{cap}_k a)^{1/k}$ exists and is equal to $\inf_k(\operatorname{cap}_k a)^{1/k}$.

Definition 1. The *capacity* of an element a of a Banach algebra $\mathcal{A}$ is defined by

$$\operatorname{cap} a = \lim_{k\to\infty} (\operatorname{cap}_k a)^{1/k} = \inf_k (\operatorname{cap}_k a)^{1/k}.$$

Clearly, there is an analogy between the capacity and the spectral radius $r(a) = \lim_{k\to\infty} \|a^n\|^{1/n} = \inf_k \|a^k\|^{1/k}$. We have $\operatorname{cap}_k a \le \|T^k\|$ for all k, and so $\operatorname{cap} a \le r(a)$.

The capacity in Banach algebras is closely related to the classical capacity of compact subsets of $\mathbb{C}$. If K is a non-empty compact subset of $\mathbb{C}$ and $k \in \mathbb{N}$, then we set $\operatorname{cap}_k K = \inf\{\|p\|_K : p \in \mathcal{P}_k^1\}$. As above it is easy to see that the limit $\lim_{k\to\infty}(\operatorname{cap}_k K)^{1/k}$ exists and is equal to the infimum. The capacity of K is defined by

$$\operatorname{cap} K = \lim_{k\to\infty} (\operatorname{cap}_k K)^{1/k} = \inf_k (\operatorname{cap}_k K)^{1/k}.$$

The use of the name capacity for Banach algebras elements is also justified by the following observation: if $C(K)$ is the algebra of all continuous functions on K with the sup-norm, and $a \in C(K)$ is defined by $a(z) = z \quad (z \in K)$, then the capacity of a in the Banach algebra $C(K)$ is equal to the classical capacity of the set K. Thus, in some sense, the capacity in Banach algebras is a generalization of the classical capacity of compact sets.

The next theorem shows that these two capacities are connected even more:

Theorem 2. *Let a be an element in a Banach algebra $\mathcal{A}$. Then:*

(i) $\operatorname{cap} a = \lim_{k\to\infty}(\operatorname{cap}'_k a)^{1/k} = \inf_k(\operatorname{cap}'_k a)^{1/k}$, *where*

$$\operatorname{cap}'_k a = \inf\{r(p(a)) : p \in \mathcal{P}_k^1\};$$

(ii) $\operatorname{cap} a = \operatorname{cap} \sigma(a)$.

Proof. (i) If $k, l \in \mathbb{N}$, $p \in \mathcal{P}_k^1$ and $q \in \mathcal{P}_l^1$, then $pq \in \mathcal{P}_{k+l}^1$ and $r((pq)(a)) \le r(p(a)) \cdot r(q(a))$. Thus $\operatorname{cap}'_{k+l} a \le \operatorname{cap}'_k a \cdot \operatorname{cap}'_l a$. As above, this implies that the limit $\lim_{k\to\infty}(\operatorname{cap}'_k a)^{1/k}$ exists and is equal to the infimum.

Obviously, $\operatorname{cap}'_k a \le \operatorname{cap}_k a$ for all k, and so $\lim_{k\to\infty}(\operatorname{cap}'_k a)^{1/k} \le \operatorname{cap} a$.

Conversely, let $p \in \mathcal{P}_k^1$. For all $m \in \mathbb{N}$ we have

$$\|(p(a))^m\|^{1/km} \ge (\operatorname{cap}_{km} a)^{1/km} \ge \operatorname{cap} a,$$

and so $r(p(a)) = \lim_{m\to\infty} \|(p(a))^m\|^{1/m} \ge (\operatorname{cap} a)^k$. Thus $r(p(a))^{1/k} \ge \operatorname{cap} a$ and $(\operatorname{cap}'_k a)^{1/k} \ge \operatorname{cap} a$ for all $k \in \mathbb{N}$. Hence $\lim_{k\to\infty}(\operatorname{cap}'_k a)^{1/k} \ge \operatorname{cap} a$.

(ii) For each $p \in \mathcal{P}_k^1$ we have

$$r(p(a)) = \max\{|\mu| : \mu \in \sigma(p(a))\} = \max\{|p(z)| : z \in \sigma(a)\} = \|p\|_{\sigma(a)},$$

and so $\operatorname{cap}'_k a = \operatorname{cap}_k \sigma(a)$. Thus

$$\operatorname{cap} \sigma(a) = \lim_{k\to\infty} (\operatorname{cap}'_k a)^{1/k} = \operatorname{cap} a. \qquad \square$$

We next generalize the concept of capacity to commuting n-tuples of Banach algebra elements.

A polynomial of degree $\le k$ in n variables can be written as

$$p(z_1, \dots, z_n) = \sum_{\substack{\alpha \in \mathbb{Z}_+^n \\ |\alpha| \le k}} c_\alpha z^\alpha,$$

where, as usually, $z^\alpha = z_1^{\alpha_1} \cdots z_n^{\alpha_n}$ and $c_\alpha \in \mathbb{C}$. We say that p is monic of degree k if $\sum_{|\alpha|=k} |c_\alpha| = 1$. The set of all polynomials of degree $\le k$ in n variables will be denoted by $\mathcal{P}_k(n)$. Denote by $\mathcal{P}_k^1(n)$ the set of all monic polynomials of degree k.

Let $a = (a_1, \dots, a_n)$ be mutually commuting elements of a Banach algebra $\mathcal{A}$. As in the one variable case, for $k \in \mathbb{N}$ set

$$\operatorname{cap}_k a = \inf\{\|p(a)\| : p \in \mathcal{P}_k^1(n)\}.$$

For $n \ge 2$ there is a technical difficulty since the product of two monic polynomial need not be monic. Therefore the sequence $\operatorname{cap}_k a$ is not submultiplicative and it is not clear that the limit $\lim_{k\to\infty}(\operatorname{cap}_k a)^{1/k}$ exists (we will see later that this is still true). Therefore we define:

Definition 3. Let $a = (a_1, \dots, a_n)$ be a commuting n-tuple of elements in a Banach algebra $\mathcal{A}$. The capacity of the n-tuple a is defined by

$$\operatorname{cap} a = \limsup_{k\to\infty} (\operatorname{cap}_k a)^{1/k}.$$

For a compact subset $K \subset \mathbf{C}^n$ define the corresponding capacity (sometimes called the Tshebyshev constant) by

$$\operatorname{cap} K = \limsup_{k\to\infty} (\operatorname{cap}_k K)^{1/k},$$

where

$$\operatorname{cap}_k K = \inf\{\|p\|_K : p \in \mathcal{P}_k^1(n)\}.$$

We show that the capacity can be expressed in another, more convenient way. Denote by $Q_k(n)$ the set of all polynomials $p(z) = \sum_{|\mu|\le k} c_\mu z^\mu \in \mathcal{P}_k(n)$ such that

$$\sup\left\{\left|\sum_{|\nu|=k} c_\mu z^\nu\right| : z \in \mathbb{T}^n\right\} = 1,$$

where $\mathbb{T}^n = \{z = (z_1, \dots, z_n) \in \mathbf{C}^n : |z_1| = \cdots = |z_n| = 1\}$ is the n-dimensional torus.

Theorem 4. *Let $x_1, \dots, x_n$ be mutually commuting elements of a Banach algebra $\mathcal{A}$. Then:*

(i) $\operatorname{cap}(x_1, \dots, x_n) = \lim_{k\to\infty} \operatorname{cap}_k(x_1, \dots, x_n)^{1/k} = \inf_k \inf\{\|p(x)\|^{1/k} : p \in Q_k(n)\}$;

(ii) $\operatorname{cap}(x_1, \dots, x_n) = \inf_k \inf\{(\operatorname{cap} p(x_1, \dots, x_n))^{1/k} : p \in Q_k(n)\}$;

(iii) $\operatorname{cap}(x_1, \dots, x_n) = \operatorname{cap} \sigma_H(x_1, \dots, x_n)$.

Proof. (i) Let $p = \sum_{|\nu|\le k} c_\nu z^\nu \in \mathcal{P}_k(n)$. By the Cauchy formulas, for each $\mu \in \mathbb{Z}_+^n$ with $|\mu| = k$ we have

$$|c_\mu| \le \max\left\{\left|\sum_{|\nu|=k} c_\nu z^\nu\right| : z \in \mathbb{T}^n\right\} = \left\|\sum_{|\nu|=k} c_\nu z^\nu\right\|_{\mathbb{T}^n}.$$

Further,

$$\left\|\sum_{|\nu|=k} c_\nu z^\nu\right\|_{\mathbb{T}^n} \le \sum_{|\mu|=k} |c_\mu| \le \binom{k+n-1}{n-1} \left\|\sum_{|\nu|=k} c_\nu z^\nu\right\|_{\mathbb{T}^n},$$

where $\binom{k+n-1}{n-1}$ is the number of coefficients c_μ with $|\mu| = k$. Write

$$\alpha_k = \inf\{\|p(x_1, \dots, x_n)\| : p \in Q_k(n)\}.$$

Then

$$\operatorname{cap}_k(x_1, \dots, x_n) \le \alpha_k \le \binom{k+n-1}{n-1} \operatorname{cap}_k(x_1, \dots, x_n). \tag{1}$$

Let $p \in Q_k(n)$ and let m, s be non-negative integers, $0 \le s \le k-1$. Then $q = p^m \cdot z_1^s \in Q_{mk+s}(n)$. Thus $\alpha_{mk+s} \le \alpha_k^m \cdot \|x_1\|^s$, $\alpha_{mk+s}^{1/mk+s} \le \alpha_k^{\frac{m}{mk+s}} \cdot \max\{1, \|x_1\|^{k-1}\}^{1/mk+s}$ and $\limsup_{r\to\infty} \alpha_r^{1/r} \le \alpha_k^{1/k}$. So the limit $\lim_{k\to\infty} \alpha_k^{1/k}$ exists and is equal to $\inf_k \alpha_k^{1/k}$.

By (1), the limit $\lim_{k\to\infty} \big(\operatorname{cap}_k(x_1, \dots, x_n)\big)^{1/k}$ also exists and

$$\begin{aligned}\operatorname{cap}(x_1, \dots, x_n) &= \lim_{k\to\infty} \big(\operatorname{cap}_k(x_1, \dots, x_n)\big)^{1/k} = \lim_{k\to\infty} \alpha_k^{1/k} \\ &= \inf_k \alpha_k^{1/k} = \inf_k \inf\{\|p(x_1, \dots, x_n)\|^{1/k} : p \in Q_k(n)\}.\end{aligned}$$

(ii) Let $p \in Q_k(n)$ and let $q = z^s + \sum_{i=0}^{s-1} a_i z^i \in \mathcal{P}_s^1(1) = Q_s(1)$. Then $q \circ p \in Q_{sk}(n)$, and so, by (i),

$$\operatorname{cap}(x_1, \dots, x_n) \le \|(q \circ p)(x_1, \dots, x_n)\|^{1/sk} \qquad (q \in Q_s(1)). \tag{2}$$

Hence

$$\begin{aligned}\operatorname{cap}(x_1,\dots,x_n) &\le \inf_s \inf\{\|q(p(x_1,\dots,x_n))\|^{1/sk} : q \in Q_s(1)\}\\ &= (\operatorname{cap} p(x_1,\dots,x_n))^{1/k}\end{aligned}$$

and

$$\operatorname{cap}(x_1,\dots,x_n) \le \inf_k \inf\{(\operatorname{cap} p(x_1,\dots,x_n))^{1/k} : p \in Q_k(n)\}.$$

On the other hand, $\operatorname{cap} p(x_1,\dots,x_n) \le \|p(x_1,\dots,x_n)\|$ for every $p \in Q_k(n)$. Together with (i) this gives

$$\operatorname{cap}(x_1,\dots,x_n) = \inf_k \inf\{(\operatorname{cap} p(x_1,\dots,x_n))^{1/k} : p \in Q_k(n)\}.$$

(iii) Let $p \in Q_k(n)$. By (ii), we have

$$\begin{aligned}(\operatorname{cap}(x_1,\dots,x_n))^k &\le \operatorname{cap} p(x_1,\dots,x_n) \le r(p(x_1,\dots,x_n))\\ &= \max\{|p(z)| : z \in \sigma_H(x_1,\dots,x_n)\} = \|p\|_{\sigma_H(x_1,\dots,x_n)}.\end{aligned}$$

So

$$\begin{aligned}\operatorname{cap}(x_1,\dots,x_n) &\le \inf_k \inf\{\|p\|^{1/k}_{\sigma_H(x_1,\dots,x_n)} : p \in Q_k(n)\}\\ &\le \inf_k \binom{k+n-1}{n-1}^{1/k} (\operatorname{cap}_k \sigma_H(x_1,\dots,x_n))^{1/k}.\end{aligned}$$

Hence $\operatorname{cap}(x_1,\dots,x_n) \le \operatorname{cap} \sigma_H(x_1,\dots,x_n)$.

On the other hand,

$$\|p(x_1,\dots,x_n)\| \ge r(p(x_1,\dots,x_n)) = \|p\|_{\sigma_H(x_1,\dots,x_n)}$$

for each polynomial $p \in \mathcal{P}_k(n)$, so

$$\operatorname{cap}_k(x_1,\dots,x_n) \ge \operatorname{cap}_k \sigma_H(x_1,\dots,x_n)$$

and

$$\operatorname{cap}(x_1,\dots,x_n) \ge \operatorname{cap} \sigma_H(x_1,\dots,x_n). \qquad \square$$

Theorem 5. *Let $\mathcal{A}$ be a Banach algebra and let $\tilde\sigma$ be a compact-valued spectral system satisfying $\operatorname{cap} \tilde\sigma(x_1) = \operatorname{cap} \sigma(x_1)$ for each $x_1 \in \mathcal{A}$. Then*

$$\operatorname{cap} \tilde\sigma(x_1,\dots,x_n) = \operatorname{cap} \sigma_H(x_1,\dots,x_n) = \operatorname{cap}(x_1,\dots,x_n)$$

for all n-tuples $x_1,\dots,x_n$ of mutually commuting elements of $\mathcal{A}$.

Proof. Let $x_1,\dots,x_n$ be mutually commuting elements of $\mathcal{A}$. Consider the algebra $C(K)$ of all continuous functions on the compact set $K = \tilde\sigma(x_1,\dots,x_n) \subset \mathbf{C}^n$ with the sup-norm on K and let $z_1,\dots,z_n$ be the independent variables.

As $\|q\|_K = \|q(z_1,\dots,z_n)\|_{C(K)}$ for each polynomial q, it is easy to see that $\operatorname{cap} K = \operatorname{cap}(z_1,\dots,z_n)$ and $\operatorname{cap} p(K) = \operatorname{cap} p(z_1,\dots,z_n)$ for all polynomials p.

Thus

$$\begin{aligned}\operatorname{cap}(x_1,\dots,x_n) &= \inf_k \inf\{\left(\operatorname{cap} p(x_1,\dots,x_n)\right)^{1/k} : p\in Q_k(n)\}\\ &= \inf_k \inf\{\left(\operatorname{cap} \sigma(p(x_1,\dots,x_n))\right)^{1/k} : p\in Q_k(n)\}\\ &= \inf_k \inf\{\left(\operatorname{cap} \tilde\sigma(p(x_1,\dots,x_n))\right)^{1/k} : p\in Q_k(n)\}\\ &= \inf_k \inf\{\left(\operatorname{cap} p(\tilde\sigma(x_1,\dots,x_n))\right)^{1/k} : p\in Q_k(n)\}\\ &= \inf_k \inf\{\left(\operatorname{cap} p(z_1,\dots,z_n)\right)^{1/k} : p\in Q_k(n)\}\\ &= \operatorname{cap}(z_1,\dots,z_n) = \operatorname{cap}\tilde\sigma(x_1,\dots,x_n).\end{aligned}$$

□

Corollary 6. *Let $\tilde\sigma$ be a spectral-radius-preserving spectral system in a Banach algebra $\mathcal{A}$. Then $\operatorname{cap}(a_1,\dots,a_n) = \operatorname{cap}\tilde\sigma(a_1,\dots,a_n)$ for all commuting n-tuples $(a_1,\dots,a_n)\in\mathcal{A}$.*

Proof. By Theorem 7.22, $\partial\sigma(a_1)\subset\tilde\sigma(a_1)\subset\sigma(a_1)$ for all $a_1\in\mathcal{A}$. By the maximum principle, $\operatorname{cap}\tilde\sigma(a_1) = \operatorname{cap}\sigma(a_1)$, so the statement follows from the previous theorem. □

Another important example for which we can use Theorem 5 is the essential spectrum.

Lemma 7. *Let X be a Banach space. Then $\operatorname{cap}\sigma_e(T) = \operatorname{cap}\sigma(T)$ for all $T\in\mathcal{B}(X)$.*

Proof. Clearly, $\operatorname{cap}\sigma_e(T)\le\operatorname{cap}\sigma(T)$.

Conversely, let $\varepsilon>0$. There exists $k\in\mathbb{N}$ and $p\in\mathcal{P}_k^1$ such that $\sup\{|p(z)| : z\in\sigma_e(T)\} < (\operatorname{cap}\sigma_e(T)+\varepsilon)^k$. Let $U = \{z\in\mathbb{C} : |p(z)| < (\operatorname{cap}\sigma_e(T)+\varepsilon)^k\}$. Then $\sigma(T)\setminus U$ is a finite set, $\sigma(T)\setminus U = \{\lambda_1,\dots,\lambda_r\}$. Let $n\in\mathbb{N}$ and set $q(z) = (p(z))^n(z-\lambda_1)\cdots(z-\lambda_r)$. Then $q\in\mathcal{P}^1_{nk+r}$ and

$$\max\{|q(z)| : z\in\sigma(T)\}\le\max\{|p(z)| : z\in U\}^n\cdot m\le(\operatorname{cap}\sigma_e(T)+\varepsilon)^{kn}\cdot m,$$

where $m = \max\{|(z-\lambda_1)\cdots(z-\lambda_r)| : z\in\sigma(T)\}$. Consequently, we have $\operatorname{cap}\sigma(T)\le(\operatorname{cap}\sigma_e(T)+\varepsilon)^{\frac{kn}{kn+r}}\cdot m^{\frac{1}{kn+r}}$. If $n\to\infty$, then we get $\operatorname{cap}\sigma(T)\le\operatorname{cap}\sigma_e(T)+\varepsilon$. Letting $\varepsilon\to0$ yields $\operatorname{cap}\sigma(T)\le\operatorname{cap}\sigma_e(T)$. □

Corollary 8. *Let X be a Banach space, let $\tilde\sigma$ be a spectral system in $\mathcal{B}(X)$ satisfying $\partial\sigma_e(T)\subset\tilde\sigma(T)$ for all $T\in\mathcal{B}(X)$. Then*

$$\operatorname{cap}\tilde\sigma(T_1,\dots,T_n) = \operatorname{cap}\sigma_H(T_1,\dots,T_n) = \operatorname{cap}(T_1,\dots,T_n)$$

for all commuting n-tuples $T_1,\dots,T_n$ of operators on X.

Proof. For all $T\in\mathcal{B}(X)$ we have $\operatorname{cap}\sigma_e(T)\le\operatorname{cap}\tilde\sigma(T)\le\operatorname{cap}\sigma(T) = \operatorname{cap}\sigma_e(T)$, and so we can use Theorem 5. □

In fact, the condition $\partial\sigma_e(T)\subset\tilde\sigma(T)$ for all $T\in\mathcal{B}(X)$ was satisfied by all spectral systems considered in this monograph.

37 Invariant subset problem and large orbits

By an *orbit* of $T \in \mathcal{B}(X)$ we mean a sequence of the form $(T^n x)_{n=0}^{\infty}$, where $x \in X$ is a fixed vector.

Orbits of operators are closely connected with the most important open problem of operator theory – the invariant subspace/subset problem.

Let $T \in \mathcal{B}(X)$. A non-empty subset $M \subset X$ is called *invariant* for T if $TM \subset M$. The set M is non-trivial if $\{0\} \neq M \neq X$ (the trivial subsets $\{0\}$ and X are always invariant for any operator $T \in \mathcal{B}(X)$). The invariant subspace problem may be formulated as follows:

Problem 1. Let T be an operator on a Hilbert space H of dimension ≥ 2. Does there exist a non-trivial closed subspace invariant for T?

It is easy to see that the problem has sense only for separable infinite-dimensional spaces. Indeed, if H is non-separable and $x \in H$ any non-zero vector, then the vectors $x, Tx, T^2x, \dots$ span a non-trivial closed subspace invariant for T.

If $\dim H < \infty$, then T has at least one eigenvalue and the corresponding eigenvector generates an invariant subspace of dimension 1. Note that the existence of eigenvalues is equivalent to the fundamental theorem of algebra that each non-constant complex polynomial has a root. Thus the invariant subspace problem is non-trivial even for finite-dimensional spaces.

Examples of Banach space operators without non-trivial closed invariant subspaces were given by Enflo [En2], Beuzamy [Bea1] and Read [Re1]. Read [Re6] also gave an example of an operator T (acting on ℓ^1) with a stronger property that T has no non-trivial closed invariant subset.

It is not known whether such an operator exists on a Hilbert space. The following "invariant subset problem" may be easier than Problem 1.

Problem 2. (invariant subset problem) Let T be an operator on a Hilbert space H. Does there exist a non-trivial closed subset invariant for T?

Both Problems 1 and 2 are also open for operators on reflexive Banach spaces. More generally, the following problem is open:

Problem 3. Let T be an operator on a Banach space X. Does T^* have a non-trivial closed invariant subset/subspace?

The existence of non-trivial invariant subspaces/subsets is closely connected with the behaviour of orbits. It is easy to see that an operator $T \in B(X)$ has no non-trivial closed invariant subspace if and only if all orbits corresponding to non-zero vectors span all the space X (i.e., each non-zero vector is cyclic). Similarly, $T \in B(X)$ has no non-trivial closed invariant subset if and only if all orbits corresponding to non-zero vectors are dense, i.e., all non-zero vectors are hypercyclic.

Thus orbits provide the basic information about the structure of an operator.

Typically, the behaviour of an orbit $(T^n x)$ depends essentially on the choice of the initial vector x. This can be illustrated by the following simple example:

Example 4. Let S be the backward shift on a Hilbert space H, i.e., $Se_0 = 0$ and $Se_i = e_{i-1} \quad (i \geq 1)$, where $\{e_i : i = 0, 1, 2, \dots\}$ is an orthonormal basis in H. Let $T = 2S$. Then:

(i) there is a dense subset of points $x \in H$ such that $\|T^n x\| \to 0$;
(ii) there is a dense subset of points $x \in H$ such that $\|T^n x\| \to \infty$;
(iii) there is a residual subset of points $x \in H$ such that the set $\{T^n x : n = 0, 1, \dots\}$ is dense in H.

Statement (i) is easy to show: any vector x which is a finite linear combination of the basis vectors e_i satisfies $\|T^n x\| \to 0$. Statements (ii) and (iii) are not so obvious. They will follow from the subsequent general results.

Note that it is very simple to find an operator such that almost all (up to a set of the first category) vectors are hypercyclic, but it is very difficult to find an operator such that all non-zero vectors are hypercyclic.

In this section we study orbits that are "large" in some sense (e.g., of type (ii)). Note that orbits satisfying $\|T^n x\| \to \infty$ provide a simple example of a non-trivial closed invariant subset.

It is an easy consequence of the Banach-Steinhaus theorem that an operator $T \in \mathcal{B}(X)$ has unbounded orbits if and only if $\sup \|T^n\| = \infty$.

Another result of this type was proved in Section 14. For each operator $T \in \mathcal{B}(X)$ there are points $x \in X$ with the property that the local spectral radius $r_x(T) = \limsup_{n\to\infty} \|T^n x\|^{1/n}$ is equal to the spectral radius $r(T)$. In particular, for such points x there are infinitely many powers such that $\|T^n x\|$ is "large" (asymptotically, $\|T^n x\| \sim r(T)^n$).

More precisely, it is possible to prove the following stronger result:

Theorem 5. *Let $T \in \mathcal{B}(X)$, let $(a_n)_{n\geq 0}$ be a sequence of positive numbers such that $a_n \to 0$. Then the set of all $x \in X$ with the property that*

$$\|T^n x\| \geq a_n \|T^n\| \qquad \text{for infinitely many powers } n$$

is residual.

Consequently, the set $\{x \in X : r_x(T) = r(T)\}$ is residual.

Proof. For $k \in \mathbb{N}$ set

$$M_k = \{x \in X : \text{ there exists } n \geq k \text{ such that } \|T^n x\| > a_n \|T^n\|\}.$$

Clearly, M_k is an open set. We prove that M_k is dense. Let $x \in X$ and $\varepsilon > 0$. Choose $n \geq k$ such that $a_n \varepsilon^{-1} < 1$. There exists $z \in X$ of norm 1 such that $\|T^n z\| > a_n \varepsilon^{-1} \|T^n\|$. Then

$$2a_n \|T^n\| < \|T^n(2\varepsilon z)\| \leq \|T^n(x + \varepsilon z)\| + \|T^n(x - \varepsilon z)\|,$$

and so either $\|T^n(x+\varepsilon z)\| > a_n\|T^n\|$ or $\|T^n(x-\varepsilon z)\| > a_n\|T^n\|$. Thus either $x+\varepsilon z \in M_k$ or $x-\varepsilon z \in M_k$, and so $\operatorname{dist}\{x, M_k\} \le \varepsilon$. Since x and ε were arbitrary, the set M_k is dense.

By the Baire category theorem, the intersection $\bigcap_{k=1}^\infty M_k$ is a dense G_δ-set, hence it is residual. Clearly, each $x \in \bigcap_{k=1}^\infty M_k$ satisfies $\|T^n x\| > a_n\|T^n\|$ for infinitely many powers n.

In particular, for $a_n = n^{-1}$ we obtain

$$r_x(T) = \limsup_{n\to\infty} \|T^n x\|^{1/n} \ge \limsup_{n\to\infty}\left(\frac{\|T^n\|}{n}\right)^{1/n} = r(T)$$

for all x in a residual subset of X. □

In fact, a much stronger result is also true: there are points $x \in X$ such that all powers $T^n x$ are "large" in the norm.

The following lemma is a useful technical tool in many constructions.

Lemma 6. *Let E be a finite-dimensional subspace of a Banach space X, let $\varepsilon > 0$. Then there exists a closed subspace $Y \subset X$ of finite codimension such that*

$$\|e+y\| \ge (1-\varepsilon)\max\bigl\{\|e\|, \|y\|/2\bigr\}$$

for all $e \in E$ and $y \in Y$.

Proof. We can assume that $\varepsilon < 1$. The unit sphere in E is compact, therefore there exists a finite subset $D \subset \{e \in E : \|e\| = 1\}$ with the property that $\operatorname{dist}\{e, D\} \le \varepsilon$ for all $e \in E$, $\|e\| = 1$. For each $d \in D$ there exists a functional $f_d \in X^*$ such that $\langle d, f_d\rangle = 1 = \|f_d\|$. Set $Y = \bigcap_{d\in D} \operatorname{Ker} f_d$. Clearly, $\operatorname{codim} Y < \infty$.

To prove the required inequality, let $e \in E$ and $y \in Y$. We can assume that $\|e\| \ne 0$ since the assertion is clear for $e = 0$. Find $d \in D$ with $\left\|d - \frac{e}{\|e\|}\right\| \le \varepsilon$. Then

$$\begin{aligned}\|e+y\| &\ge |\langle e+y, f_d\rangle| = \bigl|\langle e - \|e\|d, f_d\rangle + \langle \|e\|d, f_d\rangle\bigr| \\ &\ge \|e\| - \bigl\|e - \|e\|d\bigr\| \ge \|e\|(1-\varepsilon).\end{aligned}$$

Furthermore,

$$\begin{aligned}\|e+y\| &\ge \frac{1}{2}(1-\varepsilon)\frac{2-\varepsilon}{1-\varepsilon}\|e+y\| = \frac{1}{2}(1-\varepsilon)\Bigl(\|e+y\| + \frac{1}{1-\varepsilon}\|e+y\|\Bigr) \\ &\ge \frac{1}{2}(1-\varepsilon)\bigl(\|y\| - \|e\| + \|e\|\bigr) = \frac{1}{2}(1-\varepsilon)\|y\|.\end{aligned}$$

□

If X is a Hilbert space, then we can take $Y = E^\perp$. Thus the constructed subspace $Y \subset X$ substitutes the role of the orthogonal complement of a finite-dimensional subspace for general Banach spaces.

Lemma 7. *Let $T \in \mathcal{B}(X)$, $r_e(T) = 1$, $x \in X$ and let $\{a_j\}_{j=0}^{\infty}$ be a sequence of positive numbers satisfying $\lim_{j\to\infty} a_j = 0$. Let m_0, m_1, m_2 be integers, $0 \le m_0 \le m_1 \le m_2$ and let $\delta > 0$ satisfy $\sup\{a_j : j \ge m_1 + 1\} < \delta/3$. Suppose that $\|T^j x\| > a_j \quad (j = m_0+1, \dots, m_1)$. Then there exists $y \in X$ such that $\|y - x\| \le \delta$ and $\|T^j y\| > a_j \quad (j = m_0 + 1, \dots, m_2)$.*

Proof. Let $\lambda \in \sigma_e(T)$ satisfy $|\lambda| = 1$. Then $\lambda \in \partial\sigma_e(T) \subset \sigma_{\pi e}(T)$, and so

$$\inf\{\|(T - \lambda)u\| : u \in M, \|u\| = 1\} = 0$$

for each subspace $M \subset X$ of finite codimension.

Let $E = \bigvee\{T^j x : j = m_0 + 1, \dots, m_1\}$. Choose $\varepsilon > 0$ such that

$$\|T^j x\|(1 - \varepsilon) - m_1\delta\varepsilon\|T\|^{j-1} > a_j \quad (j = m_0 + 1, \dots, m_1)$$

and $\frac{\varepsilon}{2} + j\|T\|^{j-1}\varepsilon < \frac{1}{6} \quad (j = m_1 + 1, \dots, m_2)$. Let Y be the subspace constructed in Lemma 6, i.e., $\operatorname{codim} Y < \infty$ and $\|e+u\| \ge (1-\varepsilon)\max\{\|e\|, \|u\|/2\}$ for all $e \in E$ and $u \in Y$. Find $z \in Y$ such that $\|z\| = 1$ and $\|(T - \lambda)z\| < \varepsilon$. Set $y = x + \delta z$. Clearly, $\|y - x\| = \delta$ and

$$\|T^j z - \lambda^j z\| = \big\|(T^{j-1} + \lambda T^{j-2} + \cdots + \lambda^{j-1})(T - \lambda)z\big\| \le j\|T\|^{j-1}\varepsilon.$$

For $j = m_0 + 1, \dots, m_1$ we have

$$\begin{aligned}\|T^j y\| &= \|T^j x + \delta T^j z\| \ge \|T^j x + \delta\lambda^j z\| - \|\delta(T^j - \lambda^j)z\| \\ &\ge (1 - \varepsilon)\|T^j x\| - \delta j\|T\|^{j-1}\varepsilon > a_j.\end{aligned}$$

Similarly, for $j = m_1 + 1, \dots, m_2$ we have

$$\begin{aligned}\|T^j y\| &= \|T^j x + \delta T^j z\| \ge \|T^j x + \delta\lambda^j z\| - \|\delta(T^j - \lambda^j)z\| \\ &\ge \frac{1}{2}(1 - \varepsilon)\delta - \delta j\|T\|^{j-1}\varepsilon \ge \frac{\delta}{2} - \frac{\delta}{6} = \frac{\delta}{3} > a_j.\end{aligned}$$

□

Theorem 8. *Let $T \in \mathcal{B}(X)$, let $(a_j)_{j=0}^{\infty}$ be a sequence of positive numbers satisfying $\lim_{j\to\infty} a_j = 0$. Then:*

(i) *for each $\varepsilon > 0$ there exists $x \in X$ such that $\|x\| \le \sup\{a_j : j = 0, 1, \dots\} + \varepsilon$ and $\|T^j x\| \ge a_j r(T^j)$ for all $j \ge 0$;*

(ii) *there is a dense subset L of X with the following property: for each $y \in L$ we have $\|T^j y\| \ge a_j r(T^j)$ for all n sufficiently large.*

Proof. We distinguish two cases:

(a) Suppose first that $r(T) > r_e(T)$.

Choose $\lambda \in \sigma(T)$ with $|\lambda| = r(T)$. Then λ is an isolated eigenvalue of T. Let x be a corresponding eigenvector, $\|x\| = s$ where $s = \sup\{a_j : j = 0, 1, \dots\}$. Then $\|T^j x\| = s \cdot r(T)^j \ge a_j r(T^j)$ for all j.

To prove (ii), let F be the spectral subspace corresponding to λ, see Corollary 1.38, and let P be the corresponding spectral projection onto F. Then $\dim F < \infty$ and $(T-\lambda)|F$ is a nilpotent operator. Set $L = \{y \in X : Py \neq 0\}$. Clearly, L is a dense subset of X. Let $y \in L$. Write $z = Py$. Then $PT^j y = T^j z$, and so

$$\|T^j y\| \geq \|P\|^{-1} \|T^j z\| \tag{1}$$

for all j. Let k be the integer such that $(T-\lambda)^k z = 0$ and $(T-\lambda)^{k-1} z \neq 0$. Let $Q \in \mathcal{B}(F)$ be a projection such that $Qz = z$ and $Q\operatorname{Ker}(T-\lambda)^{k-1} = \{0\}$. Then $Q(T-\lambda)T^{j-1}z = 0$, and so $QT^j z = \lambda Q T^{j-1} z$ for all $j \geq 1$. Thus, by induction, $QT^j z = \lambda^j z$ and

$$\|T^j z\| \geq \|Q\|^{-1} r(T^j) \|z\| \quad (j = 0, 1, 2, \dots).$$

Using (1), this gives

$$\|T^j y\| \geq \frac{r(T^j)\|Py\|}{\|Q\| \cdot \|P\|} \qquad (j = 0, 1, 2, \dots).$$

Hence $\|T^j y\| \geq a_j r^j$ for all j sufficiently large.

(b) Let $r(T) = r_e(T)$. Replacing a_j by $\sup\{a_i : i \geq j\}$, we can assume that $a_0 \geq a_1 \geq a_2 \geq \cdots$. Also, we can assume that $r(T) = 1$.

(i) For $i = 0, 1, \dots$ let n_i be the smallest index such that $a_{n_i} < \frac{\varepsilon}{3 \cdot 2^{i+2}}$.

Find $\lambda \in \sigma_e(T)$ with $|\lambda| = 1$. Then $\lambda \in \partial\sigma_e(T) \subset \sigma_{\pi e}(T)$, so there is an approximate eigenvector $x_0 \in X$ satisfying $\|x_0\| = a_0 + \varepsilon/2$ and $\|(T^j - \lambda^j)x_0\| < \varepsilon/2 \quad (j = 0, 1, \dots, n_0)$. For $j = 0, 1, \dots, n_0$ we have

$$\|T^j x_0\| \geq \|\lambda^j x_0\| - \|(T^j - \lambda^j)x_0\| > a_0 + \varepsilon/2 - \varepsilon/2 = a_0 \geq a_j.$$

Using the previous lemma repeatedly, we construct a sequence (x_k) of vectors in X such that $\|x_{k+1} - x_k\| \leq \frac{\varepsilon}{2^{k+2}}$ and $\|T^j x_k\| > a_j \quad (j = 0, 1, \dots, n_k)$. Denote by x the limit of the Cauchy sequence (x_k). Then $\|T^j x\| = \lim_{k\to\infty} \|T^j x_k\| \geq a_j$ for all $j \geq 0$ and $\|x\| \leq \|x_0\| + \|x_1 - x_0\| + \cdots \leq a_0 + \frac{\varepsilon}{2} + \frac{\varepsilon}{4} + \cdots = a_0 + \varepsilon$.

(ii) Let $x \in X$ and $\delta > 0$. For $i = 0, 1, \dots$ let n_i be the smallest index such that $a_{n_i} < \frac{\delta}{3 \cdot 2^{i+1}}$. Set $y_0 = x$. Using Lemma 7 repeatedly we construct a sequence (y_k) of vectors in X such that $\|y_{k+1} - y_k\| \leq \frac{\varepsilon}{2^{k+1}}$ and $\|T^j y_k\| > a_j$ for $j = n_0 + 1, \dots, n_k$. Let $y = \lim_{k\to\infty} y_k$. Then $\|y - x\| \leq \frac{\varepsilon}{2} + \frac{\varepsilon}{4} + \cdots = \varepsilon$ and $\|T^j y\| \geq a_j$ for all $j \geq n_0 + 1$. This completes the proof. □

Taking $a_n = n^{-1}$ in the previous theorem yields the following corollary:

Corollary 9. *The set* $\{x \in X : \limsup_{n\to\infty} \|T^n x\|^{1/n} = r(T)\}$ *is residual for each* $T \in \mathcal{B}(X)$. *The set* $\{x \in X : \liminf_{n\to\infty} \|T^n x\|^{1/n} = r(T)\}$ *is dense.*

In particular, there is a dense subset of points $x \in X$ *with the property that the limit* $\lim_{n\to\infty} \|T^n x\|^{1/n}$ *exists and is equal to* $r(T)$.

Corollary 10. *Let $T \in \mathcal{B}(X)$. Then*

$$\sup_{\substack{x\in X\\ \|x\|=1}} \inf_{n\geq 1} \|T^n x\|^{1/n} = \inf_{n\geq 1} \sup_{\substack{x\in X\\ \|x\|=1}} \|T^n x\|^{1/n} = r(T).$$

Proof. The statement is clear if $r(T) = 0$. Let $r(T) > 0$ and let ε be a positive number, $\varepsilon < r(T)$. By Theorem 8, there is an $x \in X$ of norm 1 such that $\|T^n x\| \geq (r(T) - \varepsilon)^n$ for all n. Thus $\inf_n \|T^n x\|^{1/n} \geq r(T) - \varepsilon$. Letting $\varepsilon \to 0$ gives $\sup_{\|x\|=1} \inf_{n\geq 1} \|T^n x\|^{1/n} \geq r(T)$.

The second inequality is clear. □

In general it is not possible to replace the word "dense" in Corollary 9 by "residual".

Example 11. Let H be a separable Hilbert space with an orthonormal basis $\{e_j : j = 0, 1, \dots\}$ and let S be the backward shift, $Se_j = e_{j-1}$ $(j \geq 1)$, $Se_0 = 0$. Then $r(S) = 1$ and the set $\{x \in H : \liminf_{n\to\infty} \|S^n x\|^{1/n} = 0\}$ is residual.

In particular, the set $\{x \in H :$ the limit $\lim_{n\to\infty} \|S^n x\|^{1/n}$ exists$\}$ is of the first category (but it is always dense by Corollary 9).

Proof. For $k \in \mathbb{N}$ let

$$M_k = \{x \in X : \text{ there exists } n \geq k \text{ such that } \|S^n x\| < k^{-n}\}.$$

Clearly, M_k is an open subset of X. Further, M_k is dense in X. To see this, let $x \in X$ and $\varepsilon > 0$. Let $x = \sum_{j=0}^{\infty} \alpha_j e_j$ and choose $n \geq k$ such that $\sum_{j=n}^{\infty} |\alpha_j|^2 < \varepsilon^2$. Set $y = \sum_{j=0}^{n-1} \alpha_j e_j$. Then $\|y - x\| < \varepsilon$ and $S^n y = 0$. Thus $y \in M_k$ and M_k is a dense open subset of X.

By the Baire category theorem, the intersection $M = \bigcap_{k=0}^{\infty} M_k$ is a dense G_δ-subset of X, hence it is residual.

Let $x \in M$. For each $k \in \mathbb{N}$ there is an $n_k \geq k$ such that $\|S^{n_k} x\| < k^{-n_k}$, and so $\liminf_{n\to\infty} \|S^n x\|^{1/n} = 0$.

Since the set $\{x \in H : \limsup_{n\to\infty} \|S^n x\|^{1/n} = r(S) = 1\}$ is also residual, we see that the set $\{x \in H :$ the limit $\lim_{n\to\infty} \|S^n x\|^{1/n}$ exists$\}$ is of the first category.

Remark 12. If $r(T) = 1$ and $a_n > 0$, $a_n \to 0$, then Theorem 8 says that there exists x such that $\|T^n x\| \geq a_n$ for all n. This is the best possible result since the previous example $S \in \mathcal{B}(H)$ satisfies $S^n x \to 0$ for all $x \in H$. By Theorem 8, there are orbits converging to 0 arbitrarily slowly.

Theorem 8 implies that there is always a dense subset of points x satisfying $\sum_j \frac{\|T^j x\|}{r(T^j)} = \infty$. In fact, the set of all points with this property is even residual.

Theorem 13. *Let $T \in \mathcal{B}(X)$, $r(T) \neq 0$ and let $0 < p < \infty$. Then the set*

$$\left\{x \in X : \sum_{j=0}^{\infty}\left(\frac{\|T^j x\|}{r(T^j)}\right)^p = \infty\right\}$$

is residual.

Proof. For $k \geq 1$ set

$$M_k = \left\{x \in X : \sum_{j=0}^{\infty}\left(\frac{\|T^j x\|}{r(T^j)}\right)^p > k\right\}.$$

Clearly, M_k is an open subset of X. It is sufficient to show that M_k is dense. Indeed, by the Baire category theorem, the intersection

$$\bigcap_k M_k = \left\{x \in X : \sum_j\left(\frac{\|T^j x\|}{r(T^j)}\right)^p = \infty\right\}$$

is a dense G_δ set.

Fix $x \in X$, $\varepsilon > 0$ and $k \in \mathbb{N}$. We show that there is a $u \in X$ such that $\|u - x\| \leq \varepsilon$ and $u \in M_k$.

By Theorem 8, there is a $v \in X$ of norm 1 such that $\|T^j v\| \geq \frac{1}{(j+2)^{1/p}} r(T^j)$ for all $j \geq 0$. We have

$$\begin{aligned}
&\|T^j(x+\varepsilon v)\|^p + \|T^j(x-\varepsilon v)\|^p \geq \max\{\|T^j(x+\varepsilon v)\|, \|T^j(x-\varepsilon v)\|\}^p \\
&\geq \left(\frac{\|T^j(x+\varepsilon v)\| + \|T^j(x-\varepsilon v)\|}{2}\right)^p \geq \left(\frac{\|T^j(2\varepsilon v)\|}{2}\right)^p \\
&= \varepsilon^p \|T^j v\|^p \geq \frac{\varepsilon^p r(T^j)^p}{j+2}
\end{aligned}$$

and

$$\sum_{j=0}^{\infty}\left(\frac{\|T^j(x+\varepsilon v)\|}{r(T^j)}\right)^p + \sum_{j=0}^{\infty}\left(\frac{\|T^j(x-\varepsilon v)\|}{r(T^j)}\right)^p \geq \sum_{j=0}^{\infty}\frac{\varepsilon^p}{j+2} = \infty.$$

Thus either $y = x + \varepsilon v$ or $y = x - \varepsilon v$ satisfies $\|y - x\| = \varepsilon$ and $\sum_{j=0}^{\infty}\left(\frac{\|T^j y\|}{r(T^j)}\right)^p = \infty$, so $y \in M_k$. Hence M_k is dense and $\bigcap_k M_k$ is residual. □

In Theorem 5 we proved an estimate of $\|T^n x\|$ in terms of the norm $\|T^n\|$. It is also possible to construct points $x \in X$ with $\|T^n x\| \geq a_n \cdot \|T^n\|$ for all n; in this case, however, there is a restriction on the sequence (a_n).

Theorem 14. *Let X, Y be Banach spaces, let $(T_n) \subset \mathcal{B}(X,Y)$ be a sequence of operators. Let (a_n) be a sequence of positive numbers such that $\sum_{n=1}^{\infty} a_n < \infty$. Then there exists $x \in X$ such that $\|T_n x\| \geq a_n \|T_n\|$ for all n.*

Moreover, it is possible to choose such an x in each ball in X of radius greater than $\sum_{n=1}^{\infty} a_n$.

Proof. Let $u \in X$ and $\varepsilon > 0$. We find $x \in X$ such that $\|x-u\| \le \sum_{n=1}^\infty a_n + \varepsilon$ and $\|T_n x\| \ge a_n \|T_n\|$ for all n.

Without loss of generality we can assume that $T_n \neq 0$ for all n.

Let $s = \sum_{n=1}^\infty a_n$ and $a_n' = \frac{a_n}{s+\varepsilon/2}$. Then $\sum_{n=1}^\infty a_n' < 1$. For each n find $y_n^* \in Y^*$ such that $\|y_n^*\| = 1$ and $\|T_n^* y_n^*\| > \|T_n^*\| \cdot \frac{s+\varepsilon/2}{s+\varepsilon} = \|T_n\| \cdot \frac{s+\varepsilon/2}{s+\varepsilon}$. Let $x_n^* = \frac{T_n^* y_n^*}{\|T_n^* y_n^*\|}$. Then $\|x_n^*\| = 1$ for all n.

By Theorem A.5.1, there exists $x' \in X$ such that $\|x' - u/(s+\varepsilon)\| \le 1$ and $|\langle x', x_n^*\rangle| \ge a_n'$ for all n. Let $x = (s+\varepsilon)x'$. Then $\|x-u\| = (s+\varepsilon)\|x' - u/(s+\varepsilon)\| \le s+\varepsilon$. Furthermore,

$$\begin{aligned}\|T_n x\| &\ge |\langle T_n x, y_n^*\rangle| = (s+\varepsilon)|\langle T_n x', y_n^*\rangle| = (s+\varepsilon)|\langle x', T_n^* y_n^*\rangle| \\ &= (s+\varepsilon)|\langle x', x_n^*\rangle| \cdot \|T_n^* y_n^*\| \ge (s+\varepsilon) a_n' \frac{s+\varepsilon/2}{s+\varepsilon}\|T_n\| = a_n\|T_n\|\end{aligned}$$

for all n. □

Lemma 15. *Let $a_n > 0 \quad (n = 1, 2, \dots)$, $\varepsilon > 0$ and let $\sum_{n=1}^\infty a_n < \varepsilon$. Then there exist positive numbers $\beta_n \quad (n \in \mathbb{N})$ such that $\beta_n \to \infty$ and $\sum_{n=1}^\infty \beta_n a_n < \varepsilon$.*

Proof. Let $\delta = \varepsilon - \sum_{n=1}^\infty a_n$. For each $k \in \mathbb{N}$ let n_k be the smallest number such that $\sum_{i=n_k+1}^\infty a_i < 2^{-2k}\delta$.

Let $\beta_i = 1$ for $1 \le i \le n_1$ and $\beta_i = 2^k \quad (n_k < i \le n_{k+1})$. Then $\beta_i \to \infty$ and

$$\begin{aligned}\sum_{i=1}^\infty \beta_i a_i &= \sum_{i=1}^{n_1} a_i + \sum_{k=1}^\infty 2^k \sum_{i=n_k+1}^{n_{k+1}} a_i \le \sum_{i=1}^\infty a_i + \sum_{k=1}^\infty \sum_{i=n_k+1}^\infty 2^k a_i \\ &< \varepsilon - \delta + \sum_{k=1}^\infty 2^{-k}\delta = \varepsilon.\end{aligned}$$

□

Corollary 16. *Let $T \in \mathcal{B}(X)$ satisfy $\sum_{n=1}^\infty \|T^n\|^{-1} < \infty$. Then there exists a dense subset of points $x \in X$ such that $\|T^n x\| \to \infty$.*

Proof. Let $u \in X$ and $\varepsilon > 0$. Find n_0 such that $\sum_{n=n_0}^\infty \|T^n\|^{-1} < \varepsilon$.

Find positive numbers β_n such that $\beta_n \to \infty$ and $\sum_{n=n_0}^\infty \frac{\beta_n}{\|T^n\|} < \varepsilon$. Let $a_n = \frac{\beta_n}{\|T^n\|}$. Then $\sum_{n=n_0}^\infty a_n < \varepsilon$. By Theorem 14, there exists $x \in X$ such that $\|x-u\| < \varepsilon$ and $\|T^n x\| \ge a_n \|T^n\| = \beta_n$ for all $n \ge n_0$. Hence $\|T^n x\| \to \infty$. □

Better results can be obtained for Hilbert space operators.

Theorem 17. *Let H, K be Hilbert spaces and let $T_n \in \mathcal{B}(H, K)$ be a sequence of operators. Let a_n be a sequence of positive numbers such that $\sum_{n=1}^\infty a_n^2 < \infty$. Let $\varepsilon > 0$. Then:*

(i) *there exists* $x \in H$ *such that* $\|x\| \le \left(\sum_{n=1}^\infty a_n^2\right)^{1/2} + \varepsilon$ *and* $\|T_n x\| \ge a_n \|T_n\|$ *for all* n;

(ii) *there is a dense subset of vectors* $x \in H$ *such that* $\|T_n x\| \ge a_n \|T_n\|$ *for all* n *sufficiently large.*

Proof. Without loss of generality we may assume that all operators T_n are non-zero.

(i) Let $s = \left(\sum_{n=1}^\infty a_n^2\right)^{1/2}$. For each n we find $y_n \in K$ with $\|y_n\| = 1$ and $\|T_n^* y_n\| > \frac{s+\varepsilon/2}{s+\varepsilon}\|T_n\|$. By Theorem A.5.2, there is a $u \in H$ with $\|u\| = 1$ and $\left|\left\langle u, \frac{T_n^* y_n}{\|T_n^* y_n\|}\right\rangle\right| \ge \frac{a_n}{s+\varepsilon/2}$. Let $x = (s+\varepsilon)u$. Then $\|x\| = s+\varepsilon$ and

$$\begin{aligned}\|T_n x\| &= (s+\varepsilon)\|T_n u\| \ge (s+\varepsilon)|\langle T_n u, y_n\rangle| = (s+\varepsilon)|\langle u, T_n^* y_n\rangle| \\ &\ge a_n \frac{s+\varepsilon}{s+\varepsilon/2}\|T_n^* y_n\| \ge a_n \|T^n\|\end{aligned}$$

for all $n \in \mathbb{N}$.

(ii) Let $y \in H$, $y \neq 0$ and $\varepsilon > 0$. We show that there is an $x \in H$ such that $\|x - y\| < \varepsilon$ and $\|T_n x\| \ge a_n \|T_n\|$ for all n sufficiently large.

Without loss of generality we may assume that $\|y\| = 1$. Indeed, for general non-zero $y \in H$ replace y by $\frac{y}{\|y\|}$ and the numbers a_n by $\frac{a_n}{\|y\|}$.

Let $\|y\| = 1$, $0 < \varepsilon < 1$ and set $\delta = 1 - \varepsilon^2/2$. Find n_0 such that

$$\sum_{n=n_0}^\infty (2a_n)^2 < 1 - \delta^2.$$

For each $n \ge n_0$ find $y_n \in K$ such that $\|y_n\| = 1$ and $\|T_n^* y_n\| \ge \frac{1}{2}\|T_n\|$. By Theorem A.5.2, there is a $u \in H$ such that $\|u\| = 1$, $|\langle u, y\rangle| \ge \delta$ and

$$\left|\left\langle u, \frac{T_n^* y_n}{\|T_n^* y_n\|}\right\rangle\right| \ge 2a_n \qquad (n \ge n_0).$$

Set $x = \frac{\langle y,u\rangle}{|\langle y,u\rangle|} \cdot u$. Then $\|x\| = 1$ and $\langle x, y\rangle = |\langle u, y\rangle| \ge \delta$. Therefore $\|x - y\|^2 = \|x\|^2 + \|y\|^2 - 2\,\mathrm{Re}\langle x, y\rangle \le 2 - 2\delta = \varepsilon^2$, and so $\|x - y\| \le \varepsilon$. Finally, for $n \ge n_0$ we have

$$\|T_n x\| = \|T_n u\| \ge |\langle T_n u, y_n\rangle| = |\langle u, T_n^* y_n\rangle| \ge 2a_n \|T_n^* y_n\| \ge a_n \|T_n\|. \qquad \square$$

Corollary 18. *Let* $T \in \mathcal{B}(H)$ *be a Hilbert space operator and* $\sum_{n=1}^\infty \|T^n\|^{-2} < \infty$. *Then there exists a dense subset of points* $x \in H$ *such that* $\|T^n x\| \to \infty$.

Proof. By Lemma 15, there are positive numbers β_n such that $\beta_n \to \infty$ and $\sum_{n=1}^{\infty} \frac{\beta_n}{\|T^n\|^2} < \infty$. By Theorem 17, there is a dense subset of vectors $x \in H$ such that

$$\|T^n x\| \geq \frac{\beta_n^{1/2}}{\|T^n\|} \cdot \|T^n\| = \beta_n^{1/2}$$

for all n sufficiently large. Hence $\|T^n x\| \to \infty$. □

Corollary 19. *Let $T \in \mathcal{B}(X)$ satisfy $\sum_{n=1}^{\infty} \|T^n\|^{-1} < \infty$. Then T has a non-trivial closed invariant subset.*

If X is a Hilbert space, then it is sufficient to assume that $\sum_{n=1}^{\infty} \|T^n\|^{-2} < \infty$.

Proof. By Corollaries 16 and 18, there exists $x \in X$ such that $\|T^n x\| \to \infty$. Hence the set $\{T^x : n = 0, 1, \dots\}$ is a non-trivial closed subset invariant for T. □

Corollary 20. *Let $T \in \mathcal{B}(X)$ satisfy $r(T) \neq 1$. Then T has a non-trivial closed invariant subset.*

Proof. If $r(T) > 1$, then there exists an $x \in X$ with $\|T^n x\| \to \infty$. If $r(T) < 1$, then $\|T^n x\| \to 0$ for each $x \in X$. In both cases there are non-trivial closed invariant subsets. □

The previous results are in some sense the best possible.

Example 21. There exists a Banach space X and an operator $T \in \mathcal{B}(X)$ such that $\|T^n\| = n + 1$ for all n, but there is no vector $x \in X$ with $\|T^n x\| \to \infty$.

There exists a Hilbert space operator T such that $\|T^n\| = \sqrt{n+1}$ for all n and there in no vector x with $\|T^n x\| \to \infty$.

Proof. We construct both operators simultaneously. Let either $p = 1$ or $p = 2$. Let $(e_k)_{k=1}^{\infty}$ be the standard basis in the space $X = \ell^p$. Let $T \in \mathcal{B}(X)$ be the weighted backward shift defined by

$$Te_k = \begin{cases} \left(\frac{k}{k-1}\right)^{1/p} e_{k-1} & \text{for } k > 1, \\ 0 & \text{for } k = 1. \end{cases}$$

Hence

$$\|T^n\| = \prod_{k=2}^{n+1} \left(\frac{k}{k-1}\right)^{1/p} = (n+1)^{1/p}$$

for all n. Suppose on the contrary that there is an $x = \sum_{k=1}^{\infty} c_k e_k \in \ell^p$ such that $\|x\| = \left(\sum_{k=1}^{\infty} |c_k|^p\right)^{1/p} = 1$ and $\|T^n x\| \to \infty$ as $n \to \infty$. Consequently,

$$\frac{1}{n} \sum_{j=n}^{2n-1} \|T^j x\|^p \to \infty \qquad \text{as } n \to \infty.$$

Let us estimate the above arithmetic mean. We have

$$\|T^j x\|^p = \left\| \sum_{k=j+1}^{\infty} \left(\frac{k}{k-j}\right)^{1/p} c_k e_{k-j} \right\|^p$$
$$= \sum_{k=j+1}^{2j} |c_k|^p \frac{k}{k-j} + \sum_{k=2j+1}^{\infty} |c_k|^p \frac{k}{k-j},$$

where the second sum can be estimated by $2\|x\|^p \le 2$ since $\frac{k}{k-j} < 2$ for $k > 2j$. We have

$$\sum_{j=n}^{2n-1} \|T^j x\|^p \le 2n + \sum_{j=n}^{2n-1} \sum_{k=j+1}^{2j} |c_k|^p \frac{k}{k-j}$$
$$\le 2n + \sum_{k=n+1}^{4n} |c_k|^p \sum_{i=1}^{k} \frac{k}{i} \le 2n + \sum_{k=n+1}^{4n} |c_k|^p 4n(1+\log 4n),$$

and so

$$2 + 4(1+\log 4n) \sum_{k=n+1}^{4n} |c_k|^p \ge \frac{1}{n} \sum_{j=n}^{2n-1} \|T^j x\|^p \to \infty.$$

Hence, for all n large enough, the left-hand side is greater than 6, i.e., if we write $s_n = \sum_{k=n+1}^{4n} |c_k|^p$, then

$$s_n \ge \frac{1}{1+\log 4n}.$$

But this is a contradiction since for such an n we have

$$1 = \sum_{k=1}^{\infty} |c_k|^p \ge s_n + s_{4n} + s_{4^2 n} + s_{4^3 \cdot n} + \dots$$
$$\ge \sum_{j=1}^{\infty} \frac{1}{1+\log 4^j n} = \sum_{j=1}^{\infty} \frac{1}{1+\log n + j\log 4} = \infty.$$

□

It is also possible to consider orbits that are large in the sense of $\sum \frac{\|T^n x\|}{\|T^n\|}$.

Theorem 22. *Let X, Y be Banach spaces, let $(T_j) \subset \mathcal{B}(X,Y)$ be a sequence of non-zero operators and let $0 < p < 1$. Then the set*

$$\left\{ x \in X : \sum_{j=1}^{\infty} \left(\frac{\|T_j x\|}{\|T_j\|} \right)^p = \infty \right\}$$

is residual in X.

Proof. For $k \geq 1$ set

$$M_k = \left\{ x \in X : \sum_{j=1}^{\infty} \left(\frac{\|T_j x\|}{\|T_j\|} \right)^p > k \right\}.$$

Clearly, M_k is an open subset of X. It is sufficient to show that M_k is dense. Indeed, the Baire theorem then implies that the intersection $\bigcap_k M_k = \Big\{ x \in X : \sum_j \left(\frac{\|T_j x\|}{\|T_j\|} \right)^p = \infty \Big\}$ is a dense G_δ set.

Fix $x \in X$, $\delta > 0$ and $k \in \mathbb{N}$. We show that there is a $u \in X$ such that $\|u - x\| \leq \delta$ and $u \in M_k$.

Let $\varepsilon = \frac{1}{p} - 1$. Then $\varepsilon > 0$ and $s = \sum_{j=1}^{\infty} \frac{1}{j^{1+\varepsilon}} < \infty$. For $i = 1, 2, \dots$ set $\varepsilon_i = \frac{\delta}{s i^{1+\varepsilon}}$. So $\sum_{j=1}^{\infty} \varepsilon_i = \delta$ and $\sum_{i=1}^{\infty} \varepsilon_i^p = \left(\frac{\delta}{s}\right)^p \sum_{i=1}^{\infty} \frac{1}{i} = \infty$. Fix n such that $\frac{3}{8^{p+1}} \sum_{i=1}^{n} \varepsilon_i^p > k$.

For $i = 1, \dots, n$ find $u_i \in X$ of norm 1 such that $\|T_i u_i\| \geq \|T_i\|/2$. Set

$$\Lambda = \big\{ \lambda = (\lambda_1, \dots, \lambda_n) \in \mathbb{C}^n : |\lambda_i| \leq \varepsilon_i \quad (i = 1, \dots, n) \big\}.$$

For $\lambda \in \Lambda$ set $x_\lambda = x + \sum_{i=1}^{k} \lambda_i u_i$. Clearly, $\|x_\lambda - x\| \leq \sum_{i=1}^{n} \varepsilon_i \leq \delta$ for each $\lambda \in \Lambda$.

For $i = 1, \dots, n$ let $\Lambda_i = \Big\{ \lambda \in \Lambda : \left(\frac{\|T_i x_\lambda\|}{\|T_i\|} \right)^p < \frac{1}{2} \left(\frac{\varepsilon_i}{8} \right)^p \Big\}$.

Let $1 \leq i \leq n$ and suppose that $\lambda, \lambda' \in \Lambda_i$ where $\lambda = (\lambda_1, \dots, \lambda_n)$ and $\lambda' = (\lambda_1, \dots, \lambda_{i-1}, \lambda_i', \lambda_{i+1}, \dots, \lambda_n)$. Then

$$\begin{aligned}
\left(\frac{\varepsilon_i}{8} \right)^p &\geq \left(\frac{\|T_i x_\lambda\|}{\|T_i\|} \right)^p + \left(\frac{\|T_i x_{\lambda'}\|}{\|T_i\|} \right)^p \geq \max \left\{ \frac{\|T_i x_\lambda\|}{\|T_i\|}, \frac{\|T_i x_{\lambda'}\|}{\|T_i\|} \right\}^p \\
&\geq \left(\frac{\|T_i x_\lambda\| + \|T_i x_{\lambda'}\|}{2\|T_i\|} \right)^p \geq \left(\frac{\|T_i (x_\lambda - x_{\lambda'})\|}{2\|T_i\|} \right)^p \\
&= |\lambda - \lambda'|^p \left(\frac{\|T_i u_i\|}{2\|T_i\|} \right)^p \geq \frac{|\lambda - \lambda'|^p}{4^p}.
\end{aligned}$$

Thus $|\lambda - \lambda'| \leq \frac{\varepsilon_i}{2}$.

Consequently, for fixed $\lambda_1, \dots, \lambda_{i-1}, \lambda_{i+1}, \dots, \lambda_n$, the set

$$\big\{ \nu \in \mathbb{C} : (\lambda_1, \dots, \lambda_{i-1}, \nu, \lambda_{i+1}, \dots, \lambda_n) \in \Lambda_i \big\}$$

is contained in a ball of radius $\frac{\varepsilon_i}{2}$.

Let m be the Lebesgue measure in $\mathbb{C}^n$. Then $m(\Lambda) = \prod_{i=1}^{n} (\pi \varepsilon_i^2)$ and, by the Fubini theorem,

$$m(\Lambda_i) \leq \pi \frac{\varepsilon_i^2}{4} \prod_{\substack{1 \leq j \leq n \\ j \neq i}} (\pi \varepsilon_j^2) = \frac{m(\Lambda)}{4}.$$

Set $f(\lambda)=\sum_{i=1}^n\left(\frac{\|T_ix_\lambda\|}{\|T_i\|}\right)^p$. Then

$$\frac{1}{m(\Lambda)}\int_\Lambda f(\lambda)\,\mathrm{d}m=\frac{1}{m(\Lambda)}\sum_{i=1}^n\int\left(\frac{\|T_ix_\lambda\|}{\|T_i\|}\right)^p\mathrm{d}m\geq\sum_{i=1}^n\frac{\varepsilon_i^p}{2\cdot 8^p}\frac{3}{4}>k.$$

Thus there exists $\lambda\in\Lambda$ such that $f(\lambda)>k$, and so $x_\lambda\in M_k$. Hence M_k is dense and the proof is complete. □

Corollary 23. *Let $T\in\mathcal{B}(X)$ be a non-nilpotent operator and $0<p<1$. Then the set*

$$\left\{x\in X:\sum_{j=0}^\infty\left(\frac{\|T^jx\|}{\|T^j\|}\right)^p=\infty\right\}$$

is residual.

The previous result is not true for $p=1$.

Example 24. There are a Banach space X and a non-nilpotent operator $T\in\mathcal{B}(X)$ such that $\sum_{n=0}^\infty\frac{\|T^nx\|}{\|T^n\|}<\infty$ for all $x\in X$.

Proof. Let X be the ℓ^1 space with the standard basis $\{e_0,e_1,\dots\}$. Let $T\in\mathcal{B}(X)$ be the weighted backward shift defined by $Te_0=0$ and $Te_k=(\frac{k+1}{k})^2e_{k-1}\quad(k\geq 1)$. For $n\in\mathbb{N}$ we have

$$T^ne_k=\begin{cases}0 & (n>k),\\ \frac{(k+1)^2}{(k-n+1)^2}e_{k-n} & (n\leq k)\end{cases}$$

and $\|T^n\|=(n+1)^2$.

Let $x\in X$, $x=\sum_{k=0}^\infty\alpha_ke_k$ where $\sum_{k=0}^\infty|\alpha_k|<\infty$. Then

$$\sum_{n=0}^\infty\frac{\|T^nx\|}{\|T^n\|}=\sum_{n=0}^\infty\sum_{k=n}^\infty\frac{|\alpha_k|(k+1)^2}{(n+1)^2(k-n+1)^2}=\sum_{k=0}^\infty|\alpha_k|\sum_{n=0}^k\frac{(k+1)^2}{(n+1)^2(k-n+1)^2}.$$

We have

$$\begin{aligned}&\sum_{n=0}^k\frac{(k+1)^2}{(n+1)^2(k-n+1)^2}\\ &\quad=\sum_{n=0}^{[k/2]}\frac{(k+1)^2}{(n+1)^2(k-n+1)^2}+\sum_{n=[k/2]+1}^k\frac{(k+1)^2}{(n+1)^2(k-n+1)^2}\\ &\quad\leq\sum_{n=0}^{[k/2]}\frac{4}{(n+1)^2}+\sum_{n=[k/2]+1}^k\frac{4}{(k-n+1)^2}\leq 8\sum_{j=1}^\infty\frac{1}{j^2}=\frac{4\pi^2}{3}.\end{aligned}$$

Thus

$$\sum_{n=0}^\infty\frac{\|T^nx\|}{\|T^n\|}\leq\sum_{k=0}^\infty|\alpha_k|\cdot\frac{4\pi^2}{3}=\frac{4\|x\|\pi^2}{3}<\infty.$$

□

It is also possible to construct vectors x in a prescribed infinite-dimensional subspace with all powers $\|T^n x\|$ large.

Recall the quantity $j_\mu(T) = \sup\{j(T|M) : M \subset X, \operatorname{codim} M < \infty\}$, which was studied in Section 24.

Lemma 25. *Let $T_1, \dots, T_k \in \mathcal{B}(X, Y)$, let $X_1 \subset X$ be an infinite-dimensional subspace. Let $\varepsilon > 0$. Then there exists $x \in X_1$ of norm 1 such that $\|T_i x\| > j_\mu(T_i) - \varepsilon \quad (i = 1, \dots, k)$.*

Proof. For $i = 1, \dots, k$ there is a subspace $M_i \subset X$ of finite codimension such that $j(T_i|M_i) > j_\mu(T_i) - \varepsilon$. Let x be any vector of norm 1 in $X_1 \cap \bigcap_{i=1}^k M_i$. Then

$$\|T_i x\| \geq j(T|M_i) > j_\mu(T_i) - \varepsilon$$

for all $i = 1, \dots, k$. □

Theorem 26. *Let X, Y be Banach spaces, let $T_n \in \mathcal{B}(X, Y) \quad (n = 1, 2, \dots)$, let (a_n) be a sequence of positive numbers such that $\lim_{i \to \infty} a_i = 0$ and let $X_1 \subset X$ be a closed infinite-dimensional subspace. Let $\delta > 0$. Then there exists a vector $x \in X_1$ with $\|x\| \leq \sup_i a_i + \delta$ and $\|T_n x\| \geq a_n \cdot j_\mu(T_n)$ for all $n \in \mathbb{N}$.*

Moreover, there is a subset X_2 dense in X_1 with the property that for each $x \in X_2$, $\|T_n x\| \geq a_n j_\mu(T_n)$ for all n sufficiently large.

Proof. Without loss of generality we can assume that $a_1 \geq a_2 \geq \cdots$. Let $\varepsilon > 0$ satisfy $(1-\varepsilon)^2(a_1 + \frac{\delta}{2}) > a_1$. For each $k = 0, 1, \dots$ find r_k with $a_{r_k} < \frac{(1-\varepsilon)^3\delta}{2^{k+3}}$. Find $z_0 \in X_1$ such that $\|z_0\| = a_1 + \delta/2$ and $\|T_n z_0\| > (1-\varepsilon)(a_1 + \delta/2) j_\mu(T_n) \quad (n \leq r_0)$.

Let $k \geq 0$ and suppose that $z_0, \dots, z_k$ have already been constructed. Let $E_k = \bigvee\{T_n z_i : 0 \leq i \leq k, 1 \leq n \leq r_{k+1}\}$. Let M_k be a subspace of X of finite codimension such that

$$\|e + m\| \geq (1-\varepsilon)\max\{\|e\|, \|m\|/2\} \qquad (e \in E, m \in M).$$

Since the space $L_k = \bigcap_{i=1}^k \bigcap_{n=1}^{r_{k+1}} T_n^{-1} M_i < \infty$ is of finite codimension, we can choose $z_{k+1} \in X_1 \cap L_k$ such that $\|z_{k+1}\| = \delta 2^{-(k+2)}$ and

$$\|T_n z_{k+1}\| \geq (1-\varepsilon)\delta 2^{-(k+2)} j_\mu(T_n) \qquad (1 \leq n \leq r_{k+1}).$$

Set $z = \sum_{i=0}^\infty z_i$. Then $z \in X_1$ and

$$\|z\| \leq \sum_{i=0}^\infty \|z_i\| \leq a_1 + \delta/2 + \sum_{i=1}^\infty \delta 2^{-(i+1)} = a_1 + \delta.$$

For $n = 1, \dots, r_0$ we have

$$\|T_n z\| = \left\| T_n z_0 + \sum_{i=1}^\infty T_n z_i \right\| \geq (1-\varepsilon)\|T_n z_0\| > a_1 j_\mu(T_n) \geq a_n j_\mu(T_n).$$

Let $k \geq 0$ and $r_k < n \leq r_{k+1}$. Then

$$\|T_n z\| = \left\|\sum_{i=0}^{\infty} T_n z_i\right\| \geq (1-\varepsilon)\left\|\sum_{i=0}^{k+1} T_n z_i\right\|$$
$$\geq \frac{(1-\varepsilon)^2}{2}\|T_n z_{k+1}\| \geq \frac{(1-\varepsilon)^3}{2} \cdot \frac{\delta}{2^{k+2}} j_\mu(T_n) \geq a_n \cdot j_\mu(T_n).$$

Thus $\|T_n x\| \geq a_n j_\mu(T_n)$ for all $n \in \mathbb{N}$.

To show the second statement, let $u \in X_1$ and $\varepsilon > 0$. Find n_0 such that $a_n < \varepsilon$ for all $n \geq n_0$. As in the first part, taking $z_0 = u$, construct a vector $x \in X_1$ with $\|x - u\| \leq \varepsilon$ and $\|T_n x\| \geq a_n j_\mu(T_n) \quad (n \geq n_0)$. □

Corollary 27. *Let $T_n \in \mathcal{B}(X, Y)$ satisfy that $j_\mu(T_n) \to \infty$. Let $M \subset X$ be a closed infinite-dimensional subspace. Then there exists a dense subset of vectors $x \in M$ such that $\|T_n x\| \to \infty$.*

Proof. There exists a sequence (β_n) of positive numbers such that $\beta_n \to 0$ and $\beta_n j_\mu(T_n) \to \infty$. □

Corollary 28. *Let $T \in \mathcal{B}(X)$, let $\inf\{|\lambda| : \lambda \in \sigma_e(T)\} > 1$. Let $M \subset X$ be a closed infinite-dimensional subspace. Then there exists a dense subset of vectors $x \in M$ such that $\|T^n x\| \to \infty$.*

Proof. By Theorem 24.14, Corollary 17.10 and Theorem 9.25, $\lim j_\mu(T^n)^{1/n} = \lim_{n\to\infty} j(\tilde{T}^n)^{1/n} = \inf\{|\lambda| : \lambda \in \sigma_{\pi e}(T)\} > 1$. Thus $j_\mu(T^n) \to \infty$ and we can apply Corollary 27. □

38 Hypercyclic vectors

In the last section we studied vectors x with "very regular" orbits $(T^n x)$. In this section we will study the opposite extreme – vectors with very irregular orbits.

Definition 1. Let $T \in \mathcal{B}(X)$ be an operator. A vector $x \in X$ is called *hypercyclic* for T if the set $\{T^n x : n \in \mathbb{N}\}$ is dense in X. An operator T is called hypercyclic if there is a vector hypercyclic for T.

The notion has sense only in separable Banach spaces. Clearly, in non-separable Banach spaces there are no hypercyclic operators.

It is easy to find an operator that has no hypercyclic vectors. For example, if $\|T\| \leq 1$, then all orbits are bounded, and therefore not dense. On the other hand, if T is hypercyclic, then almost all vectors are hypercyclic for T.

Theorem 2. *Let $T \in \mathcal{B}(X)$ be a hypercyclic operator. Then the set of all vectors $x \in X$ that are hypercyclic for T is a dense G_δ set, and hence residual in X.*

Proof. Let $x \in X$ be a vector hypercyclic for T. For any $n \in \mathbb{N}$, the vector $T^n x$ is also hypercyclic for T and therefore the set of all vectors hypercyclic for T is dense in X.

Note that the space X is separable. Let (U_j) be a countable base of open sets in X. A vector $u \in X$ is hypercyclic for T if and only if it belongs to the set $\bigcap_{j=1}^{\infty} \left(\bigcup_{n=0}^{\infty} T^{-n} U_j \right)$, which is a G_δ set. □

Lemma 3. *Let $T \in \mathcal{B}(X)$ be a hypercyclic operator. Then $\sigma_p(T^*) = \emptyset$.*

Proof. Suppose on the contrary that $\lambda \in \mathbb{C}$ belongs to the point spectrum of T^*. Let $x^* \in X^*$ be a corresponding eigenvector, i.e., $x^* \neq 0$ and $T^* x^* = \lambda x^*$.

Let $x \in X$ be a vector hypercyclic for T. Then the set $\{\langle T^n x, x^* \rangle : n = 0, 1, \dots\}$ is dense in $\mathbb{C}$. We have

$$\begin{aligned} \{\langle T^n x, x^* \rangle : n = 0, 1, \dots\} &= \{\langle x, T^{*n} x^* \rangle : n = 0, 1, \dots\} \\ &= \{\lambda^n \langle x, x^* \rangle : n = 0, 1, \dots\}. \end{aligned}$$

The last set is bounded if either $|\lambda| \leq 1$ or $\langle x, x^* \rangle = 0$. If $|\lambda| > 1$ and $\langle x, x^* \rangle \neq 0$, then $|\lambda^n \langle x, x^* \rangle| \to \infty$. So the set $\{\lambda^n \langle x, x^* \rangle : n = 0, 1, \dots\}$ cannot be dense in $\mathbb{C}$, which is a contradiction. □

Corollary 4. *Let $\dim X < \infty$. Then there are no hypercyclic operators acting in X.*

Theorem 5. *Let $T \in \mathcal{B}(X)$ be a hypercyclic operator. Then there exists a dense linear manifold $M \subset X$ such that each non-zero vector $x \in M$ is hypercyclic for T.*

Proof. Let $x \in X$ be a vector hypercyclic for T. Let

$$M = \{q(T)x : q \text{ a polynomial}\}.$$

Clearly M is a dense linear manifold since it contains the orbit $\{T^n x : n = 0, 1, \dots\}$. We show that $q(T)x$ is hypercyclic for T for each non-zero polynomial q.

Write $q(z) = \beta(z - \alpha_1) \cdots (z - \alpha_n)$, where $n \geq 0$, $\alpha_1, \dots, \alpha_n$ are the roots of q and $\beta \neq 0$. Since $\sigma_p(T^*) = \emptyset$, the operators $T - \alpha_i$ have dense ranges. Hence $q(T)$ has also dense range. We have

$$\{T^n q(T)x : n = 0, 1, \dots\} = q(T)\{T^n x : n = 0, 1, \dots\},$$

which is dense in X, since $\{T^n x : n = 0, 1, \dots\}$ is dense in X. □

Hypercyclic vectors seem to be very strange and exceptional but in fact they are quite common. The next theorem provides a criterion for hypercyclicity of an operator.

Theorem 6. *Let X be a separable Banach space. Let $T \in \mathcal{B}(X)$. Suppose that there exists an increasing sequence of positive integers (n_k) such that the following two conditions are satisfied:*

(i) *there exists a dense subset* $X_0 \subset X$ *such that* $\lim_{k\to\infty} T^{n_k}x = 0 \quad (x \in X_0)$;

(ii) $\overline{\bigcup_k T^{n_k} B_X} = X$.

Then T *is hypercyclic.*

Proof. To simplify the notation we write for short $T_k = T^{n_k} \quad (k \in \mathbb{N})$. We first show that for all $c > 0$ and $k_0 \in \mathbb{N}$ we have $\overline{\bigcup_{k\geq k_0} T_k(cB_X)} = X$. Let $u \in X$, $u \neq 0$ and $\varepsilon > 0$, $\varepsilon < \min\{c, \|u\|\}$. Choose $s > \varepsilon^{-1}\|u\| + \max\{\|T_i\| : 1 \leq i < k_0\}$. Find $v \in B_X$ and $k \in \mathbb{N}$ such that $\left\|T_k v - \frac{su}{\|u\|}\right\| < 1$. For $k < k_0$ we have $\left\|T_k v - \frac{su}{\|u\|}\right\| \geq s - \|T_k v\| > \varepsilon^{-1}\|u\| > 1$. Thus $k \geq k_0$ and $\left\|T_k\left(\frac{v\|u\|}{s}\right) - u\right\| < \frac{\|u\|}{s} < \varepsilon$. Further, $\frac{v\|u\|}{s} \in cB_X$.

We now show that T is hypercyclic. Let $(x_i)_{i=1}^\infty$ be a sequence dense in X; without loss of generality we can assume that each member of the sequence appears in the sequence infinitely many times.

We construct vectors $u_k \in X_0$ and an increasing sequence $s_k \quad (k \in \mathbb{N})$ of positive integers. Set formally $s_0 = 0$.

Let $k \geq 1$ and suppose that the vectors $u_1, \dots, u_{k-1} \in X_0$ and numbers $s_0 < s_1 < \cdots < s_{k-1}$ have already been constructed. Since $u_1, \dots, u_{k-1} \in X_0$, there exists $m \in \mathbb{N}$ such that $\|T_j u_i\| < \frac{1}{2^{i+k}} \qquad (j \geq m, i = 1, \dots, k-1)$. Find $s_k > \max\{m, s_{k-1}\}$ and $u_k \in X_0$ such that

$$\|u_k\| < \frac{1}{2^k \max\{1, \|T_{s_1}\|, \dots, \|T_{s_{k-1}}\|\}}$$

and

$$\|T_{s_k} u_k - x_k\| < \frac{1}{2^k}.$$

Set $u = \sum_{i=1}^\infty u_i$. Clearly, the series is convergent. Further,

$$\begin{aligned}
\|T_{s_k} u - x_k\| &\leq \sum_{i=1}^{k-1} \|T_{s_k} u_i\| + \|T_{s_k} u_k - x_k\| + \sum_{i=k+1}^{\infty} \|T_{s_k} u_i\| \\
&< \sum_{i=1}^{k-1} \frac{1}{2^{i+k}} + \frac{1}{2^k} + \sum_{i=k+1}^{\infty} \frac{1}{2^i} \leq 3 \cdot 2^{-k}.
\end{aligned}$$

Since each x_k is contained in the sequence (x_i) infinitely many times, u is a hypercyclic vector. □

The last criterion is usually easy to apply. For example, it implies easily the hypercyclicity of the operator $T = 2S$, where S is the backward shift, see Example 37.4.

In the same way it is possible to obtain the hypercyclicity of any weighted backward shift with weights w_i which satisfy $\sup_n(w_1 \cdots w_n) = \infty$.

It was a longstanding open problem whether there are hypercyclic operators that do not satisfy the conditions of Theorem 6. The problem has several equivalent formulations. The simplest formulation is whether there exists a hypercyclic operator $T \in \mathcal{B}(X)$ such that $T \oplus T \in \mathcal{B}(X \oplus X)$ is not hypercyclic. The problem was solved recently by [DeR], see also [BaM], where such an operator was constructed in any space $\ell^p \quad (1 \le p < \infty)$ or c_0.

Theorem 7. *Let X be a separable Banach space and let $T \in \mathcal{B}(X)$. Then T is hypercyclic if and only if, for all non-empty open sets $U, V \subset X$, there exists $n \in \mathbb{N}$ such that $T^n U \cap V \ne \emptyset$.*

Proof. Suppose that T is hypercyclic and let U, V be non-empty open subsets of X. Since the set of all hypercyclic vectors is dense, there is an $x \in U$ hypercyclic for T. Therefore there is an $n \in \mathbb{N}$ such that $T^n x \in V$, and so $T^n U \cap V \ne \emptyset$.

Conversely, suppose that $T^n U \cap V \ne \emptyset$ for all non-empty open subsets U, V of X. Let (U_j) be a countable basis of open subsets of X. For each j let $M_j = \bigcup_{n\in\mathbb{N}} T^{-n} U_j$. Clearly, M_j is open. We show that it is also dense.

Let $y \in X$ and $\varepsilon > 0$. By assumption, there are $n \in \mathbb{N}$ and $x \in X$, $\|x-y\| < \varepsilon$ such that $T^n x \in U_j$. Thus $x \in M_j$ and M_j is dense.

By the Baire category theorem, $\bigcap_j M_j$ is non-empty and clearly each vector in $\bigcap_j M_j$ is hypercyclic for T. □

For $T \in \mathcal{B}(X)$ and $x \in X$ write $\operatorname{Orb}(T, x) = \{T^n x : n = 0, 1, \dots\}$. For $y \in X$ and $\varepsilon > 0$ denote by $B(y, \varepsilon) = \{u \in X : \|u - y\| < \varepsilon\}$ the open ball with center y and radius ε.

Theorem 8. *Let $T \in \mathcal{B}(X)$, $d > 0$ and let $x \in X$ satisfy that for each $y \in X$ there is an $n \in \mathbb{N}$ with $\|T^n x - y\| < d$. Then T is hypercyclic.*

Proof. Clearly, X is separable. Further, $\dim X = \infty$ (if $\dim X < \infty$, then consider the Jordan form of T). Let U, V be non-empty open subsets of X. We show that $T^n U \cap V \ne \emptyset$ for some $n \in \mathbb{N}$.

Choose $u \in U$, $v \in V$ and $\varepsilon > 0$ such that $\{y \in X : \|y - u\| < \varepsilon\} \subset U$ and $\{y \in X : \|y - v\| < \varepsilon\} \subset V$.

Let $x' = \frac{\varepsilon x}{3d}$. We show first that the set $\operatorname{Orb}(T, x')$ intersects each open ball with radius $\varepsilon/3$. If $y \in X$, then there is an n such that $\|T^n x - \frac{3dy}{\varepsilon}\| < d$. Therefore $\|T^n x' - y\| < \frac{\varepsilon}{3d}\|T^n x - \frac{3dy}{\varepsilon}\| < \frac{\varepsilon}{3}$.

Next we show that $\operatorname{Orb}(T, x')$ intersects each ball with radius ε in an infinite set. Suppose on the contrary that there is a $y \in X$ such that the set $\{n : \|T^n x' - y\| < \varepsilon\}$ is finite. Since the ball $B(y, \frac{2\varepsilon}{3})$ cannot be covered by a finite number of balls of radii $\frac{\varepsilon}{3}$ by Proposition 24.4, there is a $y_1 \in B(y, \frac{2\varepsilon}{3})$ such that $\operatorname{dist}\{y_1, \operatorname{Orb}(T, x')\} \ge \frac{\varepsilon}{3}$. Thus $\operatorname{Orb}(T, x') \cap B(y_1, \varepsilon/3) = \emptyset$, a contradiction.

Therefore there exist $n_1, n_2 \in \mathbb{N}$ such that $T^{n_1} x' \in B(u, \varepsilon) \subset U$, $n_2 > n_1$ and $T^{n_2} x' \in B(v, \varepsilon) \subset V$. Hence $T^{n_2 - n_1} T^{n_1} x' \in V$, and so $T^{n_2 - n_1} U \cap V \ne \emptyset$. By Theorem 7, T is hypercyclic. □

Theorem 9. *Let $T \in \mathcal{B}(X)$ and $x \in X$. Suppose that $\overline{\mathrm{Orb}(T,x)}$ has non-empty interior. Then x is hypercyclic for T.*

Proof. Write $F = \overline{\mathrm{Orb}(T,x)}$ and let U be the interior of F. Without loss of generality we can assume that $x \in U$. Indeed, there exists n_0 such that $T^{n_0}x \in U$. Clearly, the interiors of the sets $\overline{\mathrm{Orb}(T,T^{n_0}x)}$ and $\overline{\mathrm{Orb}(T,x)}$ are equal and the hypercyclicity of x is equivalent to the hypercyclicity of $T^{n_0}x$.

We prove the statement in several steps.

(a) $\sigma_p(T^*) = \emptyset$ and $q(T)$ has dense range for each non-zero polynomial q.

Proof. Suppose on the contrary that $\lambda \in \sigma_p(T^*)$. Let x^* be a corresponding eigenvector of T^*. Then $\{\langle T^n x, x^*\rangle : n = 0,1,\dots\}^-$ has non-empty interior in $\mathbb{C}$. However,

$$|\langle T^n x, x^*\rangle| = |\langle x, T^{*n}x^*\rangle| = |\lambda^n| \cdot |\langle x, x^*\rangle|$$

and this sequence either converges to zero (for $|\lambda| < 1$), or is constant (for $|\lambda| = 1$ or $\langle x, x^*\rangle = 0$), or tends to infinity (for $|\lambda| > 1$ and $\langle x, x^*\rangle \neq 0$). So the closure $\{\langle T^n x, x^*\rangle : n = 0,1,\dots\}^-$ cannot have non-empty interior. Hence $\sigma_p(T^*) = \emptyset$ and $T - \lambda$ has dense range for each $\lambda \in \mathbb{C}$.

If q is a non-zero polynomial, then $q(T)$ can be written as a product of a finite number of operators with dense range. Hence $q(T)$ has also dense range.

(b) The set $\{q(T)x : \{q$ a polynomial$\}$ is dense in X.

Proof. The set $\{q(T)x : q \text{ a polynomial}\}^-$ is a closed subspace of X containing $\overline{\mathrm{Orb}(T,x)}$ which has non-empty interior. Note that the only closed subspace of X with non-empty interior is the whole space X.

(c) $T(X \setminus U) \subset X \setminus U$.

Proof. Suppose on the contrary that there is a $y \in X \setminus U$ such that $Ty \in U$. Without loss of generality we can assume that $y \notin F$. Indeed, if $y \in F \setminus U$, then $y \in \partial F$ and we can replace y by a vector $y' \notin F$ close enough to y such that still $Ty' \in U$.

Since the set $\{q(T)x : q \text{ a polynomial}\}$ is dense in X, we can change y' slightly to obtain a polynomial q such that $q(T)x \notin F$ and $Tq(T)x \in U$.

Since $U \subset F$ and $TF \subset F$, we have $\overline{\mathrm{Orb}(T,Tq(T)x)} \subset F$. However, $\mathrm{Orb}(T,Tq(T)x) = \{q(T)T^n x : n = 1,2,\dots\}$. For $n \geq 1$ we have $q(T)T^n x = T^n q(T)x \in F$. Since x is a limit point of $\{T^n x : n \geq 1\}$, by continuity of $q(T)$ we get $q(T)x \in F$, a contradiction.

(d) Let q be a non-zero polynomial. Then $q(T)x \notin \partial U$.

Proof. Suppose on the contrary that $q(T)x \in \partial U \subset F$. Let $\mathcal{P}$ be the set of all polynomials p such that $p(T)x \in X \setminus F$. Then $\{p(T)x : p \in \mathcal{P}\}^- = \overline{X \setminus F} = X \setminus U$. Let $M = U \cup \{p(T)x : p \in \mathcal{P}\}$. Clearly $\overline{M} = X$.

Since $q(T)$ has dense range, we have $\overline{q(T)M} = X$. Each point in U is a limit of $\{T^n x : n = 0, 1, \dots\}$. We have $q(T)T^n x = T^n q(T)x \subset T^n(X \setminus U) \subset X \setminus U$. Thus $q(T)U \subset X \setminus U$.

Let $p \in \mathcal{P}$. We have

$$q(T)p(T)x = p(T)q(T)x \in p(T)F = p(T)\{T^n x : n = 0, 1, \dots\}^- \\ \subset \{T^n p(T)x : n = 0, 1, \dots\}^- \subset X \setminus U.$$

Hence $q(T)M \subset X \setminus U$, a contradiction.

(e) x is hypercyclic for T.

Proof. The set $\{p(T)x : p \text{ a non-zero polynomial}\}$ is connected, contains points of U (e.g., x) and contains no boundary points of U. Therefore

$$\{p(T)x : p \text{ a non-zero polynomial}\} \subset U$$

and

$$F = \overline{U} \supset \{p(T)x : p \text{ a non-zero polynomial}\}^- = X.$$

Hence x is hypercyclic for T. □

Corollary 10. *Let $T \in \mathcal{B}(X)$, $k \in \mathbb{N}$. Suppose that T is k-hypercyclic, i.e., there are vectors $x_1, x_2, \dots, x_k \in X$ such that $\bigcup_{i=1}^k \operatorname{Orb}(T, x_i)$ is dense in X. Then T is hypercyclic.*

Proof. We have $X = \overline{\bigcup_{i=1}^k \operatorname{Orb}(T, x_i)} = \bigcup_{i=1}^k \overline{\operatorname{Orb}(T, x_i)}$. By the Baire category theorem, there is a j, $1 \le j \le k$ such that $\overline{\operatorname{Orb}(T, x_j)}$ has non-empty interior. By Theorem 9, x_j is hypercyclic for T. □

Corollary 11. *Let $T \in \mathcal{B}(X)$ be a hypercyclic operator. Then for every positive n, the operator T^n is also hypercyclic. Moreover, T and T^n share the same collection of hypercyclic vectors.*

Proof. Let x be hypercyclic for T. We have $\operatorname{Orb}(T, x) = \bigcup_{j=0}^{n-1} \operatorname{Orb}(T^n, T^j x)$ and as above, there is a k, $0 \le k \le n-1$ such that $\overline{\operatorname{Orb}(T^n, T^k x)}$ has non-empty interior. By Theorem 9, $T^k x$ is hypercyclic for T^n. Since T has dense range, the set $T^{n-k} \operatorname{Orb}(T^n, T^k x) = \operatorname{Orb}(T^n, T^n x)$ is also dense. So $T^n x$ is hypercyclic for T^n, and so x is also hypercyclic for T^n. □

Let $T \in \mathcal{B}(X)$ and let $x \in X$ be a hypercyclic vector for T. Then x is also hypercyclic for $-T$. Indeed,

$$\operatorname{Orb}(-T, x) \supset \{(-T)^{2n} x : n = 0, 1, \dots\} = \operatorname{Orb}(T^2, x),$$

which is dense by Corollary 11.

Similarly, one can show that x is hypercyclic for each operator λT, where $\lambda = e^{2\pi i t}$ with t rational, $0 \le t < 1$. The next result shows that in fact x is hypercyclic for each operator λT with $|\lambda| = 1$.

It is easy to show that in general x is not hypercyclic for λT if $|\lambda| \ne 1$.

Theorem 12. *Let $T \in \mathcal{B}(X)$ be a hypercyclic operator and let $\lambda \in \mathbb{C}$, $|\lambda| = 1$. Then λT is also hypercyclic. Moreover, T and λT share the same collection of hypercyclic vectors.*

Proof. For $u, v \in X$ set

$$F_{u,v} = \{\mu \in \mathbb{T} : \mu v \in \overline{\mathrm{Orb}(\lambda T, u)}\}.$$

Clearly $F_{u,v}$ is a closed subset of the unit circle $\mathbb{T} = \{\mu \in \mathbb{C} : |\mu| = 1\}$.

The proof will be done in several steps:

(a) Let $u \in X$ be hypercyclic for T. Then $F_{u,v} \neq \emptyset$ for all $v \in X$.

Proof. There is a sequence (n_k) of positive integers such that $T^{n_k} u \to v$. Passing to a subsequence if necessary, we can assume that (λ^{n_k}) is convergent, $\lambda^{n_k} \to \mu$ for some $\mu \in \mathbb{T}$. Then

$$\|(\lambda T)^{n_k} u - \mu v\| \leq \|(\lambda T)^{n_k} u - \lambda^{n_k} v\| + \|(\lambda^{n_k} - \mu) v\| \to 0.$$

Thus $\mu \in F_{u,v}$.

(b) Let $u, v, w \in X$, $\mu_1 \in F_{u,v}$ and $\mu_2 \in F_{v,w}$. Then $\mu_1\mu_2 \in F_{u,w}$.

Proof. Let $\varepsilon > 0$. There exist $n_1 \in \mathbb{N}$ with $\|(\lambda T)^{n_1} v - \mu_2 w\| < \varepsilon/2$ and $n_2 \in \mathbb{N}$ such that $\|(\lambda T)^{n_2} u - \mu_1 v\| < \frac{\varepsilon}{2\|T^{n_1}\|}$. Then

$$\left\|(\lambda T)^{n_1+n_2} u - \mu_1\mu_2 w\right\| \leq \left\|(\lambda T)^{n_1}\left((\lambda T)^{n_2} u - \mu_1 v\right)\right\| + \left\|\mu_1\left((\lambda T)^{n_1} v - \mu_2 w\right)\right\| < \varepsilon.$$

Hence $\mu_1\mu_2 \in F_{u,w}$.

Let $x \in X$ be a fixed vector hypercyclic for T. By (a) and (b), $F_{x,x}$ is a non-empty closed subsemigroup of the unit circle $\mathbb{T}$.

Suppose first that $F_{x,x} = \mathbb{T}$. Then (a) and (b) imply that $F_{x,y} = \mathbb{T}$ for each $y \in X$. In particular, $1 \in F_{x,y}$, and so $y \in \overline{\mathrm{Orb}(\lambda T, x)}$ for all $y \in X$. Hence x is hypercyclic for λT.

In the following we shall assume that $F_{x,x} \neq \mathbb{T}$. We show that this assumption leads to a contradiction.

(c) There exists $k \in \mathbb{N}$ such that $F_{x,x} = \{e^{2\pi i j/k} : j = 0, 1, \dots, k-1\}$.

Proof. Let $s = \inf\{t > 0 : e^{2\pi i t} \in F_{x,x}\}$. Clearly, $s > 0$, since otherwise $F_{x,x}$ would be dense in $\mathbb{T}$. We have $e^{2\pi i s} \in F_{x,x}$. Let $k = \min\{n \in \mathbb{N} : ns \geq 1\}$. If $ks > 1$, then $e^{2\pi i(ks-1)} \in F_{x,x}$ and $0 < ks - 1 < s$, a contradiction with the definition of s. Hence $ks = 1$ and

$$F_{x,x} \supset \{e^{2\pi i j/k} : j = 0, 1, \dots, k-1\}.$$

If there is an $\mu \in F_{x,x} \setminus \{e^{2\pi i j/k} : j = 0, 1, \dots, k-1\}$, then $\mu = e^{2\pi i t}$ and $j_0/k < t < (j_0+1)/k$ for some j_0, $0 \leq j_0 \leq k-1$. Then $\mu \cdot e^{-2\pi i j_0/k} = e^{2\pi i(t - j_0/k)} \in F_{x,x}$ where $0 < t - j_0/k < 1/k = s$, which is again a contradiction with the definition of s.

Thus $F_{x,x} = \{e^{2\pi i j/k} : j = 0, 1, \dots, k-1\}$.

(d) Let $y \in X$ be a vector hypercyclic for T. Then there exists $\mu_y \in \mathbb{T}$ such that $F_{x,y} = \{\mu_y e^{2\pi ij/k} : j = 0, 1, \dots, k-1\}$.

Proof. By (a), there are $\mu \in F_{x,y}$ and $\alpha \in F_{y,x}$. By (b), we have $\mu F_{x,x} \subset F_{x,y}$ and $\alpha F_{x,y} \subset F_{x,x}$. In particular, $\operatorname{card} F_{x,y} = \operatorname{card} F_{x,x}$ and $F_{x,y} = \mu F_{x,x} = \{\mu e^{2\pi ij/k} : j = 0, 1, \dots, k-1\}$.

By the proof of Theorem 5, each non-zero vector y in the subspace generated by x and Tx is hypercyclic for T. For such a y define $f(y) = \mu^k$ where μ is any element of $F_{x,y}$. Clearly the function f is well defined by (d).

(e) f is a continuous function.

Proof. Suppose on the contrary that there exist non-zero vectors $u_n, u \in \bigvee\{x, Tx\}$ such that $u_n \to u$ and $f(u_n) \not\to f(u)$. Without loss of generality we can assume that the sequence $(f(u_n))$ converges to some $\alpha \in \mathbb{T}$, $\alpha \neq f(u)$. Let $\mu_n \in F_{x,u_n}$. Passing to a subsequence if necessary we can assume that $\mu_n \to \mu$ for some $\mu \in \mathbb{T}$. Then $\mu_n u_n \in \overline{\operatorname{Orb}(\lambda T, x)}$ and $\mu_n u_n \to \mu u$. So we have $\mu u \in \overline{\operatorname{Orb}(\lambda T, x)}$ and $\mu \in F_{x,u}$. Hence $\alpha = \lim f(u_n) = \lim \mu_n^k = \mu^k = f(u)$, a contradiction. Hence f is continuous on the set $\bigvee\{x, Tx\} \setminus \{0\}$.

Proof of Theorem 12. Since x is hypercyclic for T, the vectors x and Tx are linearly independent.

Let $g : \overline{\mathbb{D}} \to \mathbb{T}$ be the function defined by $g(z) = f\big(zx + (1 - |z|)Tx\big)$. Clearly g is continuous. For all z satisfying $|z| = 1$ we have $F_{x,zx} = z^{-1}F_{x,x}$ and $g(z) = f(zx) = z^{-k}f(x) = z^{-k}$. It is well known that such a function g cannot exist, see, e.g., [R], Theorem 10.40. Indeed, the function g would provide a homotopy between the constant path $\gamma_1 : \langle 0, 2\pi\rangle \to \mathbb{T}$ defined by $\gamma_1(t) = g(0)$ and the path $\gamma_2 : \langle 0, 2\pi\rangle \to \mathbb{T}$ given by $\gamma_2(t) = g(e^{it}) = e^{-kit}$, which has the winding number $-k$.

Thus $F_{x,x} = \mathbb{T}$ and the set $\operatorname{Orb}(\lambda T, x)$ is dense in X. □

39 Weak orbits

By a *weak orbit* of an operator $T \in \mathcal{B}(X)$ we mean a sequence of the form $(\langle T^j x, x^*\rangle)_{j=0}^\infty$, where $x \in X$ and $x^* \in X^*$.

Some results concerning orbits also remain true for weak orbits. An example is the statement of Theorem 37.5.

Theorem 1. *Let T be an operator on a Banach space X, let $(a_n)_{n\geq 0}$ be a sequence of positive numbers such that $a_n \to 0$. Then the set of all pairs $(x, x^*) \in X \times X^*$ such that*

$$|\langle T^n x, x^*\rangle| \geq a_n \|T^n\| \qquad \textit{for infinitely many powers } n$$

is residual in $X \times X^*$.

In particular, the set $\{(x,x^*)\in X\times X^* : \limsup_{n\to\infty}|\langle T^nx,x^*\rangle|^{1/n}=r(T)\}$ *is residual in* $X\times X^*$.

Proof. For $k\in\mathbb{N}$ set

$$M_k=\{(x,x^*)\in X\times X^* :\ \text{there exists } n\ge k \text{ such that } |\langle T^nx,x^*\rangle|>a_n\|T^n\|\}.$$

Clearly, M_k is an open subset of $X\times X^*$. We prove that M_k is dense. Let $x\in X, x^*\in X^*$ and $\varepsilon>0$. Choose $n\ge k$ such that $a_n<\varepsilon^2$. There is a vector $u\in X$ of norm 1 such that $\|T^nu\|>\frac{a_n}{\varepsilon^2}\|T^n\|$. Let $u^*\in X^*$ satisfy $\|u^*\|=1$ and $\langle T^nu,u^*\rangle=\|T^nu\|$. We have

$$\begin{aligned}
&|\langle T^n(x+\varepsilon u),x^*+\varepsilon u^*\rangle|+|\langle T^n(x+\varepsilon u),x^*-\varepsilon u^*\rangle|\\
&\qquad+|\langle T^n(x-\varepsilon u),x^*+\varepsilon u^*\rangle|+|\langle T^n(x-\varepsilon u),x^*-\varepsilon u^*\rangle|\\
&\ge\Big|\langle T^n(\varepsilon u+x),\varepsilon u^*+x^*\rangle+\langle T^n(\varepsilon u+x),\varepsilon u^*-x^*\rangle\\
&\qquad+\langle T^n(\varepsilon u-x),\varepsilon u^*+x^*\rangle+\langle T^n(\varepsilon u-x),\varepsilon u^*-x^*\rangle\Big|\\
&=|4\langle T^n\varepsilon u,\varepsilon u^*\rangle|=4\varepsilon^2\|T^nu\|>4a_n\|T^n\|.
\end{aligned}$$

Thus there is a pair

$$(y,y^*)\in\{(x+\varepsilon u,x^*+\varepsilon u^*),(x+\varepsilon u,x^*-\varepsilon u^*),(x-\varepsilon u,x^*+\varepsilon u^*),(x-\varepsilon u,x^*-\varepsilon u^*)\}$$

such that $|\langle T^ny,y^*\rangle|>a_n\|T^n\|$. Hence $(y,y^*)\in M_k$ and M_k is dense in $X\times X^*$.

By the Baire category theorem, the intersection $M=\bigcap_{k=1}^\infty M_k$ is a residual subset of $X\times X^*$ and all pairs $(y,y^*)\in M$ satisfy $|\langle T^ny,y^*\rangle|>a_n\|T^n\|$ for infinitely many powers n.

In particular, for $a_n=n^{-1}$ we obtain that

$$\limsup_{n\to\infty}|\langle T^ny,y^*\rangle|^{1/n}\ge\limsup_{n\to\infty}\left(\frac{\|T^n\|}{n}\right)^{1/n}=r(T)$$

for all pairs (y,y^*) in a residual subset of $X\times X^*$. □

The analogy of Theorem 37.8 for weak orbits is not true. If $T\in\mathcal{B}(X)$ and $a_n>0$, $a_n\to 0$, then in general it is not possible to find $x\in X$ and $x^*\in X^*$ such that $|\langle T^nx,x^*\rangle|\ge a_nr(T^n)$ for all n, cf. C.39.4. However, it is possible to prove some weaker results.

We start with the following lemma:

Lemma 2. *Let* X *be a Banach space,* $T\in\mathcal{B}(X)$, $r_e(T)=1$, $n_0\in\mathbb{N}$, $\varepsilon>0$, $m\in\mathbb{N}$. *Then there are numbers* $n_0<n_1<\cdots<n_m$ *such that in each subspace* $M\subset X$ *of finite codimension there exists a vector* $x\in M$ *of norm 1 with* $\|T^{n_j}x-x\|\le\varepsilon\quad(j=1,2,\dots,m)$.

Proof. Let $\lambda \in \sigma_e(T)$, $|\lambda| = 1$. Then $\lambda \in \partial\sigma_e(T) \subset \sigma_{\pi e}(T)$. Find $s \in \mathbb{N}$ such that $s > n_0$ and $|\lambda^s - 1| \le \varepsilon/2m$. Then

$$|\lambda^{sj} - 1| = |\lambda^s - 1| \cdot |\lambda^{s(j-1)} + \lambda^{s(j-2)} + \cdots + 1| \le \varepsilon/2$$

for $j = 1, 2, \ldots, m$. Let $x \in M$ be a vector of norm 1 satisfying $\|T^{sj}x - \lambda^{sj}x\| \le \varepsilon/2 \quad (j = 1, \ldots, m)$. Then $\|T^{sj}x - x\| \le \|T^{sj}x - \lambda^{sj}x\| + \|\lambda^{sj}x - x\| \le \varepsilon$. □

Theorem 3. *Let X be a Banach space, $T \in \mathcal{B}(X)$ and let $(a_j)_{j\ge1}$ be a sequence of positive numbers with $a_j \to 0$. Then there exist $x \in X$, $x^* \in X^*$ and an increasing sequence (n_j) of positive integers such that*

$$\mathrm{Re}\langle T^{n_j}x, x^*\rangle \ge a_j \cdot r(T)^{n_j}$$

for all $j \ge 1$.

Proof. Without loss of generality we can assume that $r(T) = 1$ and that $1 > a_0 \ge a_1 \ge \cdots$. We distinguish two cases:

(a) Suppose that there are $x \in X$, $x^* \in X^*$ and $c > 0$ such that $|\langle T^n x, x^*\rangle| \ge c$ for infinitely many powers n (i.e., T^n does not tend to 0 in the weak operator topology).

Then

$$|\langle T^n x,x^*\rangle| \le \sqrt{2}\cdot\max\{\mathrm{Re}\langle T^n x,x^*\rangle,\mathrm{Re}\langle T^n x,ix^*\rangle,\mathrm{Re}\langle T^n x,-x^*\rangle,\mathrm{Re}\langle T^n x,-ix^*\rangle\}.$$

Thus there are $c_1 > 0$ and $x_1^* \in X^*$ such that $\mathrm{Re}\langle T^n x, x_1^*\rangle \ge c_1$ for infinitely many powers n. Hence we get the statement of Theorem 3 for a suitable multiple of x_1^*.

(b) Suppose that $\langle T^n x, x^*\rangle \to 0$ for all $x \in X, x^* \in X^*$.

Using the uniform boundedness theorem twice yields that $M := \sup\{\|T^n\| : n = 0, 1, \ldots\} < \infty$. The assumption also implies that there are no eigenvalues of modulus 1, and so $r_e(T) = 1$. Let $s = 8M$. Find numbers $m_k \in \mathbb{N}$ such that $0 = m_0 < m_1 < m_2 < \cdots$ and

$$a_j \le \frac{1}{16s^{2k}} \qquad (k \ge 0, j > m_k).$$

We construct inductively sequences $(u_k)_{k\ge0} \subset X$, $(u_k^*)_{k\ge0} \subset X^*$ and an increasing sequence of positive integers (n_j) in the following way:

Set $u_0 = 0$ and $u_0^* = 0$. Let $k \ge 0$ and suppose that vectors $u_0, \ldots, u_k \in X$, $u_0^*, \ldots, u_k^* \in X^*$ and numbers $n_1, \ldots, n_{m_k}$ have already been constructed. Write $x_k = \sum_{i=1}^k \frac{u_i}{s^{i-1}}$ and $x_k^* = \sum_{i=1}^k \frac{u_i^*}{s^{i-1}}$. Find q_k such that $|\langle T^j x_k, x_k^*\rangle| \le \frac{1}{16s^{2k}} \quad (j \ge q_k)$. Find numbers $n_{m_k+1}, \ldots, n_{m_{k+1}}$ satisfying the properties of Lemma 2 for $\varepsilon = 1/16$ such that

$$\max\{n_{m_k}, q_k\} < n_{m_k+1} < n_{m_k+2} < \cdots < n_{m_{k+1}}.$$

Let $E_k = \bigvee\{T^{n_j}u_i : 0 \le i \le k, 0 \le j \le m_{k+1}\}$. By Lemma 37.6, there exists a subspace Y_k of finite codimension such that

$$\|e + y\| \ge \max\{\|e\|/2, \|y\|/4\} \qquad (e \in E_k, y \in Y_k).$$

Let $u_{k+1} \in Y_k \cap {}^\perp\Big(\bigvee\{T^{*n_j}u_i^* : 1 \le i \le k, 0 \le j \le m_{k+1}\}\Big)$ be a vector of norm 1 such that

$$\|T^{n_j}u_{k+1} - u_{k+1}\| < 1/16 \qquad (m_k < j \le m_{k+1}).$$

Find $u_{k+1}^* \in E_k^\perp$ such that $\|u_{k+1}^*\| = 1$ and

$$\langle u_{k+1}, u_{k+1}^*\rangle = \operatorname{dist}\{u_{k+1}, E_k\} \ge 1/4.$$

Note that $\langle T^{n_j}u_i, u_{k+1}^*\rangle = 0$ and $\langle T^{n_j}u_{k+1}, u_i^*\rangle = 0$ for all $i \le k$ and $j \le m_{k+1}$.

Continue the inductive construction, and set finally $x = \sum_{i=1}^\infty \frac{u_i}{s^{i-1}}$ and $x^* = \sum_{i=1}^\infty \frac{u_i^*}{s^{i-1}}$.

To show that x, x^* and the sequence (n_j) satisfy the properties required, let $k \ge 0$ and $m_k < j \le m_{k+1}$. We have

$$\begin{aligned}
\operatorname{Re}\langle T^{n_j}x, x^*\rangle &= \operatorname{Re}\Big\langle T^{n_j}\Big(x_k + \sum_{i=k}^\infty \frac{u_{i+1}}{s^i}\Big), x_k^* + \sum_{i=k}^\infty \frac{u_{i+1}^*}{s^i}\Big\rangle \\
&= \operatorname{Re}\langle T^{n_j}x_k, x_k^*\rangle + \frac{1}{s^{2k}}\operatorname{Re}\langle T^{n_j}u_{k+1}, u_{k+1}^*\rangle + \sum_{i=k+1}^\infty \frac{1}{s^{2i}}\operatorname{Re}\langle T^{n_j}u_{i+1}, u_{i+1}^*\rangle \\
&\ge -\frac{1}{16s^{2k}} + \frac{1}{s^{2k}}\Big(\operatorname{Re}\langle u_{k+1}, u_{k+1}^*\rangle - \operatorname{Re}\langle u_{k+1} - T^{n_j}u_{k+1}, u_{k+1}^*\rangle\Big) - \sum_{i=k+1}^\infty \frac{M}{s^{2i}} \\
&\ge \frac{1}{s^{2k}}\Big(-\frac{1}{16} + \frac{1}{4} - \frac{1}{16} - \frac{2M}{s^2}\Big) \ge \frac{1}{16s^{2k}} \ge a_j.
\end{aligned}$$

□

Corollary 4. *Let X be a Banach space, let $T \in \mathcal{B}(X)$, $0 < p < \infty$, $r(T) \ne 0$. Then the set*

$$\Big\{(x, x^*) \in X \times X^* : \sum_{n=0}^\infty \Big(\frac{|\langle T^n x, x^*\rangle|}{r(T^n)}\Big)^p = \infty\Big\}$$

is residual in $X \times X^*$.

Proof. For $k \in \mathbb{N}$ set

$$M_k = \Big\{(x, x^*) \in X \times X^* : \sum_{n=0}^\infty \Big(\frac{|\langle T^n x, x^*\rangle|}{r(T^n)}\Big)^p > k\Big\}.$$

Clearly M_k is open in $X \times X^*$. To show that M_k is dense, let $x \in X$, $x^* \in X^*$ and $\varepsilon > 0$. By the previous theorem for a suitable sequence (a_n), there are $u \in X$ and

$u^* \in X^*$ such that $\sum_{n=0}^{\infty}\Big(\frac{|\langle T^n u, u^*\rangle|}{r(T^n)}\Big)^p = \infty$. We can assume that $\|u\| < \varepsilon$ and $\|u^*\| < \varepsilon$. Since

$$\Big(\frac{|\langle T^n u, u^*\rangle|}{r(T^n)}\Big)^p = \Big|\frac{\langle T^n(u+x), u^*+x^*\rangle}{4r(T^n)} + \frac{\langle T^n(u+x), u^*-x^*\rangle}{4r(T^n)} + \frac{\langle T^n(u-x), u^*+x^*\rangle}{4r(T^n)} + \frac{\langle T^n(u-x), u^*-x^*\rangle}{4r(T^n)}\Big|^p$$
$$\le \max\Big\{\frac{|\langle T^n(u+x), u^*+x^*\rangle|}{r(T^n)}, \frac{|\langle T^n(u+x), u^*-x^*\rangle|}{r(T^n)}, \frac{|\langle T^n(u-x), u^*+x^*\rangle|}{r(T^n)}, \frac{|\langle T^n(u-x), u^*-x^*\rangle|}{r(T^n)}\Big\}^p,$$

we have that $\sum_{n=0}^{\infty}\Big(\frac{|\langle T^n y, y^*\rangle|}{r(T^n)}\Big)^p = \infty$ for at least one pair

$$(y, y^*) \in \big\{(x+u, x^*+u^*), (x+u, x^*-u^*), (x-u, x^*+u^*), (x-u, x^*-u^*)\big\}.$$

Thus M_k is dense in $X \times X^*$ and

$$M = \bigcap_k M_k = \Big\{(x, x^*) \in X \times X^* : \sum_{n=0}^{\infty}\Big(\frac{|\langle T^n x, x^*\rangle|}{r(T)^n}\Big)^p = \infty\Big\}$$

is residual in $X \times X^*$. □

The following result is analogous to Theorem 37.14.

Theorem 5. *Let X, Y be Banach spaces, let $(T_n) \subset \mathcal{B}(X,Y)$ be a sequence of operators. Let $a_n > 0$, $\sum_{n=1}^{\infty} a_n^{1/2} < \infty$. Then there are $x \in X$, $y^* \in Y^*$ such that $|\langle T^n x, y^*\rangle \ge a_n \|T^n\|$ for all n. Moreover, given balls $B \subset X$, $B' \subset Y^*$ of radii grater than $\sum_{n=1}^{\infty} a_n^{1/2}$ then it is possible to find $x \in B$ and $y^* \in B'$.*

Proof. Let $s = \sum_{n=1}^{\infty} a_n^{1/2}$. Let $u \in X$, $v^* \in Y^*$ and $\varepsilon > 0$. We find $x \in X$, $y^* \in Y^*$ such that $\|x - u\| \le s + \varepsilon$, $\|y^* - v^*\| \le s + \varepsilon$ and $|\langle T_n x, y^*\rangle| \ge a_n \|T_n\|$ for all n.

Without loss of generality we may assume that $T_n \ne 0$ for all n. For $n \in \mathbb{N}$ let $a'_n = \frac{a_n}{(s+\varepsilon/2)^2}$. Then $\sum_{n=1}^{\infty} {a'_n}^{1/2} < 1$.

For $n \in \mathbb{N}$ find $x_n \in X$ such that $\|x_n\| = 1$ and $\|T_n x\| \ge \frac{s+\varepsilon/2}{s+\varepsilon}\|T_n\|$. By Theorem A.5.1 applied to X^*, there exists $y'^* \in Y^*$ such that $\|y'^* - \frac{v^*}{s+\varepsilon}\| \le 1$ and

$$\Big|\Big\langle \frac{T_n x_n}{\|T_n x_n\|}, y'^*\Big\rangle\Big| \ge {a'_n}^{1/2}$$

for all n.

Applying Theorem A.5.1 again to the functionals $\frac{T_n^* y'^*}{\|T_n^* y'^*\|}$, we obtain $x' \in X$ such that $\|x' - \frac{u}{s+\varepsilon}\| \le 1$ and $\Big|\Big\langle x', \frac{T_n^* y'^*}{\|T_n^* y'^*\|}\Big\rangle\Big| \ge {a'_n}^{1/2}$ for all n. Set $x = (s+\varepsilon)x'$

and $y^* = (s+\varepsilon)y'^*$. Then $\|x-u\| \le s+\varepsilon$ and $\|y^*-v^*\| \le s+\varepsilon$. For $n \in \mathbb{N}$ we have

$$\begin{aligned}|\langle T_n x, y^*\rangle| &= (s+\varepsilon)^2|\langle x', T_n^* y'^*\rangle| \ge (s+\varepsilon)^2 {a'_n}^{1/2}\|T_n^* y'^*\| \\ &\ge (s+\varepsilon)^2 {a'_n}^{1/2}|\langle x_n, T_n^* y'^*\rangle| \ge (s+\varepsilon)^2 {a'_n}^{1/2}|\langle T_n x_n, y'^*\rangle| \\ &\ge (s+\varepsilon)^2 a'_n\|T_n x_n\| \ge a_n\frac{(s+\varepsilon)^2}{(s+\varepsilon/2)^2}\cdot\frac{s+\varepsilon/2}{s+\varepsilon}\|T_n\| \ge a_n\|T_n\|. \quad \square\end{aligned}$$

Corollary 6. *Let X, Y be Banach spaces, $T_n \in \mathcal{B}(X,Y)$, $\sum_{n=1}^\infty \|T_n\|^{-1/2} < \infty$. Then there exist $x \in X$ and $y^* \in Y^*$ such that $|\langle T_n x, y^*\rangle| \to \infty$. Moreover, the set of such pairs (x, y^*) is dense in $X \times Y^*$.*

Proof. Let $u \in X$, $v^* \in Y^*$, $\varepsilon > 0$. By Lemma 37.15, we can find positive numbers β_n such that $\beta_n \to \infty$ and $s = \sum_{n=1}^\infty \frac{\beta_n}{\|T_n\|^{1/2}} < \infty$.

Find n_0 such that $\sum_{n=n_0}^\infty \beta_n\|T_n\|^{-1/2} < \varepsilon$. By Theorem 5, there are $x \in X$ and $y^* \in X^*$ such that $\|x-u\| < \varepsilon$, $\|y^*-v^*\| < \varepsilon$ and

$$|\langle T^n x, y^*\rangle| \ge \frac{\beta_n^2}{\|T^n\|}\cdot\|T_n\| = \beta_n^2$$

for all $n \ge n_0$. Hence $|\langle T^n x, y^*\rangle| \to \infty$. $\square$

Corollary 7. *Let X be a Banach space, let $T \in \mathcal{B}(X)$. Then the set*

$$\{(x,x^*) \in X \times X^* : \liminf_{n\to\infty} |\langle T^n x, x^*\rangle|^{1/n} = r(T)\}$$

is dense in $X \times X^*$. In particular, there is a dense subset L of $X \times X^*$ with the property that the limit $\lim_{n\to\infty}|\langle T^n x, x^*\rangle|^{1/n}$ exists and is equal to $r(T)$ for each pair $(x,x^*) \in L$.

Proof. Let $u \in X$, $u^* \in X^*$ and $\varepsilon > 0$. Find n_0 such that $\sum_{n=n_0}^\infty n^{-3/2} < \varepsilon$. By Theorem 5, there are $x \in X$ and $x^* \in X^*$ such that $\|x-u\| < \varepsilon$, $\|x^*-u^*\| < \varepsilon$ and

$$|\langle T^n x, x^*\rangle| \ge n^{-3}\|T^n\|$$

for all n sufficiently large. Then

$$\lim_{n\to\infty} |\langle T^n x, x^*\rangle|^{1/n} \ge \lim_{n\to\infty} n^{-3/n}\|T^n\|^{1/n} = r(T).$$

So $\lim_{n\to\infty}|\langle T^n x, x^*\rangle|^{1/n} = r(T)$. $\square$

Better results are true for Hilbert space operators.

Theorem 8. *Let H, K be Hilbert spaces, let $(T_n) \subset \mathcal{B}(H,K)$ be a sequence of operators. Let $a_n > 0$, $\sum_{n=1}^\infty a_n < \infty$. Then:*

(i) *there are $x \in H$, $y \in K$ such that $|\langle T^n x, y\rangle \ge a_n\|T^n\|$ for all n.*
(ii) *there is a dense subset of pairs $(x,y) \in H \times H$ such that $|\langle T^n x, y\rangle| \ge a_n\|T_n\|$ for all n sufficiently large.*

Proof. (i) By Theorem 37.17, there exists $x \in H$ such that $\|T_n x\| \geq a_n^{1/2} \|T_n\|$ for all $n \in \mathbb{N}$. By Theorem A.5.2, there is an $y \in H$ such that

$$\left| \left\langle \frac{T_n x}{\|T_n x\|}, y \right\rangle \right| \geq a_n^{1/2}$$

for all $n \in \mathbb{N}$. Then $|\langle T_n x, y\rangle| \geq a_n^{1/2} \|T_n x\| \geq a_n \|T_n\|$ for all $n \in \mathbb{N}$.

(ii) Let $u \in H$, $v \in K$ and $\varepsilon > 0$. By Theorem 37.17 (ii), there exists $x \in H$, $\|x - u\| < \varepsilon$ such that $\|T_n x\| \geq a_n^{1/2} \|T_n\|$ for all n sufficiently large.

Consider the operators $y \mapsto \langle T_n x, y\rangle$ from K to $\mathbb{C}$. Using Theorem 37.17 (ii) again, there exists $y \in K$ such that $\|y - v\| < \varepsilon$ and

$$|\langle T_n x, y\rangle| \geq a_n^{1/2} \|T_n x\| \geq a_n \|T_n\|$$

for all n large enough. □

Corollary 9. *Let H be a Hilbert space, $T \in \mathcal{B}(H)$, $\sum_{n=1}^{\infty} \|T^n\|^{-1} < \infty$. Then there exist $x, y \in H$ such that $|\langle T^n x, y\rangle| \to \infty$. Moreover, the set of such pairs (x, y) is dense in $H \times K$.*

Proof. By Lemma 37.15, there are $\beta_n \to \infty$ such that $\sum_{n=1}^{\infty} \frac{\beta_n}{\|T^n\|} < \infty$. By Theorem 8, there is a dense subset of pairs $(x, y) \in H \times H$ such that

$$|\langle T^n x, y\rangle| \geq \frac{\beta_n}{\|T^n\|} \cdot \|T^n\| = \beta_n$$

for all n large enough. Hence $|\langle T^n x, y\rangle| \to \infty$. □

Theorem 10. *Let X, Y be Banach spaces, let $T_1, T_2, \dots$ be a sequence of non-zero operators from X to Y and let $0 < p < 1/2$. Then the set*

$$\left\{ (x, y^*) \in X \times Y^* : \sum_{j=1}^{\infty} \left(\frac{|\langle T_j x, y^*\rangle|}{\|T_j\|} \right)^p = \infty \right\}$$

is residual in $X \times Y^*$.

If X, Y are Hilbert spaces, then the same statement is true for all p, $0 < p < 1$.

Proof. For $k \geq 1$ set

$$M_k = \left\{ (x, y^*) \in X \times Y^* : \sum_{j=1}^{\infty} \left(\frac{|\langle T_j x, y^*\rangle|}{\|T_j\|} \right)^p > k \right\}.$$

Clearly, M_k is an open subset of X. It is sufficient to show that M_k is dense. Indeed, by the Baire theorem, the intersection $\bigcap_k M_k = \left\{ (x, y^*) \in X \times Y^* : \sum_j \left(\frac{|\langle T_j x, y^*\rangle|}{\|T_j\|} \right)^p = \infty \right\}$ is a dense G_δ subset of $X \times Y^*$.

Fix $u \in X$, $v^* \in Y^*$, $\delta > 0$ and $k \in \mathbb{N}$. Let $\varepsilon = \frac{1-2p}{p}$ and $a_n = \frac{1}{n^{2+\varepsilon}}$. Then $\varepsilon > 0$ and $\sum_n a_n^{1/2} < \infty$. Let n_0 satisfy $\sum_{n=n_0}^\infty a_n^{1/2} < \delta$. By Theorem 5, there exist $u \in X$, $v^* \in Y^*$ such that $\|x-u\| \le \delta$, $\|y^* - v^*\| \le \delta$ and $|\langle T_n x, y^*\rangle| \ge a_n \|T_n\|$ for all $n \ge n_0$. Then

$$\sum_{n=1}^\infty \left(\frac{|\langle T_n x, y^*\rangle|}{\|T_n\|}\right)^p \ge \sum_{n=n_0}^\infty a_n^p = \sum_{n=n_0}^\infty \frac{1}{n} = \infty.$$

Hence $(x, y^*) \in M_k$ and M_k is dense in $X \times Y^*$.

The statement for Hilbert space operators can be proved similarly. □

We finish this section with an important partial case in which the statement analogous to Theorem 37.8 is true for weak orbits. For simplicity we formulate it only for Hilbert space operators; for Banach space analogy see C.39.4.

An operator $T \in \mathcal{B}(X)$ is called *power bounded* if $\sup_n \|T^n\| < \infty$. T is of class $C_{0\cdot}$ if $\|T^n x\| \to 0$ for all $x \in X$. T is of class $C_{1\cdot}$ if there is no non-zero $x \in X$ with $\|T^n x\| \to 0$. T is of class $C_{\cdot 0}$ ($C_{\cdot 1}$) if T^* is of class $C_{0\cdot}$ ($C_{1\cdot}$). Finally, T is of class $C_{\alpha\beta}$ $(\alpha, \beta = 0, 1)$ if T is both of class $C_{\alpha\cdot}$ and $C_{\cdot\beta}$. We start with the following lemma.

Lemma 11. *Let $K \ge 1$. Then there exist positive numbers c_i $(i \in \mathbb{N})$ such that $\sum_{i=1}^\infty c_i^2 = 1$ and $\sum_{i=k+1}^\infty c_i^2 > 3K c_k$ for all $k \ge 1$.*

Proof. Note first that

$$\lim_{k\to\infty} k^{2/3} \sum_{i=k+1}^\infty i^{-4/3} = \infty. \tag{1}$$

Indeed, we have $\sum_{i=k+1}^\infty i^{-4/3} \ge \int_{k+1}^\infty x^{-4/3}\,\mathrm{d}x = 3(k+1)^{-1/3}$, and so

$$\lim_{k\to\infty} k^{2/3} \sum_{i=k+1}^\infty i^{-4/3} \ge \lim_{k\to\infty} \frac{3k^{2/3}}{(k+1)^{1/3}} = \infty.$$

By (1), there exists k_0 such that

$$k^{2/3} \sum_{i=k+1}^\infty i^{-4/3} > 3K \left(\sum_{i=1}^\infty i^{-4/3}\right)^{1/2} \tag{2}$$

for all $k \ge k_0$. For $j \in \mathbb{N}$ set

$$c_j = (j+k_0)^{-2/3} \left(\sum_{i=k_0+1}^\infty i^{-4/3}\right)^{-1/2}.$$

Then $\sum_{i=1}^\infty c_i^2 = 1$. Let $k \geq 1$. Then, by (2),

$$\sum_{i=k+1}^\infty c_i^2 = \frac{\sum_{i=k+1}^\infty (k_0+i)^{-4/3}}{\sum_{i=k_0+1}^\infty i^{-4/3}} > \frac{3K\left(\sum_{i=1}^\infty i^{-4/3}\right)^{1/2}}{\sum_{i=k_0+1}^\infty i^{-4/3}} \cdot (k+k_0)^{-2/3}$$

$$\geq \frac{3K(k+k_0)^{-2/3}}{\left(\sum_{i=k_0+1}^\infty i^{-4/3}\right)^{1/2}} = 3Kc_k. \qquad \square$$

The next result is an analogy of Theorem 37.8.

Theorem 12. *Let T be an operator of class $C_{0\cdot}$ acting on a Hilbert space H such that $1 \in \sigma(T)$. Let $(a_n)_{n=1}^\infty$ be a sequence of positive numbers such that $\lim_{n\to\infty} a_n = 0$ and $\sup a_n < 1$. Then there exists $x \in H$ of norm 1 such that $\mathrm{Re}\langle T^n x, x\rangle > a_n$ for all $n \geq 1$.*

Proof. By the Banach-Steinhaus theorem, T is power bounded. Let $K = \sup_n \|T^n\|$. Clearly $K \geq 1$ and $r(T) = 1$.

Suppose first that $1 \notin \sigma_e(T)$. Then 1 is an eigenvalue of T and there exists $x \in H$ of norm 1 such that $Tx = x$. Then $\mathrm{Re}\langle T^n x, x\rangle = 1$ for all n.

Let $1 \in \sigma_e(T)$. Then $1 \in \partial\sigma_e(T)$, and so $T - I$ is not upper semi-Fredholm, see Proposition 19.1. Consequently, for all $\varepsilon > 0$ and $M \subset H$ with $\operatorname{codim} M < \infty$ there exists $u \in M$ of norm 1 such that $\|Tu - u\| < \varepsilon$. Moreover, given $n_0 \in \mathbb{N}$, we also can find $v \in M$ of norm 1 such that $\|T^j v - v\| < \varepsilon$ for all $j \leq n_0$.

Replacing the numbers a_n by $\sup\{a_i : i \geq n\}$ we can assume without loss of generality that $1 > a_1 \geq a_2 \geq \cdots$. By Lemma 11, there are positive numbers c_i such that $\sum_{i=1}^\infty c_i^2 = 1$ and $\sum_{i=k+1}^\infty c_i^2 > 3Kc_k$ for all $k \geq 1$.

For $i = 1, 2, \ldots$ let δ_i be a positive number satisfying $\delta_i < \frac{1-a_1}{2^i}$ and $\delta_i < \min\left\{\frac{Kc_k}{i\cdot 2^{i-k+1}} : k = 1, \ldots, i+1\right\}$.

Find $m_0 \in \mathbb{N}$ such that $a_{m_0} < \sum_{i=2}^\infty c_i^2 - 3Kc_1$. We construct inductively an increasing sequence $(m_i)_{i=0}^\infty$ of positive integers and a sequence $(x_i)_{i=1}^\infty \subset H$ in the following way:

Let $k \in \mathbb{N}$ and suppose that $x_i \in H$ and m_i have already been constructed for all $i < k$. Choose $x_k \in X$ of norm 1 such that

$$x_k \perp T^j x_i \quad (i < k, 0 \leq j \leq m_{k-1}) \qquad \text{and} \qquad \|T^j x_k - x_k\| < \delta_k \quad (j \leq m_{k-1}).$$

Find $m_k > m_{k-1}$ such that

$$\|T^j x_i\| < \delta_k \qquad (i \leq k, j \geq m_k)$$

and

$$a_{m_k} < \sum_{i=k+2}^\infty c_i^2 - 3Kc_{k+1}.$$

Suppose that x_i and m_i have been constructed in the above-described way. Set $x = \sum_{i=1}^{\infty} c_i x_i$. Since (x_i) is an orthonormal sequence, we have

$$\|x\| = \Big(\sum_{i=1}^{\infty} c_i^2\Big)^{1/2} = 1.$$

For $n \le m_0$ we have

$$\begin{aligned}\operatorname{Re}\langle T^n x, x\rangle &= \operatorname{Re}\sum_{i=1}^{\infty} c_i \langle T^n x_i, x\rangle = \operatorname{Re}\sum_{i=1}^{\infty} c_i\big(\langle x_i, x\rangle - \langle x_i - T^n x_i, x\rangle\big)\\ &\ge \sum_{i=1}^{\infty} c_i^2 - \sum_{i=1}^{\infty} c_i\|x_i - T^n x_i\| \ge 1 - \sum_{i=1}^{\infty} c_i\delta_i > 1 - \sum_{i=1}^{\infty}\frac{1-a_1}{2^i} = a_1 \ge a_n.\end{aligned}$$

Let $k \ge 1$ and $m_{k-1} < n \le m_k$. Then

$$\begin{aligned}\operatorname{Re}\langle T^n x, x\rangle &= \operatorname{Re}\sum_{i=1}^{k-1} c_i\langle T^n x_i, x\rangle + \operatorname{Re} c_k\langle T^n x_k, x\rangle + \operatorname{Re}\sum_{i=k+1}^{\infty} c_i\langle T^n x_i, x\rangle\\ &\ge -\sum_{i=1}^{k-1} c_i\|T^n x_i\| - Kc_k + \operatorname{Re}\sum_{i=k+1}^{\infty} c_i\Big(\langle x_i, x\rangle - \langle x_i - T^n x_i, x\rangle\Big)\\ &\ge -\sum_{i=1}^{k-1} c_i\delta_{k-1} - Kc_k + \sum_{i=k+1}^{\infty} c_i^2 - \sum_{i=k+1}^{\infty} c_i\|x_i - T^n x_i\|\\ &\ge -(k-1)\delta_{k-1} - Kc_k + \sum_{i=k+1}^{\infty} c_i^2 - \sum_{i=k+1}^{\infty}\delta_i\\ &\ge \sum_{i=k+1}^{\infty} c_i^2 - 3Kc_k > a_{m_{k-1}} \ge a_n.\end{aligned}$$

Thus $\operatorname{Re}\langle T^n x, x\rangle > a_n$ for all $n \ge 1$. □

Corollary 13. *Let $T \in \mathcal{B}(H)$ be an operator of class $C_{0\cdot}$ satisfying $r(T) = 1$. Let $a_n > 0 \quad (n \in \mathbb{N})$ and $a_n \to 0$. Then there exists $x \in H$ such that $|\langle T^n x, x\rangle| \ge a_n$ for all $n \in \mathbb{N}$. Moreover, given $\varepsilon > 0$, it is possible to find $x \in H$ with this property such that $\|x\| < \sup_n a_n^{1/2} + \varepsilon$.*

Proof. Let $\lambda \in \sigma(T)$, $|\lambda| = 1$. Then $1 \in \sigma(\lambda^{-1}T)$ and the statement follows from Theorem 12. □

Let C be a subset of a Banach space X. We say that C is a *cone* if $C + C \subset C$ and $tC \subset C$ for all $t \ge 0$.

Theorem 14. *Let T be an operator a Hilbert space H of class $C_{0\cdot}$ such that $1 \in \sigma(T)$. Then T has a non-trivial closed invariant cone.*

Proof. By Theorem 12, there is a non-zero vector $x \in H$ with $\mathrm{Re}\langle T^n x, x\rangle \geq 0$ for all n. Let $C = \left\{\sum_{i=1}^n \alpha_i T^i x : n \in \mathbb{N}, \alpha_i \geq 0 \text{ for } 1 \leq i \leq n\right\}^-$. Clearly C is a closed cone invariant for T. We have $x \in C$, so $C \neq \{0\}$. For each $u \in C$ we have $\mathrm{Re}\langle u, x\rangle \geq 0$, so $C \neq H$ (for example, $-x \notin C$). □

The assumption that T is of class $C_{0\cdot}$ is not essential. It can be omitted by a standard reduction technique.

Let X, Y be Banach spaces, $T \in \mathcal{B}(X)$ and $S \in \mathcal{B}(Y)$. We write $T \prec S$ if there is a one-to-one operator $A : X \to Y$ with dense range such that $AT = SA$. We say that T is quasisimilar to S if $T \prec S$ and $S \prec T$. It is easy to see that $T \prec S$ implies $S^* \prec T^*$. Furthermore, if $T \prec S$ and $S \prec V$, then $T \prec V$.

Theorem 15. *Let T be a power bounded operator of class C_{11} acting on a Hilbert space H. Then T is quasisimilar to a unitary operator.*

Proof. Fix a Banach limit – a linear functional $LIM : \ell^\infty \to \mathbb{C}$ such that, for all $(a_n) \in \ell^\infty$, we have $LIM(a_{n+1}) = LIM(a_n)$, $a_n \geq 0 \quad (n \in \mathbb{N}) \Rightarrow LIM(a_n) \geq 0$ and $LIM(a_n) = \lim a_n$ whenever (a_n) is a convergent sequence.

Define in H a new norm $|||\cdot|||$ by $|||x||| = LIM\|T^n x\|$. Clearly, $|||x||| \geq \inf_n \|T^n x\| > 0$ for each $x \in H$, $x \neq 0$. Then H with this norm is a pre-Hilbert space (since the norm $|||\cdot|||$ can be defined by the new scalar product $[x, y] = LIM\langle T^n x, T^n y\rangle$). Let $k = \sup \|T^n\|$. Clearly $|||x||| \leq k\|x\|$ and $|||Tx||| = |||x|||$ for all $x \in H$.

Let K_1 be the completion of the space $(H, |||\cdot|||)$ and let $U_1 \in \mathcal{B}(K_1)$ be the uniquely determined extension of the operator $x \mapsto Tx$ acting in $(H, |||\cdot|||)$. Then U_1 is an isometry.

Since T is of class C_{11}, the operator T^* is one-to-one, and so T has dense range. Consequently, U_1 has a dense range in K_1, and so it is unitary. Let $W_1 : H \to K_1$ be the operator induced by the identity on H. Clearly $\|W_1\| \leq 1$, W_1 is one-to-one and has dense range. We have $W_1 T = U_1 W_1$ and so $T \prec U_1$.

Applying the same considerations to T^* instead of T, there are a Hilbert space K_2 and a unitary operator $U_2 \in \mathcal{B}(K_2)$ such that $T^* \prec U_2$. Consequently, $U_2^* \prec T \prec U_1$ and so $U_2^* \prec U_1$. Let $W : K_2 \to K_1$ be a one-to-one operator with dense range satisfying $WU_2^* = U_1 W$. By the Fuglede-Putnam theorem, $WU_2 = U_1^* W$, which means $U_2 \prec U_1^*$. Thus $U_1 \prec U_2^*$. Hence $U_1 \prec U_2^* \prec T \prec U_1$ and T is quasisimilar to U_1. □

Theorem 16. *Let $\dim H \geq 2$ and let $T \in \mathcal{B}(H)$ be a power bounded operator of class C_{11}. Then T has a non-trivial closed invariant subspace.*

Proof. By Theorem 15, T is quasisimilar to a unitary operator $U \in \mathcal{B}(K)$. So there are one-to-one operators $W_1 : H \to K$ and $W_2 : K \to H$ with dense ranges such that $W_1 T = UW_1$ and $TW_2 = W_2 U$.

We have $(W_1 W_2)U = W_1 T W_2 = U(W_1 W_2)$.

If $U = \lambda I$ for some $\lambda \in \mathbb{C}$, then $W_1 T x = \lambda W_1 x$, and so $Tx = \lambda x$ for all $x \in H$. Hence T has non-trivial closed invariant subspaces.

If U is not a scalar multiple of the identity, then $\operatorname{card} \sigma(U) \geq 2$. Let $K_0 \subset K$ be any non-trivial spectral subspace of U. Then $UK_0 \subset K_0$ and $W_1 W_2 K_0 \subset K_0$, since $W_1 W_2$ commutes with U.

Let $H_0 = \overline{W_2 K_0}$. We have $TW_2 K_0 = W_2 U K_0 \subset W_2 K_0 \subset H_0$ and so $TH_0 \subset H_0$. Clearly $H_0 \neq \{0\}$. Further, $W_1 W_2 K_0 \subset K_0$, and so $W_2 K_0 \subset W_1^{-1} K_0$ and $H_0 \subset W_1^{-1} K_0$. Since W_1 has dense range, $W_1^{-1} K_0 \neq H$, and so H_0 is non-trivial. □

Theorem 17. *Let T be a power bounded operator on a Hilbert space H such that $1 \in \sigma(T)$. Then T has a non-trivial closed invariant cone.*

Proof. Let $H_1 = \{x \in H : T^n x \to 0\}$. Then H_1 is a closed subspace of H invariant for T. If H_1 is non-trivial, then T has even a non-trivial closed invariant subspace. If $H_1 = H$, then the statement follows from Theorem 14. Therefore we may assume that $H_1 = \{0\}$.

Let $H_2 = \{x \in H : T^{*n} x \to 0\}$. Then H_2 is a closed subspace invariant for T^*, and so $H_2^{\perp}$ is a closed subspace invariant for T. If H_2 is non-trivial, then the theorem is proved. If $H_2 = H$, then, by Theorem 12, there is a non-zero vector $x \in H$ with $\operatorname{Re}\langle T^n x, x\rangle = \operatorname{Re}\langle x, T^{*n} x\rangle \geq 0$, and so x generates a non-trivial closed cone invariant for T.

Hence we may assume that $M_1 = \{0\} = M_2$. So T is of class C_{11}. If $\dim H \geq 2$, then T has a non-trivial closed invariant subspace by Theorem 16. If $\dim H = 1$, then $T = I$ and the statement is clear. □

40 Scott Brown technique

The Scott Brown technique is an efficient way of constructing invariant subspaces. It was first used for subnormal operators but later it was adapted to contractions on Hilbert spaces and, more generally, to polynomially bounded operators on Banach spaces. Some results are also known for n-tuples of commuting operators.

We are going to give two illustrative examples showing how this method works.

The basic idea of the Scott Brown technique is to construct a weak orbit $\{\langle T^n x, x^* \rangle : n = 0, 1, \dots\}$ which behaves in a precise way. Typically, vectors $x \in X$ and $x^* \in X^*$ are constructed such that

$$\langle T^n x, x^* \rangle = \begin{cases} 0 & n \geq 1; \\ 1 & n = 0. \end{cases}$$

Equivalently,

$$\langle p(T)x, x^* \rangle = p(0) \tag{1}$$

for all polynomials p. Then T has a non-trivial closed invariant subspace. Indeed, either $Tx = 0$ (and x generates a 1-dimensional invariant subspace) or the vectors $\{T^n x : n \geq 1\}$ generate a non-trivial closed invariant subspace.

The vectors x and x^* satisfying the above-described conditions are constructed as limits of sequences that satisfy (1) approximately.

Let $\mathbb{D} = \{z \in \mathbb{C} : |z| < 1\}$ denote the open unit disc in the complex plane and $\mathbb{T} = \{z \in \mathbb{C} : |z| = 1\}$ the unit circle. Denote by $\mathcal{P}$ the normed space of all polynomials with the norm $\|p\| = \sup\{|p(z)| : z \in \mathbb{D}\}$. Let $\mathcal{P}^*$ be its dual with the usual dual norm.

Let $\phi \in \mathcal{P}^*$. By the Hahn-Banach theorem, ϕ can be extended without changing the norm to a functional on the space of all continuous function on $\mathbb{T}$ with the sup-norm. By the Riesz theorem, there exists a Borel measure μ on $\mathbb{T}$ such that $\|\mu\| = \|\phi\|$ and $\phi(p) = \int p \, \mathrm{d}\mu$ for all polynomials p. Clearly, the measure is not unique.

Let L^1 be the Banach space of all complex integrable functions on $\mathbb{T}$ with the norm $\|f\|_1 = \frac{1}{2\pi} \int_{-\pi}^{\pi} |f(e^{it})| \, \mathrm{d}t$.

Of particular interest are the following functionals on $\mathcal{P}$:

(i) Let $\lambda \in \mathbb{D}$. Denote by $\mathcal{E}_\lambda$ the evaluation at the point λ, i.e., $\mathcal{E}_\lambda$ is defined by $\mathcal{E}_\lambda(p) = p(\lambda) \quad (p \in \mathcal{P})$. Clearly, $\|\mathcal{E}_\lambda\| = 1$.

(ii) Let $f \in L^1$. Denote by $M_f \in \mathcal{P}^*$ the functional defined by

$$M_f(p) = \frac{1}{2\pi} \int_{-\pi}^{\pi} p(e^{it}) f(e^{it}) \, \mathrm{d}t \qquad (p \in \mathcal{P}).$$

Then $\|M_f\| \leq \|f\|_1$.

The evaluation functionals $\mathcal{E}_\lambda$ are also of this type. Indeed, for $\lambda \in \mathbb{D}$ we have $\mathcal{E}_\lambda = M_{P_\lambda}$, where $P_\lambda(e^{it}) = \frac{1-|\lambda|^2}{|\lambda - e^{it}|^2}$ is the Poisson kernel. In particular, if $g = 1$, then $M_g(p) = p(0)$ for all p, and so M_g is the evaluation at the origin.

(iii) Let $k > 0$ and let $T : X \to X$ a *polynomially bounded* operator with *polynomial bound* k, i.e., T satisfies the condition $\|p(T)\| \leq k\|p\|$ for all polynomials p. Fix $x \in X$ and $x^* \in X^*$. Let $x \otimes x^* \in \mathcal{P}^*$ be the functional defined by

$$(x \otimes x^*)(p) = \langle p(T)x, x^* \rangle \qquad (p \in \mathcal{P}).$$

Since T is polynomially bounded, $x \otimes x^*$ is a bounded functional and we have $\|x \otimes x^*\| \leq k\|x\| \cdot \|x^*\|$.

Of course the definition of $x \otimes x^*$ depends on the operator T but since we are going to consider only one operator T, this cannot lead to a confusion.

By the von Neumann inequality, any contraction on a Hilbert space is polynomially bounded with polynomial bound equal to 1.

Denote by L^∞ the space of all bounded measurable functions on $\mathbb{T}$ with the usual norm. Since $\mathcal{P} \subset L^\infty = (L^1)^*$, the space $\mathcal{P}$ inherits the w^*-topology from L^∞.

Of particular importance for the Scott Brown technique are those functionals on $\mathcal{P}$ that are w^*-continuous, i.e., that are continuous functions from $(\mathcal{P}, w^*)$ to $\mathbb{C}$. Equivalently, these functionals can be represented by absolutely continuous measures. For basic properties of w^*-continuous functionals on $\mathcal{P}$ see Appendix 6.

Let $T \in B(X)$ be a polynomially bounded operator such that $\|T^n u\| \to 0$ for all $u \in X$. Then all the functionals $x \otimes x^*$ can be represented by absolutely continuous measures. Equivalently, these functionals are w^*-continuous.

For each polynomially bounded operator T is is possible to extend the polynomial calculus by continuity to the norm-closure of $\mathcal{P}$, i.e., to the disc algebra $\mathcal{A}(\mathbb{D})$. If T is of class $C_{0\cdot}$, then it is even possible to extend this functional calculus to H^∞. We summarize the results in the following theorem.

Theorem 1. *Let $T \in \mathcal{B}(X)$ be a polynomially bounded operator with polynomial bound k. Suppose that $\|T^n u\| \to 0$ for all $u \in X$. Then:*

(i) *$x \otimes x^*$ can be represented by an absolutely continuous measure for all $x \in X$ and $x^* \in X^*$. Equivalently, $x \otimes x^*$ is w^*-continuous;*

(ii) *there exists an algebraic homomorphism $H^\infty \to B(X)$, $h \in H^\infty \mapsto h(T)$ such that*

$$\|h(T)\| \le k\|h\| \text{ for each } h \in H^\infty;$$
$$h(z) \equiv 1 \Rightarrow h(T) = I;$$
$$h(z) \equiv z \Rightarrow h(T) = T;$$
$$\text{if } h, h_n \in H^\infty,\ h_n \xrightarrow{w^*} h, \text{ then } \|(h_n(T)x - h(T)x\| \to 0 \text{ for each } x \in X;$$

(iii) *the set $\{h(T)x : h \in H^\infty, \|h\| \le 1\}$ is compact for all $x \in X$.*

Proof. (i) Recall that a sequence $(p_n)_n \subset \mathcal{P}$ is called Montel if $\sup \|p_n\| < \infty$ and $\lim\limits_{n\to\infty} p_n(z) = 0$ for all $z \in \mathbb{D}$. We show that $\langle p_n(T)x, x^* \rangle \to 0$ for any Montel sequence (p_n).

Without loss of generality we can assume that $\sup \|p_n\| \le 1$, $\|x\| \le 1$ and $\|x^*\| \le 1$. Let $p_n(z) = \sum_{j=0}^\infty c_{n,j} z^j$. By the Cauchy formula and the Lebesgue dominated convergence theorem, we have $\lim_{n\to\infty} c_{n,j} = 0$ for each $j \ge 0$.

Let ε be a positive number such that $\varepsilon < 2k$. Choose l such that $\|T^l x\| \le \varepsilon/4k$. There exists n_0 sufficiently large such that for every $n \ge n_0$ we have $|c_{n,j}| < \varepsilon/2lk$ $(j = 0, \dots, l)$. Fix such an n and write $g(z) = \sum_{j=0}^{l-1} c_{n,j} z^j$. Then $p_n(z) = g(z) + z^l h(z)$ for some polynomial h. Clearly $\|g\| \le \sum_{j=0}^{l-1} |c_{n,j}| \le \varepsilon/2k$ and

$\|h\| = \|p_n - g\|$. Thus

$$\begin{aligned}|\langle p_n(T)x, x^*\rangle| &\le \|p_n(T)x\| \le \|g(T)x\| + \|(p_n - g)(T)x\|\\ &\le k\|g\| + \|h(T)\| \cdot \|T^l x\| \le \frac{\varepsilon}{2} + k\|p_n - g\| \cdot \frac{\varepsilon}{4k}\\ &\le \frac{\varepsilon}{2} + \frac{\varepsilon}{4} \cdot (\|p_n\| + \|g\|) < \varepsilon.\end{aligned}$$

Thus $\langle p_n(T)x, x^*\rangle \to 0$. By Theorem A.6.3, $x \otimes x^*$ is w^*-continuous.

Note that we have proved even that $\|p_n(T)x\| \to 0$ for each Montel sequence (p_n).

(ii) By (i) and Theorem A.6.3, for all $x \in X$ and $x^* \in X^*$ there exists $f_{x,x^*} \in L^1$ such that $\|f\|_1 \le k\|x\| \cdot \|x^*\|$ and $\langle p(T)x, x^*\rangle = \frac{1}{2\pi}\int_0^{2\pi} p(e^{it})f(e^{it})\,\mathrm{d}t$ for all $p \in \mathcal{P}$.

For $h \in H^\infty$ define $F_h(x, x^*) = \frac{1}{2\pi}\int_0^{2\pi} h(e^{it})f_{x,x^*}(e^{it})\,\mathrm{d}t$. Clearly, F_h is a bilinear form and $|F_h(x, x^*)| \le k\|x\| \cdot \|x^*\|$ for all $x \in X$ and $x^* \in X^*$. Hence F_h defines a bounded linear operator $h(T) : X \to X^{**}$ by $\langle h(T)x, x^*\rangle = F_h(x, x^*)$. Clearly, $\|h(T)\| \le k\|h\|$ and the mapping $h \mapsto h(T)$ is linear.

Moreover, for each $k \in \mathbb{N}$ and $p \in \mathcal{P}$ we have

$$\begin{aligned}&\frac{1}{2\pi}\int_0^{2\pi} p(e^{it})\big(e^{ikt}f_{x,x^*}(e^{it}) - f_{T^k x,x^*}(e^{it})\big)\,\mathrm{d}t\\ &\qquad = \langle p(T)T^k x, x^*\rangle - \langle p(T)T^k x, x^*\rangle = 0,\end{aligned}$$

and so, for each $h \in H^\infty$,

$$\begin{aligned}\langle (z^k h)(T)x, x^*\rangle &= \frac{1}{2\pi}\int_0^{2\pi} h(e^{it})e^{ikt}f_{x,x^*}(e^{it})\,\mathrm{d}t\\ &= \frac{1}{2\pi}\int_0^{2\pi} h(e^{it})f_{T^k x,x^*}(e^{it})\,\mathrm{d}t = \langle h(T)T^k x, x^*\rangle.\end{aligned}$$

Hence $(z^k h)(T) = h(T)T^k$.

As in (i), we can prove that $\|h_n(T)x\| \to 0$ for every $x \in X$ and every Montel sequence $(h_n) \subset H^\infty$. For each $h \in H^\infty$ there is a sequence $(p_n) \subset \mathcal{P}$ such that $p_n \xrightarrow{w^*} h$. Hence $h(T)x = \lim_{n\to\infty} p_n(T)x \in X$ and $h(T) \in \mathcal{B}(X)$.

We show that the mapping $h \mapsto h(T)$ is multiplicative. Let $h, h' \in H^\infty$. There exist sequences $(p_n), (p'_n) \subset \mathcal{P}$ such that $p_n \xrightarrow{w^*} h$ and $p'_n \xrightarrow{w^*} h'$. Then $p_n p'_n \xrightarrow{w^*} hh'$ and

$$(hh')(T)x = \lim_{n\to\infty} p_n(T)p'_n(T)x = \lim_{n\to\infty} p_n(T)h'(T)x = h(T)h'(T)x.$$

(iii) Let $x \in X$ and let $(g_n) \subset H^\infty$ be a bounded sequence. Then there exists a pointwise convergent subsequence $g_{n_k} \xrightarrow{w^*} g$ for some $g \in H^\infty$. Then $g_{n_k}(T)x \to g(T)x$, which means that the set $\{h(T)x : h \in H^\infty, \|h\| \le 1\}$ is compact. □

Proposition 2. *Let $T \in \mathcal{B}(X)$ be a polynomially bounded operator with polynomial bound k, let $\lambda \in \mathbb{C}$, $x \in X$, $x^* \in X^*$, $\|x\| = 1 = \|x^*\|$, $\varepsilon > 0$ and $\|(T-\lambda)x\| < \varepsilon$. Then*

$$\|x \otimes x^* - \mathcal{E}_\lambda \langle x, x^* \rangle\| < \frac{2k\varepsilon}{1-|\lambda|}.$$

Proof. We have

$$\big\|x \otimes x^* - \mathcal{E}_\lambda \langle x, x^*\rangle\big\| = \sup_{\|p\|=1} \big|\langle p(T)x, x^*\rangle - p(\lambda)\langle x, x^*\rangle\big|.$$

For $p \in \mathcal{P}$, $\|p\| = 1$ write $q(z) = \frac{p(z)-p(\lambda)}{z-\lambda}$. Then $\|q\| \le \frac{2\|p\|}{1-|\lambda|} = \frac{2}{1-|\lambda|}$. Thus

$$\begin{aligned}\big|\langle p(T)x, x^*\rangle - p(\lambda)\langle x, x^*\rangle\big| &= \big|\langle q(T)(T-\lambda)x, x^*\rangle\big| \\ &\le \|q(T)\| \cdot \|(T-\lambda)x\| \le \frac{2k\varepsilon}{1-|\lambda|}.\end{aligned}$$ □

The previous lemma shows that the points in the approximate point spectrum of T are of particular interest for the Scott Brown technique. Even more useful are the points of the essential approximate point spectrum. For simplicity, we formulate the next result only for Hilbert space operators. However, it can be adapted to Banach spaces easily.

Proposition 3. *Let T be a polynomially bounded operator on a Hilbert space H. Let $u_1, \dots, u_n \in H$, $\lambda \in \sigma_{\pi e}(T)$ and $\varepsilon > 0$. Then there exists $x \in H$ of norm 1 such that $x \perp \{u_1, \dots, u_n\}$ and*

$$\begin{aligned}\|x \otimes x - \mathcal{E}_\lambda\| &\le \varepsilon, \\ \|x \otimes u_i\| &\le \varepsilon \qquad (i = 1, \dots, n), \\ \|u_i \otimes x\| &\le \varepsilon \qquad (i = 1, \dots, n).\end{aligned}$$

Proof. Since the set $\{p(T)x : p \in \mathcal{P}, \|p\| \le 1\}^-$ is compact, there are vectors $v_1, \dots, v_m \in H$ such that $\operatorname{dist}\{p(T)x, \{v_1, \dots, v_m\}\} < \varepsilon$ for each polynomial p, $\|p\| \le 1$. Choose $x \in \{u_1, \dots, u_n, v_1, \dots, v_m\}^\perp$, $\|x\| = 1$ and $\|(T-\lambda)x\| < \frac{\varepsilon(1-|\lambda|)}{2k}$, where k is the polynomial bound of T. Then the first two inequalities are satisfied by Proposition 2.

For the last inequality (note that the second and third inequalities are not symmetrical!), let $i \in \{1, \dots, n\}$ and $p \in \mathcal{P}$, $\|p\| \le 1$. Find j such $\|p(T)u_i - v_j\| < \varepsilon$. Then

$$|\langle p(T)u_i, x\rangle| \le |\langle p(T)u_i - v_j, x\rangle| \le \|p(T)x - v_j\| < \varepsilon.$$ □

Now we are able to give an illustrative example how the Scott Brown technique can be applied.

A subset $\Lambda \subset \mathbb{D}$ is called dominant if $\sup_{z\in\Lambda} |f(z)| = \|f\|$ for all $f \in H^\infty$, see Appendix 6.

Theorem 4. *Let T be a polynomially bounded operator of class $C_{0\cdot}$ on a Hilbert space H such that $\sigma(T) \cap \mathbb{D}$ is dominant in $\mathbb{D}$. Then T has a non-trivial closed invariant subspace.*

Proof. Without loss of generality we can assume that neither T nor T^* has eigenvalues. In particular, $\sigma_{\pi e}(T) = \sigma(T)$.

By Proposition 3, we can approximate (with an arbitrary precision) the evaluation functionals $\mathcal{E}_\lambda$ for $\lambda \in \sigma_{\pi e}(T)$ by the functionals of the type $x \otimes x$ with $x \in H$, $\|x\| = 1$.

We show first the following result:

(a) Let $\psi \in \mathcal{P}^*$ be a w^*-continuous functional, let $x, y \in H$ and $\varepsilon > 0$. Then there are $x', y' \in H$ such that

$$\begin{aligned} \|x' \otimes y' - \psi\| &< \varepsilon, \\ \|x' - x\| &\le \|x \otimes y - \psi\|^{1/2}, \\ \|y' - y\| &\le \|x \otimes y - \psi\|^{1/2}. \end{aligned}$$

Proof. By Theorem A.6.4, there are elements $\lambda_1, \ldots, \lambda_n \in \sigma_{\pi e}(T)$ and non-zero complex numbers $\alpha_1, \ldots, \alpha_n$ such that $\sum_{i=1}^n |\alpha_i| \le \|x \otimes y - \psi\|$ and

$$\Big\| x \otimes y - \psi + \sum_{i=1}^n \alpha_i \mathcal{E}_{\lambda_i} \Big\| < \varepsilon/2.$$

Let δ be a sufficiently small positive number.

By Proposition 3, we can find inductively mutually orthogonal unit vectors $u_1, \ldots, u_n \in H$ such that

$$\begin{aligned} \|x \otimes u_i\| &< \delta, \\ \|u_i \otimes y\| &< \delta, \\ \|u_i \otimes u_j\| &< \delta \qquad (i \ne j), \\ \|u_i \otimes u_i - \mathcal{E}_{\lambda_i}\| &< \delta. \end{aligned}$$

Set $x' = x + \sum_{i=1}^n \frac{\alpha_i}{|\alpha_i|^{1/2}} u_i$ and $y' = y + \sum_{i=1}^n |\alpha_i|^{1/2} u_i$. Since the vectors $u_1, \ldots, u_n$ are orthonormal, we have $\|x' - x\|^2 = \sum_{i=1}^n |\alpha_i| \le \|x \otimes y - \psi\|$, and similarly, $\|y' - y\|^2 \le \|x \otimes y - \psi\|$. Furthermore,

$$\begin{aligned} \|x' \otimes y' - \psi\| &\le \Big\| x \otimes y - \psi + \sum_{i=1}^n \alpha_i \mathcal{E}_{\lambda_i} \Big\| + \Big\| \sum_{i=1}^n \alpha_i \big(u_i \otimes u_i - \mathcal{E}_{\lambda_i}\big) \Big\| \\ &\quad + \sum_{i=1}^n |\alpha_i|^{1/2} \|u_i \otimes y\| + \sum_{i=1}^n |\alpha_i|^{1/2} \|x \otimes u_i\| + \sum_{i \ne j} |\alpha_i|^{1/2} \cdot |\alpha_j|^{1/2} \cdot \|u_i \otimes u_j\| \\ &\le \varepsilon/2 + \delta \Big(\sum_{i=1}^n |\alpha_i| + 2 \sum_{i=1}^n |\alpha_i|^{1/2} + \sum_{i \ne j} |\alpha_i|^{1/2} |\alpha_j|^{1/2} \Big) < \varepsilon \end{aligned}$$

provided δ is sufficiently small.

(b) There are $x, y \in H$ such that $x \otimes y = \mathcal{E}_0$.

Proof. Set $x_0 = 0 = y_0$. Using (a) it is possible to construct inductively vectors $x_j, y_j \in H \quad (j \in \mathbb{N})$ such that

$$\begin{aligned} &\|x_j \otimes y_j - \mathcal{E}_0\| \le 2^{-2j}, \\ &\|x_{j+1} - x_j\| \le \|x_j \otimes y_j - \mathcal{E}_0\|^{1/2} \le 2^{-j} \quad \text{and} \\ &\|y_{j+1} - y_j\| \le 2^{-j}. \end{aligned}$$

Clearly, the sequences (x_j) and (y_j) are Cauchy. Let x and y be their limits. It is easy to verify that $x \otimes y = \mathcal{E}_0$. Indeed, for each polynomial p with $\|p\| = 1$ we have

$$\begin{aligned} \big|\langle p(T)x, y\rangle - \langle p(T)x_n, y_n\rangle\big| &\le \big|\langle p(T)x, y - y_n\rangle\big| + \big|\langle p(T)(x - x_n), y_n\rangle\big| \\ &\le k\|y - y_n\| \cdot \|x\| + k\|x - x_n\| \cdot \|y_n\|, \end{aligned}$$

and so $\langle p(T)x, y\rangle = \lim_{n\to\infty}\langle p(T)x_n, y_n\rangle = p(0)$.

It was shown above that (b) implies the existence of a non-trivial closed invariant subspace. Indeed, $M = \bigvee\{T^n x : n \ge 1\}$ is a closed subspace invariant for T. Since we assumed that $\sigma_p(T)$ is empty, we have $Tx \ne 0$. We have $y \ne 0$ and $\langle T^n x, y\rangle = 0$ for all $n \ge 1$, and so $M \ne H$. □

The condition that $T^n x \to 0$ for all $x \in H$ can be omitted by a standard reduction argument.

Theorem 5. *Let T be a polynomially bounded operator on a Hilbert space H such that the spectrum $\sigma(T) \cap \mathbb{D}$ is dominant in $\mathbb{D}$. Then T has a non-trivial closed invariant subspace.*

Proof. Let $M_1 = \{x \in H : T^n x \to 0\}$. It is easy to see that M_1 is a closed subspace of H invariant for T. If $M_1 = H$, then T has a non-trivial invariant subspace by Theorem 4. Thus we can assume without loss of generality that $M_1 = \{0\}$.

Let $M_2 = \{x \in H : T^{*n} x \to 0\}$. Then M_2 is a closed subspace invariant for T^*. Consequently, $M_2^\perp$ is invariant for T. If $M_2 = H$, then T^* has a non-trivial closed invariant subspace by Theorem 4 and so has T. Thus we can also assume that $M_2 = \{0\}$.

Hence T is of class C_{11}. By Theorem 39.15, T is quasisimilar to a unitary operator. By Theorem 39.16, T has a non-trivial closed invariant subspace. □

The assumption that the spectrum of T is dominant can be replaced by a weaker assumption that $\sigma(T) \supset \mathbb{T}$. Note that the points $\lambda \in \sigma(T)$ with $|\lambda| = 1$ cannot be used directly in the Scott Brown technique, cf. Proposition 2.

The following lemma gives the basic idea how to use the information that $\sigma(T) \supset \mathbb{T}$ for the Scott Brown technique.

Definition 6. *A subset $\Lambda \subset \mathbb{D}$ is called Apostol if*

$$\sup\{r \in [0, 1) : re^{i\theta} \in \Lambda\} = 1$$

for all but countably many numbers $\theta \in \langle -\pi, \pi\rangle$.

Theorem 7. *Let T be a polynomially bounded operator on a Banach space such that $\sigma(T) \supset \mathbb{T}$ and T has no non-trivial closed invariant subspace. Let $\varepsilon > 0$. Then the set*

$$\{\lambda \in \mathbb{D} : \text{there exists } u \in X \text{ with } \|u\| = 1 \text{ and } \|Tu - \lambda u\| \leq \varepsilon(1 - |\lambda|)^2\} \qquad (2)$$

is an Apostol set.

Proof. We can assume that $T - \lambda$ is one-to-one with dense range for each $\lambda \in \mathbb{C}$ (otherwise T has a non-trivial closed invariant subspace). Thus $T - \lambda$ is invertible for all $\lambda \in \mathbb{D} \setminus \Lambda$ and $\|(T - \lambda)^{-1}\| \leq \frac{1}{\varepsilon(1-|\lambda|)^2}$. Clearly, $\sigma(T) \subset \overline{\mathbb{D}}$ and for $|\lambda| > 1$ we have $\|(T - \lambda)^{-1}\| = \left\| -\sum_{n=0}^{\infty} \frac{T^n}{\lambda^{n+1}} \right\| \leq \frac{k}{|\lambda|-1}$, where k is the polynomial bound of T.

Suppose that the set Λ defined in (2) is not Apostol. Then there exists an uncountable set $\Omega \subset \mathbb{T}$ such that, for each $\omega \in \Omega$, there exists $r_\omega < 1$ such that $\{r\omega : r_\omega \leq r < 1\} \subset \mathbb{D} \setminus \Lambda$. Consequently, there exists a component G of the open set $\mathbb{D} \setminus \Lambda$ and mutually distinct points $\omega_1, \dots, \omega_4$ such that $\{r\omega : r_\omega \leq r < 1\} \subset G$ for $j = 1, \dots, 4$. Suppose that $\omega_1, \dots, \omega_4$ are arranged in counter clockwise order.

It is possible to find two rectifiable simple closed curves Γ, Γ' such that $\Gamma \cap \Gamma' = \emptyset$, $\Gamma \cap \mathbb{T} = \{\omega_1, \omega_2\}$, $\Gamma' \cap \mathbb{T} = \{\omega_3, \omega_4\}$, Γ and Γ' intersect $\mathbb{T}$ along radial segments and $\Gamma \cap \mathbb{D} \subset G$, $\Gamma' \cap \mathbb{D} \subset G$.

Define operators

$$A = \frac{1}{2\pi i} \int_\Gamma (z - \omega_1)^2 (z - \omega_2)^2 (z - T)^{-1} \, dz$$

and

$$A' = \frac{1}{2\pi i} \int_{\Gamma'} (z - \omega_3)^2 (z - \omega_4)^2 (z - T)^{-1} \, dz.$$

By definition, A and A' are well-defined operators commuting with T. Hence $\overline{AX}$ is a closed subspace invariant for T. To show that $\overline{AX}$ is non-trivial it is sufficient to show that $A \neq 0$, $A' \neq 0$ and $A'A = 0$.

Let $\mathcal{A} = \{T, A, A'\}''$ (the bicommutant of $\{T, A, A,\}$). Then $\mathcal{A}$ is a commutative Banach algebra and $\sigma^{\mathcal{A}}(T) = \sigma^{\mathcal{B}(X)}(T)$.

Fix a point $\eta \in \mathbb{T}$ surrounded by Γ. Then $\eta \in \sigma^{\mathcal{A}}(T)$ and there exists a multiplicative functional $\varphi \in \mathcal{M}(\mathcal{A})$ such that $\varphi(T) = \eta$. We have

$$\begin{aligned}
\varphi(A) &= \frac{1}{2\pi i} \int_\Gamma (z - \omega_1)^2 (z - \omega_2)^2 \varphi((z - T)^{-1}) \, dz \\
&= \frac{1}{2\pi i} \int_\Gamma (z - \omega_1)^2 (z - \omega_2)^2 (z - \eta)^{-1} \, dz = (\eta - \omega_1)^2 (\eta - \omega_2)^2
\end{aligned}$$

by the Cauchy theorem. Thus $\varphi(A) \neq 0$, and so $A \neq 0$.

Similarly, $A' \neq 0$.

For the proof that $AA' = 0$ we use the resolvent equality. With the notation $p(z) = (z-\omega_1)^2(z-\omega_2)^2$, $q(z) = (z-\omega_3)^2(z-\omega_4)^2$ we have

$$\begin{aligned}(2\pi i)^2 A'A &= \int_{\Gamma'}\int_{\Gamma} p(z)q(w)(z-T)^{-1}(w-T)^{-1}\,\mathrm{d}z\,\mathrm{d}w \\ &= \int_{\Gamma'}\int_{\Gamma} \frac{p(z)q(w)}{w-z}\big((z-T)^{-1}-(w-T)^{-1}\big)\,\mathrm{d}z\,\mathrm{d}w \\ &= \int_{\Gamma} p(z)(z-T)^{-1}\Big(\int_{\Gamma'}\frac{q(w)}{w-z}\,\mathrm{d}w\Big)\,\mathrm{d}z \\ &\quad - \int_{\Gamma'} q(w)(w-T)^{-1}\Big(\int_{\Gamma}\frac{p(z)}{w-z}\,\mathrm{d}z\Big)\,\mathrm{d}w = 0\end{aligned}$$

since the inner integrals are equal to 0 by the Cauchy theorem. □

Note that points λ satisfying (2) provide less information than the points of the approximate point spectrum. On the other hand, we have a larger set of such points (any Apostol set is dominant but not conversely).

For every $\lambda \in \mathbb{D}$, let $P_\lambda(t) = \frac{1-|\lambda|^2}{|\lambda-e^{it}|^2}$ $(t \in \mathbb{R})$ denote the Poisson kernel. Recall that $\int_{-\pi}^{\pi} P_\lambda\,\mathrm{d}t = 2\pi$ and $\max_t P_\lambda(t) = \frac{1+|\lambda|}{1-|\lambda|}$.

For $\lambda = re^{i\theta} \in \mathbb{D}$ with $|\lambda| \ge 3/4$, set $I_\lambda = \{e^{it} : |t-\theta| < 2(1-r)\}$ and define the 2π–periodic function Q_λ on $\mathbb{R}$ by: $Q_\lambda(t) = P_\lambda(t)$ if $e^{it} \in I_\lambda$, and $Q_\lambda(t) = 0$ otherwise. Denote by m the Lebesgue measure both on the real line $\mathbb{R}$ and on the unit circle $\mathbb{T}$.

Lemma 8. *For any $\lambda \in \mathbb{D}$ with $|\lambda| \ge 3/4$ we have $\int_{-\pi}^{\pi} Q_\lambda(t)\,\mathrm{d}t \ge \frac{7\pi}{6}$.*

Proof. Without loss of generality we can suppose that $\lambda = r \ge 3/4$. We have $\sin^2(1-r) \le \sin(1-r) \le 1-r$. If $|t| \le \pi$ and $e^{it} \in I_\lambda$, then

$$\cos t \ge \cos 2(1-r) = 1 - 2\sin^2(1-r) \ge 2r-1$$

and $|r-e^{it}|^2 = (r-\cos t)^2 + \sin^2 t \le (1-r)^2 + t^2$. Hence

$$\begin{aligned}\int_{-\pi}^{\pi} Q_\lambda(t)\,\mathrm{d}t &= \int_{-2(1-r)}^{2(1-r)} \frac{1-r^2}{|r-e^{it}|^2}\,\mathrm{d}t \\ &= 2(1-r^2)\int_0^{2(1-r)} \frac{\mathrm{d}t}{|r-e^{it}|^2} \\ &\ge 2(1-r^2)\int_0^{2(1-r)} \frac{dt}{(1-r)^2+t^2} \\ &= 2(1-r^2)\left[\frac{1}{1-r}\,\mathrm{arctg}\,\frac{t}{1-r}\right]_0^{2(1-r)} \\ &= 2(1+r)\,\mathrm{arctg}\,2 \ge \frac{7}{2}\cdot \mathrm{arctg}\,\sqrt{3} = \frac{7\pi}{6}.\end{aligned}$$

□

Corollary 9. *For each $\lambda \in \mathbb{D}$ with $|\lambda| \geq 3/4$ we have*

$$\int_{-\pi}^{\pi} (P_\lambda(t) - Q_\lambda(t))\, \mathrm{d}t \;\leq\; \frac{5}{7} \int_{-\pi}^{\pi} Q_\lambda(t)\, \mathrm{d}t.$$

Proof. By Lemma 8, we have the estimates

$$\frac{\int_{-\pi}^{\pi}(P_\lambda(t) - Q_\lambda(t))\, \mathrm{d}t}{\int_{-\pi}^{\pi} Q_\lambda(t)\, \mathrm{d}t} = \frac{\int_{-\pi}^{\pi} P_\lambda(t)\, \mathrm{d}t}{\int_{-\pi}^{\pi} Q_\lambda(t)\, \mathrm{d}t} - 1 \leq 2\pi \cdot \left(\frac{7\pi}{6}\right)^{-1} - 1 = \frac{5}{7}. \qquad \square$$

Lemma 10. *Let $\Lambda \subset \mathbb{D}$ be an Apostol set. Let $t_1, t_2 \in \mathbb{R}$ with $-\pi \leq t_1 < t_2 \leq \pi$. Let $f(t) = 1$ if $t_1 \leq t \leq t_2$, and $f(t) = 0$ otherwise. Then there is an $n_0 \geq 1$ such that for every $n \geq n_0$ there exist a finite set $F \subset \Lambda$ and positive real numbers $\alpha_\lambda \quad (\lambda \in F)$ with the following properties:*

(i) *$I_\lambda \subset \{e^{it} : t_1 < t < t_2\}$ for any $\lambda \in F$;*
(ii) *the sets $I_\lambda \quad (\lambda \in F)$ are pairwise disjoint;*
(iii) *$m(\bigcup_{\lambda \in F} I_\lambda) \geq \frac{1}{40\pi}(t_2 - t_1)$;*
(iv) *$|\lambda| \geq 3/4$ and $|\lambda^n - 1| < \frac{1}{9}$ for all $\lambda \in F$;*
(v) *$\sum_{\lambda \in F} \alpha_\lambda \leq \frac{t_2 - t_1}{7}$;*
(vi) *$\int_{-\pi}^{\pi} \left| \sum_{\lambda \in F} \alpha_\lambda \lambda^n P_\lambda(t) - f(t) \right| \mathrm{d}t \leq c_1 (t_2 - t_1)$, where $c_1 = 1 - \frac{1}{1920}$.*

Proof. For every $n \geq 1$, set $M_n = \{t \in (t_1, t_2) : |e^{int} - 1| \leq 1/10\}$. Clearly for all n sufficiently large we have

$$m(M_n) > \frac{t_2 - t_1}{10 \cdot 2\pi}. \tag{3}$$

Fix n satisfying (3). Let $\varepsilon > 0$ satisfy $m(M_n) - \varepsilon > (t_2 - t_1)/\, 20\pi$. Let $S \subset (t_1, t_2)$ be the exceptional set of the Apostol set λ, i.e., $\sup\{0 \leq r < 1 : re^{i\theta} \in \Lambda\} = 1$ for all $\theta \in (t_1, t_2) \setminus S$. Since S is at most countable, it can be covered by a countable union U of open intervals with $m(U) < \varepsilon/2$. Then the set M' defined by

$$M' = \Big(M_n \cap \langle t_1 + \varepsilon/4, t_2 - \varepsilon/4 \rangle \Big) \setminus U$$

is compact with $m(M') > (t_2 - t_1)/\, 20\pi$. For each $t \in M'$ we can find $r_t \geq 3/4$ such that $\lambda_t := r_t e^{it} \in \Lambda$, $|\lambda_t^n - 1| < 1/9$ and $I_{\lambda_t} \subset \{e^{is} : t_1 < s < t_2\}$. Then $\{e^{is} : s \in M'\} \subset \bigcup_{t \in M'} I_{\lambda_t}$. Since $\{e^{is} : s \in M'\}$ is a compact subset of the 1-dimensional set $\mathbb{T}$, there exists a finite subcover of $(I_{\lambda_t})_{t \in M'}$ such that any three of these subsets have empty intersection. Considering a cover of the minimal cardinality with this property it is easy to see that there are numbers $\lambda_1, \ldots, \lambda_k \in \Lambda$ with $\lambda_j = |\lambda_j| e^{is_j}$ such that $t_1 < s_1 < \cdots < s_k < t_2$, $\bigcup_{j=1}^{k} I_{\lambda_j} \supset \{e^{is} : s \in M'\}$ and $I_{\lambda_j} \cap I_{\lambda_{j'}} = \emptyset$ if $|j' - j| \geq 2$. Let $F_1 = \{\lambda_1, \lambda_3, \ldots\}$ and $F_2 = \{\lambda_2, \lambda_4, \ldots\}$. Let F be one of the sets F_1, F_2 such that

$$m\Big(\bigcup_{\lambda \in F} I_\lambda\Big) = \max\Big\{ m\Big(\bigcup_{\lambda \in F_1} I_\lambda\Big), m\Big(\bigcup_{\lambda \in F_2} I_\lambda\Big) \Big\}.$$

Then $I_\lambda \cap I_{\lambda'} = \emptyset$ for all distinct λ, λ' in F, and $m\Big(\bigcup_{\lambda\in F} I_\lambda\Big) \geq m(M')/2 > (t_2 - t_1)/40\pi$. For any $\lambda \in F$, set $\alpha_\lambda = (1-|\lambda|)(1+|\lambda|)^{-1}$. Then $\alpha_\lambda > 0$ and

$$\sum_{\lambda\in F} \alpha_\lambda \leq \frac{4}{7}\sum_{\lambda\in F}(1-|\lambda|) = \frac{1}{7}\sum_{\lambda\in F} m(I_\lambda) \leq \frac{t_2-t_1}{7}.$$

Finally,

$$\begin{aligned}
&\int_{-\pi}^{\pi}\Big|\sum_{\lambda\in F}\alpha_\lambda\lambda^n P_\lambda(t) - f(t)\Big|\,\mathrm{d}t \\
&\leq \int_{-\pi}^{\pi}\Big|\sum_{\lambda\in F}\alpha_\lambda\lambda^n (P_\lambda(t) - Q_\lambda(t))\Big|\,\mathrm{d}t + \int_{-\pi}^{\pi}\sum_{\lambda\in F}\alpha_\lambda|\lambda^n - 1|Q_\lambda(t)\,\mathrm{d}t \\
&\quad + \int_{t_1}^{t_2}\Big(1 - \sum_{\lambda\in F}\alpha_\lambda Q_\lambda(t)\Big)\,\mathrm{d}t \leq \sum_{\lambda\in F}\alpha_\lambda\int_{-\pi}^{\pi}(P_\lambda(t) - Q_\lambda(t))\,\mathrm{d}t \\
&\quad + \frac{1}{9}\int_{t_1}^{t_2}\sum_{\lambda\in F}\alpha_\lambda Q_\lambda(t)\,\mathrm{d}t + (t_2 - t_1) - \int_{t_1}^{t_2}\sum_{\lambda\in F}\alpha_\lambda Q_\lambda(t)\,\mathrm{d}t \\
&\leq t_2 - t_1 + \Big(\frac{5}{7} + \frac{1}{9} - 1\Big)\int_{t_1}^{t_2}\sum_{\lambda\in F}\alpha_\lambda Q_\lambda(t)\mathrm{d}t \leq t_2 - t_1 - \frac{1}{7}\int_{t_1}^{t_2}\sum_{\lambda\in F}\alpha_\lambda Q_\lambda(t)\mathrm{d}t \\
&\leq t_2 - t_1 - \frac{1}{7}\sum_{\lambda\in F}\frac{1-|\lambda|}{1+|\lambda|}\cdot\frac{7\pi}{6} \leq t_2 - t_1 - \frac{\pi}{12}\sum_{\lambda\in F}(1-|\lambda|) \\
&= t_2 - t_1 - \frac{\pi}{48}\cdot m\Big(\bigcup_{\lambda\in F} I_\lambda\Big) \leq c_1(t_2 - t_1),
\end{aligned}$$

where $c_1 = 1 - \frac{1}{1920}$. □

Corollary 11. *Let c_1 be the constant from the previous lemma and let $c_2 \in (c_1, 1)$. Let $f : (-\pi, \pi\rangle \to \langle 0, \infty)$ be an integrable function and let Λ be an Apostol set. Then for any n sufficiently large there are a finite set $F \subset \Lambda$ and positive numbers $\alpha_\lambda \quad (\lambda \in F)$ such that:*

(i) *the sets $(I_\lambda)_{\lambda\in F}$ are pairwise disjoint;*

(ii) *$|\lambda| \geq 3/4$ and $|\lambda^n - 1| \leq \frac{1}{9}$ for all $\lambda \in F$;*

(iii) *$\sum_{\lambda\in F}\alpha_\lambda \leq \frac{1}{2\pi}\int_{-\pi}^{\pi} f(t)\,\mathrm{d}t$;*

(iv) *$\int_{-\pi}^{\pi}\Big|\sum_{\lambda\in F}\alpha_\lambda\lambda^n P_\lambda(t) - f(t)\Big|\,\mathrm{d}t \leq c_2\int_{-\pi}^{\pi} f(t)\,\mathrm{d}t$.*

Proof. Let $\varepsilon > 0$ be sufficiently small ($\varepsilon < \min\{\frac{c_2-c_1}{2}, \frac{7}{2\pi} - 1\}$). Let g be a step function $g : (-\pi, \pi\rangle \to \langle 0, \infty)$ such that $\int_{-\pi}^{\pi}|f - g|\,\mathrm{d}t \leq \varepsilon\int_{-\pi}^{\pi} f(t)\,\mathrm{d}t$. By Lemma 10, applied to each interval where g is constant, we can find a finite set $F \subset \Lambda$

and positive numbers $\alpha_\lambda \quad (\lambda \in F)$ satisfying (i), (ii) and

$$\sum_{\lambda \in F} \alpha_\lambda \le \frac{1}{7}\int_{-\pi}^{\pi} g(t)\,\mathrm{d}t \le \frac{1}{7}\Big(\int_{-\pi}^{\pi} f(t)\,\mathrm{d}t + \int_{-\pi}^{\pi} |f-g|\,\mathrm{d}t\Big)$$
$$\le \frac{1}{7}(1+\varepsilon)\int f(t)\,\mathrm{d}t \le \frac{1}{2\pi}\int_{-\pi}^{\pi} f(t)\,\mathrm{d}t.$$

Further,

$$\int_{-\pi}^{\pi}\Big|\sum_{\lambda \in F}\alpha_\lambda\lambda^n P_\lambda(t) - g(t)\Big|\,\mathrm{d}t \le c_1\int_{-\pi}^{\pi} g(t)\,\mathrm{d}t.$$

Then we have

$$\int_{-\pi}^{\pi}\Big|\sum_{\lambda \in F}\alpha_\lambda\lambda^n P_\lambda(t) - f(t)\Big|\,\mathrm{d}t$$
$$\le \int_{-\pi}^{\pi}\Big|\sum_{\lambda \in F}\alpha_\lambda\lambda^n P_\lambda(t) - g(t)\Big|\,\mathrm{d}t + \int_{-\pi}^{\pi}|f(t)-g(t)|\,\mathrm{d}t$$
$$\le c_1\int_{-\pi}^{\pi} g(t)\,\mathrm{d}t + \varepsilon\int_{-\pi}^{\pi} f(t)\,\mathrm{d}t$$
$$\le (c_1+2\varepsilon)\int_{-\pi}^{\pi} f(t)\,\mathrm{d}t \le c_2\int_{-\pi}^{\pi} f(t)\,\mathrm{d}t. \qquad \square$$

Let b be the constant from the Carleson theorem, see Theorem A.6.5, i.e., if $F \subset \{z : 3/4 \le |z| < 1\}$ is a finite set with $(I_\lambda)_{\lambda\in F}$ pairwise disjoint and $c_\lambda \in \mathbb{C} \quad (\lambda \in F)$, then there exists a function $f \in H^\infty$ such that $f(\lambda) = c_\lambda \quad (\lambda \in F)$ and $\|f\| \le b \cdot \sup_{\lambda\in F}|c_\lambda|$.

Lemma 12. *Let $T \in B(X)$ be a polynomially bounded operator with polynomial bound k. Let $F \subset \mathbb{D}$ be a finite set with $(I_\lambda)_{\lambda\in F}$ pairwise disjoint and $|\lambda| \ge 3/4 \quad (\lambda \in F)$. Suppose that there are vectors $u_\lambda \in X$ and complex numbers $\mu_\lambda \quad (\lambda \in F)$ such that $\|u_\lambda\| = 1$, $\|(T-\lambda)u_\lambda\| < \frac{1}{2kb\pi}(1-|\lambda|)^2$ and $\|\sum_{\lambda\in F}\mu_\lambda u_\lambda\| = 1$. Then $|\mu_\lambda| \le 2kb$ for all $\lambda \in F$.*

Proof. Let $\lambda_0 \in F$ satisfy $|\mu_{\lambda_0}| = \max_{\lambda\in F}|\mu_\lambda|$. By Theorem A.6.5, there is a function $f \in H^\infty$ such that $\|f\| \le b$, $f(\lambda_0) = 1$ and $f(\lambda) = 0$ for $\lambda \in F \setminus \{\lambda_0\}$.

Let $u = \sum_{\lambda\in F}\mu_\lambda u_\lambda$. Then $\|f(T)u\| \le kb\|u\| = kb$.

For $\lambda \in F$ define $g_\lambda(z) = \frac{f(z)-f(\lambda)}{z-\lambda}$. Clearly g_λ is analytic on a neighbourhood of $\overline{\mathbb{D}}$ and $\|g_\lambda\| \le 2\|f\|(1-|\lambda|)^{-1} \le 2b(1-|\lambda|)^{-1}$. Hence

$$kb \ge \|f(T)u\| \ge \Big\|\sum_{\lambda\in F} f(\lambda)\mu_\lambda u_\lambda\Big\| - \Big\|\sum_{\lambda\in F}\mu_\lambda\big(f(\lambda)-f(T)\big)u_\lambda\Big\|$$
$$\ge \|\mu_{\lambda_0}u_{\lambda_0}\| - \sum_{\lambda\in F}|\mu_\lambda|\cdot\|g_\lambda(T)(T-\lambda)u_\lambda\|$$

$$\geq |\mu_{\lambda_0}| - |\mu_{\lambda_0}| \sum_{\lambda \in F} 2kb(1-|\lambda|)^{-1} \frac{1}{2kb\pi}(1-|\lambda|)^2$$

$$\geq |\mu_{\lambda_0}| \Big(1 - \sum_{\lambda \in F} \pi^{-1}(1-|\lambda|)\Big) \geq \frac{|\mu_{\lambda_0}|}{2},$$

since $\sum_{\lambda \in F}(1-|\lambda|) \leq \frac{1}{4} m\Big(\bigcup_{\lambda \in F} I_\lambda\Big) \leq \frac{\pi}{2}$. Hence $|\mu_\lambda| \leq |\mu_{\lambda_0}| \leq 2kb$ for each $\lambda \in F$. □

Let c_3 be a constant satisfying $c_2 < c_3 < 1$, where c_2 is the constant from Corollary 11.

Theorem 13. *Let $T : X \to X$ be a polynomially bounded operator with polynomial bound k, such that $\sigma(T) \supset \mathbb{T}$ and T has no non-trivial closed invariant subspace. Let $f \in L^1$ be a non-negative function with $\|f\|_1 = 1$. Then there is an n_0 such that for any $n \geq n_0$ there exist $x_n \in X$ and $x_n^* \in X^*$ with $\|x_n\| \leq 1$, $\|x_n^*\| \leq 1$ and $\|T^n x_n \otimes x_n^* - M_f\| < c_3$.*

Proof. Let $\varepsilon > 0$ satisfy $\varepsilon < \frac{c_3 - c_2}{2\pi k^2 b}$. Let

$$\Lambda = \{\lambda \in \mathbb{D} : \text{there exists } u \in X \text{ with } \|u\| = 1 \text{ and } \|(T-\lambda)u\| < \varepsilon(1-|\lambda|)^2\}$$

which is an Apostol set, by Theorem 7. Let $n \in \mathbb{N}$ be sufficiently large. By Corollary 11, there exist a finite set $F \subset \Lambda$ and numbers $\alpha_\lambda > 0 \quad (\lambda \in F)$ such that $(I_\lambda)_{\lambda \in F}$ are pairwise disjoint, $\sum_{\lambda \in F} \alpha_\lambda \leq 1$, $|\lambda| \geq 3/4$ and $|\lambda^n - 1| \leq 1/9 \quad (\lambda \in F)$ and

$$\int_{-\pi}^{\pi} \Big| \sum_{\lambda \in F} \alpha_\lambda \lambda^n P_\lambda(t) - f(e^{it}) \Big| \, dt \leq c_2 \int_{-\pi}^{\pi} f(e^{it}) \, dt = 2\pi c_2.$$

For $\lambda \in F$ write $\lambda = r_\lambda e^{i\theta_\lambda}$ with $r_\lambda \geq 0$ and $-\pi < \theta_\lambda \leq \pi$. For each $\lambda \in F$, fix $u_\lambda \in X$ with $\|u_\lambda\| = 1$ such that $\|(T-\lambda)u_\lambda\| < \varepsilon(1-|\lambda|)^2$. Slightly perturbing and renorming the vectors u_λ makes them linearly independent without affecting these inequalities. By the Zenger theorem, see Theorem A.5.3, there exist $x \in X$, $x^* \in X^*$ and numbers $\mu_\lambda \quad (\lambda \in F)$ such that $x = \sum_{\lambda \in F} \mu_\lambda u_\lambda$, $\|x\| \leq 1$, $\|x^*\| \leq 1$ and $\langle \mu_\lambda u_\lambda, x^* \rangle = \alpha_\lambda$ for every $\lambda \in F$. By Lemma 12, we have the estimates $|\mu_\lambda| \leq 2kb$. Let $g \in L^1$ be defined by $g(e^{it}) = \sum_{\lambda \in F} \alpha_\lambda \lambda^n P_\lambda(t)$.

We have

$$\begin{aligned}\|T^n x \otimes x^* - M_f\| &\leq \Big\| \sum_{\lambda \in F} \mu_\lambda T^n u_\lambda \otimes x^* - M_g \Big\| + \|M_g - M_f\| \\ &\leq \sup_{p \in \mathcal{P}, \|p\| \leq 1} \Big| \sum_{\lambda \in F} \langle \mu_\lambda T^n p(T) u_\lambda, x^* \rangle - \sum_{\lambda \in F} \alpha_\lambda \lambda^n p(\lambda) \Big| + \|g - f\|_1 \\ &\leq \sup_{p \in \mathcal{P}, \|p\| \leq 1} \Big| \sum_{\lambda \in F} \langle \mu_\lambda (T^n p(T) - \lambda^n p(\lambda)) u_\lambda, x^* \rangle \Big| + c_2.\end{aligned}$$

The supremum can be estimated in a standard way. For $\lambda \in F$ and $p \in \mathcal{P}$ with $\|p\| \leq 1$, write

$$z^n p(z) - \lambda^n p(\lambda) = (z - \lambda) q(\lambda),$$

where $q \in \mathcal{P}$. Clearly,

$$\|q\| \le 2\|p\|(1-|\lambda|)^{-1} \le 2(1-|\lambda|)^{-1}.$$

Then $\|q(T)\| \le 2k(1-|\lambda|)^{-1}$. Hence

$$\begin{aligned}\big\|\big(T^n p(T) - \lambda^n p(\lambda)\big)u_\lambda\big\| &= \|q(T)(T-\lambda)u_\lambda\| \le \|q(T)\| \cdot \|(T-\lambda)u_\lambda\| \\ &\le 2k(1-|\lambda|)^{-1}\varepsilon(1-|\lambda|)^2 = 2k\varepsilon(1-|\lambda|).\end{aligned}$$

It follows that

$$\begin{aligned}\|T^n x_n \otimes x_n^* - M_f\| &\le \sum_{\lambda\in F} |\mu_\lambda| \cdot 2k\varepsilon(1-|\lambda) + c_2 \\ &\le \sum_{\lambda\in F} 4k^2 b\varepsilon(1-|\lambda|) + c_2 = k^2 b\varepsilon \cdot m\Big(\bigcup_{\lambda\in F} I_\lambda\Big) + c_2 \\ &\le 2\pi k^2 b\varepsilon + c_2 < c_3.\end{aligned}$$

□

Theorem 14. *Let $T : X \to X$ be a polynomially bounded operator with polynomial bound k, suppose that $\|T^n x\| \to 0$ and $\|T^{*n}x^*\| \to 0$ for all $x \in X$ and $x^* \in X^*$. Suppose that $\sigma(T) \supset \mathbb{T}$ and T has no non-trivial closed invariant subspace. Let $x_1,\dots,x_k \in X$, $x_1^*,\dots,x_k^* \in X^*$, $f \in L^1$, $f \ge 0$ and $\varepsilon > 0$. Then there exist $u \in X$ and $u^* \in X^*$ such that*

(i) $\|u \otimes u^* - M_f\| \le c_3 \cdot \|f\|_1$;
(ii) $\|u \otimes x_j^*\| < \varepsilon$ *and* $\|x_j \otimes u^*\| < \varepsilon$ *for all* $j = 1,\dots,k$;
(iii) $\|u\| \le k\|f\|_1^{1/2}$ *and* $\|u^*\| \le k\|f\|_1^{1/2}$.

Proof. Choose n large enough such that $\|T^n x_j\|\cdot\|f\|_1^{1/2} < \varepsilon k^{-1}$, $\|T^{*n}x_j^*\|\cdot\|f\|_1^{1/2} < \varepsilon k^{-1}$ $(j = 1,\dots,k)$ and such that, by Theorem 13, there exist $v \in X$, $v^* \in X^*$ with $\|v\| \le \|f\|_1^{1/2}$, $\|v^*\| \le \|f\|_1^{1/2}$ and

$$\|T^{2n}v \otimes v^* - M_f\| \le c_3 \cdot \|f\|_1.$$

Set $u = T^n v$ and $u^* = T^{*n}v^*$. Then

$$\|u \otimes u^* - M_f\| = \|T^n v \otimes T^{*n}v^* - M_f\| = \|T^{2n}v \otimes v^* - M_f\| \le c_3\|f\|_1.$$

For each $j = 1,\dots,k$ we have

$$\|u \otimes x_j^*\| = \|T^n v \otimes x_j^*\| = \|v \otimes T^{*n}x_j^*\| \le k\|v\| \cdot \|T^{*n}x_j^*\| < \varepsilon$$

and similarly,

$$\|x_j \otimes u^*\| = \|x_j \otimes T^{*n}v^*\| = \|T^n x_j \otimes v^*\| \le k\|T^n x_j\| \cdot \|v^*\| < \varepsilon.$$

Finally, $\|u\| = \|T^n v\| \le k\|f\|_1^{1/2}$ and $\|u^*\| = \|T^{*n}v^*\| \le k\|f\|_1^{1/2}$. □

Fix a positive constant $C < 1$ and a positive integer N such that we have $c_3 + \pi/N < C < 1$. Set $\nu = e^{2\pi i/N}$. For any $j = 0, \dots, N-1$, let L_j be the set of all $z = re^{it}$ with $r > 0$ and $-\pi/N \le t - 2\pi j/N < \pi/N$.

Lemma 15. *For any $f \in L^1$ there are non-negative functions $g_0, \dots, g_{N-1} \in L^1$ such that $\|f\|_1 = \sum_{j=0}^{N-1} \|g_j\|_1$ and*

$$\Big\|f - \sum_{j=0}^{N-1} \nu^j g_j\Big\|_1 \le \pi N^{-1} \|f\|_1.$$

Proof. We use the above notation. Fix a representative of f and let $A_j = f^{-1}(L_j) = \{z \in \mathbb{T} : f(z) \in L_j\}$. Then the sets $A_0, \dots, A_{N-1}, f^{-1}(\{0\})$ form a measurable partition of $\mathbb{T}$. Set $g_j = |f|\chi_j$ where χ_j is the characteristic function of A_j. Let $\arg : \mathbb{C}\setminus\{0\} \to \langle 0, 2\pi)$ denote the argument. For all j and $t \in (-\pi, \pi\rangle$ with $e^{it} \in A_j$ we have

$$\begin{aligned}|f(e^{it}) - \nu^j g_j(e^{it})| &= |f(e^{it})| \cdot |e^{i \arg f(e^{it})} - e^{i\, 2\pi j/N}|\\ &\le |f(e^{it})| \cdot |\arg f(e^{it}) - 2\pi j N^{-1}| \le |f(e^{it})| \pi N^{-1}.\end{aligned}$$

Hence

$$\Big\|f - \sum_{j=0}^{N-1} \nu^j g_j\Big\|_1 = \sum_{j=0}^{N-1} \|f\chi_j - \nu^j g_j\|_1 \le \sum_{j=0}^{N-1} \|f\chi_j\|_1 \cdot \pi N^{-1} = \pi N^{-1} \|f\|_1.$$

The equality $\|f\|_1 = \sum_{j=0}^{N-1} \|g_j\|_1$ is obvious. □

Theorem 16. *Let $T : X \to X$ be a polynomially bounded operator of class C_{00} with polynomial bound k. Suppose that $\sigma(T) \supset \mathbb{T}$ and T has no non-trivial closed invariant subspace. Let $x \in X$, $x^* \in X^*$ and $h \in L^1$. Then there are $y \in X$ and $y^* \in X^*$ such that*

(i) $\|y - x\| \le Nk\|h\|_1^{1/2}$;

(ii) $\|y^* - x^*\| \le Nk\|h\|_1^{1/2}$;

(iii) $\|y \otimes y^* - x \otimes x^* - M_h\| \le C\|h\|_1$.

Proof. By Lemma 15, there are non-negative functions $g_0, \dots, g_{N-1} \in L^1$ such that $\sum_{j=0}^{N-1} \|g_j\|_1 = \|h\|_1$ and $\|h - \sum_{j=0}^{N-1} \nu^j g_j\|_1 \le \pi N^{-1} \|h\|_1$.

Let ε be a positive number such that $c_3 + \pi N^{-1} + N(N+1)\varepsilon < C$. By Theorem 14, find inductively vectors $u_0, \dots, u_{N-1} \in X$ and $u_0^*, \dots u_{N-1}^* \in X^*$ such that $\|u_j\| \le k\|g_j\|_1^{1/2}$, $\|u_j^*\| \le k\|g_j\|_1^{1/2}$,

$$\begin{aligned}\|u_j \otimes u_j^* - M_{g_j}\| &\le c_3 \cdot \|g_j\|_1,\\ \|u_j \otimes u_k^*\| &\le \varepsilon\|h\|_1 \qquad (j \ne k),\\ \|x \otimes u_j^*\| &\le \varepsilon\|h\|_1,\\ \|u_j \otimes x^*\| &\le \varepsilon\|h\|_1.\end{aligned}$$

Set $y = x + \sum_{j=0}^{N-1} \nu^j u_j$ and $y^* = x^* + \sum_{j=0}^{N-1} u_j^*$. Then

$$\|y - x\| \le \sum_{j=0}^{N-1} k\|g_j\|_1^{1/2} \le Nk\|h\|_1^{1/2}$$

and

$$\|y^* - x^*\| \le \sum_{j=0}^{N-1} k\|g_j\|_1^{1/2} \le Nk\|h\|_1^{1/2}.$$

Further,

$$\begin{aligned}
&\|y \otimes y^* - x \otimes x^* - M_h\| \\
&= \Big\|\sum_{j=0}^{N-1} \nu^j u_j \otimes u_j^* + \sum_{j=0}^{N-1} x \otimes u_j^* + \sum_{j=0}^{N-1} \nu^j u_j \otimes x^* + \sum_{j\ne k} \nu^j u_j \otimes u_k^* - M_h\Big\| \\
&\le \sum_{j=0}^{N-1} \|\nu^j u_j \otimes u_j^* - M_{\nu^j g_j}\| + \Big\|\sum_{j=0}^{N-1} \nu^j g_j - h\Big\|_1 + (2N + N(N-1))\varepsilon\|h\|_1 \\
&\le \sum_{j=0}^{N-1} c_3\|g_j\|_1 + \pi N^{-1}\|h\|_1 + N(N+1)\varepsilon\|h\|_1 \\
&\le \|h\|_1\big(c_3 + \pi N^{-1} + N(N+1)\varepsilon\big) \le C\|h\|_1.
\end{aligned}$$

□

Theorem 17. *Let $T : X \to X$ be a polynomially bounded operator of class C_{00} on a complex Banach space X. Suppose that the spectrum of T contains the unit circle. Then T has a non-trivial closed invariant subspace.*

Proof. Suppose on the contrary that T has no non-trivial closed invariant subspace. We construct inductively convergent sequences $(x_j) \subset X$ and $(x_j^*) \subset X^*$ such that $\|x_j \otimes x_j^* - M_1\| \to 0$, where 1 denotes the constant function equal to 1 on $\mathbb{T}$.

Set $x_0 = 0$ and $x_0^* = 0$. Let $\phi_0 = x_0 \otimes x_0^* - M_1$. Then $\|\phi_0\| = 1 < 2$.

Suppose that we have already constructed vectors $x_j \in X$ and $x_j^* \in X^*$ such that $\|\phi_j\| < 2C^j$ where $\phi_j = x_j \otimes x_j^* - M_1$. By Theorem 1, there exists $h \in L^1$ representing the functional ϕ_j. Moreover, we can assume that $\|h\|_1 < 2C^j$. By Theorem 16, there are $x_{j+1} \in X$ and $x_{j+1}^* \in X^*$ such that

$$\|x_{j+1} - x_j\| \le kN\|h\|^{1/2} \le \sqrt{2}kNC^{j/2},$$
$$\|x_{j+1}^* - x_j^*\| \le kN\|h\|^{1/2} \le \sqrt{2}kNC^{j/2}$$

and for $\phi_{j+1} := x_{j+1} \otimes x_{j+1}^* - M_1$ we have

$$\|\phi_{j+1}\| = \|x_{j+1} \otimes x_{j+1}^* - x_j \otimes x_j^* + \phi_j\| \le C\|h\|_1 < 2C^{j+1}.$$

Clearly (x_j) and (x_j^*) are Cauchy sequences. Let $x = \lim x_j$ and $x^* = \lim x_j^*$. For each polynomial p with $\|p\| = 1$ we have

$$\begin{aligned}
&\left|\langle p(T)x_j, x_j^*\rangle - \langle p(T)x, x^*\rangle\right| \\
&\leq \left|\langle |p(T)x_j, x_j^*\rangle - \langle p(T)x_j, x^*\rangle\right| + \left|\langle p(T)x_j, x^*\rangle - \langle p(T)x, x^*\rangle\right| \\
&\leq k\|x_j\| \cdot \|x^* - x_j^*\| + k\|x_j - x\| \cdot \|x^*\| \to 0
\end{aligned}$$

uniformly on the unit ball in $\mathcal{P}$. Thus $x \otimes x^* = \lim_{j\to\infty} x_j \otimes x_j^* = M_1$ and $\langle p(T)x, x^*\rangle = p(0)$ for each polynomial p. This implies that T has a non-trivial invariant subspace. Indeed, either $Tx = 0$ (in this case x generates a 1-dimensional invariant subspace) or the vectors $T^k x \quad (k \geq 1)$ generate a non-trivial closed invariant subspace. □

As before, the assumption that $T \in C_{00}$ can be omitted (at least for reflexive Banach spaces). By using a refined version of the Zenger theorem it is possible to replace the C_{00} condition by the weaker $C_{0\cdot}$ condition (for any Banach space).

By a standard technique (cf. Theorem 6 or 39.18), it is possible to omit then the $C_{0\cdot}$ condition. Thus it is possible to prove the following result (we omit the details).

Theorem 18. *Let $T \in \mathcal{B}(X)$ be a polynomially bounded operator satisfying $\sigma(T) \supset \mathbb{T}$. Then T^* has a non-trivial invariant subspace. In particular, if the space X is reflexive, then T itself has a non-trivial closed invariant subspace.*

Since any contraction on a Hilbert space is polynomially bounded by the von Neumann inequality, we have

Corollary 19. *Let T be a contraction on a Hilbert space satisfying $\sigma(T) \supset \mathbb{T}$. Then T has a non-trivial closed invariant subspace.*

41 Kaplansky's type theorems

Many results in operator theory connect local properties of an operator on a Banach space with its global properties. A trivial example, which is in some sense a prototype of this type of results, is the following observation: an operator $T \in \mathcal{B}(X)$ is locally nilpotent (i.e., for every $x \in X$ there exists $n \in \mathbb{N}$ such that $T^n x = 0$) if and only if it is nilpotent. This observation is an immediate consequence of the Baire category theorem.

An approximate version of this result was proved in Section 14: an operator $T \in \mathcal{B}(X)$ is quasinilpotent (i.e., $r(T) = 0$) if and only if it is locally quasinilpotent (i.e., $r_x(T) = \limsup_{k\to\infty} \|T^k x\|^{1/k} = 0$ for all $x \in X$).

The classical Kaplansky theorem considers polynomials instead of the powers:

Theorem 1. (Kaplansky) *Let $T \in \mathcal{B}(X)$. Then T is algebraic (i.e., $p(T) = 0$ for some non-zero polynomial p) if and only if it is locally algebraic (i.e., for every $x \in X$ there exists a non-zero polynomial p such that $p(T)x = 0$).*

The main result of this section is an analogous statement for n-tuples of operators. An approximate version of the Kaplansky theorem will be proved in the next section.

Theorem 2. *Let Y, Z be Banach spaces, let $(T_j)_{j\geq 1}$ be a finite or countable infinite family of operators from Y to Z. Suppose that for each $y \in Y$ the set $\{T_j y : j = 1, 2, \dots\}$ is linearly dependant. Then there exists a non-trivial linear combination of operators T_j that is a finite-rank operator.*

Proof. For each j let

$$M_j = \Big\{T_j + \sum_{i=1}^{j-1} \alpha_i T_i : \alpha_1, \dots, \alpha_{j-1} \in \mathbb{C}\Big\} \subset \mathcal{B}(Y, Z).$$

Let F be a finite-dimensional subspace of Z. For $j = 1, 2, \dots$ set

$$Y_{F,j} = \bigcup_{S \in M_j} S^{-1}F.$$

By assumption, we have $Y = \bigcup_j Y_{F,j}$, so there exists $k = k(F)$ such that $Y_{F,k}$ is of the second category and $Y_{F,l}$ is of the first category for $l < k$. Choose a finite-dimensional subspace $F \subset Z$ such that

$$k = k(F) = \min\big\{k(G) : G \subset Z, \dim G < \infty\big\}.$$

For $s \in \mathbb{N}$ let

$$M_k^{(s)} = \Big\{T_k + \sum_{i=1}^{k-1} \alpha_i T_i : \alpha_i \in \mathbb{C}, \sum_{i=1}^{k-1} |\alpha_i| \leq s\Big\} \qquad \text{and} \qquad Y_{F,k}^{(s)} = \bigcup_{S \subset M_k^{(s)}} S^{-1}F.$$

Thus $M_k = \bigcup_{s=1}^{\infty} M_k^{(s)}$, and so $Y_{F,k} = \bigcup_{s=1}^{\infty} Y_{F,k}^{(s)}$.

We prove that the set $Y_{F,k}^{(s)}$ is closed for all s. Let $y_j \in Y_{F,k}^{(s)}$ $(j = 1, 2, \dots)$ and $y_j \to y \in Y$. Then there exist operators $S_j \in M_k^{(s)}$ such that $S_j y_j \in F$. It is easy to see that there exists a subsequence S_{j_r} and an operator $S \in M_k^{(s)}$ such that

$$\lim \|S_{j_r} - S\| = 0.$$

Then

$$Sy = \lim_{r\to\infty} S_{j_r} y = \lim_{r\to\infty} \big(S_{j_r} y_{j_r} + S_{j_r}(y - y_{j_r})\big) = \lim_{r\to\infty} S_{j_r} y_{j_r} \in F,$$

and so $y \in Y_{F,k}^{(s)}$.

Thus there exist $s \in \mathbb{N}$ and a non-empty open subset $U \subset Y_{F,k}^{(s)}$. Since $\bigcup_{l<k} Y_{F,l}$ is a set of the first category, there exists $w \in U \setminus \bigcup_{l<k} Y_{F,l}$. Let $\varepsilon > 0$ satisfy

$$\{y \in Y : \|y - w\| < \varepsilon\} \subset Y_{F,k}^{(s)} \subset Y_{F,k}.$$

Let $S_1 = T_k + \sum_{i=1}^{k-1} \alpha_i T_i$ be an operator such that

$$S_1 w \in F. \tag{1}$$

Denote by F' the subspace of Z generated by F and by the elements $T_i w$ $(i = 1, \dots, k)$. Obviously, $\dim F' < \infty$. Set $V = Y_{F,k} \setminus \bigcup_{l<k} Y_{F',l}$. It follows from the choice of the subspace F that the set V is of the second category.

Let $v \in V$. Then $v \in Y_{F,k}$, and so there exists $S_2 = T_k + \sum_{i=1}^{k-1} \beta_i T_i$ such that

$$S_2 v \in F. \tag{2}$$

Since $w \in U \subset Y_{F,k}^{(s)}$ and U is open, there exists a non-zero complex number λ such that $w + \lambda v \in Y_{F,k}^{(s)}$. Thus there exists an operator $S_3 = T_k + \sum_{i=1}^{k-1} \gamma_i T_i$ such that

$$S_3(w + \lambda v) \in F. \tag{3}$$

Thus $S_3 v = \lambda^{-1}\big(S_3(w + \lambda v) - S_3 w\big) \in F'$ and, by (2), $(S_3 - S_2)v \in F'$, where $S_3 - S_2 = \sum_{i=1}^{k-1} (\gamma_i - \beta_i) T_i$. Since $v \notin \bigcup_{l<k} Y_{F',l}$, we have $\gamma_i = \beta_i$ $(i = 1, \dots, k-1)$. So $S_3 = S_2$.

Thus $S_3 v = S_2 v \in F$ a $S_3 w \in F$ by (3). Further, $(S_3 - S_1)w \in F$, where

$$S_3 - S_1 = \sum_{i=1}^{k-1} (\gamma_i - \alpha_i) T_i.$$

Since $w \notin \bigcup_{l<k} Y_{F,l}$, we get $S_3 - S_1 = 0$, and so $S_3 = S_1$. Hence $S_1 v \in F$ for all $v \in V$. The preimage $S_1^{-1} F$ is a closed subspace of Y which is of the second category since $S_1^{-1} F \supset V$. Thus $S_1^{-1} F = Y$, $S_1 Y \subset F$ and S_1 is a finite-rank operator. □

For $n \in \mathbb{N}$ denote by $\mathcal{P}(n)$ the set of all polynomials in n variables.

Definition 3. Let $A = (A_1, \dots, A_n)$ be a commuting n-tuple of operators on a Banach space X. We say that the n-tuple A is algebraic, if $p(A) = 0$ for some non-zero polynomial $p \in \mathcal{P}(n)$. We say that A is locally algebraic if, for each $x \in X$, there exists a non-zero polynomial $p_x \in \mathcal{P}(n)$ such that $p_x(A)x = 0$.

Corollary 4. *Let $A = (A_1, \dots, A_n)$ be an n-tuple of mutually commuting operators on a Banach space X. Then A is algebraic if and only if it is locally algebraic.*

Proof. Let $A = (A_1, \dots, A_n)$ be a locally algebraic n-tuple. Consider the countable set $\{A^\alpha : \alpha \in \mathbb{Z}_+^n\} \subset \mathcal{B}(X)$. By Theorem 2, there exists a non-zero polynomial $p \in \mathcal{P}(n)$ such that $p(A)$ is of finite rank. Hence $(q \circ p)(A) = 0$, where $q \in \mathcal{P}(1)$ is the characteristic polynomial of the finite-dimensional operator $p(A) | \operatorname{Ran}\big(p(A)\big)$. □

Theorem 5. *Let $A = (A_1, \dots, A_n)$ be a commuting n-tuple of operators on a Banach space X, which is not algebraic. Then there exist points $x_1, x_2, , \dots \in X$ such that $\sum_{i=1}^k p_i(A)x_i \neq 0$ for all $k \in \mathbb{N}$ and all k-tuples of polynomials $p_1, \dots, p_k \in \mathcal{P}(n)$, that are not all equal to zero.*

Proof. Suppose on the contrary that for each sequence $x_1, x_2, \dots$ of elements of X there exist k and polynomials $p_1, \dots, p_k \in \mathcal{P}(n)$, $(p_1, \dots, p_k) \neq (0, \dots, 0)$ such that $\sum_{i=1}^k p_i(A)x_i = 0$.

Denote by $Y = l^\infty(X)$ the space of all bounded sequences of elements of X with the sup-norm. For $k \in \mathbb{N}$ and $\alpha \in \mathbb{Z}_+^n$ let $S_{k,\alpha} : Y \to X$ be the operator defined by $S_{k,\alpha}(\{x_i\}_{i=1}^\infty) = A^\alpha x_k$.

By Theorem 2, there exists a finite-dimensional subspace $F \subset X$, $k \in \mathbb{N}$ and polynomials $p_1, \dots, p_k \in \mathcal{P}(n)$, $(p_1, \dots, p_k) \neq (0, \dots, 0)$ such that $\sum_{i=1}^k p_i(A)x_i \in F$ for each sequence $\{x_i\}_{i=1}^\infty \in l^\infty(X)$. Choose $j \in \{1, \dots, k\}$ such that $p_j \neq 0$. For $x_1 = \cdots = x_{j-1} = 0 = x_{j+1} = \cdots = x_k$ we have $p_j(A)x_j \in F$ for all $x_j \in X$, i.e., $p_j(A)$ is a finite-rank operator. As in the proof of Corollary 4 we get that A is an algebraic n-tuple. □

42 Polynomial orbits and local capacity

Recall that $\mathcal{P}(n)$ denotes the set of all polynomials in n variables and $\mathcal{P}_k(n)$ the set of all polynomials of degree $\leq k$ of n variables.

By a *polynomial orbit* of an operator $T \in \mathcal{B}(X)$ at $x \in X$ we mean the set $\{p(T)x : p \in \mathcal{P}(1)\}$.

In this section we generalize some results concerning orbits. We prove that there are always points $x \in X$ such that $\|p(T)x\|$ is "large" for all polynomials p. The results are also formulated for n-tuples of commuting operators. Further, we introduce the local capacity of operators and prove an approximate version of the Kaplansky theorem.

Lemma 1. *Let X be a Banach space, let $E \subset X$ and $\mathcal{M} \subset \mathcal{B}(X)$ be finite-dimensional subspaces and let $\varepsilon > 0$. Then there exists a subspace $Y \subset X$ of finite codimension such that*

$$\|S(e+y)\| \geq (1-\varepsilon)\max\{\|Se\|, \|Sy\|/2\}$$

for all $e \in E$, $y \in Y$ and $S \in \mathcal{M}$.

Proof. Let $T_1, \dots, T_k$ be a basis of $\mathcal{M} \subset \mathcal{B}(X)$. Set $E_1 = \bigvee_{i=1}^k T_i E$. By Lemma 37.6, there exists a subspace $Y_1 \subset X$ of finite codimension such that

$$\|e+y\| \geq (1-\varepsilon)\max\{\|e\|, \|y\|/2\} \qquad (e \in E_1, y \in Y_1).$$

Set $Y = \bigcap_{i=1}^k T_i^{-1} Y_1$. Then $\operatorname{codim} Y < \infty$. For $e \in E$, $y \in Y$ and $S \in \mathcal{M}$ we have $Se \in E_1$ and $Sy \in Y_1$, so

$$\|S(e+y)\| \geq (1-\varepsilon)\max\{\|Se\|, \|Sy\|/2\}. \qquad \square$$

Lemma 2. *Let $A = (A_1, \dots, A_n)$ be an n-tuple of mutually commuting operators on a Banach space X, let $(0, \dots, 0) \in \sigma_{\pi e}(A)$, $\varepsilon > 0$, let E, Y be subspaces of X such that $\dim E < \infty$ and $\operatorname{codim} Y < \infty$. Then there exists a vector $u \in Y$ of* *norm 1 such that*

$$\Big\|\Big(I+\sum_{i=1}^n \alpha_i A_i\Big)(e+u)\Big\| \geq \frac{1-\varepsilon}{2}$$

for all $e \in E$ and $\alpha_1,\dots,\alpha_n \in \mathbb{C}$.

Proof. Let $\mathcal{M}$ be the subspace of $\mathcal{B}(X)$ generated by the operators $A_1,\dots,A_n$; without loss of generality we can assume that these operators are linearly independent. Denote by $\mathcal{M}_1$ the set of all finite-rank operators in $\mathcal{M}$, $\mathcal{M}_1 = \mathcal{M}\cap\mathcal{F}(X)$. Let $\mathcal{M}_2$ be a complement of $\mathcal{M}_1$ in $\mathcal{M}$, i.e., $\mathcal{M}_1\cap\mathcal{M}_2 = \{0\}$ and $\mathcal{M}_1+\mathcal{M}_2 = \mathcal{M}$. Let $S_1, \dots, S_k$ be a basis in $\mathcal{M}_2$ and $S_{k+1},\dots,S_n$ a basis in $\mathcal{M}_1$.

Let ε' be a positive number satisfying $\frac{(1-\varepsilon')^3}{1+\varepsilon'} \geq 1-\varepsilon$ and $\varepsilon' < 1$. By Lemma 1, there exists a subspace $Y_1 \subset X$ of finite codimension such that

$$\Big\|\Big(I+\sum_{i=1}^n \alpha_i A_i\Big)(e+y)\Big\| \geq \frac{1-\varepsilon'}{2}\Big\|\Big(I+\sum_{i=1}^n \alpha_i A_i\Big)y\Big\|$$

for all $e \in E$, $y \in Y_1$ and $\alpha_1,\dots,\alpha_n \in \mathbb{C}$. Clearly, we can assume that $Y_1 \subset Y \cap \bigcap_{i=k+1}^n \operatorname{Ker} S_i$ since $\operatorname{codim}\operatorname{Ker} S_i < \infty$ for $k+1 \leq i \leq n$. Thus $S|Y_1 = 0$ for all $S \in \mathcal{M}_1$.

By Theorem 41.2, there exists $y_1 \in Y_1$ of norm 1 such that the vectors $S_1y_1,\dots,S_ky_1$ are linearly independent (otherwise there would be a non-zero operator in $\mathcal{M}_2$ of finite rank).

For $\beta_1,\dots,\beta_n \in \mathbb{C}^n$ consider the norms $\sum_{j=1}^k |\beta_j|$ and $\Big\|\sum_{j=1}^k \beta_j S_j y_1\Big\|$. These two norms are equivalent, so there exists a positive constant c such that

$$\Big\|\sum_{j=1}^k \beta_j S_j y_1\Big\| \geq c\cdot\sum_{j=1}^k |\beta_j|$$

for all $\beta_1, \dots, \beta_k \in \mathbb{C}$.

By Lemma 1, there is a subspace $Y_2 \subset X$ of finite codimension such that

$$\Big\|\Big(I+\sum_{i=1}^n \alpha_i A_i\Big)(e+y)\Big\| \geq (1-\varepsilon')\max\Big\{\Big\|\Big(I+\sum_{i=1}^n \alpha_i A_i\Big)e\Big\|, \frac{1}{2}\Big\|\Big(I+\sum_{i=1}^n \alpha_i A_i\Big)y\Big\|\Big\}$$

for all $e \in E \vee \{y_1\} \vee \{Sy_1 : S \in \mathcal{M}\}$, $y \in Y_2$ and $\alpha_1,\dots,\alpha_n \in \mathbb{C}$. We can assume that $Y_2 \subset Y_1 \subset Y$.

Since $(0,\dots,0) \in \sigma_{\pi e}(A)$ and the operators $S_1,\dots,S_n$ are linear combinations of $A_1,\dots,A_n$, we also have $(0,\dots,0) \in \sigma_{\pi e}(S_1,\dots,S_n)$ and there exists a vector $y_2 \in Y_2$ of norm 1 such that

$$\|S_i y_2\| < c\varepsilon'^2 \qquad (i=1,\dots,n).$$

Set $u = \frac{y_2+\varepsilon' y_1}{\|y_2+\varepsilon' y_1\|}$. Obviously, $u \in Y_1$ and $\|u\| = 1$.

Let $\alpha_1,\dots,\alpha_n \in \mathbb{C}$. Then $\sum_{i=1}^n \alpha_i A_i = \sum_{j=1}^n \beta_j S_j$ for some $\beta_1,\dots,\beta_n \in \mathbb{C}$. Let $e \in E$.

We distinguish two cases:

(a) Let $\sum_{j=1}^k |\beta_j| \le \frac{1}{c\varepsilon'}$. Then

$$\begin{aligned}
\Big\|\Big(I+\sum_{i=1}^n \alpha_i A_j\Big)(e+u)\Big\| &= \Big\|\Big(I+\sum_{j=1}^n \beta_j S_j\Big)\Big(e+\frac{\varepsilon' y_1+y_2}{\|\varepsilon' y_1+y_2\|}\Big)\Big\| \\
&\ge \frac{1-\varepsilon'}{2\|\varepsilon' y_1+y_2\|}\Big\|\Big(I+\sum_{j=1}^n \beta_j S_j\Big)y_2\Big\| \ge \frac{1-\varepsilon'}{2(1+\varepsilon')}\Big(\|y_2\|-\Big\|\sum_{j=1}^n \beta_j S_j y_2\Big\|\Big) \\
&\ge \frac{1-\varepsilon'}{2(1+\varepsilon')}\Big(1-\sum_{j=1}^k |\beta_j|\cdot \max\{\|S_j y_2\| : 1\le j\le k\}\Big) \ge \frac{(1-\varepsilon')^2}{2(1+\varepsilon')} \ge \frac{1-\varepsilon}{2}.
\end{aligned}$$

(b) Let $\sum_{j=1}^k |\beta_j| > \frac{1}{c\varepsilon'}$. Then

$$\begin{aligned}
\Big\|\Big(I+\sum_{i=1}^n \alpha_i A_j\Big)(e+u)\Big\| &= \Big\|\Big(I+\sum_{j=1}^n \beta_j S_j\Big)\Big(e+\frac{\varepsilon' y_1+y_2}{\|e' y_1+y_2\|}\Big)\Big\| \\
&\ge (1-\varepsilon')\Big\|\Big(I+\sum_{j=1}^n \beta_j S_j\Big)\Big(e+\frac{\varepsilon' y_1}{\|\varepsilon' y_1+y_2\|}\Big)\Big\| \\
&\ge \frac{(1-\varepsilon')^2\varepsilon'}{2\|\varepsilon' y_1+y_2\|}\Big\|\Big(I+\sum_{j=1}^n \beta_j S_j\Big)y_1\Big\| \ge \frac{(1-\varepsilon')^2\varepsilon'}{2(1+\varepsilon')}\Big(\Big\|\sum_{j=1}^k \beta_j S_j y_1\Big\|-\|y_1\|\Big) \\
&\ge \frac{(1-\varepsilon')^2\varepsilon'}{2(1+\varepsilon')}\Big(c\sum_{j=1}^k |\beta_j|-1\Big) \ge \frac{(1-\varepsilon')^2\varepsilon'}{2(1+\varepsilon')}\cdot\Big(\frac{1}{\varepsilon'}-1\Big) > \frac{1-\varepsilon}{2}.
\end{aligned}$$

□

Corollary 3. *Let $T=(T_1,\dots,T_n)$ be a commuting n-tuple of operators on a Banach space X, $\lambda\in\sigma_{\pi e}(A)$, $k\in\mathbb{N}$ and $\varepsilon>0$. Let E,Y be subspaces of X such that $\dim E<\infty$ and $\operatorname{codim} Y<\infty$. Then there exists a vector $u\in Y$ of norm 1 such that*

$$\|p(T)(e+u)\| \ge \frac{1-\varepsilon}{2}|p(\lambda)|$$

for all $e\in E$ and $p\in\mathcal{P}_k(n)$.

Proof. Without loss of generality we can assume that $\lambda=(0,\dots,0)$. Consider the tuple A formed by the operators T^α $(\alpha\in\mathbb{Z}_+^n, 1\le|\alpha|\le k)$. Then $(0,\dots,0)\in\sigma_{\pi e}(A)$ and the statement follows from the previous lemma. □

Corollary 3 gives the existence of a point $u\in X$ of norm 1 such that $\|p(T)u\|$ is "large" for all polynomials p with $\deg p\le k$. The estimate is given in terms of $|p(\lambda)|$ where λ is a given point of $\sigma_{\pi e}(T)$.

Our goal is to give an estimate of $\|p(T)u\|$ in terms of $\max\{|p(\lambda)| : \lambda \in \sigma_{\pi e}(T)\}$. By the spectral mapping theorem we have

$$\begin{aligned}\max\{|p(\lambda)| : \lambda \in \sigma_{\pi e}(T)\} &= \max\{|\mu| : \mu \in \sigma_{\pi e}(p(T))\}\\ &= \max\{|\mu| : \mu \in \sigma_e(p(T))\} = r_e(p(T)).\end{aligned}$$

The next lemma enables us to reduce a general compact set to a finite set.

Lemma 4. *Let n, k be positive integers and $K \subset \mathbb{C}^n$ a non-empty compact set. Then there exists a finite subset $K' \subset K$ with $\operatorname{card} K' = m \le \binom{n+k}{n}$ such that*

$$\|p\|_K \le m \cdot \|p\|_{K'}$$

for each polynomial $p \in \mathcal{P}_k(n)$.

Proof. Let $L = \{p \in \mathcal{P}_k(n) : \|p\|_K = 0\}$ and let M be a complementary space of L in $\mathcal{P}_k(n)$, i.e., $M \cap L = \{0\}$ and $M + L = \mathcal{P}_k(n)$. Let $m = \dim M \le \dim \mathcal{P}_k(n) = \binom{n+k}{n}$ and let $q_1, \dots, q_m \in M$ be a basis of M. For $x_1, \dots, x_m \in K$ define $V(x_1, \dots, x_m) = \det(q_i(x_j))_{i,j=1}^m$. The polynomials $q_1, \dots, q_m$ are linearly independent on K, so there exist points $x_1, \dots, x_m \in K$ such that the matrix $(q_i(x_j))_{i,j=1}^m$ is regular, i.e., $V(x_1, \dots, x_m) \ne 0$. Choose $k_1, \dots, k_m \in K$ such that

$$|V(k_1, \dots, k_m)| = \max\{|V(y_1, \dots, y_m)| : y_1, \dots, y_m \in K\}.$$

Then $V(k_1, \dots, k_m) \ne 0$. For $j = 1, \dots, m$ define polynomials $L^{(j)} \in \mathcal{P}_k(n)$ by

$$L^{(j)}(z) = V\big(k_1, \dots, k_{j-1}, z, k_{j+1}, \dots, k_m\big)/V(k_1, \dots, k_m).$$

Evidently, $|L^{(j)}(z)| \le 1$ for all $z \in K$. The polynomials $L^{(j)}$ are linear combinations of the polynomials $q_1, \dots, q_m$, and so $L^{(j)} \in M \quad (j = 1, \dots, m)$. Further, $L^{(j)}(k_i) = \delta_{ij}$ (the Kronecker symbol), so the polynomials $L^{(1)}, \dots, L^{(m)}$ are linearly independent and each polynomial $p \in M$ is a linear combination of them. Obviously,

$$p(z) = \sum_{j=1}^m p(k_j) L^{(j)}(z) \qquad (p \in M, z \in K).$$

Set $K' = \{k_1, \dots, k_m\}$. Each polynomial $p \in \mathcal{P}_k(n)$ can be written in the form $p = p_1 + p_2$ for some $p_1 \in L$ and $p_2 \in M$, and $p_2 = \sum_{j=1}^m p_2(k_j) L^{(j)}$. Hence

$$\begin{aligned}\|p\|_K = \|p_2\|_K &= \max\left\{\left|\sum_{j=1}^m p_2(k_j) L^{(j)}(z)\right| : z \in K\right\}\\ &\le \sum_{j=1}^m |p_2(k_j)| \le m \cdot \|p\|_{K'}.\end{aligned}$$

□

Theorem 5. *Let $T = (T_1, \dots, T_n)$ be an n-tuple of mutually commuting operators on a Banach space X, let $k \in \mathbb{N}$ and $\varepsilon > 0$. Let Y be a subspace of X with $\operatorname{codim} Y < \infty$. Then there exists $y \in Y$ of norm 1 such that*

$$\|p(T)y\| \geq \frac{(1-\varepsilon)}{2} \binom{n+k}{n}^{-2} r_e(p(T))$$

for all polynomials $p \in \mathcal{P}_k(n)$.

Proof. Set $K = \sigma_{\pi e}(T)$. As noted above, for each $p \in \mathcal{P}_k(n)$ we have

$$\|p\|_K = \max\{|p(z)| : z \in \sigma_{\pi e}(T)\} = \max\{|p(z)| : z \in \sigma_e(T)\} = r_e(p(T)).$$

By the previous lemma, there exist elements $\lambda_1, \dots, \lambda_m \in K$, $m \leq \binom{n+k}{n}$ such that

$$\|p\|_K \leq m \cdot \max\{|p(\lambda_i)| : i = 1, \dots, m\}$$

for all $p \in \mathcal{P}_k(n)$. We construct points $y_1, \dots, y_m \in Y$ inductively. Suppose that $0 \leq j < m$ and that the points $y_1, \dots, y_j$ have already been found. Set $E_j = \bigvee\{p(T)y_i : p \in \mathcal{P}_k(n), 1 \leq i \leq j\}$. By Lemma 1, there is a subspace $Y_j \subset X$ of finite codimension such that

$$\|p(T)(e+y)\| \geq \left(1 - \frac{\varepsilon}{2}\right) \max\left\{\|p(T)e\|, \frac{1}{2}\|p(T)y\|\right\}$$

for all $e \in E_j$, $y \in Y_j$ and $p \in \mathcal{P}_k(n)$.

By Corollary 3, there is a vector $y_{j+1} \in Y \cap \bigcap_{i=1}^{j} Y_i$ of norm 1 such that

$$\|p(T)(e + y_{j+1})\| \geq \frac{1-\varepsilon/2}{2} |p(\lambda_{j+1})|$$

for all $e \in E_j, p \in \mathcal{P}_k(n)$.

Let $y_1, \dots, y_m$ be constructed in the above-described way.

Set $y = a^{-1}\sum_{i=1}^m y_i$, where $a = \|\sum_{i=1}^m y_i\|$. We have $a \geq 1-\varepsilon/2$ since $y_1 \in E_1$ and $\sum_{i=2}^m y_i \in Y_1$. Further, $a \leq m$. Obviously, $y \in Y$ and $\|y\| = 1$. Let $p \in \mathcal{P}_k(n)$ and $1 \leq j \leq m$. We have

$$\begin{aligned} \|p(T)y\| &= a^{-1}\left\|\sum_{i=1}^m p(T)y_i\right\| \geq \left(1 - \frac{\varepsilon}{2}\right) m^{-1} \left\|\sum_{i=1}^j p(T)y_i\right\| \\ &\geq \left(1 - \frac{\varepsilon}{2}\right)^2 \cdot \frac{1}{2m} |p(\lambda_j)| \geq \frac{1-\varepsilon}{2m} |p(\lambda_j)|. \end{aligned}$$

Thus

$$\begin{aligned} \|p(T)y\| &\geq \frac{1-\varepsilon}{2m} \max\{|p(\lambda_j)| : j = 1, \dots, m\} \\ &\geq \frac{1-\varepsilon}{2m^2} \|p\|_K = \frac{1-\varepsilon}{2m^2} r_e(p(T)). \end{aligned}$$

□

It is easy to see that for Hilbert space operators it is possible to obtain a better estimate in the preceding theorem. In fact, using a Dvoretzky's type result enables us to generalize this improved estimate to Banach spaces, see C.42.1.

Theorem 6. *Let $T = (T_1, \dots, T_n)$ be an n-tuple of mutually commuting operators on a Banach space X, let $x \in X$ and $\varepsilon > 0$. Then there exists $y \in X$ and a constant $C = C(\varepsilon)$ such that $\|y - x\| < \varepsilon$ and*

$$\|p(T)y\| \geq C(1 + \deg p)^{-(2n+\varepsilon)} r_e(p(T))$$

for all polynomials $p \in \mathcal{P}(n)$.

Proof. Find $k_0 \geq 1$ such that $\sum_{i=k_0}^{\infty} \frac{1}{i^2} < \varepsilon$, $2^{k_0} \geq n$ and $k^2 \leq 2^{\varepsilon(k-1)}$ $(k \geq k_0)$. Set $C = \frac{1}{8k_0^2}(n + 2^{k_0})^{-2n}$. Choose positive numbers ε_i $(i \geq k_0)$ such that $\varepsilon_i < 1$ and $\prod_{i=k_0}^{\infty}(1 - \varepsilon_i) \geq \frac{1}{2}$. We construct inductively points $y_{k_0}, y_{k_0+1}, \dots \in X$ of norm 1. Suppose that $y_{k_0}, \dots, y_{k-1}$ have already been constructed. Set $E_k = \bigvee\{x, y_{k_0}, \dots, y_{k-1}\}$. By Lemma 1, there exists a subspace $Z \subset X$ with $\operatorname{codim} Z < \infty$ such that

$$\|p(T)(e + z)\| \geq \left(1 - \frac{\varepsilon_k}{2}\right) \max\left\{\|p(T)e\|, \frac{1}{2}\|p(T)z\|\right\}$$

for all $e \in E_k$, $z \in Z$ and $p \in \mathcal{P}_{2^k}(n)$. By Lemma 5, there exists $y_k \in Z$ of norm 1 such that

$$\|p(T)y_k\| \geq \frac{1}{2}\left(1 - \frac{\varepsilon_k}{2}\right)\binom{n + 2^k}{n}^{-2} r_e(p(T))$$

for all $p \in \mathcal{P}_{2^k}(n)$. Thus

$$\begin{aligned}\|p(T)(e + y_k)\| &\geq \left(1 - \frac{\varepsilon_k}{2}\right) \max\left\{\|p(T)e\|, \frac{1}{4}\left(1 - \frac{\varepsilon_k}{2}\right)\binom{n + 2^k}{n}^{-2} r_e(p(T))\right\} \\ &\geq (1 - \varepsilon_k) \max\left\{\|p(T)e\|, \frac{1}{4}\binom{n + 2^k}{n}^{-2} r_e(p(T))\right\}\end{aligned} \tag{1}$$

for all $e \in E_k$ and $p \in \mathcal{P}_{2^k}(n)$).

Set $y = x + \sum_{i=k_0}^{\infty} \frac{y_i}{i^2}$. Clearly, $\|y - x\| \leq \sum_{i=k_0}^{\infty} \frac{1}{i^2} < \varepsilon$. Let p be a polynomial of degree r.

We distinguish two cases:

(a) Let $r \leq 2^{k_0}$. Then, by (1), for $N \geq k_0$ we have

$$\begin{aligned}&\left\|p(T)x + \sum_{i=k_0}^{N} \frac{1}{i^2} p(T)y_i\right\| \geq (1 - \varepsilon_N)\left\|p(T)x + \sum_{i=k_0}^{N-1} \frac{1}{i^2} p(T)y_i\right\| \geq \cdots \\ &\geq \prod_{i=k_0+1}^{N}(1 - \varepsilon_i) \cdot \left\|p(T)x + \frac{1}{k_0^2} p(T)y_{k_0}\right\| \geq \prod_{i=k_0}^{N}(1 - \varepsilon_i) \cdot \frac{1}{4k_0^2}\binom{n + 2^{k_0}}{n}^{-2} r_e(p(T)) \\ &\geq \frac{1}{8k_0^2}(n + 2^{k_0})^{-2n} r_e(p(T)) \geq C \cdot r_e(p(T)).\end{aligned}$$

(b) Let $2^{k-1} < r \le 2^k$ for some $k > k_0$. For $N \ge k$ we have

$$\Big\|p(T)x + \sum_{i=k_0}^{N} \frac{1}{i^2} p(T)y_i\Big\| \ge \prod_{i=k+1}^{N} (1-\varepsilon_i) \cdot \Big\|p(T)x + \sum_{i=k_0}^{k} \frac{1}{i^2} p(T)y_i\Big\|$$

$$\ge \prod_{i=k}^{N}(1-\varepsilon_i) \cdot \frac{1}{4k^2} \binom{n+2^k}{n}^{-2} r_e(p(T)) \ge \frac{1}{8} 2^{-\varepsilon(k-1)} (n+2^k)^{-2n} r_e(p(T))$$

$$\ge \frac{1}{8} r^{-\varepsilon} (3r)^{-2n} r_e(p(T)) \ge C r^{-(2n+\varepsilon)} r_e(p(T)).$$

So for each polynomial p we have

$$\|p(T)y\| = \lim_{N\to\infty} \Big\|p(T)x + \sum_{i=k_0}^{N} \frac{1}{i^2} p(T)y_i\Big\| \ge C(1+\deg p)^{-(2n+\varepsilon)} r_e(p(T)). \qquad \square$$

Recall that $\mathcal{P}_k^1(n)$ denotes the set of all monic polynomials of degree k,

$$\mathcal{P}_k^1(n) = \Big\{p \in \mathcal{P}(n) : p = \sum_{|\alpha|\le k} a_\alpha z^\alpha, \sum_{|\alpha|=k} |a_\alpha| = 1\Big\}.$$

Let $T = (T_1, \dots, T_n)$ be an n-tuple of mutually commuting operators on a Banach space X. The joint capacity of T was defined by

$$\operatorname{cap} T = \limsup_{k\to\infty} (\operatorname{cap}_k T)^{1/k},$$

where

$$\operatorname{cap}_k(T) = \inf\{\|p(T)\| : p \in \mathcal{P}_k^1(n)\}$$

(note that the limsup in the definition of $\operatorname{cap} T$ can be replaced by limit by Theorem 36.4 (i)). For a compact subset $K \subset \mathbf{C}^n$ the corresponding capacity was defined by

$$\operatorname{cap} K = \limsup_{k\to\infty} (\operatorname{cap}_k K)^{1/k},$$

where

$$\operatorname{cap}_k K = \inf\{\|p\|_K : p \in \mathcal{P}_k^1(n)\}.$$

By Theorems 36.4 (iii) and Corollary 36.8, $\operatorname{cap} T = \operatorname{cap} \sigma_H(T) = \operatorname{cap} \sigma_{\pi e}(T)$.

The local capacity of T at a point x can be defined analogously.

Definition 7. Let $T = (T_1, \dots, T_n)$ be an n-tuple of mutually commuting operators on a Banach space X and let $x \in X$. The local capacity $\operatorname{cap}_x T$ is defined by

$$\operatorname{cap}_x T = \limsup_{k\to\infty} (\operatorname{cap}_{x,k} T)^{1/k},$$

where

$$\operatorname{cap}_{x,k} T = \inf\{\|p(T)x\| : p \in \mathcal{P}_k^1(n)\}.$$

Clearly, $\operatorname{cap}_x T \le \operatorname{cap} T$ and $\operatorname{cap}_x T \le r_x(T)$ for every $x \in X$.

Theorem 8. *Let $T = (T_1, \dots, T_n)$ be an n-tuple of mutually commuting operators on a Banach space X. Then there is a dense subset L of X with the property that $\liminf_{k\to\infty}(\operatorname{cap}_{x,k} T)^{1/k} = \operatorname{cap} T$ for all $x \in L$.*

In particular, the limit $\lim_{k\to\infty}(\operatorname{cap}_{x,k})^{1/k}$ exists and is equal to $\operatorname{cap} T$ for all $x \in L$.

Proof. Let $x \in X$ and $\varepsilon > 0$. Then there exists $y \in X$ with $\|y - x\| < \varepsilon$ and

$$\|p(T)y\| \geq C(1 + \deg p)^{-(2n+\varepsilon)} r_e(p(T))$$

for all polynomials p. Thus

$$\begin{aligned}\operatorname{cap}_k(T, y) &= \inf\{\|p(T)y\| : p \in \mathcal{P}_k^1(n)\}\\ &\geq C(1+k)^{-(2n+\varepsilon)} \inf\{r_e(p(T)) : p \in \mathcal{P}_k^1(n)\},\end{aligned}$$

where

$$r_e(p(T)) = \sup\{|p(z)| : z \in \sigma_{He}(T)\},$$

and so

$$\operatorname{cap}_k(T, y) \geq C(1+k)^{-(2n+\varepsilon)} \operatorname{cap}_k(\sigma_{He}(T)).$$

Hence, by Corollary 36.8,

$$\liminf_{k\to\infty}(\operatorname{cap}_{y,k} T)^{1/k} \geq \liminf_{k\to\infty}(\operatorname{cap}_k \sigma_{He}(T))^{1/k} = \operatorname{cap}(\sigma_{He}(T)) = \operatorname{cap} T.$$

The opposite inequality $\liminf_{k\to\infty}(\operatorname{cap}_{y,k} T)^{1/k} \leq \operatorname{cap} T$ is trivially satisfied for all $y \in X$. □

Remark 9. Example 37.11 shows that it in the previous theorem it is not possible to replace the word “dense” by “residual”.

Theorem 10. *Let $T = (T_1, \dots, T_n)$ be an n-tuple of mutually commuting operators on a Banach space X. Then the set $\{x \in X : \operatorname{cap}_x T = \operatorname{cap} T\}$ is residual in X.*

Proof. Write $K = \sigma_{He}(T)$. The statement is clear if $\operatorname{cap} T = 0$. In the following we assume that $\operatorname{cap} T = \operatorname{cap} K > 0$. In particular, $p|K \not\equiv 0$ for each non-zero polynomial p.

For $j \in \mathbb{N}$ denote by M_j the set of all $x \in X$ with the property that there exists $k \geq j$ such that

$$\|p(T)x\| > \frac{1}{4k}\binom{n+k}{n}^{-2} \|p\|_K$$

for all non-zero $p \in \mathcal{P}_k(n)$.

We first prove that M_j is open. Let $j \in \mathbb{N}$ and $x \in M_j$. Let $k \geq j$ satisfy $\|p(T)x\| > \frac{1}{4k}\binom{n+k}{n}^{-2}\|p\|_K$ for all $p \in \mathcal{P}_k(n)$. For each polynomial $p \in \mathcal{P}(n)$, $p = \sum_{\alpha\in\mathbb{Z}_+^n} a_\alpha z^\alpha$ write $|p| = \sum_{\alpha\in\mathbb{Z}_+^n} |a_\alpha|$. By a compactness argument, there is a positive number δ such that $\|p(T)x\| > \frac{1}{4k}\binom{n+k}{n}^{-2}\|p\|_K + \delta$ for all $p \in \mathcal{P}_k(n)$ with $|p| = 1$. Let $y \in X$, $\|y - x\| < \frac{\delta}{\max\{\|T^\alpha\|:|\alpha|\le k\}}$. Then

$$\begin{aligned}\|p(T)y\| &\ge \|p(T)x\| - \|p(T)(y-x)\|\\ &> \frac{1}{4k}\binom{n+k}{n}^{-2}\|p\|_K + \delta - \max\{\|T^\alpha\| : |\alpha| \le k\}\cdot\|y-x\|\\ &\ge \frac{1}{4k}\binom{n+k}{n}^{-2}\|p\|_K\end{aligned}$$

for all $p \in \mathcal{P}_k(n)$ with $|p| = 1$. Thus $y \in M_j$ and the set M_j is open.

We show that M_j is dense. Let $x \in X$ and $\varepsilon > 0$. Choose $k \ge j$ such that $\frac{4}{k} < \varepsilon$. By Lemma 1, there is a subspace $Y \subset X$ of finite codimension such that

$$\|p(T)(x+y)\| \ge \max\{\|p(T)x\|/2, \|p(T)y\|/4\}$$

for all $y \in Y$ and $p \in \mathcal{P}_k(n)$.

By Theorem 5, there is a vector $u \in Y$ of norm 1 such that

$$\|p(T)u\| > \frac{1}{4}\binom{n+k}{n}^{-2}\|p\|_K$$

for all non-zero $p \in \mathcal{P}_k(n)$. Then $\left\|(x + \frac{4u}{k}) - x\right\| < \varepsilon$ and

$$\left\|p(T)\Big(x + \frac{4u}{k}\Big)\right\| \ge \frac{1}{k}\|p(T)u\| > \frac{1}{4k}\binom{n+k}{n}^{-2}\|p\|_K$$

for all non-zero $p \in \mathcal{P}_k(n)$. Thus $x + \frac{4u}{k} \in M_j$ and M_j is dense in X.

By the Baire category theorem, the set $\bigcap_{j=1}^\infty M_j$ is residual. Let $x \in \bigcap_{j=1}^\infty M_j$. Then there are infinitely many positive integers k for which

$$\|p(T)x\| > \frac{1}{4k}\binom{n+k}{n}^{-2}\|p\|_K \qquad (p \in \mathcal{P}_k(n)).$$

For such k we have

$$\begin{aligned}\operatorname{cap}_{x,k} T &= \inf\{\|p(T)x\| : p \in \mathcal{P}_k^1(n)\}\\ &\ge \inf\left\{\frac{1}{4k}\binom{n+k}{n}^{-2}\|p\|_K : p \in \mathcal{P}_k^1(n)\right\}\\ &= \frac{1}{4k}\binom{n+k}{n}^{-2}\operatorname{cap}_k K \ge \frac{1}{4k}(n+k)^{-2n}\operatorname{cap}_k K.\end{aligned}$$

Hence

$$\operatorname{cap}_x T = \limsup_{k\to\infty}(\operatorname{cap}_{x,k} T)^{1/k} \ge \limsup_{k\to\infty}(\operatorname{cap}_k K)^{1/k} = \operatorname{cap} K = \operatorname{cap} T.$$

Since $\operatorname{cap}_x T \le \operatorname{cap} T$ for all $x \in X$, the set $\{x \in X : \operatorname{cap}_x T = \operatorname{cap} T\}$ is residual. □

Definition 11. Let $T = (T_1, \dots, T_n)$ be an n-tuple of mutually commuting operators on a Banach space X. We say that T is *quasialgebraic* if $\operatorname{cap} T = 0$. We say that T is *locally quasialgebraic* if $\operatorname{cap}_x T = 0$ for each $x \in X$.

Thus we have the following approximate version of the Kaplansky theorem:

Corollary 12. *An n-tuple of mutually commuting operators is locally quasialgebraic if and only if it is quasialgebraic.*

Comments on Chapter V

C.35.1. The joint spectral radius was first studied by Rotta and Strang [RtS], where the radius $r''_\infty(a) = \lim_n \max\{\|a^\alpha\| : |\alpha| = n\}^{1/n}$ was introduced. The radius $r'_\infty(a) = \lim_n \max\{r(a^\alpha) : |\alpha| = n\}^{1/n}$ was defined by Berger and Wang [BW].

The general r_p radii were introduced in [So5], cf. also [ChZ]. In the case of Hilbert space operators the radius r_2 was considered by Bunce [Bun].

Proposition 35.2 was proved in [ChZ], see also [So4]. The spectral radius formula for $p = \infty$ (Theorem 35.5) was proved in [So5], the corresponding result for $p < \infty$ (Theorem 35.6) in [Mü18].

C.35.2. The definition of the joint spectral radius $r_p(a) = \max\{|\lambda|_p : \lambda \in \sigma_H(a)\}$ and the formula

$$r''_p(a) = \lim_{k\to\infty} \Big(\sum_{f \in F(k,n)} \|a_{f(1)} \cdots a_{f(k)}\|^p \Big)^{1/pk} \tag{1}$$

make sense also for non-commuting n-tuples $a = (a_1, \dots, a_n)$ of elements in a Banach algebra $\mathcal{A}$.

Since the limit $\lim\limits_{k\to\infty} \Big(\sum_{f \in F(k,n)} r^p(a_{f(1)} \cdots a_{f(k)}) \Big)^{1/pk}$ in general does not exist, we set

$$r'_p(a) = \limsup_{k\to\infty} \Big(\sum_{f \in F(k,n)} r^p(a_{f(1)} \cdots a_{f(k)}) \Big)^{1/pk}.$$

For non-commuting n-tuples it is true only

$$r_p(a) \le r'_p(a) \le r''_p(a)$$

and these inequalities may be strict, see [Gu], [RsS]. However, by [BW], $r'_\infty(a) = r''_\infty(a)$ for all n-tuples (even for infinite bounded families) of matrices. This was generalized by [TS] to precompact families of weakly compact operators. In particular, $r'_\infty(T) = r''_\infty(T)$ for all n-tuples of operators on a reflexive Banach space.

C.35.3. Since the definitions of $r_p(a)$ and $r'_p(a)$ depend only on the spectrum, the equality $r_p(a) = r'_p(a)$ is true also for n-tuples $a = (a_1, \dots, a_n)$ of elements that are mutually commuting modulo rad $\mathcal{A}$.

It is an open problem whether also $r''_p(a) = r'_p(a)$ is this situation, see [RsS].

C.35.4. The previous question for $p = \infty$ is closely related to the following interesting problem, see [Tu]:
Suppose that $\mathcal{A}$ is a radical Banach algebra (i.e., a non-unital Banach algebra consisting of quasinilpotents). Is then $\mathcal{A}$ finitely quasinilpotent (in our notation, is it true that $r''_\infty(a) = 0$ for all finite tuples $a = (a_1, \dots, a_n)$, where $r''_\infty(a)$ is defined by (1))?

C.35.5. The previous question may be considered as an approximate version of the Nagata-Higman theorem:

If $\mathcal{A}$ is an algebra (without unit) which is nilpotent of order n (i.e., $a^n = 0$ for all $a \in \mathcal{A}$), then there exists $m \in \mathbb{N}$ such that $a_1 \cdots a_m = 0$ for all $a_1, \dots, a_m \in \mathcal{A}$.

The original proof of the Nagata-Higman theorem gave $m = 2^n - 1$; the best known result gives $m = n^2$. There is a conjecture that one can take $m = \binom{n}{2}$; it is known that in general m cannot be smaller. For details see [Fo].

C.35.6. Theorem 35.7 for $p = 2$ in the algebra of operators on a Hilbert space (the ℓ^2-norm seems to be natural in this setting) was conjectured by Bunce [Bun] and proved in [MS1] (for finite-dimensional Hilbert space case see [ChH]).

C.36.1. The concept of capacity for single Banach algebra elements is due to Halmos [Hal2], who also proved Theorem 36.2.

The capacity of commuting n-tuples was introduced in [Sti1], where the estimate

$$\operatorname{cap} \sigma_H(a_1, \dots, a_n) \le \operatorname{cap}(a_1, \dots, a_n) \le 2^n \operatorname{cap} \sigma_H(a_1, \dots, a_n)$$

was proved. The equality $\operatorname{cap}(a_1, \dots, a_n) = \operatorname{cap} \sigma_H(a_1, \dots, a_n)$ as well as the existence of the limit $\lim_{k\to\infty}(\operatorname{cap}_k(a_1, \dots, a_n))^{1/k}$ was proved in [Mü13].

C.36.2. For the classical capacity of compact subsets of $\mathbb{C}$ see [Ts]. The capacity of compact subsets of $\mathbb{C}^n$ is treated in [Za] and [Si2].

C.36.3. Lemma 36.7 (the equality $\operatorname{cap} \sigma_e(T) = \operatorname{cap} \sigma(T)$) was proved in [Sti2].

It also follows directly from a general result from potential theory that $\operatorname{cap}(K \cup C) = \operatorname{cap} K$ whenever $K \subset \mathbb{C}$ is compact and C countable, see [Ts, p. 53–55]. Therefore we have $\operatorname{cap} \sigma_e(T) = \operatorname{cap} \widehat{\sigma}_e(T) = \operatorname{cap} \sigma(T)$, since $\sigma(T) \setminus \widehat{\sigma}_e(T)$ is at most countable.

C.36.4. Let f be an analytic function from a domain $D \subset \mathbb{C}$ into a Banach algebra $\mathcal{A}$. Then the functions $z \mapsto \operatorname{cap} f(z)$ and $z \mapsto \log \operatorname{cap} f(z)$ are subharmonic, see [Sl3].

C.36.5. Another concept of capacity of n-tuples of commuting Banach algebra elements was introduced in [So1] and [So2]. Instead of monic polynomials $p(z) = \sum_{|\alpha|\le n} c_\alpha z^\alpha$ with $\sum_{|\alpha|=n} |c_\alpha| = 1$ it uses polynomials satisfying $\sum_{|\alpha|=n} c_\alpha = 1$.

C.37.1. The notion of orbits originated in the theory of dynamical systems. In the context of operator theory it was first used by Rolewicz [Ro]. Orbits in Hilbert spaces were studied intensely by Beauzamy; for a survey of results and relations with the invariant subspace problem see [Bea2].

For a survey of results concerning orbits in Banach spaces see [Mü22].

C.37.2. Orbits are also closely related to problems concerning the stability of semi-groups of operators. For results in this direction see Datko [Dat], Pazy [Pa] and van Neerven [Ne3].

C.37.3. It is simple to find an operator without non-trivial invariant subspaces on a real Hilbert space H with $\dim H = 2$. For example, take

$$T = \begin{pmatrix} 0 & 1 \\ -1 & 0 \end{pmatrix}.$$

It is an open problem whether each operator on a real Hilbert space H with $\dim H \ge 3$ has a non-trivial closed invariant subspace.

C.37.4. Theorem 37.8 was proved in [Mü9]. The same statement for Hilbert space operators was proved in [Bea2, p. 48].

Theorem 37.13 is due to Zabczyk [Zab].

Theorems 37.14 and 37.17 were proved in [MV]. The essential tool is the plank theorem due to K. Ball [B1], [B2], see Appendix 5.

C.37.5. Let $T \in \mathcal{B}(H)$ be a non-nilpotent Hilbert space operator. By [Bea2], there exists $x \in H$ such that $\sum \frac{\|T^n x\|}{\|T^n\|} = \infty$. This is not true for Banach space operators, see Example 37.24.

For Hilbert space operators the following result can be shown [Mü22]: let $T \in \mathcal{B}(H)$ be non-nilpotent, let $0 < p < 2$. Then the set of all $x \in H$ such that $\sum \left(\frac{\|T^n x\|}{\|T^n\|}\right)^p = \infty$ is residual.

C.38.1. The first example of a hypercyclic vector was constructed by Rolewicz [Ro].

Various versions of criterion 38.6 appeared in [GS], [Ki] and [GS], for related questions see also [GLM] and [LM].

C. 38.2. Theorem 38.8 appeared in [Fel], Theorem 38.9 was proved by Bourdon and Feldman [BF].

C.38.3. Theorem 38.11 is due to Ansari [Ans]. Theorem 38.12 was proved in [LMü].

C.39.1. Weak orbits were studied in [Ne1], [Ne2]. A version of Theorem 3 was first proved in [Ne2], where the following generalization was also shown: if $T \in \mathcal{B}(X)$, $r(T) = 1$, $1 \leq p < \infty$, $\beta_n > 0$ and $\sum_n \beta_n = \infty$, then there exist $x \in X$ and $x^* \in X^*$ such that $\sum_n \beta_n |\langle T^n x, x^* \rangle|^p = \infty$.

C.39.2. Corollary 39.4 is due to Weiss [We]. Theorems 39.5 and 39.8 are again consequences of the plank theorem, see Appendix 5. They appeared in [MV].

C.39.3. By [Mü22], there is a Hilbert space operator $T \in \mathcal{B}(H)$ such that

$$\sum \frac{|\langle T^n x, y \rangle|}{\|T^n\|} < \infty \qquad \text{for all } x, y \in H.$$

So the result of Theorem 10 is the best possible.

C.39.4. In general it is not true that for $T \in \mathcal{B}(X)$ and a sequence (a_n) of positive numbers tending to 0 there exist $x \in X$ and $x^* \in X^*$ with $|\langle T^n x, x^* \rangle| \geq a_n r(T^n)$ for all n. A simple example for real spaces (in a 2-dimensional Hilbert space) was given in [Ne2]. The complex case is more complicated. An example of a unitary operator without orbits tending to 0 arbitrarily slowly was constructed in [BMü].

In many cases, however, it is possible to find $x \in X$ and $x^* \in X^*$ with $|\langle T^n x, x^* \rangle| \geq a_n r(T^n)$ for all n. For example this is true for positive operators in Banach lattices [Ne2].

In Corollary 39.13 we stated the same result for Hilbert space operators of class $C_{0\cdot}$ with spectral radius equal to 1. It is also true for Banach space operators of the same class [Mü25], and for all completely non-unitary contractions on a Hilbert space with spectral radius equal to 1, see [BMü].

C.40.1. The Scott Brown technique was first used in [Br] for the construction of invariant subspaces for subnormal operators. Theorem 40.4 was proved for Hilbert space contractions in [BCP1].

C.40.2. Theorem 40.7 is due to Apostol [Ap5].

C.40.3. The existence of invariant subspaces for Hilbert space contractions with spectrum containing the unit circle was proved by Brown, Chevreau and Pearcy in [BCP2].

The Banach space version was proved in [AM].

C.40.4. By the von Neumann inequality, every operator on a Hilbert space which is similar to a contraction is polynomially bounded. It was a longstanding problem due to Halmos whether the opposite statement is also true. The problem was solved by Pisier [Pis], who constructed a polynomially bounded Hilbert space operator which is not similar to a construction.

C.41.1. The classical Kaplansky theorem (Theorem 41.1) was proved in [Kap]. A stronger version (Theorem 41.5 for single operators) was proved by Sinclair [Sin]. Theorem 41.2 and its corollaries 41.4 and 41.5 were proved in [Mü11].

A similar result was proved in [Au3]: if $T_1, \dots, T_n \in \mathcal{B}(X, Y)$ and the vectors $T_1 x, \dots, T_n x$ are linearly dependant for each $x \in X$, then there is a non-trivial linear combination of these operators with rank $\le n-1$.

C.41.2. By Theorem 41.2, if $T_1, T_2, \dots \in \mathcal{B}(X)$ and the vectors $T_1 x, T_2 x, \dots$ are linearly dependant for all $x \in X$, then there is a non-trivial linear combination of the operators T_i which is a finite-rank operator. In general, it is not true that the operators T_i are linearly dependant, so there in no non-trivial linear combination of the operators T_i giving 0, see [Mü11].

C.41.3. Corollary 41.4 and Theorem 41.5 are also true for countable families of operators. Also, they are true for non-commuting families (in this case it is necessary to consider polynomials in non-commuting variables in the definition of algebraic tuples). The proofs remain essentially the same, see [Mü11].

C.42.1. Lemma 42.4 uses the idea of extremal points of Fekete-Leja, see [Fe], [Si1]. Theorems 42.5 and 42.6 were proved in [Mü12] for single operators and in [MS2] for n-tuples.

Using a Dvoretzky's type result, it is possible to improve the estimate in Theorem 42.5 to

$$\|p(T)y\| \ge (1-\varepsilon)\binom{n+k}{n}^{-1} r_e(p(T))$$

for all $p \in \mathcal{P}_k(n)$; this estimate is the best possible. Similarly, in Theorem 42.6 it is possible to obtain

$$\|p(T)y\| \ge C(1+\deg p)^{-(n+k)} r_e(p(T))$$

for all polynomials p, see [Mü24].

Theorem 42.10 was proved in [Mü22].

C.42.2. An alternative definition of the local capacity of operators was studied by Vasilescu in [Va2]. For $T \in \mathcal{B}(X)$ and $x \in X$ define

$$\operatorname{cap}'(T,x) = \limsup_n \operatorname{cap}'_n(T,x)^{1/n},$$

where $\operatorname{cap}'_n(T,x) = \inf\{r_x(p(T)) : p \text{ monic of degree } n\}$.

It is easy to see that $\operatorname{cap}(T,x) \le \operatorname{cap}'(T,x)$.

By [Va2], $\operatorname{cap}\gamma_x(T) \le \operatorname{cap}'(T,x) \le \operatorname{cap}\sigma_x(T)$. In particular, the equality $\operatorname{cap}\sigma_x(T) = \operatorname{cap}'(T,x)$ holds for all operators with SVEP. This also implies that $T \in \mathcal{B}(X)$ is quasialgebraic if and only if $\operatorname{cap}'(T,x) = 0$ for all $x \in X$.

Appendix

A.1 Banach spaces

We summarize here basic notations, definitions and results from the theory of Banach spaces. The results are well known and the proofs can be found in any textbook on functional analysis. We do not state the results in the greatest generality but only in the form which is relevant for spectral theory.

The reader may wish to use this section merely for reference regarding notation.

As usual, we denote by $\mathbb{N}, \mathbb{Z}, \mathbb{R}$ and $\mathbb{C}$ the set of all positive integers, integers, real and complex numbers, respectively.

As throughout this monograph we consider only complex Banach spaces (however, all results in this section are true also for real spaces).

Let X, Y be Banach spaces. By an *operator* $T : X \to Y$ we mean a bounded linear mapping. It is well known that a linear mapping is bounded if and only if it is continuous. The set of all operators from X to Y will be denoted by $\mathcal{B}(X, Y)$.

The norm of $T \in \mathcal{B}(X, Y)$ is defined by $\|T\| = \sup\{\|Tx\| : x \in X, \|x\| \le 1\}$. With this norm $\mathcal{B}(X, Y)$ becomes a Banach space. If $Y = X$ then we write briefly $\mathcal{B}(X)$ instead of $\mathcal{B}(X, X)$.

For $Y = \mathbb{C}$ we obtain the *dual* $X^* = \mathcal{B}(X, \mathbb{C})$, i.e., the space of all bounded linear functionals on X. For $x \in X$ and $f \in X^*$ it is sometimes more convenient to denote the value of f at x by $\langle x, f\rangle$ instead of $f(x)$. We use alternatively both notations.

The duality between X and X^* defines the *weak topology* on X and the *w^*-topology* on X^*.

The w^*-topology on X^* is the weakest topology for which the mappings $f \mapsto \langle x, f\rangle$ are continuous for every $x \in X$. In other words, the w^*-topology is the topology of pointwise convergence of functionals, i.e., if (f_α) is a net of elements of X^* and $f \in X^*$ then $f_\alpha \to f$ in the w^*-topology if and only if $\langle x, f_\alpha\rangle \to \langle x, f\rangle$ for all each $x \in X$.

Analogously, the weak topology on X is the weakest topology on X for which all functionals $f \in X^*$ are continuous. Equivalently, $x_\alpha \to x$ weakly if and only if $\langle x_\alpha, f\rangle \to \langle x, f\rangle$ for each $f \in X^*$.

When we are speaking about convergence, closure etc. without any specification it is always meant the convergence in the norm topology.

It is well known that X can be isometrically embedded into the second dual X^{**}, so X is usually regarded as a closed subspace of X^{**}.

We list the basic results from the theory of Banach spaces.

Theorem 1. (Hahn-Banach) *Let M be a closed subspace of a Banach space X. Let $f \in M^*$. Then there exists $g \in X^*$ such that $\|g\| = \|f\|$ and $g|M = f$.*

The following two results are variants of the Hahn-Banach theorem:

Theorem 2. *Let M be a closed subspace of X and let $x \in X$. Then there exists $g \in X^*$ such that $\|g\| = 1$, $\langle x, g\rangle = \operatorname{dist}\{x, M\}$ and $g|M = 0$.*

In particular, if $x \in X$ then there exists $g \in X^*$ such that $\|g\| = 1$ and $\langle x, g\rangle = \|x\|$.

Theorem 3. *Let L be a closed absolutely convex subset of X (i.e., L is convex and $c \in L, \alpha \in \mathbb{C}, |\alpha| \le 1 \Rightarrow \alpha c \in L$). Let $x \in X \setminus C$. Then there exists $f \in X^*$ such that $|\langle c, f\rangle| \le 1$ for all $c \in L$ and $\langle x, f\rangle > 1$.*

An easy consequence of the Hahn-Banach theorem is that a convex set (in particular a subspace) is closed if and only if it is closed in the weak topology.

A w^*-closed subspace is closed but a closed subspace is not necessarily closed in the w^*-topology.

Denote by $B_X = \{x \in X : \|x\| \le 1\}$ the *closed unit ball* in a Banach space X.

Theorem 4. *B_X is compact if and only if X is finite dimensional.*

Theorem 5. (Banach-Alaoglu) *The closed unit ball $B_{X^*} = \{f \in X^* : \|f\| \le 1\}$ is compact in the w^*-topology.*

The unit ball in X is not always weakly compact. In fact, B_X is weakly compact if and only if X is reflexive (i.e., $X^{**} = X$).

Theorem 6. (Banach open mapping theorem) *Let T be an operator from a Banach space X to a Banach space Y. The following conditions are equivalent:*

(i) *T is onto;*

(ii) *there exists a positive constant k such that $TB_X \supset k \cdot B_Y$;*

(iii) *there exists a positive constant k such that $\overline{TB_X} \supset k \cdot B_Y$.*

Corollary 7. *If an operator $T \in \mathcal{B}(X,Y)$ is one-to-one and onto, then T^{-1} is bounded.*

Another consequence of the open mapping theorem is:

Corollary 8. *Let $T \in \mathcal{B}(X,Y)$ and $TX \ne Y$. Then the range TX is of the first category in Y.*

Theorem 9. (closed graph theorem) *A linear mapping $T : X \to Y$ is bounded if and only if its graph $\{(x, Tx) : x \in X\}$ is closed in $X \times Y$.*

In particular, suppose that a linear mapping T satisfies the following condition: if $(x_i)_{i=1}^{\infty} \subset X$, $x_i \to 0$, $y \in Y$ and $Tx_i \to y$, then $y = 0$. Then T is bounded.

Theorem 10. (Banach-Steinhaus uniform boundedness theorem) *Let $\mathcal{M}$ be a subset of $\mathcal{B}(X,Y)$ such that $\sup\{\|Tx\| : T \in \mathcal{M}\} < \infty$ for every $x \in X$. Then $\sup\{\|T\| : T \in \mathcal{M}\} < \infty$.*

In particular, any weakly converging sequence of elements of X is bounded.

Let M be a subset of a Banach space X. Its *annihilator* $M^{\perp}$ is defined by

$$M^{\perp} = \{f \in X^* : \langle x, f \rangle = 0 \text{ for all } x \in M\}.$$

Clearly, $M^{\perp}$ is a w^*-closed (and so closed) subspace of X^*.

Similarly, if L is a subset of X^* then the *preannihilator* ${}^{\perp}L$ is defined by

$${}^{\perp}L = \{x \in X : \langle x, f \rangle = 0 \text{ for all } f \in L\}.$$

Clearly, ${}^{\perp}L$ is a closed subspace of X.

Theorem 11. *If M is a subspace of X then ${}^{\perp}(M^{\perp}) = \overline{M}$. If L is a subspace of X^* then $({}^{\perp}L)^{\perp}$ is the w^*-closure of L.*

Theorem 12. *If $\{M_\alpha\}_\alpha$ is any family of subsets of X then*

$$\Big(\bigcup_\alpha M_\alpha\Big)^{\perp} = \bigcap_\alpha M_\alpha^{\perp}.$$

If $\{L_\alpha\}_\alpha$ is any family of subsets of X^* then

$${}^{\perp}\Big(\bigcup_\alpha L_\alpha\Big) = \bigcap_\alpha {}^{\perp}L_\alpha.$$

In particular, if M_1, M_2 are closed subspaces then $(M_1 + M_2)^{\perp} = M_1^{\perp} \cap M_2^{\perp}$, where $M_1 + M_2 = \{m_1 + m_2 : m_1 \in M_1, m_2 \in M_2\}$.

The dual relation $(M_1 \cap M_2)^{\perp} = M_1^{\perp} + M_2^{\perp}$ is not always true; it is true under the assumption that $M_1 + M_2$ is closed.

Theorem 13. *Let M_1, M_2 be closed subspaces of a Banach space X. The following assertions are equivalent:*

(i) *$M_1 + M_2$ is closed;*
(ii) *$M_1^{\perp} + M_2^{\perp}$ is closed;*
(iii) *$(M_1 \cap M_2)^{\perp} = M_1^{\perp} + M_2^{\perp}$;*
(iv) *${}^{\perp}(M_1^{\perp} \cap M_2^{\perp}) = M_1 + M_2$.*

Let $T : X \to Y$ be an operator. Its *adjoint* T^* is the uniquely determined operator $T^* : Y^* \to X^*$ satisfying

$$\langle Tx, g\rangle = \langle x, T^*g\rangle \qquad \text{for all } x \in X \text{ and } g \in Y^*.$$

Then

$$\|T^*\| = \|T\| = \sup\{|\langle Tx, g\rangle| : x \in X, g \in Y^*, \|x\| \le 1, \|g\| \le 1\}.$$

If X, Y and Z are Banach spaces, $S \in \mathcal{B}(X,Y)$ and $T \in \mathcal{B}(Y,Z)$ then $(TS)^* = S^*T^*$ and $\|TS\| \le \|T\| \cdot \|S\|$. It is easy to check that $T = T^{**}|X$.

For $T \in \mathcal{B}(X,Y)$ denote by $\operatorname{Ker} T$ its *kernel*, $\operatorname{Ker} T = \{x \in X : Tx = 0\}$, and by $\operatorname{Ran} T$ its *range*, $\operatorname{Ran} T = TX = \{Tx : x \in X\}$. Clearly, $\operatorname{Ker} T$ is a closed subspace of X and $\operatorname{Ran} T$ is a (not necessarily closed) subspace of Y. It is easy to check that $\operatorname{Ker} T^*$ is even a w^*-closed subspace of Y^*.

Theorem 14. *Let $T \in \mathcal{B}(X,Y)$. Then*

$$\operatorname{Ker} T^* = (\operatorname{Ran} T)^\perp \quad \textit{and} \quad \operatorname{Ker} T = {}^\perp(\operatorname{Ran} T^*) = \operatorname{Ker} T^{**} \cap X.$$

Furthermore, ${}^\perp \operatorname{Ker} T^* = \overline{\operatorname{Ran} T}$ *and* $(\operatorname{Ker} T)^\perp = \overline{\operatorname{Ran} T^*}^{w^*}$ *(the w^*-closure of* $\operatorname{Ran} T^*$*).*

Corollary 15. *Let $T \in \mathcal{B}(X,Y)$. Then T^* is one-to-one if and only if* $\operatorname{Ran} T$ *is dense. Similarly, T is one-to-one if and only if* $\operatorname{Ran} T^*$ *is w^*-dense.*

Theorem 16. *Let $T \in \mathcal{B}(X,Y)$. Then the following statements are equivalent:*

(i) $\operatorname{Ran} T$ *is closed;*
(ii) $\operatorname{Ran} T^*$ *is closed;*
(iii) $\operatorname{Ran} T^*$ *is w^*-closed.*

Corollary 17. *If* $\operatorname{Ran} T$ *is closed then* $\operatorname{Ran} T = {}^\perp(\operatorname{Ker} T^*)$ *and* $\operatorname{Ran} T^* = (\operatorname{Ker} T)^\perp$. *Furthermore,* $\operatorname{Ran} T = \operatorname{Ran} T^{**} \cap Y$.

Corollary 18. *Let $T \in \mathcal{B}(X,Y)$. Then T is one-to-one and onto if and only if T^* is one-to-one and onto.*

Theorem 19. *Let M be a closed subspace of a Banach space X. Then M^* can be identified with $X^*/M^\perp$.*

More precisely, let $J : M \to X$ be the natural embedding of M into X. Then $J^* : X^* \to M^*$ assigns to a functional $f \in X^*$ the restriction $J^*f = f|M \in M^*$. Clearly, $\operatorname{Ker} J^* = M^\perp$*, so J^* defines an operator $S : X^*/M^\perp \to M^*$ by $S(f + M^\perp) = J^*f$. A straightforward calculation shows that S is an isometry. So M^* can be identified with $X^*/M^\perp$ and J^* with the canonical projection $X^* \to X^*/M^\perp$.*

Theorem 20. *Let M be a closed subspace of X. Then $(X/M)^*$ can be identified with $M^\perp$.*

More precisely, let $Q : X \to X/M$ *be the canonical projection. Then* $Q^* : (X/M)^* \to X^*$ *is an isometrical operator and* $\operatorname{Ran} Q^* = (\operatorname{Ker} Q)^\perp = M^\perp$*. Thus* Q^* *defines an isometrical operator from* $(X/M)^*$ *onto* $M^\perp$*.*

Theorem 21. *Let* $T \in \mathcal{B}(X,Y)$*. Then* T *decomposes in the following way:*

$$X \xrightarrow{Q} X/\operatorname{Ker} T \xrightarrow{T_0} \overline{\operatorname{Ran} T} \xrightarrow{J} Y,$$

where $T = JT_0Q$, $Q : X \to X/\operatorname{Ker} T$ *is the canonical projection, the operator* $T_0 : X/\operatorname{Ker} T \to \overline{\operatorname{Ran} T}$ *induced by* T *is one-to-one with dense range, and* $J : \overline{\operatorname{Ran} T} \to Y$ *is the natural embedding.*

The corresponding decomposition for T^* *is* $T^* = Q^* T_0^* J^*$*.*

Let M be a subset of a Banach space X. The *span* of M, denoted by $\bigvee M$, is the smallest closed subspace of X containing M (= the intersection of all closed subspaces containing M). Similarly, we write $M \vee M' = \bigvee(M \cup M')$.

If M, M' are closed subspaces of X then $M + M'$ is not necessarily closed.

Theorem 22. *Let* M *be a closed subspace of* X *and let* F *be a finite-dimensional subspace of* X*. Then* $M + F$ *is closed. In particular, a finite-dimensional subspace is closed.*

Similarly, if $M' \subset X^*$ *is a* w^**-closed subspace and* F' *is a finite-dimensional subspace of* X^**, then* $M' + F'$ *is* w^**-closed.*

Theorem 23. (Auerbach lemma) *Let* X *be a finite-dimensional Banach space. Then there exist bases* $\{x_1, \dots, x_n\}$ *in* X *and* $\{f_1, \dots, f_n\}$ *in* X^* *such that* $\|x_i\| = 1 = \|f_j\|$ *and* $\langle x_i, f_j \rangle = \delta_{ij}$ *(the Kronecker symbol) for all* $i, j = 1, \dots, n$*.*

An operator $P \in \mathcal{B}(X)$ is called a *projection* if $P^2 = P$. If $P \in \mathcal{B}(X)$ is a projection then $I - P$ is also a projection (here I denotes the identity operator on X). Further, $\operatorname{Ran} P = \operatorname{Ker}(I - P)$ and $\operatorname{Ker} P = \operatorname{Ran}(I - P)$. In particular, the range of a projection is closed.

Let M be a closed subspace of a Banach space X. Then M is called *complemented* in X if there exists a closed subspace $M' \subset X$ such that $X = M \oplus M'$ (i.e., $M \cap M' = \{0\}$ and $M + M' = X$).

Theorem 24. *Let* M *be a closed subspace of a Banach space* X*. The following statements are equivalent:*

(i) *M is complemented;*

(ii) *there exists a projection* $P \in \mathcal{B}(X)$ *such that* $\operatorname{Ran} P = M$*;*

(iii) *there exists a projection* $Q \in \mathcal{B}(X)$ *such that* $\operatorname{Ker} Q = M$*.*

It is well known that every closed subspace of a Hilbert space is complemented.

Theorem 25.

(i) *Let F be a finite-dimensional subspace of a Banach space X. Then F is complemented. More precisely, there exists a projection $P \in \mathcal{B}(X)$ such that* $\operatorname{Ran} P = F$ *and* $\|P\| \le (\dim M)^{1/2}$.

(ii) *Let M be a closed subspace of finite codimension in X. Then M is complemented. More precisely, for every $\varepsilon > 0$ there exists a projection $Q \in \mathcal{B}(X)$ such that* $\operatorname{Ker} Q = M$ *and* $\|Q\| \le \operatorname{codim} M + \varepsilon$.

(iii) *Let L be a complemented subspace of X, let $L' \subset X$ be a closed subspace. Suppose that either $L' \subset L$,* $\dim L/L' < \infty$, *or* $L' \supset L$, $\dim L'/L < \infty$. *Then L' is complemented.*

We finish this section with the following deep result from the algebraic topology that will be used frequently:

Theorem 26. (Borsuk antipodal theorem) *Let M, L be finite-dimensional Banach spaces with* $\dim M > \dim L$, *let $S_M = \{m \in M : \|m\| = 1\}$ be the unit sphere in M and let $g : S_M \to L$ be a continuous mapping satisfying $g(-m) = -g(m)$ for all $m \in S_M$. Then there exists $m \in S_M$ with $g(m) = 0$.*

A.2 Analytic vector-valued functions

Many well-known results for complex functions are true also for vector-valued functions.

Let K be a compact Hausdorff space, let μ be a finite complex Borel measure on K and let f be a continuous function from K into a Banach space X. Then, exactly as in the case of scalar functions, it is possible to define $\int f \,\mathrm{d}\mu$ as the limit of the Riemann sums. It is easy to verify that $\langle \int f \,\mathrm{d}\mu, x^* \rangle = \int \langle f(z), x^* \rangle \,\mathrm{d}\mu(z)$ for every bounded linear functional $x^* \in X^*$. More generally, if Y is another Banach space and $T \in \mathcal{B}(X, Y)$ then $T(\int f \,\mathrm{d}\mu) = \int (T \circ f) \,\mathrm{d}\mu$.

Let a_j $(j = 0, 1, 2, \dots)$ be elements of a Banach space X. Consider the power series $\sum_{j=0}^\infty a_j z^j$ $(z \in \mathbb{C})$. Its *radius of convergence* is the number defined by

$$\sup\Big\{ r \ge 0 : \sum_{j=1}^\infty a_j z^j \text{ converges for all } z \in \mathbb{C}, |z| < r \Big\}.$$

Theorem 1. *The radius of convergence of the power series $\sum_{j=0}^\infty a_j z^j$ is equal to* $\liminf_{j\to\infty} \|a_j\|^{-1/j}$.

Analytic functions with values in a Banach space play an important role in spectral theory. There are several possible definitions of analytic vector-valued functions. Fortunately, for Banach spaces all possibilities are equivalent:

Theorem 2. *Let f be a function from an open subset $G \subset \mathbb{C}$ into a Banach space X. The following conditions are equivalent:*

(i) *the function $z \mapsto \langle f(z), x^* \rangle$ is analytic for each $x^* \in X^*$;*

(ii) *the limit $\lim_{z\to w} \frac{f(z)-f(w)}{z-w}$ exists for every $w \in G$;*

(iii) *for every $w \in G$ there is a neighbourhood U of w, $U \subset G$ and elements $x_j \in X, j = 0, 1, \dots$ such that $f(z) = \sum_{j=0}^{\infty} x_j(z-w)^i \quad (z \in U)$.*

A function satisfying any of the conditions of Theorem 2 will be called *analytic* (on G). The set of all X-valued functions analytic on G will be denoted by $H(G, X)$.

Condition (i) of Theorem 2 enables us to generalize most of the properties of analytic scalar-valued functions to the vector-valued case. In particular, the Cauchy and Liouville theorems remain true.

If $G \subset \mathbb{C}$ is an open set and K a compact subset of G, then it is always possible to find a *contour* Γ that surrounds K in G. By this we mean that Γ is a finite union of disjoint, piecewise smooth simple closed curves in $G \setminus K$ such that the winding number

$$\mathrm{Ind}_\Gamma(w) = \frac{1}{2\pi i} \int_\Gamma \frac{1}{z-w}\, \mathrm{d}z = \begin{cases} 1 & (w \in K) \\ 0 & (w \notin G). \end{cases}$$

For details see [Co].

Theorem 3. (Cauchy) *Let $f \in H(G, X)$, let $w \in G$, and let Γ be a contour in G that surrounds $\{w\}$. Then $\int_\Gamma f(z)\,\mathrm{d}z = 0$ and $f(w) = \frac{1}{2\pi i}\int_\Gamma f(z)(z-w)^{-1}\,\mathrm{d}z$.*

Moreover, the coefficients x_j in condition (iii) *of Theorem* 2 *can be expressed as $x_j = \frac{1}{2\pi i}\int_\Gamma f(z)(z-w)^{-j-1}\,\mathrm{d}z$.*

Corollary 4. *If $f \in H(G, X)$, $w \in G$, $r > 0$, $\{z : |z-w| \le r\} \subset G$, $M = \max\{\|f(z)\| : |z-w| = r\}$ and $f(z) = \sum_{j=0}^{\infty} x_j(z-w)^j$ for all z with $|z-w| < r$, then*

$$\|x_j\| \le \frac{M}{r^j} \qquad (j = 0, 1, \dots).$$

Theorem 5. (Liouville) *Let $f : \mathbb{C} \to X$ be a bounded analytic function. Then f is a constant.*

Theorem 6. (residue theorem) *Let $x_j \quad (j \in \mathbb{Z})$ be elements of a Banach space X, $r > 0$ and $w \in \mathbb{C}$. Let*

$$f(z) = \sum_{-\infty}^{\infty} x_j(z-w)^j$$

be convergent in $\{z \in \mathbb{C} : 0 < |z-w| < r\}$. Then

$$\frac{1}{2\pi i} \int_{|z|=r/2} f(z)\,\mathrm{d}z = x_{-1}.$$

Theorem 7. (maximum principle) *Let G be an open subset of $\mathbb{C}$, let $f \in H(G,X)$, and let K be a compact subset of G. Then*

$$\max\{|f(z)| : z \in K\} = \max\{|f(z)| : z \in \partial K\}.$$

The analogy between the properties of scalar analytic functions and the analytic functions with values in a Banach space remains true also for analytic functions of n variables.

We use the standard multi-index notation. Denote by $\mathbb{Z}_+$ the set of all non-negative integers. For $\alpha = (\alpha_1, \dots, \alpha_n) \in \mathbb{Z}_+^n$ and $z = (z_1, \dots, z_n) \in \mathbb{C}^n$ we write $|\alpha| = \alpha_1 + \cdots + \alpha_n$, $\alpha! = \alpha_1! \cdots \alpha_n!$ and $z^\alpha = z_1^{\alpha_1} \cdots z_n^{\alpha_n}$.

Theorem 8. *Let G be an open subset of $\mathbb{C}^n$ and let f be a function from G to a Banach space X. The following properties are equivalent:*

(i) *the function $z \mapsto \langle f(z), x^* \rangle$ is analytic for each $x^* \in X^*$;*

(ii) *for every $w \in G$ there exists a neighbourhood U of w and elements $x_\alpha \in X \quad (\alpha \in \mathbb{Z}_+^n)$ such that*

$$f(z) = \sum_{\alpha \in \mathbb{Z}_+^n} x_\alpha (z-w)^\alpha \qquad (z \in U);$$

(iii) *f is analytic in each variable separately.*

A function satisfying the conditions of Theorem 8 is called analytic on G. The coefficients x_α from condition (ii) can be expressed by means of a multiple Cauchy integral. In particular, if $w \in G$, $r > 0$, $\{z = (z_1, \dots, z_n) : |z_k - w_k| \le r, k = 1, \dots, n\} \subset G$ and $M = \max\{\|f(z)\| : |z_k - w_k| \le r, k = 1, \dots, n\}$, then

$$f(w) = \frac{1}{(2\pi i)^n} \int_{|z_1 - w_1| = r} \cdots \int_{|z_n - w_n| = r} f(z)(z_1 - w_1)^{-1} \cdots (z_n - w_n)^{-1} \mathrm{d}z_1 \cdots \mathrm{d}z_n$$

and $\|x_\alpha\| \le \frac{M}{r^{|\alpha|}}$ for all $\alpha \in \mathbb{Z}_+^n$.

If X, Y are Banach spaces then $\mathcal{B}(X,Y)$ is also a Banach space. So we can consider operator-valued analytic functions.

Theorem 9. *Let G be an open subset of $\mathbb{C}^n$ and let $T : G \to \mathcal{B}(X,Y)$ be a function. The following conditions are equivalent:*

(i) *$T : G \to \mathcal{B}(X,Y)$ is analytic on G;*

(ii) *the function $z \mapsto T(z)x$ is analytic for each $x \in X$;*

(iii) *the function $z \mapsto \langle T(z)x, y^* \rangle$ is analytic for all $x \in X$ and $y^* \in Y^*$.*

A.3 C^∞-functions

The existence of partitions of unity is a standard tool in complex analysis.

Theorem 1. (partition of unity) *Let M be a metric space, let $\mathcal{O}$ be an open cover of M. Then there exist continuous functions $f_\alpha : M \to \langle 0,1\rangle \quad (\alpha \in \Lambda)$ such that:*

(i) $\sum_{\alpha\in\Lambda} f_\alpha(z) = 1 \qquad (z \in M)$;

(ii) *for every $\alpha \in \Lambda$ there exists an open set $U \in \mathcal{O}$ such that the support of f_α, $\operatorname{supp} f_\alpha = \{z \in M : f_\alpha(z) \neq 0\}^-$, is contained in U;*

(iii) *for every $z \in M$ there is a neighbourhood U of z such that U intersects $\operatorname{supp} f_\alpha$ for only a finite number of indices α.*

The functions f_α with the properties of Theorem 1 are called a *partition of unity subordinate to the cover $\mathcal{O}$.*

Let G be an open subset of $\mathbb{C}^n$. Then we can consider G as a subset of $\mathbb{R}^{2n}$ and denote by $C^\infty(G)$ the set of all functions $f : G \to \mathbb{R}$ that have continuous (real) partial derivatives of all orders.

In $\mathbb{C}^n$ there exist partitions of unity consisting of smooth functions. Moreover, any partition of unity is countable.

Theorem 2. *Let G be a subset of $\mathbb{C}^n$, let $\mathcal{O}$ be an open cover of G. Then there exist C^∞-functions $f_i : G \to \langle 0,1\rangle \quad (i = 1,2,\dots)$ with the properties of Theorem 1.*

Let G be an open subset of $\mathbb{C}^n$ and let X be a Banach space. Write $z_j = x_j + iy_j$ and denote by $C^\infty(G,X)$ the set of all functions $f : G \to X$ that have continuous partial derivatives of all orders with respect to the real variables $x_1,\dots,x_n,y_1,\dots,y_n$.

Instead of the derivatives $\frac{\partial f}{\partial x_j}$ and $\frac{\partial f}{\partial y_j}$ of a function $f \in C^\infty(G,X)$ it is usual to consider the formal derivatives

$$\frac{\partial f}{\partial z_j} = \frac{1}{2}\Big(\frac{\partial f}{\partial x_j} - i\frac{\partial f}{\partial y_j}\Big), \quad \frac{\partial f}{\partial \bar z_j} = \frac{1}{2}\Big(\frac{\partial f}{\partial x_j} + i\frac{\partial f}{\partial y_j}\Big).$$

By the Cauchy-Riemann conditions, a function $f \in C^\infty(G,X)$ is analytic if and only if $\frac{\partial f}{\partial \bar z_j} = 0 \quad (j = 1,\dots,n)$. In this case $\frac{\partial f}{\partial z_j}$ coincide with the familiar complex derivatives of f.

Let $0 \le p,q \le n$. A *differential form* of bidegree (p,q) with coefficients in $C^\infty(G,X)$ can be expressed as

$$\eta_{p,q} = \sum_{0\le i_1<\cdots<i_p\le n}\ \sum_{0\le j_1<\cdots<j_q\le n} f_{i_1,\dots,i_p,j_1,\dots,j_q} \mathrm{d}z_{i_1}\wedge\cdots\wedge\mathrm{d}z_{i_p}\wedge\mathrm{d}\bar z_{j_1}\wedge\cdots\wedge\mathrm{d}\bar z_{j_q} \quad (1)$$

where $f_{i_1,\dots,i_p,j_1,\dots,j_q} \in C^\infty(G,X)$.

Functions in $C^\infty(G,X)$ can be identified with differential forms of bidegree $(0,0)$.

A differential form of degree r can be written as $\eta = \sum_{p+q=r} \eta_{p,q}$, where $\eta_{p,q}$ are differential forms of bidegree (p,q). Denote by $\Lambda^r[\mathrm{d}z, \mathrm{d}\bar{z}, C^\infty(G,X)]$ the linear space of all differential forms of degree r with coefficients in $C^\infty(G,X)$. Further, denote by $\Lambda[\mathrm{d}z, \mathrm{d}\bar{z}, C^\infty(G,X)] = \bigoplus_{r=0}^{2n} \Lambda^r[\mathrm{d}z, \mathrm{d}\bar{z}, C^\infty(G,X)]$ the linear space of all differential forms.

On $\Lambda[\mathrm{d}z, \mathrm{d}\bar{z}, C^\infty(G,X)]$ we consider the wedge product defined by the following properties (for all $\omega, \eta, \tau \in \Lambda[\mathrm{d}z, \mathrm{d}\bar{z}, C^\infty(G,X)]$, $f, g \in C^\infty(G,X)$):

(i) $\omega \wedge (\eta + \tau) = \omega \wedge \eta + \omega \wedge \tau$;

(ii) $(\omega + \eta) \wedge \tau = \omega \wedge \tau + \eta \wedge \tau$;

(iii) $\omega \wedge (\eta \wedge \tau) = (\omega \wedge \eta) \wedge \tau$;

(iv) $(f\omega) \wedge (g\eta) = fg\omega \wedge \eta$;

(v) if $\omega \in \Lambda^k[\mathrm{d}z, \mathrm{d}\bar{z}, C^\infty(G,X)]$, $\eta \in \Lambda^l[\mathrm{d}z, \mathrm{d}\bar{z}, C^\infty(G,X)]$ then $\omega \wedge \eta = (-1)^{kl} \eta \wedge \omega$.

In particular, $s \wedge t = -t \wedge s$ for all $s, t \in \{\mathrm{d}z_1, \dots, \mathrm{d}z_n, \mathrm{d}\bar{z}_1, \dots, \mathrm{d}\bar{z}_n\}$.

For a function $f \in C^\infty(G,X)$ we set

$$\partial f = \sum_{j=1}^{n} \frac{\partial f}{\partial z_j} \mathrm{d}z_j, \quad \bar{\partial} f = \sum_{j=1}^{n} \frac{\partial f}{\partial d\bar{z}_j} \mathrm{d}\bar{z}_j \quad \text{and} \quad \mathrm{d}f = \partial f + \bar{\partial} f.$$

For a differential form $\eta_{p,q}$ of type (1) we set

$$\bar{\partial}\eta = \sum_{0 \le i_1 < \cdots < i_p \le n} \ \sum_{0 \le j_1 < \cdots < j_q \le n} \bar{\partial} f_{i_1,\dots,i_p,j_1\dots,j_q} \wedge \mathrm{d}z_{i_1} \wedge \cdots \wedge \mathrm{d}z_{i_p} \wedge \mathrm{d}\bar{z}_{j_1} \wedge \cdots \wedge \mathrm{d}\bar{z}_{j_q}.$$

Similarly we define $\partial\eta$ and $\mathrm{d}\eta = \partial\eta + \bar{\partial}\eta$. This defines linear mappings $\partial, \bar{\partial}$ and d acting in the space $\Lambda[\mathrm{d}z, \mathrm{d}\bar{z}, C^\infty(G,X)]$ which map $\Lambda^r[\mathrm{d}z, \mathrm{d}\bar{z}, C^\infty(G,X)]$ into $\Lambda^{r+1}[\mathrm{d}z, \mathrm{d}\bar{z}, C^\infty(G,X)]$ for all $r, 0 \le r \le 2n-1$.

It is easy to verify that $\mathrm{d}(\mathrm{d}\eta) = 0$ for every form η.

We interpret

$$(2i)^{-n} \mathrm{d}\bar{z}_1 \wedge \cdots \mathrm{d}\bar{z}_n \wedge \mathrm{d}z_1 \wedge \cdots \wedge \mathrm{d}z_n$$

as the Lebesgue measure in $\mathbb{R}^{2n}$. Then we can define integrals of differential forms of degree $2n$ on subsets of $\mathbb{C}^n$. Analogously we define integrals of differential forms on submanifolds of $\mathbb{C}^n$. For details see [Spi].

Theorem 3. (Stokes) *Let G be an open subset of $\mathbb{C}^n$, let η be a differential form with coefficients in $C^\infty(G,X)$ of degree $2n-1$. Let Δ be an open bounded subset such that $\overline{\Delta} \subset G$ and $\partial\Delta$ is a piecewise smooth surface. Then*

$$\int_{\partial\Delta} \eta = \int_{\Delta} \mathrm{d}\eta.$$

Corollary 4. *Let τ be a differential form of degree $2n-1$ with compact support. Then*

$$\int_{\mathbb{C}^n} \mathrm{d}\tau = 0.$$

Theorem 5. *Let $U \subset \mathbb{C}^n$ be a polydisc and let $j : H(U,X) \to C^\infty(U,X)$ be the natural embedding. Then the sequence*

$$0 \to H(U,X) \xrightarrow{j} C^\infty(U,X) \xrightarrow{\bar\partial} \Lambda^1[\mathrm{d}\bar z, C^\infty(U,X)] \xrightarrow{\bar\partial} \cdots \xrightarrow{\bar\partial} \Lambda^n[\mathrm{d}\bar z, C^\infty(U,X)] \to 0$$

is exact.

A.4 Semicontinuous set-valued functions

Definition 1. Let X, Y be metric spaces, let Φ be a mapping which assigns to each point $x \in X$ a closed subset of Y. Let $x_0 \in X$. Then Φ is called *upper semicontinuous* at x_0 if for every neighbourhood U of $\Phi(x_0)$ there exists $\varepsilon > 0$ such that $\Phi(x) \subset U$ for all $x \in X$ with $\operatorname{dist}\{x_0, x\} < \varepsilon$.

The mapping Φ is called *lower semicontinuous* at x_0 if for every open set U with $U \cap \Phi(x_0) \neq \emptyset$ there exists $\varepsilon > 0$ such that $U \cap \Phi(x) \neq \emptyset$ for all $x \in X$ with $\operatorname{dist}\{x, x_0\} < \varepsilon$.

The mapping Φ is called *continuous* at x_0 if it is both upper and lower semicontinuous at x_0. Φ is called upper (lower) semicontinuous or continuous if it has the corresponding property at every point $x \in X$.

Definition 2. Let M, L be subsets of a metric space Y. Write

$$\Delta(M,L) = \sup_{m \in M} \operatorname{dist}\{m, L\}$$

(supremum of an empty set is considered to be equal to 0). The *Hausdorff distance* $\widehat{\Delta}(M,L)$ is defined by

$$\widehat{\Delta}(M,L) = \max\bigl\{\Delta(M,L), \Delta(L,M)\bigr\}.$$

The most important set-valued functions for our purpose are those with compact values. The semicontinuity of such mappings can be characterized easily:

Theorem 3. *Let X, Y be metric spaces, Y locally compact, $x \in X$ and let Φ be a mapping which assigns to every point of X a compact subset of Y. The following statements are equivalent:*

(i) *Φ is upper semicontinuous at x;*

(ii) *if $x_n \in X$ $(n = 1, 2, \ldots)$ and $x_n \to x$ then $\lim_{n\to\infty} \Delta\bigl(\Phi(x_n), \Phi(x)\bigr) = 0$;*

(iii) *if $x_n \in X, y_n \in \Phi(x_n)$ $(n = 1, 2, \ldots)$ and $y \in Y$ satisfy $x_n \to x$ and $y_n \to y$, then $y \in \Phi(x)$.*

Theorem 4. *Let X, Y be metric spaces, $x \in X$ and let Φ be a mapping which assigns to every point of X a compact subset of Y. The following statements are equivalent:*

(i) *Φ is lower semicontinuous at x;*

(ii) *if $x_n \in X \quad (n = 1, 2, \dots)$ and $x_n \to x$ then $\lim_{n\to\infty} \Delta(\Phi(x), \Phi(x_n)) = 0$;*

(iii) *if $x_n \in X \quad (n = 1, 2, \dots)$, $x_n \to x$ and $y \in \Phi(x)$, then there exist $y_n \in \Phi(x_n) \quad (n = 1, 2, \dots)$ such that $y_n \to y$.*

Theorem 5. (Michael's selection theorem) *Let X be a metric space, let Φ be a lower semicontinuous mapping which assigns to each point $x \in X$ a closed convex subset of a Banach space Y. Then there exists a continuous mapping $f : X \to Y$ such that $f(x) \in \Phi(x)$ for all $x \in X$.*

A.5 Some geometric properties of Banach spaces

The first result is the solution to the so-called plank problem which is due to Ball [B1].

Proposition 1. Let X be a (real or complex) Banach space, $y \in X$ any vector and $f_1, f_2, \dots \in X^*$ unit functionals. For each $n \in \mathbb{N}$, let a_n be positive numbers such that $\sum_{n=1}^{\infty} a_n < 1$. Then there is a point $x \in X$ such that $\|x - y\| \leq 1$ and $|\langle x, f_n \rangle| \geq a_n$ for every n.

A stronger result is known for operators on a complex Hilbert space [B2]. It is worth to note that it is not true for real Hilbert spaces.

Proposition 2. Let H be a complex Hilbert space and $f_1, f_2, \dots \in H$ unit vectors. For each $n \in \mathbb{N}$, let $a_n > 0$ be such that $\sum_{n=1}^{\infty} a_n^2 < 1$. Then there is a point $x \in H$ such that $\|x\| = 1$ and $|\langle x, f_n \rangle| \geq a_n$ for every n.

The following Zenger theorem can be found in [BD], p. 18–20.

Theorem 3. *Let X be a (real or complex) Banach space, let $u_1, \dots, u_n \in X$ be linearly independent. Let $\alpha_j \quad (j = 1, \dots, n)$ be positive numbers with $\sum_{j=1}^{n} \alpha_j = 1$. Then there exist complex numbers $w_1, \dots, w_n$ and $f \in X^*$ such that*

$$\Big\|\sum_{j=1}^{n} w_j u_j\Big\| \leq 1, \ \|f\| \leq 1 \quad \text{and} \quad \langle w_j u_j, f \rangle = \alpha_j$$

for all $j = 1, \dots, n$.

A.6 Basic properties of H^∞

Denote by $\mathcal{P}$ the normed space of all complex polynomials with the norm $\|p\| = \sup\{|p(z)| : z \in \mathbb{D}\}$. Recall that $\mathcal{A}(\mathbb{D})$ denotes the disc algebra of all functions analytic on $\mathbb{D}$ and continuous on $\overline{\mathbb{D}}$ and H^∞ the Hardy space of all bounded analytic functions on $\mathbb{D}$.

Let L^∞ be the space of all complex bounded measurable function on $\mathbb{T}$ with the usual norm. Then $\mathcal{P} \subset \mathcal{A}(\mathbb{D}) \subset H^\infty \subset L^\infty$ and $\mathcal{A}(\mathbb{D})$ is the norm closure of $\mathcal{P}$. The inclusion $H^\infty \subset L^\infty$ follows from the following theorem.

Theorem 1. (Fatou) *Let $f \in H^\infty$. Then the limit $\lim_{r\to 1-} f(re^{it})$ exists for almost all $t \in \langle 0, 2\pi)$ (with respect to the Lebesgue measure). If we write $f(e^{it}) = \lim_{r\to 1-} f(re^{it})$, then $f(e^{it}) \in L^\infty$ and* ess sup $|f(e^{it})| = \sup_{z\in\mathbb{D}} |f(z)|$.

Let L^1 be the space of all integrable functions on $\mathbb{T}$ with the norm $\|f\|_1 = \frac{1}{2\pi}\int |f(e^{it})|\,\mathrm{d}t$. Since $L^\infty = (L^1)^*$, the space H^∞ inherits the w^*-topology from L^∞.

The w^*-convergence of sequences in H^∞ is easy to describe as the bounded pointwise convergence.

Theorem 2.

(i) *Let $(f_n) \subset H^\infty$ be a sequence of bounded analytic functions. Then $f_n \xrightarrow{w^*} 0$ if and only if (f_n) is a Montel sequence, i.e., $\sup_n \|f_n\| < \infty$ and $f_n(z) \to 0 \quad (z \in \mathbb{D})$;*

(ii) *H^∞ is a w^*-closed subspace of L^∞ and $\overline{\mathcal{P}}^{w^*} = H^\infty$. Moreover, every function $f \in H^\infty$ is a w^*-limit of a sequence of polynomials.*

Of particular interest are the functionals on $\mathcal{P}$ that are w^*-continuous, i.e., the continuous linear mappings from $(\mathcal{P}, w^*)$ to $\mathbb{C}$.

Theorem 3. *Let $\psi \in \mathcal{P}^*$. Then the following statements are equivalent:*

(i) *$\psi \in \mathcal{P}^*$ is w^*-continuous;*

(ii) *$\psi(p_n) \to 0$ for each Montel sequence (p_n) of polynomials;*

(iii) *there is a w^*-continuous functional ψ_0 on H^∞ such that $\psi = \psi_0|\mathcal{P}$ and $\|\psi_0\| = \|\psi\|$;*

(iv) *there is an absolutely continuous measure μ on $\mathbb{T}$ such that $\psi(p) = \int p\,\mathrm{d}\mu$ $(p \in \mathcal{P})$. By the F. and M. Riesz theorem, in this case each measure representing ψ is absolutely continuous;*

(v) *there exists $f \in L^1$ such that $\|f\|_1 = \|\psi\|$ and $\psi(p) = \int_0^{2\pi} p(e^{it})f(e^{it})\,\mathrm{d}t$ for all $p \in \mathcal{P}$.*

A subset $\Lambda \subset \mathbb{D}$ is called dominant if $\sup_{\lambda\in\Lambda} |f(\lambda)| = \|f\|$ for all $f \in H^\infty$. An example of a dominant set is an annulus $\{z : 1-\varepsilon < |z| < 1\}$. However, a dominant set can be much smaller, even countable. Clearly $\overline{\Lambda} \supset \mathbb{T}$ for each dominant set. The next result is an application of the Hahn-Banach theorem.

Theorem 4. *Let $\psi \in \mathcal{P}^*$ be w^*-continuous. Let $\Lambda \subset \mathbb{D}$ be a dominant subset. Let $\varepsilon > 0$. Then there are numbers $\lambda_1, \dots, \lambda_n \in \Lambda$ and $\alpha_1, \dots, \alpha_n \in \mathbb{C}$ such that $\sum_{i=1}^n |\alpha_i| \leq \|\psi\|$ and $\left\|\psi - \sum_{i=1}^n \alpha_i \mathcal{E}_{\lambda_i}\right\| < \varepsilon$, where $\mathcal{E}_\lambda \in \mathcal{P}^*$ is the evaluation functional defined by $\mathcal{E}_\lambda(p) = p(\lambda) \quad (\lambda \in \mathbb{D}, p \in \mathcal{P})$.*

For $\lambda = re^{i\theta} \in \mathbb{D}$ let $I_\lambda = \{e^{it} : |t - \theta| < 2(1-r)\}$. The following result is a particular case of a more general interpolation theorem of Carleson.

Theorem 5. *There is a constant b with the following property: if $F \subset \mathbb{D}$ is a finite set such that the sets I_λ are pairwise disjoint and $|\lambda| \geq 3/4 \quad (\lambda \in F)$, and $c_\lambda \in \mathbb{C} \quad (\lambda \in F)$ are given, then there exists $f \in H^\infty$ such that $f(\lambda) = c_\lambda \quad (\lambda \in F)$ and $\|f\| \leq b \cdot \sup_{\lambda \in F} |c_\lambda|$.*

Bibliography

[Alb] D.W. ALBRECHT, Integral formulae for special cases of Taylor's functional calculus, Studia Math. 105 (1993), 51–68.

[Al1] E. ALBRECHT, Funktionalkalküle in mehreren Veränderlichen für stetige lineare Operatoren auf Banachräumen, Manuscripta Math. 14 (1974), 1–40.

[Al2] E. ALBRECHT, On joint spectra, Studia Math. 64 (1979), 263–271.

[AV] E. ALBRECHT, F.-H. VASILESCU, Stability of the index of a semi-Fredholm complex of Banach spaces, J. Funct. Anal. 66 (1986), 141–172.

[All1] G.R. ALLAN, A spectral theory for locally convex algebras, Proc. London Math. Soc. 15 (1965), 399–421.

[All2] G.R. ALLAN, On one-sided inverses in Banach algebras of holomorphic vector-valued functions, J. London Math. Soc. 42 (1967), 463–470.

[All3] G.R. ALLAN, Holomorphic vector-valued functions on a domain of holomorphy, J. London Math. Soc. 42 (1967), 509–513.

[Am1] C.-G. AMBROZIE, Stability of index of a Fredholm symmetrical pair, J. Operator Theory 25 (1991), 61–77.

[Am2] C.-G. AMBROZIE, The Euler characteristic is stable under compact perturbations, Proc. Amer. Math. Soc. 124 (1996), 2041–2050.

[Am3] C.-G. AMBROZIE, Noncompactness measure invariance of the index, J. Operator Theory 38 (1997), 225–242.

[AM] C.-G. AMBROZIE, V. MÜLLER, Invariant subspaces for polynomially bounded operators, J. Funct. Anal. 213 (2004), 321–345.

[AV] C.-G. AMBROZIE, F.-H. VASILESCU, Semi-Fredholm pairs, Proceedings of the 16th OT Conference, Timisoira, 1997.

[AL] S.A. AMITSUR, J. LEVITZKI, Minimal identities for algebras, Proc. Amer. Math. Soc. 1 (1950), 449–463.

[An] M. ANDERSSON, Taylor's functional calculus for commuting operators with Cauchy-Fantappie-Leray formulas, Int. Math. Res. Not. 6 (1997), 247–258.

[AS] M. ANDERSSON, S. SANDBERG, A constructive proof of the composition rule for Taylor's functional calculus, Studia Math. 142 (2000), 65–69.

[Ans] S.I. ANSARI, Hypercyclic and cyclic vectors, J. Funct. Anal. 128 (1995), 372–383.

[Ap1] C. APOSTOL, Spectral decompositions and functional calculus, Rev. Roumaine Math. Pures Appl. 13 (1968), 1481–1528.

[Ap2] C. APOSTOL, The correction by compact perturbation of the singular behavior of operators, Rev. Roumaine Math. Pures Appl. 21 (1976), 155–175.

[Ap3] C. APOSTOL, Inner derivations with closed range, Rev. Roumaine Math. Pures Appl. 21 (1976), 249–265.

[Ap4] C. APOSTOL, The spectrum and the spectral radius as functions in Banach algebras, Bull. Acad. Pol. Sci., Ser. Sci. Mat. Astronom. Phys. 26 (1978), 975–978.

[Ap5] C. APOSTOL, Ultraweakly closed operator algebras, J. Operator Theory 2 (1979), 49–61.

[Ap6] C. APOSTOL, The reduced minimum modulus, Michigan Math. J. 32 (1985), 275–294.

[ApC1] C. APOSTOL, K. CLANCEY, Generalized inverses and spectral theory, Trans. Amer. Math. Soc. 215 (1976), 293–300.

[ApC2] C. APOSTOL, K. CLANCEY, On generalized resolvents, Proc. Amer. Math. Soc. 58 (1976), 163–168.

[Ar1] R. ARENS, The space L_p and convex topological rings, Bull. Amer. Math. Soc. 52 (1946), 931–935.

[Ar2] R. ARENS, Inverse-producing extensions of normed algebras, Trans. Amer. Math. Soc. 88 (1958), 536–548.

[Ar3] R. ARENS, Dense inverse limit rings, Michigan Math. J. 5 (1958), 166–182.

[Ar4] R. ARENS, Extensions of Banach algebras, Pacific J. Math 10 (1960), 1–16.

[Ar5] R. ARENS, Ideals in Banach algebra extensions, Studia Math. 31 (1968), 29–34.

[AC] R. ARENS, A.P. CALDERON, Analytic functions of several Banach algebra elements, Ann. of Math. 62 (1955), 204–216.

[AZ] H. ARIZMENDI, W. ŻELAZKO, A B_0-algebra without topological divisors of zero, Studia Math. 82 (1985), 191–198.

[AT] K. ASTALA, H.-O. TYLLI, On the bounded compact approximation property and measures of noncompactness, J. Funct. Anal. 70 (1987), 388–401.

[At] F.V. Atkinson, The normal solvability of linear equations in normed spaces, Mat. Sbornik 28 (1951), 3–14.

[Au1] B. Aupetit, Propriétés Spectrales des Algèbres de Banach, Lecture Notes in Math. 735, Springer-Verlag, Berlin 1979.

[Au2] B. Aupetit, Charactérisation spectrale des algèbres de Banach commutatives, Pacific J. Math. 63 (1976), 23–35.

[Au3] B. Aupetit, An improvement of Kaplansky's lemma on locally algebraic operators, Studia Math. 88 (1988), 275–278.

[Au4] B. Aupetit, A Primer on Spectral Theory, Springer-Verlag, New York, 1991.

[BM] C. Badea, M. Mbekhta, Compressions of resolvents and maximal radius of regularity, Trans. Amer. Math. Soc. 351 (1999), 2949–2960.

[BMü] C. Badea, V. Müller, On weak orbits of operators, to appear.

[B1] K.M. Ball, The plank problem for symmetric bodies, Invent. Math. 10 (1991), 535–543.

[B2] K.M. Ball, The complex plank problem, Bull. London Math. Soc. 33 (2001), 433–442.

[Ba1] B.A. Barnes, A generalized Fredholm theory for certain maps in the regular representation of an algebra, Canad. J. Math. 20 (1968), 495–504.

[Ba2] B.A. Barnes, The Fredholm elements of a ring, Canad. J. Math. 21 (1969), 84–95.

[Ba3] B.A. Barnes. Common operator properties of the linear operators RS and SR, Proc. Amer. Math. Soc. 126 (1998), 1055–61.

[BMSW] B.A. Barnes, G.J. Murphy, M.R.F. Smyth, T.T. West, Riesz and Fredholm Theory in Banach Algebras, Research Notes in Mathematics 67, Pitman (Advanced Publishing Program), Boston, Mass.-London 1982.

[BG] R.G. Bartle, L.M. Graves, Mappings between function spaces, Trans. Amer. Math. Soc. 72 (1952), 400–413.

[BaM] F. Bayart, É. Matheron, Hypercyclicity operators which do not satisfy the hypercyclicity criterion, to appear.

[Bea1] B. Beauzamy, Un opérateur sans sous-espace invariant non-trivial: simplification de l'example de P. Enflo, Integral Equations Operator Theory 8 (1985), 314–384.

[Bea2] B. Beauzamy, Introduction to Operator Theory and Invariant Subspaces, North-Holland Math. Library 42, North-Holland, Amsterdam, 1988.

[Ber] S.K. Berberian, Approximate proper vectors, Proc. Amer. Math. Soc. 13 (1962), 111–114.

[BW] M.A. Berger, Y. Wang, Bounded semigroups of matrices, Linear Algebra Appl. 166 (1992), 21–27.

[Be1] M. Berkani, On a class of quasi-Fredholm operators, Integral Equations Operator Theory 34 (1999), 244–249.

[Be2] M. Berkani, Restriction of an operator to the range of its powers, Studia Math. 140 (2000), 163–175.

[BO] M. Berkani, A. Ouahab, Opérateurs essentiellement réguliers dans les espaces de Banach, Rend. Circ. Mat. Palermo 46 (1997), 131–160.

[BS] R. Berntzen, A. Sołtysiak, The Harte spectrum is not contained in the Taylor spectrum, Comment. Math. Prace Mat. 38 (1998), 29–35.

[Boe] B. den Boer, Linearization of operator functions on arbitrary open sets, Integral Equations Opererator Theory 1 (1978), 19–27.

[Bo1] B. Bollobás, Adjoining inverses to commutative Banach algebras, Trans. Amer. Math. Soc. 181 (1973), 165–181.

[Bo2] B. Bollobás, Normally subregular systems in normed algebras, Studia Math. 49 (1974), 263–266.

[Bo3] B. Bollobás, Adjoining inverses to commutative Banach algebras, Algebras in analysis (J.H. Williamson ed.), Academic Press, New York, 1975, pp. 256–257.

[Bo4] B. Bollobás, To what extend can the spectrum of an operator be diminished under an extension, in: Linear and Complex Analysis Problem Book, V.P. Havin, S.V. Hruščev and N.K. Nikol'skij (eds.), Lecture Notes in Math. 1043, Springer-Verlag, Berlin 1984, p. 210.

[BD] F.F. Bonsal, J. Duncan, Complete Normed Algebras, Springer - Verlag, Berlin 1973.

[BF] P.S. Bourdon, N. Feldman, Somewhere dense orbits are everywhere dense, Indiana Univ. Math. J. 52 (2003), 811–819.

[Br] S.W. Brown, Some invariant subspaces for subnormal operators, Integral Equations Operator Theory 1 (1978), 310–333.

[BCP1] S. Brown, B. Chevreau, C. Pearcy, Contractions with rich spectrum have invariant subspaces, J. Operator Theory 1 (1979), 123–136.

[BCP2] S. Brown, B. Chevreau, C. Pearcy, On the structure of contraction operators II., J. Funct. Anal. 76 (1988), 30–55.

[Bun] J.W. Bunce, Models for n-tuples of noncommuting operators, J. Funct. Anal. 57 (1984), 21–30.

[BHW] J.J. Buoni, R. Harte, T. Wickstead, Upper and lower Fredholm spectra, Proc. Amer. Math. Soc. 66 (1977), 309–314.

[Bu1] L. Burlando, Continuity of spectrum and spectral radius in algebras of operators, Ann. Fac. Sci. Toulouse 9 (1988), 5–54.

[Bu2] L. BURLANDO, Continuity of spectrum and spectral radius in Banach algebras, in: Functional Analysis and Operator Theory, Banach Center Publications, vol. 30, Warszawa 1994, pp. 53–100.

[Ca] J.W. CALKIN, Two-sided ideals and congruences in the ring of bounded operators in Hilbert spaces, Ann. of Math. 42 (1941), 839–873.

[CPY] S.R. CARADUS, W.E. PFAFFENBERGER, B. YOOD, Calkin Algebras of Operators on Banach Spaces, Lecture Notes in Pure and Applied Mathematics, Marcel Dekker, Inc., New York, 1974.

[ChH] M. CHŌ, T. HURUYA, On the joint spectral radius, Proc. Roy. Irish Acad. Sect. A 91 (1991), 39–44.

[ChT] M. CHŌ, M. TAKAGUCHI, Boundary of Taylor's joint spectrum for two commuting operators, Sci. Rep. Hirosaki Univ. 28 (1981), 1–4.

[ChZ] M. CHŌ, W. ŻELAZKO, On geometric spectral radius of commuting n-tuples of operators, Hokkaido Math. J. 21 (1992), 251–258.

[ChD] M.-D. CHOI, C. DAVIS, The spectral mapping theorem for joint aproximate spectrum, Bull. Amer. Math. Soc. 80 (1974), 317–321.

[CF] I. COLOJOARA, C. FOIAS, Theory of generalized spectral operators, Gordon and Breach, New York, 1968.

[Co] J.B. CONWAY, Functions of One Complex Variable, Springer-Verlag, New York, 1978.

[CM] J.B. CONWAY, B. MORREL, Operators that are points of spectral continuity, Integral Equations Operator Theory 2 (1979), 174–198.

[Cr] R.W. CROSS, On the continuous linear image of a Banach space, J. Austral. Math. Soc. A29 (1980), 219–234.

[Cu1] R.E. CURTO, Fredholm and invertible n-tuples of operators. The deformation problem, Trans. Amer. Math. Soc. 266 (1981), 129–159.

[Cu2] R.E. CURTO, Spectral permanence for joint spectra, Trans. Amer. Math. Soc. 270 (1982), 659–665.

[Cu3] R.E. CURTO, Connections between Harte and Taylor spectrum, Rev. Roumaine Math. Pures Appl. 31 (1986), 203–215.

[Cu4] R.E. CURTO, Applications of Several Complex Variables to Multiparameter Spectral Theory, in: Surveys of some recent results in operator theory, vol. II (J.B. Conway, B.B. Morrel eds.), pp. 25–90, Pitman Research Notes in Mathematics Series 192, 1988.

[Cu5] R.E. CURTO, Spectral theory of elementary operators (Blaubeuren 1991), 3–52, World Sci. Publishing, River Edge, NJ. 1992.

[CD] R.E. CURTO, A.T. DASH, Browder spectral systems, Proc. Amer. Math. Soc. 103 (1988), 407–413.

[CF] R.E. CURTO, L. FIALKOW, The spectral picture of (L_A, R_B), J. Funct. Anal. 71 (1987), 371–392.

[Dal] H.G. DALES, Banach Algebras and Automatic Continuity, London Math. Soc. Monographs, New Series 24, Oxford Science Publications, The Clarence Press, Oxford University Press, New York 2000.

[Dan] J. DANEŠ, On local spectral radius, Čas. Pěst. Mat. 112 (1987), 177–178.

[Das] A.T. DASH, Joint essential spectra, Pacific J. Math. 64 (1976), 119–128.

[Dat] R. DATKO, Extending a theorem of A.M. Liapunov to Hilbert space, J. Math. Anal. Appl. 32 (1970), 610–616.

[DR] C. DAVIS, P. ROSENTHAL, Solving linear operator equations, Canad. Math. J. 26 (1974), 1384–1389.

[DeR] M. DE LA ROSA, C. READ, A hypercyclic operator whose direct sum is not hypercyclic, to appear.

[Di] P. DIXON, A Jacobson semi-simple Banach algebra with dense nil subalgebra, Colloq. Math. 37 (1977), 81–82.

[Do] R.G. DOUGLAS, Banach Algebra Techniques in Operator Theory, Academic Press, New York 1972.

[Du1] N. DUNFORD, Spectral Theory I., Convergence to projections, Trans. Amer. Math. Soc. 54 (1943), 185–217.

[Du2] N. DUNFORD, Spectral operators, Pacific J. Math. 4 (1954), 321–354.

[DS] N. DUNFORD, J.T. SCHWARTZ (with the assistance of W.G. Bade and R.G. Bartle), Linear Operators, Part I: General Theory, Pure and Applied Mathematics, vol 7, Interscience Publishers, Inc., New York, 1958.

[Em] M.R. EMBRY, Factorization of operators on Banach spaces, Proc. Amer. Math. Soc. 38 (1973), 587–590.

[En1] P. ENFLO, A conterexample to the approximation property in Banach spaces, Acta Math. 130 (1973), 309–317.

[En2] P. ENFLO, On the invariant subspace problem in Banach spaces, Acta Math. 158 (1987), 213–313.

[Es1] J. ESCHMEIER, Local peoperties of Taylor's analytic functional calculus, Invent. Math. 68 (1982), 103–116.

[Es2] J. ESCHMEIER, On two notions of the local spectrum for several commuting operators, Michigan Math. J. 30 (1983), 245–248.

[Es3] J. ESCHMEIER, Analytic spectral mapping theorems for joint spectra, in: Operator Theory in Indefinite Metric Spaces, Scattering Theory and Other Topics (Bucharest 1985), H. Helson, B. Sz-Nagy, F.-H. Vasilescu and D. Voiculescu (eds.), Oper. Theory Adv. Appl. 24, Birkhäuser, Basel, 1987, 167–181.

[EP] J. ESCHMEIER, M. PUTINAR, Spectral decompositions and analytic sheaves, London Math. Soc. Monographs New Series 10, Oxford University Press, Oxford, 1996.

[Fa1] A.S. FAINSHTEIN, On the joint essential spectrum of a family of linear operators, Funct. Anal. Appl. 14 (1980), 83–84 (Russian).

[Fa2] A.S. FAINSHTEIN, A stability of Fredholm complexes of Banach spaces with respect to perturbations small in q-norm, Izv. Akad. Nauk Azerbaidzanskoi SSR, Ser. Fiz.-Tekn. Mat. Nauk 1 (1980), 3–8 (Russian).

[Fa3] A.S. FAINSHTEIN, On the gap between subspaces of a Banach space, Inst. Mat. Mech. Akad. Azerbaidzanskoi SSR, Baku 1983 (Russian).

[Fa4] A.S. FAINSHTEIN, Measures of noncompactness of linear operators and analogues of the minimum modulus for semi-Fredholm operators, in: Spectral Theory of Operators 6, 182–195, Elm, Baku, 1985 (Russian).

[FS1] A.S. FAINSHTEIN, V.S. SHUL'MAN, On Fredholm complexes of Banach spaces, Funktsional. Anal. i Prilozhen. 14:4 (1980), 87–88 (Russian).

[FS2] A.S. FAINSHTEIN, V.S. SHUL'MAN, Stability of index of a short Fredholm complex of Banach spaces with respect to perturbations small in a measure of noncompactness, in: Spectral Theory of Operators 4, 189–198, Elm, Baku 1982 (Russian).

[Fe] M. FEKETE, Über die Verteilung der Wurzeln bei gewissen algebraischen Gleichungen mit ganzzahligen Koefficienten, Math. Z. 17 (1923), 228–249.

[Fel] N.S. FELDMAN, Perturbations of hypercyclic vectors, J. Math. Anal. Appl. 27 (2002), 67–74.

[FM] A. FERNÁNDEZ, V. MÜLLER, Renormalization of Banach and locally convex algebras, Studia Math. 96 (1990), 237–242.

[Fi1] L. FIALKOW, Spectral properties of elementary operators, Acta Sci. Math. (Szeged) 46 (1983), 269–282.

[Fi2] L. FIALKOW, Spectral properties of elementary operators II, Trans. Amer. Math. Soc. 290 (1985), 415–429.

[FW] P.A. FILLMORE, J.P. WILLIAMS, On operator ranges, Advances in Math. 7 (1971), 254–281.

[FS1] C.-K. FONG, A. SOŁTYSIAK, Existence of a multiplicative functional and the joint spectra, Studia Math. 81 (1985), 213–220.

[FS2] C.-K. FONG, A. SOŁTYSIAK, On the left and right joint spectra, Studia Math. 97 (1990), 151–156.

[Fo] E. FORMANEK, The Nagata-Hingman theorem, Acta Appl. Math. 21 (1990), 185–192.

[FK] K.-H. FÖRSTER, M.A. KAASHOEK, The asymptotic behaviour of the reduced minimum modulus of a Fredholm operator, Proc. Amer. Math. Soc. 49 (1975), 123–131.

[FL] K.-H. FÖRSTER, E.-O. LIEBENTRAU, Semi-Fredholm operators and sequence conditions, Manuscripta Math. 44 (1983), 35–44.

[Fr] S. FRUNZĂ, The Taylor spectrum and spectral decompositions, J. Funct. Anal. 19 (1975), 390–421.

[Ga] T.W. GAMELIN, Uniform Algebras, Prentice Hall, Inc., Englewood Cliffs, N. J., 1969.

[Ge] I.M. GELFAND, Normierte Ringe, Rec. Mat. (Mat. Sbornik) N.S. 51 (1941), 3–24.

[GS] R.M. GETHNER, J.H. SHAPIRO, Universal vectors for operators on spaces of holomorphic functions, Proc. Amer. Math. Soc. 100 (1987), 281–288.

[Gle] A.M. GLEASON, A characterization of maximal ideals, J. Anal. Math. 19 (1967), 171–172.

[GS] G. GODEFROY, J.H. SHAPIRO, Operators with dense, invariant, cyclic vector manifolds, J. Funct. Anal. 98 (1991), 229–269.

[Goh1] I.C. GOHBERG, On linear equations in normed spaces, Dokl. Akad. Nauk SSSR 76 (1951), 477–480 (Russian).

[Goh2] I.C. GOHBERG, On linear equations depending analytically on a parameter, Dokl. Akad. Nauk SSSR 78 (1951), 629–632 (Russian).

[GKL] I.C. GOHBERG, M.A. KAASHOEK, D.C. LAY, Equivalence, linearization, and decomposition of holomorphic operator functions, J. Funct. Anal. 28 (1978), 102–144.

[GhK1] I.C. GOHBERG, M.G. KREIN, The basic propositions on defect numbers, root numbers, and indices of linear operators, Uspekhi Mat. Nauk 12 (1957), 43–118 (Russian); English translation: Amer. Math. Soc. Transl. 13 (1960), 185–265.

[GhK2] I.C. GOHBERG, M.C. KREIN, Introduction to the Theory of Linear Nonselfadjoint Operators, Translations of Math. Monographs vol. 18, A.M.S., Providence, 1969.

[GhM] I.C. GOHBERG, A.S. MARKUS, Two theorems on the opening of subspaces of Banach space, Uspekhi Mat. Nauk 14 (1959), 135–140 (Russian).

[GMF] I.C. GOHBERG, A.S. MARKUS, I.A. FEL'DMAN, Normally solvable operators and ideals associated with them, Bul. Akad. Stiince RSS Moldoven 10 (1960), 51–70 (Russian). English translation: Amer. Math. Soc. Transl. 61 (1967), 63–84.

[Gol] S. GOLDBERG, Unbounded linear operators with applications, McGraw-Hill, New York, 1966.

[GGM] L.S. GOLDENSTEIN, I.C. GOHBERG, A.S. MARKUS, Investigation of some properties of bounded linear operators in connection with their q-norms, Uch. Zap. Kishinev Gos. Univ. 29 (1957), 29–36 (Russian).

[GlM] L.S. GOLDENSTEIN, A.S. MARKUS, On a measure of non-compactness of bounded sets and linear operators, in: Studies in Algebra and Mathematical Analysis, pp. 45–54, Izdat. Karta Moldovenjaski, Kishinev, 1965, (Russian).

[GlK] M.A. Gol'dman, S.N. Krachkovskiĭ, On the stability of some properties of a closed linear operator, Dokl. Akad. Nauk SSSR 209 (1973), 769–772 (Russian); English translation: Soviet Math. Dokl. 14 (1973), 502–505.

[GLM] M. González, F. Leon-Saavedra, A. Montes-Rodriguez, Semi-Fredholm theory: hypercyclic and supercyclic subspaces, Proc. London Math. Soc. 81 (2000), 169–186.

[GnM1] M. González, A. Martinón, Operational quantities derived from the norm and measures of noncompactness, Proc. Roy. Irish Acad. Sect. A 91 (1991), 63–70.

[GnM2] M. González, A. Martinón, Operational quantities characterizing semi-Fredholm operators, Studia Math. 114 (1995), 13–27.

[GwM] W.T. Gowers, B. Maurey, The unconditional basis sequence problem, J. Amer. Math. Soc. 6 (1993), 851–874.

[Gr1] S. Grabiner, Ascent, descent, and compact perturbations, Proc. Amer. Math. Soc. 71 (1978), 79–80.

[Gr2] S. Grabiner, Uniform ascent and descent of bounded operators, J. Math. Soc. Japan, 34 (1982), 317–337.

[GL] B. Gramsch, D. Lay, Spectral mapping theorems for essential spectra, Math. Ann. 192 (1971), 17–32.

[Gro] C.W. Groetsch, Generalized Inverses of Linear Operators, Representation and Approximation, Marcel Dekker, Inc. New York and Basel, 1977.

[Gu] P.S. Guinand, On quasinilpotent semigroups of operators, Proc. Amer. Math. Soc. 86 (1982), 485–486.

[GR] R.C. Gunning, H. Rossi, Analytic Functions of Several Complex Variables, Prentice Hall, Englewood Cliffs, N.J. 1965.

[Hal1] P.R. Halmos, A Hilbert Space Problem Book, D. van Nostrand Company, Inc., Toronto, 1967.

[Hal2] P.R. Halmos, Capacity in Banach algebras, Indiana Univ. Math. J. 20 (1971), 855–863.

[Ha1] R. Harte, The spectral mapping theorems in several variables, Bull. Amer. Math. Soc. 78 (1972), 871–875.

[Ha2] R. Harte, Spectral mapping theorems, Proc. Roy. Irish Acad. Sect. A 72 (1972), 89–107.

[Ha3] R. Harte, Tensor products, multiplication operators and the spectral mapping theorem, Proc. Roy. Irish Acad. Sect. A 73 (1973), 285–302.

[Ha4] R. HARTE, The exponential spectrum in Banach algebras, Proc. Amer. Math. Soc. 58 (1976), 114–118.

[Ha5] R. HARTE, Berberian-Quigley and the ghost of a spectral mapping theorem, Proc. Roy. Irish Acad. Sect. A 78 (1978), 63–68.

[Ha6] R.E. HARTE, Invertibility, singularity and Joseph L. Taylor, Proc. Roy. Irish Acad. Sect. A 81 (1981), 71–79.

[Ha7] R.E. HARTE, Fredholm theory relative to a Banach algebra homomorphism, Math. Z. 179 (1982), 431–438.

[Ha8] R. HARTE, Invertibility and Singularity for Bounded Linear Operators, Marcel Dekker, Inc., New York and Basel, 1988.

[HW] R. HARTE, T. WICKSTEAD, Upper and lower Fredholm spectra II, Math. Z. 154 (1977), 253–256.

[Hel1] A. YA. HELEMSKII, Homological methods in the holomorphic calculus of several operators in Banach spaces, after Taylor, Uspekhi Mat. Nauk 36 (1981), 127–172 (Russian); English translation: Russian Math. Surveys 36 (1981).

[Hel2] A. YA. HELEMSKII, The Homology of Banach and Topological Algebras (translated from Russian), Mathematics and its Applications (Soviet Series) 41. Kluwer, Academic Publishers Group, Dordrecht 1989.

[He] H. HEUSER, Über Operatoren mit endlichen Defekten, Inaug. Diss., Tübingen, 1956.

[Hi] D. HILBERT, Zur Variationsrechnung, Math. Ann. 62 (1906), 351–370.

[IF] G.A. ISAEV, A.S. FAINSHTEIN, Joint spectra of finite commutative families, in: Spectral Theory of Operators No. 3, pp. 222–257, Elm, Baku 1980 (Russian).

[Ja] N. JACOBSON, The radical and semisimplicity for arbitrary rings, Amer. J. Math. 67 (1945), 300–320.

[Jan] R. JANTZ, Stetige und holomorphe Scharen von Teilräumen und Operatoren in Banachräumen, Dissertation, Konstanz, 1986.

[J] B.E. JOHNSON, The uniqueness of the (complete) norm topology, Bull. Amer. Math. Soc. 73 (1967), 537–539.

[Ka1] M.A. KAASHOEK, Stability theorems for closed linear operators, Indag. Math. 27 (1965), 452–466.

[Ka2] M.A. KAASHOEK, Ascent, descent, nullity and defect, A note on a paper by A.E. Taylor, Math. Ann. 172 (1967), 105–115.

[Kab1] W. KABALLO, Projectoren und relative Inversion holomorpher Semi-Fredholmfunktionen, Math. Anal. 219 (1976), 85–96.

[Kab2] W. KABALLO, Holomorphe Semi-Fredholmfunktionen ohne komplementierte Kerne bzw. Bilder, Math. Nachr. 91 (1979), 327–335.

[KZ] J.P. KAHANE, W. ŻELAZKO, A characterization of maximal ideals in commutative Banach algebras, Studia Math. 29 (1968), 339–343.

[Kap] I. KAPLANSKY, Infinite Abelian Groups, University of Michigan Press, Ann Arbor, 1954.

[Kat1] T. KATO, Perturbation theory for nullity, deficiency and other quantities of linear operators, J. Analyse Math. 6 (1958), 261–322.

[Kat2] T. KATO, Perturbation Theory for Linear Operators, Springer-Verlag, Berlin, 1966.

[Ki] C. KITAI, Invariant closed sets for linear operators, thesis, University of Toronto, 1982.

[Kol] J. KOLIHA, A generalized Drazin inverse, Glasgow Math. J. 38 (1996), 367–381.

[KMMP] J. KOLIHA, M. MBEKHTA, V. MÜLLER, P.W. POON, Corrigendum and Addendum: "On the axiomatic theory of spectrum II", Studia Math. 130 (1998), 193–198.

[Ko1] V. KORDULA, Spectrum in Banach algebras, thesis, Charles University, Prague, 1995.

[Ko2] V. KORDULA, The essential Apostol spectrum and finite dimensional perturbations, Proc. Roy. Irish Acad. Sect. A 96 (1996), 105–109.

[KM1] V. KORDULA, V. MÜLLER, Vasilescu-Martinelli formula for operators in Banach spaces, Studia Math. 113 (1995), 127–139.

[KM2] V. KORDULA, V. MÜLLER, On the axiomatic theory of spectrum, Studia Math. 119 (1996), 109–128.

[KM3] V. KORDULA, V. MÜLLER, The distance from the Apostol spectrum, Proc. Amer. Math. Soc. 124 (1996), 3055–3061.

[KM4] V. KORDULA, N. MÜLLER, A continuous semicharacter, Czechoslovak Math. J. 46 (1996), 133–139. 127–139.

[KMR] V. KORDULA, V. MÜLLER, V. RAKOČEVIĆ, On the semi-Browder spectrum, Studia Math. 123 (1997), 1–13.

[KS] S. KOWALSKI, Z. SŁODKOWSKI, A characterization of multiplicative linear functionals in Banach algebras, Studia Math 67 (1980), 215–223.

[KKM] M.G. KREIN, M.A. KRASNOSEL'SKIJ, D.C. MIL'MAN, On the defect numbers of linear operators in Banach space and on some geometric problems, Sbornik Trud. Inst. Mat. Akad. Nauk Ukr. SSR 11 (1948), 97–112.

[KV] H. KROH AND P. VOLKMANN, Störungssätze für Semifredholmoperatoren, Math. Z. 148 (1976), 295–297.

[Kr] N.YA. KRUPNIK, Banach Algebras with Symbol and Singular Integral Operators, Birkhäuser, Basel, 1987.

[Lab] J.P. LABROUSSE, Les opérateurs quasi-Fredholm: une généralisation des opérateurs semi-Fredholm, Rend. Circ. Mat. Palermo 29 (1980), 161–258.

[LW] T.J. LAFFEY, T.T. WEST, Fredholm commutators, Proc. Roy. Irish Acad. Sect. A 82 (1982), 129–140.

[LN] K.B. LAURSEN, M.M. NEUMANN, An Introduction to Local Spectral Theory, London Math. Soc. Monographs New Series 20, Oxford University Press, Oxford 2000.

[La] D.C. LAY, Spectral analysis using ascent, descent, nullity and defect, Math. Ann. 184 (1970), 197–214.

[LS] A. LEBOW, M. SCHECHTER, Semigroups of operators and measures of noncompactness, J. Funct. Anal. 7 (1971), 1–26.

[Le] J. LEITERER, Banach coherent analytic Fréchet sheaves, Math. Nachr. 85 (1978), 91–109.

[LM] F. LEON-SAAVEDRA, A. MONTES-RODRIGUEZ, Spectral theory of hypercyclic and supercyclic subspaces, Trans. Amer. Math. Soc. 252 (2001), 247–267.

[LMü] F. LEON-SAAVEDRA, V. MÜLLER, Rotations of hypercyclic and supercyclic vectors, Integral Equations Operator Theory 50 (2004), 385–391.

[Lev1] R. LEVI, Notes on the Taylor joint spectrum of commuting operators, Spectral Theory, Banach Center Publications, Vol. 8, PWN – Polish Scientific Publishers, Warsaw, 1982, 321–332.

[Lev2] R. LEVI, Cohomological invariants for essentially commuting systems of operators, Funktsional. Anal. Prilozhen. 17 (1983), 79–80 (Russian).

[LvS] S. LEVI, Z. SŁODKOWSKI, Measurability properties of spectra, Proc. Amer. Math. Soc. 98 (1986), 225–231.

[Li] J.A. LINDBERG, Extension of algebra norms and applications, Studia Math. 40 (1973), 35–39.

[LT] J. LINDENSTRAUSS, L. TZAFRIRI, On complement subspace problem, Israel J. Math. 9 (1971), 262–269.

[Liv] A. JA. LIVČAK, Operators with finite dimensional salient of zeros on the Riesz kernel: duality and non-commuting perturbations, VINITI, Voronezh, 1983 (in Russian).

[LR] G. LUMER, M. ROSENBLUM, Linear operator equations, Proc. Amer. Math. Soc. 10 (1959), 32–41.

[MZ] E. MAKAI, J. ZEMÁNEK, The surjectivity radius, packing numbers and boundedness below, Integral Equations Operator Theory 6 (1983), 372–384.

[Man] F. MANTLIK, Linear equations depending differentiably on a parameter, Integral Equations Operator Theory 13 (1990), 231–250.

[Mar] A.S. Markus, On some properties of linear operators connected with the notion of gap, Kishinev Gos. Uchen. Zap. 39 (1959), 265–272 (Russian).

[Ma] S. Mazur, Sur les anneaux linéaires, C.R. Acad. Sci. Paris 207 (1938), 1025–1027.

[Mb1] M. Mbekhta, Résolvant généralisé et théorie spectrale, J. Operator Theory 21 (1989), 69–105.

[Mb2] M. Mbekhta, Formules de distance au spectre généralisé et au spectre semi-Fredholm, preprint.

[Mb3] M. Mbekhta, Fonctions perturbation et formules du rayon spectral et de distance au spectre, preprint.

[MM] M. Mbekhta, V. Müller, On the axiomatic theory of spectrum II., Studia Math. 119 (1996), 129–147.

[MO1] M. Mbekhta, A. Ouahab, Opérateurs s-régulier dans un espace de Banach et théorie spectrale, Pub. Irma, Lille – 1990, Vol. 22, No XII.

[MO2] M. Mbekhta, A. Ouahab, Contribution à la théorie spectrale généralisé dans les espaces de Banach, C.R. Acad. Sci. Paris 313 (1991), 833–836.

[MP] M. Mbekhta, R. Paul, Sur la conorme essentielle, Studia Math. 117 (1996), 243–252.

[MW] A.M. Meléndez, A. Wawrzynczyk An approach to joint spectra, Ann. Polon. Math. 72 (1999), 131–144.

[Mi] E. Michael, Locally multiplicatively-convex topological algebras, Mem. Amer. Math. Soc. 11 (1952).

[Mo] E.H. Moore, Abstract, Bull. Amer. Math. Soc. 26 (1920), 394–395.

[Mü1] V. Müller, On discontinuity of the spectral radius in Banach algebras, Comment. Math. Univ. Carolinae 18 (1977), 591–598.

[Mü2] V. Müller, Non-removable ideals in commutative Banach algebras, Studia Math. 74 (1982), 97–104.

[Mü3] V. Müller, On domination and extensions of Banach algebras, Studia Math. 73 (1982), 75–80.

[Mü4] V. Müller, The inverse spectral radius formula and removability of spectrum, Čas. Pěst. Mat. 108 (1983), 412–415.

[Mü5] V. Müller, Inverse elements in extensions of Banach algebras, Studia Math. 80 (1984), 191–195.

[Mü6] V. Müller, Removability of ideals in commutative Banach algebras, Studia Math. 78 (1984), 297–307.

[Mü7] V. Müller, On quasialgebraic operators in Banach spaces, J. Operator Theory 17 (1987), 291–300.

[Mü8] V. MÜLLER, Adjoining inverses to non-commutative Banach algebras and extensions of operators, Studia Math. 91 (1988), 73–77.

[Mü9] V. MÜLLER, Local spectral radius formula for operators in Banach spaces, Czechoslovak Math. J. 38 (1988), 726–729.

[Mü10] V. MÜLLER, Adjoining one-sided inverses to noncommutative Banach algebras, Bull. Pol. Acad. Sci. Math. 37 (1989), 415–419.

[Mü11] V. MÜLLER, Kaplansky's theorem and Banach PI-algebras, Pacific J. Math. 141 (1990), 355–361.

[Mü12] V. MÜLLER, Local behaviour of the polynomial calculus of operators, J. Reine Angew. Math. 430 (1992), 61–68.

[Mü13] V. MÜLLER, A note on joint capacities in Banach algebras, Czechoslovak Math. J. 43 (1993), 367–372.

[Mü14] V. MÜLLER, Słodkowski spectra and higher Shilov boundaries, Studia Math. 105 (1993), 69–75.

[Mü15] V. MÜLLER, On the regular spectrum, J. Operator Theory 31 (1994), 363–380.

[Mü16] V. MÜLLER, Local behaviour of operators, in: Functional Analysis and Operator Theory, Banach Center Publications, vol. 30, PWN Warsaw 1994, 251–258.

[Mü17] V. MÜLLER, The splitting spectrum differs from the Taylor spectrum, Studia Math. 123 (1997), 291–294.

[Mü18] V. MÜLLER, Stability of index for semi-Fredholm chains, J. Operator Theory 37 (1997), 247–261.

[Mü19] V. MÜLLER, On the joint spectral radius, Annales Polonici Mathematici 66 (1997), 173–182.

[Mü20] V. MÜLLER, On the topological boundary of the one-sided spectrum, Czechoslovak Math. J. 49 (1999), 561–568.

[Mü21] V. MÜLLER, Axiomatic theory of spectrum III -Semiregularities, Studia Math. 142 (2000),159–169.

[Mü22] V. MÜLLER, Orbits, weak orbits and local capacity of operators, Integral Equations Operator Theory 41 (2001), 230–253.

[Mü23] V. MÜLLER, On the punctured neighbourhood theorem, J. Operator Theory 43 (2000), 83–95.

[Mü24] V. MÜLLER, Dvoretzky's type result for operators on Banach spaces, Acta Sci. Math. (Szeged) 66 (2000), 697–709.

[Mü25] V. MÜLLER, Power bounded operators and supercyclic vectors II., Proc. Amer. Math. Soc. 13 (2005), 2997–3004.

[MS1] V. MÜLLER, A. SOŁTYSIAK, Spectral radius formula for commuting Hilbert space operators, Studia Math. 103 (1992), 329–333.

[MS2] V. MÜLLER, A. SOŁTYSIAK, On local joint capacities of operators, Czechoslovak Math. J. 43 (1993), 743–751.

[MV] V. MÜLLER, J. VRŠOVSKÝ, Orbits of linear operators tending to infinity, Rocky Mountain J. Math., to appear.

[Ne1] J.M.A.M. VAN NEERVEN, Exponential stability of operators and operator semigroups, J. Funct. Anal. 130 (1995), 293–309.

[Ne2] J.M.A.M. VAN NEERVEN, On the orbits of an operator with spectral radius one, Czechoslovak Math. J. 45 (1995), 495–502.

[Ne3] J.M.A.M. VAN NEERVEN, The Asymptotic Behaviour of Semigroups of Linear Operators, Operator Theory: Advances and Applications 88, Birkhäuser, Basel, 1996.

[Neu1] G. NEUBAUER, Zur Spektraltheorie in lokalkonvexen Algebren I, Math. Ann. 142 (1961), 131–134.

[Neu2] G. NEUBAUER, Zur Spektraltheorie in lokalkonvexen Algebren II, Math. Ann. 143 (1961), 251–263.

[New] J.D. NEWBURGH, The variation of spectra, Duke Math. J. 18 (1951), 165–176.

[Ni] N.K. NIKOL'SKIJ, Treatise on the Shift Operator, Nauka, Moskva, 1980 (Russian). English translation: Grundlehren der Mathematischen Wissenschaften 273, Springer-Verlag, Berlin 1986.

[Ob1] K.K. OBERAI, On the Weyl spectrum, Illinois J. Math. 18 (1974), 208–212.

[Ob2] K.K. OBERAI, Spectral mapping theorem for essential spectra, Rev. Roumaine Math. Pures Appl. 25 (1980), 365–373.

[Pal] T.W. PALMER, Banach Algebras and the General Theory of $*$-algebras, volume I: Algebras and Banach Algebras, Encyklopedia of Mathematics and its applications, vol. 49, Cambridge University Press 1994.

[Pa] A. PAZY, Semigroups of linear operators and applications to partial differential equations, Springer-Verlag, Berlin, 1983.

[Pe] R. PENROSE, A generalised inverse for matrices, Proc. Cambridge Philos. Soc. 51 (1955), 406–413.

[Pi1] A. PIETSCH, Zur Theorie der σ-Tranformationen in localconvexen Vektorräumen, Math. Nachr. 21 (1960), 347–369.

[Pi2] A. PIETSCH, s-numbers of operators in Banach spaces, Studia Math. 51 (1974), 201–223.

[Pi3] A. PIETSCH, Operator Ideals, Mathematische Monographien, Band 16, VEB Deutscher Verlag der Wissenschaften, Berlin, 1978.

[Pis] G. PISIER, A polynomially bounded operator on Hilbert space which is not similar to a contraction, J. Amer. Math. Soc. 10 (1997), 351–369.

[PZ] V. Pták, J. Zemánek, On uniform continuity of the spectral radius in Banach algebras, Manuscripta Math. 20 (1977), 177–189.

[Pu1] M. Putinar, The superposition principle for Taylor's functional calculus, J. Operator Theory 7 (1982), 149–155.

[Pu2] M. Putinar, Uniqueness of Taylor's functional calculus, Proc. Amer. Math. Soc. 89 (1983), 647–650.

[Pu3] M. Putinar, Functional calculus and the Gelfand transformation, Studia Math. 79 (1984), 83–86.

[Pu4] M. Putinar, Base change and the Fredholm index, Integral Equations Operator Theory 8 (1985), 674–692.

[Ra1] V. Rakočević, Measures of non-strict-singularity of operators, Mat. Vestnik 35 (1983), 79–82.

[Ra2] V. Rakočević, Approximate point spectrum and commuting compact perturbations, Glasgow Math. J. 28 (1986), 193–198.

[Ra3] V. Rakočević, On the essential spectrum, Zbornik radova Filozofskog fakulteta u Nišu, Ser. Mat. 6 (1992), 39–48.

[Ra4] V. Rakočevič, Generalized spectrum and commuting compact perturbations, Proc. Edinb. Math. Soc. 36 (1993), 197–209.

[Ra5] V. Rakočević, Semi-Fredholm operators with finite ascent or descent and perturbations, Proc. Amer. Math. Soc. 123 (1995), 3823–3825.

[Ra6] V. Rakočević, Semi-Browder operators and perturbations, Studia Math 122 (1996), 131–137.

[RZ] V. Rakočević, J. Zemánek, Lower s-numbers and their asymptotic behaviour, Studia Math. 91 (1988), 231–239.

[RZi] V. Rakočević, S. Živković, Lower analogues of Gelfand and Kolgomorov numbers, Indian J. Pure Appl. Math. 30 (1999), 777–785.

[Rn] T.J. Ransford, Generalised spectra and analytic multivalued functions, J. London Math. Soc. 29 (1984), 306–322.

[Re1] C. Read, A solution to the invariant subspace problem, J. London Math. Soc. 16 (1984), 337–401.

[Re2] C. Read, Inverse producing extension of a Banach algebra which eliminates the residual spectrum of one element, Trans. Amer. Math. Soc. 286 (1986), 715–725.

[Re3] C. Read, A short proof concerning the invariant subspace problem, J. London Math. Soc. 34 (1986), 335–348.

[Re4] C. Read, Extending an operator from a Hilbert space to a larger Hilbert space so as to reduce its spectrum, Israel J. Math. 57 (1987), 375–380.

[Re5] C.J. Read, Spectrum reducing extension for one operator on a Banach space, Trans. Amer. Math. Soc. 308 (1988), 413–429.

[Re6] C.J. READ, The invariant subspace problem for a class of Banach spaces II. Hypercyclic operators, Israel J. Math. 63 (1988), 1–40.

[Ric] C.E. RICKERT, General Theory of Banach Algebras, The University Series in Higher Mathematics, D. Van Nostrand, Co. Princeton, N.J., Toronto-London-New York 1960.

[Ri1] F. RIESZ, Sur certains systèmes singuliers d'équations intégrales, Ann. École Norm. Sup. 28 (1911), 33–62.

[Ri2] F. RIESZ, Über lineare Funktionalgleichungen, Acta Math. 41 (1918), 71–98.

[RN] F. RIESZ, B. SZ.-NAGY, Functional Analysis, Ungar, New York, 1955.

[RoS] S. ROCH, B. SILBERMANN, Continuity of generalized inverses in Banach algebras, Studia Math. 136 (1999), 197–227.

[Ro] S. ROLEWICZ, On orbits of elements, Studia Math. 32 (1969), 17–22.

[RsS] P. ROSENTHAL, A. SOŁTYSIAK, Formulas for the joint spectral radius of non-commuting Banach algebra elements, Proc. Amer. Math. Soc. 123 (1995), 2705–2708.

[RtS] G.-C. ROTTA, W.G. STRANG, A note on the joint spectral radius, Indag. Math. 22 (1960), 379–381.

[Sa] B.N. SADOVSKII, Limit-compact and condensing operators, Uspekhi Mat. Nauk 27 (1972), 81–146 (Russian), English transl. Russian Math. Surveys 27 (1972), 85–155.

[Sap] P. SAPHAR, Contribution à l'étude des applications linéaires dans un espace de Banach, Bull. Soc. Math. France 92 (1964), 363–384.

[Sd] J. SCHAUDER, Über lineare, vollstetige Funktionaloperationen, Studia Math. 2 (1930), 183–196.

[Sch1] M. SCHECHTER, On the essential spectrum of an arbitrary operator, J. Math. Anal. Appl. 13 (1966), 205–215.

[Sch2] M. SCHECHTER, Riesz operators and Fredholm perturbations, Bull. Amer. Math. Soc. 74 (1968), 1139–1144.

[Sch3] M. SCHECHTER, On perturbations of essential spectra, J. London Math. Soc. 1 (1969), 343–347.

[Sch4] M. SCHECHTER, Quantities related to strictly singular operators, Indiana Univ. Math. J. 21 (1972), 1061–1071.

[SW] M. SCHECHTER, R. WHITLEY, Best Fredholm perturbation theorems, Studia Math. 90 (1988), 175–190.

[Sm1] CH. SCHMOEGER, Ein Spektralabbildungssatz, Arch. Math. 55 (1990), 484–489.

[Sm2] CH. SCHMOEGER, Relatively regular operators and a spectral mapping theorem, J. Math. Anal. Appl. 175 (1993), 315–320.

[Se] Ó. SEARCÓID, Economical finite-rank perturbations of semi-Fredholm operators, Math. Z. 198 (1988), 431–434.

[Seg] G. SEGAL, Fredholm complexes, Quart. J. Math. Oxford Ser. 21 (1970), 385–402.

[Shl] A.L. SHIELDS, Weighted shift operators, in: Topics in Operator Theory (ed. C. Pearcy), Mathematical Surveys No. 13, American Mathematical Society, Providence, Rhode Island 1974, pp. 49–128.

[Sh1] G.E. SHILOV, Sur la théorie des idéaux dans les anneaux normés de fonctions, C.R. (Doklady) Acad. Sci. URSS (N.S.) 27 (1940), 900–903.

[Sh2] G.E. SHILOV, On decomposition of a commutative normed ring in a direct sums of ideals, Mat. Sbornik N.S. 74 (1953), 352–364 (Russian); English translation: Amer. Math. Soc. Transl. 1 (1955), 37–48.

[Shu] M.A. SHUBIN, On holomorphic families of subspaces of a Banach space, Mat. Issled. (Kishinev) 5, (1970), 153–165 (Russian). English translation: Integral Equations Operator Theory 2 (1979), 407–420.

[Shm] V.S. SHUL'MAN, Invariant subspaces and spectral mapping theorems, in: Functional Analysis and Operator Theory, ed. J. Zemánek, Banach Center Publ. vol. 30, Warsaw 1994.

[Si1] J. SICIAK, Extremal points in the space $\mathbb{C}^n$, Colloq. Math. 11 (1964), 157–163.

[Si2] J. SICIAK, Extremal plurisubharmonic functions and capacities in $\mathbb{C}^n$, Sophia Kokyuroku in Mathematics 14, Sophia University, Tokyo, 1982.

[Sin] A. M. SINCLAIR, Automatic Continuity of Linear Operators, London Math. Soc., Lect. Notes Series 21, Cambridge University Press, Cambridge, 1976.

[Sl1] Z. SŁODKOWSKI, On ideals consisting of joint topological divisors of zero, Studia Math. 48 (1973), 83–88.

[Sl2] Z. SŁODKOWSKI, An infinite family of joint spectra, Studia Math. 61 (1977), 239–255.

[Sl3] Z. SŁODKOWSKI, On subharmonicity of the capacity of the spectrum, Proc. Amer. Math. Soc. 81 (1981), 243–249.

[Sl4] Z. SŁODKOWSKI, Operators with closed ranges in spaces of analytic functions, J. Funct. Anal. 28 (1986), 155–177.

[SZ1] Z. SŁODKOWSKI, W. ŻELAZKO, On joint spectra of commuting families of operators, Studia Math. 50 (1974), 127–148.

[SZ2] Z. SŁODKOWSKI, W. ŻELAZKO, A note on semicharacters, in: Spectral Theory, Banach Center Publications vol. 8, PWN, Warsaw, 1982, pp. 397–402.

[So1] A. SOŁTYSIAK, Capacity of finite systems of elements in Banach algebras, Commentat. Math. 19 (1977), 381–387.

[So2] A. SOŁTYSIAK, Some remarks on the joint capacities in Banach algebras, Comment. Math. Univ. Carolinae 20 (1977), 197–204.

[So3] A. SOŁTYSIAK, Approximate point joint spectra and multiplicative functionals, Studia Math. 86 (1987), 277–286.

[So4] A. SOŁTYSIAK, On a certain class of subspectra, Comment. Math. Univ. Carolinae 32 (1991), 715–721.

[So5] A. SOŁTYSIAK, On the joint spectral radii of commuting Banach algebra elements, Studia Math. 105 (1993), 93–99.

[Spi] M. SPIVAK, Calculus on Manifolds, Mathematics Monograph Series, W.A. Benjamin Inc., New York, 1965.

[St] J.G. STAMPFLI, Compact perturbations, normal eigenvalues and a problem of Salinas, J. London Math. Soc. 9 (1974), 165–175.

[Sti1] P.S.G. STIRLING, The joint capacity of elements of Banach algebras, J. London Math. Soc. 10 (1975), 212–218.

[Sti2] P.S. STIRLING, Perturbations of operators which leave capacity invariant, J. London Math. 10 (1975), 75–78.

[Tay] A.E. TAYLOR, Theorems on ascent, descent, nullity and defect of linear operators, Math. Ann. 163 (1966), 18–49.

[Ta1] J.L. TAYLOR, A joint spectrum for several commuting operators, J. Funct. Anal. 6 (1970), 172–191.

[Ta2] J.L. TAYLOR, Analytic-functional calculus for several commuting operators, Acta Math. 125 (1970), 1–38.

[Ts] M. TSUJI, Potential Theory in Modern Function Theory, Maruzen, Tokyo, 1959.

[Tu] YU.V. TUROVSKIĬ, Spectral properties of certain Lie subalgebras and the spectral radius of subsets of a Banach algebra, in: Spectral Theory of Operators and its Applications 6, Elm, Baku, 1985, 144–181 (Russian).

[TS] YU.V. TUROVSKIĬ, V.S. SHUL'MAN, Joint spectral radius and invariant subspaces, Funktsional. Anal. i Prilozhen. 34 (2000), 91–94 (Russian); English translation: Transl. Funct. Anal. Appl. 34 (2000), 156–158.

[Ty1] H.-O. TYLLI, Lifting non-topological divisors of zero modulo the compact operators, J. Funct. Anal. 124 (1994), 389–415.

[Ty2] H.-O. TYLLI, The essential norm of an operator is not self-dual, Israel J. Math. 91 (1995), 93–110.

[Va1] F.-H. VASILESCU, Analytic functions and some residual properties, Rev. Roumaine Math. Pures Appl. 15 (1970), 435–451.

[Va2] F.-H. VASILESCU, Local capacity of operators, Indiana Univ. Math. J. 21 (1972), 743–749.

[Va3] F.-H. VASILESCU, A Martinelli type formula for the analytic functional calculus, Rev. Roumaine Math. Pures Appl. 23 (1978), 1587–1605.

[Va4] F.-H. VASILESCU, Analytic Functional Calculus and Spectral Decompositions, Editura Academiei Republicii Socialiste România, Bucuresti, 1979 (Roumanian); English translation: D. Reidel Publishing Company, Dordrecht 1982.

[Va5] F.-H. VASILESCU, Stability of index of a complex of Banach spaces, J. Operator Theory 2 (1979), 247–275.

[Va6] F.-H. VASILESCU, A multidimensional spectral theory in C^*-algebras, in: Spectral Theory, Banach Center Publications, Vol. 8, PWN – Polish Scientific Publishers, Warsaw, 1982, 471–491.

[Va7] F.-H. VASILESCU, Nonlinear objects in the linear analysis, in: Spectral theory of linear operators and the related topics (Timisoira/Herculane 1983), Operator Theory, Adv. Appl. 14, Birkhäuser, Basel 1984, pp. 265–278.

[Ve1] E. VESENTINI, On the subharmonicity of the spectral radius, Boll. Un. Mat. Ital. 4 (1968), 427–429.

[Ve2] E. VESENTINI, Maximum theorems for spectra, in: Essays on topology and related topics, 111–117, Springer-Verlag, New York 1970.

[Vr] P. VRBOVÁ, On local spectral properties of operators in Banach spaces, Czechoslovak Math. J. 23 (1973), 483–492.

[Wa] L. WAELBROECK, Le calcul symbolique dans les algèbres commutatives, J. Math. Pures Appl. 33 (1954), 147–186.

[Waw] A. WAWRZYNCZYK, On ideals consisting of topological zero divisors, Studia Math. 142 (2000), 245–251.

[We] G. WEISS, Weakly ℓ^p-stable linear operators are power stable, Int. J. Systems Sci. 20 (1989), 2323–2328.

[Wes1] T.T. WEST, The decomposition of Riesz operators, Proc. London Math. Soc. 16 (1966), 737–752.

[Wes2] T.T. WEST, A Riesz-Schauder theorem for semi-Fredholm operators, Proc. Roy. Irish Acad. Sect. A 87 (1987), 137–146.

[Wr] V. WROBEL, The boundary of Taylor's joint spectrum for two commuting Banach space operators, Studia Math. 84 (1986), 105–111.

[Yo1] B. YOOD, Properties of linear transformations preserved under addition of a completely continuous transformation, Duke Math. J. 18 (1951), 599–612.

[Yo2] B. YOOD, Difference algebras of linear transformations on a Banach space, Pacific J. Math. 4 (1954), 615–636.

[Zab] J. ZABCZYK, Remarks on the control of discrete-time distributed parameter systems, SIAM J. Control 12 (1974), 721–735.

[ZKKP] M.G. ZAIDENBERG, S.G. KREIN, P.A. KUČMENT, A.A. PANKOV, Banach bundles and linear operators, Uspekhi Mat. Nauk 30 (1975), 101–

157 (Russian); English translation: Russian Math. Surveys 30 (1975) 115–175.

[Za] V.P. ZAKHARYUTA, Transfinite diameter Tschebyshev constant and a capacity of a compact set in $\mathbb{C}^n$, Mat. Sb. 96 (1975), 374–389 (Russian).

[Zam] W.R. ZAME, Existence uniqueness and continuity of functional calculus homomorphisms, Proc. London Math. Soc. 39 (1979), 73–92.

[Zel1] W. ŻELAZKO, Metric generalizations of Banach algebras, Dissertationes Math. (Rozprawy Mat.) 47 (1965).

[Zel2] W. ŻELAZKO, A characterization of multiplicative linear functionals in complex Banach algebras, Studia Math. 30 (1968), 83–85.

[Zel3] W. ŻELAZKO, A characterization of Shilov boundary in function algebras, Comment. Math. 14 (1970), 63–68.

[Zel4] W. ŻELAZKO, Selected Topics in Topological Algebras, Aarhus University Lecture Notes Series 31, 1971.

[Zel5] W. ŻELAZKO, On a certain class of non-removable ideals in Banach algebras, Studia Math. 44 (1972), 87–92.

[Zel6] W. ŻELAZKO, Banach Algebras, PWN and Elsevier, Amsterdam 1973.

[Zel7] W. ŻELAZKO, Axiomatic approach to joint spectra I., Studia Math. 64 (1979), 249–261.

[Ze1] J. ZEMÁNEK, Concerning spectral characterizations of the radical in Banach algebras, Comment. Math. Univ. Carolinae 17 (1976), 689–691.

[Ze2] J. ZEMÁNEK, Spectral radius characterizations of commutativity in Banach algebras, Studia Math. 61 (1977), 2571–268.

[Ze3] J. ZEMÁNEK, A survey of recent results on the spectral radius in Banach algebras, in: Proc. of the Fourth Prague Symp. on General Topology and its Relations to Modern Analysis and Algebra, August 1976, Part B, Prague 1977, pp. 531–540.

[Ze4] J. ZEMÁNEK, A note on the radical of a Banach algebras, Manuscripta Math. 20 (1977), 191–196.

[Ze5] J. ZEMÁNEK, One-sided spectral calculus, Problem 2.10 in: Linear and Complex Analysis problem Book (V.P. Havin, N.K. Nikolski, eds.), Lect. Notes Math. 1573 (1994), 102–104.

[Ze6] J. ZEMÁNEK, The stability radius of a semi-Fredholm operator, Integral Equations Operator Theory 8 (1985), 137–144.

[Ze7] J. ZEMÁNEK, Approximation of the Weyl spectrum, Proc. Roy. Irish Acad. Sect. A 87 (1987), 177–180.

Index

List of Symbols

Operator Theory: Advances and Applications (OT)

Edited by
Israel Gohberg, Tel Aviv University, Israel

BIRKHÄUSER

This series is devoted to the publication of current research in operator theory, with particular emphasis on applications to classical analysis and the theory of integral equations, as well as to numerical analysis, mathematical physics and mathematical methods in electrical engineering.

OT 160: Kaashoek, M.A. / Seatzu, S. / van der Mee, C. (Eds.), Recent Advances in Operator Theory and its Applications. The Israel Gohberg Anniversary Volume (2005). ISBN 3-7643-7290-7

OT 159: Reissig, M. / Schulze, B.-W. (Eds.), New Trends in the Theory of Hyperbolic Functions (2005). Subseries **A**dvances in **P**artial **D**ifferential **E**quations. ISBN 3-7643-7283-4

OT 158: Eiderman, V.Ya. / Samokhin, M.V. (Eds.), Selected Topics in Complex Analysis (2005). ISBN 3-7643-7251-6

OT 157: Alpay, D. / Vinnikov, V. (Eds.), Operator Theory, Systems Theory and Scattering Theory: Multidimensional Generalizations (2005). ISBN 3-7643-7212-5

OT 156: Ebenfelt, P. / Gustafsson, B. / Khavinson, D. / Putinar, M. (Eds.), Quadrature Domains and Their Applications. The Harold S. Shapiro Anniversary Volume (2005). ISBN 3-7643-7145-5

OT 155: Ashino, R. / Boggiatto, P. / Wong, M.W. (Eds.), Advances in Pseudo-Differential Operators (2004). ISBN 3-7643-7140-4

OT 154: Janas, J. / Kurasov, P. / Naboko, S. (Eds.), Spectral Methods for Operators of Mathematical Physics (2004). ISBN 3-7643-7133-1

OT 153: Gaspar, D. / Gohberg, I. / Timotin, D. / Vasilescu, F.H. / Zsido, L. (Eds.), Recent Advances in Operator Theory, Operator Algebras, and their Applications (2004). ISBN 3-7643-7127-7

OT 152: Eidelman, S.D. / Ivasyshen, S.D. / Kochubei, A.N., Analytic Methods in the Theory of Differential and Pseudo-differential Equations of Parabolic Type (2004). ISBN 3-7643-7115-3

OT 151: Gil, J.B. / Krainer, T. / Witt, I. (Eds.), Aspects of Boundary Problems in Analysis and Geometry (2004). Subseries **A**dvances in **P**artial **D**ifferential **E**quations. ISBN 3-7643-7069-6

OT 150: Rabinovich, V. / Roch, S. / Silbermann, B., Limit Operators and their Applications in Operator Theory (2004). ISBN 3-7643-7081-5

OT 149: Ball, J.A. / Helton, J.W. / Klaus, M. / Rodman, L. (Eds.), Current Trends in Operator Theory and its Applications (2004). ISBN 3-7643-7067-X

OT 148: Ashyralyev, A. / Sobolevskii, P.E., New Difference Schemes for Partial Differential Equations (2004). ISBN 3-7643-7054-8

OT 147: Gohberg, I. / dos Santos, A.F. / Speck, F.-O. / Teixeira, F.S. / Wendland, W. (Eds.), Operator Theoretical Methods and Applications to Mathematical Physics. The Erhard Meister Memorial Volume (2004). ISBN 3-7643-6634-6

OT 146: Kopachevsky, N.D. / Krein, S.G., Operator Approach to Linear Problems of Hydrodynamics. Volume 2: Nonself-adjoint Problems for Viscous Fluids (2003). ISBN 3-7643-2190-3

OT 145: Albeverio, S. / Demuth, M. / Schrohe, E. / Schulze, B.-W. (Eds.), Nonlinear Hyperbolic Equations, Spectral Theory, and Wavelet Transformations (2003). Subseries **A**dvances in **P**artial **D**ifferential **E**quations. ISBN 3-7643-2168-7

OT 144: Belitskii, G. / Tkachenko, V., One-dimensional Functional Equations (2003). ISBN 3-7643-0084-1

OT 143: Alpay, D. (Ed.), Reproducing Kernel Spaces and Applications (2003). ISBN 3-7643-0068-X

MIX
Papier aus verantwortungsvollen Quellen
Paper from responsible sources
FSC® C105338

If you have any concerns about our products,
you can contact us on
ProductSafety@springernature.com

In case Publisher is established outside the EU,
the EU authorized representative is:
Springer Nature Customer Service Center GmbH
Europaplatz 3, 69115 Heidelberg, Germany

Printed by Libri Plureos GmbH
in Hamburg, Germany